Superlubricity

Superlubricity

Second Edition

Edited by

Ali Erdemir
Texas A&M University, J. Mike Walker '66 Department of Mechanical Engineering, College Station, TX, United States

Jean Michel Martin
University of Lyon, Ecole Centrale de Lyon, Laboratory of Tribology and System Dynamics UMR CNRS 5513, Lyon, France

Jianbin Luo
State Key Laboratory of Tribology, Tsinghua University, Beijing, P.R. China

ELSEVIER

Elsevier
Radarweg 29, PO Box 211, 1000 AE Amsterdam, Netherlands
The Boulevard, Langford Lane, Kidlington, Oxford OX5 1GB, United Kingdom
50 Hampshire Street, 5th Floor, Cambridge, MA 02139, United States

British Library Cataloguing-in-Publication Data
A catalogue record for this book is available from the British Library

Library of Congress Cataloging-in-Publication Data
A catalog record for this book is available from the Library of Congress

ISBN: 978-0-444-64313-1

For Information on all Elsevier publications
visit our website at https://www.elsevier.com/books-and-journals

Publisher: Susan Dennis
Acquisitions Editor: Anneka Hess
Editorial Project Manager: Lena Sparks
Production Project Manager: Debasish Ghosh
Cover Designer: Greg Harris

Typeset by MPS Limited, Chennai, India

Contents

List of Contributors

Koshi Adachi Department of Mechanical Systems Engineering, Tohoku University, Sendai, Japan

Maria-Isabel De Barros Bouchet Université de Lyon, Ecole Centrale de Lyon, LTDS, CNRS, Ecully, France

Mehmet Z. Baykara Department of Mechanical Engineering, University of California Merced, Merced, CA, United States

Michel Belin Université de Lyon, Ecole Centrale de Lyon, LTDS, CNRS, Ecully, France

Juliette Cayer-Barrioz Laboratoire de Tribologie et Dynamique des Systèmes, CNRS UMR5513, Ecole centrale de Lyon, Ecully, France

Xinchun Chen State Key Laboratory of Tribology, Tsinghua University, Beijing, P.R. China

Zhe Chen Department of Chemical Engineering and Materials Research Institute, Pennsylvania State University, University Park, PA, United States

Alexia Crespo Laboratoire de Tribologie et Dynamique des Systèmes, CNRS UMR5513, Ecole centrale de Lyon, Ecully, France

Nicole Dörr AC2T research GmbH, Wiener Neustadt, Austria

Ali Erdemir Texas A&M University, J. Mike Walker '66 Department of Mechanical Engineering, College Station, TX, United States

Hélène Fay Laboratoire de Tribologie et Dynamique des Systèmes, CNRS UMR5513, Ecole centrale de Lyon, Ecully, France; Centre de Recherche Paul Pascal, CNRS UMR5031, Université de Bordeaux, Pessac, France; Now at Solvay, LOF, CNRS UMR5258, Université de Bordeaux, Pessac, France

Xiangyu Ge School of Mechanical Engineering, Beijing Institute of Technology, Beijing, P.R. China

Thilo Glatzel Institute of Physics, University of Basel, Basel, Switzerland

Enrico Gnecco Otto Schott Institute of Materials Research (OSIM), Friedrich Schiller University Jena, Jena, Germany

Hanjuan Gong State Key Laboratory of Tribology, Department of Mechanical Engineering, Tsinghua University, Beijing, China

Dan Guo State Key Laboratory of Tribology, Tsinghua University, Beijing, P.R. China

Motohisa Hirano Department of Mechanical Engineering, Faculty of Science and Engineering, Hosei University, Koganei, Tokyo, Japan

Takahisa Kato Surface Science and Tribology Laboratory, The University of Tokyo, Tokyo, Japan

Seong H. Kim Department of Chemical Engineering and Materials Research Institute, Pennsylvania State University, University Park, PA, United States

Jinjin Li State Key Laboratory of Tribology, Tsinghua University, Beijing, P.R. China

Ke Li Intelligent Transport Systems Research Center, Wuhan University of Technology, Wuhan, P.R. China

Qunyang Li State Key Laboratory of Tribology, Tsinghua University, Beijing, P.R. China; AML, CNMM, Department of Engineering Mechanics, Tsinghua University, Beijing, P.R. China

Chenxu Liu State Key Laboratory of Tribology, Tsinghua University, Beijing, P.R. China

Dameng Liu State Key Laboratory of Tribology, Tsinghua University, Beijing, P.R. China

Yanmin Liu State Key Laboratory of Tribology, Tsinghua University, Beijing, P.R. China

Yuhong Liu State Key Laboratory of Tribology, Tsinghua University, Beijing, P.R. China

Yun Long Université de Lyon, Ecole Centrale de Lyon, LTDS, CNRS, Ecully, France

Jianbin Luo State Key laboratory of Tribology, Tsinghua University, Beijing, P.R. China; State Key Laboratory of Tribology, Department of Mechanical Engineering, Tsinghua University, Beijing, China

Liran Ma State Key laboratory of Tribology, Tsinghua University, Beijing, P.R. China

Ming Ma Center for Nano and Micro Mechanics, Tsinghua University, Beijing, P.R. China; State Key Laboratory of Tribology, Tsinghua University, Beijing, P.R. China; Department of Mechanical Engineering, Tsinghua University, Beijing, P.R. China

Tianbao Ma State Key Laboratory of Tribology, Tsinghua University, Beijing, P.R. China

Stefan Makowski Fraunhofer Institute for Material and Beam Technology (IWS), Dresden, Germany

Jean Michel Martin Université de Lyon, Ecole Centrale de Lyon, LTDS, CNRS, Ecully, France

Denis Mazuyer Laboratoire de Tribologie et Dynamique des Systèmes, CNRS UMR5513, Ecole centrale de Lyon, Ecully, France

Yonggang Meng State Key Laboratory of Tribology, Tsinghua University, Beijing, P.R. China

Ernst Meyer Institute of Physics, University of Basel, Basel, Switzerland

Nazario Morgado Laboratoire de Tribologie et Dynamique des Systèmes, CNRS UMR5513, Ecole centrale de Lyon, Ecully, France

Tasuku Onodera Functional Materials Research Department, Center for Technology Innovation – Materials, Research & Development Group, Hitachi, Ltd., Hitachi, Japan

Rémy Pawlak Institute of Physics, University of Basel, Basel, Switzerland

Xiaoyong Ren State Key Laboratory of Tribology, Department of Mechanical Engineering, Tsinghua University, Beijing, China

M. Clelia Righi Department of Physics and Astronomy, University of Bologna, Bologna, Italy

Andjelka Ristic AC2T research GmbH, Wiener Neustadt, Austria

Shuai Shi State Key Laboratory of Tribology, Tsinghua University, Beijing, P.R. China

Xinfeng Tan State Key Laboratory of Tribology, Tsinghua University, Beijing, P.R. China

Hongdong Wang State Key Laboratory of Tribology, Tsinghua University, Beijing, P.R. China

Wei Wang State Key Laboratory of Tribology, Department of Mechanical Engineering, Tsinghua University, Beijing, China

Yongfu Wang State Key Laboratory of Solid Lubrication, Lanzhou Institute of Chemical Physics, Chinese Academy of Sciences, Lanzhou, P.R. China; Center of Materials Science and Optoelectronics Engineering, University of Chinese Academy of Sciences, Beijing, P.R. China

Volker Weihnacht Fraunhofer Institute for Material and Beam Technology (IWS), Dresden, Germany

Shuai Wu State Key Laboratory of Tribology, Department of Mechanical Engineering, Tsinghua University, Beijing, China

Guoxin Xie State Key Laboratory of Tribology, Department of Mechanical Engineering, Tsinghua University, Beijing, China

Qunfeng Zeng Key Laboratory of Education Ministry for Modern Design and Rotor-Bearing System, Xi'an Jiaotong University, Xi'an, P.R. China

Chenhui Zhang State Key Laboratory of Tribology, Tsinghua University, Beijing, P.R. China

Junyan Zhang State Key Laboratory of Solid Lubrication, Lanzhou Institute of Chemical Physics, Chinese Academy of Sciences, Lanzhou, P.R. China; Center of Materials Science and Optoelectronics Engineering, University of Chinese Academy of Sciences, Beijing, P.R. China

Quanshui Zheng Center for Nano and Micro Mechanics, Tsinghua University, Beijing, P.R. China; State Key Laboratory of Tribology, Tsinghua University, Beijing, P.R. China; Department of Engineering Mechanics, Tsinghua University, Beijing, P.R. China

Introduction

Among all physical phenomena, friction perhaps poses one of the greatest challenges to the scientific and industrial communities and has a direct linkage to energy efficiency and environmental cleanliness of all moving mechanical systems. In our everyday life, we rarely think about friction or appreciate its importance, but there is no doubt that it is a major cause of lost energy and hazardous emissions to our environment with dire consequences for a sustainable future. Hence, the prospect of vanishing friction in all types of mechanical systems has real-life implications for not only preserving our limited energy resources, but also saving our planet from hazardous emissions for generations to come. Considering that in most industrialized nations, the annual cost of friction- and wear-related energy and material losses is estimated to account for a significant fraction of their gross national products, the further reduction (or even elimination) of friction in all kinds of moving mechanical systems would be extremely beneficial to the societal and economic well-being of all nations. In short, reducing friction further is extremely important for conserving not only our ever-diminishing energy resources but also preventing our planet from catastrophic health and environmental consequences.

Recent advances in the high-power computational, analytical, and experimental capabilities for more advanced tribological studies have paved the way for a much better understanding and hence control of friction across the force, length, and time scales. In particular, dedicated fundamental friction studies at the subnanometer scale using a wide range of advanced computational and experimental tools (i.e., ab initio and molecular dynamic simulations, atomic and friction force microscopy, scanning tunneling microscopy, surface force apparatus, in situ and real-time probing capabilities, and so on) have enabled scientists to not only more accurately predict/simulate, but also to identify the specific types of materials and test conditions under which near-zero friction or superlubricity can be attained in a number of nano- to macroscale sliding systems. In particular, high-power supercomputers with huge processing capabilities now enable simulation and real-time visualization of atomic-scale friction and allow predictions of those conditions that can lead to superlubricity. In combination with fast-developing artificial intelligence and machine learning capabilities, prospects for more precisely predicting those materials and/or operating conditions that can lead to sustainable superlubricity are very exciting and will hopefully be within our reach in coming years.

The word "superlubricity" was first used by Motohisa Hirano to describe a theoretical sliding regime in which friction or resistance to sliding completely vanishes (see his chapter in this book). Historically, the earliest studies on superlubricity started in mid-1980s, but the real progress occurred during the 1990s. In particular, theoretical studies by Professors Jeffrey Sokoloff and Motohisa Hirano predicted the existence of such superlubricious states between weakly interacting, atomically smooth surfaces. Their original work has formed the foundation of more in-depth studies since then and the superlubricity research has now become one of the most popular research topics in our tribology field as manifested by thousands of publications within the last two decades.

Due to the strict structural requirements (i.e., optimal imcommensurability and/or misalignment) for superlubricity, some scientists who worked in this field during the early 2000s often preferred to use the term "structural superlubricity" rather than "superlubricity," mainly because "superlubricity" seemed related to such well-known physical phenomena as superconductivity, superfluidity, and so on. Because superlubricious situations were predicted and observed only in ultrahigh vacuum in the 1990s, there was a tendency to see an analogy with superconductivity, in which electrical resistance vanishes at very low temperatures. However, the term superlubricity is quite appropriate from a tribological standpoint in the sense that the prefix "super" means "extreme"; hence, superlubricity means extreme lubricity but in no way it suggests "zero" friction. In addition, we are presently unable to measure friction coefficients at values below 10^{-4}. In the case of superconductivity, electric resistance completely vanishes or it is "zero" so that the use of the term "super" may not be totally appropriate here, either.

Apart from structural superlubricity, it now appears that some tribological systems lubricated with specific liquids could also approach the superlow friction regime (friction coefficients below 0.01) and several examples are including in this book. The term "liquid superlubricity" is used to describe these recent discoveries. Since the publication of first book on Superlubricity in 2007, tens of new kinds of liquids providing superlubricity have been found, that is, water-based liquids, oil-based lubricants, and liquids combined with new additives of 2D materials. In the past 13 years, the friction coefficient has been further decreased by tenfold to reach the 0.0001 level and the sustained contact pressures have been increased by more than an order of magnitude, reaching GPa level.

The main objective of this book is to bring together leading researchers who work in superlubricity and other relevant fields and to provide a concise state-of-the-art overview of the most recent developments and new discoveries. Because this field has become rather large, complex, and multidisciplinary in nature, the chapters in this book may not sufficiently cover everything that relates to superlubricity; however, these editors have done their best to bring out some of the most important new developments in superlubricity since the publication of last Superlubricity book in 2007.

This second edition consists of 27 chapters providing a wealth of new information ranging from theoretical and practical aspects of friction in general to superlubricity in particular. Each chapter is written by a group of leading experts who are well known for their invaluable contributions to the field. The book starts with a section covering the theoretical aspects of superlubricity by Prof. Hirano. This is followed by other sections that deal with the observation and/or measurement of superlubricity in a wide range of sliding systems including 2D or layered materials. Specifically, Chapters 1–10 are devoted to the theoretical and simulations aspects of structural superlubricity, while Chapter 11 and 12 provide fundamental insight into superlubricity of various sliding systems involving carbon-based coatings. Chapters 13 and 14 cover the liquid superlubricity in the presence of glycerol and Chapters 15–20 discuss the liquid superlubricity with water-based systems and biomaterials. Chapters 21–26 highlight liquid superlubricity of sliding surfaces under boundary lubricated sliding regimes as well as new approaches using black phosphorous for example as a new means for achieving superlubricity (Chapter 22).

Readers will appreciate that recent advances in superlubricity research have indeed been phenomenal. However, they will also realize that there remain several key challenges for future scientists and engineers to overcome for large-scale industrial applications. These editors are truly indebted to the authors of each chapter in this book. Without their enthusiasm and eagerness, we could not have put this book together. The editors also acknowledge the support of their institutions (Texas A&M University, USA; Ecole Centrale de Lyon, France and CNRS; and SKLT Tsinghua University, China) and their funding agencies. Last but not the least, the editors thank their families for their support and understanding during the preparation of this book.

Ali Erdemir[1], Jean Michel Martin[2] and Jianbin Luo[3]

[1]J. Mike Walker '66 Department of Mechanical Engineering, Texas A&M University, College Station, TX, United States, [2]Laboratory of Tribology and System Dynamics LTDS, Ecole Centrale de Lyon, University of Lyon, Lyon, France, [3]State Key Laboratory of Tribology, Tsinghua University, Beijing, P.R. China

Chapter 1

Atomistics of superlubricity

Motohisa Hirano
Department of Mechanical Engineering, Faculty of Science and Engineering, Hosei University, Koganei, Tokyo, Japan

1.1 Atomistics in tribology

The work done by friction has very different nature from the work done by gravity. Work by gravity happens when objects are moved against gravity, which is always acting on objects. In contrast, static friction force appears as a reaction force in response to the tangential force applied to the contacting surface under the operation of the gravitational force perpendicular to the tangential force. The relative sliding is not always necessary in static friction. The dynamic friction force is the drag against relative sliding. From the atomistic point of view, no work by friction is generated as long as the interatomic force between surfaces is perpendicular to the sliding direction. Some researchers did not agree with atomistic idea proposed by Desaguliers [1], and others claimed that friction force stems from molecular interaction at contacting surfaces [2,3]. Tomlinson [4] was the first to explain the energy dissipation stemming from molecular interaction by introducing the concept of "mechanical adiabaticity."

Very little research on the atomistics of friction followed because of the difficulty of handling the complexity of actual non-well-defined surfaces. A solution to the problems of "friction coefficients" in real systems has been discussed by studying the elastoplastic deformation of the interfacial junctions, which are formed by the adhesive interaction between real-contacting surface asperities under the combined stress field in which both of compressive and shearing stresses operate [5]. It has been considered that the friction force observed in real systems is attributed to the shearing resistance for the deformation and growth of the junction against the applied tangential force along contacting surfaces. The friction data usually measured in real systems contain lots of unknown factors such as surface asperities and contaminants which should be the extrinsic factors to cause friction. It had been thus difficult for a very long time to study the atomistic origin of friction force intrinsically generated by the interatomic interactions at sliding interfaces.

Superlubricity. DOI: https://doi.org/10.1016/B978-0-444-64313-1.00001-6

Friction research has been renewed by much progress in modern surface physics, nanotechnology, and its related experimental techniques. These have provided us the atomically well-defined tribological surfaces, whose atomic structures and superstructures are specified by various surface analytical tools such as scanning probe microscopy and field electron microscopy in ultrahigh vacuum. We thus have obtained "ideal friction experiments" [6,7] that are directly comparable to the theoretical models. Theory can investigate in detail the fundamental properties of interatomic interactions and the mechanism for the appearance of friction using empirical computer studies [8,9] such as molecular dynamics simulation.

1.2 Atomistic origin of friction and superlubricity

Superlubricity, which is the state of vanishing friction, is closely related to the origin for the appearance of friction. The atomistic models explaining the physical origins of the static and the dynamic friction forces are presented [10]. This study has clarified the atomistic origin of the static friction that is generated by intrinsic factors, such as interatomic interactions between constituent atoms, not by extrinsic factors such as surface asperities or surface contaminants. The origin of the dynamic friction force is formulated as the problem of how the given translational kinetic energy dissipates into the internal motions of constituent atoms during sliding. From studying that, the available phase-space volume of the translational motion becomes negligibly small, compared with that of the internal motions, thus concluding that the energy dissipation occurs irreversibly from the translational motion to the internal motions in which recurrence phenomena, in other words, almost periodic oscillation of mass center velocity of the friction model occurs.

1.2.1 Friction model

Let us consider the case in which upper surface very slowly, that is, quasistatically slides against lower surface [11]. The adiabatic potential of the frictional system is defined by the total energy when two contacting surfaces slide against each other. The upper surface has N^u atoms and the lower one N^l atoms, and the constituent atoms belonging to both surfaces interact with each other. The position coordinates of the atoms are denoted by $\mathbf{r}_i = (r_i^x, r_i^y, r_i^z)$, where $i = 1,2, \ldots, (N^u + N^l)$. Total energy is a function of the position vectors r_i of all the atoms as: $V(\mathbf{Q}, \{\mathbf{r}_i(\mathbf{Q})\})$, where $\mathbf{Q}$ stands for the displacement vector expressed by Eq. (1.1) of the upper surface against the stationary lower surface. An r_i coordinate set satisfies the relationship given by

$$\mathbf{Q} = \sum_i^{N^u} \frac{\mathbf{r}_i}{N^u} \quad \text{and} \quad \mathbf{0} = \sum_i^{N^l} \frac{\mathbf{r}_i}{N^l}. \tag{1.1}$$

The friction model involves two contacting surfaces having simple symmetry. Each atom belonging to the upper (or lower) surface is denoted by a (or b). $V_{ab}(r)$ describes the interaction potential energy between atoms a and b as: $V_{aa}(r)$, $V_{ab}(r)$, and $V_{bb}(r)$, where r is the interatomic distance between two atoms.

The adiabatic potential is obtained by

$$V(Q,\{\mathbf{r}_i(Q)\}) = \sum_i^{N^u}\sum_j^{N^l} V_{ab}(|\mathbf{r}_i - \mathbf{r}_j|) + \frac{1}{2}\sum_{i,j}^{N^u,N^l} V_{aa}(|\mathbf{r}_i - \mathbf{r}_j|). \tag{1.2}$$

Here the summation of j in the first term of the right-hand side in Eq. (1.2) is expressed by

$$V(\mathbf{r}_i(Q)) = \sum_j^{N^l} V_{ab}(|\mathbf{r}_i - \mathbf{r}_j|). \tag{1.3}$$

$V_l(r_i(Q))$ is the interaction energy that the atoms of the upper surface receive from the atoms of the lower surface. The terms $V_{bb}(|\mathbf{r}_i - \mathbf{r}_j|)$ is dropped, since it has no $\mathbf{Q}$ dependence by assuming that an upper surface moves quasistatically against a stationary lower body. $V_l(\mathbf{r}_i(\mathbf{Q}))$ has a periodicity characterized by the primitive vectors of the lower surface. The summation of j in the second term of the right-hand side in Eq. (1.2) is expressed as

$$V^u(\mathbf{r}_i(Q) - \mathbf{r}_j(Q)) = \sum_j^{N^u} V_{aa}(|\mathbf{r}_i - \mathbf{r}_j|). \tag{1.4}$$

Equation (1.2) can be rewritten as

$$V(Q,\{\mathbf{r}_i(Q)\}) = \sum_i^{N^u}\left\{\frac{1}{2}\sum_{i\neq j}^{N^u} V^u(\mathbf{r}_i(Q) - \mathbf{r}_j(Q)) + V^l(\mathbf{r}_i(Q))\right\}, \tag{1.5}$$

The following frictional system will be considered to be given by

$$H(\{\mathbf{p}_i\},\{\mathbf{r}_i\}) = \sum_i^{N^u}\frac{|\mathbf{p}_i|^2}{2} + \sum_i^{N}\left\{\frac{1}{2}\sum_{j(\neq i)}^{N^u} V^u(\mathbf{r}_i - \mathbf{r}_j) + V^l(\mathbf{r}_i)\right\}, \tag{1.6}$$

which is obtained by adding the kinetic energy term of each atom to the model given in Eq. (1.5). The first, the second, and the third terms of the right-hand side stand for the kinetic energy of the ith atom, the mutual interaction potential energy $V^u(\mathbf{r}_i - \mathbf{r}_j)$ between the atoms of the upper surface, and the friction potential energy given by $V^l(\mathbf{r}) = \sum_j v_a(\mathbf{r} - \mathbf{r}_j)$, where $v_a(\mathbf{r} - \mathbf{r}_j)$ is the interaction energy from the jth atom of the lower surface.

In the frictional process, it is convenient to distinguish the translational degree of freedom, that is, the mass center of the upper surface from the other degrees of freedom concerning with the internal relative motions. We then introduce the following notations as:

$$\mathbf{P} = \sum_{i}^{N^u} \frac{\mathbf{p}_i}{N^u}, \tag{1.7}$$

$$\mathbf{Q} = \sum_{i}^{N^u} \frac{\mathbf{r}_i}{N^u}, \tag{1.8}$$

and

$$\overline{\mathbf{p}}_i = \mathbf{p}_i - \mathbf{P}, \qquad \overline{\mathbf{r}}_i = \mathbf{r}_i - \mathbf{Q}, \tag{1.9}$$

where $i = 1, 2, \ldots, 3(N^u - 1)$. In the aforementioned equations, $\mathbf{P}$ and $\mathbf{Q}$ are the momentum and the coordinate of the mass center, specifying the transnational motion, and $\overline{\mathbf{p}}_i$ and $\overline{\mathbf{r}}_i$ are the momentum and the position coordinate of the ith atom, specifying the internal relative motions. By using these notations, the frictional system in Eq. (1.6) can be rewritten as:

$$H(\{\overline{\mathbf{p}}_i\}, \{\overline{\mathbf{r}}_i\}; \mathbf{P}, \mathbf{Q}) = N\frac{|\mathbf{P}|^2}{2} + \sum_i V^l(\overline{\mathbf{r}}_i + \mathbf{Q}) + H_0(\{\overline{\mathbf{p}}_i\}, \{\overline{\mathbf{r}}_i\}), \tag{1.10}$$

$$H(\{\overline{\mathbf{p}}_i\}, \{\overline{\mathbf{r}}_i\}) = \sum_{i}^{N-1} \frac{|\overline{\mathbf{p}}_i|^2}{2} + \frac{1}{2}\sum_{i \neq j}^{N-1} V^u(\overline{\mathbf{r}}_i - \overline{\mathbf{r}}_j). \tag{1.11}$$

$H_0(\{\overline{\mathbf{p}}_i\}, \{\overline{\mathbf{r}}_i\})$ involves only the internal degrees of freedom of the upper body, and the translational motion $(\mathbf{P}, \mathbf{Q})$ is connected with the internal motions $(\overline{\mathbf{p}}_i, \overline{\mathbf{r}}_i)$ by the second term, that is, the frictional term in the right-hand side in Eq. (1.10). The equation of motion for the mass center of the upper surface is given from Eqs. (1.10) and (1.11):

$$\frac{d\mathbf{P}}{dt} = \mathbf{F}(\{\overline{\mathbf{r}}_i\}; \mathbf{Q}), \tag{1.12}$$

$$\frac{d\mathbf{Q}}{dt} = \mathbf{P},$$

where $\mathbf{F}(\{\overline{\mathbf{r}}_i\}; \mathbf{Q})$ is a force acting on the mass center of the upper surface, and is defined by

$$\mathbf{F}(\{\overline{\mathbf{r}}_i\}; \mathbf{Q}) = -\frac{1}{N^u}\sum_i \frac{\partial V^l(\overline{\mathbf{r}}_i(t) + \mathbf{Q})}{\partial \mathbf{Q}}. \tag{1.13}$$

1.2.2 Static friction

A picture for the one-dimensional frictional systems is shown in Fig. 1.1 [10,11]. The upper surface is simply expressed by the linear atomic chain where the atoms interact with each other. Each atom of the upper surface

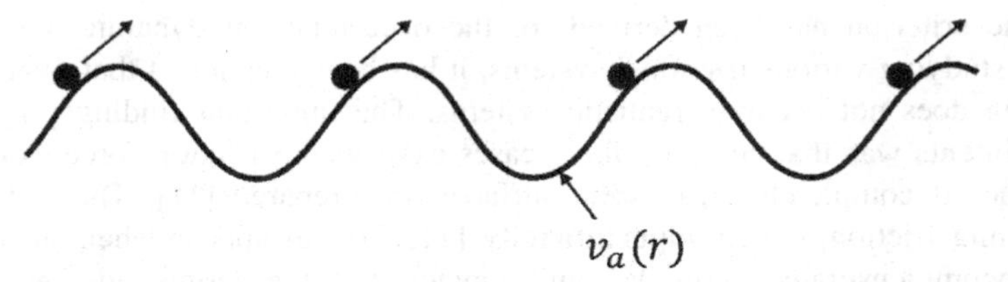

FIGURE 1.1 The atomistic picture for the origin of the static friction force, shown for the one-dimensional friction systems [11]. The upper surface is simply expressed by the linear chain where each atom interacts with each other. Each atom of the upper surface feels the interaction potential $v_a(r) = V^l(\overline{\mathbf{r}}_i(\mathbf{Q}))$ from the lower surface, which is represented by the potential curve. When we apply an external force to push the linear chain in the right direction, each atom rises the mountain of the interaction potential coherently or cooperatively. The drag against the applied force is the sum of the forces along the chain which each atom feels from the lower surface expressed by $v_2(r)$.

feels the interaction potential from the lower surface. Here, imagine we apply an external force to push the linear chain in a sliding direction. Each atom climbs the mountain of the friction potential energy. During sliding, the interdistances between two adjacent atoms change. The drag can be obtained by calculating the total friction potential energy for each sliding distance **Q**, and by taking its first derivative with respect to **Q**. Alternatively, the drag against the applied force can be the sum of the forces along the chain which each atom feels from the lower surface. The drag from each atom can become positive or negative. Then we have the static friction force as:

$$\mathbf{F_s}(Q) = -\sum_i \frac{\partial V^l(\overline{\mathbf{r}}_i(\mathbf{Q}) + \mathbf{Q})}{\partial \mathbf{Q}}, \tag{1.14}$$

or using Eq. (1.13) to obtain

$$\mathbf{F_s}(Q) = -N^u \mathbf{F}(\overline{\mathbf{r}}_i(\mathbf{Q}), \mathbf{Q}). \tag{1.15}$$

The positive part of this $\mathbf{F_s}(\mathbf{Q})$ gives the static friction force. The static friction force stems from the cooperative motion of atoms when the upper surface lattice is commensurate with the lower surface lattice along the sliding direction. This picture is called atomistic locking [11]. We have clarified the two atomistic origins of friction such as atomistic locking and dynamic locking [11] based on the concept of mechanical adiabaticity. When all the constituent atoms move adiabatically or continuously, the atomistic locking occurs for an arbitrary strength of interatomic potential. While the classical mechanical locking model, the nonflat potential surface that the upper body feels from the lower body spans on a large scale, in atomistic locking the nonflat potential surface spans on an atomic scale. Therefore the static friction force appears when contacting surfaces are commensurate. Dynamic locking occurs when the surface atomic configuration discontinuously changes due to the dynamic process and if the interatomic potential is stronger than a specific given value [11].

The criterion has been derived for the occurrence of dynamic locking. From studying various frictional systems, it has been concluded that dynamic locking does not occur in realistic systems. The important finding prior to experiments was that certain unique cases exist where friction force exactly vanishes if completely clean solid surfaces are prepared [11]. The state of vanishing friction, called **superlubricity** [11,12], can appear when surfaces are incommensurately brought into contact and the atoms adiabatically change their equilibrium positions during sliding in infinite system.

1.2.3 Incommensurate and commensurate contacts

The static friction force appears when contacting surfaces are commensurate. A state of vanishing friction, that is, superlubricity appears when surfaces are brought into incommensurate contact and the atoms adiabatically change their equilibrium positions in quasistatic sliding. Equation (1.15) shows the static friction force **Fs** (Q), which is the sum of the forces, expressed by $\mathbf{F}(\{\bar{\mathbf{r}}_i\};\mathbf{Q})$ shown in Eq. (1.13), acting on each atom, and superlubricity appears when the sum becomes zero. The conclusion that this sum becomes vanishing does not depend on the type of atomic interaction or crystal structure of the solids and is not based on quantum-mechanical effects, which means that it holds in general for either classical or quantum theory. Whether this sum is zero or finite depends on the atomic arrangement of the two solid surfaces that are in contact with each other. If the ratio of the lattice spacing of two solid surfaces is a rational number, that sum will be finite and friction will appear. In order for a solid to slide, all atoms must climb over the potential peak of the other solid in phase as shown in Fig. 1.1. Here each atom receives a force, that is, resistance to motion, in a direction opposite the external pulling force. In order for all of the atoms to move together in this situation, this external force must be larger than the sum total of each atom's resistance. This mechanism is called atomistic locking mentioned in the previous section.

In contrast, for contact between two solid surfaces where their lattice-spacing ratio is an irrational number the magnitude and direction of the forces received by the atoms never coincide with each other along the sliding direction. Thus for an infinite system, the sum total of the forces received by the atoms, that is, the friction forces, becomes zero. In other words, the loss and gain of each atom's interaction energy compensate with each other out. As a consequence, the total energy of the system at ground state becomes invariant with respect to sliding distance of the mass center, that is, the friction force exactly vanishes. Although a real surface has a finite size, a simple calculation tells us that the friction forces are sufficiently small for a surface area about 100×100 Å^2 in size. To produce solid contact in which forces acting on each atom cancel each other out, one way would be to bring two crystalline surfaces in contact with each other and then rotate either one by an appropriate angle so that the orientations of their lattices do not match.

Such modulated atomic structures with periods which are incommensurate with the basic lattices are quite common in solid state physics. The structure can be another lattice, a periodic lattice distortion, a helical or sinusoidal magnetic structure, or a charge density wave (CDW) in one, two, or three dimensions. The model having an array of atoms connected with harmonic springs interacts with a periodic sinusoidal potential was originally introduced by Frenkel and Kontorova [13]. The Frenkel–Kontorova model, denoted as FK model, in solid friction is related to the CDW model, which is a physical system that describes electronic sliding motion against ions in crystals. It is well known that the CDW appearing from interactions between electrons and ionic structures in crystals result in free sliding of electrons when the interactions become lower than the specific threshold value. Such sliding phenomenon without energy dissipation may be familiar in the field of solid-state physical systems with two interacting periodicities [14]. In friction models with such properties, when the surfaces in contact are incommensurate, or when the ratio of periodicity along sliding surfaces is an irrational number, the two surfaces are found to be able to slide without energy dissipation. Sokoloff showed that the FK model for CDWs can reproduce phenomena such as stick slip in friction, thus demonstrating its usefulness as a model for friction in solids [15]. The idea of commensurability in solid surfaces in contact is leading to new developments in recent theoretical and experimental research in nanotribology [16,17] and nanoengineering [18].

1.2.4 Energy dissipation in dynamic friction

The origin of the dynamic friction force is presented. Consider that the upper surface is pushed to slide against the lower surface at a constant velocity $\mathbf{P}(0)$ and $\overline{\mathbf{p}}_i = 0$ initially. If the translational momentum $\mathbf{P}(t)$ subsequently decreases during sliding, the force must be applied to push the upper surface to keep a sliding velocity constant. This applied force corresponds to the dynamic friction force. The origin of the dynamic friction force is thus reformulated as the problem of how the translational kinetic energy of the mass center decreases. The energy dissipation rate $R(t)$ at time t is given as the reduction rate of the translational kinetic energy is written as

$$R(t) = -N^u \frac{d|\mathbf{P}(t)|^2}{dt} = -N^u \mathbf{P}(t) \cdot \mathbf{F}(\{\mathbf{r}_i\}, \mathbf{Q}) \tag{1.16}$$

or using Eq. (1.13) as

$$R(t) = \mathbf{P}(t) \cdot \sum_i \frac{\partial V^l(\overline{\mathbf{r}}_i + \mathbf{Q})}{\partial \mathbf{Q}}, \tag{1.17}$$

where a symbol · stands for the inner product between two vectors. The dynamic friction force $\mathbf{F}_d(t)$ can be obtained from the relation $R(t) = \mathbf{P}(t) \cdot {}^*\mathbf{F}_d(t)$. From Eq. (1.17), we have

$$\mathbf{F}_d(t) = -\sum_i \frac{\partial V^l(\bar{\mathbf{r}}_i + \mathbf{Q})}{\partial \mathbf{Q}}, \tag{1.18}$$

which is equal to the N^u times of the force acting on the mass center, as seen in Eq. (1.18).

If $\sum V^l(\bar{\mathbf{r}}_i + \mathbf{Q})$ has a $\mathbf{Q}$ dependence, namely, $\mathbf{F}(\{\mathbf{r}_i\}, \mathbf{Q}) \neq 0$, the translational kinetic energy can be transformed into the kinetic energies of the internal relative motions. If the transformed energy does not turn again to the translational kinetic energy, this energy transfer occurs irreversibly. In order to examine the possibility of this irreversibility to occur, the phase-space volume concerning with the translational motion is estimated to have an order of $N^u|\mathbf{P}|^2/2$ since the translational kinetic energy, being less than or equal to $N^u|\mathbf{P}|^2/2$, is available. There are many combinations of how this kinetic energy $N^u|\mathbf{P}|^2/2$ may be distributed on many degrees of freedom for the internal relative motions. The number of combinations increases with an exponential order of the total number of the internal degrees of freedom. So the available phase-space volume of the internal relative motions may be estimated to have an order of $e^{\gamma(N^u-1)}$, where the value of $\gamma(>0)$ depends on details of the model with internal variables, for an example, $\gamma = 3\ln(|\mathbf{P}|^2/2\ \nu)$ when H_0 in Eq. (1.11) is described as an ensemble of many independent harmonic oscillators with frequency ν. From this study, the phase-space volume for the translational motion becomes negligibly small for a large N^u, compared with that for the internal motions. Thus the energy transfer from the translational motion to the internal motion occurs irreversibly, and so it has been concluded that the energy is dissipated from the translational motion to the internal motions.

The frictional system is the dynamic one conserving the total energy. The irreversible energy dissipation indicates that the internal relative motions are excited, and hence the adiabaticity does not hold true for the internal relative motions. However in order that the present idea is adequate, the system, described by $H_0(\{\bar{\mathbf{p}}_i\}, \{\bar{\mathbf{r}}_i\})$, must work as a host system absorbing energy. In other words, the dynamic system given by Eq. (1.10) has the ergodic property. If the system energy is sufficiently low, one participates finding energy surface, which is filled by the nonergodic torus with the Kolmogorov–Arnolod–Moser (KAM) stability. As the energy increases, the system recovers the ergodic property. The number of empirical computer studies supports this picture. When the energy surface is filled by the KAM torus, the system is well known to show the recurrence phenomenon where the energy repeats decreasing and increasing [19], and hence the energy does not diffuse. The number of studies, for example, the first computational experiment by Ferm, Ulam, and Pasta, has been made to find the border, the critical energy, where the system becomes from nonergodic to ergodic [20].

The simulations show that the energy seems to be distributed over the entire degrees of freedoms even for the case of the weak adhesion. The FK frictional system [13], where the potential terms in Eq. (1.6) is replaced by the pure FK model has been studied. The Hamiltonian of the one-dimensional FK friction model with the kinetic term can be written as

$$H(\{p_i\},\{x_i\}) = \sum_i^N \frac{p_i^2}{2m} + \sum_i^N \left\{ \frac{1}{2} k(x_{i+1} - x_i - \ell)^2 + \frac{f}{2\pi} \sin\left(\frac{2\pi x_i}{L}\right) \right\}, \quad (1.19)$$

where p_i is momentum, k is the spring constant, ℓ is the natural length of a spring, L is the potential period of the solid below, and f is the amplitude of friction potential energy. The subsequent calculations assume $\ell = (1+\sqrt{5})/2$, which is called golden mean number and $L = 1$. The number of atoms N is set 100. The periodic boundary condition is applied to the FK model. The momentum $\mathbf{P}(t)$ as a function of $\mathbf{Q}(t)$, started from the initial $\mathbf{P}(0)$, is shown in Fig. 1.2(A). The momentum $\overline{\mathbf{p}}_i(t)$ as a function of the coordinate $\overline{\mathbf{r}}_i(t)$ is shown in Fig. 1.2(B). Fig. 1.2(B) implies that the system is ergodic, and so the host system works as an energy absorber, as seen in Fig. 1.2.

1.2.5 Friction-phase diagram

The dynamic FK friction system written by Eq. (1.19) given in the previous section is studied. The periodicity length L of the sinusoidal potential in Eq. (1.19) is taken as unity, while the natural length of the FK chain ℓ is assumed to be equal to the golden mean number $\ell = (\sqrt{5} + 1)/2$. To examine the frictional property including superlubricity, the dynamics has been studied after the upper solid surface at the ground state is pushed with initial sliding velocity $\mathbf{P}(0)$ under the condition of $\overline{\mathbf{p}}_i = 0$ for any i with the Hamiltonian dynamics conserving the energy. The friction dynamics is studied by examining quantities such as $\mathbf{P}(t)$, $\mathbf{Q}(t)$, $\overline{\mathbf{q}}_i(t)$, $\overline{\mathbf{p}}_i(t)$, and the sliding distance $Ls(t)$ defined as the distance over which the upper solid surface slides during time t. These quantities are obtained by numerically solving Eq. (1.19) by using the velocity Verlet algorithm [21]. Here the speed of sound of longitudinal waves can be written as $v_s = \sqrt{E/\rho} = \sqrt{k/m\ell}$, and is a unit of velocity. The speed of sound in solids is around $v \approx 5000\ \text{ms}^{-1}$. Therefore if a solid slides at a constant speed of 0.1 for 10^5 time, the sliding speed is around $500\ \text{ms}^{-1}$ and the sliding distance is about 5 μm $(=0.1 \times 5000\ \text{ms}^{-1} \times 10^5 \times 0.1\ \text{ps})$.

Fig. 1.3 illustrates the frictional properties of the FK model, showing how the mass center velocity Vc of the model changes with time after the upper body is pushed with the initial sliding velocity $\mathbf{P}(0) = 0.1$ at three different friction potential amplitudes of $f = 0.005$, 0.031, and 0.069, respectively. At the small friction potential amplitude of $f = 0.005$, superlubricity appears, resulting

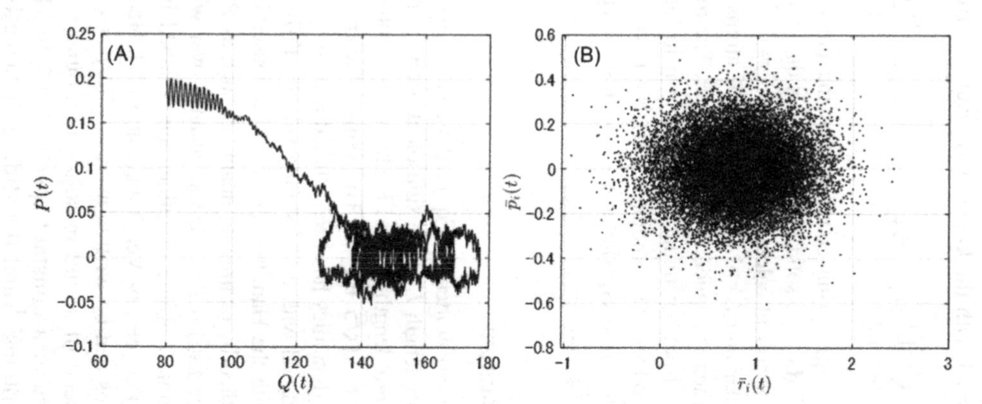

FIGURE 1.2 Translational momentum $\mathbf{P}(t)$ as a function of $\mathbf{Q}(t)$ in (A) and the internal momentum $\overline{\mathbf{p}}_i(t)$ as a function of $\overline{\mathbf{r}}_i(t)$ in (B) for the Frenkel–Kontorova frictional system. The dots stand for their values at every 1000 unit time intervals. The magnitude of k of the spring describing the interaction between the upper body and the magnitude f describing the adhesion is taken equal to 1 and 0.1, respectively. (B) It implies that the system is ergodic, and so the host system works as an energy absorber, as seen from (A).

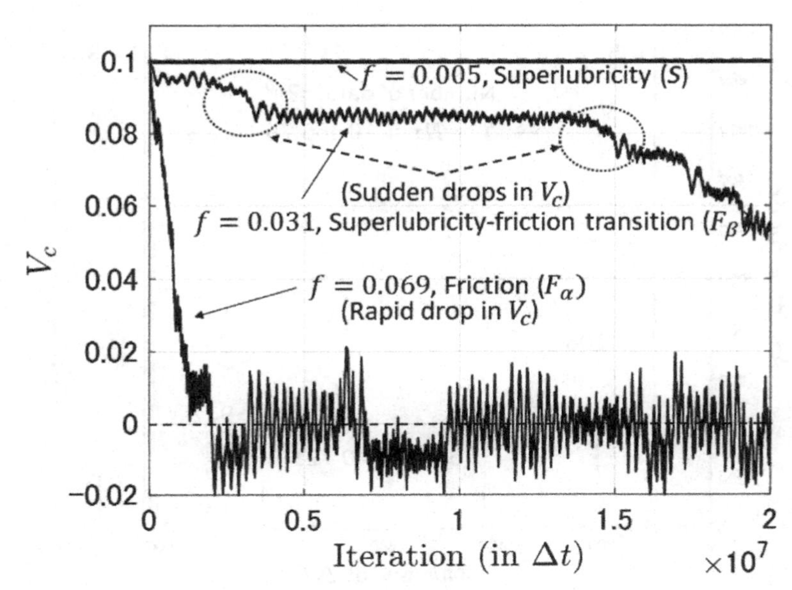

FIGURE 1.3 Mass center velocity changing with time (t = iteration $\times$ Δt). The initial velocity is **P**(0) = 0.1, and the friction potential amplitudes are $f = 0.005$, $f = 0.031$, and $f = 0.069$, respectively, showing superlubricity, superlubricity-friction transition, and friction regimes which will be illustrated in Fig. 1.5.

in vanishing friction and the upper body moves at a constant velocity during the time. The atomic oscillation amplitude spectrum has been obtained as shown in Fig. 1.4 by Fourier transforming the time-series data of the relative displacement of the specific atom with respect to the mass center. The spectrum shows the fundamental harmonic wave and its higher-order harmonic waves, and the frequency of the fundamental one was confirmed to almost equal to the fundamental frequency calculated by the lattice vibration model of one-dimensional atomic chain.

When subsequently increasing the friction potential amplitude up to the maximum value of $f = 0.069$, friction appears, demonstrating the mass center velocity **P**(t) = Vc rapidly decreases as soon as an upper body begins to slide, as shown in Fig. 1.3. At intermediate regime at $f = 0.031$, **P**(t) = Vc gradually decreases and it suddenly drops at unexpected time, which is called **superlubricity-friction transition**. Fig. 1.4 shows the spectrum for the relative displacement between the displacement of the specific 50th atoms from the origin, which is actually next to the mass center with respect to the mass center position. As seen in Fig. 1.4, almost harmonic oscillation with the fundamental mode at 105 of the number of Δf and the multiple higher-order modes are observed. The frequency of the fundamental mode calculated by

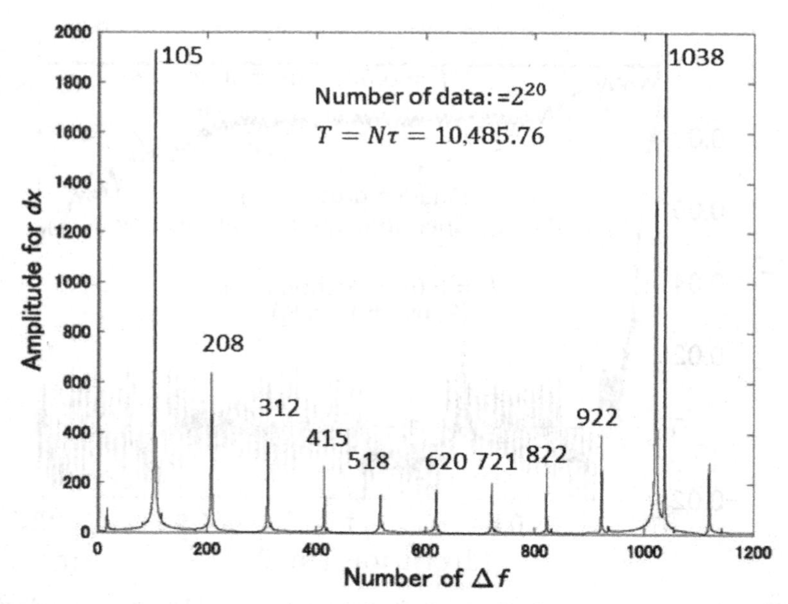

FIGURE 1.4 The Fourier spectrum of the atomic oscillation appeared in superlubricity regime.

this numerical method corresponds the value solved by the dynamical matrix of the FK chain with periodic boundary condition. This implies that the physical picture of superlubricity is the smooth sliding without friction while atomic chain almost harmonically oscillating. When increasing the initial sliding velocity, the simple harmonic mode shown in Fig. 1.4 turns to be more complex one when increasing the friction potential amplitude f, in which multiple oscillation peaks beside the harmonic one emerged and the frequency components are continuously distributed for the wide range of frequencies. The role of nonlinear dynamics by combining the fast Fourier transform analysis and the phase-space investigation will be studied.

Fig. 1.5 shows the friction-phase diagram determined by the frictional properties representing how the mass center velocity changes with calculation time of 2×10^4, shown in Fig. 1.3. Two regimes such as friction regime (F) and superlubricity regime (S) appear. Friction regime has two subarea of F_α and F_β. It seems that the boundaries between F area and S area look like irregular. It will need to perform a large number of calculations to determine more precise boundaries for successive study.

In this friction diagram, atoms continuously move in equilibrium positions in the area $f < 0.14$ and $\mathbf{P}(0) = v_0 = 0$, resulting in zero friction, that is, superlubricity. On the other hand, the friction regime (F_α) extends if $0 < f < 0.14$ and the initial velocity is finite ($\mathbf{P}(0) \neq 0$). In summary, superlubricity is likely to occur when the interface interaction parameter is small because of dynamic effects of kinetic energy. Interestingly, superlubricity is specifically unlikely to appear at

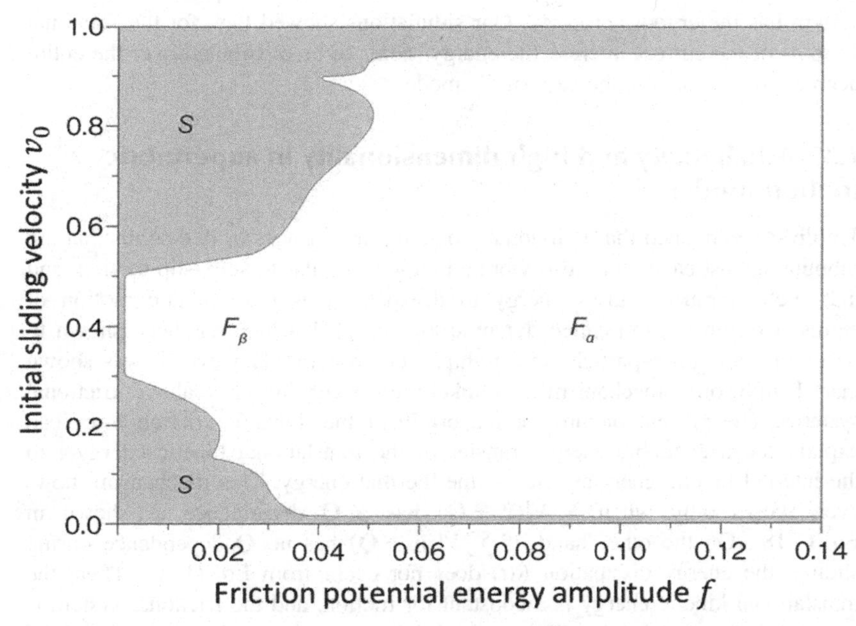

FIGURE 1.5 Friction-phase diagram of one-dimensional dynamic Frenkel–Kontorova model. The boundary for S, F_α, and F_β regimes is determined as shown in Fig. 1.3.

the range of intermediate velocity (0.3–0.5) even if the interface interaction is small, and a friction regime with different friction properties (F_β).

The friction system of multiparticles, which are coupled with each other while feeling the sinusoidal potential when mass center moves, is essentially nonlinear dynamical system. In superlubricity case, it is easier to evaluate the vibration mode, which looks like simple harmonic oscillation mode, where a primary vibration peak and the associated higher-order peaks have been verified. When sliding speed approaches to the region of F_β area at small f, simple almost harmonic mode turns to be nonharmonic mode when multivibration peaks have appeared. The mechanism for the appearance of the multipeaks is complex problem; it will need to elaborately examine vibration mode signals in terms of nonlinear system analysis. A scenario for the transition superlubricity (S) mode to friction (F_β) could be described in terms of several nonlinear vibrations such as subharmonic oscillation, summed-and-differential harmonic oscillation, superharmonic oscillation, and internal resonance phenomena.

In the case of F_α mode, the dynamics of the system has shown to be chaotic. The occurrence of irreversible energy dissipation indicates that the internal relative motions are excited, and hence the adiabaticity does not hold true for the internal relative motions. In order that the present idea is adequate, the system must work as a host system absorbing energy. In other words, the dynamical system has the ergodic property. Our simulations showed that, for the frictional systems of the current interest, the energy seems to be distributed over the entire degrees of freedoms in the case of F_α mode.

1.3 Adiabaticity and high dimensionality in superlubric friction model

Tomlinson explained that individual atoms on the surfaces of two solids that are rubbing against each other also vibrate in a way similar to stick-slip motion, and that such vibration causes energy to dissipate. This nonadiabatic motion of atoms is origin friction called dynamic locking [11], which has been shown to occur in both single-particle and multiparticle systems. However it was shown that Tomlinson's mechanism is unlikely to occur in the realistic frictional systems. The present picture for the origin of the dynamic friction force can explain the irreversible energy transfer of the translational kinetic energy into the internal kinetic energies, that is, the thermal energy. This mechanism, however, works only when $\sum V^l(\bar{\mathbf{r}}_i + \mathbf{Q})$ has a $\mathbf{Q}$ dependence as shown in Eq. (1.18). On the other hand, if $\sum V^l(\bar{\mathbf{r}}_i + \mathbf{Q})$ has no $\mathbf{Q}$ dependence during sliding, the energy dissipation $R(t)$ does not occur from Eq. (1.16). Then, the translational kinetic energy is a constant for motion, and the frictional system is in a state of superlubricity. The superlubricity can appear when the sum of the forces of each atom vanishes. It has been proved that the superlubric state occurs when the atoms have their equilibrium positions for each $\mathbf{Q}$ and, at the same time, the frictional system satisfies some conditions [11]. The condition is satisfied when two solid surfaces are incommensurate. The superlubricity has been theoretically discussed for the weak adhesion. It was argued that the system of incommensurately contacting surfaces has zero dynamic threshold for sliding when two contacting surfaces form a perfect periodic lattice. The state of superlubricity is not associated with the energy dissipation. Thus the concept of superlubricity contradicts with Tomlinson's mechanism.

The problems will be interesting of how the superlubricity is influenced by the dynamic effect [22] when two bodies moves relatively so fast that $\bar{\mathbf{r}}_i$ may not be in the equilibrium position $\bar{\mathbf{r}}_i(\mathbf{Q})$ and by the existence of the surface roughness and the imperfections such as the defects and the dislocations. It has been considered that the superlubricity may be stable, as is different from the one-dimensional case of the CDW pinned easily by the defects.

Friction was investigated essentially on the basis of one-dimensional models, as shown in Fig. 1.6(A). In such a one-dimensional system, the degree of freedom in the motion of an atom is low. This means that if unstable areas, where the open areas in Fig. 1.6(A) in which atoms cannot stably exist, appear, an atom will undergo nonadiabatic motion as it passes through those areas. Such an unstable area, which corresponds to the area in the aforementioned chalk example where the chalk does not stick, appears as

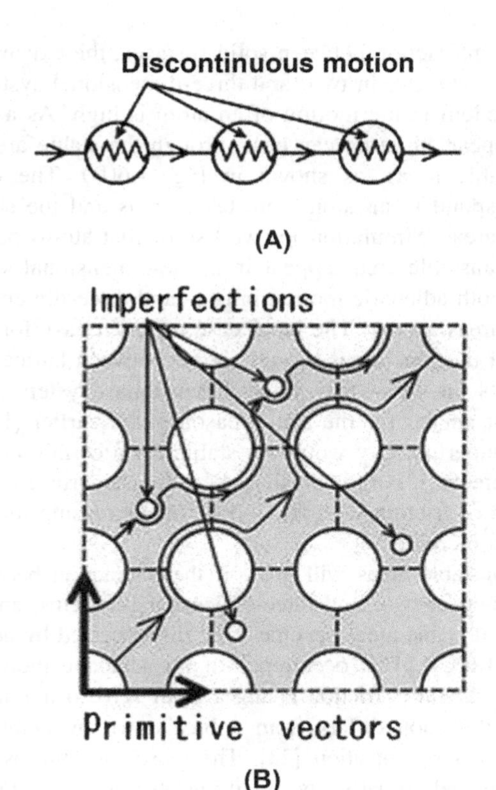

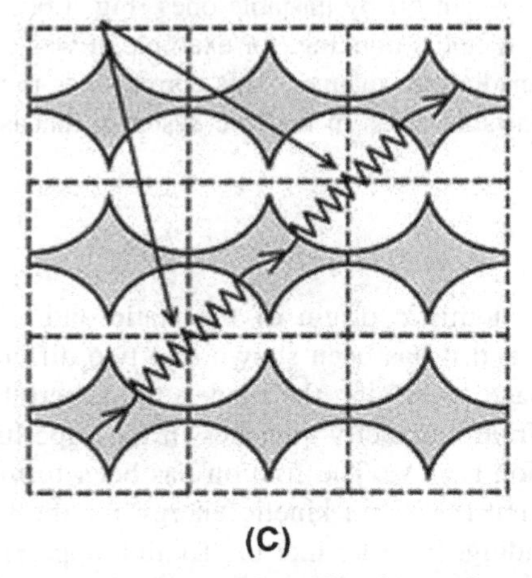

FIGURE 1.6 Motion of atoms at a contact surface. The *white* sections represent unstable areas in which atoms cannot stably exist and the shaded sections to stable areas in which they can stably exist. (A) One-dimensional system. [(B) and (C)] Two- and three-dimensional systems [10].

a result of strong interaction between solid surfaces, the existence of impurities and lattice defects, etc. In two- and three-dimensional systems, however, the degree of freedom in the motion of an atom is high. As a result, even if unstable areas appear, an atom can pass through the stable areas by moving around the unstable areas, as shown in Fig. 1.6(B). The open areas in Fig. 1.6(B) correspond to an atom's unstable areas and the shaded areas to an atom's stable areas. Simulations as well show that atoms perform nonadiabatic motion if unstable areas appear in a one-dimensional system but that they perform smooth adiabatic motion in two- and three-dimensional systems even if unstable areas appear. The latter case makes it easy for superlubricity to appear. Even if unstable areas should appear between lattices due to impurities and defects in two- and three-dimensional systems, nonadiabatic motion would not appear for the same reason given earlier [Fig. 1.6(B)]. In this case, while superlubricity would be stable for a certain concentration of impurities and defects, it is thought that exceeding a certain value will result in the appearance of friction with friction forces increasing monotonically as that concentration increases.

Incidentally, unstable areas will grow if the interaction between solid surfaces is made strong in two- and three-dimensional systems, and nonadiabatic motion will occur if stable areas become to be disconnected by unstable areas at some time [Fig. 1.6(C)]. This occurrence of nonadiabatic motion results in a friction transition in which friction changes from zero to a finite value. Note that this friction transition differs from a breakdown in analytical properties referred to as an Aubry transition [23]. The reason for this is as follows. A breakdown in analytical properties occurs if unstable areas for atoms appear in a one-dimensional system [Fig. 1.6(A)]. A friction transition, on the other hand, does not occur simply because unstable areas appear [Fig. 1.6(B)] but does occur if stable areas are cut off by unstable ones [Fig. 1.6(C)]. In a realistic system characterized by metallic bonding, for example, it was concluded that atoms do not move in a nonadiabatic manner. This corresponds to no disconnectedness of stable areas by unstable areas in realistic systems, that is, friction transitions do not occur.

1.4 Summary

The picture of the atomistic origin of the static and the dynamic friction forces is presented, and it has been shown that two different regimes appear in the parameter space specifying the model: the superlubricity and the friction regimes. The friction exactly vanishes in the superlubricity regime and appears in the friction regime. The friction has been formulated as the problem of whether or not the given kinetic energy for the translational motion dissipates during sliding. In order that the friction appears, the given translational kinetic energy must be absorbed by the host system describing the internal motions.

We have studied the dynamic property in friction from an atomistic point of view. Let us compare the results with those obtained by assuming the quasistatically sliding where the upper surface slides with very low velocity. From the study of the FK model with kinetic energy terms, it was found that superlubricity appears even for the case with finite sliding velocity as well as for the quasistatically sliding case. Superlubricity occurs due to the persistent recurrence phenomenon where the translational kinetic energy repeatedly increases and decreases with time. It has been emphasized that for high-dimensional systems, superlubricity is a generic phenomenon, appearing for a wide class of (strong or weak) adhesion such as metallic bonding and van der Waals interaction.

We would like to make a comment on the friction. What causes the friction? We shall discuss it based on the maximal entropy principle that the dynamics proceeds forward to increase the entropy. The phase space is spanned by a set of $\{\overline{\mathbf{p}}_i, \overline{\mathbf{q}}_i\}$ and $\{\mathbf{P}, \mathbf{Q}\}$, and its volume is expressed by the multiplication of $\{\overline{\mathbf{p}}_i, \overline{\mathbf{q}}_i\}$ and $\{\mathbf{P},\mathbf{Q}\}$, that is, $\Omega = \int_\omega \prod_{i=1}^{N-1} d\overline{\mathbf{p}}_i d\overline{\mathbf{q}}_i d\mathbf{P} d\mathbf{Q}$, where ω stands for the region being integrated. We can focus only on the phase volume, Ω_h, for the internal motions for the given energy, say, $(N-1)\varepsilon$ (ε stands for the mean energy for the internal motion), since the number, $N-1$, of the degrees of freedoms for the internal motions is much larger than that for the translational motion and so the Ω is dominated by the internal motions. The maximal entropy principle implies that when the Ω_h is a decreasing (increasing) function of ε, the dynamics goes forward to decrease (increase) ε. The phase-space volume grows large if the mixing property holds, and its size can be estimated. Suppose that the $(N-1)\varepsilon$ is distributed on the host system. The translational kinetic energy may be considered to have $(NP(0)^2/2 - (N-1)\varepsilon)$. Under the mixing property, $(N-1)\varepsilon$ may be distributed on the $(N-1)$ degrees of freedom for the internal motions in the various ways. Then, the phase-space volume for the internal motions have an order of $\exp[\gamma(\varepsilon)\varepsilon(N-1)]$, where the $\gamma(\varepsilon)$ (>0) depends on details of the model and is generally an increasing function of ε: for an example, $\gamma(\varepsilon) = 3\ln(\varepsilon/\nu)$ when H_0 in Eq. (1.10) is described as an ensemble of many independent harmonic oscillators with frequency ν. The phase-space volume size is the increasing function of ε. Therefore based on the maximal entropy principle, we may conclude that the dynamics proceeds forward to increase ε; in other words, the $\mathbf{P}(t)$ decreases, and so the dissipation occurs.

The superlubricity phenomenon is interesting from both theoretical and applicable points of view [24]. To confirm an existence of superlubric state, many experiments have been made to observe the frictional anisotropy by using various materials [7,18,25–30]. The high-dimensionality yields another remarkable property in the friction. For the case of two-dimensional FK model, the friction becomes anisotropic with respect to the lattice misfit angle between the upper and the lower solid surfaces. In fact, lots of anisotropic frictional systems have been observed at various scales [26]. For the applicable viewpoint, the anisotropy implies a feasibility for controlling friction and designing friction systems by taking the lattice misfit angle as a controlling parameter.

References

[1] J.T. Desaguliers, Phil. Trans. Roy. Soc. (London) 33 (1725) 345–347.

[2] J. Leslie, An Experimental Inquiry into the Nature and Propagation of Heat, The Giller Printer, Poultry, London, 1804.

[3] F. Jenkin, J.A. Ewing, On friction between surfaces moving at low speeds, Phil. Trans. Roy. Soc. (London) 167 (1877) 509–528.

[4] G.A. Tomlinson, Phil. Mag 7 (46) (1929) 905–939.

[5] D. Tabor, Junction growth in metallic friction: the role of combined stresses and surface contamination, Proc. Roy. Soc. Lond A251 (1959) 378–393.

[6] S. Kawai, A. Benassi, E. Gnecco, H. Sode, R. Pawlak, X. Feng, et al., Superlubricity of graphene nanoribbons on gold surfaces, *Science* 351 (6276) (2016) 957–961.

[7] Y. Kobayashi, T. Taniguchi, K. Watanabe, Y. Maniwa, Y. Miyata, Slidable atomic layers in van der Waals heterostructures, Appl. Phys. Express 10 (2017) 045201.

[8] L. Verlet, Computer "experiments" on classical fluids. I. Thermodynamical properties of Lennard-Jones molecules, Phys. Rev. 159 (1) (1967) 98–103.

[9] L. Verlet, Computer "experiments" on classical fluids. II. Equilibrium correlation functions, Phys. Rev. 165 (1) (1968) 201–214.

[10] M. Hirano, Friction at the Atomic Level: Atomistic Approaches in Tribology, Wiley-VCH, Weinheim, Germany, 2018.

[11] M. Hirano, K. Shinjo, Atomistic locking and friction, Phys. Rev B41 (1990) 11837–11851.

[12] K. Shinjo, M. Hirano, Dynamics of friction: superlubric state, Surf. Sci. 283 (1993) 473–478.

[13] Y.I. Frenkel, T. Kontorowa, On the theory of plastic deformation and doubling, Zh. Eksp. Teor. Fiz. 8 (1938) 1340–1349.

[14] P. Bak, Commensurate phases, incommensurate phase, and the devil's staircase, Rep. Prog. Phys. 45 (1982) 587–629.

[15] J.B. Sokoloff, Theory of dynamical friction between idealized sliding surfaces, Surf, Sci. 144 (1984) 267–272.

[16] J. Krim, Friction at the atomic scale, Sci. Am. 275 (4) (1996) 74–80.

[17] M.Z. Baykara, M.R. Vazirisereshk, A. Martini, Emerging superlubricity: a review of the state of the art and perspectives on future research, Appl. Phys. Rev. 5 (2018) 041102. Available from: https://doi.org/10.1063/1.5051445.

[18] J. Cumings, A. Zettl, Low-friction nanoscale linear bearing realized from multiwall carbon nanotubes, Science 289 (5479) (2000) 602–604.

[19] V.I. Arnold, A. Avez, Problemes Ergodiques de la Mecatiique Classique, Ganthier-Villas, Paris, 1967.

[20] E. Fermi, J. Pasta, S. Ulam, Nonlinear wave motion, Lect. Appl. Math. 15 (1974) 143–155.

[21] D.W. Heermann, Computer Simulation Methods in Theoretical Physics, second Ed., Springer-Verlag, 1990.

[22] J.B. Sokoloff, Theory of energy dissipation in sliding crystal surfaces, Phys. Rev. B 42 (1990) 760–765.

[23] S. Aubry, Devil's staircase and order without periodicity in classical condensed matter, J. Phys. (Paris) 44 (1983) 147–162.
[24] J.M. Martin, A. Erdemir, Superlubricity: friction's vanishing act, Phys. Today 71 (4) (2018) 40–46. Available from: https://doi.org/10.1063/PT.3.3897.
[25] J.M. Martin, C. Donnet, Th. Le Mogne, Th Epicier, Superlubricity of molybdenum disulphide, Phys. Rev B48 (2004) 10583–10586.
[26] M. Dienwiebel, G.S. Verhoeven, N. Pradeep, J.W.M. Frenken, Superlubricity of graphite, Phys. Rev. Lett 92 (2004) 126101.
[27] Z. Liu, J. Yang, F. Grey, J.Z. Liu, Y. Liu, Y. Wang, et al., Observation of microscale superlubricity in graphite, Phys. Rev. Lett. 108 (2012) 205503.
[28] D. Berman, S.A. Deshmukh, S.K.R.S. Sankaranarayanan, A. Erdemir, A.V. Sumant, Macroscale superlubricity enabled by graphene nanoscroll formation, Science 348 (2015) 1118–1122.
[29] A. Bylinskii, D. Gangloff, V. Vuletic, Tuning friction atom-by-atom in an ion-crystal simulator, Science 348 (2015) 1115–1118.
[30] J.J. Li, T.Y. Gao, J.B. Luo, Superlubricity of graphite induced by multiple transferred graphene nanoflakes, Adv. Sci. 5 (2018) 1700616. URL: <https://doi.org/10.1002/advs.201700616>.

Chapter 2

Ab initio insights into graphene lubricity

M. Clelia Righi
Department of Physics and Astronomy, University of Bologna, Bologna, Italy

2.1 Introduction

The massive waste of energy and environmental costs associated to friction and wear has triggered active research activity in the search for novel lubricant materials. Among them two-dimensional (2D) materials have recently attracted increasingly attention as environmental-friendly alternative to lubricants additives in base oil [1] and atomically thin coatings [2,3]. Thanks to its mechanical strength [4,5] and chemical inertness [6] graphene is emerging as an ideal material in this context, as revealed by the extremely low coefficients of friction (COF) measured at the nanoscales and microscales [7–14]. However, the tribological performances of graphene as solid lubricant for macroscale contacts are much less impressive, with friction coefficients typically increasing by two orders of magnitudes [15–21], with only few exceptions where lower values have been obtained by realizing a full interfacial coverage by graphene [22,23]. Furthermore the effectiveness of graphene as solid lubricant at the macroscale highly depends on the presence of passivating species in the surrounding environment, in particular on air humidity [17,24–27]. This behavior is common to graphite, the bulk counterpart of graphene, which has long been used as solid lubricant.

The aforementioned macroscopic behaviors are rooted in molecular-level mechanisms occurring at the buried interface that are extremely difficult to monitor in real time by experiments. Here we show that atomistic simulations represent a very powerful tool in this context, especially those based on an ab initio approach, which is essential to accurately describe the surface interactions and the chemical processes in conditions of enhanced reactivity, as those imposed by the applied mechanical stresses.

First we apply ab initio calculations to describe the interlayer sliding in graphene. A wide set of applications of few layer graphene (FLG) films,

Superlubricity. DOI: https://doi.org/10.1016/B978-0-444-64313-1.00002-8

such as their use in nanoelectromechanical systems and in solid lubrication, involve the relative displacement of the layers composing the film. It has also been demonstrated that the electronic properties in FLG can be tuned by changing the layer stacking [28], with very promising consequences for a graphene-based electronics. The effects of shear forces constitute, thus, an important aspect of FLG physics, which has been recently investigated by different experiments [7,8,13,15,29–35].

At the fundamental level, the mobility of one graphene layer onto another is governed by the shape of the potential energy surface (PES), which describes the interlayer binding energy as a function of the relative position of two layers, $\gamma(x, y, z)$. This energy variation gives rise to frictional forces [36], and the PES corrugation determines the maximum energy that can be dissipated during sliding. Through the PES knowledge, one can, thus, understand the sliding properties of graphene on graphene and predict how they will change in the presence of an applied load. We construct the ab initio PES and then identify an analytical expression that describes the PES in excellent agreement with first principles calculations [37]. Thanks to its formal simplicity, the proposed model allows for an immediate interpretation of the interactions, and of their contribution to the PES corrugation at different applied loads. From the PES shape, we derive the ideal shear strength of graphene and compare it with other layered materials used as solid lubricants [38].

In the second part of the chapter, we apply ab initio calculations to provide an explanation for recent tribological experiments that showed that graphene is able to lubricate macroscale steel-on-steel sliding contacts very effectively [18]. This effect has been attributed to a *mechanical action* of graphene related to its load-carrying capacity [39]. However this explanation does not account for the Raman spectra recorded after the tribological test. On the basis of ab initio calculations, we provide a different explanation based on graphene *chemical action*. We show, in fact, that graphene binds strongly to native iron surfaces highly reducing their surface energy. Thanks to a passivating effect, the metal surfaces coated by graphene become almost inert and present very low adhesion and shear strength when mated in a sliding contact [6].

The final part of the chapter aims at providing insights into the dependence of graphene/graphite lubricity on the presence of humidity. The fact that graphite lubricity is not an intrinsic property of graphite, but depends on the environment, was discovered in the 1930s, when it was found that devices lubricated by graphite used in airplanes stopped to properly function due to the low humidity at high altitude [40]. Since then, several experiments have investigated the role of humidity in graphite tribology [41,42], and different hypotheses have been made. The firstly proposed mechanism assumes that water physisorbed on graphite surface forms a hydration layer that lubricates the contact [43]. Another early considered mechanism is based on the idea that water intercalates between graphite planes increasing their spacing and consequently favors the interlayer slipperiness [44,45]. Later it was proposed that the saturation of reactive sites by passivating species present in moist air is the key process for friction reduction [46–50]. However a validation of this hypothesis through a direct observation of the molecular mechanisms is still lacking and the effects of edges-water interaction on graphene/graphite friction are still debated [51–53]. We show that significant progress can be made by the possibility, given by the so-called quantum mechanics/molecular mechanics (QM/MM) approach, to combine the computational advantages of classical molecular dynamics (MD) with the accuracy of QM. The QM/MM simulations we performed [54,55] turn out to be of paramount importance to understand the effects of humidity in graphitic materials, as they provide an accurate description of the collective behavior of confined water molecules that lead to efficient edge passivation. A close comparison with O_2 molecules in similar tribological conditions highlighted the relevance of hydrogen bond networks in favoring the formation of hydrated lubricant media [55].

2.2 Interlayer shear strength of graphene films

We compare the most widely used numerical methods within the density functional theory (DFT) to describe the interlayer binding in graphene and identified the semiempirical functional proposed by Grimme, which includes the van der Waals (vdW) interactions [56], as the most suitable one to describe not only the strength of interlayer binding, but also its variation as a function of the layer stacking [37]. A 2D representation of the PES is reported in Fig. 1 for two different interlayer distances. The stacking configurations corresponding to the PES stationary points can be figured out by shifting the unit cell origin of the superimposed layer (not shown in Fig. 1) into the labeled positions: the PES maximum corresponds to the AA stacking with all the carbon atoms of the upper layer on-top those of the lower ones; the PES minimum corresponds to AB stacking with half of the atoms on-top, the other half at the hollow sites directly above the centers of the hexagonal rings; the saddle stacking presents two connected carbon atoms falling within every hexagonal ring. The path with the highest statistical weigh is the minimum energy path (MEP) that connects the PES minima passing through the saddle points. Scanning tunneling microscope images corresponding to the saddle configuration has been acquired during the tip-induced horizontal shift of the top layer in a graphite sample [29,57]. In agreement with our findings, this observation indicates that the saddle configuration (defined as "no overlap" configuration in ref. [29]) is the configuration the top layer goes through in passing from one PES minima to another, that is, from ABA to ABC stacking on graphite.

The graphene-graphene interaction can be described as a sum of two contributions, one repulsive due to Pauli principle and one attractive of vdW nature:

$$\gamma(x,y,z) = C_0(x,y)e^{-zC_1(x,y)} - \frac{C_2(x,y)}{z^4}, \tag{2.1}$$

the repulsive term is described as an exponential decay that mimics the decay of the surface charge density into the vacuum, and the attractive term is described as the inverse of a power law. As in the model we developed for rare gas adsorbed on metals [58,59], the *Ci* coefficients are bidimensional functions with the same periodicity of the hexagonal lattice [37]. This model provides an excellent description of the DFT data. Moreover it allows for a direct interpretation of the physical interactions, in particular of the potential corrugation.

Upon bilayer compression, the location of the PES stationary points might change [59], this is not the case for bilayer graphene. As seen in Fig. 2.1, the PES shape at the two selected nonequilibrium distances is the same, a part from the enhanced corrugation at the lower separation. In Fig. 2.2, the PES corrugation as a function of the interlayer separation z is reported for the layer stackings corresponding to the PES maxima (red) and saddle points (blue). It can be seen that the ab initio data (dots) and the model of Eq. (2.1) (lines) are in excellent agreement. The inset shows the contributions at the corrugation arising from the repulsive (dashed-dot line)

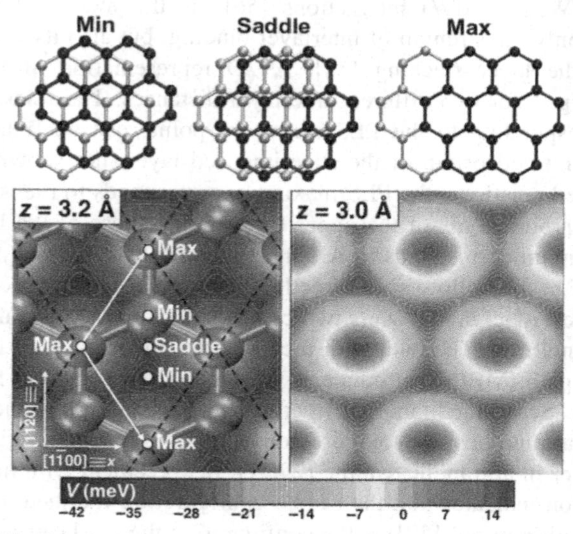

FIGURE 2.1 The function $Y(x,y, z)$ [Eq. (2.1)], which describes the potential energy surface for bilayer graphene, is represented in two dimensions for two different values of the interlayer separation, z.

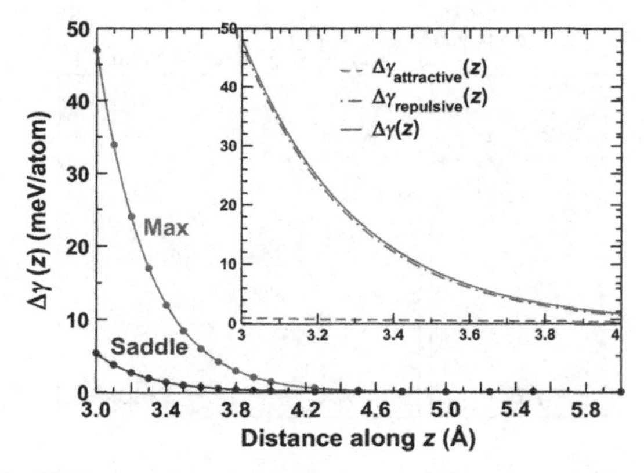

FIGURE 2.2 The PES corrugations, $\Delta\gamma$, for the layer stackings corresponding to the PES maxima (*red*) and saddle points (*blue*) are calculated for different layer separations. The excellent agreement between the ab initio data (dots) and the model (lines) can be appreciated. The inset shows the contributions at the corrugation arising from the repulsive (dashed-dot line) and attractive (dashed line) parts of the potential. The considered stacking corresponds to the potential energy surface maxima. (*PES*: Potential energy surface).

and attractive (dashed line) parts of the interaction. It appears evident that the corrugation is almost completely determined by the repulsive energy, while the long-range attraction produces a negligible corrugation. This clearly indicates that the vdW adhesion does not produce significant effects on the intrinsic resistance to sliding of the system, which is almost completely determined by the Pauli repulsion existing between the two layers. This repulsion rapidly increases with load causing a corresponding increase of the frictional forces.

The differences in the repulsive character of the different stacking configurations, which mostly contribute to the PES corrugation, are rooted in a different distribution of the electronic charge. In Fig. 2.3A, we show the charge displacements occurring when a compressed bilayer is formed from two isolated layers, that is, we plot $\Delta\rho = \rho_{12} - (\rho_1 + \rho_2)$, where ρ_{12} is the charge density of the bilayer, and ρ_1 and ρ_2 are the charge densities of single graphene layers at $z = 0$ Å and $z = 3$ Å, respectively. In Fig. 2.3A, a slice of $\Delta\rho$ is taken in the yz plane with origin (0,0,0). It can be seen that the bilayer compression induces a charge depletion from the region between the two layers (indicated by vertical lines). This charge depletion is much more consistent in the max than in the min and saddle configurations. The depleted electronic charge gathers in the regions around the nuclei, where the electrostatic potential is lower, with a consequent increase of Pauli repulsion. In Fig. 2.3B, the profile of the charge displacement along the z direction, obtained by integration in two dimensions, clearly reveals a difference in the amount of charge which is present in the region between the two layers for the different stackings. We have recently shown that the amount of charge

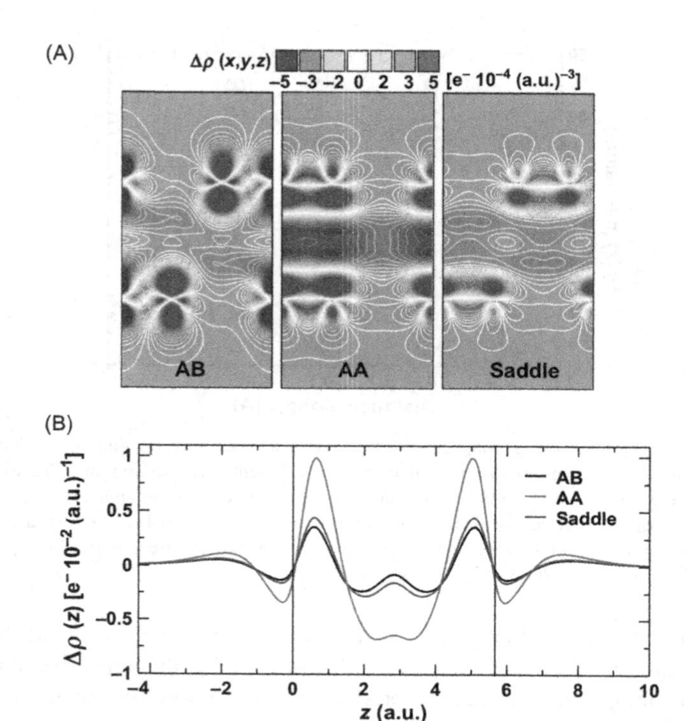

FIGURE 2.3 Electronic charge displacement $\Delta\rho$ occurring when a bilayer is formed from isolated graphene layers. A compressed bilayer, $z = 3.0$ Å, is considered. The same scale is used for different stacking configurations to highlight the differences (A). The profile of $\Delta\rho$ along the z direction is obtained by two-dimensional integration (B).

accumulated at the middle of an interface is directly proportional to the interfacial adhesion. Moreover the differences in the interfacial charge at different lateral positions is directly proportional to the PES corrugation [60]. With this information at hand, we can fully understand the microscopic origin of the PES features and their evolution as a function of load.

The forces acting on the system during sliding can be obtained as derivatives of the PES profile along the sliding path. We define the shear strength, τ, along that path as the maximum resistance force per unit area [36]. In Table 2.1, we report τ calculated for bilayer graphene along the MEP and the y direction (see Fig. 2.1) and compare it with the values obtained for transition metal dicalcogenides, calculated within the same computational scheme [38]. The geometrical parameters, interlayer adhesion, and maximum value of the potential corrugation are also reported. The comparison highlights the superior tribological properties of graphene over MoS_2, which is a well-known solid lubricant, and all the other layered materials here considered. It is interesting to notice that this theoretical prediction has been recently verified by atomic force microscopy (AFM) measurements [61].

TABLE 2.1 Calculated values for the lattice constant (a), interfacial separation (d_{sep}), interlayer adhesion (γ), maximum sliding corrugation ($\Delta\gamma_{max}$), ideal shear strengths along the minimum energy path (τ_{MEP}) and along the y (τ_y) for each considered bilayer system.

	a (Å)	d_{sep} (Å)	γ (J m^{-2})	$\Delta\gamma_{max}$ (J m^{-2})	τ_{MEP} (GPa)	τ_y (GPa)
Graphene	2.46	3.31	−0.26	0.06	0.21	0.63
MoS_2	3.19	3.10	−0.27	0.10	0.50	0.65
$MoSe_2$	3.33	3.19	−0.37	0.14	0.71	1.00
$MoTe_2$	3.54	3.44	−0.50	0.18	0.98	1.04

2.3 Why graphene coating makes iron slippery? An answer from first principles

Recent experimental findings have revealed that graphene possess a great potential as solid lubricant not only for nanoscale, but also for macroscale applications. It has, in fact, been shown that graphene flakes, delivered at the sliding interface thorough an ethanol solution, are able to decrease the wear of steel surfaces by four orders of magnitudes and their coefficient of friction by a factor of six [18]. The lifetime of the slippery regime decreased with the applied load [19] as also observed in other works [13–15].

A possible explanation for these results has been provided by Klements et al. [39], who combined classical MD simulations and AFM experiments on graphene-covered Pt(111), and proposed that the ability of graphene to reduce the COF resides on its ability to increase the load-carrying capacity of the surface. According to the authors, this reduces the penetration depth of the tip and consequently the friction and wear. Once the graphene has ruptured, the tribological behavior of the bare metal substrate is recovered. Therefore the authors conclude that graphene can be an excellent coating for low friction and wear as far as it is not damaged. However this explanation is not fully consistent with the Raman spectra recorded inside the wear tracks after the tribological experiment, which shows that defective graphene can provide low friction, while the high-friction regime is associated to the situation where graphene has been actually removed from the wear track, not just ruptured [18]. Therefore we consider another hypothesis to explain the lubricating properties of graphene, which is related to its chemical action, rather than its mechanical action. In particular, we show by means of first principles calculations that graphene can bind strongly to native iron surfaces highly reducing their surface energy. Thanks to a passivating effect, the metal surfaces coated by graphene become almost inert and present very low adhesion and shear strength when mated in a sliding contact.

We consider iron, which is different from steel, but it may constitute a suitable model for the native metal surfaces exposed during scratching [62]. The binding energy of a graphene layer on the (110) surface of iron, which is the most stable surface for this material, is $E_{ad} = 0.89\ \mathrm{J\,m^{-2}}$ according to our calculation within the DFT scheme including the vdW interaction [6]. Such value indicates a relatively strong binding due to the hybridization of the p_z orbitals of graphene with the partially occupied d states of the metal, in agreement with X-ray photoelectron spectroscopy (XPS) and near-edge X-ray absorption fine structure (NEXAFS) spectroscopy observations that show that the electronic structure of graphene adsorbed on iron is significantly disturbed by the substrate [63]. The presence of unsaturated graphene edges highly increase the adhesion of graphene on the metal substrate, anchoring pristine flakes and patches on the substrate. The latters can highly reduce the adhesion between the substrate and a countersurface, as can be seen from Fig. 2.4, where the adhesion of iron-on-iron is reported as a function of graphene coverage. We can see that graphene intercalation has a dramatic effect on the interfacial adhesion, which consistently decreases even at low coverage, the reduction then reaches the 88% when both the surfaces in contact are fully coated by graphene. This last situation is representative of a condition often occurring during pin-on-disc experiments, where part of the powder lubricant is transferred from the substrate to the sliding pin.

We can see from Fig. 2.5 that the interaction of two surfaces fully covered by graphene (black curve) closely resembles the interlayer interaction in graphite (in red). The graphene coating changes, thus, the nature of the

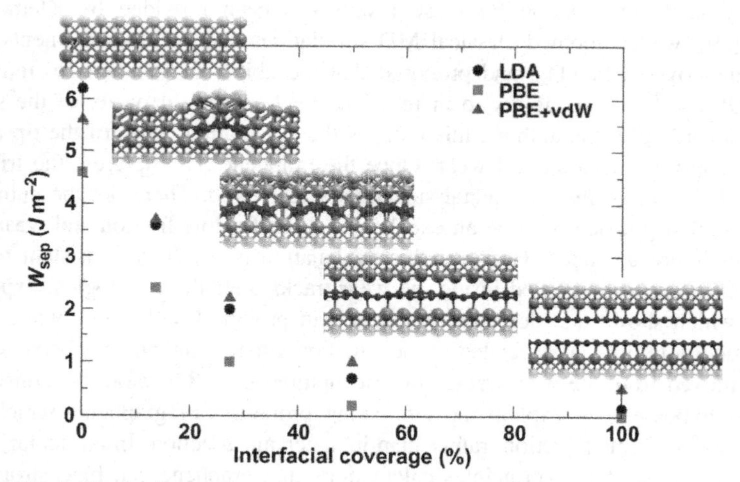

FIGURE 2.4 Work of separation of iron interfaces as a function of graphene coverage calculated with different approximations for the exchange correlation functional. Partial coverage (<50%) is modeled by intercalated graphene ribbons, while the 50% (100%) coverage corresponds to a situation where a graphene layer (bilayer) is present at the interface.

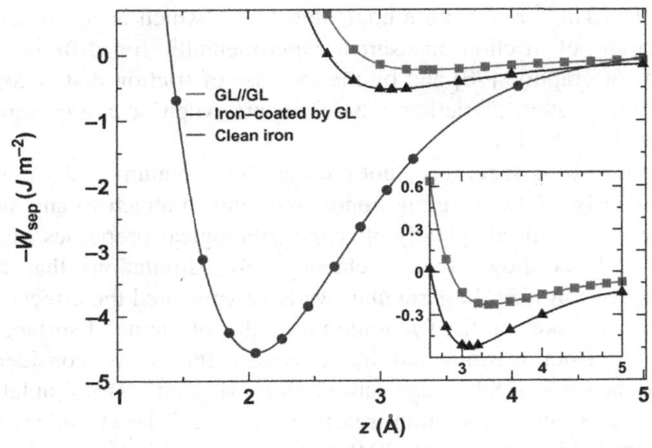

FIGURE 2.5 Interaction energy as a function of separation for clean (*blue line*) and graphene-covered (black line) iron surfaces. The binding energy curve for an isolated graphene bilayer is also displayed for comparison (*red line*).

surface-surface interaction from chemical to physical, as can be seen by comparing the black curve with the blue one, representing the iron-iron interaction. Graphene is able to passivate the iron surfaces very effectively, screening almost completely the metal-metal interaction at the interface. This may suggest that sliding microasperities fully covered by graphene may show similar friction anysotropy and "superlubricity" as observed in graphite [64,65].

The adhesion reduction is accompanied by a reduction of adhesive friction, as demonstrated by the values of the shear strength that we calculated following the same procedure as described in the previous section. We obtain $\tau_{MEP} = 9.20$ GPa for the clean iron interface, in agreement with previous DFT results [66,67], and $\tau_{MEP} = 0.17$ GPa for the iron interface covered by graphene, indicating that the intrinsic resistance to sliding of iron dramatically decreases in the presence of graphene.

2.4 Reactive defects destroy graphene lubricity, but humidity can recover it

When graphene is used in macroscale applications or subject to severe mechanical stresses, it can tear with the consequent exposure of edges. Graphene/graphite flakes with nonpassivated edges are anything but lubricious because graphene edges are extremely reactive [68]. To give an idea of that, we calculate from first principles the work of adhesion of self-mating (1120) surfaces, which expose armchair edges: it corresponds to 10.0 J m^{-2}, a value that almost doubles the adhesion of the strongest elements in the periodic table, such as chromium and tungsten [69]. The adhesion energy of self-mating (0001) surfaces, corresponding to the basal planes of graphene, is instead 0.24 $J\,m^{-2}$ revealing a huge anisotropy, which is confirmed by the large variation of friction measured experimentally for different surface orientations of graphite [70] and by the increase of friction that is commonly recorded during atomic friction measurements when scanning across graphene edges [51–53,71].

When passivating species are not present, as in vacuum or dry conditions, the tremendously sticky dangling bonds will tend to attach to any surrounding surface. This will completely alter the tribological properties of the graphitic material, as shown by the classical MD simulations that we have recently carried out [55]. In particular, we have simulated the effects of interfacial graphene flakes on the frictional properties of diamond surfaces in relative motion. Both pristine and hydrogenated flakes are considered, the latters with the 49% or 89% edge saturation by H atoms. The simulations are based on the second-generation reactive empirical bond order potential [72,73], composed of more than 20,000 atoms, and their time scale is of the order of half a nanosecond. In Fig. 2.6, the system atomic structures in the final relaxation stage are reported and the observed time behavior for the relative velocity of the two mated surfaces lubricated by pristine and hydrogenated flakes are compared. It appears evident that even relatively small pristine flakes are able to attach simultaneously to both the surfaces producing a high static friction, while when passivated flakes are present the two surfaces can slide with relatively low friction.

It is well known that humidity plays a key role in enabling the lubricating capability of graphitic materials. To understand the molecular mechanisms underlying this effect, we perform QM/MM simulations of a tribological interface containing graphene edges and water molecules. In the QM/MM

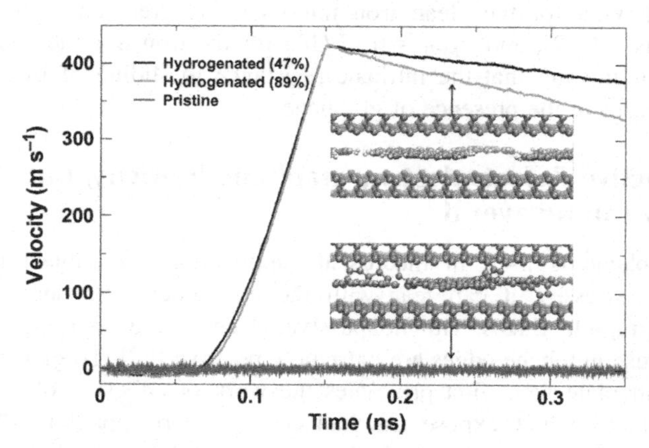

FIGURE 2.6 Snapshots of the final atomic configuration for the diamond interface lubricated by pristine (lower image) and hydrogenated (upper image) graphene flakes. The time evolution of the relative diamond slab velocity is also displayed.

approach, the chemically active region is treated by quantum mechanics, while the rest of the system is described by classical MD. This method thus combines the computational advantages of classical MD with the accuracy of quantum mechanics. Since its first introduction [74], QM/MM has been successfully applied to several systems, especially the bimolecular ones, but are not aware of any previous application to tribochemistry. The QM/MM simulations we performed [55] turn out to be of paramount importance to understand the effects of humidity in graphitic materials, as they highlighted numerous collective processes of confined water molecules that lead to efficient edge passivation. Some examples are: the edge passivation mediated by Grotthus-like proton diffusion (Fig. 2.7A and B), the concerted H diffusion along the dimer that favors the formation of an epoxide group (Fig. 2.7C and D), the hydrogen bond networks established among water molecules and passivated edges (Fig. 2.7E). Collective mechanisms involving critical numbers of water molecules are key in determining the evolution of multimolecular systems in confined conditions, as those present at tribological interfaces.

Graphene edges or vacancies are not the only reactive sites that can be present in graphitic materials under tribological conditions. An interesting outcome of a fully ab initio MD simulation that we performed considering single graphene layers instead of bilayers, reveals, in fact, that the thinner graphene films can undergo large out-of-plane deformations under at the effect of mechanical stresses, and that the curved graphene regions are very reactive. The C atoms at the convexly curved graphene regions change, in fact, their hybridization from sp^2 to sp^3 and expose dangling bonds that can interact with edges, forming strong C–C bonds, or catalyze water splitting. This observation is in agreement with the site-selective adsorption of atomic hydrogen on convexly curved regions of monolayer graphene grown on SiC (0001), observed by scanning tunneling microscopy [75].

Out-of-plane deformations, or puckers, have been previously introduced to explain the friction enhancement observed in AFM experiments when passing from thicker to thinner films of 2D materials [8]. However the role

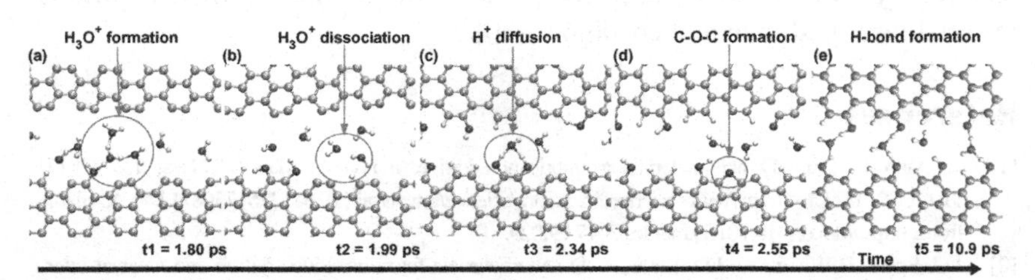

FIGURE 2.7 Snapshots, subsequent in time from left to right, showing cooperative mechanisms of water dissociation along graphene edges. (a, b) The edge passivation is mediated by the formation and dissociation of a H_3O^+ cation. (c, d) The concerted H diffusion along the dimer favors the formation of an epoxide group. (e) Free standing water molecules allow the interaction of fully passivated edges by means of hydrogen bonds.

of puckers in enhancing friction was attributed to an increase of contact area [8,76,77] or an increase of the contact commensurability [9]. We add a further insight, which is based on the observation that curved graphene is much more adhesive than flat graphene. Out-of-plane deformations are more easily formed in single-layer graphene than on multilayer graphene due to the lower stiffness [78], and the C dangling bonds appearing at the curved regions can be responsible for the friction enhancement observed in the experiments.

The earlier discussion highlights the importance of having passivating molecules all around the graphitic media to reduce the reactivity of wrinkles and edges that can be formed during the tribological process. However it does not provide any specific clue for interpreting the superior effect of humidity on graphite lubricity with respect to other molecular species present in air. Our tribological simulations, where H_2O molecules have replaced the O_2 ones address this issue [55]. The results indicate that oxygen is as efficient as water in passivating the carbon dangling bonds, but it lacks the peculiar property of water molecules to form hydrogen bonds, which can be the key process to interpret the experimental observations. Indeed, the physysorption energies that we calculated with DFT indicate high attraction of water molecules by chemisorbed water fragments. This suggests that the (curved) basal plane, and edges of graphene that have been passivated by water are highly hydrophilic. On the contrary, oxygenated graphene is weakly interacting with surrounding O_2 molecules. Therefore in humid conditions one can imagine the existence of a reservoir of passivating molecules surrounding the graphitic media able to continuously passivate the reactive sites that are formed during rubbing. Such reservoir of passivating molecules is not expected to be present in dry air. Moreover, the exfoliation of graphite oxide is more easy than that of graphite [79]. In the specific case of graphene, a hydrated media could favor the flake mobility, making them to more easily reach reactive regions, such as native metallic surfaces, where to adsorb and reduce the interfacial shear strength [20]. It has also been proposed that a boundary layer of water formed on hydroxylated carbon surfaces can produce low friction coefficients of this material measured in humid conditions [80].

References

[1] H. Xiao, S. Liu, 2D nanomaterials as lubricant additive: a review, Mater. & Des. 135 (2017) 319–332. URL: <http://www.sciencedirect.com/science/article/pii/S0264127517308717>. Available from: https://doi.org/10.1016/j.matdes.2017.09.029.

[2] J.C. Spear, B.W. Ewers, J.D. Batteas, 2D-nanomaterials for controlling friction and wear at interfaces, Nano Today 10 (2015) 301–314. URL: <http://www.sciencedirect.com/science/article/pii/S174801321500050X>. Available from: https://doi.org/10.1016/j.nantod.2015.04.003.

[3] D. Berman, A. Erdemir, A.V. Sumant, Approaches for achieving superlubricity in two-dimensional materials, ACS Nano 12 (2018) 2122–2137. URL: <https://pubs.acs.org/doi/10.1021/acsnano.7b09046. Available from: https://doi.org/10.1021/acsnano.7b09046

[4] K.S. Novoselov, V.I. Falko, L. Colombo, P.R. Gellert, M.G. Schwab, K. Kim, A roadmap for graphene, Nature 490 (2012) 192–200. URL: <https://doi.org/10.1038/nature11458>.

[5] C. Lee, X. Wei, Q. Li, R. Carpick, J.W. Kysar, J. Hone, Elastic and frictional properties of graphene. Phys. Status Solidi B. 246 (2009) 2562–2567. URL: <https://onlinelibrary.wiley.com/doi/abs/10.1002/pssb.200982329>. Available from: https://doi.org/10.1002/pssb.200982329.

[6] P. Restuccia, M. Righi, Tribochemistry of graphene on iron and its possible role in lubrication of steel, Carbon 106 (2016) 118–124. URL: <http://www.sciencedirect.com/science/article/pii/S0008622316303797>. Available from: https://doi.org/10.1016/j.carbon.2016.05.025.

[7] T. Filleter, J.L. McChesney, A. Bostwick, E. Rotenberg, K.V. Emtsev, T. Seyller, et al., Friction and dissipation in epitaxial graphene films, Phys. Rev. Lett. 102 (2009) 086102. URL: <https://link.aps.org/doi/10.1103/PhysRevLett.102.086102>. Available from: https://doi.org/10.1103/PhysRevLett.102.086102.

[8] C. Lee, Q. Li, W. Kalb, X.-Z. Liu, H. Berger, R.W. Carpick, et al., Frictional characteristics of atomically thin sheets, Science 328 (2010) 76–80. URL: <http://science.sciencemag.org/content/328/5974/76>. Available from: https://doi.org/10.1126/science.1184167.

[9] S. Li, Q. Li, R.W. Carpick, P. Gumbsch, X.Z. Liu, X. Ding, et al., The evolving quality of frictional contact with graphene, Nature 539 (2016) 541–545. URL: <https://doi.org/10.1038/nature20135>.

[10] Q. Zheng, B. Jiang, S. Liu, Y. Weng, L. Lu, Q. Xue, et al., Self-retracting motion of graphite microflakes, Phys. Rev. Lett 100 (2008) 067205. URL: <https://link.aps.org/doi/10.1103/PhysRevLett.100.067205>. Available from: https://doi.org/10.1103/PhysRevLett.100.067205.

[11] X. Feng, S. Kwon, J.Y. Park, M. Salmeron, Superlubric sliding of graphene nanoflakes on graphene, ACS Nano 7 (2013) 1718–1724. URL: <https://doi.Org/10.1021/nn305722d>. Available from: https://doi.org/10.1021/nn305722d.

[12] S.-W. Liu, H.-P. Wang, Q. Xu, T.-B. Ma, G. Yu, C. Zhang, et al., Robust microscale superlubricity under high contact pressure enabled by graphene-coated microsphere, Nat. Commun. 8 (2017) 14029. URL: <https://doi.org/10.1038/ncomms14029>.

[13] Y.J. Shin, R. Stromberg, R. Nay, H. Huang, A.T. Wee, H. Yang, et al., Frictional characteristics of exfoliated and epitaxial graphene, Carbon 49 (2011) 4070–4073. URL: <http://www.sciencedirect.com/science/article/pii/S000862231100409X>. Available from: https://doi.org/10.1016/j.carbon.2011.05.046.

[14] D. Marchetto, C. Held, F. Hausen, F. Wahlisch, M. Dienwiebel, R. Bennewitz, Friction and wear on single-layer epitaxial graphene in multi-asperity contacts, Tribol. Lett. 48 (2012) 77–82. URL: <https://doi.org/10.1007/s11249-012-9945-4>.

[15] K.-S. Kim, H.-J. Lee, C. Lee, S.-K. Lee, H. Jang, J.-H. Ahn, et al., Chemical vapor deposition-grown graphene: the thinnest solid lubricant, ACS Nano 5 (2011) 5107–5114. URL: <https://doi.org/10.1021/nn2011865>.

[16] M.-S. Won, O.V. Penkov, D.-E. Kim, Durability and degradation mechanism of graphene coatings deposited on Cu substrates under dry contact sliding, Carbon 54 (2013) 472–481. URL: <http://www.sciencedirect.com/science/article/pii/S0008622312009712>. Available from: https://doi.org/10.1016/j.carbon.2012.12.007.

[17] D. Berman, S.A. Deshmukh, S.K.R.S. Sankaranarayanan, A. Erdemir, A.V. Sumant, Extraordinary macroscale wear resistance of one atom thick graphene layer, Adv. Funct. Mater. 24 (2014) 6640–6646. URL: <https://onlinelibrary.wiley.com/doi/abs/10.1002/adfm.201401755>.

[18] D. Berman, A. Erdemir, A.V. Sumant, Few layer graphene to reduce wear and friction on sliding steel surfaces, Carbon 54 (2013) 454–459. URL: <http://www.sciencedirect.com/science/article/pii/S0008622312009529>. Available from: https://doi.org/10.1016/j.carbon.2012.11.061.

[19] D. Berman, A. Erdemir, A.V. Sumant, Reduced wear and friction enabled by graphene layers on sliding steel surfaces in dry nitrogen, Carbon 59 (2013) 167–175. URL: <http://www.sciencedirect.com/science/article/pii/S0008622313002108>. Available from: https://doi.org/10.1016/j.carbon.2013.03.006.

[20] D. Marchetto, P. Restuccia, A. Ballestrazzi, M. Righi, A. Rota, S. Valeri, Surface passivation by graphene in the lubrication of iron: a comparison with bronze, Carbon 116 (2017) 375–380. URL: <http://www.sciencedirect.com/science/article/pii/S000862231730129X>. Available from: https://doi.org/10.1016/j.carbon.2017.02.011.

[21] J. Ou, J. Wang, S. Liu, B. Mu, J. Ren, H. Wang, et al., Tribology study of reduced graphene oxide sheets on silicon substrate synthesized via covalent assembly, Langmuir 26 (2010) 15830–15836. URL: <https://doi.org/10.1021/la102862d>. Available from: https://doi.org/10.1021/la102862d.

[22] D. Berman, S.A. Deshmukh, S.K.R.S. Sankaranarayanan, A. Erdemir, A.V. Sumant, Macroscale superlubricity enabled by graphene nanoscroll formation, Science 348 (2015) 1118–1122. URL: <http://science.sciencemag.org/content/348/6239/1118>. Available from: https://doi.org/10.1126/science.1262024.

[23] P. Wu, X. Li, C. Zhang, X. Chen, S. Lin, H. Sun, et al., Self-assembled graphene film as low friction solid lubricant in macroscale contact, ACS Appl. Mater. Inter. 9 (2017) 21554–21562. URL: <https://doi.org/10.1021/acsami.7b04599>. Available from: https://doi.org/10.1021/acsami.7b04599.

[24] Y. Huang, Q. Yao, Y. Qi, Y. Cheng, H. Wang, Q. Li, et al., Wear evolution of monolayer graphene at the macroscale, Carbon 115 (2017) 600–607. URL: <http://www.sciencedirect.com/science/article/pii/S0008622317300660>. Available from: https://doi.org/10.1016/j.carbon.2017.01.056.

[25] Z. Li, W. Yang, Y. Wu, S. Wu, Z. Cai, Role of humidity in reducing the friction of graphene layers on textured surfaces, Appl. Surf. Sci. 403 (2017) 362–370. URL: <http://www.sciencedirect.com/science/article/pii/S0169433217302490>. Available from: https://doi.org/10.1016/j.apsusc.2017.01.226.

[26] S. Bhowmick, A. Banerji, A.T. Alpas, Role of humidity in reducing sliding friction of multilayered graphene, Carbon 87 (2015) 374–384. URL: <http://www.sciencedirect.com/science/article/pii/S0008622315000780>. Available from: https://doi.org/10.1016/j.carbon.2015.01.053.

[27] Z. Yang, S. Bhowmick, F.G. Sen, A. Banerji, A.T. Alpas, Roles of sliding-induced defects and dissociated water molecules on low friction of graphene, Sci. Rep. 8 (2018) 121. URL: <https://www.nature.com/articles/s41598-017-17971-1>. Available from: https://doi.org/10.1038/s41598-017-17971-1.

[28] W. Bao, L. Jing, J. Velasco, Y. Lee, G. Liu, D. Tran, et al., Stacking-dependent band gap and quantum transport in trilayer graphene, Nat. Phys. 7 (2011) 948.

[29] P. Xu, Y. Yang, D. Qi, S.D. Barber, M.L. Ackerman, J.K. Schoelz, et al., A pathway between bernal and rhombohedral stacked graphene layers with scanning tunneling microscopy, Appl. Phys. Lett. 100 (2012) 201601.

[30] P.H. Tan, W.P. Han, W.J. Zhao, Z.H. Wu, K. Chang, H. Wang, et al., The shear mode of multilayer graphene, Nat. Mater. 11 (2012) 294–300.

[31] X. Liu, T.H. Metcalf, J.T. Robinson, B.H. Houston, F. Scarpa, Shear modulus of monolayer graphene prepared by chemical vapor deposition, Nano Lett. 12 (2012) 1013–1017.

[32] A.P.M. Barboza, H. Chacham, C.K. Oliveira, T.F.D. Fernandes, E.H.M. Ferreira, B.S. Archanjo, et al., Dynamic negative compressibility of few-layer graphene, h-BN, and MoS_2, Nano Lett. 12 (2012) 2313–2317.

[33] Z. Liu, J. Yang, F. Grey, J.Z. Liu, Y. Liu, Y. Wang, et al., Observation of microscale superlubricity in graphite, Phys. Rev. Lett. 108 (2012) 205503.

[34] Z. Liu, J.Z. Liu, Y. Cheng, Z. Li, L. Wang, Q. Zheng, Interlayer binding energy of graphite: a mesoscopic determination from deformation, Phys. Rev. B 85 (2012) 205418.

[35] B. Li, J. Yin, X. Liu, H. Wu, J. Li, X. Li, et al., Probing van der Waals interactions at two-dimensional heterointerfaces, Nat. Nanotechnol. 14 (2019) 567.

[36] G. Zilibotti, M.C. Righi, Ab initio calculation of the adhesion and ideal shear strength of planar diamond interfaces with different atomic structure and hydrogen coverage, Langmuir 27 (2011) 6862–6867.

[37] M. Reguzzoni, A. Fasolino, M. Molinari, M.C. Righi, Phys. Rev. B Condens. Matter 86 (2012) 245434-1–245434-7. URL: <https://journals.aps.org/prb/abstract/10.1103/PhysRevB.86.245434>.

[38] G. Levita, E. Molinari, T. Polcar, M.C. Righi, First-principles comparative study on the interlayer adhesion and shear strength of transition-metal dichalcogenides and graphene, Phys. Rev. B Condens. Matter. 92 (2015) 1–8. URL: <http://harvest.aps.org/bagit/articles/10.1103/PhysRevB.92.085434/apsxml>. Available from: https://doi.org/10.1103/PhysRevB.92.085434.

[39] A. Klemenz, L. Pastewka, S.G. Balakrishna, A. Caron, R. Bennewitz, M. Moseler, Atomic scale mechanisms of friction reduction and wear protection by graphene, Nano Lett. 14 (2014) 7145–7152. URL: <https://doi.org/10.1021/nl5037403>. Available from: https://doi.org/10.1021/nl5037403.

[40] E.F. Bracken, Humidity control prevents AC brush disintegration, Electr. World 102 (1933) 410.

[41] J.R. Felts, A.J. Oyer, S.C. Hernandez, K.E. Whitener Jr, J.T. Robinson, S.G. Walton, et al., Direct mechanochemical cleavage of functional groups from graphene, Nat. Commun. 6 (2015) 6467. URL: <https://www.nature.com/articles/ncomms7467>. Available from: https://doi.org/10.1038/ncomms7467.

[42] Z. Chen, X. He, C. Xiao, S.H. Kim, Effect of humidity on friction and wear—a critical review, Lubricants 6 (2018) 74. URL: <http://www.mdpi.com/2075-4442/6/3/74>. Available from: https://doi.org/10.3390/lubricants6030074.

[43] R.H. Savage, Graphite lubrication, J. Appl. Phys. 19 (1948) 1–10. URL: <https://doi.org/10.1063/1.1697867>.

[44] G. Rowe, Some observations on the frictional behaviour of boron nitride and of graphite, Wear 3 (1960) 274–285. URL: <http://www.sciencedirect.com/science/article/pii/0043164860902921>. Available from: https://doi.org/10.1016/0043-1648(60)90292-1.

[45] P. Bryant, P. Gutshall, L. Taylor, A study of mechanisms of graphite friction and wear, Wear 7 (1964) 118–126. URL: <http://www.sciencedirect.com/science/article/pii/0043164864900833?via%3Dihub>. Available from: https://doi.org/10.1016/0043-1648(64)90083-3.

[46] J.K. Lancaster, J.R. Pritchard, The influence of environment and pressure on the transition to dusting wear of graphite, J. Phys. D: Appl. Phys. 14 (1981) 747. URL: <http://stacks.iop.org/0022-3727/14/i = 4/a = 027>.

[47] N. Kumar, S. Dash, A. Tyagi, B. Raj, Super low to high friction of turbostratic graphite under various atmospheric test conditions, Tribol. Int. 44 (2011) 1969–1978. URL: <http://www.sciencedirect.com/science/article/pii/S0301679X11002313>. Available from: https://doi.org/10.1016/j.triboint.2011.08.012.

[48] H. Zaidi, H. Nery, D. Paulmier, Stability of lubricating properties of graphite by orientation of the crystallites in the presence of water vapour, Appl. Surf. Sci. 70-71 (1993) 180–185. URL: <http://www.sciencedirect.com/science/article/pii/0169433293904239>. Available from: https://doi.org/10.1016/0169-4332(93)90423-9.

[49] J.-C. Rietsch, P. Brender, J. Dentzer, R. Gadiou, L. Vidal, C. Vix-Guterl, Evidence of water chemisorption during graphite friction under moist conditions, Carbon 55 (2013) 90–97. URL: <http://www.sciencedirect.com/science/article/pii/S0008622312009803>. Available from: https://doi.org/10.1016/j.carbon.2012.12.013.

[50] J. Xiao, L. Zhang, K. Zhou, J. Li, X. Xie, Z. Li, Anisotropic friction behaviour of highly oriented pyrolytic graphite, Carbon 65 (2013) 53–62. URL: <http://www.sciencedirect.com/science/article/pii/S0008622313007483>. Available from: https://doi.org/10.1016/j.carbon.2013.07.101.

[51] P. Egberts, Z. Ye, X.Z. Liu, Y. Dong, A. Martini, R.W. Carpick, Environmental dependence of atomic-scale friction at graphite surface steps, Phys. Rev. B 88 (2013) 035409. URL: <https://link.aps.org/doi/10.1103/PhysRevB.88.035409>. Available from: https://doi.org/10.1103/PhysRevB.88.035409.

[52] H. Lang, Y. Peng, X. Zeng, X. Cao, L. Liu, K. Zou, Effect of relative humidity on the frictional properties of graphene at atomic-scale steps, Carbon 137 (2018) 519–526. URL: <http://www.sciencedirect.com/science/article/pii/S0008622318305475>. Available from: https://doi.org/10.1016/j.carbon.2018.05.069.

[53] Y. Qi, J. Liu, Y. Dong, X.-Q. Feng, Q. Li, Impacts of environments on nanoscale wear behavior of graphene: edge passivation vs. substrate pinning, Carbon 139 (2018) 59–66. URL: <http://www.sciencedirect.com/science/article/pii/S0008622318305852>. Available from: https://doi.org/10.1016/j.carbon.2018.06.029.

[54] P. Restuccia, M. Ferrario, M.C. Righi, Quantum mechanics/molecular mechanics (QM/MM) applied to tribology: real-time monitoring of tribochemical reactions of water at graphene edges, Comput. Mater. Sci. 173 (2020) 109400.

[55] P. Restuccia, M. Ferrario, M.C. Righi, Monitoring water and oxygen splitting at graphene edges and folds: insights into the lubricity of graphitic materials, Carbon 156 (2019) 93–103. URL: <http://www.journals.elsevier.com/carbon/>. Available from: https://doi.org/10.1016/j.carbon.2019.09.040.

[56] S. Grimme, Semiempirical gga-type density functional constructed with a long-range dispersion correction, J. Comput. Chem. 27 (2006) 1787–1799.

[57] Y. Wang, Y. Ye, K. Wu, Simultaneous observation of the triangular and honeycomb structures on highly oriented pyrolytic graphite at room temperature: an STM study, Surf. Sci. 600 (2006) 729–734.

[58] M.C. Righi, M. Ferrario, Potential energy surface for rare gases adsorbed on Cu(111): parameterization of the gas/metal interaction potential, J. Phys. Condens. Matter 19 (2007) 305008–305018.

[59] M.C. Righi, M. Ferrario, Pressure induced friction collapse of rare gas boundary layers sliding over metal surfaces, Phys. Rev. Lett. 99 (2007) 176101.

[60] M. Wolloch, G. Levita, P. Restuccia, M.C. Righi, Interfacial charge density and its connection to adhesion and frictional forces, Phys. Rev. Lett. 121 (2018)N/A-N/A. URL: <http://harvest.aps.org/v2/bagit/articles/10.1103/PhysRevLett.121.026804/apsxml>. Available from: https://doi.org/10.1103/PhysRevLett.121.026804.

[61] M.R. Vazirisereshk, H. Ye, Z. Ye, A. Otero-de-la Roza, M.-Q. Zhao, Z. Gao, et al., Origin of nanoscale friction contrast between supported graphene, MoS2, and a graphene/MoS2 heterostructure, Nano Lett. 19 (2019) 5496–5505. URL: doi:10.1021/acs.nanolett.9b02035.arXiv: https://doi.org/10.1021/acs.nanolett.9b02035>. Available from: https://doi.org/10.1021/acs.nanolett.9b02035. pMID: 31267757.

[62] D. Philippon, M.-I. De Barros-Bouchet, T. Le Mogne, O. Lerasle, A. Bouffet, J.-M. Martin, Role of nascent metallic surfaces on the tribochemistry of phosphite lubricant additives,

Tribol. Int. 44 (6) (2011) 684–691. URL: http://linkinghub.elsevier.com/retrieve/pii/S0301679X0900351X. Available from: https://doi.org/10.1016/j.triboint.2009.12.014.

[63] N.A. Vinogradov, A.A. Zakharov, V. Kocevski, J. Rusz, K.A. Simonov, O. Eriksson, et al., Formation and structure of graphene waves on Fe(110), Phys. Rev. Lett. 109(2012) 026101. URL: <https://doi.org/10.1103/PhysRevLett.109.026101>.

[64] O. Hod, Interlayer commensurability and superlubricity in rigid layered materials, Phys. Rev. B 86 (2012) 075444. URL: <https://doi.org/10.1103/PhysRevB.86.075444>.

[65] M. Dienwiebel, G.S. Verhoeven, N. Pradeep, J.W.M. Frenken, J.A. Heimberg, H.W. Zandbergen, Superlubricity of graphite, Phys. Rev. Lett. 92 (2004) 126101. URL: <https://doi.org/10.1103/PhysRevLett.92.126101>.

[66] S. Ogata, J. Li, N. Hirosaki, Y. Shibutani, S. Yip, Ideal shear strain of metals and ceramics, Phys. Rev. B 70 (2004) 104104. URL: <https://doi.org/10.1103/PhysRevB.70.104104>.

[67] D.M. Clatterbuck, D.C. Chrzan, J.W. Morris Jr., The ideal strength of iron in tension and shear, Acta Mater. 51 (8) (2003) 2271–2283. Available from: http://dx.doi.org/10.1016/S1359-6454(03)00033-8. URL: <http://www.sciencedirect.com/science/article/pii/S1359645403000338>.

[68] G. Levita, P. Restuccia, M.C. Righi, Graphene and MoS_2 interacting with water: a comparison by ab initio calculations, Carbon 107 (2016) 878–884. URL: <http://www.sciencedirect.com/science/article/pii/S0008622316305292?via%3Dihub>. Available from: https://doi.org/10.1016/j.carbon.2016.06.072.

[69] M. Wolloch, G. Losi, M. Ferrario, M.C. Righi, High-throughput screening of the static friction and ideal cleavage strength of solid interfaces, Sci. Rep. 9 (2019) 17062–17072. URL: <www.nature.com/srep/index.html>. Available from: https://doi.org/10.1038/s41598-019-49907-2.

[70] J.W. Midgley, D.G. Teer, Surface orientation and friction of graphite, graphitic carbon and non-graphitic carbon, Nature 189 (1961) 735–736. URL: <https://www.nature.com/articles/189735a0>. Available from: https://doi.org/10.1038/189735a0.

[71] D.P. Hunley, T.J. Flynn, T. Dodson, A. Sundararajan, M.J. Boland, D.R. Strachan, Friction, adhesion, and elasticity of graphene edges, Phys. Rev. B 87 (2013) 035417. URL: <https://link.aps.org/doi/10.1103/PhysRevB.87.035417>. Available from: https://doi.org/10.1103/PhysRevB.87.035417.

[72] D.W. Brenner, O.A. Shenderova, J.A. Harrison, S.J. Stuart, B. Ni, S.B. Sinnott, A second-generation reactive empirical bond order (REBO) potential energy expression for hydrocarbons, J. Phys. Condens. Matter 14 (2002) 783.

[73] S.J. Stuart, A.B. Tutein, J.A. Harrison, A reactive potential for hydrocarbons with intermolecular interactions, J. Chem. Phys. 112 (2000) 6472–6486. URL: <http://scitation.aip.org/content/aip/journal/jcp/112/14/10.1063/1.481208>. Available from: https://doi.org/10.1063/1.481208.

[74] A. Warshel, M. Levitt, Theoretical studies of enzymic reactions: dielectric, electrostatic and steric stabilization of the carbonium ion in the reaction of lysozyme, J. Mol. Biol. 103 (1976) 227–249. URL: <http://www.sciencedirect.com/science/article/pii/0022283676903119>. Available from: https://doi.org/10.1016/0022-2836(76)90311-9.

[75] S. Goler, C. Coletti, V. Tozzini, V. Piazza, T. Mashoff, F. Beltram, et al., Influence of graphene curvature on hydrogen adsorption: toward hydrogen storage devices, J. Phys. Chem. C 117 (2013) 11506–11513. URL: <https://doi.org/10.1021/jp4017536>. Available from: https://doi.org/10.1021/jp4017536.

[76] Z. Ye, C. Tang, Y. Dong, A. Martini, Role of wrinkle height in friction variation with number of graphene layers, J. Appl. Phys. 112 (2012) 116102. URL: <https://doi.org/10.1063/1.4768909>.

[77] Q. Li, C. Lee, R.W. Carpick, J. Hone, Substrate effect on thickness-dependent friction on graphene. Phys. Status Solidi B. 247 (2010) 2909–2914. URL: <https://onlinelibrary.wiley.com/doi/abs/10.1002/pssb.201000555>. Available from: https://doi.org/10.1002/pssb.201000555.
[78] M. Reguzzoni, A. Fasolino, E. Molinari, M.C. Righi, Friction by shear deformations in multilayer graphene, J. Phys. Chem. C. 116 (2012) 21104–21108. URL: <https://pubs.acs.org/doi/10.1021/jp306929g>. Available from: https://doi.org/10.1021/jp306929g.
[79] M. Cai, D. Thorpe, D.H. Adamson, H.C. Schniepp, Methods of graphite exfoliation, J. Mater. Chem. 22 (2012) 24992–25002. URL: <https://doi.org/10.1039/C2JM34517J>. Available from: https://doi.org/10.1039/C2JM34517J.
[80] S. Kajita, M.C. Righi, A fundamental mechanism for carbon-film lubricity identified by means of ab initio molecular dynamics, Carbon 103 (2016) 193–199. URL: <http://www.sciencedirect.com/science/article/pii/S0008622316301713>. Available from: https://doi.org/10.1016/j.carbon.2016.02.078.

Chapter 3

Molecular simulation of superlow friction provided by molybdenum disulfide

Tasuku Onodera[1] and Jean Michel Martin[2]
[1]Functional Materials Research Department, Center for Technology Innovation – Materials, Research & Development Group, Hitachi, Ltd., Hitachi, Japan, [2]Université de Lyon, Ecole Centrale de Lyon, LTDS, CNRS, Ecully, France

3.1 Tribological aspect of molybdenum disulfide

Since the Industrial Revolution, solid-state lubricants, such as graphite, polytetrafluoroethylene (PTFE), hexagonal boron nitride (h-BN), and hexagonal molybdenum disulfide (h-MoS_2), have made it possible to construct high-performance and reliable industrial products. Especially, h-MoS_2 is recognized as an excellent solid lubricant that reduces friction extremely well without being helped by any liquid-state lubricants. According to a referential definition [1], the friction coefficient of h-MoS_2 is always at the "ultralow" ($0.01 < \mu < 0.1$) or "superlow" friction level ($\mu < 0.01$), although it depends on the friction conditions, namely, load, speed, working environment, and so on.

A prime example of tribological applications of h-MoS_2 is the automotive industry. In thermal engines, controlling friction by lubricants is one of the most important issues concerning fuel economy. Accordingly, a large number of research works have been devoted to understanding friction, wear, and lubrication [2–4]. High friction at a rubbing interface significantly increases fuel consumption while decreasing durability of mechanical systems. To overcome these problems, an organic chemical compound called molybdenum dithiocarbamate (MoDTC) has been used as an additive for engine oils because of its excellent performance in regard to friction reduction. It is widely accepted that the reduction of friction by MoDTC is mainly attributed to a solid lubricant film of h-MoS_2 formed by complex tribochemical reactions [5–7].

The crystalline structure of h-MoS_2 is shown in Fig. 3.1. A layer of molybdenum atoms is sandwiched between two sulfur layers, and the whole structure can be seen as lamellar. It is important to characterize such an h-MoS_2 coating

Superlubricity. DOI: https://doi.org/10.1016/B978-0-444-64313-1.00003-X

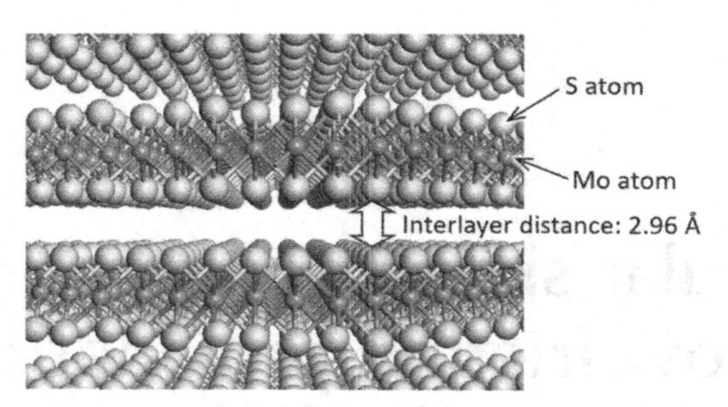

FIGURE 3.1 Crystal structure of h-MoS_2. Distance between two sheets is 2.96 Å.

(or one derived from MoDTC tribochemical reactions) in detail because it has been shown that its chemical composition and crystal structure are strongly correlated with its friction properties and wear resistance [8]. To understand the tribological mechanism of an h-MoS_2 coating, its chemical composition and crystal structure have been investigated in the following researches.

- Chemical composition: Raman spectroscopy, X-ray photoelectron spectroscopy, and Auger electron spectroscopy [9]
- Long-range crystal structure: X-ray diffraction and selected-area electron diffraction in a transmission electron microscope [10]
- Short-range crystal structure: high-resolution transmission electron microscope (HRTEM) and extended X-ray absorption fine structure [11]

These researches on surface chemistry and microstructure revealed that h-MoS_2 crystals exhibit many imperfections, including faults, kinks, and curvature, unlike the perfect crystal structure shown in Fig. 3.1. As an example, a sputtered h-MoS_2 coating includes significant amounts of oxygen (10–20 at.%) because of its sulfur defects and existence of environmental water and oxygen molecules. Under friction conditions with engine oils, an MoDTC-derived MoS_2 film also contains many impurities [7], such as carbon species and oxygen, which are formed through decomposition of molybdenum compounds or by dissolved chemical species in the oil. This oxygen incorporation affects the crystalline structure and interaction between MoS_2 sheets. It thus eventually affects the friction property of the film.

To investigate the mechanism of superlow friction by an h-MoS_2 film, it is necessary to understand the atomic and molecular interactions between the MoS_2 sheets. As well as the superlow-friction property of an h-MoS_2 film, it is also important to understand the formation mechanism of the h-MoS_2 film through decomposition of MoDTC molecules. However, obtaining knowledge at atomistic and molecular levels is often difficult by using conventional experimental and analytical techniques. Accordingly, approaches based on molecular simulation enabling electronic and atomistic analyses have been applied extensively in the field of nanotribology [12–20].

In present work, molecular-simulation methods [namely, molecular dynamics (MD) and quantum-chemistry simulations] were applied to investigate the dynamic behavior and superlow-friction mechanism of an h-MoS_2 film. Prior to those applications, a series of chemical reactions of the MoDTC molecule (namely, the source of an h-MoS_2 solid lubricant film applied inside automotive engines) was investigated.

3.2 Decomposition reaction of MoDTC molecule

As described earlier, an important industrial compound concerning lubrication by h-MoS_2 is MoDTC, which is dispersed in engine oils. It is chemically adsorbed on metallic substrates like a piston and cylinder, where it decomposes to form solid-state MoS_2 [5]. The mechanism of this decomposition reaction was theoretically investigated by a tribochemical simulator called "Hybrid-Colors," which is based on a hybrid quantum chemical/classical-MD method [21]. The simulator hybridizes two methods: one based on tight-binding quantum chemistry and one based on molecular mechanics. The quantum-chemistry-based method calculates the central part of the chemical reaction dynamics, while the molecular-mechanics-based method calculates the remaining part. The combination of these two methods enables simulation of the chemical reaction dynamics of a large complex system such as the friction process occurring at contacting interfaces. This tribochemical simulator is described in detail elsewhere [21]. The dynamic behavior of MoDTC molecules inside a heated oil phase was simulated first [22,23]. The model shown in Fig. 3.2A consists of 1 MoDTC molecule and 22 molecules of polyalphaolefine (PAO) as a model compound of a synthetic-base oil used for automotive-engine oils. As shown in Fig. 3.3, the structure of the MoDTC molecule changes distinctly. That is, several chemical bonds change as described in the figure caption. The simulation results shown in Fig. 3.4B suggest that the MoDTC molecule (whose original configuration is shown in Fig. 3.4A) thermally reacts with itself to form its linkage isomer (LI-MoDTC) in heated PAO phase. Decomposition through this unique chemical reaction (namely, forming a linkage isomer) is more energetically predominant than that of MoDTC in its original configuration [22].

It is expected that the LI-MoDTC molecule adsorbs onto a metallic substrate in engine parts such as a piston and cylinder (Fig. 3.4C). Further chemical reaction of the LI-MoDTC molecule on the metallic surface was investigated by using the Hybrid-Colors simulator [23,24]. The simulation model includes a LI-MoDTC molecular layer sandwiched between two nascent iron surfaces in a manner that mimics boundary friction due to metallic contact. Infriction process, extension of two Mo–O bonds can be observed in Fig. 3.4. It was also found that the LI-MoDTC molecule was negatively charged, meaning that electrons are donated from the iron substrate to the LI-MoDTC molecule. A nascent iron surface thus strongly

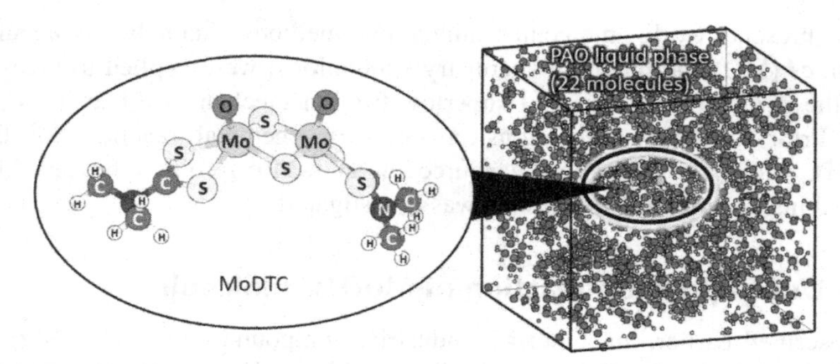

FIGURE 3.2 Calculation model including MoDTC and PAO molecules. Each MoDTC molecule includes four methyl groups. Size of the simulation cell is $30 \times 30 \times 30$ Å^3. Temperature was set to 425K to mimic the condition of heated engine oil, and external pressure and sliding velocity were not introduced. *Credit: T. Onodera, Y. Morita, A. Suzuki, R. Sahnoun, M. Koyama, H. Tsuboi et al., A theoretical investigation on the dynamic behavior of molybdenum dithiocarbamate molecule in the engine oil phase, Tribol. Online, 3 (2008) 80–85.*

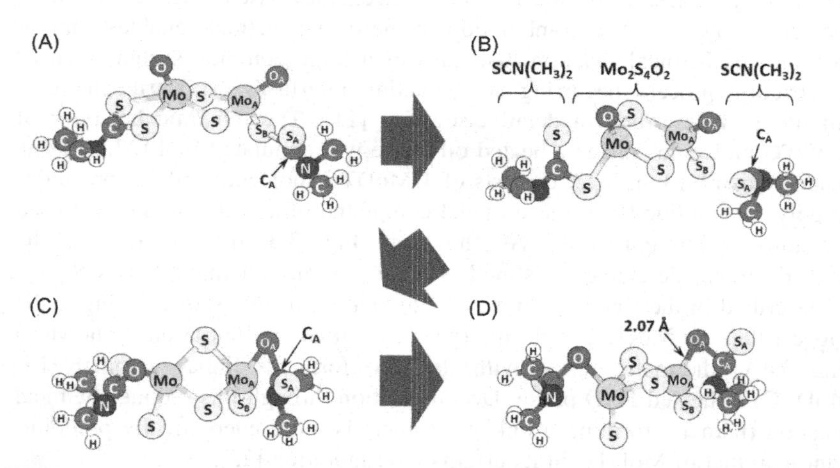

FIGURE 3.3 Molecular structures of MoDTC in heated PAO phase at simulation times of (A) 100, (B) 200, (C) 400, and (D) 500 fs. Four interatomic interactions in the molecule were focused on: Mo_A-S_A, C_A-S_B, C_A-O_A, and Mo_A-O_A. It can be observed that (A) an Mo_A-S_A bond is first dissociated and (B) a C_A-S_B bond is subsequently dissociated. Liberation of unsaturated carbon species, $SCN(CH_3)_2$, from the $Mo_2S_4O_2$ central core can be also observed in (B). In (C), the liberated $SCN(CH_3)_2$ parts are rapidly bound to the $Mo_2S_4O_2$ central core via formation of a C_A-O_A bond, which is a different type of connection from the initial structure with a C_A-S_B bond. Simultaneously, in (D), the Mo_A-O_A bond distance significantly increases; namely, it is elongated to 2.07 Å, which is 0.36 Å longer than that in the initial structure (1.71 Å). *Credit: T. Onodera, Y. Morita, A. Suzuki, R. Sahnoun, M. Koyama, H. Tsuboi et al., A theoretical investigation on the dynamic behavior of molybdenum dithiocarbamate molecule in the engine oil phase, Tribol. Online, 3 (2008) 80–85.*

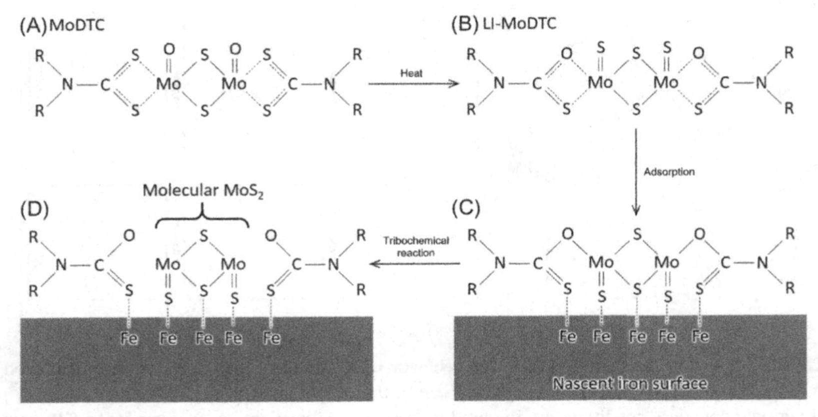

FIGURE 3.4 Schematic illustrations of molecular structure of (A) MoDTC and (B) LI-MoDTC, (C) adsorbed structure of LI-MoDTC on nascent iron surface, and (D) formation of molecular molybdenum disulfide by tribochemical effects. *R* denotes alkyl groups.

influences bond dissociation of the molecule. Further simulation without sliding was done, and the simulation results showed that the Mo–O bond was not clearly elongated. This simulation suggests that the dissociation reaction of LI-MoDTC is initiated by the electron donation by the nascent iron surface and accelerated by a dynamic friction. Subsequent chemical reaction probably takes place as schematically shown in Fig. 3.4D. Formation of molecular molybdenum disulfide (Mo_2S_4) and two thiocarbamic acid ($SOCN(CH_3)_2$) molecules are the main reaction products in the initial stage of formation of the MoS_2 film. In addition, an h-MoS_2 film derived from MoDTC usually contains oxygen impurities [6,25–27]. Our suggested aforementioned tribochemical reaction, however, generates molecular Mo_2S_4 without any oxygen atoms. It is hence inferred that oxygen content in an actual h-MoS_2 film comes from dissolved molecular oxygen in the hydrocarbon oil phase, that is, not from the film itself.

3.3 Formation process of crystalline h-MoS_2

The simulation results obtained by the hybrid quantum chemical/classical-MD simulation described in the previous section suggest that the MoDTC molecule first changes into its linkage isomer and then forms a molecular molybdenum disulfide. Indeed, as shown in Fig. 3.1, an h-MoS_2 film usually exists in the form of a lamellar crystal structure. Accordingly, the process by which a crystalline h-MoS_2 molecule is formed from a molecular state was investigated next [28]. This investigation used an original classical MD method that considers chemical reaction stochastically. Chemical reactions considered in the simulation were specified a priori, and certain interatomic distances in a reaction pair were monitored. The simulation model used,

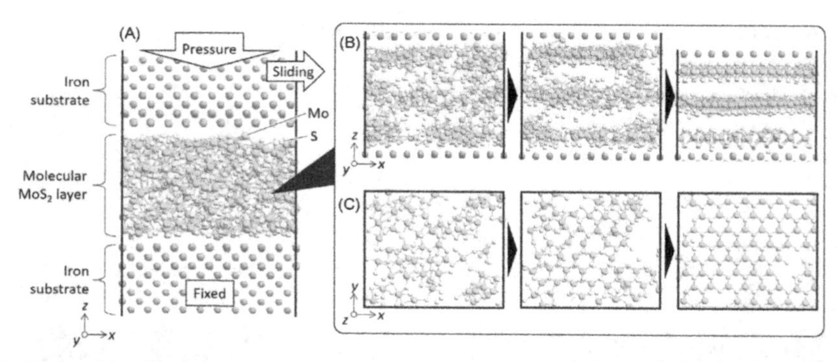

FIGURE 3.5 (A) Calculation model and behavior of molecular MoS_2 layer during simulation (B) in the *z-x* plane and (C) in the *x-y* plane. Size of the simulation cell is $27.5 \times 23.5 \times 100.0$ Å^3, and periodic boundary conditions are imposed in the *x* and *y* directions. A pressure of 0.5 GPa was vertically exerted on the upper iron substrate, and the bottom iron substrate was fixed. A sliding velocity of 100 m/s was applied horizontally. *Credit: Y. Morita, T. Onodera, A. Suzuki, R. Sahnoun, M. Koyama, H. Tsuboi, N. Hatakeyama, et al., Development of a new molecular dynamics method for tribochemical reaction and its application to formation dynamics of MoS_2 tribofilm, Appl. Surf. Sci., 254 (2008) 7618–7621.*

shown in Fig. 3.5, included 180 MoS_2 molecules between two iron substrates, forming an amorphous layer. Friction was generated by applying external pressure and sliding. The figure depicts the dynamic behavior of the MoS_2 molecular layer observed on the (B) *z-x* and (C) *x-y* planes. As a result of chemical reaction between MoS_2 molecules, the amorphous layer self-organized its structure and formed lamellar crystalline (Fig. 3.5B). An MoS_2 sheet was formed first on the iron surface, and an intermediate MoS_2 sheet was then generated. Three layers of MoS_2 were consequently formed between the two iron substrates. In Fig. 3.5C, a hexagonal crystal structure is gradually formed by the chemical reaction between MoS_2 molecules. The influence of the formation of the lamellar crystalline structure on the frictional property of MoS_2 was studied by analyzing the friction coefficient during the sliding simulation. In the initial stage of the simulation, the friction coefficient significantly decreased as the layered structure of MoS_2 formed. Finally, the friction coefficient stabilized at a value below 0.01. This simulation result suggests that the self-organization phenomenon caused by external shear is necessary to form the lamellar crystal structure of h-MoS_2 that achieves superlow friction.

3.4 Atomistic mechanism of superlow friction

To explain the superlow-friction property, it is necessary to understand the interlayer interaction between the MoS_2 crystalline sheets formed from the molecular state of MoS_2. One of the advantages of a molecular simulation is that atomistic interaction can be analyzed in detail. Here, the hybrid quantum-chemical/classical-MD simulation by Hybrid-Colors was adopted again [21] for the model used in the previous section that the lamellar h-MoS_2 layer has been finally formed between iron substrates. To investigate atomistic interactions between the h-MoS_2 sheets during the friction process, several contents of diatomic interaction energy between two sulfur atoms in different layers (one in the bottom sulfur layer of the top sheet, and the other in the top sulfur layer of the middle sheet) were analyzed. The sulfur–sulfur interaction energy between different MoS_2 sheets and their interatomic distance during the friction simulation is shown in Fig. 3.6. In the figure, individual interaction energies are shown separately as "atomic orbitals," "Coulombic interaction," and "exchange repulsive interaction" energies. Each energy component reaches its maximum value when the sulfur atom in the top MoS_2 sheet is just passing on the sulfur atom in the middle MoS_2 sheet. Furthermore, the Coulombic interaction energy is significantly larger than the atomic-orbital energy and exchange-repulsive-interaction energies. These results suggest that the dominant interaction between sulfur atoms in different MoS_2 sheets is Coulombic repulsion in nature.

Another interesting insight from this simulation is charge transfer on the interface between the top MoS_2 sheet and the iron substrate; that is, an electron is transferred from the substrate to the sheet. This charge transfer significantly affects the friction property of h-MoS_2. It is explained in detail as

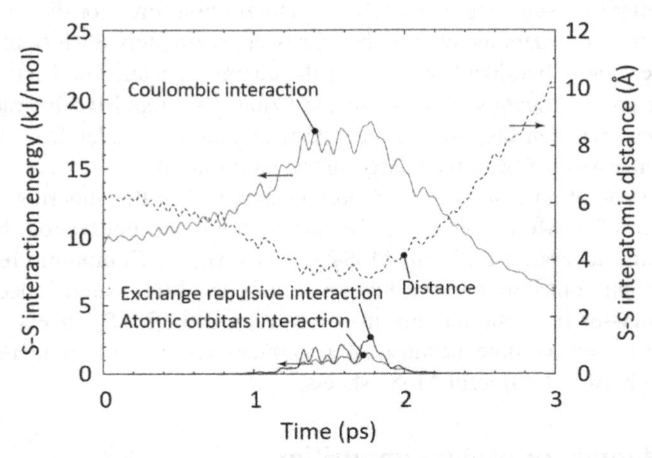

FIGURE 3.6 Interaction energy and interatomic distance between a sulfur atom in the top MoS_2 sheet and a sulfur atom in the middle MoS_2 sheet. In the simulation, a pressure of 0.5 GPa was vertically exerted on the upper iron substrate, and the bottom iron substrate was fixed. A sliding velocity of 100 m/s was horizontally exerted. The temperature was set at 300K. *Credit: T. Onodera, Y. Morita, A. Suzuki, M. Koyama, H. Tsuboi, N. Hatakeyama, et al., A computational chemistry study on friction of h-MoS_2. Part I. Mechanism of single sheet lubrication, J Phys. Chem. B, 113 (2009) 16526–16536.*

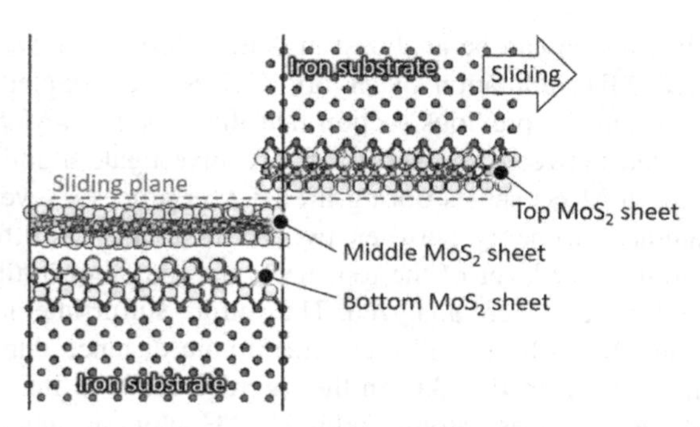

FIGURE 3.7 Result of a classical MD simulation on lamellar MoS_2 layers (considering charge transfer effect obtained by quantum-chemistry calculation). The figure ignores the periodic boundary for a clear understanding of friction behavior. The applied contact pressure was 0.5 GPa, and sliding speed was set as 100 m/s. Temperature was maintained at 300K. *Credit: T. Onodera, Y. Morita, A. Suzuki, M. Koyama, H. Tsuboi, N. Hatakeyama, et al., A computational chemistry study on friction of h-MoS_2. Part I. Mechanism of single sheet lubrication, J. Phys. Chem. B, 113 (2009) 16526–16536.*

follows. In the simulation, it was firstly observed that the top MoS_2 sheet tracks the sliding iron substrate, while the middle and bottom MoS_2 sheets almost keep their initial positions (see Fig. 3.7). Therefore sliding took place mainly between the top and the middle MoS_2 sheets. To quantitatively represent the effect of charge transfer on this sliding phenomenon, frictional force on the sliding iron substrate was analyzed. The friction force for the simulation that explicitly considers the charge transfer is approximately 88% smaller than that for the model that does not consider the charge transfer. The friction force was lower so significantly to increase the Coulombic repulsive interaction at the interface between the MoS_2 sheets; that is, charge transfer from the iron substrate enlarged the negative charge of the sulfur atoms.

These simulation results give important insights on the superlow-friction mechanism of h-MoS_2. That is, the most dominant interaction between sulfur-atomic layers on different MoS_2 sheets involves Coulombic repulsive energy, which contributes to the low friction of the MoS_2 single sheets. The electron transfer from the nascent iron substrate to the MoS_2 sheets makes a much lower friction state because of it induces large Coulombic repulsive interaction between different MoS_2 sheets.

3.5 Influence of oxygen impurities

Realistic MoS_2 solid lubricant films, including MoDTC-derived ones, applied in engine systems typically include many oxygen impurities [6,25–27]. The next point to be discussed is the influence of oxygen impurities on the friction property of crystalline h-MoS_2. The friction property of $MoS_{2-x}O_x$ was studied by a classical MD method. Although the model used is similar to that

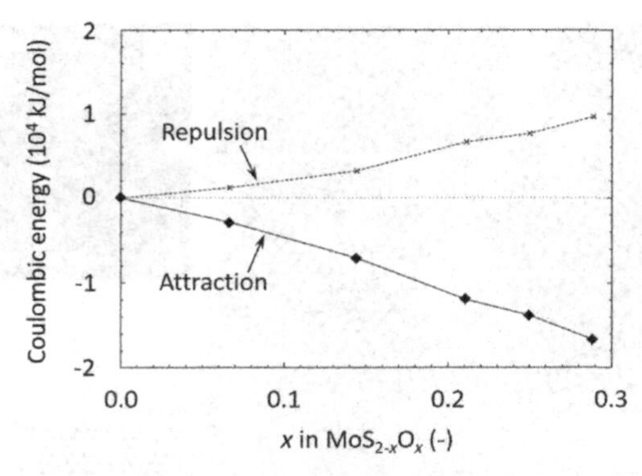

FIGURE 3.8 Coulombic attractive (Mo–S and Mo–O interactions) and repulsive (Mo–Mo, S–S, S–O, and O–O interactions) energies in $MoS_{2-x}O_x$ structures. The attractive and repulsive energies are relative to themselves for the structure at $x = 0$. *Credit: T. Onodera, Y. Morita, A. Suzuki, M. Koyama, H. Tsuboi, N. Hatakeyama, et al., A computational chemistry study on friction of h-MoS_2. Part I. Mechanism of single sheet lubrication, J. Phys. Chem. B, 113 (2009) 16526–16536.*

shown in Fig. 3.7, several sulfur atoms were randomly replaced by the same number of oxygen atoms in order to construct $MoS_{2-x}O_x$ structures. According to the result of a sliding simulation for these models, the sliding took place mainly between two $MoS_{2-x}O_x$ sheets. This finding agrees with the simulation results for standard MoS_2 sheets presented in the previous section. A different point is the frictional force. The models with oxygen impurities showed much larger frictional force than that of pure MoS_2. The contributions of Coulombic attractive and repulsive energies to the interlayer interaction are shown in Fig. 3.8. It is clear from the figure that the decrease rate of attractive energy is higher than that of repulsive energy. This result suggests that oxygen impurities in the h-MoS_2 crystal structure moderately affect its interlayer interactions. The simulation result simply indicates that the larger friction due to oxygen impurities is basically attributed to the decrement of Coulombic repulsive interaction between MoS_2 sheets.

3.6 Friction anisotropy

The results of HRTEM observation of the stoichiometric h-MoS_2 coatings showed that the superlubricity of an h-MoS_2 coating can be attributed to "structural anisotropy" between MoS_2 sheets [29]. Misfit angles between the nanocrystals composing MoS_2 wear debris were observed in the superlow friction regime. An example of structural anisotropy in the superlow friction regime in a part of a wear particle coming from the h-MoS_2 coatings is shown in Fig. 3.9. The material here is composed of only five MoS_2 single sheets in thickness, while a mosaic of nanometer-scale domains is depicted

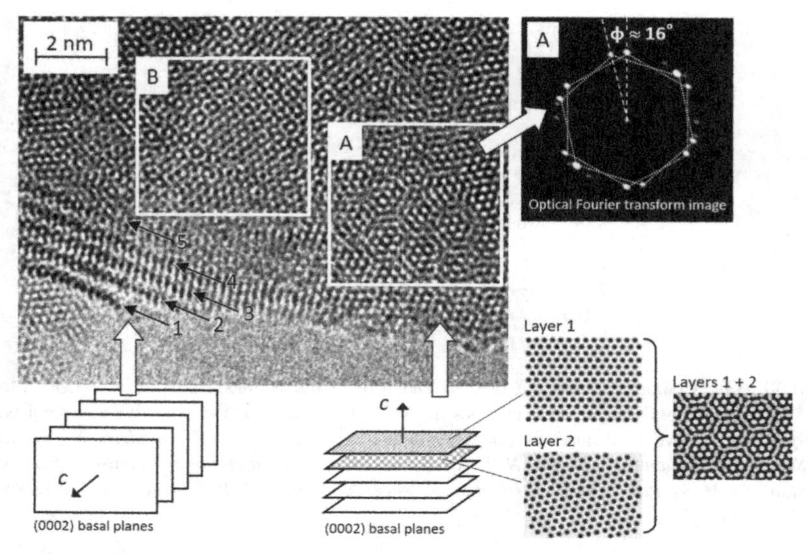

FIGURE 3.9 HRTEM image for wear debris of MoS_2 coatings. *Credit: T. Onodera, Y. Morita, R. Nagumo, R. Miura, A. Suzuki, H. Tsuboi, et al., A computational chemistry study on friction of h-MoS_2. Part II. Friction anisotropy, J. Phys. Chem. B, 114 (2010) 15832–15838.*

with different rotation angles between the sheets: (A) is about 15 degrees and (B) is 30 degrees corresponding to the incommensurate situation. This result indicates that a rotational disorder might play an important role in the superlow friction of h-MoS_2.

Here, a classical MD method was adopted to investigate the detailed atomistic mechanism on superlow friction through formation of a misfit angle between MoS_2 single sheets [30]. The model consisted of an MoS_2 flake on an MoS_2 plate with specific misfit angle φ. Motions of the sliding MoS_2 flake for $\varphi = 0°$ and $\varphi = 30°$ are, respectively, illustrated in Fig. 3.10. Different motions related to the formation of a misfit angle are shown. The MoS_2 flake moved in a zigzag way with respect to the sliding direction for $\varphi = 0°$, while it moved in a straight line for $\varphi = 30°$. The reason for this difference in motion is explained by the potential surface of MoS_2 sheet, as also shown in Fig. 3.10. Since potential barriers exist at the center of sulfur-atomic sites on the plate, the MoS_2 flake moves to avoid sulfur–sulfur overlapping. The relationship between misfit angle and calculated frictional force is shown in Fig. 3.11. The figure shows that the friction for $\varphi = 0°$ and 60° is a hundred times larger (10^{-2} pN/atom order) than that for the other angles (10^{-4} pN/atom order). A slight misfit angle between the MoS_2 sheets greatly reduces friction, and it is thought to be a source of superlubricity as well as the interlayer Coulombic repulsive interaction, as discussed in the previous section.

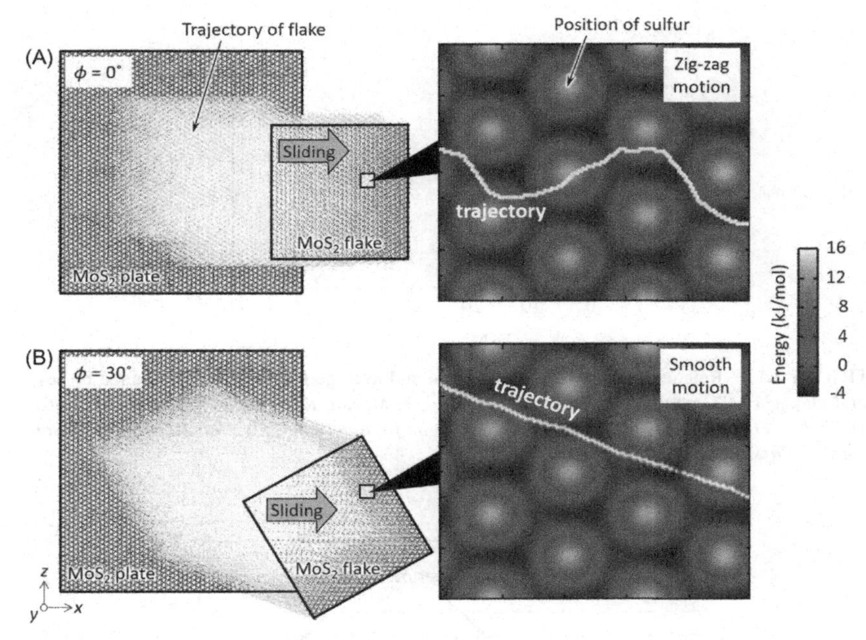

FIGURE 3.10 Trajectory of a MoS_2 flake sliding on a MoS_2 plate. Potential surface of the plate is also shown. The angle between one side of the MoS_2 flake and the x-axis is defined as misfit angle φ, where (A) $\varphi = 0°$ and (B) $\varphi = 30°$. Size of a periodic cell is $15.4 \times 15.8 \times 10.0\ nm^3$. Contact pressure of 0.5 GPa and sliding speed of 100 m/s were applied simultaneously. System temperature was maintained at 300K. *Credit: T. Onodera, Y. Morita, R. Nagumo, R. Miura, A. Suzuki, H. Tsuboi, et al., A computational chemistry study on friction of h-MoS₂. Part II. Friction anisotropy,* J. Phys. Chem. *B, 114 (2010) 15832–15838.*

The effects of the sliding direction on the frictional properties of full-commensurate h-MoS_2 were recently investigated by using the/a MD method [31]. In that investigation, different types of dynamics were obtained with different sliding directions; in particular, a stick-slip dynamics was characterized by a highly dissipative behavior. As well as formation of a misfit angle, the sliding direction also contributes to the reduction of friction by h-MoS_2.

3.7 Superlubricity controlled by structure and property

The relationships between parameters studied in this simulation work are illustrated schematically in Fig. 3.12. To achieve superlubricity, the quality of the h-MoS_2 films or coatings is an important factor. Atomistic structure (i.e., lamellar crystal, oxygen insertion, and misfit angle) and interface property (interlayer interaction and charge transfer) are also key factors in controlling its friction. In the case MoDTC is used in an oil phase, molecular reactivity

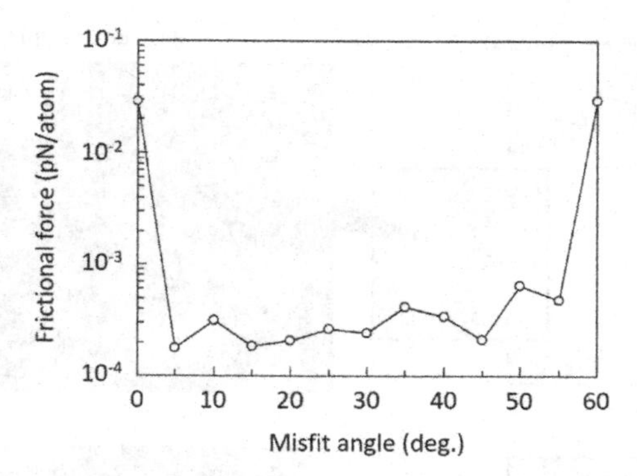

FIGURE 3.11 Relationship between misfit angle and average frictional force obtained by several classical MD simulations. *Credit: T. Onodera, Y. Morita, R. Nagumo, R. Miura, A. Suzuki, H. Tsuboi, et al., A computational chemistry study on friction of h-MoS_2. Part II. Friction anisotropy, J. Phys. Chem. B, 114 (2010) 15832–15838.*

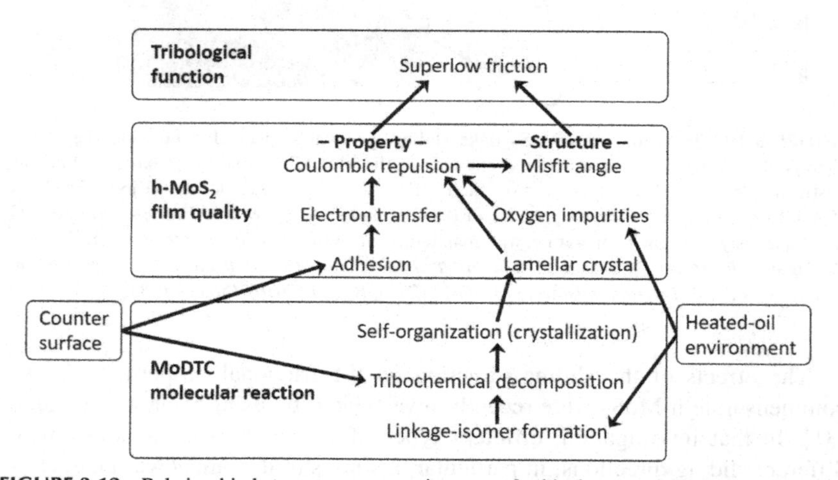

FIGURE 3.12 Relationship between parameters in terms of achieving superlubricity by h-MoS_2.

determined by friction condition, environmental temperature, and condition of the counter face are also essential factors. Controlling these parameters is one way to achieve a superlow-friction regime with h-MoS_2 solid lubricants.

References

[1] J.M. Martin, Superlubricity of molybdenum disulfide, in: A. Erdemir, J.M. Martin (Eds.), Superlubricity, Elsevier B. V, Oxford, 2005.

[2] J.M. Martin, Antiwear mechanisms of zinc dithiophosphate: a chemical hardness approach, Tribol. Lett. 6 (1999) 1–8.

[3] C. Grossiord, J.M. Martin, K. Varlot, B. Vacher, T. Le Mogne, Y. Yamada, Tribochemical interactions between ZnDTP, MoDTC and calcium borate, Tribol. Lett. 8 (2000) 203–212.

[4] G. Pereira, A. Lachenwitzer, M. Kasrai, P.R. Norton, T.W. Capehart, T.A. Perry, et al., A multi-technique characterization of ZDDP antiwear films formed on Al (Si) alloy (A383) under various conditions, Tribol. Lett. 26 (2007) 103–117.

[5] C. Grossiord, K. Varlot, J.M. Martin, T. Le Mogne, C. Esnouf, K. Inoue, MoS_2 single sheet lubrication by molybdenum dithiocarbamate, Tribol. Intern. 31 (1998) 737–743.

[6] J.M. Martin, C. Grossiord, T. Le Mogne, J. Igarashi, Transfer films and friction under boundary lubrication, Wear 245 (2000) 107–115.

[7] M.I. De Barros Bouchet, J.M. Martin, T. Le Mogne, P. Bilas, B. Vacher, Y. Yamada, Mechanisms of MoS_2 formation by MoDTC in presence of ZnDTP: effect of oxidative degradation, Wear 258 (2005) 1643–1650.

[8] C. Donnet, Problem-solving methods in tribology with surface-specific techniques, in: J. C. Rivière, S. Myhra (Eds.), Handbook of Surface and Interface Analysis, Marcel Dekker, Inc, 2000.

[9] J.R. Lince, P.F. Frantz, Anisotropic oxidation of MoS_2 crystallites studies by XPS, Tribol. Lett. 9 (2000) 211–218.

[10] M.R. Hilton, P.D. Fleischauer, TEM lattice imaging of the nanostructure of early-growth sputter-deposited MoS_2 solid lubricant films, J. Mater. Res. 5 (1990) 406–420.

[11] J.R. Lince, M.R. Hilton, A.S. Bommannavar, EXAFS of sputter-deposited MoS_2 films, Thin Solid. Film. 264 (1995) 120–134.

[12] H. Tamura, M. Yoshida, K. Kusakabe, C. Young-Mo, R. Miura, M. Kubo, et al., Molecular dynamics simulation of friction of hydrocarbon thin films, Langmuir 15 (1999) 7816–7821.

[13] D. Kamei, H. Zhou, K. Suzuki, K. Konno, S. Takami, M. Kubo, et al., Computational chemistry study on the dynamics of lubricant molecules under shear conditions, Tribol. Int. 36 (2003) 297–303.

[14] A. Rajendran, Y. Takahashi, M. Koyama, M. Kubo, A. Miyamoto, Tight-binding quantum chemical molecular dynamics simulation of mechano-chemical reactions during chemical-mechanical polishing process of SiO_2 surface by CeO_2 particle, Appl. Surf. Sci. 244 (2005) 34–38.

[15] M. Koyama, J. Hayakawa, T. Onodera, K. Ito, H. Tsuboi, A. Endou, et al., Tribochemical reaction dynamics of phosphoric ester lubricant additive by using a hybrid tight-binding quantum chemical molecular dynamics method, J. Phys. Chem. B 110 (2006) 17507–17511.

[16] G.T. Gao, P.T. Mikulski, J.A. Harrison, Molecular-scale tribology of amorphous carbon coatings: effects of film thickness, adhesion, and long-range interactions, J. Am. Chem. Soc. 124 (2002) 7202–7209.

[17] G.T. Gao, P.T. Mikulski, G.M. Chateauneuf, J.A. Harrison, The effects of film structure and surface hydrogen on the properties of amorphous carbon films, J. Phys. Chem. B 107 (2003) 11082–11090.

[18] N.J. Mosey, T.K. Woo, Finite temperature structure and dynamics of zinc dialkyldithiophosphate wear inhibitors: a density functional theory and ab initio molecular dynamics study, J. Phys. Chem. A 107 (2003) 5058–5070.

[19] N.J. Mosey, M.H. Müser, T.K. Woo, Molecular mechanisms for the functionality of lubricant additives, Science 307 (2005) 1612–1615.

[20] N.J. Mosey, T.K. Woo, M. Kasrai, P.R. Norton, G.M. Bancroft, M.H. Müser, Interpretation of experiments on ZDDP anti-wear films through pressure-induced cross-linking, Tribol. Lett. 24 (2006) 105–114.

[21] T. Onodera, Y. Morita, A. Suzuki, M. Koyama, H. Tsuboi, N. Hatakeyama, et al., A computational chemistry study on friction of h-MoS_2. Part I. Mechanism of single sheet lubrication, J. Phys. Chem. B 113 (2009) 16526–16536.

[22] T. Onodera, Y. Morita, A. Suzuki, R. Sahnoun, M. Koyama, H. Tsuboi, et al., A theoretical investigation on the dynamic behavior of molybdenum dithiocarbamate molecule in the engine oil phase, Tribol. Online 3 (2008) 80–85.

[23] T. Onodera, R. Miura, A. Suzuki, H. Tsuboi, N. Hatakeyama, A. Endou, et al., Development of a quantum chemical molecular dynamics tribochemical simulator and its application to tribochemical reaction dynamics of lubricant additives, Model. Simul. Mater. Sci. Eng. 18 (2010) 034009.

[24] T. Onodera, Y. Morita, A. Suzuki, R. Sahnoun, M. Koyama, H. Tsuboi, et al., Tribochemical reaction dynamics of molybdenum dithiocarbamate on nascent iron surface: a hybrid quantum chemical/classical molecular dynamics study, J. Nanosci. Nanotechnol. 10 (2010) 2495–2502.

[25] A. Morina, A. Neville, M. Priest, J.H. Green, ZDDP and MoDTC interactions and their effect on tribological performance –tribofilm characteristics and its evolution, Tribol. Lett. 24 (2006) 243–256.

[26] A. Morina, A. Neville, M. Priest, J.H. Green, ZDDP and MoDTC interactions in boundary lubrication – the effect of temperature and ZDDP/MoDTC ratio, Tribol. Int. 39 (2006) 1545–1557.

[27] H. Tanaka, M. Hashimoto, S. Gondo, Y. Yamamoto, Friction and wear characteristics of molybdenum dithiocarbamate with alloy steels, Tribol. Online 1 (2006) 9–13.

[28] Y. Morita, T. Onodera, A. Suzuki, R. Sahnoun, M. Koyama, H. Tsuboi, et al., Development of a new molecular dynamics method for tribochemical reaction and its application to formation dynamics of MoS_2 tribofilm, Appl. Surf. Sci. 254 (2008) 7618–7621.

[29] J.M. Martin, C. Donnet, T. Le Mogne, T. Epicier, Superlubricity of molybdenum disulphide, Phys. Rev. B 48 (1993) 10583–10586.

[30] T. Onodera, Y. Morita, R. Nagumo, R. Miura, A. Suzuki, H. Tsuboi, et al., A computational chemistry study on friction of h-MoS_2. Part II. Friction anisotropy, J. Phys. Chem. B 114 (2010) 15832–15838.

[31] V.E.P. Claerbout, T. Polcar, P. Nicolini, Superlubricity achieved for commensurate sliding: MoS_2 frictional anisotropy in silico, Comput. Mater. Sci. 163 (2019) 17–23.

Chapter 4

Vibration-induced superlubricity

Dan Guo, Shuai Shi, Xinfeng Tan and Jianbin Luo
State Key Laboratory of Tribology, Tsinghua University, Beijing, P.R. China

4.1 Introduction

Friction is one of the most common physical phenomena encountered in our daily life [1]. It is one of the greatest unsolved mysteries of science and has attracted many researchers' interest. On one hand, it plays an important role in many instruments playing or the writing to improve quality of life. On the other hand, some useless or harmful friction will cause harm to human life, such as the serious accidents of construction machinery caused by friction failure. Therefore the method of reducing and controlling friction is of great significance for saving useful energy, improving the life of mechanical equipment, and reducing environmental impact.

Friction is a complicated function of interface structure, physical, and chemical properties, as well as the environmental conditions including temperature, humidity, and vibration. Despite the difficulty of studying so many factors at once, over the past three decades, researchers have found out some novel ways to significantly reduce friction under certain conditions [2–4].

For the first time in 1983, Peyrard and Aubry theoretically predicted such a state of vanishing static friction in crystalline interfaces for infinite incommensurate contacts [5]. The superlubricity was firstly proposed by Hirano and Shinjo in the 1990s [6]. They predicted the friction would exactly vanish in the superlubric regime from the study of the Frenkel–Kontorova model by molecular dynamics simulation calculation [7]. The effect, also called structural lubricity, was verified between two graphite surfaces in 2004 by Dienwiebel et al. [8]. They showed that the origin of the ultralow friction of graphite lied in the incommensurability between rotated graphite layers. A new mechanism called thermolubricity was proposed by Krylov et al. [9] in 2005 to show that the temperature contributed to the superlubric behavior observed by Dienwiebel et al. Friction is expected to vanish at high temperatures, low sliding velocities, or low corrugations of the interaction potential between the surfaces, based on the emergence of a thermal diffusion regime [10]. In 2012, the microscale superlubricity phenomenon "self-retraction"

Superlubricity. DOI: https://doi.org/10.1016/B978-0-444-64313-1.00004-1

was firstly observed in graphite flakes on highly oriented pyrolytic graphite (HOPG) substrates by Liu et al. [11]. In 2015, the macroscopic superlubricity was observed between diamond-like carbon films and nanoscrolls formed from graphene flakes and nanodiamond particles by Erdemir et al. [12].

In the research of the microscopic mechanisms, Liu et al. [13] have obtained robust microscale superlubricity between 2D materials (both graphene/graphene and graphene/hexagonal boron nitride) under applied normal loads by using a graphene-coated microsphere. This new discovery breaks rotational angle dependence rule in the ultralow friction, which is attributed to the sustainable overall incommensurability due to the multiasperity contact covered with randomly oriented graphene nanograins. Chen et al. [14] have revealed the mechanisms governing superlubricity in a-C:H is generally dependent on the formation of interfacial nanostructures by different carbon rehybridization pathways. Liu et al. [15] first fabricated various 2D flake-wrapped atomic force microscope (AFM) tips and directly measured the interlayer friction coefficient as low as 10^{-4} between 2D flakes in single-crystalline contact. In appropriate liquid environment, the ultralow friction sliding can also be achieved [16]. The superlubricity (friction coefficient $0.004 \sim 0.006$) has been found between glass and Si_3N_4 surfaces lubricated by mixtures of acid solutions and glycerol solutions by Luo's group [17]. Not only for the mixed liquid, but also for the pure ionic liquid, when it is used to lubricate the silica-graphite interface, the superlubricity can be obtained by applying a potential [18]. A new theoretical concept called quantum lubricity was discovered in a laser-cooled, one-dimensional crystal of ions sliding over an optical lattice, which can enable the measurement of friction at the individual atom scale [19]. Superlubricity in various environments has attracted the significant attention from the fundamental science and engineering communities [20–23], because of its great potential in application.

The aforementioned mechanism in the solid interface contact is called structural superlubricity, which can be achieved by the nature of the material. A different mechanism of superlubricity due to external conditions was firstly proposed in 2006 by Socoliuc et al. [24], who added voltage to conducting probe when scanning ionic crystal by AFM. They found that friction energy consumption almost disappeared when the voltage frequency was applied. Riedo et al. [25] applied lateral vibrations to the cantilever holder while sliding on mica and the average friction force reduced around the resonance frequency when the sliding velocity below the critical value. Adjusting sliding speed and thermal excitation can also be used to reduce friction force. Shi et al. [26] have observed a new superlubricity phenomenon. They added the sinusoidal voltage signal to the scanning ceramic tube to achieve the superposition of tip vibration and the tube movement, thus realized a controllable superlubricity. This new method has been named vibration-induced superlubricity (VIS).

4.2 Measured methods

Friction is pervasive in the macroworld, but the explanation of the origin is still scarce and controversial [27,28]. Therefore how to measure the friction force and the energy dissipation in the friction and wear process is an important issue in terms of the origin of friction and its control. AFM, an innovative technique invented in 1986, has become a multifunctional and powerful apparatus for the application of micronanoscale imaging and force detection, including topography image, modulus of elasticity, viscoelastic properties, friction force, and energy dissipation in microscale and nanoscale worlds in various environments [29–34].

4.2.1 Dynamic atomic force microscope

There are two major dynamic AFM modes, amplitude modulation AFM (AM-AFM) (also known as tapping mode or semicontact mode) and frequency modulation AFM (FM-AFM) (also called noncontact AFM, NC-AFM) in the dynamic measurement methods. The difference is the feedback signal in the control system. In AM-AFM, the feedback signal is the vibration amplitude from the microcantilever, which is excited at or near its resonance frequency. The fundamental application is to measure the topography of a sample surface. In addition, the other physical quantity, such as the phase shift between the driving signal and the vibration signal, can be used to map the material properties variations, such as the adhesion distribution and the modulus of elasticity. The amplitude and phase can also be integrated to calculate the energy dissipation. Due to its low quality factor, AM mode is often carried out in air or the liquid to obtain rapid feedback responses. Schematic diagram of AM-AFM is shown in Fig. 4.1.

In FM-AFM, the cantilever is kept vibrating with a fixed amplitude at its resonance frequency. The interaction between the tip and the sample can be reflected in the varying resonance frequency. The frequency shift which is the feedback signal to control the piezo tube to follow the sample topography can be also integrated to calculate the tip-sample interaction in the attractive or repulsive regime in a real noncontact mode, where the noncontact friction energy dissipation can be explored in a pendulum configuration [28]. This mode is usually employed to obtain the high-resolution image for its excellent performance, however, the experimental environments are in liquids or in ultrahigh vacuum. Schematic diagram of FM-AFM is shown in Fig. 4.2.

4.2.2 Excitation methods

There are three main excitation methods developed in dynamic AFM. The acoustic and magnetic excitation methods are conventional and some disadvantages appear [35]. In the acoustic excitation method, the piezoelectric

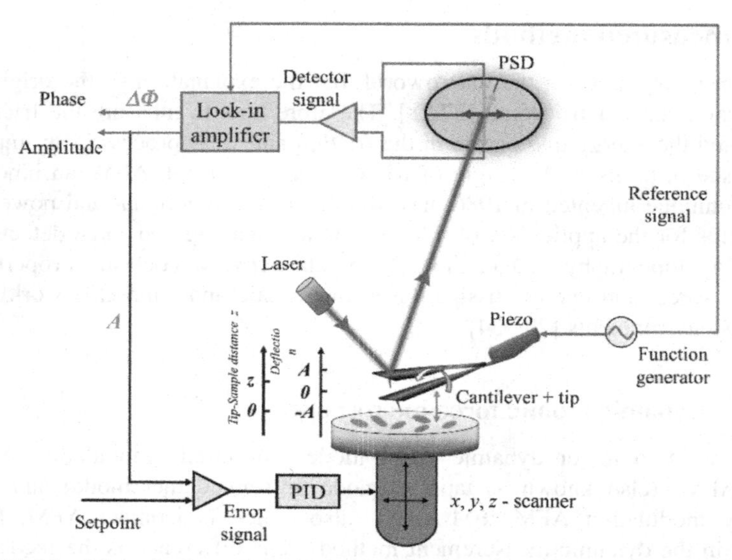

FIGURE 4.1 Schematic diagram of AM-AFM. The phase and amplitude signals of the microcantilever are processed by the lock-in amplifier. Setpoint is the percentage of the free vibration amplitude. (*AM-AFM*: Amplitude modulation-atomic force microscope).

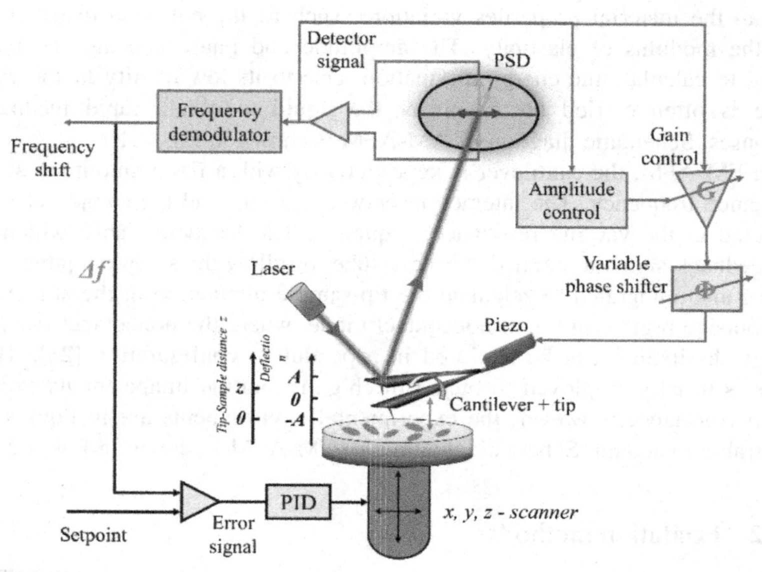

FIGURE 4.2 Schematic diagram of FM-AFM. The frequency shift signal of the microcantilever is processed by the frequency demodulator. (*FM-AFM*: Frequency modulation-atomic force microscope).

ceramic stacks will introduce the considerable parasitic resonance peaks, making the cantilever spectrum difficult to distinguish [36]. Due to the natural impedance of the magnetic coil, it is hard to achieve high-frequency excitation and detection in the higher vibration mode or multifrequency AFM by the magnetic method [37].

The photothermal excitation (PTE) method first reported in the dynamic scanning attractive force microscope in 1991 [38]. Tan et al. [39] modified the PTE method to make it excite the flexural and torsional vibration mode simultaneously. In Fig. 4.3, a schematic of the PTE experimental setup is displayed. The average excitation laser power is modulated by a signal generator to produce a variable power laser beam. After passing through the optical system, the blue laser beam sparkles on the base of the cantilever to excite the cantilever. The two-dimensional photoelectric position-sensitive detector is used to detect the red laser reflected from the back of the cantilever.

The frequency spectra (0–1000 kHz) of the cantilever AC-240-TM is shown in Fig. 4.4. The red line represents the flexural signal and the blue line represents the torsional signal of the cantilever response. A few spurious resonance peaks in the sweeping frequency spectrum by piezoelectric excitation method appears in Fig. 4.4B while the PTE sweeping frequency spectrum is clean and clear in Fig. 4.4A, the reason of which is that the piezoelectric excitation signal is transmitted from the dither piezo stacks to the cantilever through the cantilever holder and substrate, which makes the vibration transmission process more complicated and the coupling effect is larger. But for the PTE method, the laser position can be adjusted to the

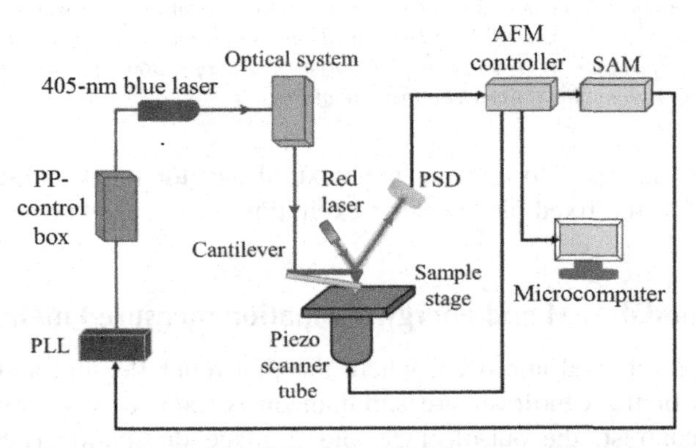

FIGURE 4.3 Schematic description of the experimental setup and optical path for the PTE method. (*PTE*: Photothermal excitation). *Reproduced with permission from X. Tan, S. Shi, D. Guo, J. Luo, Dynamical characterization of micro cantilevers by different excitation methods in dynamic atomic force microscopy. Rev. Sci. Instrum., 89(11): 115109, 2018, Copyright 2018, American Institute of Physics.*

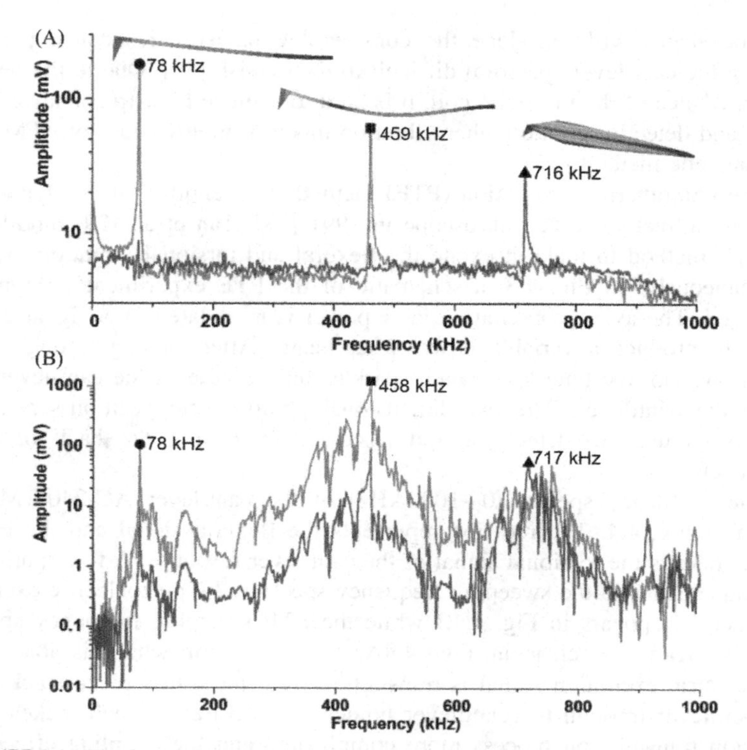

FIGURE 4.4 The frequency spectra (0–1000 kHz) of the cantilever AC-240-TM. (A) The flexural signal spectral response when the laser sparkles on the middle of the cantilever beam root *(red line)* and torsional signal spectral response when the laser sparkle on the edge of the cantilever beam root *(blue line)* and (B) the flexural signal spectral response (*red* line) and torsional signal spectral response *(blue line)* under fixed piezoelectric excitation. *Reproduced with permission from X. Tan, S. Shi, D. Guo, J. Luo, Dynamical characterization of micro cantilevers by different excitation methods in dynamic atomic force microscopy. Rev. Sci. Instrum., 89(11): 115109, 2018, Copyright 2018, American Institute of Physics.*

edge of the cantilever to activate the flexural and torsional vibration mode together unlike the fixed piezoelectric excitation.

4.2.3 Bimodal AFM and energy dissipation measured method

In Fig. 4.5, a bimodal approach, where the first-order flexural and torsional eigenmodes of the cantilever are simultaneously excited, was developed to detect the contrast, the out-of-plane and in-plane dissipation (flexural and torsional dissipation) between the tip and the polymer blend of polystyrene (PS) and low-density polyethylene (LDPE) [40]. The flexural and torsional signal spectra were obtained by sweeping frequency and the resonance peaks were distinguished. Typically the approaching tip first senses the attractive

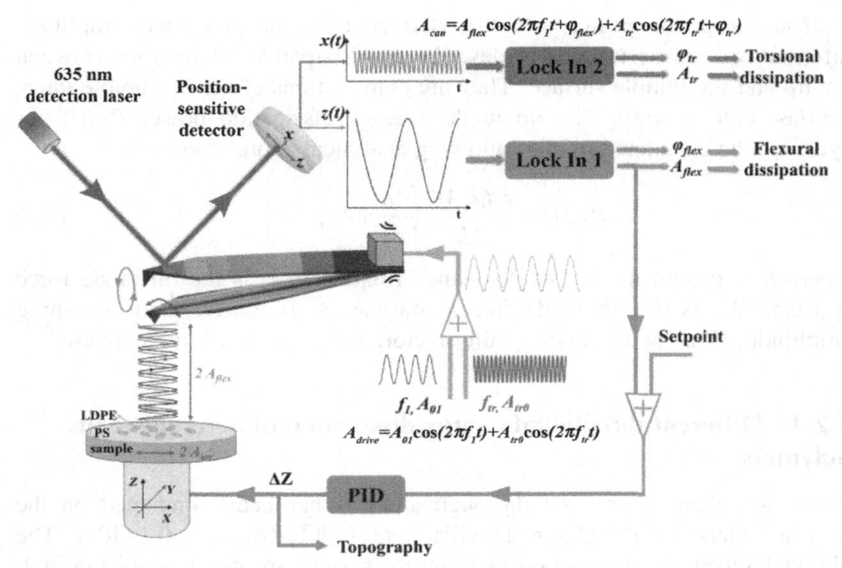

FIGURE 4.5 Bimodal schematic description and control system. *Reproduced with permission from X. Tan, D. Guo, J. Luo, Different directional energy dissipation of heterogeneous polymers in bimodal atomic force microscopy. RSC Adv., 9(47): 27464, 2019, Copyright 2019, Royal Society of Chemistry.*

force, and then the mixture of the attractive force acting on the tip body and the repulsive force between the tip and sample surface. The amplitude and phase of the flexural and torsional response signals have been recorded and analyzed to calculate the energy dissipation power at various setpoint ratios.

In bimodal dynamic force microscopy, the rectangular cantilever beam can be approximately modeled by the Euler–Bernoulli partial differential equation to describe the dynamics of the cantilever-tip system. The analytical expressions are simplified as follows [40]:

$$\frac{k_i}{{\omega_i}^2}\ddot{z}_i(t) = -\frac{k_i}{Q_i\omega_i}\dot{z}_i(t) - k_i z_i + F_{0i}\cos(\omega_i t) + F_{ts} \tag{4.1}$$

$$\frac{k_{tri}}{{\omega_{tri}}^2}\ddot{x}_i(t) = -\frac{k_{tri}}{Q_{tri}\omega_{tri}}\dot{x}_i(t) - k_{tri}x_i + F_{tr0i}\cos(\omega_{tri}t) + F_{trts} \tag{4.2}$$

where i represents the ith mode; k_i, $\omega_i = 2\pi f_i$, and Q_i are the modal stiffness, resonance frequency, and quality factor of the ith mode, respectively. $F_{0i} = k_i A_{0i}/Q_i$ is the magnitude of the ith mode external driving force. A_{0i} and z_i are the free amplitude and deflection of the cantilever for the ith mode. F_{ts} and F_{trts} are the out-of-plane and in-plane tip-sample interaction, respectively. The subscript tr means the torsional vibration mode.

The dissipation power, which is derived from the phase and amplitude information, can be utilized to describe the dissipative interactions between the tip and the sample surface. They are convolutions of the tip sample interactions with velocity. The *i*th mode average dissipation power $P_{dis}(i)$ per cycle can be calculated by the following analytical expressions:

$$P_{dis}(i) = \frac{\pi f_i k_i A_i^2}{Q_i}\left[\frac{A_{0i}}{A_i}\sin\varphi_i - 1\right] \tag{4.3}$$

where f_i is the *i*th mode free resonance frequency, k_i is the *i*th mode force constant, A_{0i} is the *i*th mode free amplitude, A_i is the *i*th mode scanning amplitude, Q_i is the *i*th mode quality factor, and φ_i is the *i*th mode phase.

4.2.4 Different directional energy dissipation of heterogeneous polymers

As is described earlier, the thorough analysis has been conducted on the polymer blend of PS ($E_{PS} = 2.0$ GPa) and LDPE ($E_{LDPE} = 0.1$ GPa). The bimodal schematic description and control system are designed in Fig. 4.4. When simultaneously driving the piezo shaker at both the first-order flexural and torsional frequencies, the cantilever responses are not usually only in the flexural mode, but also the torsional mode. The resonant spectra of the two signals were displayed in Fig. 4.6.

The distance dependence of the tip-sample (PS/LDPE) energy dissipation has been recorded at different setpoint ratios to capture not only the spatial variation in the *X-Y* plane, but also as a function of the tip-sample distance to some extent. Fig. 4.7 shows the sectional energy dissipation distribution at

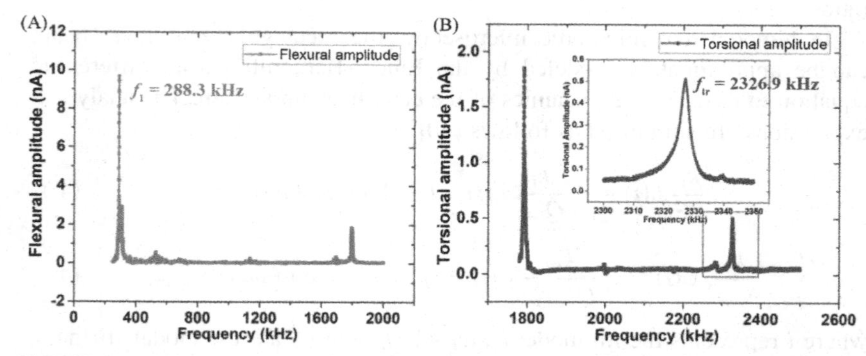

FIGURE 4.6 The frequency spectra (200–2500 kHz) of the cantilever PPP-NCH. (A) The flexural signal spectral response *(red line)* and the first-order flexural vibration frequency $f_I = 288.3$ kHz. (B) The torsional signal spectral response *(blue line)* and the first-order torsional vibration frequency $f_{tr} = 2326.9$ kHz. *Reproduced with permission from X. Tan, D. Guo, J. Luo, Different directional energy dissipation of heterogeneous polymers in bimodal atomic force microscopy. RSC Adv., 9(47): 27464, 2019, Copyright 2019, Royal Society of Chemistry.*

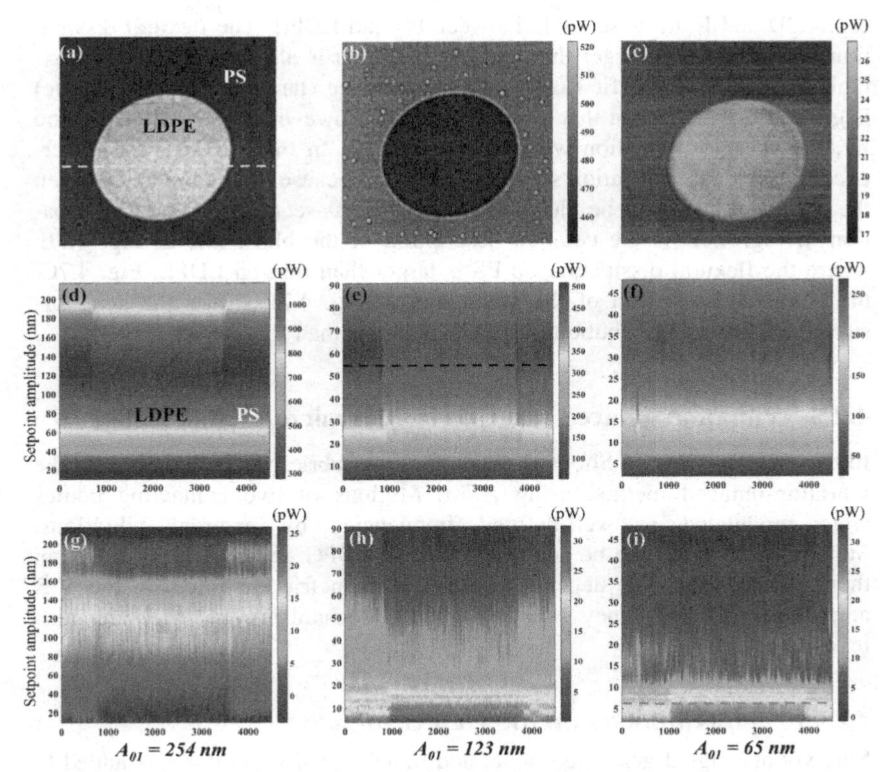

FIGURE 4.7 The sectional energy dissipation distribution on PS/LDPE. (LDPE: Low-density polyethylene; PS: polystyrene). (A) Phase image of the PS/LDPE. (B) Cross sectional flexural dissipation distribution at the setpoint amplitude 54 nm (A_{01} = 123 nm) as shown by the black line in (E). (C) Cross sectional torsional dissipation distribution at the setpoint amplitude 6.5 nm (A_{01} = 65 nm) as shown by the green line in (I). The longitudinal sectional flexural dissipation distribution at (D) A_{01} = 254 nm (E) A_{01} = 123 nm and (F) A_{01} = 65 nm. The longitudinal sectional torsional dissipation distribution at (G) A_{01} = 254 nm (H) A_{01} = 123 nm and (J) A_{01} = 65 nm. The side length is 4.5 micrometer and the longitudinal sectional position of (D)∼(I) is the yellow line in (A). *Reproduced with permission from X. Tan, D. Guo, J. Luo, Different directional energy dissipation of heterogeneous polymers in bimodal atomic force microscopy. RSC Adv., 9(47): 27464, 2019, Copyright 2019, Royal Society of Chemistry.*

different setpoint ratios. Fig. 4.7A shows the phase image of PS (the dark brown area) and LDPE (the bright circular area). In Fig. 4.7D–I, the tip-sample interaction is different on PS and LDPE, especially in the torsional dissipation. The visualization is not a strict way to distinguish the two materials, but some differences can be intuitively seen. For example, the flexural dissipation power of the two materials is close to each other especially within the low setpoint range at a small free amplitude 65 nm. It does show the absent difference in Fig. 4.7F, however, the small differences in

Fig. 4.7D and E are observable between PS and LDPE. The flexural dissipation power on PS is larger than that on LDPE. It is also found that the flexural dissipation of the first order at long distance (large setpoint amplitude) is generally greater than that at short distance, however, if the distance is too far, the energy dissipation will also be reduced. In Fig. 4.7G–I, the difference in torsional dissipation seems a bit better because the boundary between the PS and LDPE can be clearly seen. The cross-sectional flexural dissipation in Fig. 4.7B is the detailed description of the black line in Fig. 4.7E, where the flexural dissipation on PS is larger than that on LDPE. Fig. 4.7C, the detailed description of the green line in Fig. 4.7I, shows the near torsional dissipation distribution on LDPE is larger than that on PS.

4.2.5 Vibration-induced superlubricity in micro/atomic scale

In the recently work of Shi et al. [26], the superlubricity was achieved by the vibration-induced method using AFM. Motions of two contacting bodies were modulated at well-defined frequencies by external vibrations. Superlubricity state can be achieved both on HOPG and polymer surfaces in their experiments. The dependence of the mean friction force on the VIS amplitude and frequency was explored. Mechanisms of VIS are also investigated.

4.2.5.1 Superlubricity on the HOPG surface

Sine voltage signal generated by an additional signal generator was added to the scanning ceramic tube to achieve the superposition of movement in a commercial AFM. Piezoelectric ceramic vibration drives the probe to perform simple harmonic vibration, thus realizing the vibration during the friction process.

A commercial AFM probe was used to conduct the friction experiments on HOPG surface. The movement of frictional model is displayed in Fig. 4.8.

Firstly the influence of amplitude and frequency on the friction on the graphite surface is researched. Frequencies were 1, 2.5, 5, 10, 30, 61.7, and 100 kHz, and the amplitude range was 0 ~ 17 nm. Fig. 4.9 shows the curves of friction caused by the probe sliding on the HOPG surface with different amplitudes and frequencies. It can be seen that friction decreases rapidly with the increase of amplitude and tends to be flat at 0.5 nN in the case of each modulated frequency. When it is 5 kHz, the friction decreases fastest.

The influence of scanning speed on the friction was also studied on the graphite surface. In the experiments, the scanning speed changed (0.5, 1, and 2 Hz), the vibration frequency was 1 kHz, and the amplitude range was 0 ~ 16 nm. As shown in Fig. 4.10, the friction results of the probe on HOPG surface at different scanning speeds, friction force suddenly decreases with

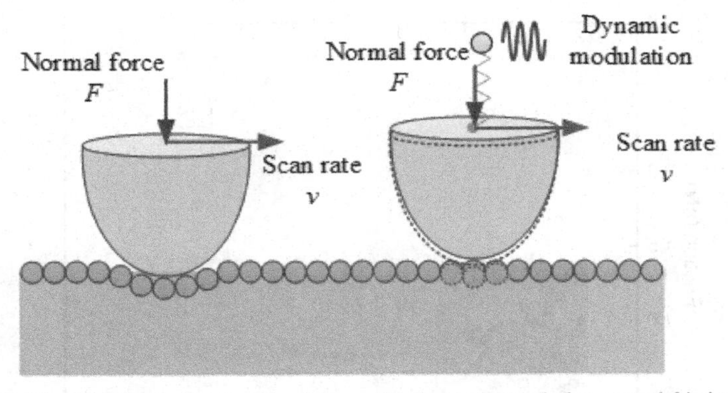

FIGURE 4.8 Additional vibration signal forces the tip to vibrate during normal friction movement [26].

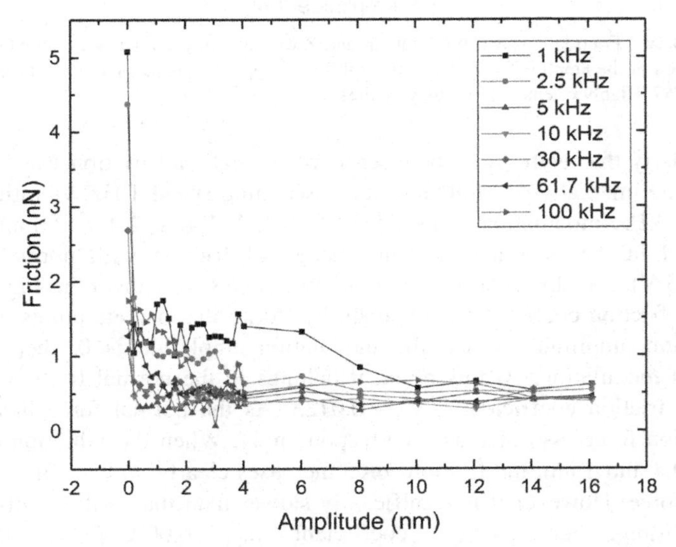

FIGURE 4.9 Friction detected by scanning and Z exciting on the microscale HOPG surface as a function of the amplitude (0 ~ 16 nm) for different applied frequencies (1, 2.5, 5, 10, 30, 61.7, and 100 kHz) [26]. (*HOPG*: Highly oriented pyrolytic graphite).

the increase of amplitude for each scanning speed. It reduced from 0.99 to 0.2 ~ 0.3 nN for 0.5 Hz, and reduced from 1.75 to 0.3 ~ 0.4 nN for 1 Hz, and reduced from 2.61 to 0.7 ~ 0.8 nN for 2 Hz. It is obvious that the friction decreases significantly with vibration. When the scanning speed is large, the friction in the stationary state is large. It indicates that scanning speed has a great influence on the stability of the friction.

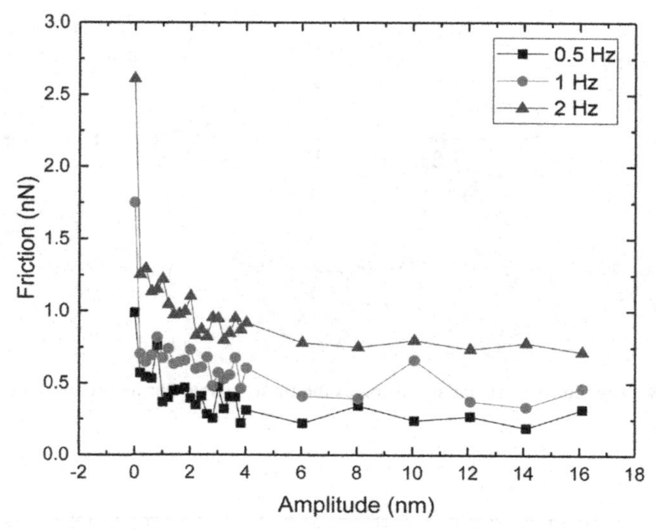

FIGURE 4.10 Friction detected by scanning and Z exciting on the microscale HOPG surface as a function of the amplitudes (0 ~ 17 nm) for different applied scanning rate (0.5, 1, and 2 Hz) [26]. (*HOPG*: Highly oriented pyrolytic graphite).

Similarly the relationship between normal force and friction was investigated experimentally on HOPG surface. Scanning speed 1 Hz, vibration frequency 1 kHz, modulation amplitude of 0, 0.4, 1, 1.4, 2, 2.4, and 3 nm were employed in the experiments. The change of friction with normal force (0.5 ~ 3.5 V) for different modulation amplitudes is shown in Fig. 4.11. Material friction coefficient is obtained by fitting the discrete points of each modulation amplitude. When the modulation amplitude is 0, there is no vibration modulation. At this point, it belongs to the normal friction mode. And the friction coefficient is $\mu_0 = 0.0123$. As the normal force increases, the friction force also increases correspondingly. When the vibration amplitude is 0.4 and 1 nm, the friction force increases clearly with the increase of normal force. However it is significantly slower than that without vibration. After fitting, the friction coefficient $\mu_{0.4} = 0.0078$, $\mu_{1.0} = 0.0017$, $\mu_{1.4} = 0.0004$, $\mu_2 = 0.0012$, $\mu_{2.4} = 0.0016$, and $\mu_3 = 0.0007$ were obtained for vibration amplitudes 0.4, 1.0, 1.4, 2, 2.4, and 3 nm. In comparison, friction can be significantly reduced and coefficient of friction can be greatly flattened with vibration. The friction coefficient reaches one thousandth of a magnitude, indicating that the superlubricity has been realized.

It is easier to reveal the mechanism of energy dissipation by studying atomic-scale friction. In order to explore the deep nature of the VIS mechanism, atomic-scale friction experiments were further carried out on graphite.

First of all, atomic-level friction force was obtained on the HOPG surface in the normal friction mode. Image of atomic-level friction was shown in

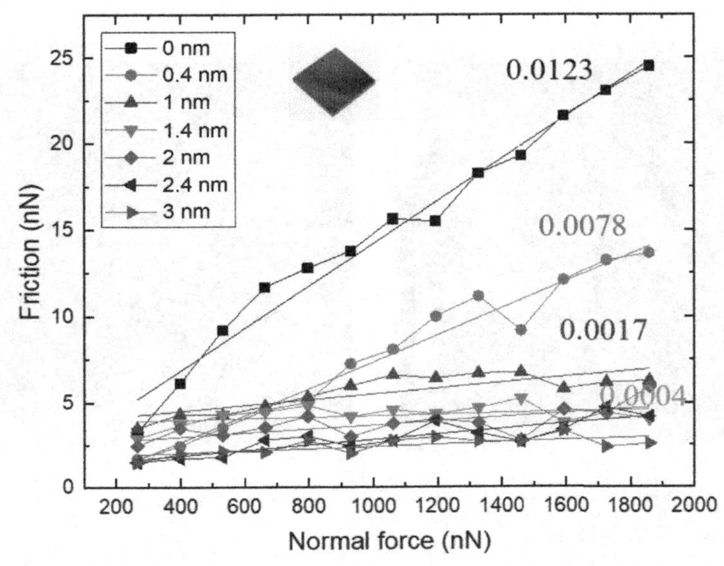

FIGURE 4.11 Friction force as a function of the applied normal load for different Z-VM amplitudes (0, 0.4, 1, 2.4, and 3 nm) on HOPG surface [26]. (*HOPG*: Highly oriented pyrolytic graphite; *Z-VM*: z-directional vibration modulation).

Fig. 4.12. In Fig. 4.12A, the real-time trace and retrace curves can be seen corresponding to the atomic-level stick-slip curves. The middle regime of the trace and retrace curves represents twice of voltage value corresponding to the friction force. Friction ring (friction energy dissipation) disappears with adding *z*-directional vibration modulation (Z-VM) and the friction ring occurs when disconnecting to vibration signal, which was shown in Fig. 4.12B. It can be seen directly from the comparison that the vibration method can achieve superlubricity.

4.2.5.2 Superlubricity on the polystyrene surface

It is found that VIS can also be realized on the polymer surface.

Fig. 4.13 shows the effect of friction caused by sliding of probe with different modulated amplitudes and frequencies on the PS surface. As shown in the figure, friction dropped fast and tended to be flat with the increase of amplitude for frequencies of 1, 2.5, 5, 10, and 30 kHz. Among these lines, there is a fastest decline of friction for 5 kHz. Enlarged detail shows friction decreases from 2000 ~ 2500 (related to the starting point) to 100 nN when the amplitude is 4 nm. Unexpectedly the friction basically did not decrease within the setting amplitude range for 61.7 and 100 kHz.

Fig. 4.14 shows that the friction decreases rapidly for the vibration frequencies of 5, 22.5, and 35 kHz. The difference is that, friction decreases

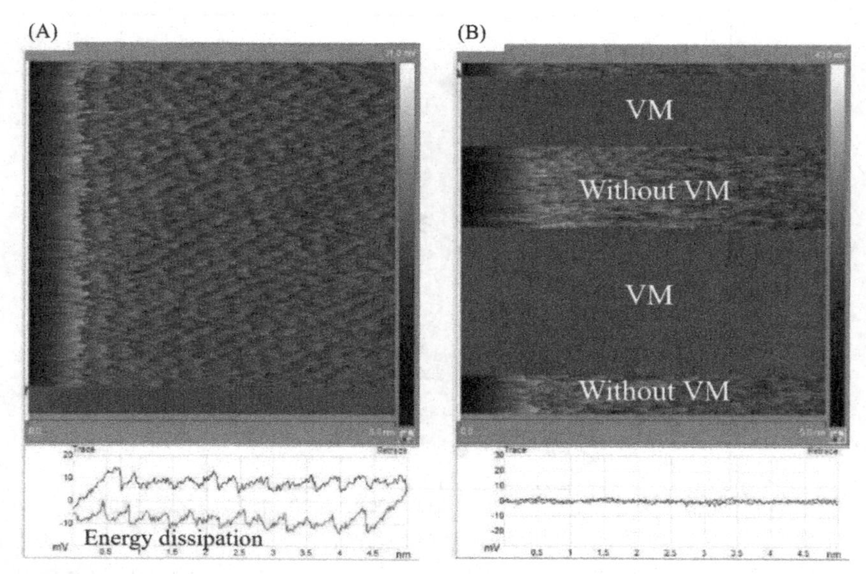

FIGURE 4.12 Frictional images force detected by scanning forward and backward on an atomically flat HOPG surface. (A) Without Z-VM and (B) a contrast between with Z-VM and without Z-VM [26]. (*HOPG*: Highly oriented pyrolytic graphite; *Z-VM*: z-directional vibration modulation).

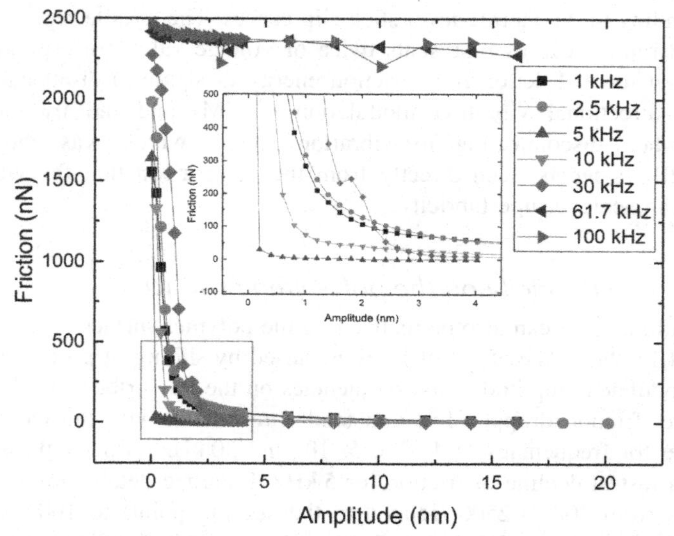

FIGURE 4.13 Friction (average frictional force) detected by scanning and Z-VM exciting on the microscale PS surface as a function of the VM amplitude (0 ~ 20 nm) for different applied VM frequencies (1, 2.5, 5, 10, 30, 61.7, and 100 kHz) [26]. (*PS*: Polystyrene; *VM*: vibration modulation; *Z-VM*: z-directional vibration modulation).

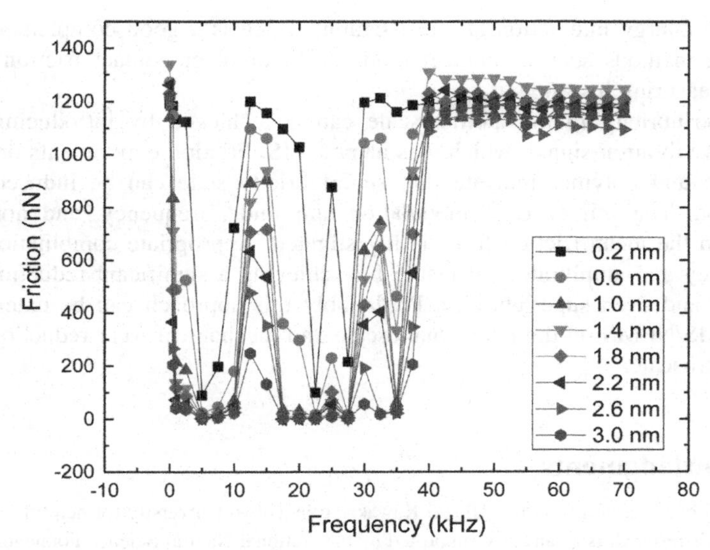

FIGURE 4.14 Friction detected by scanning and Z-VM exciting on the microscale PS surface as a function of the VM frequency (0 ~ 100 kHz) for different applied VM amplitudes (0.2, 0.6, 1.0, 1.4, 1.8, 2.2, 2.6, and 3.0 nm) [26]. (*PS*: Polystyrene; *VM*: vibration modulation; *Z-VM*: z-directional vibration modulation).

less in the frequency bands of 10–15 and 30–35 kHz which may be related to the vibration amplitude. Obviously there almost no reduction of friction for all amplitudes when the frequency is higher than 40 kHz.

In the chapter of Gnecco et al., superlubricity is achieved by modulating normal force at well-defined frequencies corresponding to mechanical resonances of the tip-surface contact on the atomically flat KBr surface, the results were supported by the theoretical basis of the Prandtl–Tomlinson model, and this is named "dynamic superlubricity." The VIS between the tip and HOPG in this chapter can also be interpreted as the vibration changes the potential energy function, the stick-slip motion is destroyed, and the energy dissipation disappears, leading to the reduction of friction. However there is no stick-slip motion in the friction of PS, and VIS also realized. The friction is complicated affected by vibration frequency, amplitude, direction, normal force, and scanning speed. The mechanisms to be revealed have potential application not only at the microscale, but also at the macroscale. So VIS was proposed to describe this superlubricity state.

4.3 Summary

The bimodal method has an excellent performance for measuring the energy dissipation near the sample surface. It is a promising technique which can act as an effective and direct strategy for the detection of out-of-plane and in-plane energy interaction and dissipation, which is a good complement to existing methods such as contact mode AFM or other contact friction and wear measuring instruments.

Superlubricity in the atomic scale can be achieved by introducing an external vibration signal, which was named VIS. Friction experiments on the graphite and polymer indicate that superlubricity state can be induced by vibration. The influence of modulation amplitude, frequency, and normal force on the mean friction force is investigated. Appropriate combination of frequency and amplitude is decisive for achieving a significant reduction of friction and even superlubricity. Predictably this approach can be extended in MEMS/NEMS on the micro-/nanoscale and mechanical wear reduction on the macroscale.

Acknowledgments

The authors are grateful to Prof. Hideki Kawakatsu in Tokyo University for helpful discussions. This research is financially supported by the National Natural Science Foundation of China (Grant No. 51527901).

References

[1] K. Holmberg, A. Erdemir, Influence of tribology on global energy consumption, costs and emissions, Friction 5 (2017) 263–284.

[2] K. Kim, H. Lee, C. Lee, S. Lee, H. Jang, J. Ahn, et al., Chemical vapor deposition-grown graphene: the thinnest solid lubricant, ACS Nano 5 (6) (2011) 5107–5114.

[3] D. Berman, A. Erdemir, A. Zinovev, A. Sumant, Nanoscale friction properties of graphene and graphene oxide, Diam. Relat. Mater. 54 (2015) 91–96.

[4] C. Lee, Q. Li, W. Kalb, X.Z. Liu, H. Berger, R. Carpick, et al., Frictional characteristics of atomically-thin sheets, Science 328 (5974) (2010) 76–80.

[5] M. Peyrard, S. Aubry, Critical behaviour at the transition by breaking of analyticity in the discrete Frenkel-Kontorova model, J. Phys. C: Solid State Phys. 16 (9) (1983) 1593–1608.

[6] M. Hirano, K. Shinjo, Atomistic locking and friction, Phys. Review. B: Condens. Matter 41 (17) (1990) 11837–11851.

[7] K. Shinjo, M. Hirano, Dynamics of friction: superlubric state, Surf. Sci. 283 (1-3) (1993) 473–478.

[8] M. Dienwiebel, G. Verhoeven, N. Pradeep, J. Frenken, J. Heimberg, H. Zandbergen, Superlubricity of graphite, Phys. Rev. Lett. 92 (12) (2004) 126101.

[9] S. Krylov, K. Jinesh, H. Valk, M. Dienwiebel, J. Frenken, Thermally induced suppression of friction at the atomic scale, Phys. Rev. E 71 (2005) 065101.

[10] K. Jinesh, S. Krylov, H. Valk, M. Dienwiebel, J. Frenken, Thermolubricity in atomic-scale friction, Phys. Rev. B 78 (2008) 155440.

[11] Z. Liu, J. Yang, F. Grey, J. Liu, Y. Liu, Y. Wang, et al., Observation of microscale superlubricity in graphite, Phys. Rev. Lett. 108 (20) (2012) 205503.

[12] D. Berman, S. Deshmukh, S. Sankaranarayanan, A. Erdemir, A. Sumant, Macroscale superlubricity enabled by graphene nanoscroll formation, Science 348 (6239) (2015) 1118–1122.

[13] S. Liu, H. Wang, Q. Xu, T. Ma, G. Yu, C. Zhang, et al., Robust microscale superlubricity under high contact pressure enabled by graphene-coated microsphere, Nat. Commun. 8 (2017) 14029.

[14] X. Chen, C. Zhang, T. Kato, X. Yang, S. Wu, R. Wang, et al., Evolution of tribo-induced interfacial nanostructures governing superlubricity in a-C:H and a-C:H:Si films, Nat. Commun. 8 (2017) 1675.

[15] Y. Liu, A. Song, Z. Xu, R. Zong, J. Zhang, W. Yang, et al., Interlayer friction and superlubricity in single-crystalline contact enabled by two-dimensional flake-wrapped atomic force microscope tips, ACS Nano 12 (8) (2018) 7638–7646.

[16] J. Klein, Hydration lubrication, Friction 1 (2013) 1–23.

[17] J. Li, C. Zhang, L. Ma, Y. Liu, J.B. Luo, Superlubricity achieved with mixtures of acids and glycerol, Langmuir 29 (1) (2013) 271–275.

[18] H. Li, R. Wood, M. Rutland, R. Atkin, An ionic liquid lubricant enables superlubricity to be "switched on" in situ using an electrical potential, Chem. Commun. 50 (33) (2014) 4368–4370.

[19] A. Bylinskii, D. Gangloff, V. Vuletic, Tuning friction atom-by-atom in an ion-crystal simulator, Science 348 (6239) (2015) 1115–1118.

[20] J. Li, J. Luo, Advancements in superlubricity, Sci. China: Technol. Sci 56 (2013) 2877–2887.

[21] E. Meyer, E. Gnecco, Superlubricity on the nanometer scale, Friction 2 (2014) 106–113.

[22] Q. Zheng, Z. Liu, Experimental advances in superlubricity, Friction 2 (2014) 182–192.

[23] D. Dietzel, U. Schwarz, A. Schirmeisen, Nanotribological studies using nanoparticle manipulation: principles and application to structural lubricity, Friction 2 (2014) 114–139.

[24] A. Socoliuc, E. Gnecco, S. Maier, O. Pfeiffer, A. Baratoff, R. Bennewitz, et al., Atomic-scale control of friction by actuation of nanometer-sized contacts, Science 313 (5784) (2006) 207–210.

[25] E. Riedo, E. Gnecco, R. Bennewitz, E. Meyer, H. Brune, Interaction potential and hopping dynamics governing sliding friction, Phys. Rev. Lett. 91 (2003) 084502.

[26] S. Shi, D. Guo, J.B. Luo, Micro/Atomic-Scale vibration induced superlubricity, Friction, to be published.

[27] M. Urbakh, E. Meyer, Nanotribology: the renaissance of friction, Nat. Mater. 9 (2010) 8–10.

[28] M. Kisiel, E. Gnecco, U. Gysin, L. Marot, S. Rast, E. Meyer, Nat. Mater. 10 (2011) 119–122.

[29] G. Binnig, C. Quate, C. Gerber, Atomic force microscope, Phys. Rev. Lett. 56 (1986) 930.

[30] R. Garcia, R. Perez, Dynamic atomic force microscopy methods, Surf. Sci. Rep. 47 (6-8) (2002) 197–301.

[31] S. Shi, D. Guo, J. Luo, Interfacial interaction and enhanced imagecontrasts in higher mode and bimodal mode atomic force microscopy, RSC Adv. 7 (87) (2017) 55121–55130.

[32] S. Shi, D. Guo, J. Luo, Enhanced phase and amplitude image contrasts of polymers in bimodal atomic force microscopy, RSC Adv. 7 (19) (2017) 11768–11776.

[33] S. Shi, D. Guo, J. Luo, Imaging contrast and tip-sample interaction of non-contact amplitude modulation atomic force microscopy with Q-control, J. Phys. D: Appl. Phys. 50 (41) (2017) 415307.

[34] D. Martin-Jimenez, E. Chacon, P. Tarazona, R. Garcia, Atomically resolved three-dimensional structures of electrolyte aqueous solutions near a solid surface, Nat. Commun. 7 (2016) 12164.

[35] D. Ramos, J. Tamayo, J. Mertens, M. Calleja, Photothermal excitation of microcantilevers in liquids, J. Appl. Phys. 99 (12) (2006) 124904.

[36] X. Xu, A. Raman, Comparative dynamics of magnetically, acoustically, and Brownian motion driven microcantilevers in liquids, J. Appl. Phys. 102 (3) (2007) 034303.

[37] M. Kageshima, T. Chikamoto, T. Ogawa, Y. Hirata, T. Inoue, Y. Naitoh, et al., Development of atomic force microscope with wide-band magnetic excitation for study of soft matter dynamics, Rev. Sci. Instrum. 80 (2) (2009) 023705.

[38] N. Umeda, S. Ishizaki, H. Uwai, Scanning attractive force microscope using photothermal vibration, J. Vac. Sci. Technol. B 9 (2) (1991) 1318.

[39] X. Tan, S. Shi, D. Guo, J. Luo, Dynamical characterization of micro cantilevers by different excitation methods in dynamic atomic force microscopy, Rev. Sci. Instrum. 89 (11) (2018) 115109.

[40] X. Tan, D. Guo, J. Luo, Different directional energy dissipation of heterogeneous polymers in bimodal atomic force microscopy, RSC Adv. 9 (47) (2019) 27464.

Chapter 5

Atomic-scale investigations of ultralow friction on crystal surfaces in ultrahigh vacuum

Enrico Gnecco[1], Rémy Pawlak[2], Thilo Glatzel[2] and Ernst Meyer[2]
[1]*Otto Schott Institute of Materials Research (OSIM), Friedrich Schiller University Jena, Jena, Germany,* [2]*Institute of Physics, University of Basel, Basel, Switzerland*

5.1 Introduction

Controlling friction on the nanometer scale is one of nowadays' challenges for scientists and engineers. Since the first observation of atomic friction reported by Mate et al. [1] for a tungsten tip sliding on graphite, a lot of progress has been made in the understanding of this phenomenon on the atomic scale. An accurate description of the motion of a sharp tip elastically driven on a crystal surface by a microcantilever was first given by Tomanek et al., who based their interpretation on the Prandtl–Tomlinson (PT) model [2]. The lateral (friction) force acting on the tip can be estimated by measuring the angle of torsion of the cantilever. The tip sticks to a given equilibrium position on the surface lattice until the driving force becomes high enough to cause a slip into the closest equilibrium position along the pull direction. The resulting stick-slip motion corresponds to a sawtooth-shaped time evolution of the lateral force with the atomic periodicity of the surface lattice. However, this scenario is observed only if a precise condition is fulfilled. The lateral stiffness of the driving spring must be lower than the curvature of the tip-surface interaction potential. If this is not the case, the tip slides on the surface without abrupt jumps, and a "superlubric" scenario is observed.

The transition from stick-slip to continuous sliding was first observed by Socoliuc et al. [3] in 2004 on a NaCl surface using a dedicated atomic force microscope (AFM) in ultrahigh vacuum (UHV) conditions. The transition occurred for normal force values in the subnanonewton range only when using brand new silicon tips, whereas superlubricity could be hardly distinguished from noise when the tips became blunt by usage. In the same year Dienwiebel et al. [4] also observed negligible friction when pulling a graphite flake out of registry on a

Superlubricity. DOI: https://doi.org/10.1016/B978-0-444-64313-1.00005-3

graphite surface. This effect arises from the incommensurability of the two surfaces in contact, and even if, following Hirano et al. [5], it is commonly known as "superlubricity," it may be better termed *structural lubricity*, as proposed by Müser et al. [6]. Friction also vanishes if the sliding velocity is very low (in the order of nm/s) due to thermally activated "back-jumps" in the stick-slip mechanism, which are more probable if the time of contact increases (*thermolubricity*) [7]. Last but not least, negligible friction can be also achieved in a *dynamic* way [8]. In this case, the contact region is vibrated in the normal direction while keeping contact with the solid surface. This leads to a periodic decrease of the energy barrier preventing the tip slip, thus favoring continuous sliding. Compared to the other methods, dynamic superlubricity is not limited to extremely low loads or velocities, or to specific orientations of the surfaces in contact. Its application to technological devices like micro- and nanoelectromechanical systems (MEMS and NEMS) may be thus easier to explore.

This chapter is organized as follows. The PT model for atomic-scale stick-slip is first discussed in the one-dimensional case at zero temperature. A simple sinusoidal potential is used, which allows to derive analytical expressions relating measurable quantities, like friction and lateral contact stiffness, to the interaction energy between the contacting surfaces. "Steady" and dynamic superlubricity are naturally introduced in this context. This part is followed by a brief presentation of the first experiments on alkali halide and metal surfaces performed at the University of Basel and demonstrating the occurrence of both forms of superlubricity. Very recent measurements of the friction force acting on a single organic molecule and on the anisotropy of friction in UHV conclude the chapter. Even if these measurements did not address superlubricity explicitly, the stick-slip observed in both cases could also be interpreted by original extensions of the PT model thus opening new scenarios for possible transitions to superlubricity. The chapter deliberately ignores chemical means to reduce friction (e.g., using standard oil lubricants), which are addressed elsewhere in this book. Solid lubricants are also not considered here.

5.2 Onset of superlubricity in a sliding point contact

5.2.1 The Prandtl–Tomlinson model for atomic scale stick-slip

In a very first approximation, a sharp tip elastically driven on a crystal surface can be represented by a point mass pulled along a periodic interaction potential. The total potential $U_{tot}(x,t)$ sensed by the tip apex is given by the sum of this potential and of the elastic potential of the composite spring system formed by the tip support (e.g., an AFM cantilever), the tip body, and the contact region. Assuming a sinusoidal profile for the first one, we can write

$$U_{tot}(x,t) = -U_0 \cos\frac{2\pi x}{a} + \frac{1}{2}k(x-vt)^2, \quad (5.1)$$

where U_0 is the amplitude of the interaction potential, a is the lattice constant of the surface lattice, k is the effective lateral spring constant of the system [9,10], and v is the velocity of the support. The minus sign in front of U_0 is introduced in order to have a minimum at $x=0$ when $t=0$. In Fig. 5.1A the total potential $U_{tot}(x,t)$ is shown at two different times t. The chosen parameter values are in the typical range for AFM measurements on alkali halide surfaces in UHV.

At a given time t the tip is located in the first position $x=x_{tip}$, where the first derivative of $U_{tot}(x,t)$ with respect to x is zero:

$$\frac{\partial U_{tot}}{\partial x} = \frac{2\pi U_0}{a}\sin\frac{2\pi x}{a} + k(x - vt) = 0. \tag{5.2}$$

Expanding the sinusoidal term to the first order in Eq. (5.2), we obtain the initial velocity of the tip,

$$v_{tip}(0) = \left.\frac{dx_{tip}}{dt}\right|_{t\to 0} = \frac{v}{1+\eta}, \tag{5.3}$$

where we have introduced the parameter

$$\eta = \frac{4\pi^2 U_0}{ka^2} \tag{5.4}$$

Fig. 5.1B shows a numerical solution of Eq. (5.2), when t increases up to the time t_c when the equilibrium becomes unstable (and the tip slips). The time dependence of the first local maximum of $U_{tot}(x,t)$, $x_{max}(t)$, is also shown in the figure. The difference between the values $U_{tot}(x_{max},t)$ and $U_{tot}(x_{min},t)$ defines the energy barrier $\Delta E(t)$ preventing the tip from slipping away. This barrier decreases with time and vanishes at $t=t_c$ ($=1.38$ s in Fig. 5.1B). The corresponding value x_c can be estimated by combining Eq. (5.2) with the condition that the second derivative of $U_{tot}(x,t)$ with respect to x be zero:

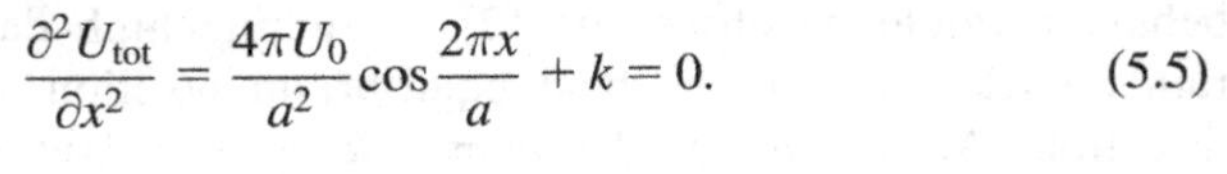

$$\frac{\partial^2 U_{tot}}{\partial x^2} = \frac{4\pi U_0}{a^2}\cos\frac{2\pi x}{a} + k = 0. \tag{5.5}$$

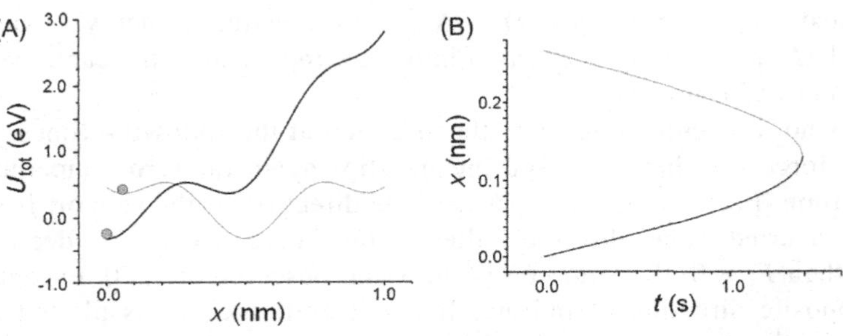

FIGURE 5.1 (A) Potential describing the AFM tip (represented by the circle) sliding over a periodic surface at $t=0$ (thick line) and $t=0.5$ s (thin line). Parameter values: $U_0=0.315$ eV, $k=1$ N/m, $a=0.5$ nm, $v=1$ nm/s. (B) First minimum (lower line) and maximum (upper line) of the potential $U_{tot}(x, t)$ as a function of time.

As a result:

$$x_c = \frac{a}{2\pi}\cos^{-1}\left(-\frac{1}{\eta}\right) \tag{5.6}$$

Note that the tip velocity, corresponding to the slope of the $x(t)$ curve in Fig. 5.1B, tends to infinity when the critical time t_c is approached [11]. The (lateral) spring force at the critical position, $F_c = k(vt\ x_c)$, can be also obtained from Eqs. (5.2) and (5.3):

$$F_c = \frac{ka}{2\pi}\sqrt{\eta^2 - 1} \tag{5.7}$$

Equations (5.6) and (5.7) clearly show that the *elastic instability* causing the slip can be reached only if $\eta > 1$, that is, if the contact is sufficiently loose and/or the tip-sample interaction is sufficiently strong. The lateral force $|F_L| = k(vt - x_{tip})$ can also be analytically evaluated when $t \to 0$. In such case, the tip velocity is given by (3), so that

$$|F_L(t)| = \frac{\eta}{\eta + 1}kvt. \tag{5.8}$$

If $\eta \gg 1$, the lateral spring constant k is approximately given by the ratio $|F_L(t)|/vt$, which can be directly measured in the experiments. On the other hand, if $\eta \to 1$, this ratio tends to *half* the value of k. Note also that the lateral force reaches its maximum value F_{max} slightly before x_c, when $x_{tip} = a/4$ [3]. The following simple relation is valid in all cases:

$$F_{max} = \frac{2\pi U_0}{a} \tag{5.9}$$

When a slip occurs, the cantilever starts to oscillate at its torsional resonance. If these oscillations were not heavily damped, a rather chaotic behavior would be observed [12]. Assuming high damping, the tip will rather stick again to the next equilibrium position x_c' in the positive x direction. At this point, the lateral force $F_c' = k(vt_c - x_c') = 0.83$ nN (as estimated also numerically). A corresponding energy amount $\Delta U = |\Delta F_L|a = 3.36$ eV is released into the sample and/or the cantilever in the form of acoustic waves.

It is not difficult to describe the situation at the following times. The lateral force $|F_L|$ increases and the tip slips again (at zero temperature) every time $|F_L| = F_c$. What happens if the direction of the support is suddenly inverted? The absolute value of the lateral force $|F_L|$ decreases and, when $F_L = 0$, the situation is the same observed at $t = 0$, except for the opposite direction of pulling. If the lateral force F_L is plotted as a function of the support position in a forward and backward scanning one gets a so-called *friction loop* similar to those measured with AFM (Fig. 5.2).

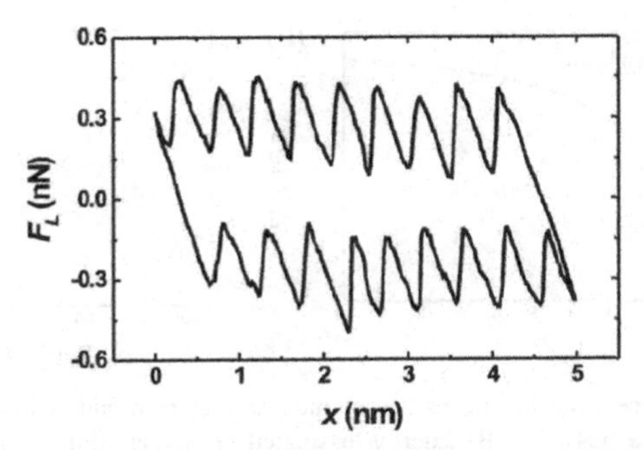

FIGURE 5.2 Friction loop observed on NaCl. The nonuniform height distribution of the peaks is due to thermal activation, as explained in the cited reference. *Reproduced with permission from [13] E. Gnecco, R. Bennewitz, T. Gyalog, Ch. Loppacher, M. Bammerlin, E. Meyer, et al., Phys. Rev. Lett. 84 (2000) 1172. © American Physical Society.*

5.2.2 The superlubric regime

The energy amount ΔU released in a slip can be approximately estimated as a function of the friction parameter η. If $\eta < 1$, the elastic instabilities are suppressed, and no abrupt release of energy occurs. Correspondingly $\Delta U = 0$, a case which may be referred to as the *superlubric regime*. As shown in Ref. [12], the transition from stick-sip to superlubricity is continuous, since the energy dissipation vanishes when $\eta \to 1$ as $\Delta U \propto (\eta - 1)^2$. In this sense, the friction parameter η behaves like the order parameter in a second-order phase transition. The relation between ΔU and η can be also estimated in the limit case $\eta \to \infty$, when the dependence $\Delta U \propto \eta$ is found [12].

In order to relate the energy dissipation to the normal force F_N acting on the tip, the relation between the friction parameter η (or the energy amplitude U_0) and F_N must be known. In the experiment on NaCl presented in Section 3.1, a linear relation was found [3]. The normal force F_N required to enter the superlubric regime was in the order of 1 nN, which is close to the threshold of the instrumental noise level in UHV. However, the restriction on the loading value could be easily overcome by oscillating F_N at a given frequency $f \gg v/a$. If the oscillation amplitude is small, this is equivalent to replace the constant U_0 in Eq. (5.1) with $U_0(1 + \alpha \cos 2\pi f t)$, where $\alpha < 1$. In this case the amplitude of the interaction potential is replaced by an effective value $U_0(1 - \alpha)$ and the condition for the occurrence of superlubricity becomes

$$\eta_{\min} = \eta(1 - \alpha) < 1 \tag{5.10}$$

We can also say that the curve $= \alpha_c$, where

$$\alpha_c = 1 - 1/\eta \tag{5.11}$$

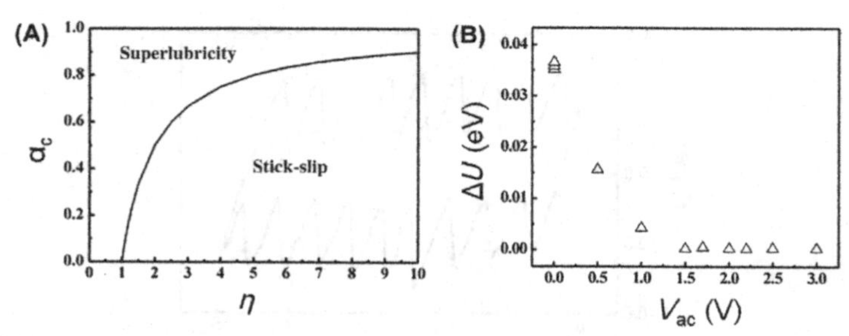

FIGURE 5.3 (A) Depending on the values of the parameters η and α introduced in the text, two sliding regimes are possible. (B) Energy dissipated per cycle while scanning a 1-mm-thick NaCl sample with a voltage V_{ac} applied between tip and surface (with frequency of 56.7 kHz). An average normal force $F_N = 2.73$ nN was kept constant by a feedback loop.

divides the (η,α) plane in two regions (Fig. 5.3A). Below this curve the normal oscillations do not suppress the instabilities leading to stick-slip and dissipation. Above it the tip slides continuously, and the friction is limited to the viscous force causing the damping in the presence of stick-slip.

5.3 Experimental evidences of superlubricity

5.3.1 Steady case

The transition from stick-slip to superlubricity was observed by Socoliuc et al. [14] with a home-built friction force microscope operated at room temperature and in UHV conditions. The experiments were performed on NaCl single crystals cleaved in UHV and heated at 150°C to remove charges produced in the cleaving process. Fig. 5.4 shows three lateral force profiles corresponding to three different values of the normal force F_N. Note that the adhesion force between tip and sample has been subtracted. This force (~0.7 nN) corresponds to the set point leading the tip out of contact. For $F_N = 4.7$ nN, the friction loop consists of two opposite sawtooth curves with the periodicity of the crystal lattice along the (100) direction corresponding to stick-slip motion (Fig. 5.4A). The area enclosed by the loop is the total energy dissipated when scanning forward and backward. If F_N is lowered to 3.3 nN, the two profiles are found to overlap partially (Fig. 5.4C) and the energy loss is consequently reduced. After reducing F_N further down to -0.47 nN, the overlap is complete and within the sensitivity of the experimental setup, no energy loss could be detected (Fig. 5.4E). The sawtooth profile is also replaced by a wavy profile still showing the periodicity of the surface lattice. The theoretical evolution of the lateral force for different values of the parameter η matching the experimental conditions is shown in Fig. 5.4B, D, and F.

According to Eq. (5.9), the amplitude U_0 of the tip-sample interaction potential is proportional to the maximum absolute value of the lateral force

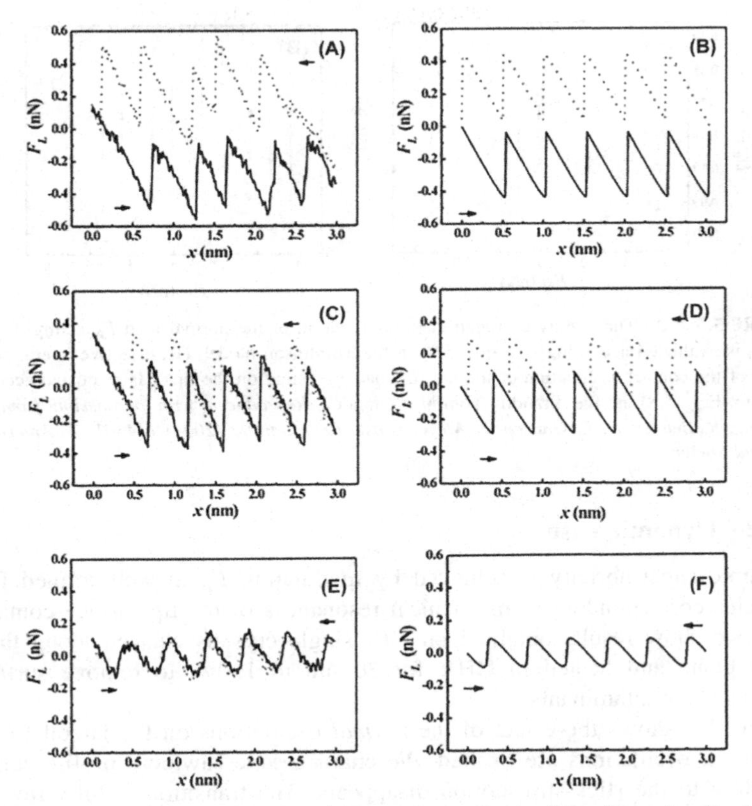

FIGURE 5.4 Measurements of the lateral force acting on the tip sliding forward and backward in the (100) direction over a NaCl(001) surface. Cross-sections through a two-dimensional scan obtained for external loads: (A) $F_N = 4.7$ nN, (C) $F_N = 3.3$ nN, and (E) $F_N = -0.47$ nN. Corresponding numerical evaluation of F_L from the PT model for (B) $\eta = 5$, (D) $\eta = 3$, and (F) $\eta = 1$. *Reproduced with permission from A. Socoliuc, R. Bennewitz, E. Gnecco, E. Meyer, Phys. Rev. Lett. 92 (2004) 134301. © American Physical Society.*

F_{max}, which can be deduced from the curves in Fig. 5.4A, C, and E. In the measurements by Socoliuc et al. [3], the dependence of U_0 on F_N turned out to be a simple proportionality, as seen in Fig. 5.5A. The increase in the potential amplitude with increasing normal load can be intuitively understood as an increase of the barrier height between adjacent atomic positions when the contacting atoms are pressed closer toward the surface lattice. A similar behavior was also reported by Riedo et al. [15] on freshly cleaved and atomically smooth muscovite mica surface in controlled humidity environment. Finally, Fig. 5.5B shows the load dependence of the lateral stiffness k, as determined from the lateral profiles using Eq. (5.8). This quantity is in the order of 1 N/m and almost independent of the load.

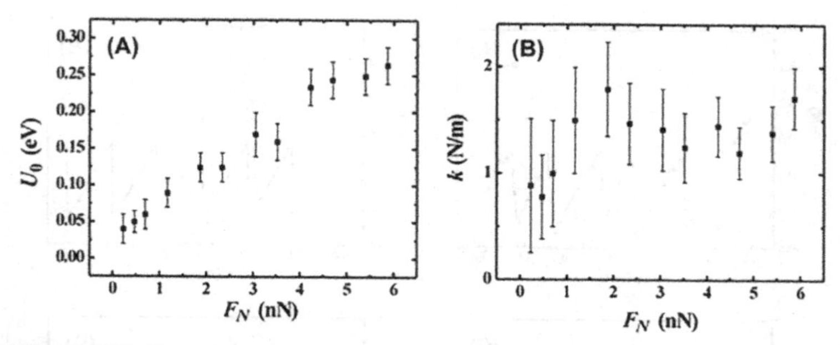

FIGURE 5.5 (A) The energy corrugation U_0 as function of the normal load F_N acting on the tip. U_0 is evaluated accordingly to Eq. (5.9) in the Tomlinson model, (B) effective lateral stiffness k of the contact as function of the normal load F_N acting on the tip. k is evaluated accordingly to Eq. (5.8) in the Prandtl–Tomlinson model. *Reproduced with permission from A. Socoliuc, R. Bennewitz, E. Gnecco, E. Meyer, Phys. Rev. Lett. 92 (2004) 134301. © American Physical Society.*

5.3.2 Dynamic case

Dynamic superlubricity is achieved by modulating F_N at well-defined frequencies corresponding to mechanical resonances of the tip-surface contact. Here we show results obtained on KBr single crystals cleaved along their (100) plane and heated in UHV for 30 min to 120°C to remove surface charges and contaminants.

Fig. 5.6 shows the effect of the *normal* oscillations on the lateral force. When the oscillations are excited, the characteristic sawtooth profile corresponding to the stick-slip motion disappears. The transition is fully reversible, as seen when the oscillations are removed backward (exactly at the same support position). In the following we discuss how the frequency for obtaining the effect is chosen and how the normal load oscillation is applied to the contact. Fig. 5.7A shows the resonance peaks of the cantilever in contact with the KBr surface. The peak values are in good agreement with the theoretical resonance frequencies of different vibration modes calculated for pinned-configuration [16]. The resonant excitation of the cantilever in contact with the surface was conveniently achieved by applying an *ac* voltage between the tip and the sample holder plate, V_{ac}. The crystal then acts as a dielectric spacer. Fig. 5.7B shows the mean lateral force as a function of the applied frequency. Friction is strongly reduced when the frequency matches one of the bending resonance frequencies of the pinned lever or half those values. Dynamic superlubricity was not achieved by exciting the torsional mode of the cantilever, although this result was possibly due to an oscillation amplitude not large enough [17]. Since the sample thickness ($\sim$1 mm) was much larger than the tip height, the capacitive interaction occurred mainly between the cantilever body and the sample holder. This interaction resulted in a capacitive force, F_C, proportional to the square of

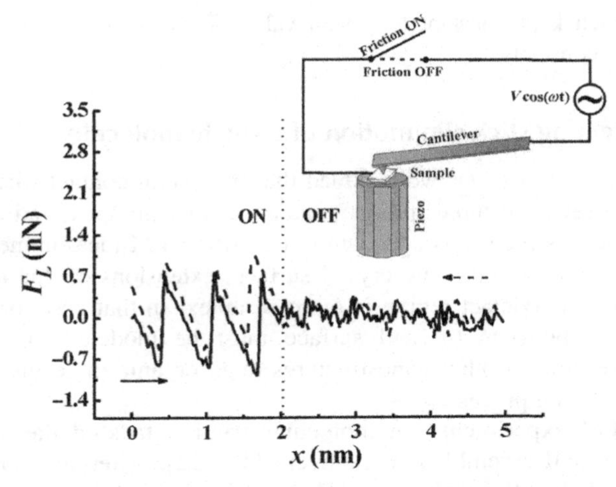

FIGURE 5.6 Frictional force detected by scanning forward (continuous line) and backward (dashed line) on the atomically flat KBr surface. An average normal load $F_N = 0.67$ nN was kept constant by a feedback loop. An *ac* voltage with $f = 41$ kHz and amplitude of 5 V was applied and later removed when the cantilever was displaced 2 nm from its initial position.

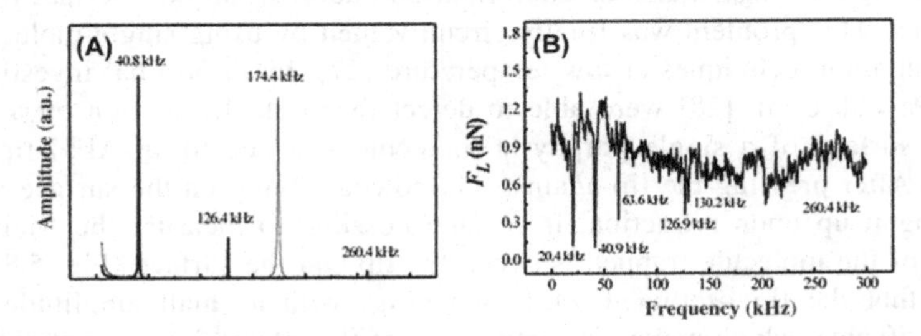

FIGURE 5.7 (A) Thermal noise spectrum of the normal (continuous line) and torsional (dashed line) oscillations of a silicon cantilever in contact with a KBr surface. (B) Average value of the frictional force recorded while applying an *ac* voltage with frequency swept between 0 and 300 kHz. Average friction force recorded with a lower sweeping rate (not shown) proves that friction falls below the sensitivity of the AFM at the resonance frequency $f_{n1} = 40.8$ kHz. *Reproduced with permission from A. Socoliuc, E. Gnecco, S. Maier, O. Pfeiffer, A. Baratoff, R. Bennewitz, et al., Science 313 (2006) 207. © American Association for the Advancement of Science.*

the applied voltage. The attractive force F_C oscillated with twice the excitation frequency f. Besides F_C, any charge trapped at the tip or any charge layer at the surface resulted in a nonzero contact potential and in an additional force F_Q, which oscillated at the actuation frequency f [18]. Thus when one of the frequencies f or $2f$ matched a bending resonance of the pinned lever, the oscillation amplitude caused the normal force F_N, and the energy corrugation U_0 to vary between two extreme values. The quick variation of the electrostatic forces could not be followed by the distance-controlling feedback, which kept constant the mean value of the total normal force over several lattice constants.

5.3.3 Detecting stick-slip motion of a single molecule

In the discussion run so far we assumed that the lateral contact stiffness corresponds to a series of three springs: contact region, tip apex, and cantilever torsion. While this picture is adequate to describe AFM measurements when a tip is sliding on an inorganic crystal surface, extensions of this model are required when the contact region is more complex. In that case, or not only the interaction between tip and surface must be modeled, but also the degrees of freedom of other nanostructures (e.g., organic molecules) playing a role in the sliding process.

Early AFM experiments on molecular friction tackled the sliding of sharp tips on self-assembled monolayers [19–22]. Alternatively the tips were covered by these layers and dragged over inorganic surfaces [23–26]. Both experimental approaches result in friction profiles similar to Fig. 5.2. However, their interpretation is hampered by the uncertainties regarding the number of interacting molecules between tip and sample as well as by their instantaneous conformation due to sliding or thermal fluctuations. This problem was finally circumvented by using single-molecule manipulation techniques at low temperature [27]. In an original investigation Pawlak et al. [28] were able to detect the stick-slip motion on a Cu (111) surface of a single porphyrin molecule attached to an AFM tip at 4.8K. After pressing the tip against the molecule lying on the surface and picking it up upon retraction, it became possible to measure the friction force of the molecule trapped between the tip and the surface (Fig. 5.8A). Note that the tip oscillated while scanning (with a small amplitude of about 50 pm), which makes the estimation of the lateral force not straightforward. Rather than the time evolution of this quantity, Fig. 5.8B shows that of the *normal* stiffness k_{ts} detected when the molecule is moved forward and backward along the Cu(111) surface lattice. The monotonic increasing followed by sudden drops of this quantity, suggests a possible stick-slip behavior of the decorated tip.

To elucidate the contribution of intramolecular mechanics to the friction pattern, numerical calculations were performed based on an extended PT model. While a porphyrin molecule contains many internal degrees of freedom, it was postulated that the relevant molecular spring dictating the friction response is the σ-bond linking the porphyrin leg in contact to the surface to its core as shown in Fig. 5.8A. The system was then numerically described as a single-point mass (symbolizing the interaction of a single nitrogen atom with the Cu(111)) dragged over the surface potential. The relevant stiffnesses corresponding to the σ-bond characteristics of the molecule were parameterized using density functional theory (DFT) calculations. The

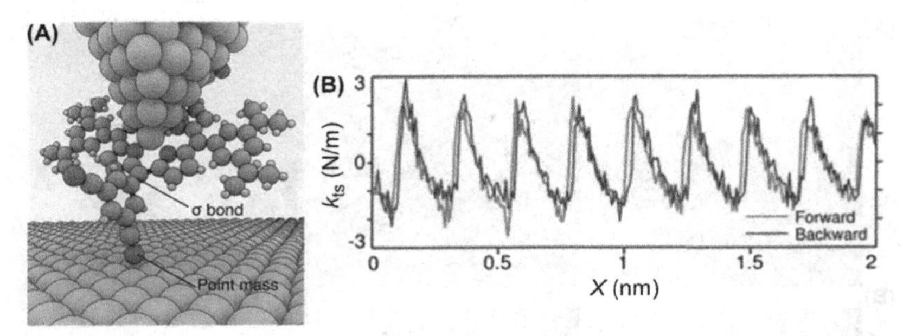

FIGURE 5.8 (A) Illustration of molecule manipulation experiment performed at low temperature to measure single-molecule friction. (B) Experimental tip-sample stiffness k_{ts} trace revealing stick-slip events. *Reprinted with permission from R. Pawlak, W. Ouyang, A.E. Filippov, L. Kalikhman-Razvozov, S. Kawai, T. Glatzel, et al. ACS Nano 10 (2016) 13–722. © 2016 American Chemical Society.*

simulations remarkably reproduced the experimental data and demonstrated the PT model's applicability down to a single-atom contact.

5.3.4 Anisotropy and asymmetry effects

Measuring anisotropy and asymmetry effects accompanying frictional processes is difficult because AFM measurements are usually restricted to a fixed scan direction perpendicular to the cantilever axis. This requires either nontrivial methods for the calibration of forces along different scan orientations or the ability to rotate the sample with respect to the probing tip via a rotatable stage [29], specially designed AFM setups [30], measurements along different scan angles without physically rotating the sample [31–33] or single measurements on domains with different crystallographic orientations [21,34].

Fig. 5.9 shows measurements on NaCl performed with a homemade AFM allowing an in situ sample rotation. In this way anisotropy measurements could be performed with very precise alignment of the crystal orientations to the coordinate system of the force sensor [35]. To interpret the experimental findings, calculations on the base of a PT model were performed using an ab initio calculated corrugation potential instead of the commonly used sinusoidal potential (introduced in Section 2 and extended to two dimensions in [36,37]). In the calculations the atomic cores were allowed to find their relaxed positions every time the position of the tip is changed with respect to the sample [35]. Friction force maps at $T = 0$ and 300 K along the [100] and [110] directions are shown in Fig. 5.9 together with the calculation results. The standard PT model with sinusoidal potential cannot indeed reproduce the experimentally observed features, namely, the asymmetric shape of the friction "diamonds," the shift of the forward and backward friction maps along the slow scan axis and as a result of it, the asymmetry of the tip path in forward and

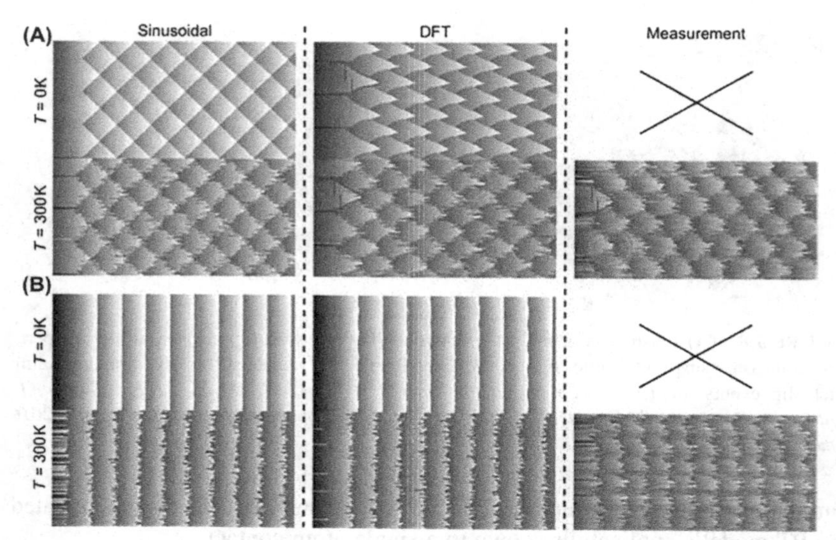

FIGURE 5.9 Friction force maps at $T=0$ and 300K along (A) the [100] and (B) the [110] directions calculated using a sinusoidal potential (left) and the density functional theory potential (middle). Measurements (right) are only available for room temperature. Typical nose-like features are marked with *green* circles, while the asymmetric feature of the initial slip is indicated by *red* and *black* marks. *Reproduced with permission from G. Fessler, A. Sadeghi, Th. Glatzel, S. Goedecker, E. Meyer, Tribol. Lett. 67 (2019) 59. © 2019 Springer Nature.*

backward direction with five different characteristic modes of tip motion. Instead the advanced DFT calculations including the more realistic silicon tip and allowing for relaxations of the atomic positions of tip and surface are able to reproduce the observations quite well. In the orientation along [110] direction, the measurements revealed a reduction of friction of almost 27% on one ionic species compared to the other. According to the simulation, the smaller friction is measured on the Cl^- sites. This ionic contrast in the [110] orientation is a material contrast, unlike in the [100] situation.

5.4 Conclusions and outlook

In conclusion, we have first presented two simple methods to achieve superlubricity on the atomic scale, and supported them with the theoretical basis of the PT model. Our results, obtained on alkali halide crystal surfaces, may not be easily applied to a general class of macroscopic bodies. It will usually be impossible to find one excitation frequency that is resonantly enhanced for a major part of the microscopic contacts between the surfaces. The situation is, however, totally different for micromechanical devices. These devices have an enormous surface-to-volume ratio and friction is a major problem in their application. On the other hand, contacting parts in MEMS often are small enough to constitute single asperity contacts, and their structure favors the development of distinct resonances. Consequently the method demonstrated here could find an interesting and important application in overcoming the problem of static friction in MEMS. New insights on the transition from stick-slip to superlubricity are expected from friction measurements based on the nanomanipulation of single molecules and on AFM with rotatable sample holders, which have recently appeared in the literature.

References

[1] C. Mate, G. McClelland, R. Erlandsson, S. Chiang, Phys. Rev. Lett. 59 (1987) 1942.
[2] D. Tomanek, W. Zhong, H. Thomas, Europhys. Lett. 15 (1991) 887.
[3] A. Socoliuc, R. Bennewitz, E. Gnecco, E. Meyer, Phys. Rev. Lett. 92 (2004) 134301.
[4] M. Dienwiebel, G. Verhoeven, N. Pradeep, J. Frenken, J. Heimberg, H. Zandbergen, Phys. Rev. Lett. 92 (2004) 126101.
[5] M. Hirano, K. Shinjo, R. Kaneko, Y. Murata, Phys. Rev. Lett. 78 (1997) 1448.
[6] M.H. Müser, L. Wenning, M.O. Robbins, Phys. Rev. Lett. 86 (2001) 1295.
[7] S. Yu Krylov, K.B. Jinesh, H. Valk, M. Dienwiebel, J.W.M. Frenken, Phys. Rev. E 71 (2005) R65101.
[8] A. Socoliuc, E. Gnecco, S. Maier, O. Pfeiffer, A. Baratoff, R. Bennewitz, et al., Science 313 (2006) 207.
[9] R.W. Carpick, D.F. Ogletree, M. Salmeron, Appl. Phys. Lett. 70 (1997) 1548.
[10] M.A. Lantz, S.J. O'Shea, M.E. Welland, K.L. Johnson, Phys. Rev. B 55 (1997) 10776.
[11] E. Gnecco, R. Bennewitz, T. Gyalog, E. Meyer, J. Phys. Condens. Matt. 13 (2001) R619.
[12] E. Gnecco, R. Roth, A. Baratoff, Phys. Rev. B 86 (2012) 035443.
[13] E. Gnecco, R. Bennewitz, T. Gyalog, Ch Loppacher, M. Bammerlin, E. Meyer, et al., Phys. Rev. Lett. 84 (2000) 1172.
[14] L. Howald, E. Meyer, R. Lüthi, H. Haefke, R. Overney, H. Rudin, et al., Appl. Phys. Lett. 63 (1993) 117.
[15] E. Riedo, E. Gnecco, R. Bennewitz, E. Meyer, H. Brune, Phys. Rev. Lett. 91 (2003) 084502.
[16] U. Rabe, et al., Rev. Sci. Instr. 67 (1996) 3281.
[17] R. Roth, O.Y. Fajardo, J.J. Mazo, E. Meyer, E. Gnecco, Appl. Phys. Lett. 104 (2014) 083013.
[18] M.R. Weaver, D. Abraham, J. Vac. Sci. Technol. B 9 (1991) 1559.
[19] E. Meyer, et al., Thin Solid. Films 220 (1992) 132–137.
[20] R. Lüthi, E. Meyer, H. Haefke, L. Howald, W. Gutmannsbauer, H.J. Güntherodt, Science 266 (1994) 1979–1981.
[21] M. Liley, D. Gourdon, D. Stamou, U. Meseth, T.M. Fischer, C. Lautz, et al., Science 280 (1998) 273–275.
[22] A.R. Burns, J.E. Houston, R.W. Carpick, T.A. Michalske, Phys. Rev. Lett. 82 (1999) 1181–1184.
[23] C.D. Frisbie, L.F. Rozsnyai, A. Noy, M.S. Wrighton, C.M. Lieber, Science 265 (1994) 2071–2074 [33].
[24] T. Ito, M. Namba, P. Bühlmann, Y. Umezawa, Langmuir 13 (1996) 4323–4332.
[25] V.V. Tsukruk, V.N. Bliznyuk, Langmuir 2 (1998) 446–455.

[26] G. Fessler, I. Zimmermann, T. Glatzel, E. Gnecco, P. Steiner, R. Roth, et al., Appl. Phys. Lett. 98 (2011) 083119.
[27] R. Pawlak, S. Kawai, T. Meier, Th Glatzel, A. Baratoff, E. Meyer, J. Phys. D App. Phys 50 (2017) 113003.
[28] R. Pawlak, W. Ouyang, A.E. Filippov, L. Kalikhman-Razvozov, S. Kawai, T. Glatzel, et al., ACS Nano 10 (2016) 13–722.
[29] F. Trillitzsch, R. Guerra, A. Janas, N. Manini, F. Krok, E. Gnecco, Phys. Rev. B 98 (2018) 165417.
[30] M. Dienwiebel, N. Pradeep, G.S. Verhoeven, H.W. Zandbergen, J.W.M. Frenken, Surf. Sci. 576 (2005) 197.
[31] Y. Namai, H. Shindo, Jpn. J. Appl. Phys. 39 (2000) 4497.
[32] S.G. Balakrishna, A.S. de Wijn, R. Bennewitz, Phys. Rev. B 89 (2014) 245440.
[33] C.M. Almeida, R. Prioli, B. Fragneaud, L.G. Cançado, R. Paupitz, D.S. Galvão, et al., Sci. Rep. 6 (2016) 31569.
[34] R.W. Carpick, D.Y. Sasaki, A.R. Burns, Tribol. Lett. 7 (1999) 79.
[35] G. Fessler, A. Sadeghi, Th Glatzel, S. Goedecker, E. Meyer, Tribol. Lett. 67 (2019) 59.
[36] P. Steiner, R. Roth, E. Gnecco, A. Baratoff, S. Maier, T. Glatzel, et al., Phys. Rev. B 79 (2009) 045414.
[37] P. Steiner, R. Roth, E. Gnecco, A. Baratoff, E. Meyer, Phys. Rev. B 82 (2010) 205417.

Chapter 6

Structural superlubricity in large-scale heterogeneous layered material junctions

Ming Ma[1,2,3] and Quanshui Zheng[1,2,4]

[1]Center for Nano and Micro Mechanics, Tsinghua University, Beijing, P.R. China, [2]State Key Laboratory of Tribology, Tsinghua University, Beijing, P.R. China, [3]Department of Mechanical Engineering, Tsinghua University, Beijing, P.R. China, [4]Department of Engineering Mechanics, Tsinghua University, Beijing, P.R. China

6.1 Introduction

During the past two decades great advances in the basic understanding of the physical processes underlying friction and wear at the atomistic level led to breakthroughs in the field of nanoscale tribology [1–4]. Nowadays, the design and fabrication of microscopic and nanoscopic electromechanical devices that present controllable mechanical and frictional properties are routinely performed in many laboratories worldwide [5–11]. With this respect, unique tribological characteristics such as superlubricity [12,13] (a state of nearly zero friction) have been demonstrated at nanoscale-layered materials junctions [14] allowing for durable wear-less operation. This is crucial when considering such miniature systems, whose surface-to-volume ratio is extremely large, especially because traditional liquid phase lubricants become too viscous when confined to nanoscale dimensions.

One of the pioneering experiments in this field demonstrated interlayer registry-dependent friction in pristine graphitic interfaces [14]. It was shown that when the lattices of the two contacting surfaces are in registry, relatively high interlayer friction values are measured, while when taken out of registry friction practically vanishes. This unique effect, termed structural superlubricity (SSL), marked layered materials as promising candidates for nanoelectromechanical systems as well as improved lubricants or lubrication additives for macroscopic devices [15–19].

Despite the great potential, this phenomenon exhibits one major obstacle to its utilization is the fact that homogeneous junctions tend to lock in their

Superlubricity. DOI: https://doi.org/10.1016/B978-0-444-64313-1.00006-5

high friction configurations thus eliminating their superlubric capabilities [20–24]. Furthermore, until very recently, observations of superlubric behavior have been limited to nano- and microscale graphitic contacts [9,23,25,26]. The main challenges today are to achieve robust superlubricity [15–19,27,28], sustainable against external load, high velocities, and varying contact geometries [29], and eventually to scale this phenomenon up to microscopic and macroscopic system dimensions [30–32]. To meet this challenge, heterogeneous layered materials junctions that exhibit surface incommensurability for all possible interlayer orientations leading to effective cancellation of interfacial interactions are among the most promising candidates. Importantly, incommensurability in these systems increases with contact size due to the appearance of Moiré patterns resulting from the intrinsic interlayer lattice constant mismatch [31,33]. In addition, external load, naturally existing in frictional contacts, is expected to reduce interfacial corrugation thus further supporting robust superlubricity [29].

In this chapter, we will summarize the state-of-the-art research results by dividing them into theoretical (including numerical simulations) and experimental aspects. For theoretical part, both one-dimensional (1D) and two-dimensional (2D) models will be covered since they all have experimental correspondence. The effects of surface adsorbates, that is, contaminations, on SSL will also be discussed. This is due to the fact that although the simplicity of the theoretical models benefits the understanding of fundamental principles underlying SSL, recent experimental observations suggest that surface adsorbates may play an important role in the tribological properties in practice [34–36]. For experimental side, we will cover various systems exhibiting structural superlubric properties, including graphite/hexagonal boron nitride (hBN) [37], graphite/MoS_2 [38], graphite/Au [39], graphite/Pt [40], and so on. In the end of the chapter, perspectives on the challenges and future directions will be discussed. Last but not the least, we want to note that the contents discussed in this chapter do not intend to cover all the literatures, but only to give the readership an overview of this field.

6.2 Structural superlubricity in homogeneous interfaces

The first convincing experimental validation of superlubricity was reported by Dienwiebel et al. in 2004 [14]. Using a lab-built scanning probe microscope, they observed a frictional anisotropy with sixfold symmetry while sliding a tip on graphite surface (Fig. 6.1). Simulation revealed that the sliding was actually happening between a 2 nm^2 graphene flake adhered to the tip and the graphite substrate [25]. The sixfold symmetry, which was attributed to commensurability and crystalline structure of graphite, along with the ultralow friction measured in incommensurate state, constituted a strong evidence of superlubricity. The underlying physical process can be described within a generalized Prandtl–Tomlinson model [25]. Recently, techniques

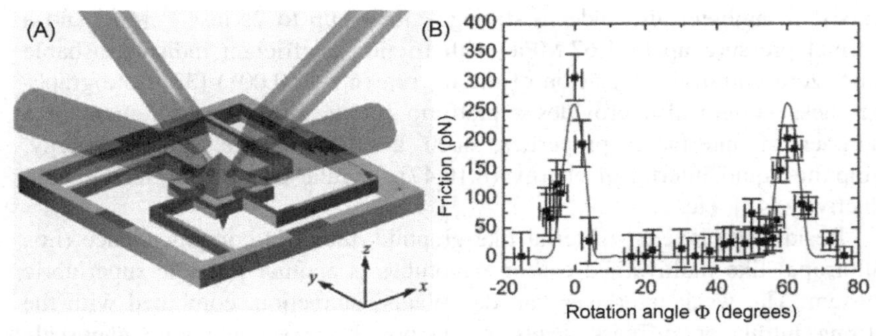

FIGURE 6.1 (A) High-resolution AFM equipment used to detect frictional force. (B) Superlubricity achieved in nanoscale graphite-graphite contacts [14]. (*AFM*: Atomic force microscope).

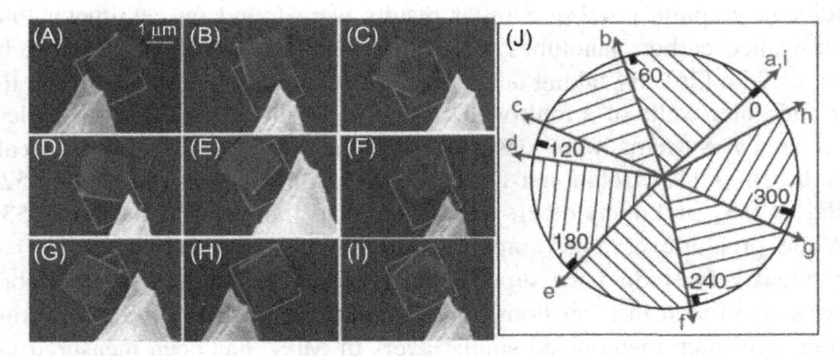

FIGURE 6.2 Microscale superlubricity achieved in graphite-graphite contacts. The commensurate states in (A)–(I) clearly indicating the 60-degree symmetry (J) [23].

were developed to directly grow or transfer graphene sheets onto an atomic force microscope (AFM) probe [38,41]. Within the nanoscale contact formed by the graphene wrapped probe and graphite, superlubricity was found to persist under ultrahigh contact pressure (~1 GPa) [42–44].

Superlubricity of homogenous interface on the microscale was firstly realized on a graphite mesa system in 2008 [45] and then confirmed in 2012 [23]. By shearing and cleaving a micrometer-sized graphite mesa using a tungsten tip, a graphite-graphite incommensurate interface with contact size up to $10\times10\ \mu m^2$ can be created. The frictional stress was measured to be 0.02–0.04 MPa [10,23,32]. Under such a low-friction condition, with the help of interfacial energy, a self-retraction motion was observed when the upper part of the mesa was sheared out and suddenly released. The self-retraction motion disappeared for narrow ranges of angle corresponding to commensurate state due to the drastically increased friction (Fig. 6.2) [23]. For incommensurate angles, the superlubricity of graphite mesa was found to persist in ambient air, under a sliding velocity up to 25 m s^{-1} [46], and a normal pressure up to 1.67 MPa, with friction coefficient indistinguishable from zero within the precision of the instruments (<0.001) [32]. The graphite mesa system also provides a platform for the direct measurement of a number of interfacial properties, such as the graphite surface energy, graphite-liquid interfacial energy [8,10,47], angular-dependent electric conductivity [48], etc.

Besides 2D layered material like graphite, their 1D correspondence (i.e., allotrope) like multiwalled carbon nanotubes is another possible superlubric system. The weak interlayer van der Waals interaction, combined with the strong intralayer stiffness, leads to the possibility of achieving nanoscale superlubricity on the interface between the inner and outer walls, when they are incommensurate. Indeed, the frictional force acting between the inner and outer shells during sliding was experimentally estimated below the measurement resolution (1.4×10^{-15} N $atom^{-1}$) [49], which resembles the superlubricity in graphite [14,23]. Similar results were found on centimeter-long double-walled carbon nanotubes, where frictional stress was measured to be as low as 9 Pa [26]. By taking advantage of the ultralow friction between the inner and outer walls of a multiwall carbon nanotube, a variety of nanoelectromechanical systems were designed, such as ultralow-friction nanoscale linear bearings [50], rotational actuators [51], tunable nanoresonators [52], oscillators [53], and many others after Zheng and Jiang's early proposal [53].

While graphitic carbon (graphite, graphene, carbon nanotubes, etc.) is most intensively studied for superlubricity in layered materials, superlubricity was also found in other homogenous contact pairs. The ultralow friction between two micrometer-sized single layers of MoS_2 has been measured via in situ scanning electron microscope technique with a Si nanowire force sensor [54]. The friction coefficient was measured on the order of 10^{-4} for incommensurate contact, comparable to the value of 10^{-3} measured in nanoscale misaligned homogeneous MoS_2 contact [55]. This method also renders a new approach to obtain SSL via assembling 2D materials. Deng et al. reported observations on microscale SSL in assembled graphite flakes under ambient condition [34]. By making use of a combination approach of swiping phenomenon [56] and enhanced diffusion [57], the amount of contamination at the interfaces was reduced during running-in process. Such an extension of the microscale fabrication/assembly method provides a key step toward practical applications for SSL.

To further increase the size of superlubricity into macroscale regime, great challenges have been met. Since SSL requires atomically flat single crystalline rigid surfaces in perfectly incommensurable contact, only very few types of materials have the potential to meet such stringent requirement. It is hardly possible to obtain ideal single-crystalline contact interface on the macroscale, considering the grain sizes of highly oriented pyrolytic graphite (HOPG), MoS_2, and hBN are approximately 10 μm, 2 mm, and 20 μm, respectively [58], not counting the intrinsic limitation of superlubricity posed by elasticity when contact scale increases [59]. Berman et al. conceived a novel approach to obtain apparent macroscale superlubricity with friction coefficients of 4×10^{-3} in diamond-like carbon (DLC)/graphene-scroll junctions [7] by partially making use of SSL between graphene layers. Given the multicontacts nature, the real contact area, however, could be orders lower than the nominal contact area.

The SSL observed in different homogeneous interfaces is essential for the function of machines from nano- to macroscale. However there are several fundamental limitations for applying superlubricity of homogenous interfaces to practical device. In previous work [14], the superlubricity between graphite flakes and an underlying HOPG substrate was found stable only for a finite time [20]. Since the commensurate state is typically energy favorable over the incommensurate one, a spontaneous rotation into the high frictional commensurate states easily leads to the failure of superlubricity and locks the interface [20,60,61]. A method was recently proposed to constrain the spontaneous rotation through interfacial energy and by careful design of the contact shape [47]. However it only ensures superlubric sliding in translational motions. Superlubric rotational bearings fabricated from homogenous contacts are still impossible, since they lock themselves in high frictional state whenever commensurate state is met. This puts a severe obstacle against the application in micromachine sphere, such as hard disk drive and durable rotational microelectromechanical systems.

Normal load was also found to induce the failure of superlubricity on graphitic homogeneous contacts. High loading forces were found to destroy the stability of superlubric orbits [22]. Breakdown of superlubricity can also occur due to normal load-induced local commensurability of contacting surfaces resulting from strong in-plane distortions [62]. Vertical forces also contributed to the out-of-plane deformation of the carbon atoms at the edge of the flake, which leads to the transition from smooth sliding to stick-slip with high-energy dissipation when the edges are saturated by hydrogen [63]. Thus for homogenous contacts, the robustness of superlubricity remains a great challenge against its scalable application.

6.3 Structural superlubricity in heterojunctions

6.3.1 Theoretical research on superlubricity of heterojunctions

Research on graphene and other 2D atomic crystals is intense and likely to remain one of the leading topics in condensed matter physics and materials science [64]. These studies deeply contribute to SSL for that 2D materials is characterized by superhigh in-plane stiffness and ultralow interlayer interaction. Compared with homojunctions, the abundance of the potentially various kinds of heterojunctions [64] could also be a driving force for the rich studies along this line.

While recently there is a surge of studies on the tribological properties of superlubric heterojunctions, interestingly, theoretical study pioneers this direction much earlier (Fig. 6.3). Based on Frenkel–Kontorova model, Peyrard et al. [65] show that for stiff crystals on a regular substrate, the static frictional behavior is determined by commensurability. Specially, for incommensurate contact the static friction does not scale with the contact area,

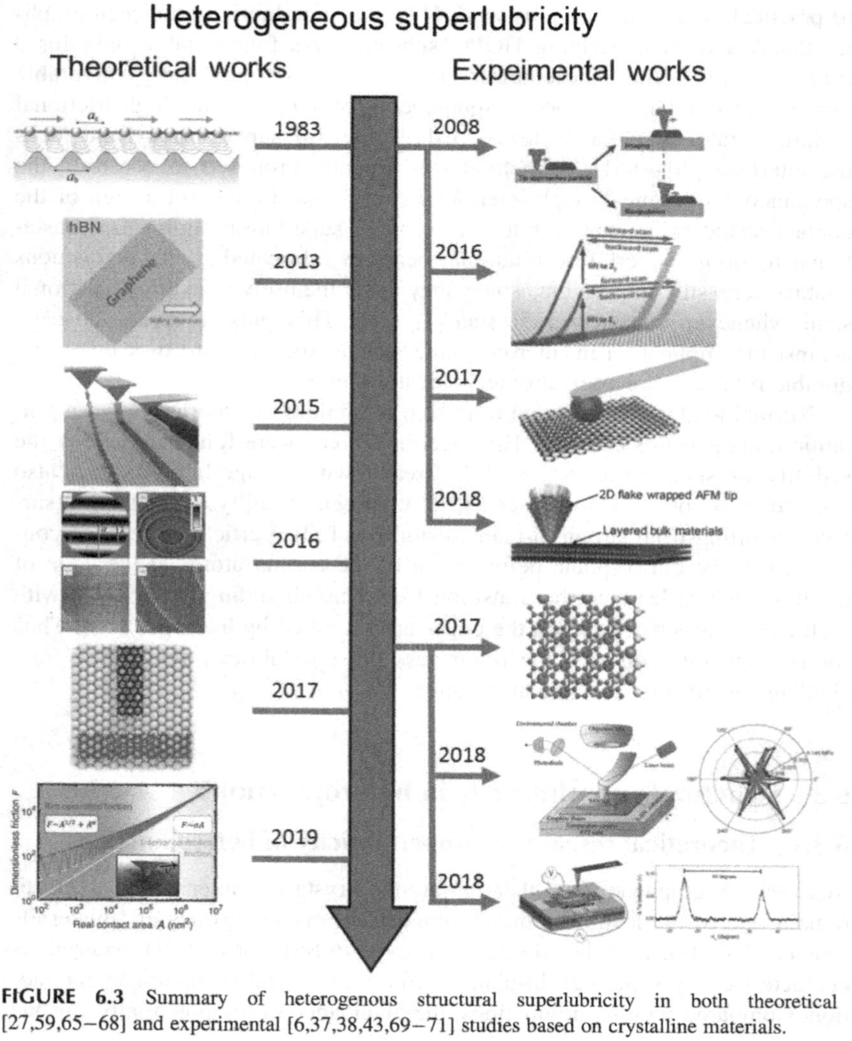

FIGURE 6.3 Summary of heterogenous structural superlubricity in both theoretical [27,59,65–68] and experimental [6,37,38,43,69–71] studies based on crystalline materials.

which is an apparent evidence for SSL [68,72]. Results of this 2D case is similar to the 1D incommensurate contacts aforementioned, that is, the Aubry transition [65], which states that static friction strictly disappear. The Aubry transition is recently validated by experiments [73]. The theoretical studies both on 1D and 2D clearly show that commensurability is the key for SSL, thus it is important to interpret this "ideal" concept into a more practical understanding. By considering a variety of hexagonal layered materials, including graphene, hBN, and molybdenum disulfide, Hod et al. show how a simple geometrical parameter, termed the "registry index" (RI) [28], can capture the interlayer sliding energy landscape as calculated using advanced electronic structure methods (Fig. 6.4). The predictive power of this method is further demonstrated by showing how the RI can fully reproduce the experimentally measured frictional behavior of a 2D structural superlubric heterojunctions.

Based on the RI, the sliding energy landscape of the heterogeneous graphene/hBN interface is studied theoretically [27]. For a graphene flake sliding on top of hBN, the anisotropy of the sliding energy corrugation with respect to the misfit angle between the two naturally mismatched lattices is found to reduce with the flake size. For sufficiently large flakes, the sliding

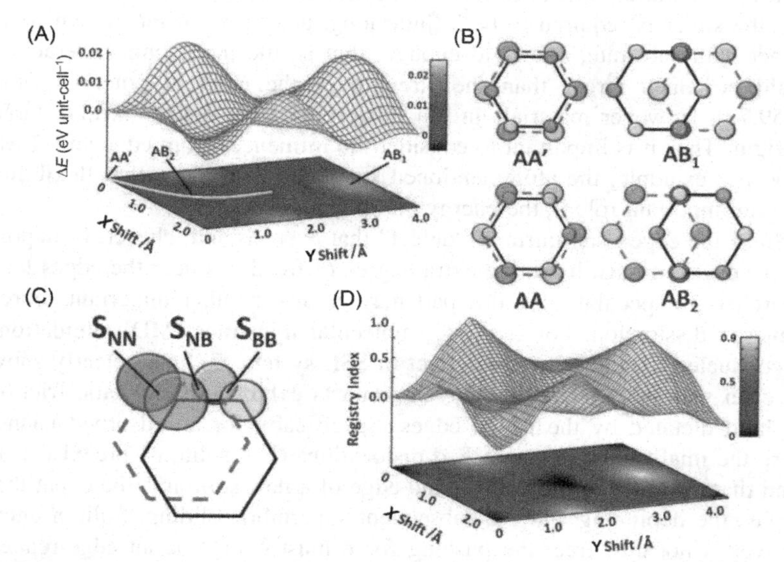

FIGURE 6.4 Definition of the registry index of hBN: [28] (A) sliding-energy landscape of bilayer hBN, as calculated by using the TS-vdW dispersion-corrected PBE exchange-correlation density functional approximation. (B) Definition of some high-symmetry stacking modes of bilayer hBN. (C) Definition of the circle overlaps that are used in the RI expression. (D) Sliding RI surface of bilayer hBN. *Blue (yellow)* circles represent nitrogen (boron) atomic sites. (*hBN*: Hexagonal boron nitride; *PBE*: perdew–burke-ernzerhof; *RI*: registry index; *vdW*: van der Waals).

energy corrugation is expected to be at least one order of magnitude lower than that obtained for matching lattices regardless of the relative interlayer orientation. Therefore, in contrast to the case of the homogeneous graphene interface where flake reorientations are known to eliminate superlubricity, a stable low-friction state is expected to occur, which is validated by later experimental studies [37].

Besides the static friction computed from the potential landscape, kinetic friction is also important for SSL [74]. One prominent property of the SSL is the ultralow kinetic friction and the coefficient of the friction (COF) is usually smaller than 0.001 [74]. Interestingly, considering the graphite-hBN heterojunctions, the COF is more complex. Negative friction coefficients, where friction is reduced upon increasing normal load, are predicted for superlubric [37] graphite-hBN heterojunctions [75]. The origin of this counterintuitive behavior lies in the load-induced suppression of the Moiré superstructure out-of-plane distortions, leading to a less dissipative interfacial dynamics. Thermally induced enhancement of the out-of-plane fluctuations leads to an unusual increase of friction with temperature. This highlighted frictional mechanism is of a general nature and is expected to appear in many layered material heterojunctions.

In the theoretical model studied by Peyard et al. [65], in order to achieve SSL, the slider is required to be infinite long, able to form incommensurate contact with substrate, and rigid enough, that is, the interatomic interaction should be much larger than the strength of the external potential field [18,59,76]. However materials in practical contact are neither infinite large nor rigid. Thus it is important to consider the influences from edge and flexibility. For example, the aforementioned studies [37,75] show that flexibility plays an important role in the energy dissipation.

Since the edge is an intrinsic "defect" that every island, cluster, or deposited nanosystem must have, the extra degree of freedoms near the edges lead researchers to speculate that this part may be also another important source of energy dissipation. For example, molecular dynamics (MD) simulations are conducted to study the edge effect in SSL system [77], and clearly show that even without defects, these weak contacts exhibit a basic static friction threshold dictated by the island edges. Specifically for an adsorbed island, where the misfit lattice-mismatched dislocations (i.e., solitons) preexist, it is found that the entry through the island edge of a new soliton is the event that initiates the depinning and the subsequent superlubric sliding. Soliton entry however is not cost free; the pushing force must overcome an edge-related energy barrier, which is thus the controlling element of the island's static friction.

Apart from the Kr/Pb(111) heterojunctions [77], the edge effect has also been studied by MD simulation in graphene nanoribbon (GNR)/Au(111) heterojunctions [67] (Fig. 6.5A), which was recently validated to be superlubric experimentally [6]. The GNR static friction is entirely edge related, whereas

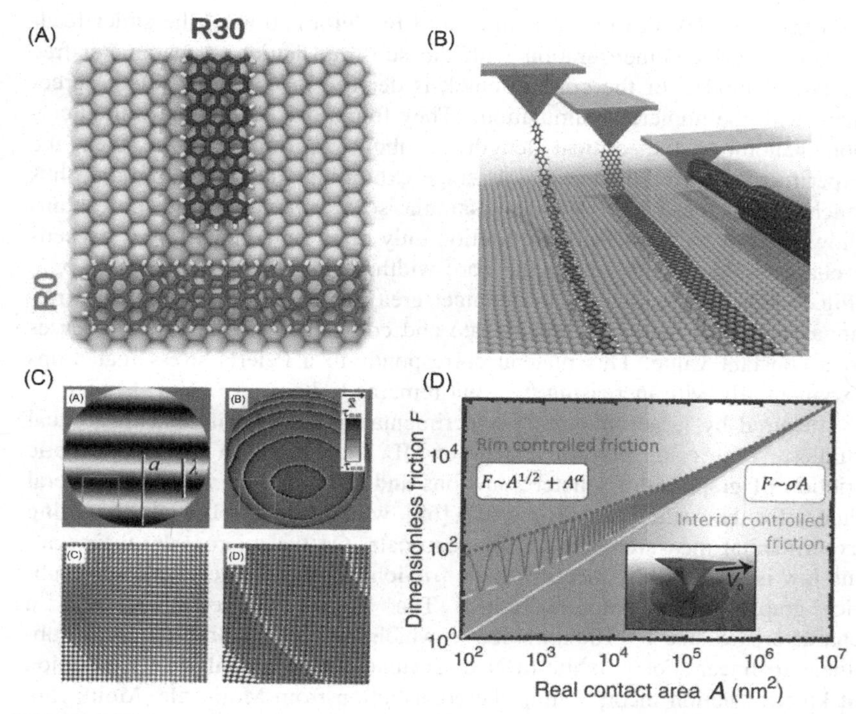

FIGURE 6.5 (A) Top-view sketch of the R0 *(orange)* and R30 *(blue)* alignments of the GNR deposited on the Au (1 1 1) substrate *(yellow)*. Broken carbon bonds at front, back, and side edges are passivated by hydrogen atoms *(white)* [67]. (B) Schematic sketch of the typical experimental setups for 1D SSL systems [66]. (C) The occurrence and sliding of dislocations at the interfaces between two atomically smooth crystalline surfaces [59]. (D) Kinetic friction versus real contact area for graphite flake sliding on hBN [68]. (*GNR*: Graphene nanoribbon; *hBN*: hexagonal boron nitride; *SSL*: structural superlubricity).

the GNR interior is superlubric. This system is similar to a finite-size Frenkel–Kontorova chain. Specifically simulations [67] allow us to correlate the periodicity of frictional oscillations to the characteristic length of the Moiré pattern, and in particular to the oscillating size of incomplete periods, namely the front and tail edges, which are responsible for the friction. This interpretation suggests that GNRs of certain "magic" lengths matching an integer number of Moiré wavelengths could be selected for ultralow-friction applications.

Flexibility effect is also considered in a few studies. By using an edge-driven Frenkel–Kontorova model (Fig. 6.5B), Ma et al. established the dependence of the critical length above which superlubricity disappears on both intrinsic material properties and experimental parameters [66]. A striking boost in dissipated energy with chain length emerges abruptly due to a high-friction stick-slip mechanism caused by deformation of the slider leading to a local commensuration with the substrate lattice. A parameter-free analytical model for the critical length is derived that is in excellent agreement with the numerical simulations. They found that for double-walled carbon nanotubes the critical length is about 50 times larger than the experimentally investigated length range exhibiting superlubric regime, thus reaching an agreement. By using atomic-scale simulations [59], Robbins shows that elasticity affects the friction only after the contact radius exceeds a characteristic length set by the core width of interfacial dislocations b_{core} (Fig. 6.5C). As the size of the contact area increases above b_{core}, the frictional stress for both incommensurate and commensurate surfaces decreases to a constant value. This plateau corresponds to a Peierls stress that drops exponentially with increasing b_{core} but remains finite.

Inspired by large amount of experimental measurements and theoretical studies, Wang et al. perform extensive MD simulations to study the kinetic friction of graphite-hBN heterojunctions and introduce an analytical general theory for the scaling law (Fig. 6.5D) [68], which could well explain existing experimental measurements on the nanoscale. On the microscale, their scaling law is validated by measuring the friction of several microscale superlubric graphite/hBN heterojunctions. The proposed theory predicts a characteristic size (~ 100 nm) above which the scaling transits from sublinear to linear. For graphite-hBN heterojunctions, the analytical expression of kinetic friction incorporating the contribution from Moiré tile, Moiré rim, and edges provides insights in the origin of friction for SSL and benefit its application on macroscale.

6.3.2 Superlubricity of heterojunctions on nanoscale

In the late 1990s, the first experiments for heterogeneous contacts with superlubric sliding were conducted by Hirano et al. in examining systems consisted of atomically clean surfaces of W and Si in ultra-high vacuum (UHV) [78]. However, since it is hard to prepare the ultraclean atomic flat surfaces and the superlubric phenomenon only exists in extreme conditions such as UHV, it is difficult to have impacts in practical applications.

After about 20 years, Dietzel et al. characterized the properties of SSL for heterojunctions by measuring sliding friction of nanoparticles (Sb and gold) on the substrates (MoS_2 and HOPG) [69,72,79,80]. According to their results, the real contact area of the particles could be accurately estimated by the morphology of the particles with AFM. By sliding nanoflake (Sb and gold) atop HOPG, SSL could persist under certain environment conditions such as UHV. For such systems, the dependence of friction on velocity was measured and agrees with the predictions derived based on thermal activation theory [81].

Another heterostructure with superlubric phenomenon is GNRs on Au (111) substrate under UHV at cryogenic temperatures. Noncontact AFM tools play a great role in the measurement of interactions between single-layer graphene and atomically smooth Au surfaces for the high sensitivity of the frequency drift [6]. With the nanoribbon attached to AFM tip and moved linearly in constant height mode, the resonance frequency shifts detected by the tuning fork indicated the total interactions from the interface. The results showed that ribbon size and elasticity are critical for the ultralow friction in such well-defined atomically smooth contacts. In the meanwhile, under ambient conditions at nanoscale interfaces, the structurally lubric sliding formed by gold islands on graphite was reported [39]. However for such metal cluster-based systems, to maintain chemically cleanness of single crystal metal surfaces is hard due to the relatively high surface activity. Therefore more stable/inert atomically smooth surfaces such as 2D materials gradually become a better choice of researchers.

With the rapid development of 2D materials synthesis [82,83], now researchers can prepare atomic flat single crystalline heterojunctions directly. Kobayashi et al. [84] reported the preparation of slidable heterostructures with atomic layers in clean, incommensurate van der Waals contact. By chemical vapor deposition, few layer WS_2 sheets can be directly grown on graphite or hBN substrates and can be moved through the manipulation of a tip. Similar process can be used in the preparation of other kinds of heterojunctions such as MoO_3 nanocrystals thermally grown on MoS_2 or $MoSe_2$ substrates [85]. For tribological test, researchers need to introduce slot on top of the MoO_3 nanocrystal to initiate sliding. It was found that the measured friction per area was far lower than for macroscale contact. To form heterojunctions, the growth method is a powerful way for the fabrication of heterostructure on nanoscale. Due to the nanometer thickness of the slider and the difficulty in loading the slider laterally, it is usually hard for the grown layer to support normal load exerting on it.

With a graphene-coated corrugated sphere sliding atop flat graphite or hBN surfaces [43], these technical problems were overcome. Merely a few separated nanoscale contact points between the sphere and the flat substrate were formed. Therefore researchers could obtain superlubric motion for both the homogeneous and heterogeneous interfaces with total real contact area on nanoscale. The friction anisotropy and the possible elastic interaction within the surfaces were effectively reduced by such multicontact approach. Nevertheless the enhanced wear would arise for the extremely high local pressures on such a small contact area under the application of large normal loads. To measure the intrinsic interlayer frictional properties between the atomic planes of 2D materials, Liu et al. coated AFM tips with graphene sheets via thermally assisted mechanical exfoliation and transfer as shown in Fig. 6.6A [38]. With different 2D materials (MoS_2, ReS_2, hBN, TaS_2, and graphene) coated tips, their friction coefficients are measured to be all within

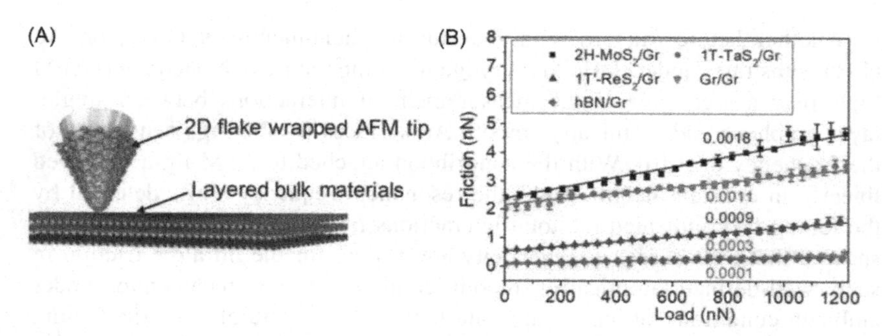

FIGURE 6.6 (A) Preparation of nanoscale heterogeneous interfaces with 2D flake-wrapped AFM tips. (B) Nanoscale structural superlubricity achieved in several heterogeneous interfaces [38]. (*AFM*: Atomic force microscopy).

the superlubric regime (Fig. 6.6B) [38]. Such graphite-coated AFM tip which can be fabricated in different ways could not only be used to measure the friction between single crystalline surfaces on nanoscale, but also the van der Waals interactions [41].

6.3.3 Superlubricity of heterojunctions on microscale

Compared to significant advances achieved experimentally on nanoscale, studies are comparatively rare on microscale. One possible reason could be that the methods suggested on nanoscale cannot be directly applied on microscale. For example, heterojunctions usually suffered from the facts that the thin layer grown via direct growth make it difficult to initiate sliding. While there are viable solutions like introducing slot on nanoscale, on microscale the low normal load capacity and large adhesion at the interfaces make the situation even worse, therefore the sample is extremely difficult to be driven laterally. For coated AFM tip, obviously it is difficult to form microscale heterogeneous contacts via such approach.

To overcome these technical problems, researchers extended the fabrication techniques of microscale graphite contact through assembly under ambient condition to form heterojunctions [34]. As shown in Fig. 6.7A, graphite/hBN contacts with contact area about $3 \times 3\ \mu m^2$ were fabricated. The fabricated contacts composed of single crystalline surfaces show an anisotropic friction with sixfold symmetry. The anisotropy ($\sim$4) is two orders lower than that for graphite/graphite homogeneous contacts ($\sim 10^3$) [37] (Fig. 6.7B). These results are in accordance with recent experimental observations for hBN on graphene which reported two prominent peaks separated by 60 degrees in lateral force measurement [71]. The friction coefficient measured is less than 0.001 for low friction states and about 0.006 for high friction state. For such heterojunctions, the highly dissipative center-of-mass stick-slip motion for homojunctions is suppressed effectively. Instead, the

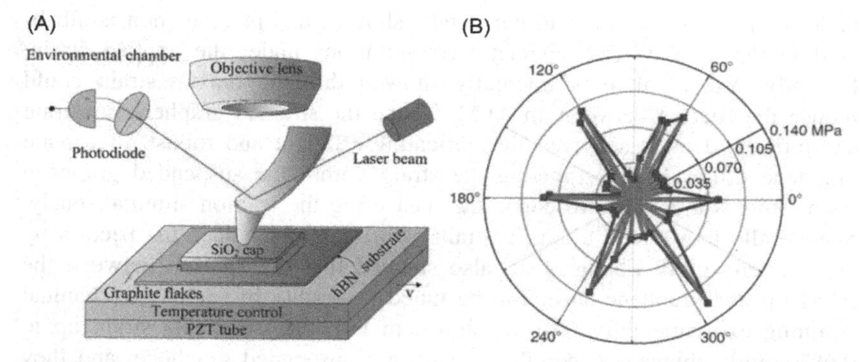

FIGURE 6.7 Microscale structural superlubricity achieved in graphite/hBN interface. (A) Schematic diagram of the experimental setup. (B) Rotational anisotropy of the measured friction for a graphite/hBN heterojunction [37]. (*hBN*: Hexagonal boron nitride).

out-of-plane motion for the atoms at the interfaces provides the main energy dissipation route. The results shed light on the fabrication of other superlubric van der Waals heterostructures.

Besides the fact that microscale graphite/hBN contacts persist the ultralow friction coefficient and small anisotropy, another interesting observation is the low friction state being energetically more favorable [33]. This is different from the graphite contact corresponding to its homojunction, where the system will tend to rotate to a high friction state. As a result, the robustness of superlubric contacts can be greatly enhanced by such intrinsic feature. What's more, the achieved SSL was verified to be stable over a wide range of normal load (0.27–11.1 MPa), environmental conditions (ambient conditions and nitrogen protected), and sliding velocities (60 nm s^{-1}–10 μm s^{-1}).

6.3.4 Strain engineering

The understanding of SSL from structural point of view makes it intuitive to control the friction via applying strain within the surfaces. In fact, recently such concept has been validated not only in layered materials, but also for system with AFM tip sliding on surfaces.

For an AFM tip sliding on strained substrate, Bai et al. carried out MD simulations of a diamond probe scanned on a suspended graphene and studied the effect of strain on the frictional properties of suspended graphene [86]. The results show that the friction coefficient could be decreased about 1 order of magnitude with the increase of the strain. The underlying mechanism is the decrease of potential energy and the fluctuation of contact region. Later, Yang et al. also using MD simulations to study the friction coefficient between a strained multilayer graphene and a diamond tip [87]. They found that the friction coefficient decreased under tensile strain while it increased under compressive strain. Further details showed this phenomenon is attributed to the atomic scale contact area variations under the applied strain. Recently, Yin et al. experimentally showed that the biaxial strain could reduce the friction between an AFM tip and the strained graphene substrate enormously. This is achieved by fabricating efficient and robust microscale graphene drums [88], estimating the strain within the suspended graphene layer using Raman spectroscope, and measuring the friction simultaneously. Specifically they found a tensile strain of 0.2% could reduce the friction by 70%. Interestingly Zhang et al. also showed that the friction between the AFM tip and graphene sheet can be tuned reversibly by simple mechanical straining experimentally [89], as shown in Fig. 6.8. A tensile strain up to 0.6% could achieve a superlubric state on a suspended graphene, and they

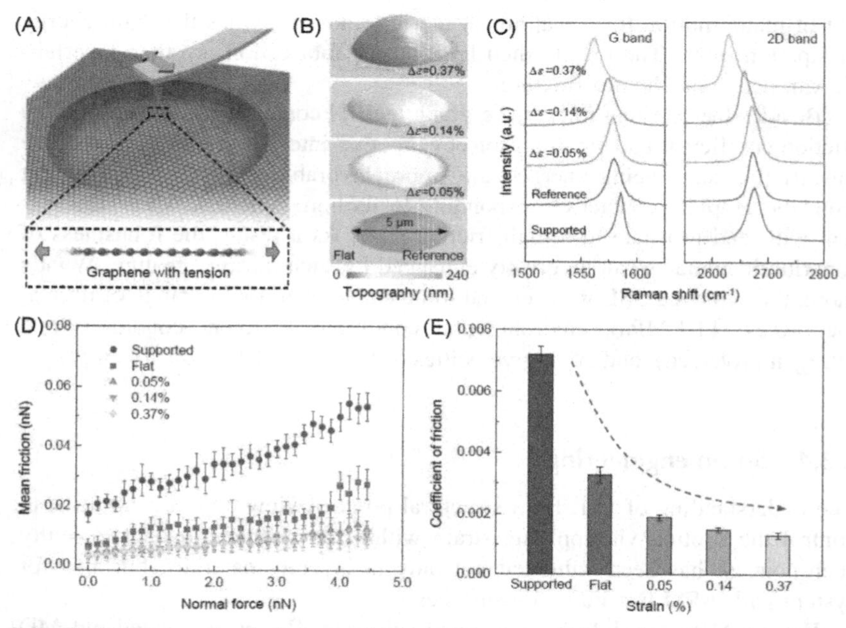

FIGURE 6.8 Preparation and characterization of graphene with varying strains [89]. (A) A schematic showing the topography and friction measurements on suspended graphene with different strains. (B) Typical 3D topographic images of pressurized monolayer graphene bubbles with different strains. The color represents the height. The heights of the bubbles are magnified (not to the ratio of the lateral size) to better illustrate the bulging. (C) Raman spectra of the suspended graphene with varying strains, measured at the center region of the graphene bubbles. (D) Friction versus normal force data measured on the supported graphene and the suspended graphene. The error bar represents the SD of the repeated measurements under the same normal load. (E) Variation of the coefficients of friction with the strain of graphene, acquired by fitting linearly the mean friction vs. normal force curves in (D). (*SD*: Standard deviation).

attributed in-plane strain effectively modulates the flexibility of graphene which in turn changing the local pinning capability of the contact interface.

For the effects of strain engineering on the interlayer friction of graphite, Wang et al. [90] studied the friction of a graphene flake sliding on top of a strained graphene substrate using MD simulation. The results show that by applying strain on the substrate, biaxial stretching is better than uniaxial stretching in terms of reducing interlayer friction. They also find that robust superlubricity can be obtained via both biaxial and uniaxial stretching. For stretching, above a critical strain which has been achieved experimentally, the friction is no longer dependent on the relative orientation between the layers. This can be used to reduce the high anisotropy effectively. Later, using MD simulation, Liu et al. [91] found that friction at the incommensurate interface is insensitive to the tensile and small compressive strains, while at the commensurate interface the strain-induced commensurability transition leads to a nonmonotonic variation of the friction with strain. Dong et al. [92] showed that as pretensile strain increases, overall softer interfacial registry and smaller Moiré patterns appear, and finally, the consequence of counteracting atomic forces results in lower friction. By studying the dependence of friction on biaxial tensile strain, it is found that friction can be reduced nonmonotonically by applying strain [93], as shown in Fig. 6.9. The critical strain needed for significant reduction in friction decreases drastically when the flake size increases. For example, for a 250 nm flake, a 0.1% biaxial strain could lead to a more than 2-order-of-magnitude reduction. This is of practical importance since graphene grown on metals generally has a biaxial strain up to 1%. The ratio between the area of the Moiré pattern and the flake size is found to play a central role in determining friction in strain engineering and possibly other scenarios of SSL as well.

6.3.5 Effects of surface adsorbates

In previous sections, theoretical works on 1D and 2D SSL have been introduced. However at ambient conditions, superlubric interfaces that are exposed to air will inevitably be contaminated by adsorbed molecules including hydrocarbons, oxygen, water, and other small molecules [94–97]. The interaction strength between adsorbate and substrate changes by orders in magnitude from weak physisorption to strong chemisorption; on the other hand, mobility of adsorbed molecules changes in a broad range as well. These two factors play key roles in the friction behavior deviating from intrinsic superlubric interfaces. Due to limited characterization means to investigate quantitatively the presence of adsorbed molecules at nanoscopic interfaces, there have been very few experimental studies on this topic. But theoretical investigation is possible thanks to mathematical and numerical methods.

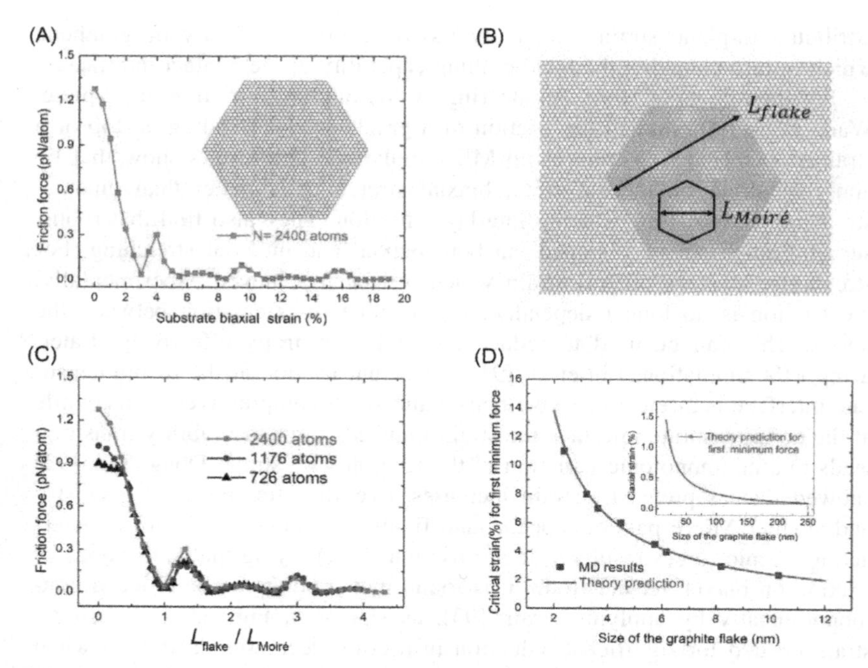

FIGURE 6.9 Size effects of the graphene flake on friction under different biaxial stretching strains in the substrate [93]. (A) Flake containing 2400 atoms. (B) The length of L_{flake} and $L_{\text{Moiré}}$. (C) Flake of different sizes is normalized by the size of Moiré pattern generated under different biaxial strains. (D) MD simulations versus theories of the critical strain for first minimum friction. (*MD*: Molecular dynamics).

Robbins et al. have studied analytically how the friction behavior of rigid crystalline interfaces are affected by adsorbed hydrocarbon molecules [94,98–100]. These "third bodies" adsorbing at local free-energy minima can arrange to lock two contacting surfaces together, thus resulting in the emergence of static friction. Collaborating with Müser, they have developed a microscopic theory on static friction coinciding with previous results, in which the mobile adsorbed layer screens direct wall-wall interaction [100]. In their subsequent work using MD to simulate sliding dynamics between rigid crystalline surfaces with a layer of adsorbed molecules, the kinetic friction is found to be consistent with Amontons' laws, that is, linear dependence of friction on normal pressure [99]. Then in a large-scale MD simulation that uses spring-chained atoms to mimic chain hydrocarbons enveloped in an elastic tip-substrate model, they systematically investigated distributions of pressure and molecules, as well as time and load dependence of friction [98].

Friction is also closely related to the adsorbate surface coverage. In the aforementioned work by Müser and Robbins, their model predicts slight dependence of static friction on coverage. An even earlier work states that size and shape of adsorbed hydrocarbons, rather than coverage is paramount in determining the magnitude of sliding friction [101]. In contrast, a strong dependence of friction on adsorbate coverage has been predicted in MD simulations mimicking rare gas quartz crystal microbalance experiments, that is, rare gas molecules sliding on a crystalline surface [77,102,103]. For the system consisting of rare gas and crystalline surface, a significant contribution of island edges to friction forces and a transition from superlubric adsorbate motion to the pinned state when the coverage approaches a full monolayer are found both in experiments and simulations. Self-assembled organic layers and thick water layers confined between flat materials are systems closer to liquid lubrication rather than superlubricity [103,104].

Friction force microscopy (FFM) is one key experiment method in studying superlubricity, but few have modelled in MD simulations a common scenario in real-world FFM experiments that considers adsorbed molecules. Simulations using opposing flat plates do not reflect the geometry or frictional dissipation of FFM experiments. Ouyang et al. have modelled the atomic sliding motion of contaminated surface relative to a crystalline spherical tip [105]. The kinetic friction is found to vary nonmonotonically with normal load. Also studied is the coverage effects of the physisorbed molecules and dependence on temperature of friction. The authors predicted three different regimes of temperature dependence.

Since it is commonly assumed that surface adsorbates may increase friction, there have been a few studies on cleaning the surfaces efficiently. Experimentally, by sliding micrometer-sized graphite flakes on HOPG and graphene, it is shown that the graphite flakes can swipe the adsorbates on the surfaces effectively, serving as an eraser [56]. For adsorbates that are initially trapped at the interface, using numerical simulations together with analytical analysis, Ma et al. suggest that mechanical oscillation of the surface could efficiently remove the contaminants. The underlying mechanism is the enhanced diffusion of particles confined at the interface due to bifurcation [57]. This theoretical prediction is partially validated by experiments based on graphite flakes exhibiting SSL [57] (Fig. 6.10).

6.3.6 Toward macroscale

SSL holds great technological promise for reducing friction and wear in practical applications. While a real contact area of 100 μm^2 has been reached experimentally, however, fulfilling this potential requires the scaling up of superlubricity to macroscale. In recent years, a few efforts have been made to achieve macroscale superlubricity.

One promising route to obtain large-scale superlubricity involves multi-contact interfacial geometries. Regarding this respect, it was recently demonstrated that superlubricity can be achieved at engineering scale when

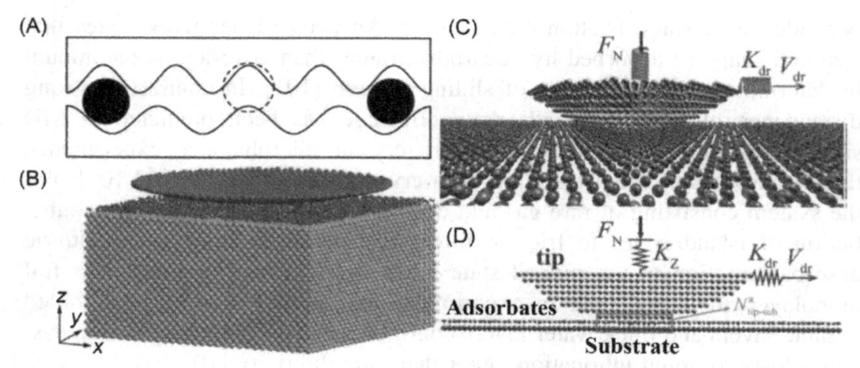

FIGURE 6.10 (A) Model proposed by Müser and Robbins, adsorbates resting at local energy minima, hindering relative motion; [100] (B) MD model used by Robbins et al., consisting of rigid tip, hydrocarbons, and elastic substrate; [98] (C) and (D) are schematic and side view of Ouyang et al.'s tip-molecule-substrate MD model, respectively [105]. (*MD*: Molecular dynamics).

graphene is used in combination with nanodiamond particles and DLC [7]. Macroscopic superlubricity originates because graphene patches at the sliding interface wrap around the nanodiamonds, and nanoscrolls are formed with reduced contact area that slide against the DLC surface (Fig. 6.11), achieving an incommensurate heterogeneous contact and substantially reduced coefficient of friction ($\sim$0.004). Large-scale MD simulation was applied to elucidate the overall mechanism. After the sliding of the interface entered a stable superlubric state, the effective contact area at the interface was reduced significantly, by $\sim$65–70%. But it has been pointed out that once the relative humidity increased to 30%, the coefficient of friction increased out of the superlubricity regime, because water occupied the defects in the graphene leading to a more ordered material that suppresses the formation of scrolls. Similar experiment was performed with 2D MoS_2 in dry nitrogen atmosphere, near-zero coefficient of friction (0.005) was observed, demonstrating a superlubric sliding [106]. The mechanism, however, is different from the aforementioned experiment. Diffusion of sulfur led to amorphization of nanodiamond and subsequent transformation to onion-like carbon structures (OLCs) under high pressure (Fig. 6.12). The in situ formation of OLCs at the sliding interface provides reduced contact area as well as incommensurate contact with respect to the hydrogenated DLC surface, thus enabling superlubricity.

In another macroscale sliding system between steel and graphite, instantaneous superlubricity was also achieved through the formation of many tribo-transferred multilayer graphene nanoflakes (MGNFs) on the steel contact zone after the initial sliding [107]. The friction coefficient could reduce to a 0.001, which randomly appeared in the sliding process. The macroscale

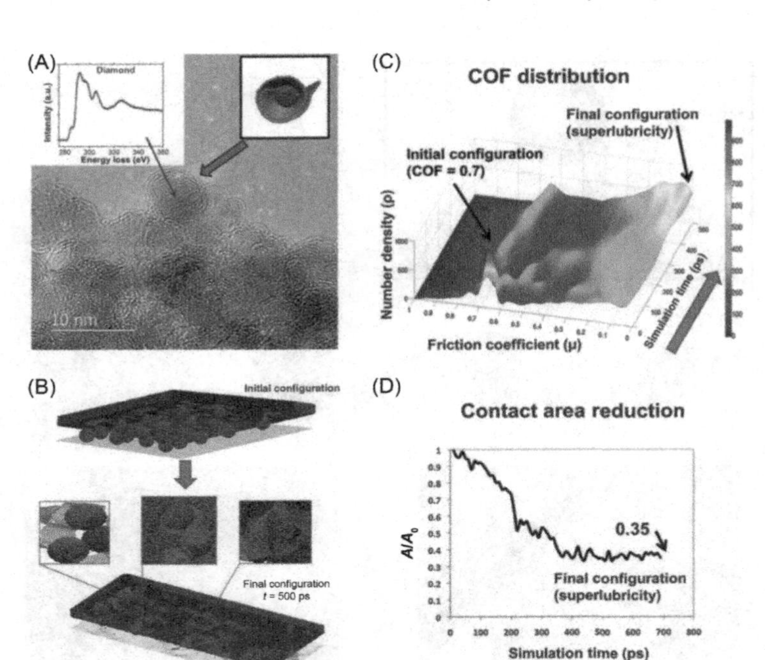

FIGURE 6.11 Graphene nanoscrolls formation [7]. (A) TEM images of the wear debris for DLC ball sliding against graphene-plus-nanodiamonds. (B) Snapshot showing the scroll formation on nanodiamonds for an ensemble of graphene patches when subjected to sliding. (C) Temporal evolution of COF distribution averaged over an ensemble of graphene patches. (D) Evolution of the corresponding contact area. (*COF*: Coefficient of the friction; *DLC*: diamond-like carbon; *TEM*: transmission electron microscope).

superlubricity results from the statistical frictional forces of multiple transferred MGNFs sliding on the graphite, which is also a multicontact approach (Fig. 6.13).

The aforementioned macroscale superlubricity are achieved with 2D layered materials. DLC film is also a frequently used material to realize macroscale superlubricity. When synthesized in hydrogen-containing environments, the DLC films may contain large amounts of hydrogen in their structures, realizing superlubricity under dry and inert tribological conditions [108]. High hydrogen concentration within the bulks, on the surfaces or at the sliding interfaces can provide ultralow friction coefficients down to 0.001 level, leading to macroscale superlubricity (Fig. 6.14).

So far, several ways have been attempted to achieve macroscale superlubricity with different materials, but they are generally limited to conditions such as dry or inert environments. Other challenges such as the edge effects, surface roughness, and contaminants [15] are remained to be resolved towards achieving macroscale superlubricity.

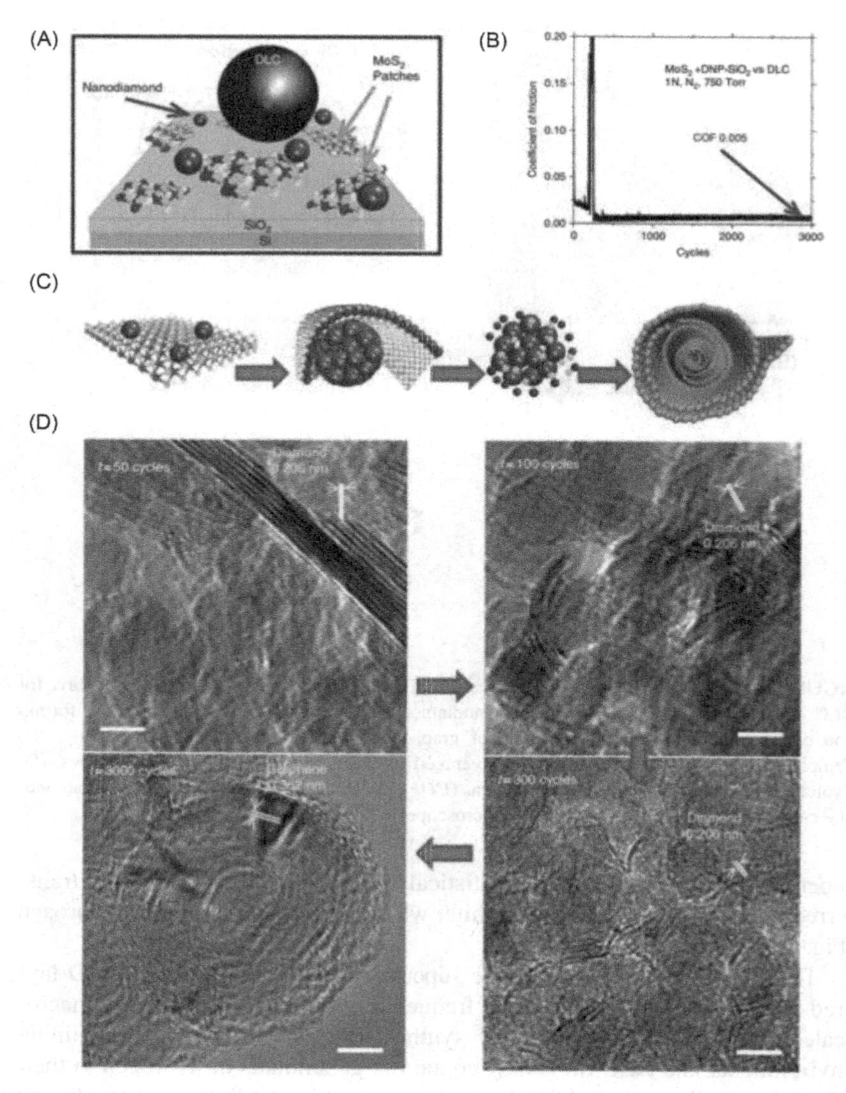

FIGURE 6.12 Tribological performance of MoS_2 layers mixed with nanodiamond [106]. (A) Schematic of the experimental setup. (B) Graphs of the coefficient of friction. (C) Schematics depicting the mechanism of OLC formation. (D) TEM images of the wear debris taken at regular intervals after interrupting the tribotest. (*OLC*: Onion-like carbon structure).

6.4 Conclusions and challenges

The study of nanoscale tribology has vastly developed over the past years providing basic understanding of the physical mechanisms underlying

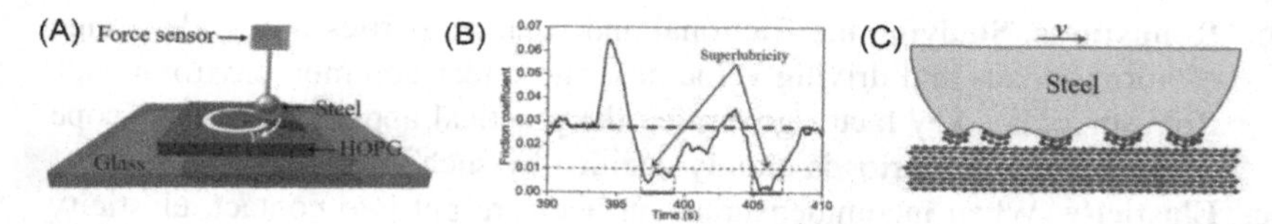

FIGURE 6.13 (A) Experimental setup of ball-on-disk friction test with a rotation radius of 1 mm under ambient conditions. (B) Evolution of friction coefficient with time under a constant load of 0.5 N and a sliding speed of 52 mm s^{-1}. (C) Illustration of the contact zone between steel ball and HOPG [107]. (*HOPG*: Highly oriented pyrolytic graphite).

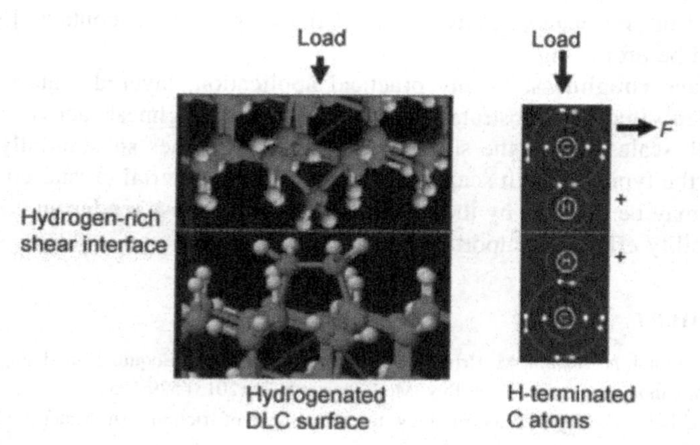

FIGURE 6.14 A mechanistic model for the superlow friction behavior of highly hydrogenated DLC surfaces [109]. (*DLC*: Diamond-like carbon).

superlubricity and clear reproducible demonstrations of its experimental realization. This has led to the identification of the necessary conditions required for superlubricity to occur and, most importantly, the limiting factors eliminating this unique phenomenon. The field has thus matured to a point where scaling up to realistic tribological contacts can be pursued. The synergic mode-of-investigation incorporating both experiments and theories plays a critical role in the achievements so far. However at present, a comprehensive understanding about the mechanisms for microscale SSL is still lacking, which further hinders both the progress toward macroscale and the practical applications of SSL. Among the many factors to be explored, we suggested the following issues which may be more significant.

a. **Defects.** Examining effects of structural interfacial defects on the frictional properties for both homogeneous and heterogeneous contacts. Thermodynamically, when the systems become sufficiently large, the presence of defects is inevitable. In this case, to estimate their effects on friction is crucial in designing mechanical systems based on superlubricity.

b. **Robustness.** Studying the frictional and wear properties in a wider range of normal loads and driving velocities, and under common environments. Robustness is a key factor governing the practical applications. The scope of application scenarios is mainly decided by such properties.
c. **Elasticity.** When incommensurate surfaces are put into contact, elasticity effects can support interlattice readjustment that leads to the formation of locally commensurate regions characterized by strong lock-in forces [59,66]. This effect grows with the contact size and hence is expected to enhance friction at macroscale junctions. Thus it would be necessary to design the superlubric contacts so that such elastic effects could be minimized or eliminated. Along this way, the usage of multicontacts [38,110] could be promising.
d. **Surface roughness.** In any practical application, layered materials will be deposited on substrates that will exhibit roughness across various length scales. When the surface corrugation becomes substantially larger than the typical length scales of the 2D layered material crystal coat, friction may be dictated by the overall surface roughness rendering commensurability effects unimportant.

References

[1] A. Vanossi, N. Manini, M. Urbakh, S. Zapperi, E. Tosatti, Colloquium: modeling friction: from nanoscale to mesoscale, Rev. Mod. Phys. 85 (2) (2013) 529–552.

[2] M. Urbakh, E. Meyer, Nanotribology: the renaissance of friction, Nat. Mater. 9 (1) (2010) 8–10.

[3] I. Szlufarska, M. Chandross, R.W. Carpick, Recent advances in single-asperity nanotribology, J. Phys. D Appl. Phys. 41 (12) (2008) 123001.

[4] O.M. Braun, A.G. Naumovets, Nanotribology: Microscopic mechanisms of friction, Surf. Sci. Rep. 60 (6-7) (2006) 79–158.

[5] P.E. Sheehan, C.M. Lieber, Nanotribology and nanofabrication of MoO_3 structures by atomic force microscopy, Science 272 (5265) (1996) 1158–1161.

[6] S. Kawai, A. Benassi, E. Gnecco, H. Sode, R. Pawlak, X.L. Feng, et al., Superlubricity of graphene nanoribbons on gold surfaces, Science 351 (6276) (2016) 957–961.

[7] D. Berman, S.A. Deshmukh, S.K.R.S. Sankaranarayanan, A. Erdemir, A.V. Sumant, Macroscale superlubricity enabled by graphene nanoscroll formation, Science 348 (6239) (2015) 1118–1122.

[8] W. Wang, S. Dai, X. Li, J. Yang, D.J. Srolovitz, Q. Zheng, Measurement of the cleavage energy of graphite, Nat. Commun. 6 (2015) 7853.

[9] S. Li, Q. Li, R.W. Carpick, P. Gumbsch, X.Z. Liu, X. Ding, et al., The evolving quality of frictional contact with graphene, Nature 539 (7630) (2016) 541–545.

[10] E. Koren, E. Loertscher, C. Rawlings, A.W. Knoll, U. Duerig, Adhesion and friction in mesoscopic graphite contacts, Science 348 (6235) (2015) 679–683.

[11] M. Imboden, P. Mohanty, Dissipation in nanoelectromechanical systems, Phys. Rep. 534 (3) (2014) 89–146.

[12] K. Shinjo, M. Hirano, Dynamics of friction: Superlubric state, Surf. Sci. 283 (1-3) (1993) 473–478.

[13] M. Hirano, K. Shinjo, Atomistic Locking and Friction, Phys. Rev. B 41 (17) (1990) 11837–11851.

[14] M. Dienwiebel, G.S. Verhoeven, N. Pradeep, J.W.M. Frenken, J.A. Heimberg, H.W. Zandbergen, Superlubricity of graphite, Phys. Rev. Lett. 92 (12) (2004) 126101.

[15] O. Hod, E. Meyer, Q. Zheng, M. Urbakh, Structural superlubricity and ultralow friction across the length scales, Nature 563 (7732) (2018) 485–492.

[16] J.M. Martin, A. Erdemir, Superlubricity: friction's vanishing act, Phys. Today 71 (4) (2018) 40–46.

[17] D. Berman, A. Erdemir, A.V. Sumant, Approaches for achieving superlubricity in two-dimensional materials, ACS Nano 12 (3) (2018) 2122–2137.

[18] M.Z. Baykara, M.R. Vazirisereshk, A. Martini, Emerging superlubricity: a review of the state of the art and perspectives on future research, Appl. Phys. Rev. 5 (4) (2018) 041102.

[19] Y. Song, C. Qu, M. Ma, Q. Zheng, Structural superlubricity based on crystalline materials, Small (2019). Available from: https://doi.org/10.1002/smll.201903018:1903018.

[20] A.E. Filippov, M. Dienwiebel, J.W.M. Frenken, J. Klafter, M. Urbakh, Torque and twist against superlubricity, Phys. Rev. Lett. 100 (4) (2008) 046102.

[21] A.S. de Wijn, A. Fasolino, A.E. Filippov, M. Urbakh, Low friction and rotational dynamics of crystalline flakes in solid lubrication, EPL 95 (6) (2011) 66002.

[22] A.S. de Wijn, C. Fusco, A. Fasolino, Stability of superlubric sliding on graphite, Phys. Rev. E 81 (4) (2010) 046105.

[23] Z. Liu, J. Yang, F. Grey, J.Z. Liu, Y. Liu, Y. Wang, et al., Observation of microscale superlubricity in graphite, Phys. Rev. Lett. 108 (20) (2012) 205503.

[24] Z. Liu, S.-M. Zhang, J.-R. Yang, J.Z. Liu, Y.-L. Yang, Q.-S. Zheng, Interlayer shear strength of single crystalline graphite, Acta Mech. Sinica-PRC 28 (4) (2012) 978–982.

[25] G.S. Verhoeven, M. Dienwiebel, J.W.M. Frenken, Model calculations of superlubricity of graphite, Phys. Rev. B 70 (16) (2004) 165418.

[26] R. Zhang, Z. Ning, Y. Zhang, Q. Zheng, Q. Chen, H. Xie, et al., Superlubricity in centimetres-long double-walled carbon nanotubes under ambient conditions, Nat. Nanotechnol. 8 (12) (2013) 912–916.

[27] I. Leven, D. Krepel, O. Shemesh, O. Hod, Robust superlubricity in graphene/h-BN heterojunctions, J. Phys. Chem. Lett. 4 (1) (2013) 115–120.

[28] O. Hod, The registry index: a quantitative measure of materials interfacial commensurability, Chemphyschem 14 (11) (2013) 2376–2391.

[29] W. Ouyang, M. Ma, Q. Zheng, M. Urbakh, Frictional properties of nanojunctions including atomically thin sheets, Nano Lett. 16 (3) (2016) 1878–1883.

[30] Q. Zheng, W. Ouyang, M. Ma, S. Zhang, Z. Zhao, H. Dong, et al., Superlubricity: a world with vanishing friction, Sci. Technol. Rev. 34 (9) (2016) 12–26.

[31] X. Zheng, L. Gao, Q. Yao, Q. Li, M. Zhang, X. Xie, et al., Robust ultra-low-friction state of graphene via Moire superlattice confinement, Nat. Commun. 7 (2016) 13204.

[32] C.C. Vu, S.M. Zhang, M. Urbakh, Q.Y. Li, Q.C. He, Q.S. Zheng, Observation of normal-force-independent superlubricity in mesoscopic graphite contacts, Phys. Rev. B 94 (8) (2016) 081405(R).

[33] D.M. Wang, G.R. Chen, C.K. Li, M. Cheng, W. Yang, S. Wu, et al., Thermally induced graphene rotation on hexagonal boron nitride, Phys. Rev. Lett. 116 (12) (2016) 126101.

[34] H. Deng, M. Ma, Y. Song, Q. He, Q. Zheng, Structural superlubricity in graphite flakes assembled under ambient conditions, Nanoscale 10 (29) (2018) 14314–14320.

[35] Y.J. Gongyang, C.Y. Qu, S.M. Zhang, M. Ma, Q.S. Zheng, Eliminating delamination of graphite sliding on diamond-like carbon, Carbon 132 (2018) 444–450.

[36] Gongyang Y., Ouyang W., Qu C., Urbakh M., Quan B., Ma M., et al., Temperature and velocity dependent friction of a microscale graphite-DLC heterostructure. Friction, 2019. DOI 10.1007/s40544-019-0288-0.
[37] Y. Song, D. Mandelli, O. Hod, M. Urbakh, M. Ma, Q. Zheng, Robust microscale superlubricity in graphite/hexagonal boron nitride layered heterojunctions, Nat. Mater. 17 (10) (2018) 894–899.
[38] Y. Liu, A. Song, Z. Xu, R. Zong, J. Zhang, W. Yang, et al., Interlayer friction and superlubricity in single-crystalline contact enabled by two-dimensional flake-wrapped atomic force microscope tips, ACS Nano 12 (8) (2018) 7638–7646.
[39] E. Cihan, S. Ipek, E. Durgun, M.Z. Baykara, Structural lubricity under ambient conditions, Nat. Commun. 7 (2016) 12055.
[40] A. Ozogul, S. Ipek, E. Durgun, M.Z. Baykara, Structural superlubricity of platinum on graphite under ambient conditions: the effects of chemistry and geometry, Appl. Phys. Lett. 111 (21) (2017) 211602.
[41] B.W. Li, J. Yin, X.F. Liu, H.R. Wu, J.D. Li, X.M. Li, et al., Probing van der Waals interactions at two-dimensional heterointerfaces, Nat. Nanotechnol. 14 (6) (2019) 567–572.
[42] J. Sun, Y. Zhang, Z. Lu, Q. Li, Q. Xue, S. Du, et al., Superlubricity enabled by pressure-induced friction collapse, J. Phys. Chem. Lett. 9 (10) (2018) 2554–2559.
[43] S.-W. Liu, H.-P. Wang, Q. Xu, T.-B. Ma, G. Yu, C. Zhang, et al., Robust microscale superlubricity under high contact pressure enabled by graphene-coated microsphere, Nat. Commun. 8 (2017) 14029.
[44] J. Li, J. Li, J. Luo, Superlubricity of graphite sliding against graphene nanoflake under ultrahigh contact pressure, Adv. Sci. 5 (11) (2018) 1800810.
[45] Q.S. Zheng, B. Jiang, S.P. Liu, Y.X. Weng, L. Lu, Q.K. Xue, et al., Self-retracting motion of graphite microflakes, Phys. Rev. Lett. 100 (6) (2008) 067205.
[46] J.R. Yang, Z. Liu, F. Grey, Z.P. Xu, X.D. Li, Y.L. Liu, et al., Observation of high-speed microscale superlubricity in graphite, Phys. Rev. Lett. 110 (25) (2013) 255504.
[47] C. Qu, S. Shi, M. Ma, Q. Zheng, Rotational instability in superlubric joints, Phys. Rev. Lett. 122 (24) (2019) 246101.
[48] E. Koren, I. Leven, E. Lortscher, A. Knoll, O. Hod, U. Duerig, Coherent commensurate electronic states at the interface between misoriented graphene layers, Nat. Nanotechnol. 11 (9) (2016) 752–757.
[49] A. Kis, K. Jensen, S. Aloni, W. Mickelson, A. Zettl, Interlayer forces and ultralow sliding friction in multiwalled carbon nanotubes, Phys. Rev. Lett. 97 (2) (2006) 025501.
[50] J. Cumings, A. Zettl, Low-friction nanoscale linear bearing realized from multiwall carbon nanotubes, Science 289 (5479) (2000) 602–604.
[51] A.M. Fennimore, T.D. Yuzvinsky, W.Q. Han, M.S. Fuhrer, J. Cumings, A. Zettl, Rotational actuators based on carbon nanotubes, Nature 424 (6947) (2003) 408–410.
[52] K. Jensen, C. Girit, W. Mickelson, A. Zettl, Tunable nanoresonators constructed from telescoping nanotubes, Phys. Rev. Lett. 96 (21) (2006) 215503.
[53] Q.S. Zheng, Q. Jiang, Multiwalled carbon nanotubes as gigahertz oscillators, Phys. Rev. Lett. 88 (4) (2002) 045503.
[54] H. Li, J. Wang, S. Gao, Q. Chen, L. Peng, K. Liu, et al., Superlubricity between MoS_2 monolayers, Adv. Mater. 29 (27) (2017) 1701474.
[55] J.M. Martin, C. Donnet, T. Lemogne, T. Epicier, Superlubricity of molybdenum-disulfide, Phys. Rev. B 48 (14) (1993) 10583–10586.
[56] Z. Liu, P. Boggild, J.-R. Yang, Y. Cheng, F. Grey, Y.-L. Liu, et al., A graphite nanoeraser, Nanotechnology 22 (26) (2011) 265706.

[57] M. Ma, I.M. Sokolov, W. Wang, A.E. Filippov, Q.S. Zheng, M. Urbakh, Diffusion through bifurcations in oscillating nano- and microscale contacts: fundamentals and applications, Phys. Rev. X 5 (3) (2015) 9.

[58] C. Lee, Q.Y. Li, W. Kalb, X.Z. Liu, H. Berger, R.W. Carpick, et al., Frictional characteristics of atomically thin sheets, Science 328 (5974) (2010) 76–80.

[59] T.A. Sharp, L. Pastewka, M.O. Robbins, Elasticity limits structural superlubricity in large contacts, Phys. Rev. B 93 (12) (2016). 121402(R).

[60] X. Feng, S. Kwon, J.Y. Park, M. Salmeron, Superlubric sliding of graphene nanoflakes on graphene, ACS Nano 7 (2) (2013) 1718–1724.

[61] Y. Liu, F. Grey, Q. Zheng, The high-speed sliding friction of graphene and novel routes to persistent superlubricity, Sci. Rep. 4 (2014) 4875.

[62] W.K. Kim, M.L. Falk, Atomic-scale simulations on the sliding of incommensurate surfaces: the breakdown of superlubricity, Phys. Rev. B 80 (23) (2009) 235428.

[63] M.M. van Wijk, M. Dienwiebel, J.W.M. Frenken, A. Fasolino, Superlubric to stick-slip sliding of incommensurate graphene flakes on graphite, Phys. Rev. B 88 (23) (2013) 235423.

[64] A.K. Geim, I.V. Grigorieva, Van der Waals heterostructures, Nature 499 (7459) (2013) 419–425.

[65] M. Peyrard, S. Aubry, Critical behavior at the transition by breaking of analyticity in the discrete Frenkel-Kontorova model, J. Phys. C: Solid State Phys. 16 (9) (1983) 1593–1608.

[66] M. Ma, A. Benassi, A. Vanossi, M. Urbakh, Critical length limiting superlow friction, Phys. Rev. Lett. 114 (5) (2015) 055501.

[67] L. Gigli, N. Manini, A. Benassi, E. Tosatti, A. Vanossi, R. Guerra, Graphene nanoribbons on gold: understanding superlubricity and edge effects, 2d Mater. 4 (4) (2017) 045003.

[68] J. Wang, W. Cao, Y. Song, C. Qu, Q. Zheng, M. Ma, Generalized scaling law of structural superlubricity, Nano Lett. 19 (11) (2019) 7735–7741.

[69] D. Dietzel, C. Ritter, T. Monninghoff, H. Fuchs, A. Schirmeisen, U.D. Schwarz, Frictional duality observed during nanoparticle sliding, Phys. Rev. Lett. 101 (12) (2008) 125505.

[70] L. Wang, X. Zhou, T. Ma, D. Liu, L. Gao, X. Li, et al., Superlubricity of a graphene/ MoS_2 heterostructure: a combined experimental and DFT study, Nanoscale 9 (30) (2017) 10846–10853.

[71] R. Ribeiro-Palau, C. Zhang, K. Watanabe, T. Taniguchi, J. Hone, C.R. Dean, Twistable electronics with dynamically rotatable heterostructures, Science 361 (6403) (2018) 690–693.

[72] D. Dietzel, M. Feldmann, U.D. Schwarz, H. Fuchs, A. Schirmeisen, Scaling laws of structural lubricity, Phys. Rev. Lett. 111 (23) (2013) 235502.

[73] T. Brazda, A. Silva, N. Manini, A. Vanossi, R. Guerra, E. Tosatti, et al., Experimental observation of the Aubry transition in two-dimensional colloidal monolayers, Phys. Rev. X 8 (1) (2018) 011050.

[74] A. Erdemir, J. Martin, Superlubricity, 1st ed., Elsevier, New York, 2007.

[75] D. Mandelli, W.G. Ouyang, O. Hod, M. Urbakh, Negative friction coefficients in superlubric graphite-hexagonal boron nitride heterojunctions, Phys. Rev. Lett. 122 (7) (2019) 076102.

[76] A.S. de Wijn, (In)commensurability, scaling, and multiplicity of friction in nanocrystals and application to gold nanocrystals on graphite, Phys. Rev. B 86 (8) (2012) 085429.

[77] N. Varini, A. Vanossi, R. Guerra, D. Mandelli, R. Capozza, E. Tosatti, Static friction scaling of physisorbed islands: the key is in the edge, Nanoscale 7 (5) (2015) 2093–2101.
[78] M. Hirano, K. Shinjo, R. Kaneko, Y. Murata, Observation of superlubricity by scanning tunneling microscopy, Phys. Rev. Lett. 78 (8) (1997) 1448–1451.
[79] D. Dietzel, T. Moenninghoff, C. Herding, M. Feldmann, H. Fuchs, B. Stegemann, et al., Frictional duality of metallic nanoparticles: influence of particle morphology, orientation, and air exposure, Phys. Rev. B 82 (3) (2010) 035401.
[80] D. Dietzel, J. Brndiar, I. Stich, A. Schirmeisen, Limitations of structural superlubricity: chemical bonds versus contact size, ACS Nano 11 (8) (2017) 7642–7647.
[81] E. Gnecco, R. Bennewitz, T. Gyalog, C. Loppacher, M. Bammerlin, E. Meyer, et al., Velocity dependence of atomic friction, Phys. Rev. Lett. 84 (6) (2000) 1172–1175.
[82] Z. Liu, L. Song, S. Zhao, J. Huang, L. Ma, J. Zhang, et al., Direct growth of graphene/hexagonal boron nitride stacked layers, Nano Lett. 11 (5) (2011) 2032–2037.
[83] W. Yang, G. Chen, Z. Shi, C.-C. Liu, L. Zhang, G. Xie, et al., Epitaxial growth of single-domain graphene on hexagonal boron nitride, Nat. Mater. 12 (9) (2013) 792–797.
[84] Y. Kobayashi, T. Taniguchi, K. Watanabe, Y. Maniwa, Y. Miyata, Slidable atomic layers in van der Waals heterostructures, Appl. Phys. Express 10 (4) (2017) 045201.
[85] P.E. Sheehan, C.M. Lieber, Friction between van der Waals solids during lattice directed sliding, Nano Lett. 17 (7) (2017) 4116–4121.
[86] Q.S. Bai, X. He, J.X. Bai, Z. Tong, An atomistic investigation of the effect of strain on frictional properties of suspended graphene, AIP Adv. 6 (5) (2016) 055308.
[87] L. Yang, Y. Guo, Q. Zhang, Frictional behavior of strained multilayer graphene: tuning the atomic scale contact area, Diam. Relat. Mater. 73 (2017) 273–277.
[88] P. Yin, M. Ma, Efficient and robust fabrication of microscale graphene drums, ACS Appl. Nano Mater. 1 (12) (2018) 6596–6602.
[89] S. Zhang, Y. Hou, S.Z. Li, L.Q. Liu, Z. Zhang, X.Q. Feng, et al., Tuning friction to a superlubric state via in-plane straining, Proc. Natl Acad. Sci. U.S.A. 116 (49) (2019) 24452–24456.
[90] K. Wang, W. Ouyang, W. Cao, M. Ma, Q. Zheng, Robust superlubricity by strain engineering, Nanoscale 11 (5) (2019) 2186–2193.
[91] X. Lin, H. Zhang, Z. Guo, T. Chang, Strain engineering of friction between graphene layers, Tribol. Int. 131 (2019) 686–693.
[92] Y. Dong, Z. Duan, Y. Tao, Z. Wei, B. Gueye, Y. Zhang, et al., Friction evolution with transition from commensurate to incommensurate contacts between graphene layers, Tribol. Int. 136 (2019) 259–266.
[93] K. Wang, C. Qu, J. Wang, W. Ouyang, M. Ma, Q. Zheng, Strain engineering modulates graphene interlayer friction by Moire pattern evolution, ACS Appl. Mater. Inter. 11 (39) (2019) 36169–36176.
[94] G. He, M.H. Muser, M.O. Robbins, Adsorbed layers and the origin of static friction, Science 284 (5420) (1999) 1650–1652.
[95] Z.T. Li, A. Kozbial, N. Nioradze, D. Parobek, G.J. Shenoy, M. Salim, et al., Water protects graphitic surface from airborne hydrocarbon contamination, ACS Nano 10 (1) (2016) 349–359.
[96] Z. Li, Y. Wang, A. Kozbial, G. Shenoy, F. Zhou, R. McGinley, et al., Effect of airborne contaminants on the wettability of supported graphene and graphite, Nat. Mater. 12 (10) (2013) 925–931.
[97] S.H. Kim, D.B. Asay, M.T. Dugger, Nanotribology and MEMS, Nano Today 2 (5) (2007) 22–29.

[98] S.F. Cheng, B.Q. Luan, M.O. Robbins, Contact and friction of nanoasperities: effects of adsorbed monolayers, Phys. Rev. E 81 (1) (2010) 016102.

[99] G. He, M.O. Robbins, Simulations of the kinetic friction due to adsorbed surface layers, Tribol. Lett. 10 (1-2) (2001) 7–14.

[100] M.H. Muser, L. Wenning, M.O. Robbins, Simple microscopic theory of Amontons's laws for static friction, Phys. Rev. Lett. 86 (7) (2001) 1295–1298.

[101] M.D. Perry, J.A. Harrison, Friction between diamond surfaces in the presence of small third-body molecules, J. Phys. Chem. B 101 (8) (1997) 1364–1373.

[102] R. Guerra, E. Tosatti, A. Vanossi, Slider thickness promotes lubricity: from 2D islands to 3D clusters, Nanoscale 8 (21) (2016) 11108–11113.

[103] A.S. de Wijn, L.G.M. Pettersson, How square ice helps lubrication, Phys. Rev. B 95 (16) (2017) 165433.

[104] M. Chandross, E.B. Webb, M.J. Stevens, G.S. Grest, S.H. Garofalini, Systematic study of the effect of disorder on nanotribology of self-assembled monolayers, Phys. Rev. Lett. 93 (16) (2004) 166103.

[105] W. Ouyang, A.S. de Wijn, M. Urbakh, Atomic-scale sliding friction on a contaminated surface, Nanoscale 10 (14) (2018) 6375–6381.

[106] D. Berman, B. Narayanan, M.J. Cherukara, S.K.R.S. Sankaranarayanan, A. Erdemir, A. Zinovev, et al., Operando tribochemical formation of onion-like-carbon leads to macroscale superlubricity, Nat. Commun. 9 (2018) 1164.

[107] J. Li, X. Ge, J. Luo, Random occurrence of macroscale superlubricity of graphite enabled by tribo-transfer of multilayer graphene nanoflakes, Carbon 138 (2018) 154–160.

[108] A. Erdemir, O. Eryilmaz, Achieving superlubricity in DLC films by controlling bulk, surface, and tribochemistry, Friction 2 (2) (2014) 140–155.

[109] A. Erdemir, The role of hydrogen in tribological properties of diamond-like carbon films, Surf. Coat. Technol. 146 (2001) 292–297.

[110] I. Barel, M. Urbakh, L. Jansen, A. Schirmeisen, Multibond dynamics of nanoscale friction: the role of temperature, Phys. Rev. Lett. 104 (6) (2010) 066104.

Chapter 7

Structural superlubricity under ambient conditions

Mehmet Z. Baykara
Department of Mechanical Engineering, University of California Merced, Merced, CA, United States

7.1 Introduction

The phenomenon of friction has been an integral part of humans' lives since before the time of written history. On the one hand, our ancestors made use of friction when starting fires. At the same time, they also tried to minimize the detrimental effects of friction on mobility, by devising new mechanical components (e.g., the wheel) or employing lubricants (e.g., animal fats) [1]. Despite thousands of years that passed since then and the associated advances in our scientific understanding of nature, humanity still struggles with friction to this day. In particular, there is no *first principles* theory that can accurately estimate the friction forces that are expected to occur at an interface formed between two arbitrary surfaces in relative motion. Likewise, despite notable improvements in lubrication technology, there is no universal scheme that can be employed to reduce friction at all interfaces under all operating conditions.

The fact that we are still unable to control friction in a robust and universal fashion in all mechanical systems is certainly not due to a lack of motivation: a tremendous portion of useful energy generated in power plants and combustion engines around the world is dissipated to overcome friction, in order to first set and then keep mechanical systems in motion [2]. While traditional friction reduction schemes based on liquid lubricants are employed to varying degrees of success in a large number of mechanical systems and processes, liquid lubrication is not always feasible (e.g., in space and biomedical medical applications), which necessitates the use of solid lubricants (e.g., graphite or MoS_2) in certain scenarios [3–5]. Despite significant effort spent toward minimizing friction via the use of liquid and solid lubricants, most of this work is of empirical and applied character; with minimal fundamental scientific understanding to base developments on, scientists and

Superlubricity. DOI: https://doi.org/10.1016/B978-0-444-64313-1.00007-7

engineers pursue improvements in lubrication technology mainly based on experience. Most of the associated difficulties arise from the fact that friction is a convoluted function of the structure and physico-chemical properties of the surfaces forming the interface (which are typically not known a priori down to the atomic level), as well as multiple factors including temperature, humidity, and the sliding speed.

Within this context, the subject of *structural superlubricity* (a state of ultralow friction that is expected to arise at molecularly clean interfaces formed by atomically flat surfaces with incommensurate structure and/or orientation; for details please see Refs. [6–8]) presents an intriguing opportunity (1) because of its promise of nearly frictionless sliding *without the use of any lubricants* and (2) because any experimental system designed to investigate it should involve atomically flat and chemically inert interfaces, eliminating most of the difficulties associated with studying friction in conventional engineering applications. Consequently, a number of milestone experiments have been performed in the last ~15 years that have been aimed at testing the basic principles of structural superlubricity that have been theorized as early as 1990 [9–12], of which three selected examples are mentioned here:

1. In 2004, Dienwiebel et al. [13] demonstrated structurally superlubric sliding between a nanoscale graphite flake and a graphite substrate under a dry nitrogen environment by utilizing a custom atomic force microscopy (AFM) setup that allowed the controlled rotation of the graphite flake with respect to the underlying substrate.
2. In 2012, Liu et al. [14] used a micromanipulator to shear micron-sized graphite flakes off microfabricated mesas on a graphite substrate in controlled fashion, whereby the rapid retraction of the flakes to their original positions on the mesas was observed in detail, and the associated retraction speeds were employed to deduce that the interlayer friction forces were in the superlubric regime.
3. In 2015, in contrast to the majority of previous experimental demonstrations of structural superlubricity, Berman et al. [15] have reported structural superlubricity on the *macroscopic scale* for a multicomponent material system under dry nitrogen.

Despite outstanding experiments in the area, structural superlubricity was generally thought to be more of a *niche* phenomenon and not feasible for conventional engineering applications by the general tribology community. The main argument behind this line of reasoning was that structural superlubricity required molecularly clean interfaces to manifest [16], which would be extremely difficult to realize under ambient conditions, where almost all conventional mechanical systems are expected to operate and all surfaces are inevitably covered by contaminant molecules (including water, various airborne hydrocarbons, etc.). Taking this into account, it is perhaps surprising that there have, in fact, been certain demonstrations of ultralow friction sliding under ambient conditions that have been attributed to structural superlubricity [14,17–22]. On the other hand, these demonstrations always involved sheets of two-dimensional (2D) materials (including graphene, hexagonal boron nitride and in one particular case, the core and shell of a double-walled carbon nanotube) sliding against each other, making a generalization of the phenomenon to mechanical systems that will comprise sliding components of dissimilar, non-2D materials difficult.

Motivated in this fashion, over the past few years, our laboratory performed AFM-based nano manipulation experiments on nano islands of noble metals (gold and platinum) situated on graphite, demonstrating (for the first time in quantitative agreement with the underlying theory) the robust occurrence of structural superlubricity under ambient conditions between dissimilar materials [23,24]. The following sections of this chapter will first describe the method employed in our work and then summarize the results of our experiments. The chapter will be concluded with a brief section on future research directions.

7.2 Nano manipulation experiments via atomic force microscopy

AFM has been used with great success to study friction on the nanometer scale as a function of various factors (including but not limited to normal load, sliding speed and temperature) over the past three decades [25–27]. On the other hand, as it is the case with every experimental approach, the method suffers from various drawbacks that can be summarized as follows:

1. Commercial AFM cantilevers featuring nanoscale tips are universally used as probes in AFM experiments. As the precise characterization of the atomic-scale structure of tip apexes (which are typically amorphous) is difficult and atomic configurations at tip ends may in fact change during the experiments, obtained AFM results cannot be correlated to specific interface structures.
2. While the sample surface to be investigated can be chosen freely to focus on a certain type of structure or chemistry, the choice of materials for commercial AFM cantilevers is very limited. This prevents the investigation of frictional properties exhibited by an interface consisting out of two freely chosen materials.
3. As tip radii remain fixed during experiments (except for extensive wear), analyzing the dependence of friction force on contact area becomes complicated. While effective contact areas may be manually changed by changing the applied load, this can be done only to a small extent; it would be impossible to study how friction evolves with contact areas changing by orders of magnitude. Moreover, contact areas in

conventional AFM experiments are not measured, but rather estimated using continuum-based contact theory models with stringent assumptions [28], which may break down at the nanometer scale [29].

4. Finally, contact areas obtained with commercial AFM cantilevers are limited to a few tens of nm^2 [30], while actual sliding components in micro- and nanoscale mechanical systems are expected to feature significantly larger contact areas. Consequently, the need for a more appropriately sized frictional contact in nanotribology experiments emerges, to be able to model such components in micro- and nanoscale mechanical systems more closely.

To overcome the majority of the limitations associated with conventional AFM-based friction experiments described earlier, Ritter et al. [31] and Dietzel et al. [30,32] have demonstrated the use of the tip in an AFM to laterally manipulate (i.e., push/slide) metallic nano islands on atomically flat substrates such as graphite or MoS_2. While the method described by Ritter et al. employs dynamic (i.e., tapping-mode) AFM and thus measures energy dissipation (rather than the friction forces experienced by the nano islands during manipulation), the contact-mode-based approach demonstrated by Dietzel et al. (Fig. 7.1) directly measures the friction force acting at the interface between the nano islands and the substrate during manipulation. The advantages associated with this manipulation approach (when compared with conventional AFM-based friction experiments) are clear:

1. In nano manipulation experiments, one can, to a large extent, freely choose the materials to use for the slider and substrate components, thus

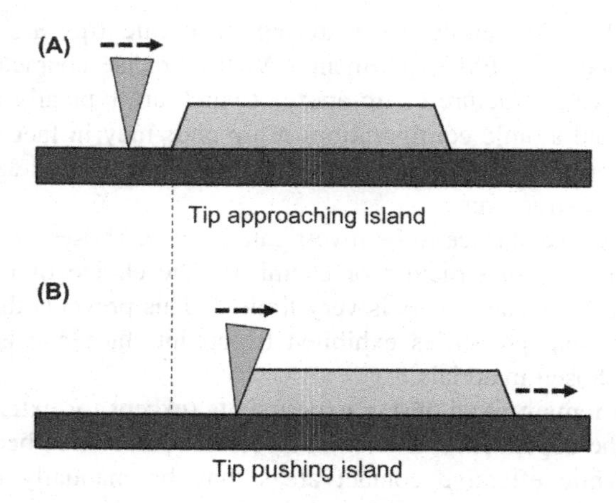

FIGURE 7.1 Schematic illustrations of the AFM tip approaching (A) and pushing (B) a nano island on a flat substrate in contact-mode manipulation. The additional twisting of the tip in (B) is due to the friction force acting at the island-substrate interface during sliding.

enabling the study of friction between material combinations that would simply not be possible to investigate via conventional AFM.
2. The dependence of friction force on contact area can be studied in a wide range in nano manipulation experiments (up to contact areas on the order of 10^5 nm^2), by performing experiments on islands of different size.
3. When islands exhibit atomically flat interfaces with the underlying substrate (as is the case in our experiments [23,24]), *true* contact areas can be accurately determined by simple topographical imaging via AFM, eliminating the need to employ continuum-based theories for contact area estimation.

Based on the advantages summarized earlier, we have exclusively employed nano manipulation experiments based on contact-mode AFM to obtain the results described in this chapter. In particular, we utilized a commercial AFM instrument (PSIA XE-100E) together with soft, contact-mode cantilevers (NanoSensors PPP-CONTR, radius of curvature of ~10 nm) under ambient conditions (with relative humidity values of 20%–30%) to both image and laterally manipulate nano islands of gold and platinum on graphite substrates. Applied normal loads were kept to a minimum (<1 nN) and the cantilevers were calibrated according to established practices in the field [33,34]. The experiments began with imaging an area of a few μm^2 on the sample via contact-mode AFM. During this initial imaging phase, the majority of nano islands in the investigated area were inadvertently manipulated by the plowing action of the tip. After only a few scans, the area under investigation featured only islands that are trapped at step edges or other defects on the substrate. Upon repeated scanning, some of these initially trapped islands would be then spontaneously manipulated. The noticeable increase in the lateral force signal (arising from the torsional twisting of the cantilever around its longitudinal axis) during the manipulation was recorded to deduce the interfacial friction force between the nano island and the substrate (see Section 7.3). Realizing that the interface formed between the noble metal islands and graphite is atomically flat, the interfacial contact area for each island was determined via topographical AFM images acquired during contact-mode scanning, whereby error bars of ±10% have been employed to account for tip convolution effects on imaged geometry [31]. The measured friction force for each nano island was then considered in conjunction with the contact area that it formed with the underlying graphite substrate.

7.3 Structural superlubricity of gold nano islands on graphite under ambient conditions

Taking into account the fact that structural superlubricity at the gold-graphite interface has been previously observed under ultrahigh vacuum (UHV) conditions [35], and the outstanding chemical inertness of gold [36], we decided to investigate the potential occurrence of structural superlubricity under ambient conditions by first focusing on gold nano islands on graphite. Our sample system was prepared via the following procedure [23,37]:

1. ZYA-quality graphite substrates were cleaved in air via adhesive tape and transferred into the vacuum chamber of a thermal evaporator system.
2. Thermal evaporation of 99.99% purity gold was performed on the freshly cleaved graphite under high vacuum at a base pressure of 5×10^{-6} Torr for a total deposition amount of 1 Å, at a typical deposition rate of 0.1 Å/s.
3. After the completion of thermal deposition, the gold-coated graphite sample was removed from the thermal evaporation system and annealed in a furnace at a temperature of 650°C for 1−2 h.

Between preparation steps and after preparation, our sample was characterized by scanning electron microscopy (SEM) and transmission electron microscopy (TEM) (Fig. 7.2). While thermal evaporation resulted in the

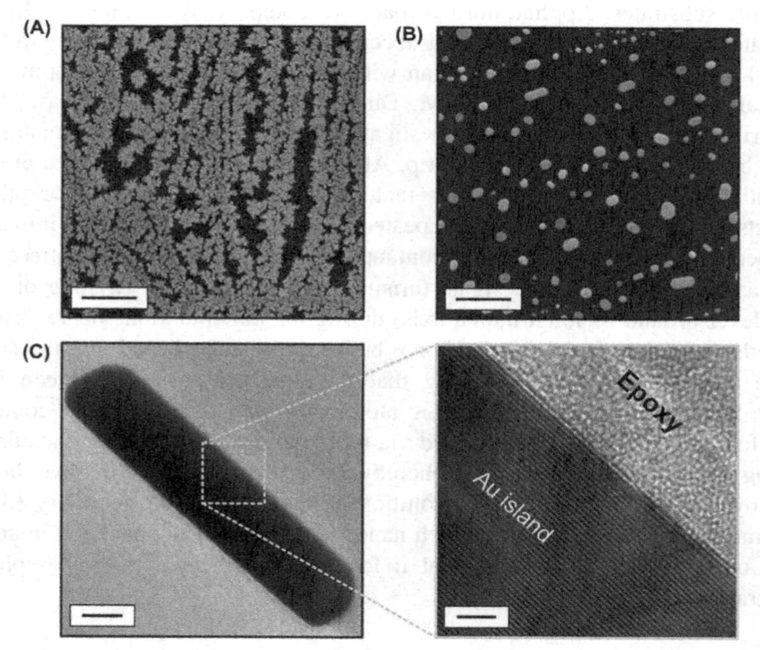

FIGURE 7.2 (A) SEM image of the graphite substrate after thermal evaporation of 1 Å gold. Scale bar: 500 nm. (B) SEM image of the graphite substrate covered with gold nano islands of various size, obtained after the annealing step. Scale bar: 500 nm. (C) Cross-sectional TEM images of a single gold island. The high-resolution image on the right confirms the crystalline character of the gold island and the absence of oxidation. Scale bars: 10 and 2 nm, respectively. *Figure reproduced from E. Cihan, et al., Nat. Commun. 7 (2016) 12055.*

creation of a thin film of gold on the graphite substrate with submonolayer coverage (Fig. 7.2A), postdeposition annealing led to the creation of individual gold islands (Fig. 7.2B). Despite the fact that one could predict the gold islands to be of crystalline character based on their well-faceted morphology in SEM images, high-resolution, cross-sectional TEM measurements rigorously confirmed their crystalline structure, with (111) planes stacked in parallel fashion to the graphite substrate (Fig. 7.2C). The TEM measurements additionally revealed the absence of any oxidation on the gold nano islands, in contrast to antimony islands that were previously the subject of manipulation experiments under ambient conditions [38]. It should be mentioned here that the wide distribution of lateral size for the gold islands (from a few tens of nm up to ~500 nm, as statistically analyzed via SEM images, see Fig. 7.3) is a major advantage for the nano manipulation experiments as it presents the opportunity to study friction as a function of contact area in a wide range that is not attainable in conventional AFM experiments.

After sample preparation and structural/morphological characterization by SEM and TEM, manipulation experiments were conducted via contact-mode AFM as described in Section 7.2. The fact that gold islands were rather easily pushed by the AFM tip during scanning indicated low resistance to motion and the potential occurrence of superlubricity. As such, friction forces experienced by the nano islands during sliding were evaluated in detailed fashion. The results of a representative manipulation experiment performed on an individual gold island are described in Fig. 7.4. In particular, during scanning, the island was slid by the AFM tip along the yellow arrow (Fig. 7.4A) and the vertical tip position (z) as well as the lateral force (F_l)

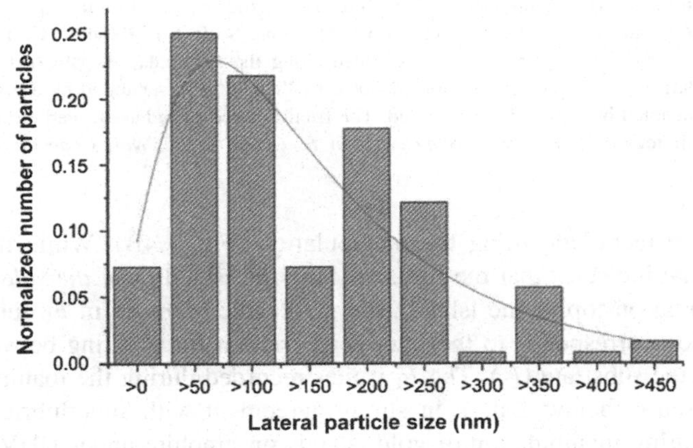

FIGURE 7.3 Lateral size distribution of gold islands (here referred to as *particles*) on graphite as determined from SEM images. Island sizes up to 500 nm are measured. The solid curve is a fit by the log-normal function. *Figure reproduced from E. Cihan, A. Ozogul, M.Z. Baykara, Appl. Surf. Sci. 354 (2015) 429.*

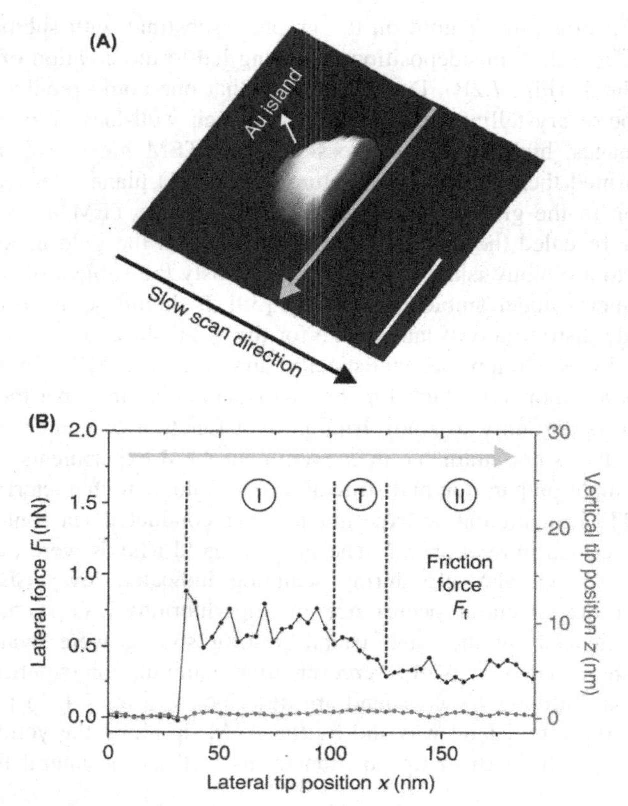

FIGURE 7.4 (A) An AFM image (shown as a three-dimensional representation) demonstrating the manipulation of a gold nano island on graphite. The island is pushed by the tip of the AFM along the arrow and thus disappears from the image frame. Scale bar: 100 nm. (C) The lateral force F_l (and vertical tip position z data recorded along the manipulation path *(arrow)*. Two regions of relatively high and low friction (denoted by "I"and "II"), separated by a brief transition regime (denoted by "T") can be observed. The friction force F_f and associated error bars are extracted from region II. *Figure reproduced from E. Cihan, et al., Nat. Commun. 7 (2016) 12055.*

signals were recorded during the manipulation (Fig. 7.4B). While the constant z signal indicates that the tip is pushing the island *from the side* (rather than climbing on top of the island), the noticeable *increase* in F_l during the manipulation corresponds to the interfacial friction force acting between the island and the substrate (F_f). The F_l values recorded during the manipulation were miniscule (below 1 nN), in strong agreement with superlubric values recorded during manipulation of gold islands on graphite under UHV conditions [35], and orders of magnitude smaller than the values measured on antimony islands under ambient conditions [39]. Interestingly, two regions involving relatively high (region I) and low (region II) lateral forces can be distinguished along the manipulation line: While the lateral force signal is measured as 0.65 ± 0.11 nN in region I, the value drops to 0.33 ± 0.05 nN in region II, following a short transition regime. A particular reason for the initially high lateral forces can be that the majority of our manipulated islands were initially located at step edges and defects (see Section 7.2), which may have increased the resistance to motion during the initial stages of motion. While this relatively clear distinction between high and low friction regions could not be made for all manipulation events in our experiments, for the ones where the distinction can be made, friction forces (F_f) were extracted from region II.

Despite the fact that the miniscule friction forces reported for the single manipulation event in Fig. 7.4 were indicative of superlubricity, multiple measurements needed to be made and the dependence of friction force on contact area (A) studied in order to quantitatively confirm the origin of the observed ultralow friction as structural superlubricity. In order to achieve this goal, manipulation experiments were repeated on multiple gold islands with contact areas from 4000–130,000 nm^2; the results of these experiments are reported in Fig. 7.5A, where F_f is plotted as a function of A for a total of 37 manipulations. All manipulation events occurred under ultralow friction forces, whereby 36 manipulations involved friction forces smaller than 1 nN, and the largest island ($\sim 130,000$ nm^2) experienced the largest friction force (2.38 nN).

The theory underlying structural superlubricity necessitates that F_f depends sublinearly on A via a power law [40], such that:

$$F_f = F_0 N^{\gamma} \tag{7.1}$$

where F_0 is the "friction force" that a single atom moving over the potential energy landscape of the substrate would experience [as deduced from the ratio of the related energy barrier to diffusion (ΔE; 50 meV for a gold atom diffusing over a graphite substrate [35,41]) and the lattice constant of the substrate (a; 0.246 nm for graphite)], N is the number of atoms on the sliding surface (as determined by multiplying the density of atoms on the (111) face of gold (14 atoms/nm^2) with A) and γ is a scaling power that is expected to be between 0 and 0.5 [hence the *sublinear* character of Eq. (7.1)], depending on the geometry of the slider and its commensurability with the underlying substrate. To test whether the friction forces experienced by our gold islands indeed scale sublinearly with contact area in accordance with Eq. (7.1), we plotted (F_f/F_0) as a function of N for all 37 manipulation events (Fig. 7.5B), by means of an approach that has been employed previously [35]. As one can clearly observe in Fig. 7.5B, all of our manipulations are comfortably in the scaling power range defined for structural superlubricity ($0 < \gamma < 0.5$, with a mean γ of 0.16), whereby the wide distribution in γ values is thought to arise from differences in the shape of the islands [24,35].

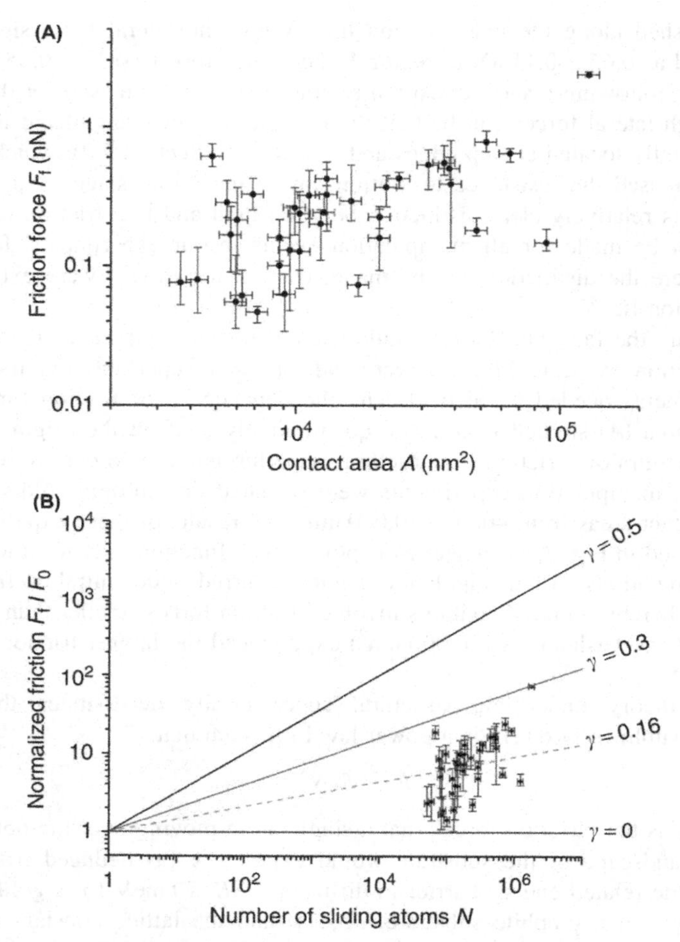

FIGURE 7.5 (A) Friction forces (F_f) for 37 gold nano island manipulations as a function of contact area (A). (B) Normalized friction (F_f/F_0) as a function of number of atoms on the sliding gold surface (N). All manipulations are within the regime of structural lubricity ($0 < \gamma < 0.5$), with a mean scaling power of $\gamma = 0.16$. *Figure reproduced from E. Cihan, et al., Nat. Commun. 7 (2016) 12055.*

The robust observation of structural superlubricity under ambient conditions for gold islands sliding on graphite was rather unexpected, mainly due to the fact that it would be highly unlikely for the gold-graphite interface to be molecularly clean under such an environment. Nevertheless, the results of our experiments were in striking agreement with those performed under UHV [35], so ab initio simulations based on density functional theory (DFT) were performed to study the interaction of the gold-graphite interface with certain contaminant molecules (such as water, O_2 and a representative hydrocarbon: propane) that would be found in an ambient setting [23]. In particular, a gold cluster of 19 atoms was situated on a three-layer graphite substrate such that its (111) planes were aligned parallel to it, which resulted in an equilibrium spacing of 3.45 Å between the two (Fig. 7.6A). Single contaminant molecules were then approached to the interface in steps of 0.5 Å and the changes in the total energy of the system (ΔE) were calculated at each step (Fig. 7.6B–G). The results revealed a rather interesting behavior: While propane was strongly repelled from the interface via rapidly increasing repulsive energy, water and oxygen faced substantial energy barriers (4.3 and 2.3 eV, respectively). If enough energy were available to the molecules to surpass the barrier, they would dissociate and the atomic/molecular constituents (O, H, and OH) would absorb at the gold-graphite interface, potentially leading to a breakdown of superlubricity [16]. On the other hand, the calculated barriers are too large for the molecules to dissociate via the thermal energy available to them at room temperature and as such, the gold-graphite interface remains uncontaminated, essentially acting as a *hermetic seal* and preserving the structural incommensurability at the interface that is required for superlubric sliding.

It should be indicated that the mechanism proposed here to explain the observation of structural superlubricity under ambient conditions has certain drawbacks, the most important of which is that it does not consider the inevitable presence of defects in the islands, which can presumably facilitate the contamination of the interface by foreign molecules via a reduction of energy barriers for absorption. Moreover, the interface may not be clean at all, and the presence of a layer of water molecules between the island and the substrate could be considered [42]. On the other hand, the close alignment of measured friction forces [23] with those obtained under UHV [35] renders this scenario rather unlikely. Within this context, one can imagine that the high temperature (650°C) annealing performed as the last step of the preparation procedure could lead to the desorption of the majority of contaminant molecules from the sample surface and thus ensure clean conditions as the gold atoms coalesce to form nano islands on the substrate.

Before concluding this section, one should think about the potential reasons behind the drastic difference in friction forces experienced by gold and antimony nano islands under ambient conditions. This is especially important as the theory of structural superlubricity should be valid for any combination of slider and substrate as long as they have atomically flat and relatively inert surfaces. As such, at first glance there is no fundamental reason why antimony islands should not undergo superlubric sliding on graphite. On the other hand, upon closer inspection via TEM, it is found that antimony nano island surfaces are covered by an amorphous oxide [38], which may, for instance, break down structural superlubricity via increased topographical roughness on the island surface [43]. Consequently, the chemical inertness of the gold nano islands studied in our experiments (and the related absence of

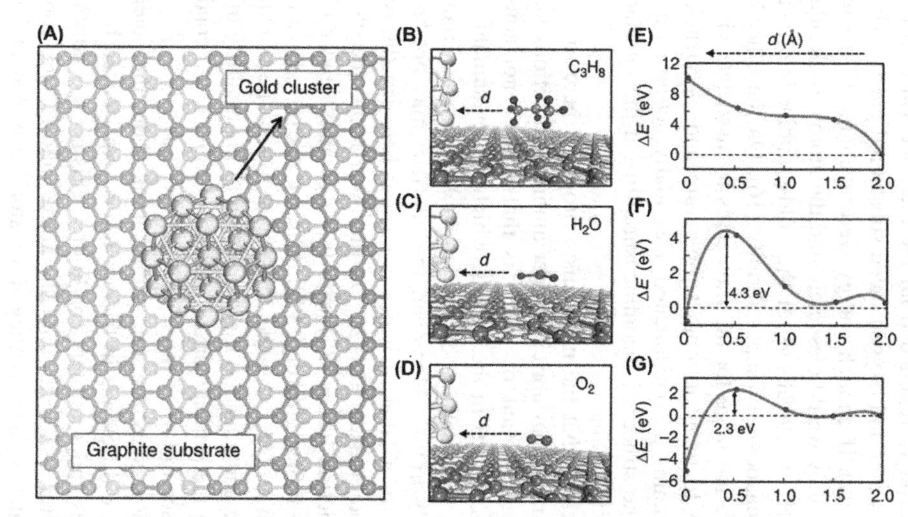

FIGURE 7.6 (A) Top view of the model system used in the ab initio simulations. (B–D) Side views of single propane (B), water (C), and oxygen (D) molecules approaching the gold-graphite interface with decreasing distance d. (E–G) Calculated change in the total energy of the system (ΔE) as a function of d for the three scenarios in (B–D). *Figure reproduced from E. Cihan, et al., Nat. Commun. 7 (2016) 12055.*

oxidation) arises as a key characteristic leading to the conservation of structural superlubricity under ambient conditions.

7.4 Structural superlubricity of platinum nano islands on graphite under ambient conditions

In order to determine whether or not the observation of structural superlubricity under ambient conditions is exclusive to the gold-graphite interface, we directed our focus on another material system: platinum nano islands on graphite. Platinum was mainly chosen as the slider material to study next due to the fact that it is also a noble metal and that we independently verified via X-ray photoelectron spectroscopy that platinum nano islands do not oxidize under ambient conditions [44].

Platinum nano islands on graphite were prepared in a similar fashion to gold nano islands on graphite (see Section 7.3) [24]. The main differences were that (1) the platinum thin film was deposited via e-beam evaporation rather than thermal evaporation and (2) the annealing was performed at a higher temperature (1000°C) for a shorter duration ($\sim$30 min). The resulting system comprised crystalline platinum islands of varying lateral size on the graphite substrate, with the absence of an oxide layer on both the top and bottom surfaces (Fig. 7.7).

Nano manipulation experiments were performed on platinum nano islands on graphite using the approach introduced in Sections 7.2 and 7.3. Fig. 7.8A shows the results of such experiments in the form of a friction force versus contact area plot, where the results obtained on gold islands are also shown for comparison purposes. While it is clear that platinum nano islands also experience ultralow friction forces ($<$5 nN) while sliding on graphite under ambient conditions, it is also evident that the mean friction force for platinum nano islands (in the same size range) is noticeably higher than the mean friction force for gold nano islands (1.41 and 0.33 nN, respectively). Nevertheless, an analysis of normalized friction (F_f/F_0) versus number of sliding atoms (N) in accordance with Eq. (7.1) (with an ΔE of 173 meV [45]) reveals that all manipulations are structurally superlubric (Fig. 7.8B), with a mean scaling power (γ) of 0.19, comparable to the one measured on the gold nano islands (0.16).

In order to confirm the origin of the noticeably higher friction experienced by platinum nano islands when compared with gold, we performed ab initio simulations utilizing DFT [24]. In particular, slabs of gold and platinum comprising three (111) layers (Fig. 7.9A) were slid in incommensurate configuration on graphite substrates of five layers. The change in the total energy of the system was calculated after each step and normalized with respect to the number of atoms on the sliding metallic surface. Fig. 7.9B shows the evolution in the normalized energy of the system as the metallic slab is moved over the graphite substrate. A comparison of the most

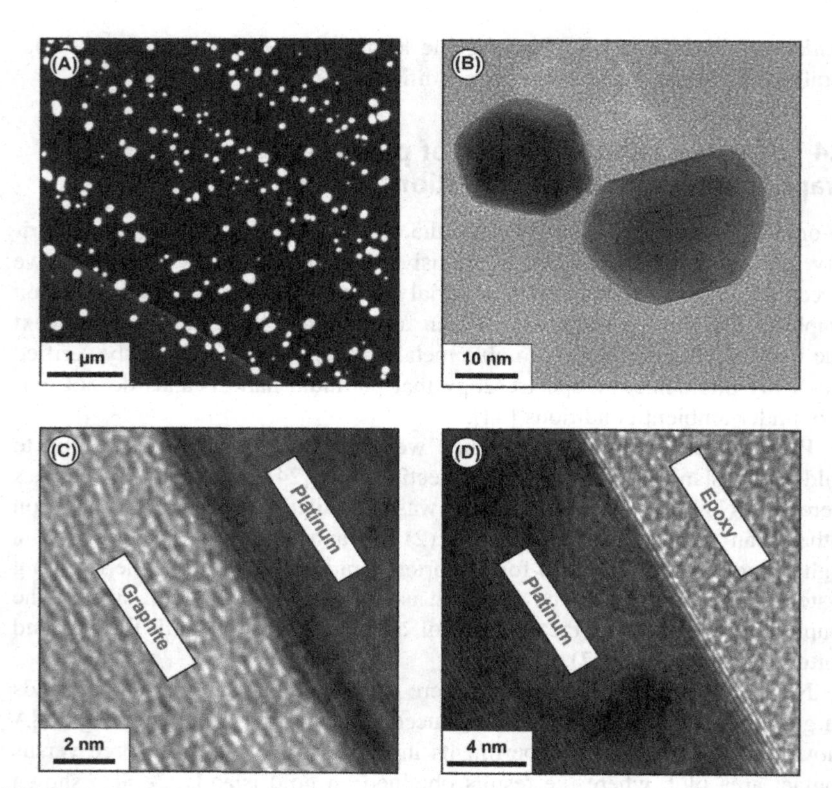

FIGURE 7.7 (A) SEM image of the graphite substrate covered with platinum nano islands of various size. (B) TEM image of two platinum nano islands with predominantly straight edges. (C) Cross-sectional TEM image of the bottom side of an individual platinum nano island on graphite. (D) Cross-sectional TEM image of the top side of an individual platinum nano island on graphite. *Figure reproduced from A. Ozogul, et al., Appl. Phys. Lett. 111 (2017) 211602.*

substantial energy barriers in the two scenarios reveals that platinum atoms experience significantly higher resistance to motion on graphite when compared with gold (1.6 meV vs 0.4 meV). Considering that similarly sized platinum and gold nano islands in the experiments will contain similar number of atoms on their sliding surfaces (due to the fact that their densities on the (111) planes are very similar: 15 and 14 atoms/nm^2, respectively), the simulations confirm the origin of the experimental results as a stronger interaction of platinum atoms with the graphite substrate when compared with gold.

7.5 Perspectives on future work

This chapter reported on AFM-based nano manipulation experiments performed on nano islands of noble metals on graphite, with the remarkable

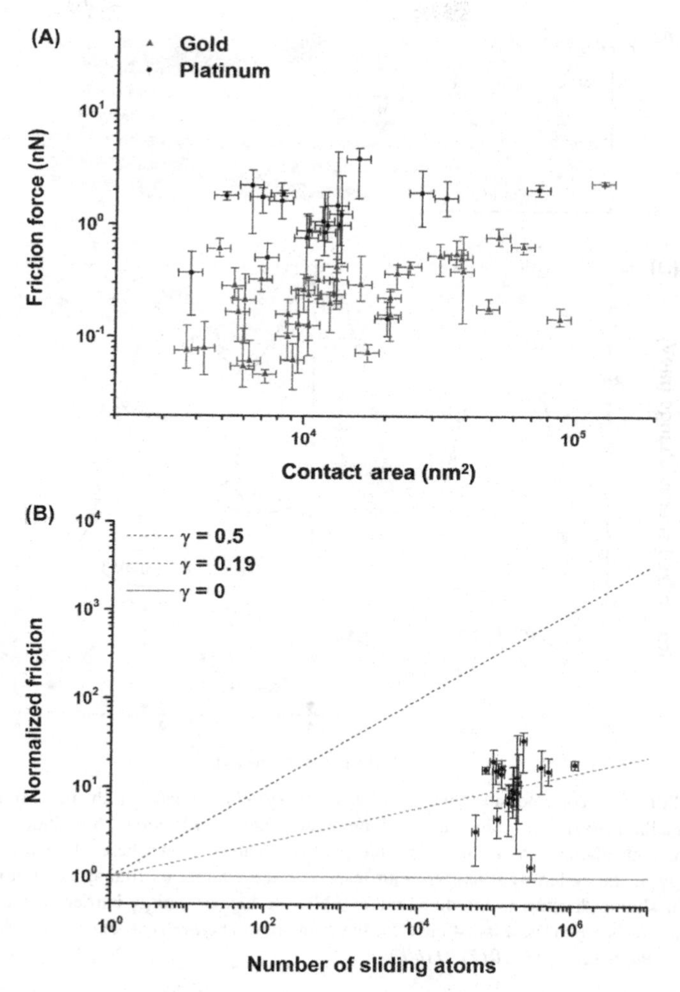

FIGURE 7.8 (A) Friction force as a function of contact area for platinum and gold nano islands. (B) Normalized friction (F_f/F_0) as a function of number of sliding atoms, for platinum nano islands. *Figure reproduced from A. Ozogul, et al., Appl. Phys. Lett. 111 (2017) 211602.*

conclusion that structural superlubricity (in quantitative agreement with the underlying theory) can indeed be observed under ambient conditions. While this discovery is certainly promising, there are important questions that need to be addressed in order to better evaluate the application potential of structural superlubricity in practical mechanical systems. In particular, the robustness of the superlubric state against changes in temperature, sliding speed, and contact size need to be carefully studied, before proof-of-concept

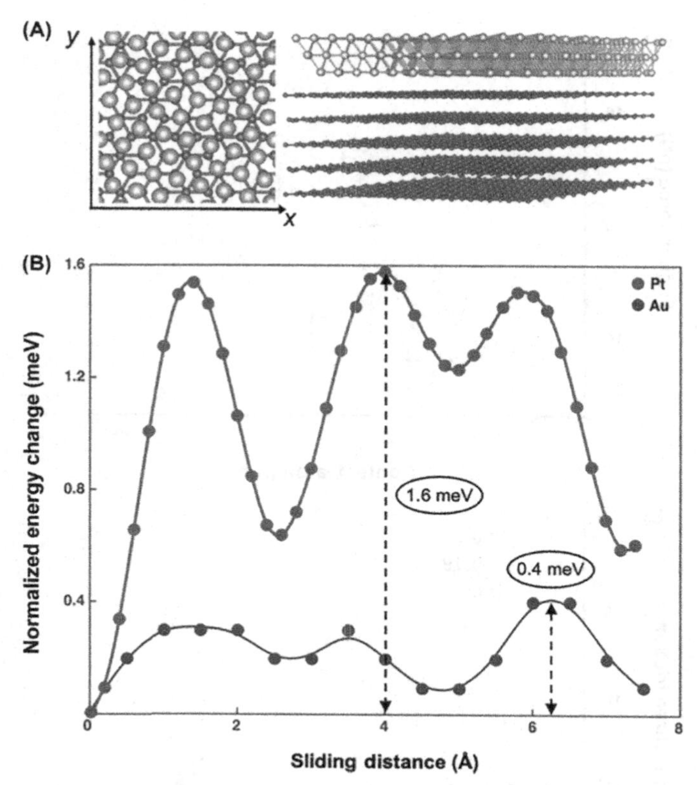

FIGURE 7.9 (A) Top and side views of the model system studied via ab initio simulations, with a metallic slab *(large atoms)* and a graphite substrate *(small atoms)*. Simulations are performed for both platinum and gold slabs; the platinum slab is shown here. (B) Change in the total energy of the system (normalized with respect to the number of sliding metal atoms) as a function of sliding distance in the direction x. While the highest energy barrier encountered by platinum atoms is 1.6 meV, it is 0.4 meV for the gold atoms. *Figure reproduced from A. Ozogul, et al., Appl. Phys. Lett. 111 (2017) 211602.*

systems on small length scales can be realistically envisioned. Within this context, it should be remembered that theoretical and experimental studies of the phenomenon point to a breakdown of the superlubric state with increasing contact size [46–49], due to the nucleation of commensurate regions at the slider-substrate interface by way of elastic deformations. A potential approach to overcome the associated limits could involve the use of amorphous, yet atomically flat sliders and/or substrates in the experiments. Recent developments in the high-precision processing of metallic glasses could provide a potential route toward the realization of such experiments [50].

References

[1] B. Bhushan, Introduction to Tribology, Wiley, New York, 2013.
[2] K. Holmberg, A. Erdemir, Friction 5 (2017) 263.
[3] A. Savan, et al., Lubrication Sci. 12 (2000) 185.
[4] C. Donnet, A. Erdemir, Surf. Coat. Technol. 180 (2004) 76.
[5] M.R. Vazirisereshk, et al., Lubricants 7 (2019) 57.
[6] M.Z. Baykara, M.R. Vazirisereshk, A. Martini, Appl. Phys. Rev. 5 (2018) 18.
[7] O. Hod, et al., Nature 563 (2018) 485.
[8] J.M. Martin, A. Erdemir, Phys. Today 71 (2018) 40.
[9] M. Hirano, K. Shinjo, Phys. Rev. B 41 (1990) 11837.
[10] J.B. Sokoloff, Phys. Rev. B 42 (1990) 760.
[11] K. Shinjo, M. Hirano, Surf. Sci. 283 (1993) 473.
[12] M. Hirano, K. Shinjo, Wear 168 (1993) 121.
[13] M. Dienwiebel, et al., Phys. Rev. Lett. 92 (2004) 126101.
[14] Z. Liu, et al., Phys. Rev. Lett. 108 (2012) 205503.
[15] D. Berman, et al., Science 348 (2015) 1118.
[16] G. He, M.H. Muser, M.O. Robbins, Science 284 (1999) 1650.
[17] R. Zhang, et al., Nat. Nanotechnol. 8 (2013) 912.
[18] J. Yang, et al., Phys. Rev. Lett. 110 (2013) 255504.
[19] E. Koren, et al., Science 348 (2015) 679.
[20] H. Deng, et al., Nanoscale 10 (2018) 14314.
[21] Y. Song, et al., Nat. Mater. 17 (2018) 894.
[22] C.C. Vu, et al., Phys. Rev. B 94 (2016). 081405(R).
[23] E. Cihan, et al., Nat. Commun. 7 (2016) 12055.
[24] A. Ozogul, et al., Appl. Phys. Lett. 111 (2017) 211602.
[25] I. Szlufarska, M. Chandross, R.W. Carpick, J. Phys. D Appl. Phys. 41 (2008) 123001.
[26] N. Manini, et al., Adv. Phys. X 2 (2017) 569.
[27] A. Vanossi, et al., Beilstein J. Nanotechnol. 9 (2018) 1995.
[28] U.D. Schwarz, J. Colloid Interface Sci. 261 (2003) 99.
[29] B.Q. Luan, M.O. Robbins, Nature 435 (2005) 929.
[30] D. Dietzel, et al., J. Appl. Phys. 102 (2007) 084306.
[31] C. Ritter, et al., Phys. Rev. B 71 (2005) 085405.
[32] D. Dietzel, U.D. Schwarz, A. Schirmeisen, Friction 2 (2014) 114.
[33] J.E. Sader, J.W.M. Chon, P. Mulvaney, Rev. Sci. Instrum. 70 (1999) 3967.
[34] M. Varenberg, I. Etsion, G. Halperin, Rev. Sci. Instrum. 74 (2003) 3362.
[35] D. Dietzel, et al., Phys. Rev. Lett. 111 (2013) 235502.
[36] B. Hammer, J.K. Norskov, Nature 376 (1995) 238.
[37] E. Cihan, A. Ozogul, M.Z. Baykara, Appl. Surf. Sci. 354 (2015) 429.
[38] C. Ritter, et al., Phys. Rev. B 88 (2013) 045422.
[39] D. Dietzel, et al., Phys. Rev. Lett. 101 (2008) 125505.
[40] A.S. de Wijn, Phys. Rev. B 86 (2012) 085429.
[41] P. Jensen, X. Blase, P. Ordejon, Surf. Sci. 564 (2004) 173.
[42] F. Hartmuth, et al., Lubricants 7 (2019) 66.
[43] U. Tartaglino, V.N. Samoilov, B.N.J. Persson, J. Phys. Condens. Matter 18 (2006) 4143.
[44] A. Ozogul, Investigation of Structural Lubricity on Platinum Nanoparticles Under Ambient Conditions, Bilkent University, 2017.
[45] X.J. Liu, et al., Phys. Chem. Chem. Phys. 14 (2012) 9157.

[46] M. Ma, et al., Phys. Rev. Lett. 114 (2015) 055501.
[47] A. Benassi, et al., Sci. Rep. 5 (2015) 16134.
[48] T.A. Sharp, L. Pastewka, M.O. Robbins, Phys. Rev. B 93 (2016) 121402.
[49] D. Dietzel, et al., ACS Nano 11 (2017) 7642.
[50] R. Li, et al., Commun. Phys. 1 (2018) 75.

Chapter 8

Toward micro- and nanoscale robust superlubricity by 2D materials

Tianbao Ma[1], Yanmin Liu[1], Qunyang Li[1,2] and Jianbin Luo[1]
[1]*State Key Laboratory of Tribology, Tsinghua University, Beijing, P.R. China,* [2]*AML, CNMM, Department of Engineering Mechanics, Tsinghua University, Beijing, P.R. China*

8.1 Introduction

Superlubricity [1] is an ultralow friction regime with exceptionally low energy dissipation and wear, which is crucial for energy savings, environmental protection, and long-life machine operation in industrial applications [2]. Two-dimensional (2D) materials have shown superior tribological behaviors with their thickness down to few atomic layers, which show intrinsic advantages, such as atomically smooth, chemically inert and weak interlayer van der Waals (vdW) interaction [3]. Researchers have made a series of achievements about excellent friction properties of 2D materials at the nanoscale, including theoretical models and experimental explorations [4]. Conventionally the nanoscale sliding was between commercial atomic force microscopy (AFM) sharp tip and various 2D materials. It is technically difficult to probe, however, the interaction between the atomic layers of 2D materials, which makes it incapable to precisely control the friction properties of 2D materials. Dienwiebel et al. [5] show an impressive periodic rotation angle dependence of friction between graphite layers, where superlubricity was achieved due to incommensurate interfacial geometry between high friction peaks every 60 degrees. However superlubricity achieved in this way was not stable due to torque-induced twist of graphite flake back to an energetically stable commensurate geometry [6]. Similar behavior has also been reported for monolayer graphene flake, where the incommensurate to commensurate transition is postulated to be driven by mechanical or thermal activation [7]. To experimentally acquire a stable interlayer sliding to avoid spontaneous rotation is a key challenge to achieve robust microscale superlubricity. Also a systematic theoretical

Superlubricity. DOI: https://doi.org/10.1016/B978-0-444-64313-1.00008-9

modeling of the interlayer coupling of 2D materials at the atomic level is necessary to reveal the fundamental mechanism of superlubricity and design of the superlubricious system.

8.2 Theoretical prediction of sustainable superlubricity by 2D heterostructures

The sliding friction between conventional AFM tip and crystalline samples including 2D materials has been modeled by the well-known Prandtl–Tomlinson (PT) model [8,9]. In this model, it is assumed that a point mass m (e.g., the AFM tip) is dragged over a one-dimensional sinusoidal potential representing the interaction between the tip and a crystalline substrate. The point tip is pulled by a spring of effective elastic constant k connected to a driving support representing the stiffness of the AFM cantilever. Also the potential has been extended to 2D, so that the AFM tip trajectory and lateral force can be modeled to better simulate the real experiments [10]. Both the amplitude of the potential corrugation and the spring stiffness contribute significantly to the friction behavior of the system. The effects of both have been carefully discussed by Socoliuc et al., where increasing normal load leads to larger potential energy corrugation and maxmium lateral force, yet also to larger effective lateral stiffness. These contradicting effects enable a modulation of nanoscale friction and superlubricity by varying the normal load [11]. Specifically for interlayer sliding friction between commensurate graphene layers, Xu et al. [12] proposed a theoretical model with regard to lateral stiffness, slip length, and maximum lateral force, which well explained the simulated thickness dependence of friction in few layer graphene. A superlubricity state is achieved when the few layer graphene is composed of two to three layers with large enough stiffness, even when the graphene layers are in commensurate contact.

In PT-model-based studies, the potential energy corrugation has been generally modeled by simple sinusoidal functions or their superimposition. This is an oversimplification when considering a specific system, where the exact magnitude of potential energy corrugation, as well as the shape of the potential corrugation plays a significant role. In this regard, first-principles calculations have been conducted by quasistatic scan of the 2D layers to establish a "interlayer sliding" potential energy surface, where the total energy variation is usually calculated by density functional theory (DFT) calculations [13,14]. Consequently, the sliding energy barrier and shear strength along certain sliding paths can be calculated [15]. The kinetic friction behavior between the 2D layers could also be simulated by a combination of PT model and a "numerical" potential energy surface interpolated by DFT results [16]. This could qualitatively guide the design of superlubricious system based on 2D layers. For example, it has been estimated that fluorographene exhibits lower friction than graphene due to the repulsive electrostatic forces between F atoms at the interfaces [17]. More importantly, this method has been adopted to predict the superlubricity between 2D heterostructure. 2D heterostructure here refers to a vertical stacking of different 2D layers, which has been proved to show novel electronic, and optical properties that differs from each composing monolayer [18]. A theoretical analysis suggests that graphene/hexagonal boron nitride (hBN) heterojunction may exhibit robust superlubricity due to the intrinsic incommensurate geometry at the interface, as shown in Fig. 8.1A [19]. When elasticity of the 2D layers is considered, however, it has been found recently, that the heterostructure composed of graphene and hBN should experience a structural reconstruction due to the similar lattice constants between the two [21]. Interlayer vdW interaction will trigger the formation of Moiré superlattice structure with commensurate regions separated by domain walls [22,23]. These local commensurate regions may contribute to increased friction of the heterostructure. So the construction of 2D heterostructure with large lattice mismatch is probably favored to eliminate the effect of sheet elasticity on superlubricity. Wang et al. give direct evidence by full DFT calculations of graphene (or fluorographene) sliding against MoS_2 monolayer, which shows more than 2-order-of-magnitude reduction in the interlayer shear strength as compared with the graphene or MoS_2 commensurate bilayers [20,24]. The lattice mismatch between the two is more than 20%, so that structural reconstruction is hard to take place [21]. Consequently, sustainable superlubricity regardless of the relative rotation angle between the layers was predicted for MoS_2/graphene 2D heterostructure.

Another important yet simple way to estimate the interlayer potential energy surface is the registry index method, proposed by Marom et al. [14]. This geometric model was firstly proposed to capture the essence of the

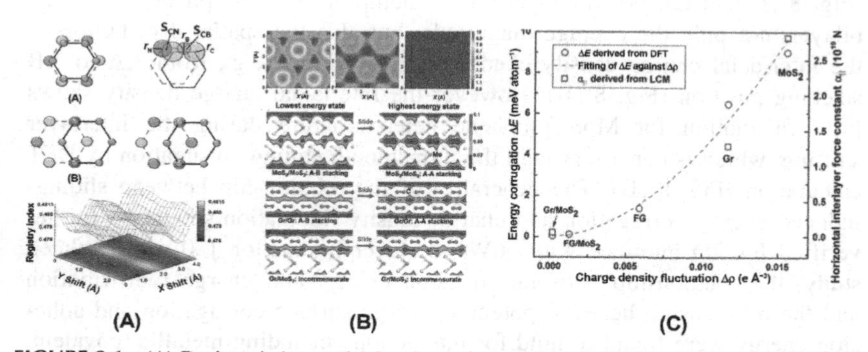

FIGURE 8.1 (A) Registry index model for simulating graphene/h-BN interlayer potential landscape [19]. (B) Interlayer sliding potential energy surface and interfacial charge density fluctuation between MoS_2 and graphene bilayers and MoS_2/graphene heterostructure. (C) Correlation between the sliding-induced energy corrugation and interfacial charge density fluctuation [20]. (*h-BN*: Hexagonal boron nitride).

DFT-calculated interlayer sliding corrugation of hBN by quantifying the registry matching between the layers. It was assumed that dispersion interactions play the role of anchoring the layers at the appropriate interlayer distance, while electrostatic forces determine the interlayer sliding corrugation. In this model, each atom in the unit cell was ascribed a circle centered around its position. Considering the projection on a plane parallel to the layers, the overlaps between the circle centered around the i atom belonging to the top layer and the j atom of the bottom layer was marked by S_{ij} (i and j being either N or B). The radii of the circles r_N and r_B were further fitted to capture the main features of the DFT-calculated sliding corrugation [14]. One successful application of this geometric model was the prediction of superlubricity between graphene/hBN heterojunctions, where a huge supercell consisting of a 56×56 graphene unit cell and a 55×55 hBN unit cell makes full DFT calculations of sliding corrugation too time consuming to be performed [19].

The mechanism of superlubricity achieved by heterostructure can be understood by the mismatched atoms at the incommensurate interface where the friction force of the atoms point in all directions and sum up to zero [25]. The explanation is consistent with the numerical simulations by Hirano et al. [1] and Frenkel–Kontorova (FK) model [26]. However both the interlayer interaction and registry matching are important in determining interlayer friction. So an in-depth physical model is needed to precisely predict and provide design criteria of superlubricious systems. In this regard, interlayer sliding corrugation has been correlated to sliding-induced charge density fluctuation at the interface by a full DFT study [20]. MoS_2, graphene, and fluorographene bilayers, as well as MoS_2/graphene and MoS_2/fluorographene heterostructures have been included as a database to fit the relation between sliding-induced energy corrugation and charge density fluctuation (Fig. 8.1B and C). For homogeneous structure such as graphene and MoS_2 bilayer, not only the average magnitude, but also the spatial distribution of the interfacial charge density changes during sliding, e.g., from AA to AB stacking position (Fig. 8.1B). However the interfacial charge density shows little fluctuation for MoS_2/graphene heterostructure during the interlayer shifting, which is consistent with the ultrasmooth sliding corrugation by DFT calculation (Fig. 8.1B). The generality of the relationship between sliding-induced energy corrugation and charge density fluctuation should be further verified for 2D interface with vdW interlayer interaction [20]. In a related study, the relationship between adhesion energy and charge redistribution and the relationship between potential energy surface corrugation and adhesion energy were found to hold for interactions including metallic, covalent, and physical bonds [27].

As discussed in previous literatures, superlubricity may breakdown even with an incommensurate interface due to the Aubry transition, especially at large length scales [28–30]. With atom-by-atom control, Bylinskii et al. [31]

observed the transition from an unpinned (superlubricity) to a pinned state (stick-slip) in finite chains of atoms as increasing the interactions between chains and a periodic substrate. Recently Brazda et al. [32] experimentally and theoretically demonstrated the occurrence of the Aubry transition in a 2D system using a colloidal monolayer on an optical periodic lattice by tuning the corrugation and the period of the substrate. Unlike the continuous 1D Aubry transition, the first-order transition with a coexistence of pinned and unpinned areas was observed in 2D system at room temperature.

From the theoretical predictions discussed earlier, sustainable ultralow interlayer friction could be achieved between 2D heterostructure with weak vdW interlayer interaction and relatively large lattice mismatch. So it is strongly favorable to establish an experimental platform to precisely measure the interlayer friction between various 2D layers under a certain normal load as will be discussed in details in the next section.

8.3 Achieving robust superlubricity at the nano-, micro-, and macroscale

To achieve interlayer superlubricity requires both surfaces in relative sliding are made of or covered with 2D materials. AFM is a widely accepted platform for sub-nN precision intermolecular and surface force measurement. What is especially useful in tribology area is the lateral force mode, or called friction force microscopy, which has been widely adopted to learn the tribological behaviors of well-defined surfaces, for example, atomic-scale friction on atomically thin sheets like 2D materials [33]. Conventional tips used for nanofriction measurement are fabricated from silicon nitride or silicon. Liu et al. developed a thermally assisted mechanical exfoliation and transfer method to fabricate various 2D flake-wrapped AFM tips (Fig. 8.2A) [34]. Take graphite flake-wrapped tip, for example, the fabrication procedure includes a tip fracture stage to expose the fresh tip surface with larger radius of curvature, a heating stage to drive off the water adsorbed on the surface of the tip and the sample as well as promote the adhesion of the fractured tip onto the graphite substrate, and a scanning-transfer stage to roll up and exfoliate the graphite flake from the substrate and consequently, the graphite flake is transferred and wrapped firmly onto the fractured tip. Through the same procedure, graphite, MoS_2, hBN, 1T-TaS_2, and 1T'-ReS_2 flake-wrapped AFM tips were fabricated, as confirmed by high-resolution transmission electron microscopy images of the tip apexes (MoS_2 wrapped tip in Fig. 8.2B and others in ref. [34]), all showing extremely low friction as well as friction coefficient. The lowest friction coefficient obtained in ambient air, as defined by slope of friction-load curve, was 1.0×10^{-4} for graphite/hBN interface, under high local contact pressure in the order of 1.0 GPa. Notably the tribological behavior has been measured recently under ultrahigh vacuum (UHV) at a base pressure of about 4.5×10^{-10} Torr, where the

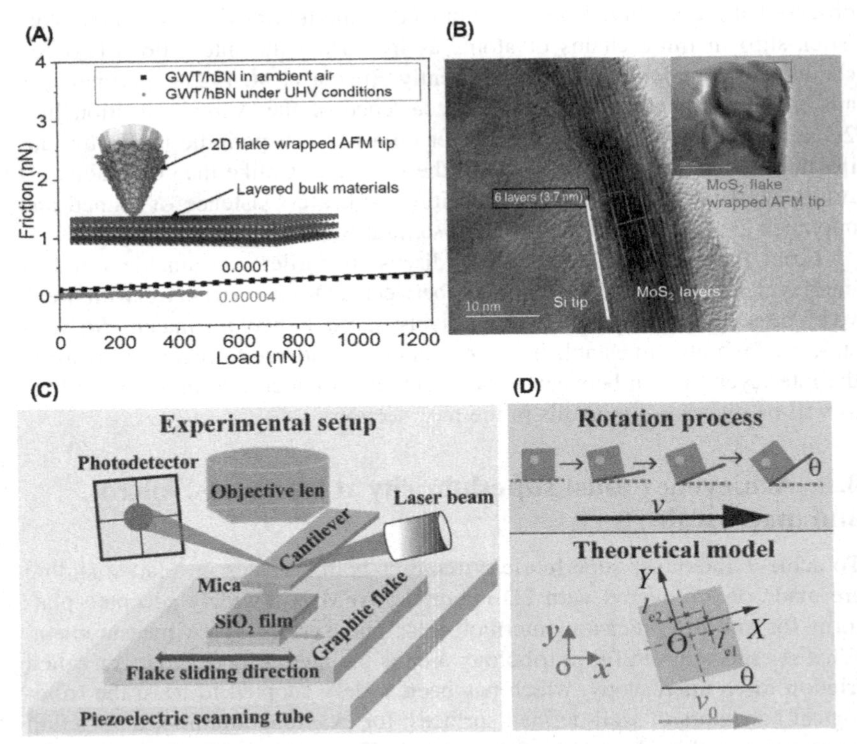

FIGURE 8.2 Superlubricity by 2D heterostructures. (A) Friction characteristics by graphite-wrapped AFM tip against h-BN in ambient air or ultrahigh vacuum. (B) HRTEM of the tip apex of the MoS_2-wrapped AFM tip [34]. [(C) and (D)] Direct fabrication of graphite-mica heterostructure and in situ control of their relative orientation [35]. (*AFM*: Atomic force microscopy; *h-BN*: hexagonal boron nitride; *HRTEM*: high-resolution transmission electron microscopy).

friction coefficient can be as low as 4.0×10^{-5} and friction force is in the order of pN in the load range of the UHV-AFM (RHK technology). Rotational angle-independent superlubricity was achieved between graphite-wrapped tip and hBN substrate, while the wear life was also significantly prolonged to 40,000 cycles with no sign of breakdown of superlubricity, which verifies the predicted sustainable superlubricity between 2D heterostructure. Besides, the atomically resolved lateral force image between ReS_2-wrapped tip and ReS_2 substrate was obtained. The ReS_2 interlayer lateral force image shows a novel hexagonal pattern, which is a reflection of the integrated effect of both upper and lower ReS_2 layers and is distinct with the chain-like pattern corresponding to the characteristic surface structure of 1T'-ReS_2 crystal. The pattern of lateral force image varies by changing the relative orientation angle between the upper and lower ReS_2 layers, which is consistent with theoretical simulations [34].

Interestingly another dry transfer method has been independently proposed by Li et al. [36] to prepare graphite-wrapped tip to measure the critical adhesion force between graphite and graphite; graphite and hBN; and graphite and MoS_2. The results showed that the critical adhesion forces between hBN and graphite and MoS_2 and graphite were, respectively, 0.953 and 1.028 times that between graphite and graphite, which were consistent with the prediction based on Lifshitz theory, implying an important role of material dielectric function in the vdW interactions at heterointerfaces [36]. Both the method of Liu et al. [34] and Li et al. [36] provide a way to quantitatively study interlayer couplings and friction between 2D flakes and layered bulk materials, including 2D heterostructures.

Another important category of method to measure the interlayer adhesion and shear strength has been developed by Liu et al. [37] and Koren et al. [38], inspired by the self-retraction motion of micrometer-sized graphite mesas [39]. Recently, Song et al. [40], Liu et al. [35], and Gongyang et al. [41] measured the sliding friction between heterogenous materials, for example, graphite and hBN, graphite and mica, graphite and diamond-like carbon (DLC) films, respectively. All the upper surfaces in these experiments were graphite mesas prepared by the self-retraction motion, where the graphite mesa was attached to the AFM tungsten tip. External normal loads were applied with a maximum contact pressure of about 11 MPa. And the friction coefficient between graphite flake and hBN surface was measured to be as low as 1.4×10^{-4}, which accomplished the microscale superlubricity with a nominal contact area was 9 μm^2. Interestingly the graphite/hBN heterostructure also shows frictional anisotropy, where less than fourfold enhancement of friction was observed when rotating every 60 degrees, corresponding to an aligned configuration. This anisotropy for the heterostructure, however, was orders of magnitude smaller than that measured for homogeneous graphitic interface. This anisotropy for heterostructure originates mainly from the enhanced out-of-plane atomic undulations in the aligned configuration, which rapidly decay upon rotation to an orientationally misaligned configuration [40]. A further theoretical study predicts a negative friction coefficient for graphite/hBN heterojunction, which was attributed to the load-induced suppression of the Moiré superstructure out-of-plane distortions [42]. Sun et al. theoretically reported a pressure-induced friction collapse between 2D materials by first-principles calculations, which was attributed to an abnormal transition of the sliding potential energy surface from corrugated to substantially flattened [43].

In addition to the nanoscale superlubricity achieved by construction of 2D heterostructure, people strive to find more solutions to scale up the superlubricity system [30]. One attempt was made by Liu et al. to achieve microscale superlubricity by a sliding contact between a 10-μm-diameter graphene-coated microsphere (GMS) and atomically flat graphite surface [44]. They propose to utilize the multiasperity contact geometry and polycrystalline structure of the graphene coating. In their study, multilayer graphene coatings were grown on silica microsphere by a catalyst-free chemical vapor deposition method [44]. The GMS was then attached firmly onto the AFM cantilever with UV light solidify glue. A series of characterizations of the microsphere show that the graphene coating was polycrystalline with an average grain size less than 160 nm, with a thickness of 8–10 atomic layers, not atomically smooth with a standard deviation of height distribution 21.75 nm. The friction coefficient between bare silica microsphere and SiO_2/Si surface was as high as 0.6. If only one side of the sliding interface was covered with graphene coating, the friction coefficient can be reduced significantly to 0.1–0.2. If both sides of the sliding interface were covered with graphene, especially when highly oriented pyrolytic graphite or mechanically exfoliated graphene was used as the lower surface, the friction coefficient can be further lowered to 0.003, entering the superlubricity regime. The superlubricity state achieved between GMS and graphite preserved in both dry nitrogen and humid air, which held a much longer wear life as compared with that without graphene coating. The absence of rotation angle dependence of the ultralow friction suggests that the mechanism of superlubricity should differ from the previously reported registry-induced high friction where single-crystalline contact between graphite was presumed [5,37]. Since the graphene coating was polycrystalline, it is postulated that graphene nanograins exhibit random orientation, so that the nanograins cannot be in registry simultaneously with the counterface. This was verified by theoretical simulations. Firstly, the topography of GMS was measured by a conventional AFM tip, which was then transferred to contact mechanics calculation to obtain the areal contact area. It shows that only several highest asperities were in direct contact with the lower surface, while the distances between contacting asperities were at least 200 nm, which is larger than the grain size (<160 nm) of the graphene coating. This implies that the asperities were covered by different graphene grains with random orientations. Under this assumption, molecular dynamics simulations were conducted to reveal the friction properties of multiasperity contact covered with randomly oriented graphene nanograins, which avoids the concerted motion in commensurate cases, and shows much lower friction and friction coefficient. Thus the superlubricity was attributed to the sustainable overall incommensurability due to the multiasperity contact covered with randomly oriented graphene nanograins. Besides, the proposed mechanism was believed to be one of the future directions to solve the challenges facing scaling-up of superlubricity [30].

Despite the excellent tribological behaviors of 2D materials at the nanoscale under precisely controlled test condition, it is always a huge challenge to achieve superlubricity at macroscale, where the performance of 2D materials tends to degrade at large scale [45]. Berman et al. devised a method to utilize a large amount of graphene flakes instead of a continuous graphene layer to achieve a multicontact geometry together with nanodiamond particles [46]. In their experiments, DLC balls were slid against graphene-flakes-coated SiO_2 surface with nanodiamond as the additive. During the sliding process, the few-layer graphene can wrap the nanodiamond forming the nanoscrolls, which forms incommensurate contact with the amorphous DLC surface. In a late work from the same group, MoS_2 flakes were used instead of graphene flakes [47]. Graphitization of the nanodiamond surface and formation of onion-like carbon (OLC) structures by the catalysis effect of MoS_2 dissociated during the tribochemical processes (Fig. 8.3C and D). The graphitic layers of OLCs form incommensurate contact with the DLC surface, leading to superlubricity at the macroscale. Both the findings by Liu et al. [44] and Berman et al. [46,47] provide promising solutions to achieve micro- and macroscale superlubricity by forming the multicontact interfacial geometry [30].

8.4 Suppression of nanoscale wear

Another important factor hindering the possible application of 2D materials is its vulnerability to wear failure in two ways: (1) as atomically thin sheets, the exfoliation of the 2D materials from the supporting substrate when the interface adhesion is weak; (2) imperfect surface structures unavoidably generated during preparation. It has been reported that for graphene/glass substrate architecture, the mechanically transferred graphene shows a poor adhesion to the glass substrate, however, a direct chemical vapor deposition of graphene on glass significantly strengthens the interface interaction which shows potential application in touch screen industries [48].

The step edges, wrinkles, grain boundaries (GBs), vacancy defects, etc. act as weak points of graphene and other 2D materials subject to nanoscale friction (Fig. 8.4). Qi et al. [49] found that although graphene is strong in the interior region, the step edge is easily damaged by either atom-by-atom adhesive wear or peeling-induced rupture. Vasić et al. [50] showed that graphene tearing during the AFM tip scanning process tend to happen at the wrinkles. Xu et al. [51] suggested that vacancy defects may act as chemically active sites with dangling bonds, increasing the risk of tribochemical wear. Zhang et al. [52] showed that low-angle GBs deteriorate wear resistance of graphene, due to the existence of "long" bond at the GBs, while the wear resistance of graphene with large-angle GBs is just slightly lower than that of pristine graphene. These mechanisms explain why 2D materials are nanoscopically strong yet macroscopically weak. It is favorable to prepare large area, atomically flat 2D materials with low defect density. Also by increasing the interfacial strength between graphene and substrate, the wear resistance can be markedly enhanced [52]. Besides, it was suggested that to construct a 2D/2D contact architecture, that is, to prepare 2D layers on both sides of the sliding interface, could significantly increase the wear resistance, which was attributed to the suppression of the local contact pressure fluctuations and atomic interlocking by forming atomically smooth contact interface with 2D materials beyond certain number of layers [51].

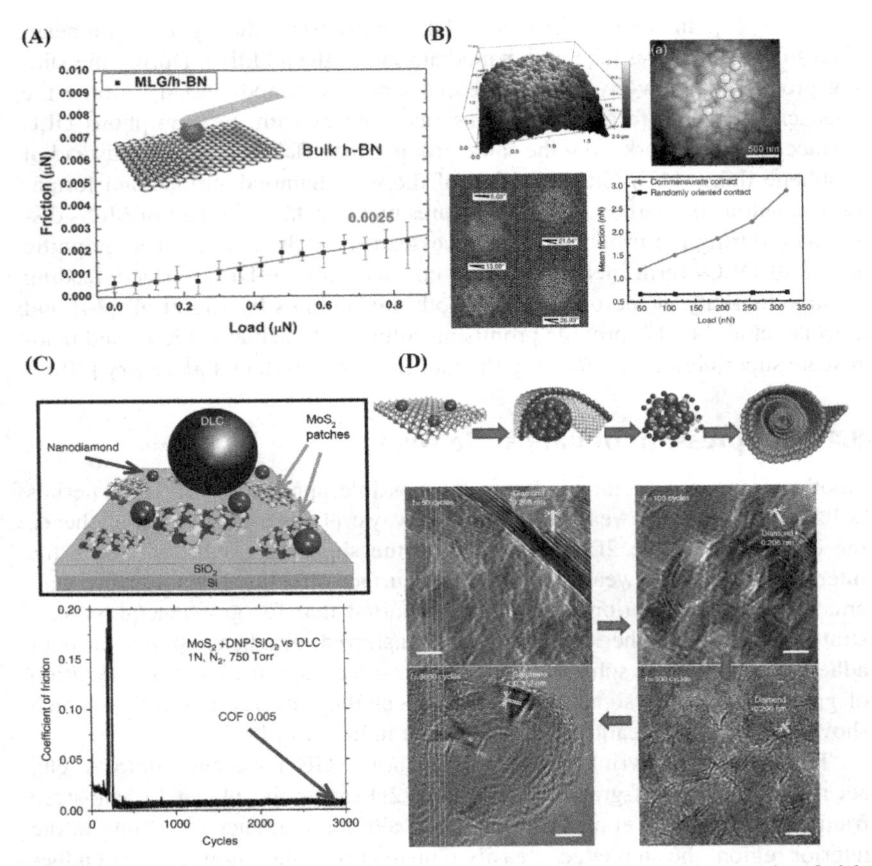

FIGURE 8.3 Micro- and macroscale superlubricity by forming the multicontact interfacial geometry. (A) Superlubricity regime enabled by graphene-coated microsphere sliding against various 2D materials (e.g., bulk hexagonal boron nitride with a friction coefficient of 0.0025), the inset is a schematic of the experimental method. (B) Upper: 3D and 2D surface topography of the graphene-coated microsphere measured by a conventional AFM tip. The five *red*-dashed circles denote the contacting asperities with atomically flat counterface by contact mechanics calculations. Lower: Molecular dynamics simulations of multiasperity contact model with randomly orientated graphene flakes covered asperities and calculated friction at different normal loads [44]. [(C) and (D)] Macroscale superlubricity enabled by in situ formation of onion-like carbon structures by tribochemical reactions involving molybdenum disulfide flakes [47]. (*AFM*: Atomic force microscopy).

8.5 Perspectives

In this chapter, an up-to-date overview is presented in terms of the mechanisms and schemes to achieve robust superlubricity. Despite the exciting progresses, the wonderland of superlubricity is still largely unexplored. Firstly, the fundamental question of friction energy dissipation remains unresolved. Theoretically, general descriptions of the energy dissipation processes [53]

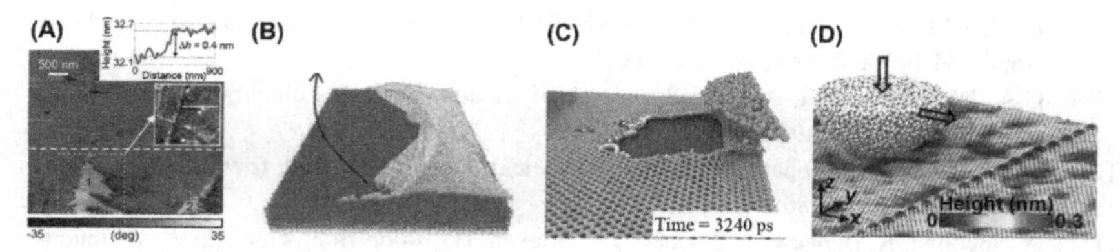

FIGURE 8.4 Wear failure of graphene at weak points of (A) wrinkles [50], (B) step edges [49], (C) vacancy defects [51], and (D) grain boundaries [52].

together with modeling of nonequilibrium phonon excitation and interfacial transmission from atomistic levels [54], are urgently needed to understand the physical origin of ultralow friction. Experimentally, quantitative measurements of different dissipation processes at the superlubricity state, including phonon, electron, and photon excitation, are extremely challenging but necessary to reveal the superlubricity mechanism. Secondly, novel 2D materials should be developed, either to possess extremely low friction coefficient at specific conditions or to exhibit excellent combined performance of tribology and other functionalities. An example is the black phosphorous, which shows high potential as lubrication additives and fillers in self-lubricating composite materials [55,56]. Thirdly, there is still a gap between laboratory-level fundamental studies and industrial applications in tribology of 2D materials. For example, in space technology, especially under harsh conditions of high vacuum, atomic oxygen environment, high/low temperature, etc., solid lubrication using 2D materials shows unique advantages where low friction of the tribopairs is highly desired [57]. However the wear life, reliability, and generation of wear debris are the most important factors needed to be solved.

References

[1] M. Hirano, K. Shinjo, Atomistic locking and friction, Phys. Rev. B 41 (17) (1990) 11837–11851.

[2] J.J. Li, J.B. Luo, Advancements in superlubricity. Sci. China Technol, Sci. 56 (12) (2013) 2877–2887.

[3] S. Zhang, T. Ma, A. Erdemir, Q. Li, Tribology of two-dimensional materials: from mechanisms to modulating strategies, Mater. Today 26 (2019) 67–86.

[4] M.R. Vazirisereshk, H. Ye, Z. Ye, A. Otero-de-la-Roza, M.-Q. Zhao, Z. Gao, et al., Origin of nanoscale friction contrast between supported graphene, MoS_2, and a graphene/MoS_2 heterostructure, Nano Lett. 19 (2019) 5496–5505.

[5] M. Dienwiebel, G.S. Verhoeven, N. Pradeep, J.W.M. Frenken, J.A. Heimberg, H.W. Zandbergen, Superlubricity of graphite, Phys. Rev. Lett. 92 (12) (2004) 126101.

[6] A.E. Filippov, M. Dienwiebel, J.W.M. Frenken, J. Klafter, M. Urbakh, Torque and twist against superlubricity, Phys. Rev. Lett. 100 (4) (2008) 046102.

[7] X. Feng, S. Kwon, J.Y. Park, M. Salmeron, Superlubric sliding of graphene nanoflakes on graphene, ACS Nano 7 (2) (2013) 1718–1724.

[8] L. Prandtl, Ein Gedankenmodell zur kinetischen Theorie der festen Körper, ZAMM-Z Angew Math Me 8 (2) (1928) 85–106.

[9] G.A. Tomlinson, CVI. A molecular theory of friction, London, Edinburgh Dublin Philos. Mag. J. Sci. 7 (46) (1929) 905–939.

[10] Y. Dong, A. Vadakkepatt, A. Martini, Analytical models for atomic friction, Tribol. Lett. 44 (3) (2011) 367–386.

[11] A. Socoliuc, R. Bennewitz, E. Gnecco, E. Meyer, Transition from stick-slip to continuous sliding in atomic friction: Entering a new regime of ultralow friction, Phys. Rev. Lett. 92 (13) (2004) 134301.

[12] L. Xu, T.-B. Ma, Y.-Z. Hu, H. Wang, Vanishing stick-slip friction in few-layer graphenes: the thickness effect, Nanotechnology 22 (28) (2011) 285708.

[13] T. Liang, W.G. Sawyer, S.S. Perry, S.B. Sinnott, S.R. Phillpot, First-principles determination of static potential energy surfaces for atomic friction in MoS_2 and MoO_3, Phys. Rev. B 77 (10) (2008) 104105.

[14] N. Marom, J. Bernstein, J. Garel, A. Tkatchenko, E. Joselevich, L. Kronik, et al., Stacking and registry effects in layered materials: the case of hexagonal boron nitride, Phys. Rev. Lett. 105 (4) (2010) 046801.

[15] L.-F. Wang, T.-B. Ma, Y.-Z. Hu, H. Wang, Atomic-scale friction in graphene oxide: an interfacial interaction perspective from first-principles calculations, Phys. Rev. B 86 (12) (2012) 125436.

[16] R.-Y. Shi, L.-F. Wang, L. Gao, A.-S. Song, Y.-M. Liu, Y.-Z. Hu, et al., Quantitative calculation of atomic-scale frictional behavior of a two-dimensional material based on sliding potential energy surface, Acta Phys. Sinca 66 (2017) 196803.

[17] L.-F. Wang, T.-B. Ma, Y.-Z. Hu, H. Wang, T.-M. Shao, Ab initio study of the friction mechanism of fluorographene and graphane, The J. Phys. Chem. C. 117 (24) (2013) 12520–12525.

[18] A.K. Geim, I.V. Grigorieva, Van der Waals heterostructures, Nature 499 (7459) (2013) 419–425.

[19] I. Leven, D. Krepel, O. Shemesh, O. Hod, Robust superlubricity in graphene/h-BN heterojunctions, J. Phys. Chem. Lett. 4 (1) (2013) 115–120.

[20] L. Wang, X. Zhou, T. Ma, D. Liu, L. Gao, X. Li, et al., Superlubricity of a graphene/MoS_2 heterostructure: a combined experimental and DFT study, Nanoscale 9 (30) (2017) 10846–10853.

[21] K.S. Novoselov, A. Mishchenko, A. Carvalho, A.H. Castro Neto, 2D materials and van der Waals heterostructures, Science 353 (6298) (2016) aac9439.

[22] C.R. Woods, L. Britnell, A. Eckmann, R.S. Ma, J.C. Lu, H.M. Guo, et al., Commensurate-incommensurate transition in graphene on hexagonal boron nitride, Nat. Phys. 10 (6) (2014) 451–456.

[23] M.M. van Wijk, A. Schuring, M.I. Katsnelson, A. Fasolino, Moiré patterns as a probe of interplanar interactions for graphene on h-BN, Phys. Rev. Lett. 113 (13) (2014) 135504.

[24] L.-F. Wang, T.-B. Ma, Y.-Z. Hu, Q. Zheng, H. Wang, J. Luo, Superlubricity of two-dimensional fluorographene/MoS_2 heterostructure: a first-principles study, Nanotechnology 25 (38) (2014) 385701.

[25] D.E. Kim, Investigation of the microscopic mechanisms of friction. Massachusetts Institute of Technology, Department of Mechanical Engineering, 1991.

[26] T. Kontorova, J. Frenkel, On the theory of plastic deformation and twinning, II. Zh. Eksp. Teor. Fiz. (1938) 81340–81348.

[27] M. Wolloch, G. Levita, P. Restuccia, M.C. Righi, Interfacial charge density and its connection to adhesion and frictional forces, Phys. Rev. Lett. 121 (2) (2018) 026804.
[28] M. Peyrard, S. Aubry, Critical behaviour at the transition by breaking of analyticity in the discrete Frenkel-Kontorova model, J. Phys. C: Solid. State Phys. 16 (9) (1983) 1593–1608.
[29] D. Dietzel, J. Brndiar, I. Štich, A. Schirmeisen, Limitations of structural superlubricity: chemical bonds versus contact size, ACS Nano 11 (8) (2017) 7642–7647.
[30] O. Hod, E. Meyer, Q. Zheng, M. Urbakh, Structural superlubricity and ultralow friction across the length scales, Nature 563 (7732) (2018) 485–492.
[31] A. Bylinskii, D. Gangloff, I. Counts, V. Vuletic, Observation of Aubry-type transition in finite atom chains via friction, Nat. Mater. 15 (7) (2016) 717–722.
[32] T. Brazda, A. Silva, N. Manini, A. Vanossi, R. Guerra, E. Tosatti, et al., Experimental observation of the Aubry transition in two-dimensional colloidal monolayers, Phys. Rev. X 8 (1) (2018) 011050.
[33] C. Lee, Q. Li, W. Kalb, X.-Z. Liu, H. Berger, R.W. Carpick, et al., Frictional characteristics of atomically thin sheets, Science 328 (5974) (2010) 76–80.
[34] Y. Liu, A. Song, Z. Xu, R. Zong, J. Zhang, W. Yang, et al., Interlayer friction and superlubricity in single-crystalline contact enabled by two-dimensional flake-wrapped atomic force microscope tips, ACS Nano 12 (8) (2018) 7638–7646.
[35] B. Liu, J. Wang, X. Peng, C. Qu, M. Ma, Q. Zheng, Direct fabrication of graphite-mica heterojunction and in situ control of their relative orientation, Mater. Des. 160 (2018) 371–376.
[36] B. Li, J. Yin, X. Liu, H. Wu, J. Li, X. Li, et al., Probing van der Waals interactions at two-dimensional heterointerfaces, Nat. Nanotechnol. 14 (2019) 567–572.
[37] Z. Liu, J. Yang, F. Grey, J.Z. Liu, Y. Liu, Y. Wang, et al., Observation of microscale superlubricity in graphite, Phys. Rev. Lett. 108 (20) (2012) 205503.
[38] E. Koren, E. Lörtscher, C. Rawlings, A.W. Knoll, U. Duerig, Adhesion and friction in mesoscopic graphite contacts, Science 345 (6235) (2015) 679–683.
[39] Q. Zheng, B. Jiang, S. Liu, Y. Weng, L. Lu, Q. Xue, et al., Self-retracting motion of graphite microflakes, Phys. Rev. Lett. 100 (6) (2008) 067205.
[40] Y. Song, D. Mandelli, O. Hod, M. Urbakh, M. Ma, Q. Zheng, Robust microscale superlubricity in graphite/hexagonal boron nitride layered heterojunctions, Nat. Mater. 17 (10) (2018) 894–899.
[41] Y. Gongyang, W. Ouyang, C. Qu, M. Urbakh, B. Quan, M. Ma, et al., Temperature and velocity dependent friction of a microscale graphite-DLC heterostructure, Friction (2019) 1–9.
[42] D. Mandelli, W. Ouyang, O. Hod, M. Urbakh, Negative friction coefficients in superlubric graphite-hexagonal boron nitride heterojunctions, Phys. Rev. Lett. 122 (7) (2019) 076102.
[43] J. Sun, Y. Zhang, Z. Lu, Q. Li, Q. Xue, S. Du, et al., Superlubricity enabled by pressure-tnduced friction collapse, J. Phys. Chem. Lett. 9 (10) (2018) 2554–2559.
[44] S.W. Liu, H.P. Wang, Q. Xu, T.B. Ma, G. Yu, C. Zhang, et al., Robust microscale superlubricity under high contact pressure enabled by graphene-coated microsphere, Nat. Commun. 8 (1) (2017) 14029.
[45] H. Song, L. Ji, H. Li, J. Wang, X. Liu, H. Zhou, et al., Self-forming oriented layer slip and macroscale super-low friction of graphene, Appl. Phys. Lett. 110 (2017) 073101.
[46] D. Berman, S.A. Deshmukh, S.K.R.S. Sankaranarayanan, A. Erdemir, A.V. Sumant, Macroscale superlubricity enabled by graphene nanoscroll formation, Science 348 (6239) (2015) 1118–1122.

[47] D. Berman, B. Narayanan, M.J. Cherukara, S.K.R.S. Sankaranarayanan, A. Erdemir, A. Zinovev, et al., Operando tribochemical formation of onion-like-carbon leads to macroscale superlubricity, Nat. Commun. 9 (1) (2018) 1164.

[48] J. Sun, Z. Chen, L. Yuan, Y. Chen, J. Ning, S. Liu, et al., Direct chemical-vapor-deposition-fabricated, Large-scale graphene glass with high carrier mobility and uniformity for touch panel applications, ACS Nano 10 (2016) 11136–11144.

[49] Y. Qi, J. Liu, J. Zhang, Y. Dong, Q. Li, Wear resistance limited by step edge failure: the rise and fall of graphene as an atomically thin lubricating material, ACS Appl. Mater. Inter. 9 (2017) 1099–1106.

[50] B. Vasić, A. Zurutuza, R. Gajić, Spatial variation of wear and electrical properties across wrinkles in chemical vapour deposition graphene, Carbon 102 (2016) 304–310.

[51] Q. Xu, X. Li, J. Zhang, Y. Hu, H. Wang, T. Ma, Suppressing nanoscale wear by graphene/graphene interfacial contact architecture: a molecular dynamics study, ACS Appl. Mater. Inter. 9 (2017) 40959–40968.

[52] J. Zhang, X. Chen, Q. Xu, T. Ma, Y. Hu, H. Wang, et al., Effects of grain boundary on wear of graphene at the nanoscale: a molecular dynamics study, Carbon 143 (2019) 578–586.

[53] A. Buldum, D.M. Leitner, S. Ciraci, Model of phononic energy dissipation in friction, Phys. Rev. B 59 (1999) 16042–16046.

[54] Z. Wei, Y. Kan, Y. Zhang, Y. Chen, The frictional energy dissipation and interfacial heat conduction in the sliding interface, AIP Adv. 8 (2018) 115321.

[55] W. Wang, G. Xie, J. Luo, Black phosphorus as a new lubricant, Friction 6 (2018) 116–142.

[56] S. Wu, F. He, G. Xie, Z. Bian, J. Luo, S. Wen, Black phosphorus: degradation favors lubrication, Nano Lett. 18 (2018) 5618–5627.

[57] L. Liu, M. Zhou, L. Jin, L. Li, Y. Mo, G. Su, et al., Recent advances in friction and lubrication of graphene and other 2D materials: mechanisms and applications, Friction 7 (3) (2019) 199–216.

Chapter 9

Energy dissipation through phonon and electron behaviors of superlubricity in 2D materials

Dameng Liu and Jianbin Luo
State Key Laboratory of Tribology, Tsinghua University, Beijing, P.R. China

9.1 Introduction

Friction is a typical energy dissipation process, transforming the mechanical kinetic energy of the sliding friction pairs into other forms of energy, for example, heat. When two contacting surfaces move relatively, localized vibration is excited by stick-slip process and translated into collective vibrations including acoustic and optical phonons. Besides, in metallic and semiconducting materials electrons are excited to unoccupied levels, forming electron–hole pairs (excitons), and finally decay via excitation of phonons, fluorescence, or rupture of chemical bonds [1]. Fig. 9.1 shows the well-acknowledged model of friction energy dissipation, in which phonon and

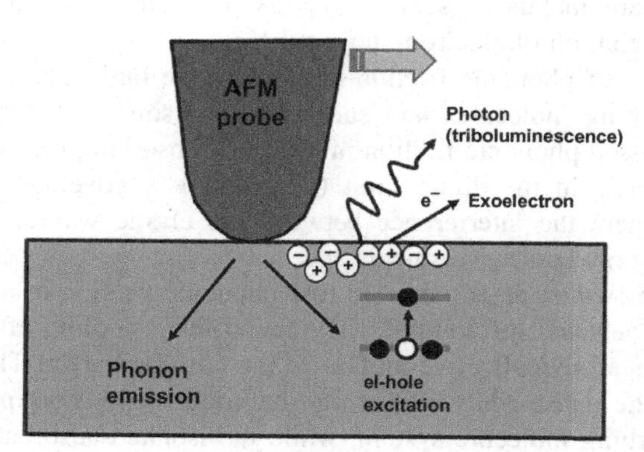

FIGURE 9.1 General view of friction energy dissipation process [1]. *Reprinted with permission from J.Y. Park, M. Salmeron, Fundamental aspects of energy dissipation in friction, Chem. Rev. 114(1) (2013) 677–711. Copyright 2013, American Chemical Society.*

Superlubricity. DOI: https://doi.org/10.1016/B978-0-444-64313-1.00009-0

electron–hole excitation are the dominated dissipation channels. Beyond that, emission of light (triboluminescence) or electrons (exoelectron emission) can also arise from electronic excitations.

Understanding of the interaction between these excitations and energy dissipation is very important for exploring the atomistic origins of friction. But it is still challenging because many factors, like mechanical property, electronic property, and surface topography (roughness or surface defects) of the sliding pairs may all contribute to friction. Moreover, it is difficult to measure and modulate these excitations.

Another obstacle for studying the frictional energy dissipation process is that most basic excitations are in play simultaneously. Fortunately, under special experimental conditions, some of the dissipating channels can be suppressed, and some specific channels can be studied. Therefore, this chapter reviews recent experimental and theoretical studies of these energy dissipating channels in various systems.

9.2 Phonon channel of friction energy dissipation

9.2.1 Theory of phononic dissipation

The processes of phononic energy dissipation are the excitations of lattice vibrations during sliding. According to Prandtl–Tomlinson model, in relative sliding process, friction is constructed by periodic stick-slip process, as shown in Fig. 9.2 [2,3]. In atomic force microscopy(AFM) experiments, the atomic motions are observed as stick-slip events with lattice periodicity in crystalline materials [4]. In this stick-slip process, energy released in the slip process leads to the vibration of surface atoms. These vibrations of surface atoms spread and transfer the energy to lattice of the bulk material. When friction pairs are metals or semiconductors, this vibration can also damp through the excitation of electron–hole pairs.

The theories of phononic friction dissipation are built mainly on friction between absorbing molecules and substrates. Persson et al. [5] and Lewis et al. [6] discussed phononic friction in model of absorbing molecules sliding on bulk materials in the dilute limit (i.e., with low coverage of adsorbed molecules), where the interference between the elastic waves, that emitted from absorbing molecules, was neglected.

However, Lewis et al. [6] showed that phononic friction of a dense layer of adsorbates behaved differently. They found that coupling efficiency was severely enhanced by collective motion of the adsorbed layer. They also discovered that the dense-adlayer limit can be confirmed according to a large range of adsorbing molecule system, while the isolated-adsorbate limit cannot unless coverages well below that which can be observed by current experimental techniques.

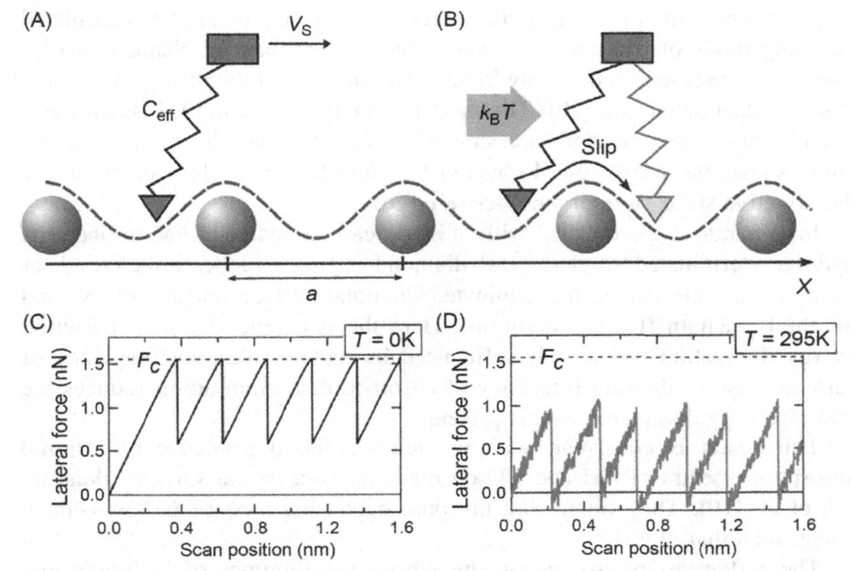

FIGURE 9.2 (A) Schematic diagram of Prandtl–Tomlinson model in stick process. (B) Schematic diagram of Prandtl–Tomlinson model in slip process. (C) Theoretical lateral force of stick-slip process at $T = 0K$. (D) Predicted lateral force at higher temperatures with influence of thermal activation [2]. *Reprinted with permission from U.D. Schwarz, H. Hölscher, Exploring and explaining friction with the Prandtl–Tomlinson model, ACS Nano 10(1) (2016) 38–41. Copyright 2016, American Chemical Society.*

9.2.2 Experimental studies of phononic friction

The phononic channel of energy dissipation in friction is mainly studied for cases involving adsorbate molecules and metallic substrates, where phonon mode can be easily measured and modulated. For example, Krim et al. [7,8] measured the friction forces between condensing xenon (Xe) gas and (111) surface of silver film at a temperature of 77.4K with resonant frequencies of $f \approx 8\text{MHz}$ and quality factors of $Q \approx 10,000$.

Molecular dynamics (MD) simulation results by Tomassone et al. [8] have shown that the slip time τ, defined as the time it takes for the sliding speed to fall to $1/e$ of the original value, is about $2.45 \times 10^9\ s^{-1}$. In their study, the slip time is determined through both thermal equilibrium autocorrelation and velocity decay after driving of vibration turned off, and results in the same damping coefficient of $2.45 \times 10^9\ s^{-1}$ for model of Xe molecule sliding on Ag(111) surface. This conforming result of experimental measurement and theoretical simulation, which only includes the phonon contribution to frictional energy dissipation, leads to the conclusion that phononic channel of energy dissipation dominates in the Xe/Ag(111) friction system.

To better simulate real friction process, Cannara et al. [9] modulated vibrating mode of friction pairs using isotopes. Pronounced changes in friction were observed, and the hydrogenated surface exhibited higher friction than the deuterated one [10]. The model of experiments and measured average effective shear strength are showed in Fig. 9.3, in which the solid red circle shows the results on H-covered Si while the open blue square shows the effective shear strength on D-covered Si.

In separate experiments, with a hydrocarbon-coated AFM probe, the hydrogen-terminated single-crystal diamond led to a higher effective shear strength than the deuterium-terminated diamond surface under both N_2 and ultrahigh vacuum (UHV) conditions. Here the phononic channel of friction energy dissipation is intensely influenced by natural vibrating frequency of surface. Lower vibrating frequency of absorbed deuterium atoms reduces the rate of energy dissipation in this system.

This result is consistent with the temperature dependence of infrared absorption spectra of CH and CD on diamond nanocrystal surfaces done by Lin et al. [10]. They found that the pure dephasing time of CD was much longer than that of CH.

The influences of isotope on the vibrational lifetimes of hydrogen and deuterium are also proved by theoretical calculations. Sakong et al. [11] calculated the vibrational lifetime of H and D molecules absorbed on Ge surface, and found that lifetime was 1.56 ns at 400K when surface was adsorbed by pure H, while this lifetime was reduced by a factor of up to 6 when Ge surface was adsorbed by a mixture of H and D [12].

9.3 Electronic channel of friction energy dissipation

As mentioned earlier, kinetic energy of friction pairs dissipates only through phonon channel when the two surfaces are insulator. However when friction

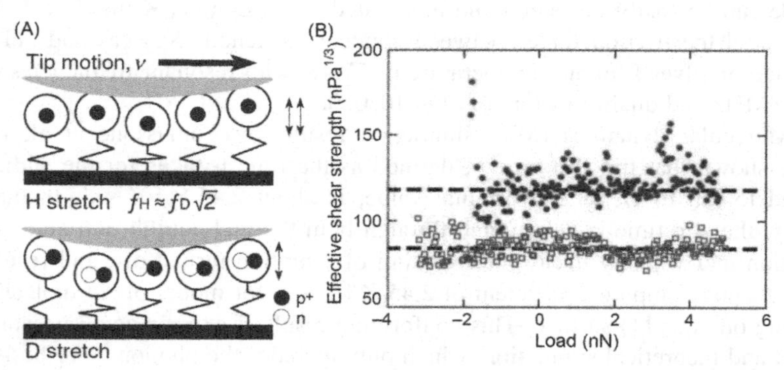

FIGURE 9.3 (A) A schematic of the frictional interface with H and D; (B) effective shear strength measured with tip rubbing on Si(111) terminated with hydrogen *(solid red circle)* or deuterium atoms *(open blue square)* [9].

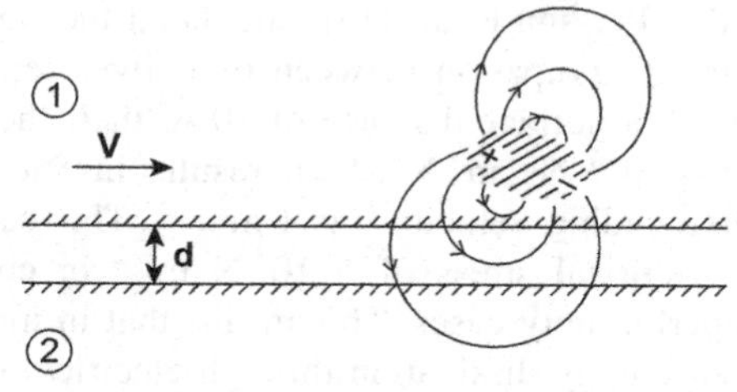

FIGURE 9.4 Schematic drawing of the origin of electrical friction dissipation [13]. *Reprinted with permission from B. Persson, Z. Zhang, Theory of friction: Coulomb drag between two closely spaced solids, Phys. Rev. B 57(12) (1998) 7327. Copyright 1998, American Physical Society.*

pairs are metals or semiconductors, energy dissipation also occurred through conduction electron coupling in friction pairs, which is electronic dissipation. In general, moving atoms exerts dragging force on surface electrons and leads to energy dissipation by resistive Ohmic heat via creation of electron–hole pairs that eventually decay into phonons.

9.3.1 Theory of electronic friction

Persson gave a widely acknowledged theoretical model [13], in which electrical channel of dissipation occurred between ideal smooth metal surfaces separated in vacuum. Though no electric field exists between surfaces on average, local charge imbalance still happens in metals and semiconductors because of thermal or quantum fluctuations. This will lead to electric field patches extending from one solid into the other solid, as showed in Fig. 9.4. The fluctuating electric field will induce electric currents in the solids, which are damped due to, for example, normal "Ohmic" processes such as scattering of conduction electrons against imperfections.

In static systems that no relative movement occurs, such electronic energy transfer takes place in both directions, resulting in energy equilibrium with no net energy transfer between contacting surfaces. However, when relative sliding happens, a net energy and momentum transfer will occur from the sliding body to the stationary body, which finally leads to friction force. Detailed studies have shown that the results of quantum and thermal fluctuation are completely different. The contribution of thermal fluctuations makes up the lowest order in the (electric field) coupling between the solids, while quantum fluctuations only contribute to the linear (in v) sliding friction in the second (or higher) order of perturbation theory. Furthermore, while thermal contribution to the friction force requires that both solids are metallic, the case in which only one of the solids is metallic satisfies the requirements of the quantum fluctuations.

To test their model, Persson et al. [13] calculated the noncontact friction that caused by electrical dissipation between two silver semiinfinite surface based on their theory. For surface distance of 10 Å, their theory gives damping coefficient of $\gamma \sim 10^{-7}$ Ns m^{-3}, which results in friction stress of $F/A \sim 10^{-7}$ N m^{-2} when sliding velocity v is 1 m s^{-1}. This calculated stress is much smaller than frictional stress of $\sim 10^{8}$ N m^{-2} in contacting friction process, even for superlubricity cases. This means that in most real frictional process, proportion of energy dissipation through electrical channel is nearly negligible. But in some special situations, such as noncontact friction, electrical channel of energy dissipation may be dominated.

9.3.2 Electronic friction in absorbing membrane

Similar to the study of the phononic dissipation channels, the sliding between adsorbing molecular film and substrate, as a relatively simple nanoscale friction system, is typically used to study the friction energy dissipation through the electronic channels. By measuring the Q-factor change of quartz crystal microbalance (QCM), the role of electrical channels in friction dissipation was earlier proposed by Krim et al. [14,15] who found that the slip time for chemisorbed oxygen/silver surfaces was longer than that for clean silver, implying that electronic contributions to friction should be considered [16].

Krim's team measured the frictional dissipation of N_2 molecules sliding on a Pb film using a QCM, as shown in Fig. 9.5. A sharp decrease of friction was found when Pb substrate crossed the superconducting transition temperature ($T_c \approx 5$K) [17,18]. They attributed it to the suppression of electronic excitations in the substrate [19] when it entered superconducting state.

However, Persson et al. [20] questioned this explanation. They pointed out that the abrupt disappearance of normal nonpaired electrons in the superconducting transition could not lead to this friction variation. The normal electron density decreases steadily as the temperature drops down to 0K, so contrary to observation, the electronic contribution from electron–hole pair formation should also decrease smoothly.

A similar experiment was conducted by Renner et al. [21,22]. They found that N_2 film was pinned to the substrate, and could not slip at low temperature. Aiming at solving this problem, superconductivity-dependent friction was studied by many researches [20,23–26]. Bruch calculated electronic friction of N_2 and Xe molecules when sliding on Pb and Ag substrates, and gave a damping coefficient of $\eta_{el} = 5 \times 10^{6}$ s^{-1} for Xe/Ag and 5×10^{7} s$^{-1} < \eta_{el} < 5 \times 10^{8}$ s^{-1} for N_2/Pb [26].

Highland and Krim [19] exerted similar superlubricating friction measurements on N_2, H_2O, and He films adsorbed on Pb(111) substrates. Sharp decrease of friction was observed again when substrate joined

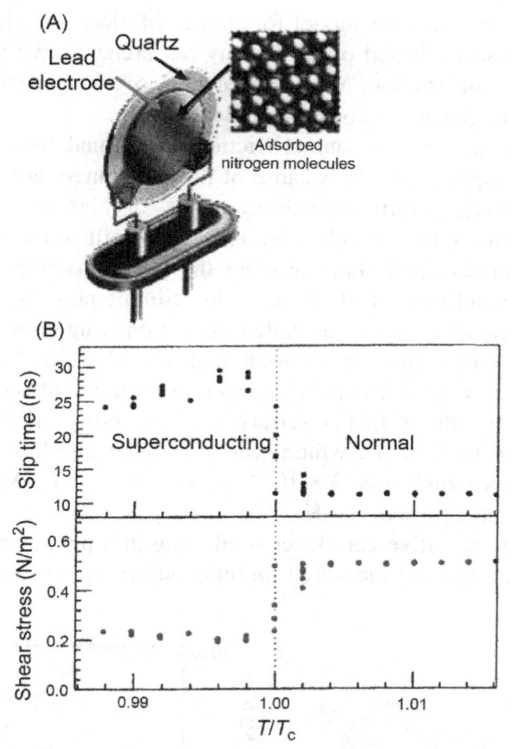

FIGURE 9.5 (A) Scheme of friction measurements using quartz crystal microbalance (QCM). (B) Slip time and corresponding shear stress across the superconducting transition temperature (7.2K) [1]. *Reprinted with permission from J.Y. Park, M. Salmeron, Fundamental aspects of energy dissipation in friction, Chem. Rev. 114(1) (2013) 677–711. Copyright 2014, American Chemical Society.*

superconducting state, with even bigger extent than He. These results are consistent with Bruch's theory that connects electronic friction to adsorbate polarizability [26].

9.3.3 Electronic friction in noncontact friction

Phononic dissipating channel of nanoscale friction is induced by the vibration of interface atoms in stick-slip process, while electrical dissipating channel is originated from electrons dragged by the moving atom. This means that when friction pairs do not contact, but separated by a little distance (several nanometer or less), phononic channel of friction dissipation will disappear as stick-slip process do not exist in noncontact friction. However, electrical energy dissipation channel decrease only a little as electrical interaction is a relative long-range interaction. Thus, in a noncontact friction situation, phononic friction can be avoided, and electrical friction can be isolated, offering an excellent model for studies of electrical channel of friction energy dissipation. Based on this, many researchers have tried to reveal electrical channel of friction, which is always overlapped with phononic energy dissipation channel in contacting friction.

But for researches on noncontact friction, traditional friction measuring methods are no longer available because of limited contact area. Focusing on this problem, several efforts have been made for high-sensitivity friction detection apparatus. One possible way is by controlling the tip oscillation parallel to the surface, and characterizing the lateral friction by measuring frequency and amplitude drift of the tip. Simultaneously, the distance between tip and sample can be controlled using tunneling current [27].

Using cantilevers with spring constant of $k = 3.3 \times 10^{-4}$ N m^{-1} and a resonance frequency of 3.86 kHz, Stipe et al. studied noncontact friction between silicon tip and Au(111) surface [28] as shown in Fig. 9.6A. The measured viscous friction, Γ, which was gained by amplitude damping of the cantilever, was found to be 3×10^{-12} kg s^{-1} at bias of 1 V bias, distance of 20 nm, and temperature of 300K.

Using a highly sensitive cantilever oscillating in a pendulum geometry in UHV, Kisiel et al. [29,30] measured the noncontact friction between tip and

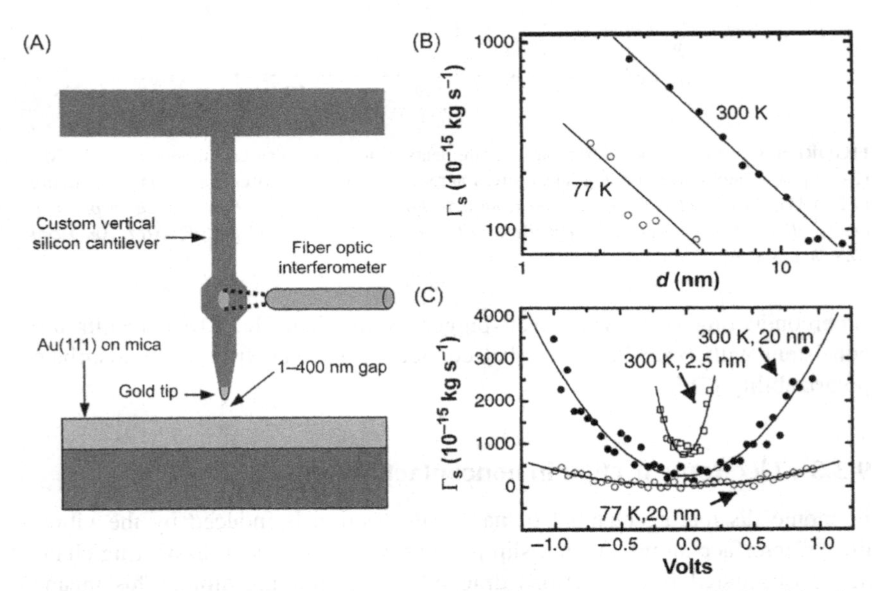

FIGURE 9.6 (A) Schematic diagram of noncontact friction measurement. (B) Zero-bias tip-sample friction measured as a function of distance for $T = 300$ and 77K. (C) Friction measured as a function of bias voltage for three different combinations of tip-sample spacing and temperature [28]. *Reprinted with permission from B. Stipe, H. Mamin, T. Stowe, T. Kenny, D. Rugar, Noncontact friction and force fluctuations between closely spaced bodies, Phys. Rev. Lett. 87(9) (2001) 096801. Copyright 2001, American Physical Society.*

Nb films across the critical superconducting temperature. In their measurements, noncontact friction coefficient was found to decrease by a factor of 3 during the process of Nb surface entering the superconducting state, which implied that friction had an electronic nature in the metallic state and in the superconducting state the phononic friction was in charge. They also found that under metallic state, friction coefficient increased gradually with decreasing tip-sample distance, while friction coefficient kept ultralow until distance decrease to less than 0.3 nm under superconducting state. This proved their previous explanation that under superconducting state, electronic channel of friction dissipation disappeared, leading to ultralow friction until phononic channel of energy dissipation began to dominate (Fig. 9.7).

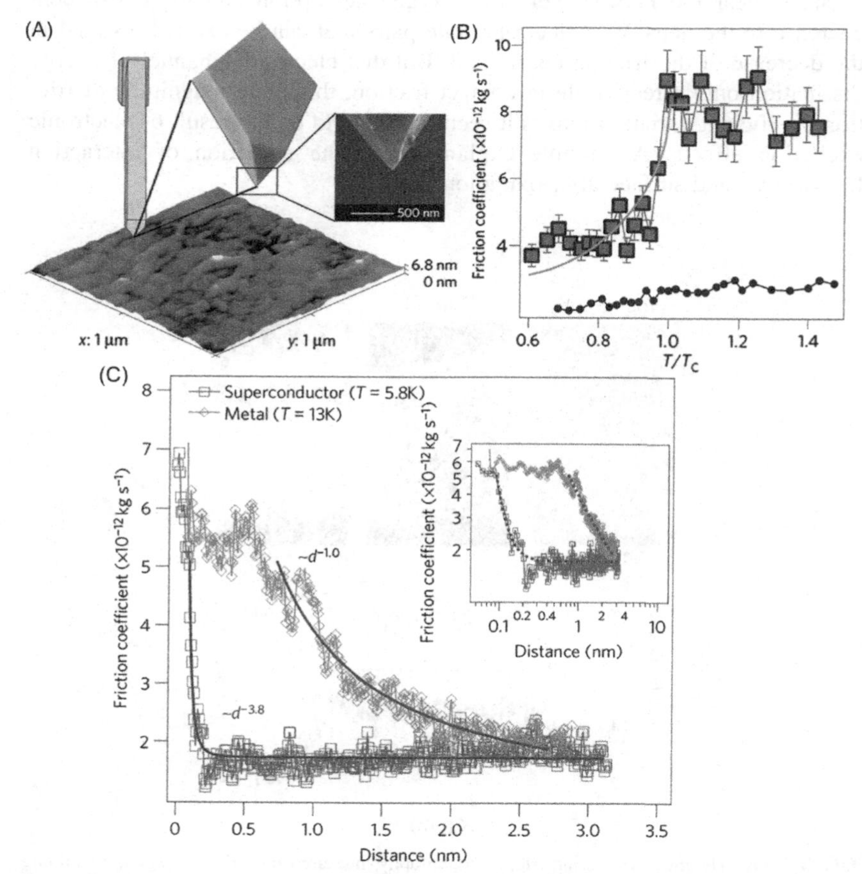

FIGURE 9.7 (A) AFM topography image of the Nb film studied in the experiment. (B) Temperature variation of the friction coefficient across the critical point $T_c = 9.2K$ of Nb. (C) Distance dependence of the friction coefficient in the metallic and in the superconducting state of Nb [29]. (*AFM*: Atomic force microscopy).

9.3.4 Electronic friction in contacting friction

Electronic channel of friction energy dissipation was shown to have only weak influence in friction, and can be observed only when contributions of phononic friction is suppressed. The question then arises that whether electronic channel of friction energy dissipation can obviously affect conventional contacting friction. An early study of electronic friction was performed using AFM to measure friction on vanadium carbide (100) in UHV by Merrill et al. [31]. The friction coefficient between vanadium and silicon nitride probe tip was measured to be 0.52 ± 0.04. However, this friction coefficient decreased about 40% to 0.32 ± 0.05 after the surface of vanadium carbide was oxidized. They believe the decrease in the *d*-band density of states near the Fermi level after oxygen adsorption and the consequent decrease in the density of electron–hole pairs that can be excited resulted in the decrease of the friction coefficient. But that electronic channel of energy dissipation only affects little in contact friction, this large magnitude of friction coefficient variation makes it seem unlikely to be the result of electronic excitation effects. A possible explanation is the reduction of interaction between tip and surface after oxidation.

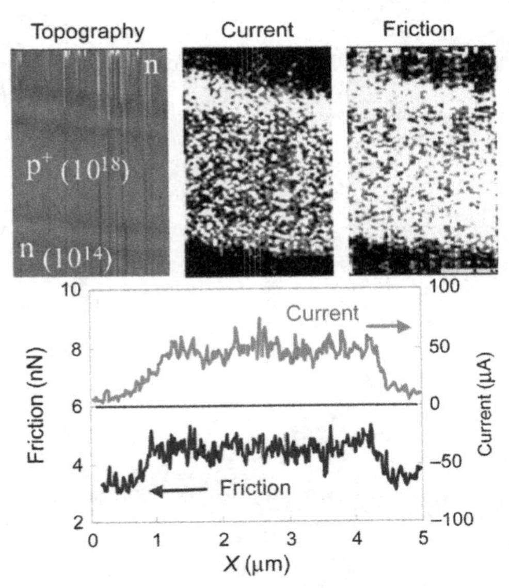

FIGURE 9.8 Height, current, and friction topography measured by AFM on region containing a p-type stripe under bias of +4 V. (*AFM*: Atomic force microscopy) [32]. *Reprinted with permission from J.Y. Park, Y. Qi, D.F. Ogletree, P.A. Thiel, M. Salmeron, Influence of carrier density on the friction properties of silicon pn junctions, Phys. Rev. B 76(6) (2007). Copyright 2007, American Physical Society.*

Park et al. [32,33] used doped Si samples to change electron density in contact area by several orders of magnitude via application of a bias voltage. Fig. 9.8 shows topographic, current, and friction images at a sample bias of +4 V. The surface of the doped Si samples was covered with an oxide layer about 9 Å thick. This oxide layer was served as a passivation layer to make the chemical interaction keep same in *n* and *p* regions, and prevented Fermi-level pinning at the same time. They found that friction greatly increased in the highly doped *p* regions when the region was forward biased and in strong electron accumulation, which led to high carrier concentration at surface. At the same bias voltage, the *n* region was reverse-biased, causing depletion or weak inversion. No increase in friction was observed there.

Similar study on *n*-type GaAs(100) substrates with a donor concentration of 1.2×10^{18} cm^{-3} was carried out by Qi et al. [34]. The GaAs surface was also covered by an approximately 1-nm-thick oxide layer. Substantial increase in friction in accumulation state was observed, as shown in Fig. 9.9. This excess friction on forward biased *n*-GaAs corresponds to the early results of Park [33].

Wu et al. [35] measured the friction and current signals simultaneously on VO_2 surface in UHV when thermal evolution of the specimen was heated from 31°C to 80°C, as shown in Fig. 9.10. After temperature was increased to 70°C, insulator-metal transition began to occur and high current area was shown, as bright area in current topography. In friction topography, higher

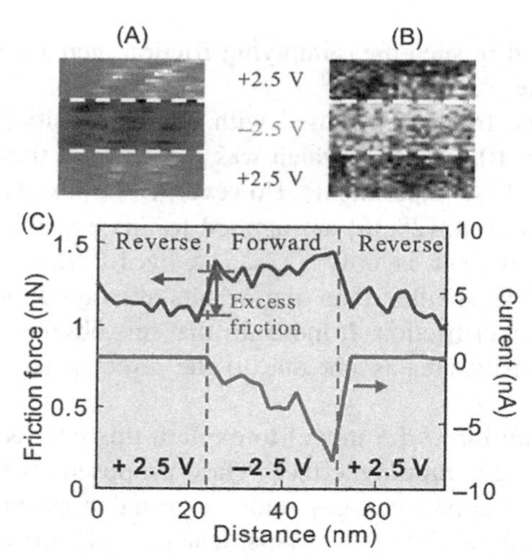

FIGURE 9.9 (A) current and (B) friction images with sample biases of +2.5 and −2.5 V. (C) Averaged line profiles of the vertical cross sections in the current image shown in (A) and in the friction image shown in (B) [34]. *Reprinted with permission from Y. Qi, J.Y. Park, B.L.M. Hendriksen, D.F. Ogletree, M. Salmeron, Electronic contribution to friction on GaAs: an atomic force microscope study, Phys. Rev. B 77(18) (2008). Copyright 2008, American Physical Society.*

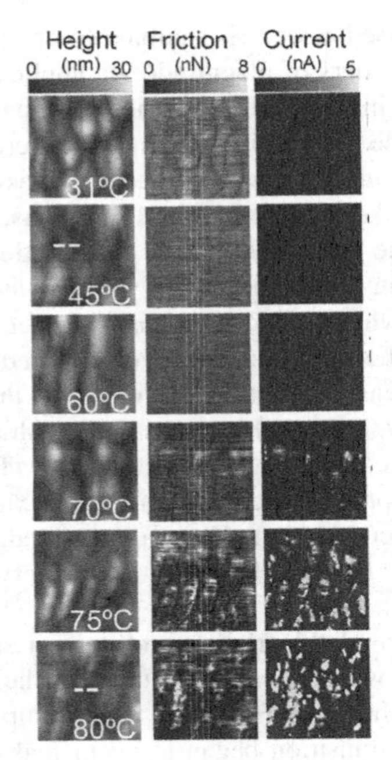

FIGURE 9.10 Height topography, friction topography, and current topography measured simultaneously under different temperature [35].

friction also showed in such area, implying friction increase after transforming to metallic state.

In these studies, friction increased with carrier density increasing. And coefficient was 1×10^{-5} kg s^{-1}, which was gained from the velocity dependence observed in these experiments. However, if the results of Stippe et al. [28,36] and Kuehn et al. [28,36] are applied to this experimental condition, the calculated coefficient is only $\Gamma_{\text{extrapol}} \approx 1 \times 10^{-8}$ kg s^{-1}, which is 3 orders of magnitude smaller than the results measured in corresponding experiments of contact friction. It indicates that this observed excess friction in high carrier density area is not due to the direct generation of electron–hole pairs.

Qi et al. [34] introduced a model to explain this excess friction, which was called as charge trapping effect. They proposed that the tip would leave a "trail" of trapped charges under external bias during sliding on sample, as showed in Fig. 9.11. This leads to electrostatic interaction between tip and surface and shows as resistance in measurements, which finally results in friction increase. The model also explains the observed velocity dependence.

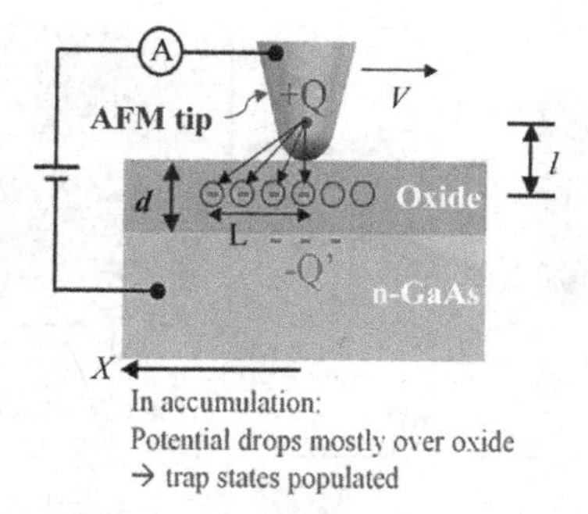

FIGURE 9.11 Model of charge trapping effect in friction measuring under bias [1]. *Reprinted with permission from J.Y. Park, M. Salmeron, Fundamental aspects of energy dissipation in friction, Chem. Rev. 114(1) (2013) 677–711. Copyright 2014, American Chemical Society.*

9.4 Energy dissipation affected by electron–phonon coupling

Except for phononic channel and electronic channel of energy dissipation, another mechanism that may influence the process of energy dissipation is electron–phonon coupling. When friction pairs are metallic or semiconducting, in which electron plays a nonnegligible role in heat transfer, excited phonon may be scattered by electron and thus influences damping process of phonon. This phenomenon makes electron–phonon coupling to become another channel for energy dissipation. When electron–phonon coupling strength is stronger, energy can dissipate through electron–phonon coupling and thus increase the speed of energy dissipation, leading to bigger friction. Filleter et al. measured friction on monolayer and bilayer graphene grown on SiC, and found that friction on monolayer was much higher than bilayer, with a factor of 2 [37,38]. At the same time, they found that strength of electron-phonon coupling was higher on monolayer than bilayer graphene using angle-resolved photoemission spectroscopy. Thus they explained that higher friction on monolayer graphene was due to its higher electron–phonon coupling strength. An interesting point of the Filleter's results is that it shows another aspect of the electronic contribution to friction, not by the direct excitation of electron–hole pairs, but rather through their efficient contribution to phonon decay. In other words, friction is still due primarily to phonon excitations, while electrons contribute to their decay, a necessary part of the energy dissipation process (Fig. 9.12).

Using MD simulation methods, Dong [39] calculated the influence of electron–phonon coupling on friction of graphene. Since conventional MD simulation is not suitable for the model of electron–phonon coupling in atomic friction, they introduced the two-temperature method into MD simulation, where electrons are treated as continuum materials and energy transport through electrons. Accuracy of the the two-temperature method depends

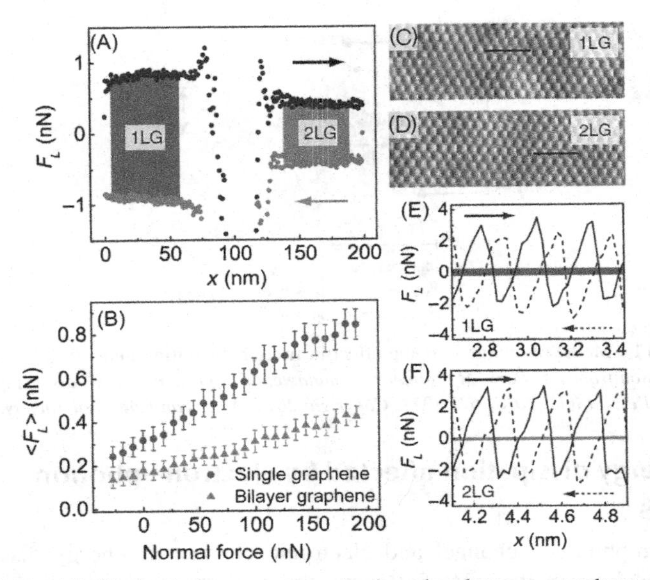

FIGURE 9.12 (A) Friction loop measured on the boundary between monolayer and bilayer region of grapheme. (B) Average friction force as a function of normal load on 1LG and 2LG. [(C) and (D)] Atomic stick-slip friction force maps measured on 1LG and 2LG films. [(E) and (F)] Corresponding lateral force line profiles [38]. *Reprinted with permission from T. Filleter, J. L. McChesney, A. Bostwick, E. Rotenberg, K.V. Emtsev, T. Seyller, et al., Friction and dissipation in epitaxial graphene films, Phys. Rev. Lett. 102(8) (2009) 086102. Copyright 2009, American Physical Society.*

on the electron–phonon coupling parameter g_{ep}, which is still an ongoing research topic. They made a rough estimation based on the values of metals. Since the electron density of graphene was much smaller than that of metals, they anticipated that g_{ep} was smaller than 10^{16} W m^{-3} K^{-1}. In their calculation, g_{ep} varied from 10^{14} to 10^{18} W m^{-3} K^{-1}, and calculated mean friction is present in Fig. 9.13. One can clearly see that friction varies little with electron–phonon coupling. Compared with the friction variation with thicknesses, its effects are negligible. At the same time, it must be noted that they have used a wide range of g_{ep} with an upper limit greatly exceeding the value estimated for graphene. But at the same time, it should be noticed that this study only simulates the energy dissipation in graphene, while the energy dissipation of the tip is not included, which may have a noticeable influence.

9.5 Energy dissipation from frictional interface

Processes happening on the surface and interface are pivotal to the understanding of energy dissipation. On one hand, they are relatively easier to

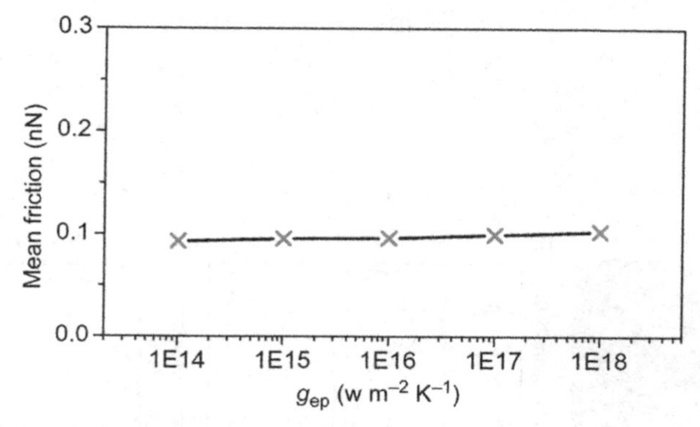

FIGURE 9.13 Calculated mean friction as a function of electron–phonon coupling parameter using the two-temperature method [39].

detect and can provide more direct and effective proofs to the existing theories. On the other hand, they are closely related to the traditional concept of friction, which is a kind of special process taking place between two contacting surfaces in relative motion. Researches concerning these processes have to answer several fundamental questions. First, certain substances must be selected properly as the object of a study. Next, we have to figure out how to detect them effectively. And on this basis, factors which may have influences on the whole thing should be deeply thought over.

Two-dimensional (2D) materials have long been the star of investigation concerning various fields. Their special layered structures can provide well-defined surfaces and interfaces. Besides, regular atomic arrangement in 2D materials makes it possible, as well as easier, to quantitatively discuss some physical phenomena such as phonon, excitation, defects, etc.

9.5.1 Interlayer interaction

Interaction at sliding interfaces is one of the most important factors that affect the value of friction force, which has been focused and demonstrated by different simulation researches [40,41]. However, measuring such interactions under nanoscale is not that easy. Luckily, indirect measurement of this property can be realized through Raman spectroscopy, since the frequency of characteristic peaks in Raman spectrum is determined by the strength of interaction. By some simplification of the considered material, in which the molecular layers of 2D materials are taken as rigid balls connected by springs, an analytic formula between interlayer interaction and vibrational frequency can be derived [42,43]. The whole model is called the linear chain model (LCM), and with its help, the only difficulty left is to effectively acquire Raman spectrum. Usually, preanalysis of the structure of 2D materials can help reveal some of the features of the spectrum, and such analysis

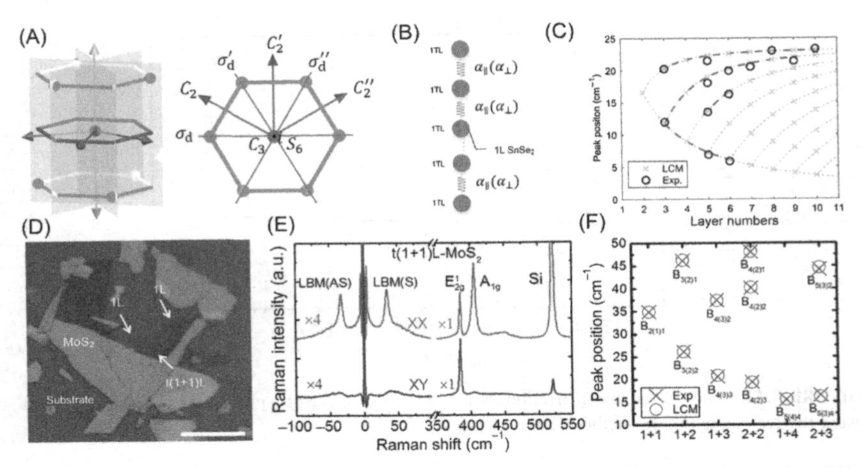

FIGURE 9.14 Low-frequency Raman spectroscopy and detection of interlayer interaction. (A) Group theory analysis of $SnSe_2$ to obtain symmetry of the material and thus help Raman measurement. (B) Schematic diagram of LCM and (C) the corresponding fitting result of low-frequency Raman peaks. (D) Artificially fabricated twisted MoS_2 sample and its (E) low-frequency breathing vibrational mode. (F) Comparison of the measured and calculated peak positions in twisted MoS_2 structures [44,45]. *Reprinted from X. Zhou, J. Li, Y. Leng, X. Cong, D. Liu, J. Luo, Exploring interlayer interaction of SnSe2 by low-frequency Raman spectroscopy, Phys. E Low Dimens. Syst. Nanostruct. 105 (2019) 7–12; X. Zhou, K. Jin, X. Cong, Q. Tan, J. Li, D. Liu, et al., Interlayer interaction on twisted interface in incommensurate stacking MoS2: a Raman spectroscopy study, J. Colloid Interface Sci. 538 (2019) 159–164. Copyright 2019, with permission from Elsevier.*

may need some knowledge of group theory, as conducted in the investigation of $SnSe_2$ (Fig. 9.14) [44]. The same strategy, using the combination of Raman measurement and LCM, can be used not only in natural interface in 2D materials, but also in artificially fabricated incommensurate interface in homostructure of heterostructures [45,46].

9.5.2 Electron energy dissipation and dynamics issues

In another research, attention has been focused on electron energy dissipations in few-layer MoS_2, WS_2, and WSe_2 by way of the fluorescence lifetime, which is an important parameter to describe the electron transition process. It also contains information about the specific mode of electron radiation transition, such as direct band gap transition, indirect band gap transition, and so on. As for WS_2, when the number of layer increases from a monolayer to four-layer, lifetimes of direct transition excitons (electron–hole pair) and trions (a negative charged exciton) tend to increase over 10 times and 2.5 times, respectively, while the lifetime of indirect transition excitons tends to be reduced by nearly 2.5 times (Fig. 9.15B and E). This layer-dependent signature is ascribed to the reduced binding energy in thicker WS_2

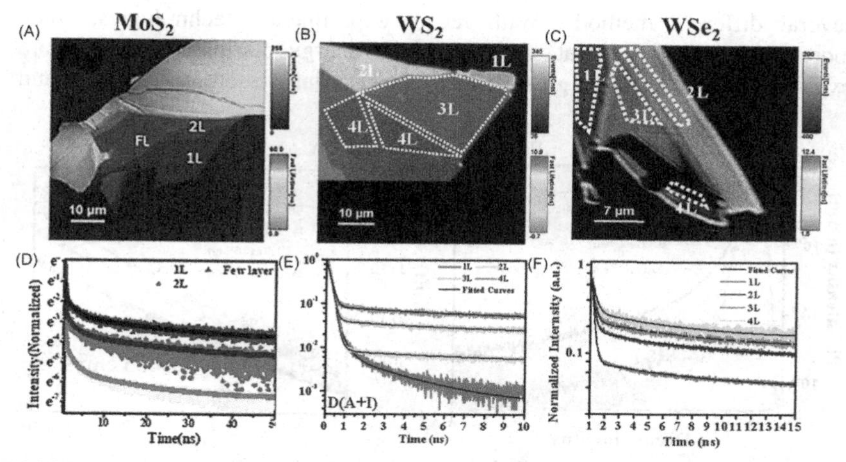

FIGURE 9.15 FLIM images of monolayer to multilayer MoS_2 [47], WS_2 [48], and WSe_2 [49] reflected distinctive fluorescent lifetime (ns) in different fluorescence colors as shown in (A)–(C). Correspondingly, (D)–(F) indicated the lifetime decay curves. To better quantify different decay processes, a biexponential lifetime equation $I = I_0 + A_1 e^{t/t_1} + A_2 e^{t/t_2}$ is used to fit the decay curves. Among the equation, t_1 and t_2 represent the different exciton decay lifetime [47,48]. (*FLIM*: fluorescence lifetime imaging microscopy). *Reprinted with permission from T. Wang, Y. Zhang, Y. Liu, J. Li, D. Liu, J. Luo, et al., Layer-number-dependent exciton recombination behaviors of MoS_2 determined by fluorescence-lifetime imaging microscopy, J. Phys. Chem. C. 122 (32) (2018) 18651–18658. Copyright 2018, American Chemical Society; reprinted with permission from Y. Liu, X. Hu, T. Wang, D. Liu, Reduced binding energy and layer-dependent exciton dynamics in monolayer and multilayer WS2, ACS Nano, 2019. Copyright 2019, American Chemical Society; reprinted with permission from Y. Liu, H. Li, C. Qiu, X. Hu, D. Liu, Layer-dependent signatures for exciton dynamics in monolayer and multilayer WSe2 revealed by fluorescence lifetime imaging measurement, Nano Res. (2020) 1–6. Copyright 2020, Springer.*

at room temperature. As for WSe_2, two specific kinds of excitons are considered: the direct transition neutral excitons and trions. Compared with the lifetime of neutral excitons (<0.3 ns within four-layer), trions possess a longer lifetime (~6.6 ns within four-layer) which increases with the number of layers (Fig. 9.15C and F). The longer-lived lifetime is attributed to the increasing number of trions as well as the varieties of trion configurations in thicker WSe_2. Besides, the whole average lifetime increases over 10% when WSe_2 flakes increase from a monolayer to four-layer. Layer-dependent method provides a novel tunable method to control the exciton dynamics as well as the electron energy dissipation process.

9.5.3 Defect-induced exciton dynamics in 2D materials

Defects play a very important role in the context of surface energy dissipation, and excitons are main carriers for the energy transports. But measurement of the interaction between defects and energy transports must use several different methods. With recent experimental technologies, some more profound issues that may affect the energy dissipation process are explored. Defects have great effects on crystal structure, energy band

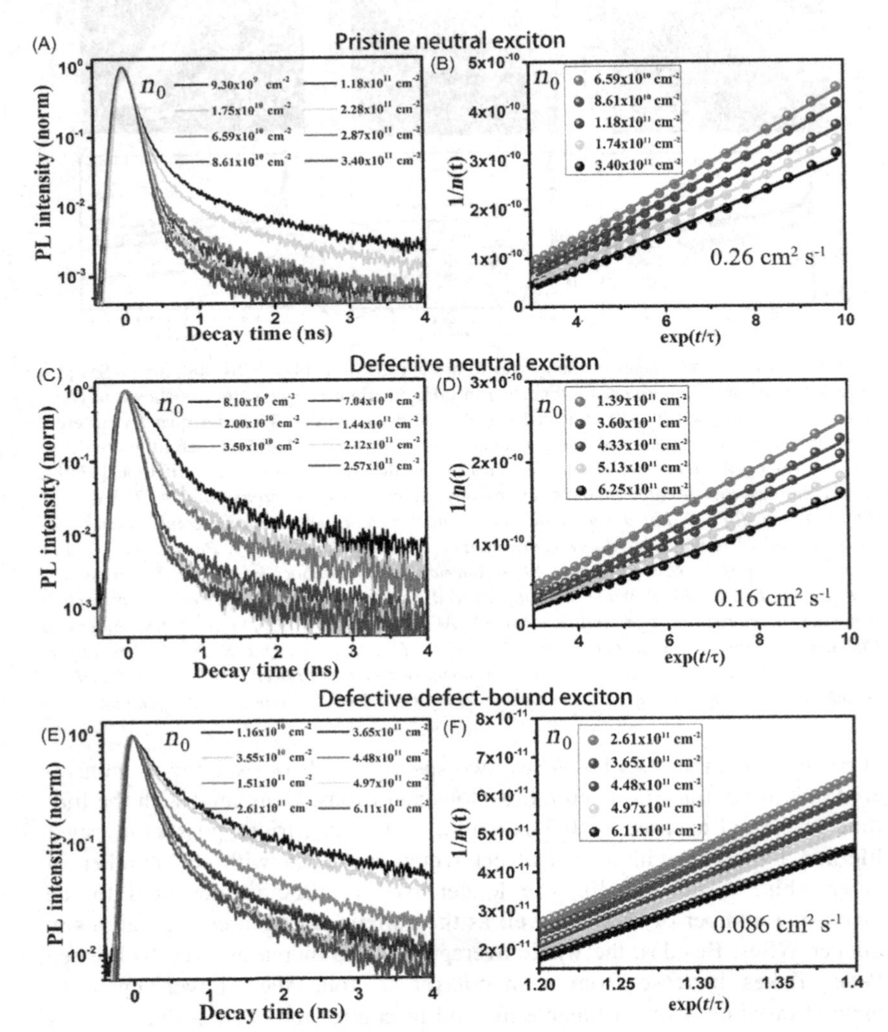

FIGURE 9.16 EEA rates of neutral exciton and defected-bound exciton. [(A), (C), and (E)] Time-resolved PL decay profiles of neutral and defect-bound exciton at various initial exciton density n_0; [(B), (D), and (F)] linearized data of TRPL curves for neutral exciton in (B) and (D) and defect-bound exciton in (F). The solid lines are linear fits [53]. (*TRPL*: time-resolved photoluminescence; *EEA*: exciton-exciton annihilation; *PL*: photoluminescence). *Reproduced with permission from H. Liu, C. Wang, D. Liu, J. Luo, Neutral and defect-induced exciton annihilation in defective monolayer WS2, Nanoscale 11(16) (2019) 7913–7920. Copyright 2019, The Royal Society of Chemistry.*

structure, and exciton transport [50,51]. Particularly, for 2D transition metal dichalcogenides (TMDCs), the interfacial defects have huge influence on exciton dynamics, for only several atom layers are on the surface, which enhance the probability of defects scattering process compared with bulk crystals [52]. By plasma treatment, defects can be introduced in WS_2 and a new defect-induced photoluminescence spectral feature is observed. When excitation intensity is increased, owing to exciton–exciton annihilation (EEA) as an additional relaxation channel that competes with radiative exciton recombination, photoluminescence decay rates of both neutral excitons and defect-bound excitons become significantly faster in pristine and defective monolayer WS_2. What's more, the EEA rate of neutral excitons in defective monolayer WS_2 has a significant decrease compared with pristine sample (Fig. 9.16), resulting from the localized nature of defect-bound excitons and the impeded exciton diffusion. These results provide insight into the effect of defects on exciton dynamics and exciton–exciton annihilation in monolayer WS_2 and guidance for designing light-emitting devices based on monolayer TMDCs [53].

9.6 Conclusion

In conclusion, recent research progresses of phononic and electronic channels of friction energy dissipation are introduced in this chapter. Phononic energy dissipation generally occurs on the insulator frictional pairs. While electronic dissipation occurs when friction pairs are metals or semiconductors. The processes of phononic energy dissipation are the excitations of lattice vibrations during sliding. The theories of phononic friction dissipation are built mainly on friction between adsorbing molecules and substrates. And it is experimentally proved that the phononic channel of friction energy dissipation is intensely influenced by natural vibrating frequency of surface. Electronic dissipation channel is originated from electrons dragged by the moving atom, and it dominates in noncontact friction. The fluorescence lifetime is an important parameter to describe the electron transition process, electron energy dissipation can be investigated by the fluorescence lifetime of 2D materials. These new insights on energy dissipation mechanism help to explore the origin of friction and control it, servicing for human production activities.

Acknowledgments

This work was financially supported by the National Natural Science Foundation of China (Grant Nos. 51527901 and 51575298).

References

[1] J.Y. Park, M. Salmeron, Fundamental aspects of energy dissipation in friction, Chem. Rev. 114 (1) (2013) 677–711.

[2] U.D. Schwarz, H. Hölscher, Exploring and explaining friction with the Prandtl–Tomlinson model, ACS Nano 10 (1) (2016) 38–41.

[3] G. Tomlinson, CVI. A molecular theory of friction, London, Edinburgh, Dublin Philos. Mag. J. Sci. 7 (46) (1929) 905–939.

[4] C. Lee, Q. Li, W. Kalb, X.-Z. Liu, H. Berger, R.W. Carpick, et al., Frictional characteristics of atomically thin sheets, Science 328 (5974) (2010) 76–80.

[5] B. Persson, R. Ryberg, Brownian motion and vibrational phase relaxation at surfaces: CO on Ni(111), Phys. Rev. B 32 (6) (1985) 3586.

[6] S.P. Lewis, M. Pykhtin, E. Mele, A.M. Rappe, Continuum elastic theory of adsorbate vibrational relaxation, J. Chem. Phys. 108 (3) (1998) 1157–1161.

[7] C. Daly, J. Krim, Sliding friction of solid xenon monolayers and bilayers on Ag(111), Phys. Rev. Lett. 76 (5) (1996) 803.

[8] M. Tomassone, J. Sokoloff, A. Widom, J. Krim, Dominance of phonon friction for a xenon film on a silver(111) surface, Phys. Rev. Lett. 79 (24) (1997) 4798.

[9] R.J. Cannara, M.J. Brukman, K. Cimatu, A.V. Sumant, S. Baldelli, R.W. Carpick, Nanoscale friction varied by isotopic shifting of surface vibrational frequencies, Science 318 (5851) (2007) 780–783.

[10] J.C. Lin, K.H. Chen, H.C. Chang, C.S. Tsai, C.E. Lin, J.K. Wang, The vibrational dephasing and relaxation of CH and CD stretches on diamond surfaces: an anomaly, J. Chem. Phys. 105 (10) (1996) 3975–3983.

[11] X. Han, T. Balgar, E. Hasselbrink, Vibrational dynamics of hydrogen on Ge surfaces, J. Chem. Phys. 130 (13) (2009) 134701.

[12] S. Sakong, P. Kratzer, X. Han, T. Balgar, E. Hasselbrink, Isotope effects in the vibrational lifetime of hydrogen on germanium (100): theory and experiment, J. Chem. Phys. 131 (12) (2009) 124502.

[13] B. Persson, Z. Zhang, Theory of friction: Coulomb drag between two closely spaced solids, Phys. Rev. B 57 (12) (1998) 7327.

[14] J. Krim, QCM tribology studies of thin adsorbed films, Nano Today 2 (5) (2007) 38–43.

[15] J. Krim, D. Solina, R. Chiarello, Nanotribology of a Kr monolayer: a quartz-crystal microbalance study of atomic-scale friction, Phys. Rev. Lett. 66 (2) (1991) 181.

[16] C. Mak, C. Daly, J. Krim, Atomic-scale friction measurements on silver and chemisorbed oxygen surfaces, Thin Solid Films 253 (1-2) (1994) 190–193.

[17] N. Manini, G. Mistura, G. Paolicelli, E. Tosatti, A. Vanossi, Current trends in the physics of nanoscale friction, Adv. Phys. X 2 (3) (2017) 569–590.

[18] B. Mason, S. Winder, J. Krim, On the current status of quartz crystal microbalance studies of superconductivity-dependent sliding friction, Tribol. Lett. 10 (1-2) (2001) 59–65.

[19] M. Highland, J. Krim, Superconductivity dependent friction of water, nitrogen, and superheated He films adsorbed on Pb(111), Phys. Rev. Lett. 96 (22) (2006) 226107.

[20] B. Persson, E. Tosatti, The puzzling collapse of electronic sliding friction on a superconductor surface, Surf. Sci. 411 (1-2) (1998) L855–L857.

[21] R. Renner, J. Rutledge, P. Taborek, Quartz microbalance studies of superconductivity-dependent sliding friction, Phys. Rev. Lett. 83 (6) (1999) 1261.

[22] R. Renner, P. Taborek, J. Rutledge, Friction and pinning of nitrogen films on lead substrates near the superconducting transition, Phys. Rev. B 63 (23) (2001) 233405.

[23] S.S. Rekhviashvili, The phenomenon of decrease in the friction force upon transition to the superconducting state, Tech. Phys. Lett. 30 (1) (2004) 4–5.
[24] T. Novotný, B. Velický, Electronic sliding friction of atoms physisorbed at superconductor surface, Phys. Rev. Lett. 83 (20) (1999) 4112.
[25] J. Sokoloff, M. Tomassone, A. Widom, Strongly temperature dependent sliding friction for a superconducting interface, Phys. Rev. Lett. 84 (3) (2000) 515.
[26] L. Bruch, Ohmic damping of center-of-mass oscillations of a molecular monolayer, Phys. Rev. B 61 (23) (2000) 16201.
[27] O. Pfeiffer, R. Bennewitz, A. Baratoff, E. Meyer, P. Grütter, Lateral-force measurements in dynamic force microscopy, Phys. Rev. B 65 (16) (2002) 161403.
[28] B. Stipe, H. Mamin, T. Stowe, T. Kenny, D. Rugar, Noncontact friction and force fluctuations between closely spaced bodies, Phys. Rev. Lett. 87 (9) (2001) 096801.
[29] M. Kisiel, E. Gnecco, U. Gysin, L. Marot, S. Rast, E. Meyer, Suppression of electronic friction on Nb films in the superconducting state, Nat. Mater. 10 (2) (2011) 119–122.
[30] U. Gysin, S. Rast, M. Kisiel, C. Werle, E. Meyer, Low temperature ultrahigh vacuum noncontact atomic force microscope in the pendulum geometry, Rev. Sci. Instrum. 82 (2) (2011) 023705.
[31] P.B. Merrill, S.S. Perry, Fundamental measurements of the friction of clean and oxygen-covered VC (100) with ultrahigh vacuum atomic force microscopy: evidence for electronic contributions to interfacial friction, Surf. Sci. 418 (1) (1998) 342–351.
[32] J.Y. Park, Y. Qi, D.F. Ogletree, P.A. Thiel, M. Salmeron, Influence of carrier density on the friction properties of silicon pn junctions, Phys. Rev. B 76 (6) (2007).
[33] J.Y. Park, Electronic control of friction in silicon pn junctions, Science 313 (5784) (2006) 186.
[34] Y. Qi, J.Y. Park, B.L.M. Hendriksen, D.F. Ogletree, M. Salmeron, Electronic contribution to friction on GaAs: an atomic force microscope study, Phys. Rev. B 77 (2008) 18.
[35] J.H. Kim, D. Fu, S. Kwon, K. Liu, J. Wu, J.Y. Park, Crossing thermal lubricity and electronic effects in friction: vanadium dioxide under the metal-insulator transition, Adv. Mater. Interfaces 3 (2) (2016).
[36] S. Kuehn, R.F. Loring, J.A. Marohn, Dielectric fluctuations and the origins of noncontact friction, Phys. Rev. Lett. 96 (15) (2006) 156103.
[37] T. Filleter, R. Bennewitz, Structural and frictional properties of graphene films on SiC (0001) studied by atomic force microscopy, Phys. Rev. B 81 (2010) 15.
[38] T. Filleter, J.L. McChesney, A. Bostwick, E. Rotenberg, K.V. Emtsev, T. Seyller, et al., Friction and dissipation in epitaxial graphene films, Phys. Rev. Lett. 102 (8) (2009) 086102.
[39] Y. Dong, Effects of substrate roughness and electron–phonon coupling on thickness-dependent friction of graphene, J. Phys. D: Appl. Phys. 47 (5) (2014) 055305.
[40] A. Buldum, S. Ciraci, Atomic-scale study of dry sliding friction, Phys. Rev. B 55 (4) (1997) 2606–2611.
[41] Z.M. Xu, P. Huang, Study on the energy dissipation mechanism of atomic-scale friction with composite oscillator model, Wear 262 (7-8) (2007) 972–977.
[42] P.H. Tan, W.P. Han, W.J. Zhao, Z.H. Wu, K. Chang, H. Wang, et al., The shear mode of multilayer graphene, Nat. Mater. 11 (4) (2012) 294–300.
[43] X. Zhang, W.P. Han, J.B. Wu, S. Milana, Y. Lu, Q.Q. Li, et al., Raman spectroscopy of shear and layer breathing modes in multilayer MoS_2, Phys. Rev. B 87 (11) (2013) 115413.

[44] X. Zhou, J. Li, Y. Leng, X. Cong, D. Liu, J. Luo, Exploring interlayer interaction of $SnSe_2$ by low-frequency Raman spectroscopy, Physica E Low Dimens. Syst. Nanostruct. 105: (2019) 7–12.

[45] X. Zhou, K. Jin, X. Cong, Q. Tan, J. Li, D. Liu, et al., Interlayer interaction on twisted interface in incommensurate stacking MoS_2: a Raman spectroscopy study, J. Colloid Interface Sci. 538: (2019) 159–164.

[46] L. Wang, X. Zhou, T. Ma, D. Liu, L. Gao, X. Li, et al., Superlubricity of a graphene/ MoS_2 heterostructure: a combined experimental and DFT study, Nanoscale 9 (30) (2017) 10846–10853.

[47] T. Wang, Y. Zhang, Y. Liu, J. Li, D. Liu, J. Luo, et al., Layer-number-dependent exciton recombination behaviors of MoS_2 determined by fluorescence-lifetime imaging microscopy, J. Phys. Chem. C. 122 (32) (2018) 18651–18658.

[48] Y. Liu, X. Hu, T. Wang, D. Liu, Reduced binding energy and layer-dependent exciton dynamics in monolayer and multilayer WS2, ACS Nano (2019).

[49] Y. Liu, H. Li, C. Qiu, X. Hu, D. Liu, Layer-dependent signatures for exciton dynamics in monolayer and multilayer WSe2 revealed by fluorescence lifetime imaging measurement, Nano Res. (2020) 1–6.

[50] J. Lu, A. Carvalho, X.K. Chan, H. Liu, B. Liu, E.S. Tok, et al., Atomic healing of defects in transition metal dichalcogenides, Nano Lett. 15 (5) (2015) 3524–3532.

[51] H. Nan, Z. Wang, W. Wang, Z. Liang, Y. Lu, Q. Chen, et al., Strong photoluminescence enhancement of MoS_2 through defect engineering and oxygen bonding, ACS Nano 8 (6) (2014) 5738–5745.

[52] L. Yuan, T. Wang, T. Zhu, M. Zhou, L. Huang, Exciton dynamics, transport, and annihilation in atomically thin two-dimensional semiconductors, J. Phys. Chem. Lett. 8 (14) (2017) 3371–3379.

[53] H. Liu, C. Wang, D. Liu, J. Luo, Neutral and defect-induced exciton annihilation in defective monolayer WS2, Nanoscale 11 (16) (2019) 7913–7920.

Chapter 10

Liquid superlubricity with 2D material additives

Hongdong Wang, Jinjin Li, Yuhong Liu and Jianbin Luo
State Key Laboratory of Tribology, Tsinghua University, Beijing, P.R. China

10.1 Introduction

Friction is widespread in human daily life and industrial production, but unnecessary friction will cause huge losses in material wear and low energy efficiency. Therefore it has become an urgent problem to increase energy efficiency in the development of modern industrialization [1–4]. Nowadays, various lubricating additives have been commonly utilized in industrial production process to reduce friction and wear, thereby improving production efficiency, extending component life, and increasing energy transfer efficiency. In recent years, two-dimensional (2D) nanomaterials have drawn great attention due to their unique physical structure and special chemical properties, so that they have been widely investigated in tribological research [5,6]. With continuously innovative synthesis method and characterization technique, most 2D nanomaterials have been successfully applied as lubricant additives. Among them, carbon nanomaterials (e.g., graphene) and transition metal disulfides (e.g., molybdenum disulfide) are the most typical ones [7–9]. In addition, other 2D materials such as layered double hydroxide (LDH), hexagonal boron nitride, metal or covalent organic framework materials, and graphite-like carbon nitride have also been reported to show excellent tribological performance due to their special structure and various chemical compositions [10–14]. As compared with ordinary lubricant nanoadditives, 2D nanomaterials are more likely to enter the contact area of the lubricant because of their relatively small longitudinal dimension, thereby showing synergistic effects and taking sufficient tribochemical reaction with the sliding solid surfaces. Due to the strong mechanical properties of laminates, the direct collision between asperities of solid surfaces can be effectively prevented during friction process. The probability of severe wear will be greatly reduced, thereby enhancing the wear resistance of solid surfaces.

Superlubricity. DOI: https://doi.org/10.1016/B978-0-444-64313-1.00010-7

As 2D nanomaterials with perfect crystallinity are utilized as solid lubricants, the ultralow friction coefficient (COF) (less than 0.01) can be achieved under certain atmosphere. The "solid superlubricity" can be achieved between two sliding surfaces with ultralow interaction under certain conditions [15–21]. However the applied scale is still limited to the micro- or nanometer due to the requirement of environment and nondefective material. It is still difficult to achieve continuous solid superlubricity and stable ultralow COF in the macroscopic scale in ambient environment [22,23].

In addition to the solid lubricants, ultralow COF can also be obtained by liquid lubricants due to hydration effect, electric double layer, and hydrodynamic lubrication [24–26]. In our group, a novel liquid superlubricity system (ultralow COF of 0.004) was obtained with phosphoric acid solution between sliding solid surfaces of Si_3N_4 and glass [27–30]. Then, the superlubricity system was further promoted to the mixture of polyhydroxy alcohol and acid solution [31,32]. Moreover, ionic liquids were also utilized as lubricant additives to develop the excellent superlubricity property [33]. Since "liquid superlubricity" can be realized in a relatively large scale in the ambient environment, it is quite suitable for practical production in industry. Currently, the water content in liquid lubricant and the sensitivity of hydrogen ions during running-in process have great influence on the reliability and stability of lubricating system [34]. According to the statistical results in Table 10.1, it is still quite necessary to significantly improve the load-carrying capacity and meanwhile shorten the indispensable running-in period of liquid superlubricity system. Therefore it comes a great challenge to combine both liquid superlubricity of water-based lubricant and solid superlubricity of 2D nanomaterials to deeply explore the superlubricity materials as well as their mechanism, regulation, and application in the lubricating system.

10.2 Lubricant additives of 2D nanomaterials

Nowadays, 2D nanomaterials have been widely utilized as lubricating additives to synergize lubricants to improve tribological properties, thereby effectively reducing the energy loss and wear condition caused by violent friction process. Chen et al. [36] dispersed 1 wt.% ultrathin MoS_2 nanosheets in liquid paraffin, and the highest load capacity at 120°C got tremendously improved from less than 50 N to more than 2000 N. As shown in Fig. 10.1, it was considered that the ultrathin shape of 2D nanosheets made it much easier to enter the contact area of sliding surfaces as compared to three-dimensional nanoparticles, and then the protective film was formed to prevent seizure. Additionally, ultrathin MoS_2 sheets with various lateral sizes were compared, and the sheets with a larger size were found to show better lubrication performance [37]. As long as the contact pressure between asperities was below the breaking strength of ultrathin MoS_2, then the localized

TABLE 10.1 Contact pressures for various lubrication systems during the liquid-superlubricity period at the macroscale.

Initial (MPa)	Final (MPa)	Lubricant	Frictional pairs	WSD (μm)
1170	385	PAO + BN	Si_3N_4/DLC	182
280	280	Castor oil	TiNi60/HSS	–
270	≤270	Glycerol (GL)	ta-C/ta-C	–
480	255	Water	Si_3N_4/a-C:H:Si	100
1600	200	GO + $EG_{(aq)}$	Si_3N_4/Si_3N_4	–
700	170	$Salt_{(aq)}$	Si_3N_4/Sapphire	150
1600	132	$LiPF_6$-based $IL_{(aq)}$	Si_3N_4/Si_3N_4	170
750	111	PEG + boric $acid_{(aq)}$	Si_3N_4/SiO_2	200
2570	95	$H_3PO_{4(aq)}$	Ruby/Sapphire	200
63.5	63.5	Phosphate-buffered saline	PTFE/PVPA	–
1680	60	$PAG_{(aq)}$	Si_3N_4/Al_2O_3	252
668	40	$GL_{(aq)}$	Steel/Steel	310
750	37	$IL_{(aq)}$	Si_3N_4/SiO_2	320
683	20	1,3-diketone	Steel/Steel	568
700	16	GL + nanodiamonds	Steel/Steel	300
440	15	$PEG_{(aq)}$	Steel/Steel	500
–	<10	Water	Si_3N_4/Si_3N_4	–
–	2.5	IL + carbon quantum dots	Al_2O_3/Si	–
0.48	≤0.48	Polymer	Metal or ceramic	–
700	–	PAO + GMO	Steel/ta-C	–

Notes: "Initial" is the initial contact pressure (Hertzian pressure) and "Final" is the final contact pressure (dividing the normal load by the contact area) during the superlubricity period; (aq) means the aqueous solution. *WSD*, worn scar diameter, ta-C, hydrogen-free amorphous DLC. *BN*, boron nitride; *EG*, ethylene glycol; *EDS*, energy disperse spectroscopy; *PAO*, poly alpha olefin; *PEG*, olyethylene glycol; *GMO*, glycerol mono-oleate.
Source: Adapted from X. Ge, J. Li, H. Wang, C. Zhang, Y. Liu, J. Luo, Macroscale superlubricity under extreme pressure enabled by the combination of graphene-oxide nanosheets with ionic liquid, Carbon 151 (2019) 76–83. Copyright 2019, Elsevier Ltd.

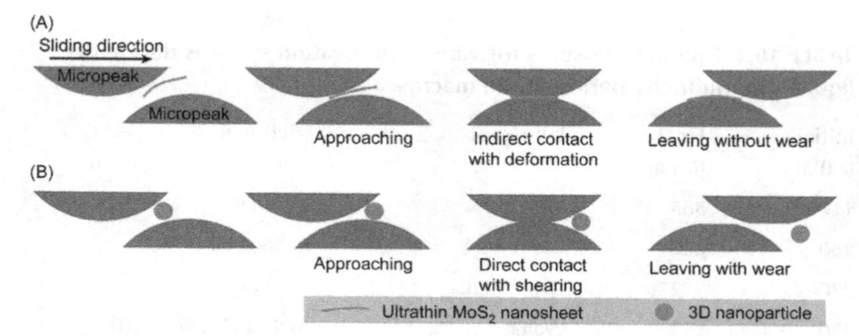

FIGURE 10.1 Schematics of the lubrication mechanism of ultrathin 2D nanosheets and 3D nanoparticles. *Adapted from Z. Chen, X. Liu, Y. Liu, S. Gunsel, J. Luo, Ultrathin MoS2 nanosheets with superior extreme pressure property as boundary lubricants, Sci. Rep. 5 (2015) 12869. Copyright 2015, Springer Nature Limited.*

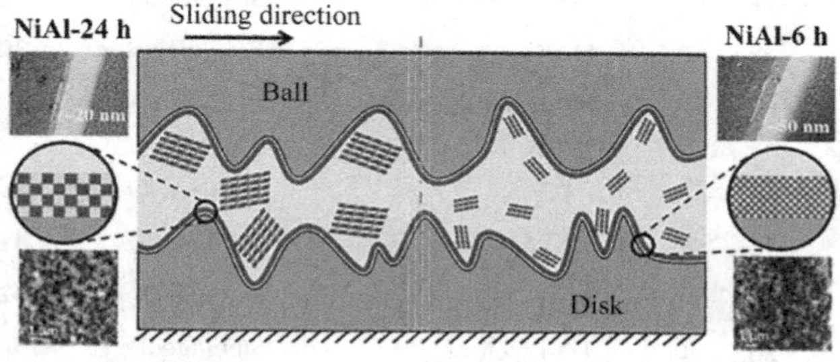

FIGURE 10.2 Schematic illustration of proposed lubrication model with different nanosized LDH additives in base oil. (*LDH*: Layered double hydroxide) *Adapted from H. Wang, Y. Liu, W. Liu, R. Wang, J. Wen, H. Sheng, et al., Tribological behavior of NiAl-layered double hydroxide nanoplatelets as oil-based lubricant additives, ACS Appl. Mater. Interfaces 9(36) (2017) 30891–30899. Copyright 2017, American Chemical Society.*

seizure and further wear can be prevented because of no direct metal-to-metal contact.

Wang et al. [38] prepared three different Ni-Al LDH nanoplatelets, confirmed their layered structure and evaluated their tribological properties as additives in base oil. It was revealed as shown in Fig. 10.2 that the large additives, rather than the small ones, showed the best and most stable tribological performance because their high degree of crystallinity was conductive to the formation of a tribofilm with superior mechanical properties.

2D nanomaterials can also be uniformly dispersed in the water-based lubricant with successful surface modification. Wang et al. [39] prepared NiAl-LDH nanoplatelets with few ordered layers in a microemulsion by a hydrothermal method, dispersed them in water and obtained a stable translucent solution. A lubricating layer was formed under high contact pressure (about 1.5 GPa initially) due to their small size and excellent

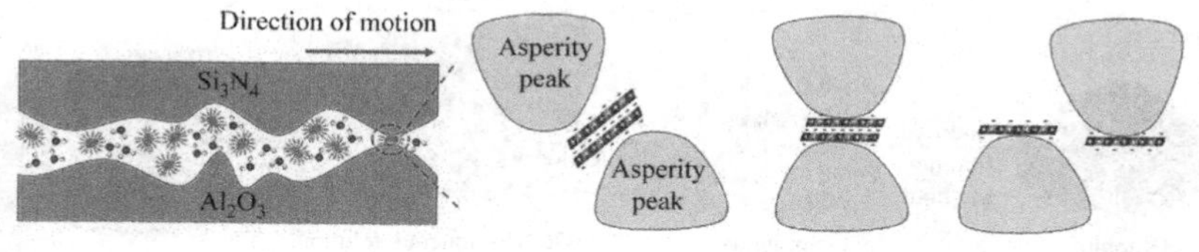

FIGURE 10.3 Schematic lubrication model of the aqueous solution with LDH nanoplatelet additives. (*LDH*: Layered double hydroxide). *Adapted from H. Wang, Y. Liu, Z. Chen, B. Wu, S. Xu, J. Luo, Layered double hydroxide nanoplatelets with excellent tribological properties under high contact pressure as water-based lubricant additives, Sci. Rep. 6(1) (2016) 22748. Copyright 2016, Springer Nature Limited.*

dispersion. In comparison with the tribological results tested by pure water, the aqueous sample with 0.5 wt.% NiAl-LDH nanoplatelets exhibited superior lubricating performance. The COF, wear scar diameter, depth, and width of wear track decreased by 83.1%, 43.2%, 88.5%, and 59.5%, respectively. As shown in Fig. 10.3, the exfoliated sheets absorbing on the sliding solid surfaces in friction period were able to prevent direct collision of asperities because of their layered structure, high specific surface area, and laminates with high carrying capacity.

Kinoshita et al. [40] investigated the application and tribological properties of graphene oxide (GO) monolayer sheets (the lateral size distributed from 10 to 50 μm) as additives in water-based lubricants. After 60,000 cycles of friction testing, no obvious surface wear was found and low COF (around 0.05) was obtained. GO sheets in water provided the adsorption layer on both sliding solid surfaces, which may behave as protective coatings. Song et al. [41] investigated the tribological properties of GO nanosheets and oxide multiwall carbon nanotubes (CNTs-COOH) as water-based lubricant additives. Excellent dispersion and superior property of GO nanosheets were considered to result in better friction-reducing ability while CNTs-COOH presented poor dispersion because of the large CNTs-COOH bundles easily formed in water, which resulted in the COF increasing. The physical tribofilm formed on the metal substrate under high contact pressure can not only improve the load-bearing capacity but also reduce the probability of direct collision of asperities on sliding solid surfaces.

10.3 Superlubricity enabled by the synergy of graphene oxide and ethanediol

Ge et al. [42] reported a robust macroscale superlubricity state ($\mu = 0.0037$) with the synergy effect of GO nanoflakes (GONFs) and ethanediol (EDO) between sliding Si_3N_4-SiO_2 solid surfaces. The preparation process was shown in Fig. 10.4A. The as-prepared GONFs with an oxygen content of 30%–40% were obtained via a modified Hummers method. The lateral

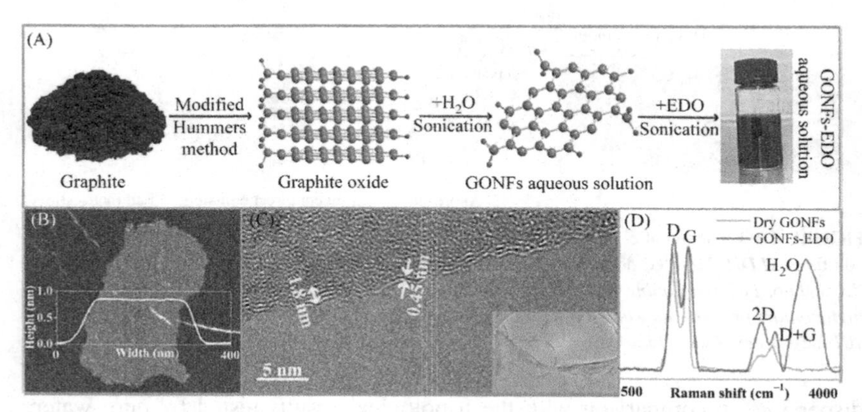

FIGURE 10.4 (A) Preparation of aqueous GONFs-EDO solution; (B) AFM image of GONFs dried on the mica substrate; inset depicts the structural profile of GONFs; (C) HRTEM images of dry GONFs; inset depicts an HRTEM image at low magnification; (D) Raman spectra of dry GONFs and aqueous GONFs-EDO solution. (*AFM*: Atomic force microscopy; *EDO*: ethanediol; GONFs: graphene-oxide nanoflakes; *HRTEM*: high-resolution transmission electron microscopy). *Adapted from X. Ge, J. Li, R. Luo, C. Zhang, J. Luo, Macroscale superlubricity enabled by the synergy effect of graphene-oxide nanoflakes and ethanediol, ACS Appl. Mater. Interfaces 10(47) (2018) 40863–40870. Copyright 2018, American Chemical Society.*

dimension of 200–400 nm and the thickness of 0.8 nm were detected using an AFM in Fig. 10.4B. The interlayer spacing of 0.45 nm was demonstrated with a high-resolution transmission electron microscopy (HRTEM) image shown in Fig. 10.4C. It is a bit larger than that of graphite (0.335 nm) due to the existence of oxygen-containing groups intercalated among the laminates of graphene. Raman spectroscopy was utilized to explore the interactions between GONFs and EDO. As shown in Fig. 10.4D, a slight shift in the Raman spectrum of GONFs-EDO solution, as compared to dry GONFs, indicated that the surface of solid additives could be covered by water molecules due to the hydrophilic oxygen-containing groups and therefore GONFs can be well dispersed in the aqueous solution.

After that, the aqueous solutions of GONFs, EDO, and GONFs-EDO were prepared to investigate the synergy effect of GONFs and EDO. Their tribological properties were evaluated and shown in Fig. 10.5. The high COF of water can be attributed to the inability of pure water to form an effective lubricating film; the COF of GONFs was found to reduce from 0.35 to 0.013; the COF of EDO showed gradual reduction initially and then increased to 0.012 at the end of test, as shown in Fig. 10.5A. Herein, all these samples were unable to achieve superlubricity directly. However the COF of GONFs-EDO aqueous solution gradually reduced to the value less than 0.01 after the running-in period of 600 s. After test, GONFs were found to be adsorbed on the contact surfaces, and thereby they were considered to avoid the direct contact between asperities. Meanwhile, the wear volume of

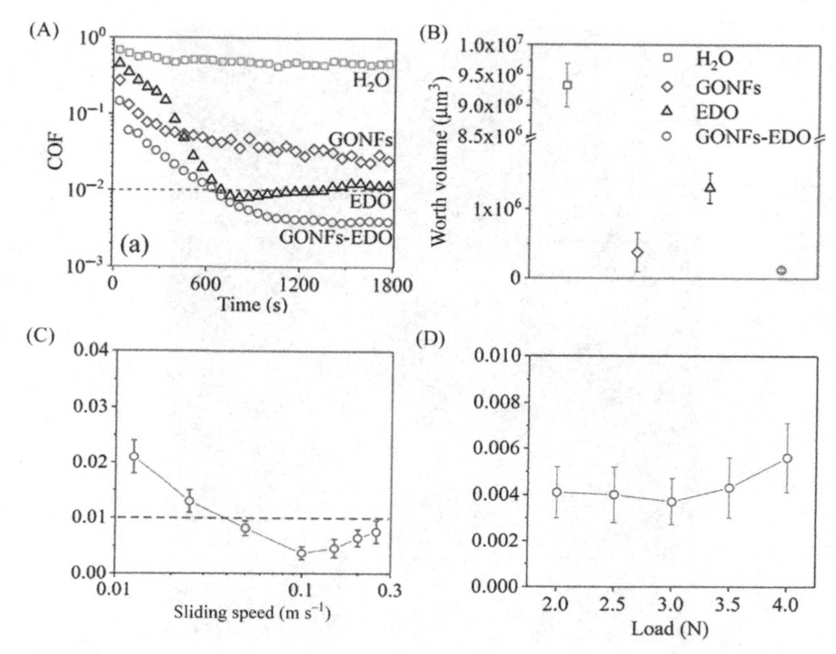

FIGURE 10.5 (A) COF evolution and (B) wear volume of water and aqueous solutions of GONFs, EDO, and GONFs-EDO. Average COF values for GONFs-EDO during the superlubricity period (C) at different sliding speeds from 12.5 to 250 mm s^{-1} under an applied load of 3 N and (D) under different applied loads from 2 to 4 N at a sliding speed of 0.1 m s^{-1}. (*COF*: Friction coefficient; *EDO*: ethanediol; *GONFs*: graphene oxide nanoflakes). *Adapted from X. Ge, J. Li, R. Luo, C. Zhang, J. Luo, Macroscale superlubricity enabled by the synergy effect of graphene-oxide nanoflakes and ethanediol, ACS Appl. Mater. Interfaces 10(47) (2018) 40863–40870. Copyright 2018, American Chemical Society.*

the ball lubricated by GONFs-EDO (5.1×10^4 μm^3) was only 5% of that lubricated by EDO (1.3×10^6 μm^3), as shown in Fig. 10.5B. The extremely low shear stresses between asperities were considered to result from the formation of hydrodynamic boundary condition at the GONFs-EDO interface, which effectively stabilized superlubricity and leaded to negligible wear. In Fig. 10.5C, the average COF values reduced from 0.021 to 0.0037 with the increasing sliding speed from 12.5 to 100 mm s^{-1}. At the point of 50 mm s^{-1}, it started achieving superlubricity, in which COF approached 0.008. With further increasing sliding speed, the COF values were still in the superlubricity regime. Therefore it was considered that in a proper velocity range (50–250 mm s^{-1}), the robust superlubricity of GONFs-EDO lubricating system can be realized. This work also applied various loads from 2 to 4 N, and recorded their average COFs during the sliding period in Fig. 10.5D. According to different applied condition, the superlubricity can finally be realized as the contact pressure was no more than 111 MPa.

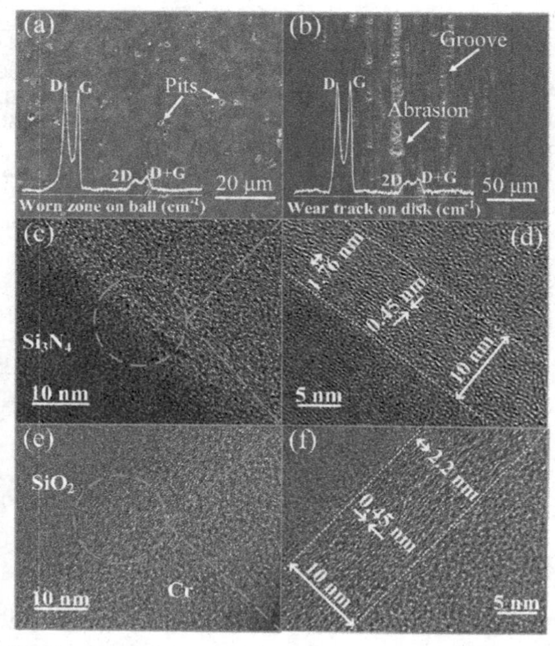

FIGURE 10.6 SEM images of the contact area in (A) Si_3N_4 ball and (B) SiO_2 disk, lubricated by aqueous GONFs-EDO solution; insets depict Raman spectra of adsorbed GONFs in the contact area. Cross-sectional structures of GONF in the worn-out zone, as obtained by HRTEM; (C) and (E) represent low-magnification images; (D) and (F) represent high-magnification images. (*EDO*: ethanediol; *GONF*: graphene oxide nanoflake; *HRTEM*: high-resolution transmission electron microscopy; *SEM*: scanning electron microscopy) *Adapted from X. Ge, J. Li, R. Luo, C. Zhang, J. Luo, Macroscale superlubricity enabled by the synergy effect of graphene-oxide nanoflakes and ethanediol, ACS Appl. Mater. Interfaces 10(47) (2018) 40863–40870. Copyright 2018, American Chemical Society.*

The topographies of contact area lubricated by GONFs-EDO were detected and analyzed using scanning electron microscope and Raman spectrum in Fig. 10.6A and B, which demonstrated that GONFs were adsorbing on the surfaces of both ball and disk. With further observation of HRTEM images in Fig. 10.6C–F, it was interesting to find that the thicknesses of adsorbed GONFs after friction test (corresponding film thickness of 10 nm) was much larger than that of original GONFs but the interlayer structure (corresponding space of 0.45 nm) made no obvious change, thereby indicating that GONFs in EDO solution might gather on sliding solid surfaces subjected to rubbing motion, which provided better protection to surfaces against direct contact.

As illustrated in Fig. 10.7, the mechanism was proposed for underlying the realization of superlubricity. In asperity contact region shown in Fig. 10.7B, the low shear between GONF laminates resulted in friction

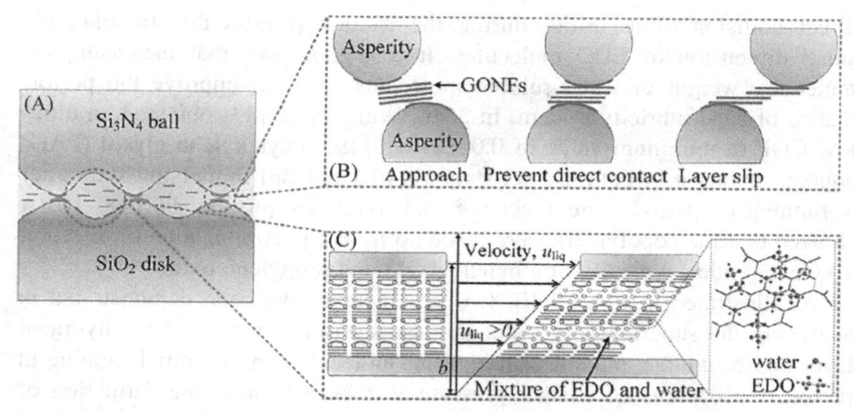

FIGURE 10.7 Proposed mechanism underlying the realization of superlubricity through use of GONFs-EDO (A) overall view; (B) asperity contact region; and (C) liquid-contact region. The extremely low shear stresses of the liquid lubricating film is caused by the formation of partial-slip hydrodynamic boundary condition at the GONFs − EDO interface and the formation of hydrated GONFs − EDO networks through hydrogen-bond interactions. Inset in (C) depicts schematic of hydrated GONFs-EDO networks; the GONF planar is depicted in the stick model, wherein *red* sticks correspond to the oxygen-containing group; hydrogen bond interactions are marked by *red*-dotted circles. (*EDO*: Ethanediol; *GONF*: graphene oxide nanoflake). *Adapted from X. Ge, J. Li, R. Luo, C. Zhang, J. Luo, Macroscale superlubricity enabled by the synergy effect of graphene-oxide nanoflakes and ethanediol, ACS Appl. Mater. Interfaces 10(47) (2018) 40863–40870. Copyright 2018, American Chemical Society.*

reduction and ultralow wear condition during the running-in period. After that, it is much easy for the adsorbed GONF layers to slide between approaching asperities, thereby leading to the achievement of low friction in the boundary lubrication regime during the superlubricity period. As shown in Fig. 10.7C, the relatively low shear stress between hydrated GONF and EDO networks resulted in the low friction in the liquid-contact region. Finally, the superlubricity was achieved by the synergy effect of GONFs and EDO. Besides, different kinds of ceramic solid surfaces were also selected for friction test and proved that the GONFs-EDO aqueous solution was an excellent lubricant for achieving superlubricity with ceramic surfaces. The obtained results in this study demonstrated the potential of such a solution in achieving robust superlubricity in practical applications.

10.4 Superlubricity enabled by ultrathin layered double hydroxide nanosheets

In GONFs-EDO lubricating system, it was clear that neither GONFs nor EDO can achieve robust superlubricity. However the superlubricity can be achieved with the synergy effect of GONFs and EDO. It was difficult for EDO aqueous solution to form a thick enough lubricating film to prevent the direct collision of asperities during the friction process due to relatively small dimension of EDO molecules. It is a good idea that increasing the molecular weight of water-soluble molecules so as to improve the performance of superlubricity system. In 2016, Wang et al. [43] obtained an ultra-low COF (a minimum value of 0.0023) with the polyalkylene glycol (PAG) aqueous solution in both droplet state (40 μL) and full immersion state after a running-in period. The reactants and synthesis process in Fig. 10.8A showed that the copolymers were made up from any combination of different alkylene oxides, typically of ethylene oxide and propylene oxide.

As illustrated in Fig. 10.8B, two key factors have been demonstrated in achieving the superlubricity state: the low shearing strength of the hydrated layer and a suitable amount of free water molecules. In the initial running-in period, the decreasing contact pressure was beneficial to the formation of elastohydrodynamic lubrication of fluid. The hydrated PAG chains provided low shear strength between two sliding solid surfaces. Meanwhile, contained free water molecules effectively weakened interactions between polymer chains. Thus the superlubricity can be stably achieved over a wide range of concentration (30–60 wt.%) and relative linear velocity (no less than 6 mm s^{-1}). However a relatively long running-in period was still an indispensable

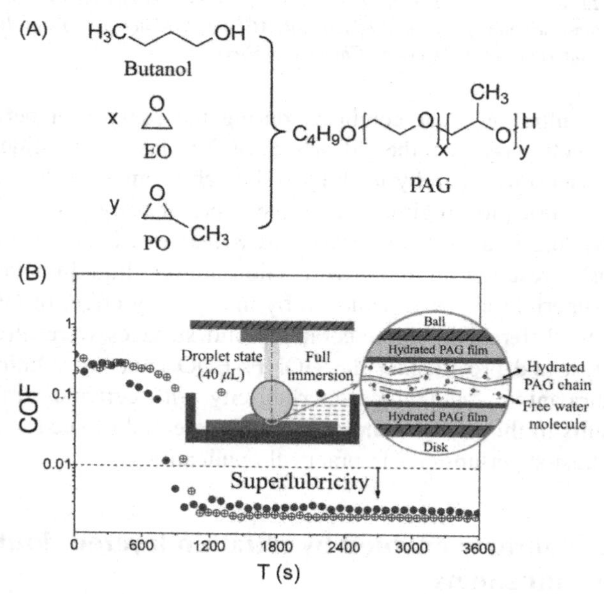

FIGURE 10.8 (A) Reactants and synthesis process leading to random copolymer PAG. (B) Schematic illustration of PAG aqueous solutions in both droplet and full immersion states. (*PAG*: Polyalkylene glycol). *Adapted from H. Wang, Y. Liu, J. Li, J. Luo, Investigation of superlubricity achieved by polyalkylene glycol aqueous solutions, Adv. Mater. Interfaces 3(19) (2016) 1600531. Copyright 2016, Wiley-VCH Verlag GmbH & Co.*

process for achieving ultralow COF. After a long period of storage, the PAG aqueous solution would gradually decompose and turn acidic, which may cause severe corrosion of the contact surfaces and thereby limited their widespread applications.

Since the LDH is able to improve the tribological performance and suppress acidification as a water-based lubricant additive, it is logical to disperse nanosized LDHs into the PAG aqueous solution to explore whether superlubricity behavior can be further boosted. In 2019, Wang et al. [44] prepared two different nano-LDHs and dispersed them in the PAG aqueous solution for tribological experiments: ultrathin LDH nanosheets (ULDH-NS) and LDH nanoparticles (LDH-NPs). The ULDH-NS in Fig. 10.9A with a lateral dimension of about 80 nm and a longitudinal dimension of about 1 nm can

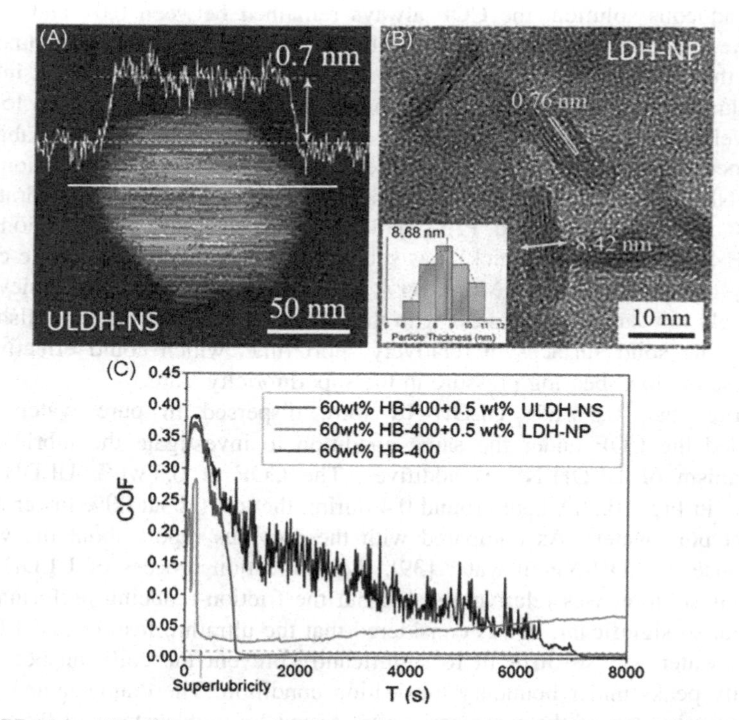

FIGURE 10.9 (A) AFM image and cross-sectional profile of ULDH-NS. (B) High-resolution TEM image of LDH-NP and its thickness distribution. (C) Friction coefficient of 60 wt.% PAG aqueous solution and those with 0.5 wt.% LDH additives (ULDH-NS and LDH-NP). (*AFM*: Atomic force microscopy; *LDH*: layered double hydroxide; *LDH-NP*: layered double hydroxide nanoparticle; *PAG*: polyalkylene glycol; *TEM*: transmission electron microscopy; *ULDH-NS*: ultrathin LDH nanosheets). *Adapted from H. Wang, Y. Liu, W. Liu, Y. Liu, K. Wang, J. Li, et al., Superlubricity of polyalkylene glycol aqueous solutions enabled by ultrathin layered double hydroxide nanosheets, ACS Appl. Mater. Interfaces 11(22) (2019) 20249–20256. Copyright 2019, American Chemical Society.*

be considered as a single or double layer in the aqueous solution. The LDH-NP in Fig. 10.9B possessed a lateral dimension of about 20 nm and a longitudinal dimension of about 10 nm. Then, 0.5% LDH nanoadditives were added to HB-400 PAG aqueous solution for tribological test. As shown in Fig. 10.9C, it took the original PAG aqueous solution about 7000 s to finally enter the superlubricity regime, but the COF obviously fluctuated during the initial running-in process. Through the analysis by three-dimensional topography, the diameter of wear scar was about 400 μm (corresponding contact pressure was 24.5 MPa) with the surface roughness *Sa* of 3.5 nm. With the addition of 0.5 wt.% ULDH-NS, the running-in process (about 1000 s) was greatly shortened. The diameter of wear scar was about 200 μm (corresponding contact pressure was 92.7 MPa) with the surface roughness *Sa* (roughness average) of 3.9 nm. However, with the addition of 0.5 wt.% LDH-NP in the same aqueous solution, the COF always remained between 0.04 and 0.05, and the superlubricity state could not be achieved at all. It is easily understood that more ULDH-NS sample would be dragged into the contact interface during the sliding process. As for the sample of LDH-NP, due to its relatively round and granular shape, it can effectively reduce the probability of asperity-asperity collusion. However the relatively large dimension of LDH-NP in contrast may disturb the ultralow shear between the hydration layers among the hydrated PAG molecular chains while the addition of ULDH-NS (only ~1 nm thick) was substantially inconsequential to the continuity of fluid layer [43]. Moreover it held the great potential to achieve a relatively low surface roughness with the addition of ULDH-NS by polishing the sliding solid surfaces in relatively short time, which could effectively increase the load-bearing pressure in the superlubricity state.

After that, 0.5 wt.% ULDH-NS were dispersed in pure water and recorded the COF under the same condition to investigate the lubrication mechanism of ULDH-NS as additives. The COF of 0.5 wt.% ULDH-NS sample in Fig. 10.10A kept around 0.4 during the test (about 20% lower than that of pure water). As compared with the previous report about the wear resistance of LDH-NP in water [39], the lubrication process of ULDH-NS aqueous solution was relatively stable but the friction-reducing performance was not so significant. It was considered that the ultrathin feature of ULDH-NS in water was so difficult to significantly prevent the collision between asperity peaks under boundary lubrication condition. The diameter and surface morphology of the wear scars after tested by pure water and 0.5 wt.% ULDH-NS were analyzed and compared in Fig. 10.10B–E. As lubricated by pure water, the diameter of wear scar was about 377 μm, and the surface roughness *Sa* was about 12.4 nm. As lubricated by 0.5 wt.% ULDH-NS, the diameter of wear scar was about 343 μm, and the surface roughness *Sa* was about 3.9 nm. Obviously, the surface roughness turned nearly 68.5% lower than the sample lubricated by pure water, though the diameter got only about 9% smaller. The analysis of sliding solid surfaces demonstrated that the

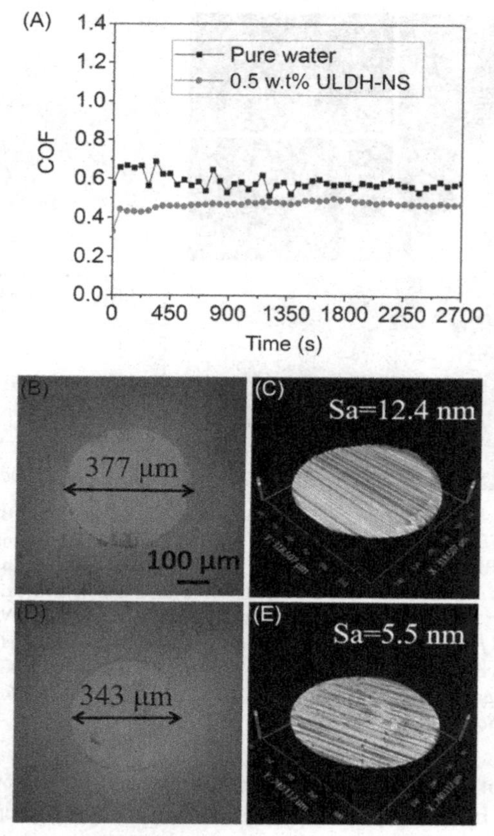

FIGURE 10.10 (A) Friction coefficient of pure water and 0.5 wt.% ULDH-NS aqueous solution. The optical image and three-dimensional topography of the wear scar on the ball, which was lubricated by [(B) and (C)] pure water and [(D) and (E)] 0.5 wt.% ULDH-NS aqueous solution. (*ULDH-NS*: Ultrathin layered double hydroxide nanosheets). *Adapted from H. Wang, Y. Liu, W. Liu, Y. Liu, K. Wang, J. Li, et al., Superlubricity of polyalkylene glycol aqueous solutions enabled by ultrathin layered double hydroxide nanosheets, ACS Appl. Mater. Interfaces 11(22) (2019) 20249–20256. Copyright 2019, American Chemical Society.*

physically adsorbed ULDH-NS enabled the sliding solid surfaces to be polished so as to realize the quick access to superlubricity.

Then, the AFM microscale friction test in the mode of lateral force measurement was conducted to simulate the interactions between asperity peaks and stacked ULDH-NS additives in water so as to further investigate the lubrication mechanism of ULDH-NS as additives. As the scanning progressed, it was interesting to find that the stacked ULDH-NS were exfoliated with the reduction thickness of one laminate (0.6–0.7 nm) because of relatively weak interaction between laminates. After scanning several LDH flake

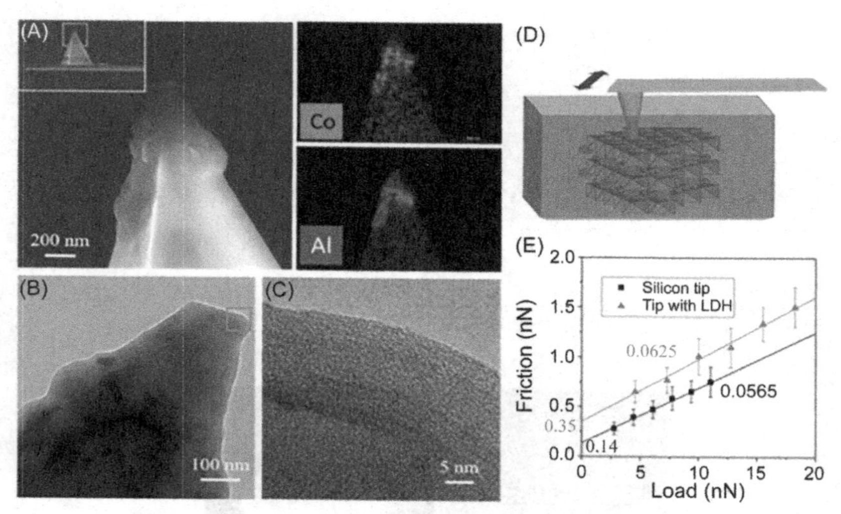

FIGURE 10.11 (A) SEM image and EDS mapping analysis (Co and Al); (B) TEM image; and (C) high-resolution TEM image of the AFM tip with ULDH-NS. (D) Schematic illustration of AFM friction test. (E) Friction between silicon tip versus mica and the tip with LDH versus mica in water. (*AFM*: Atomic force microscopy; *LDH*: layered double hydroxide; *SEM*: scanning electron microscopy; *TEM*: transmission electron microscopy; *ULDH-NS*: ultrathin LDH nanosheets). *Adapted from H. Wang, Y. Liu, W. Liu, Y. Liu, K. Wang, J. Li, et al., Superlubricity of polyalkylene glycol aqueous solutions enabled by ultrathin layered double hydroxide nanosheets, ACS Appl. Mater. Interfaces 11(22) (2019) 20249–20256. Copyright 2019, American Chemical Society.*

samples with an AFM tip, the morphology of the tip was shown in Fig. 10.11A. Its EDS analysis and transmission electron microscopy (TEM) image in Fig. 10.11B confirmed absorbed LDH nanosheets. The high-resolution TEM image Fig. 10.11C showed ULDH-NS stacking effect during scanning process as illustrated in Fig. 10.11D, which was quite similar to the adsorption condition of macroscopic experiment.

Then, the tip with LDH nanosheets was applied to a friction test against a fresh mica substrate. The COF obtained by the tip with LDH nanosheets ($\sim$0.0625) in Fig. 10.11E showed no significant difference from that of the fresh silicon tip ($\sim$0.0565) but the adhesion (0.35 nN) got significantly improved, which was beneficial to the adsorption of PAG molecules on the solid surfaces. Because there were a large amount of hydroxyl groups and some ethylene glycol molecules on the surface of LDH nanosheets, the solid surfaces will turn extremely hydrophilic and their adhesion force got effectively improved. Thus the violent collision can be better prevented during the sliding process so as to accelerate the realization of superlubricity.

From aforementioned experimental results and analysis, it was clearly found that the addition of ULDH-NS did not show obvious friction-reducing and antiwear performance during the friction process. However a significant

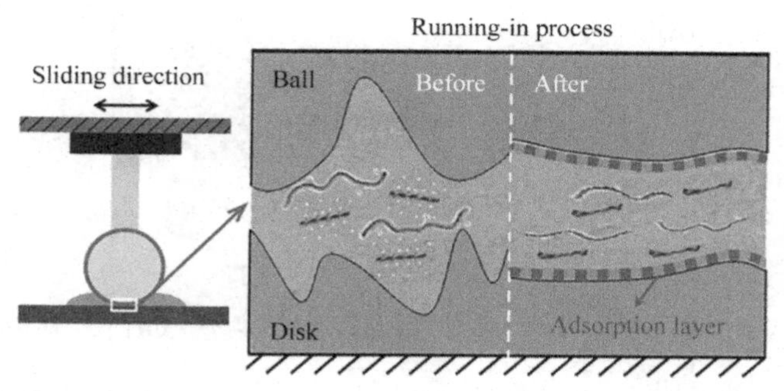

FIGURE 10.12 Schematic illustration of the proposed lubrication model with ULDH-NS additives in PAG aqueous solution. (*PAG*: Polyalkylene glycol; *ULDH-NS*: ultrathin layered double hydroxide nanosheets). *Adapted from H. Wang, Y. Liu, W. Liu, Y. Liu, K. Wang, J. Li, et al., Superlubricity of polyalkylene glycol aqueous solutions enabled by ultrathin layered double hydroxide nanosheets, ACS Appl. Mater. Interfaces 11(22) (2019) 20249–20256. Copyright 2019, American Chemical Society.*

contribution to the polishing, planarization, and protective effect was made, so that it was easy to obtain the contact area with lower surface roughness. As illustrated in Fig. 10.12, the adsorption layer provided PAG aqueous solution with favorable conditions for shortening the running-in process. Therefore the violent collision can be better prevented so as to shorten the running-in time (within 1000 s) for achieving superlubricity by 85% and improve load-bearing capacity ($\sim$92.7 MPa) by four times.

10.5 Superlubricity enabled by the combination of graphene oxide and ionic liquid

In order to further realize the widespread application of superlubricity, it is quite necessary to improve the load-bearing capacity of superlubricity to the level of actual working condition. As for the friction test condition of superlubricity liquid, the initial contact pressure between sliding solid surfaces was relatively high (even to the order of GPa), but the final contact pressure during the superlubricity period was quite low, which was mainly caused by the mechanical collision during the running-in period. Recently it was reported that a robust macroscale superlubricity was achieved by the combination of GO and ionic liquid (IL), in which the corresponding extreme contact pressure reached 600 MPa between Si_3N_4/sapphire sliding solid surfaces [35].

The composite aqueous solution IL + $GO_{(aq)}$ was prepared by mixing as-synthesized $[Li(EG)]PF_6$ with GO aqueous solution, as shown in Fig. 10.13. Then, the friction tests were conducted under a load of 3 N and relative sliding velocity of 100 mm s^{-1} with 30 μL $IL_{(aq)}$ or IL + $GO_{(aq)}$ (0.03%), as shown in Fig. 10.14A. The COF of $IL_{(aq)}$ increased from 0.04 to 0.1 and

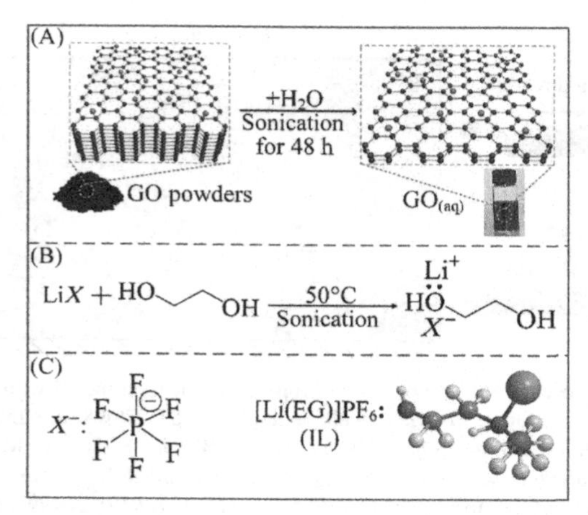

FIGURE 10.13 (A) Preparation of $GO_{(aq)}$ and the structure of the GO nanosheets, where the *black* orbs represent carbon atoms and *red* orbs represent oxygen-containing groups. (B) Preparation of $[Li(EG)]PF_6$. (C) Structure of PF_6^- anion and the possible model of $[Li(EG)]PF_6$. The cation $[Li(EG)]^+$ could be formed by the donation of lone pairs on the oxygen of an EG molecule to Li^+. Thus, the weakly Lewis-acidic cation could form an IL ($[Li(EG)]PF_6$) with the weakly Lewis-basic anion (PF_6^-). (*GO*: Graphene oxide). *Adapted from X. Ge, J. Li, H. Wang, C. Zhang, Y. Liu, J. Luo, Macroscale superlubricity under extreme pressure enabled by the combination of graphene-oxide nanosheets with ionic liquid, Carbon 151 (2019) 76–83. Copyright 2019, Elsevier Ltd.*

finally reduced to 0.033. When 30 μL IL + $GO_{(aq)}$ was applied, the COF dramatically dropped from 0.05 to 0.002 within a period of 900 s, which was regarded as the wearing-in period. Subsequently, the COF slightly increased from 0.002 to 0.005 and finally stabilized at around 0.005 for 2 h (Fig. 10.14B). In addition, the IL + $GO_{(aq)}$ prepared after one month (old IL + $GO_{(aq)}$) still exhibited the same lubrication performance as the fresh IL + $GO_{(aq)}$, demonstrating its stability under natural conditions. As the sliding velocity was in the range of 25 − 250 mm s^{-1}, it was demonstrated in Fig. 10.14C that the COFs were all less than 0.01 (indicating the achievement of superlubricity).

It was shown in Fig. 10.14D that the influence of lubricant volume on both average COFs and contact pressure during the superlubricity period. With the increase in lubricant volume from 2 to 50 μL, the contact pressure fluctuated in the range of 517–628 MPa, indicating that the variation of lubricant volume had no significant influence on the improvement of final contact pressure. Notably it also implies that the superlubricity can be achieved as the contact pressure was less than 597 MPa. However the COF would exceed the range of superlubricity when the contact pressure was more than 600 MPa. To further investigate the influence of contact pressure on the value of COF, friction tests lubricated by IL + $GO_{(aq)}$ (30 μL) under various normal loads were conducted in Fig. 10.14E. When the normal load

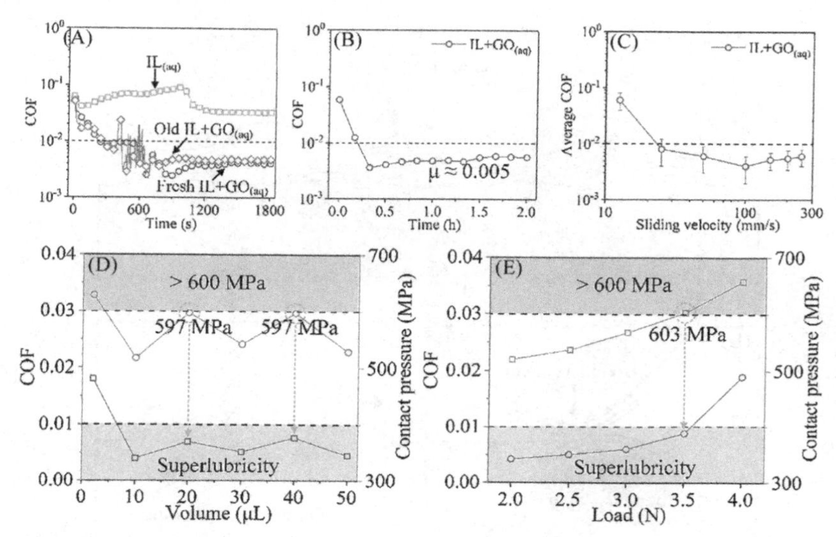

FIGURE 10.14 (A) COFs versus time with the lubrication of 30 μL $IL_{(aq)}$, fresh and old IL + $GO_{(aq)}$ (3 N and 100 mm s^{-1}). (B) COF for long-run testing with the lubrication of 30 μL IL + GO(aq) (3 N and 100 mm s^{-1}). (C) The influence of sliding velocity on the COF during the superlubricity period (3 N). (D) COFs and corresponding contact pressures during the superlubricity period with the lubrication of various volumes of IL + $GO_{(aq)}$ (3 N and 100 mm s^{-1}). (E) COFs and corresponding contact pressures during the superlubricity period under various normal loads with the lubrication of 30 μL IL + $GO_{(aq)}$ (100 mm s^{-1}). (*COFs*: Friction coefficients; *GO*: graphene oxide; *IL*: ionic liquid). *Adapted from X. Ge, J. Li, H. Wang, C. Zhang, Y. Liu, J. Luo, Macroscale superlubricity under extreme pressure enabled by the combination of graphene-oxide nanosheets with ionic liquid, Carbon 151 (2019) 76–83. Copyright 2019, Elsevier Ltd.*

increased from 2 to 4 N, the corresponding contact pressure increased from 520 to 658 MPa and meanwhile the COF also increased. When the contact pressure increased from 520 to 603 MPa (2–3.5 N), although the corresponding COF increased from 0.0042 to 0.0088, they were all still in the superlubricity level ($\mu < 0.01$). It also implied the upper limit of contact pressure (600 MPa) for achieving superlubricity with IL + $GO_{(aq)}$.

According to the analysis, a schematic lubrication model was proposed in Fig. 10.15. The lubricating process was considered to be divided into two parts—a running-in period and a superlubricity period. During the running-in period, a flat contact surface was formed, which gradually caused an obvious decrease of contact pressure occurred between sliding solid asperities. Meanwhile, the tribochemical reactions took place so that a thin composite boundary layer containing phosphates was formed on the contact surfaces. It can contribute to the surface protection and low wear but not enough to achieve superlubricity. Notably, the asperities between contact surfaces were also absorbed by GO nanosheets. The adsorption layer in Fig. 10.15B, which possessed extremely low shear stress between contact areas, could further

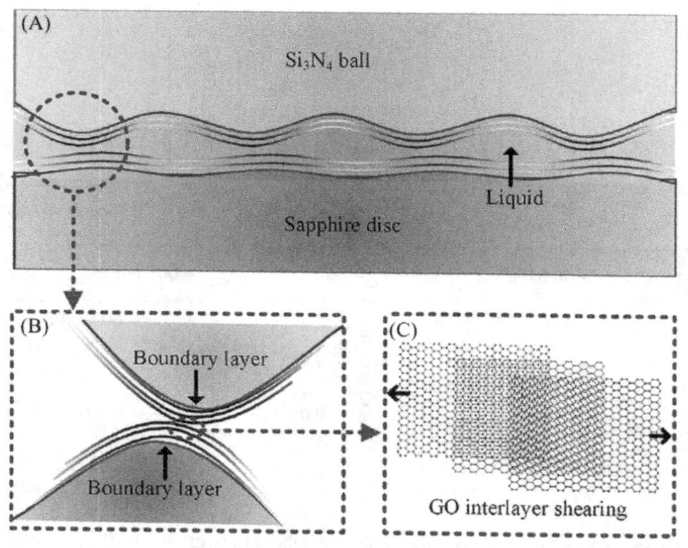

FIGURE 10.15 Superlubricity model for IL + $GO_{(aq)}$ at the macroscale under extreme pressure: (A) contact surfaces at low magnification; (B) the protection of asperity contact derived from the adsorption layer of GO nanosheets and boundary layer at high magnification; and (C) the shearing between GO interlayers. (*GO*: Graphene oxide; *IL*: ionic liquid). *Adapted from X. Ge, J. Li, H. Wang, C. Zhang, Y. Liu, J. Luo, Macroscale superlubricity under extreme pressure enabled by the combination of graphene-oxide nanosheets with ionic liquid, Carbon 151 (2019) 76–83. Copyright 2019, Elsevier Ltd.*

reduce the friction to a superlubricity level. Besides, the extremely low shear stress was also obtained between GO interlayers in Fig. 10.15C. Above all, the formation of boundary layer during the running-in period resulted in the low-wear condition to achieve extreme pressure state, and meanwhile the adsorption layer of GO nanosheets resulted in the achievement of superlubricity because of their extreme pressure property and low shear stress between GO interlayers. The combination of GO nanosheets and IL effectively improved the load-bearing capacity for liquid superlubricity to a high level of 600 MPa, which was beneficial for the achievement of superlubricity under extreme pressures.

10.6 Conclusion

Based on all aforementioned researches, it can be concluded as follows. The "solid superlubricity," achieved by homogenous or heterogeneous 2D nanomaterials due to their unique crystal structure, is usually limited to the micro-nano-scale and sometimes requires a specific atmosphere. Suitable 2D nanomaterials as additives, which are uniformly dispersed in lubricant, are able to prevent severe wear condition on the solid surfaces and obviously improve the tribological properties. Based on the previously studies of "liquid superlubricity," it was clearly found that the tribochemical reaction and hydrodynamic effect took key effects. However there are still many problems (e.g., low load-bearing pressure, dependence on hydrogen ions, and relatively long running-in process) that need to be further explored and solved. Here, various 2D nanomaterials, such as LDH and GO, were prepared and dispersed in water-based lubricants for optimizing the applied conditions of superlubricity. Until now, the "liquid superlubricity" has been increasingly controllable and adjustable. The load-bearing capacity has been effectively improved to 600 MPa; the running-in period from initially high contact pressure to the realization of superlubricity was controlled within 1000 s; the corrosion problem caused by the pH value of the lubricant has been effectively suppressed.

These work successfully combined 2D solid materials with water-based lubricants in the field of superlubricity. Their synergistic effect was deeply analyzed and the lubricating mechanism was explored. It holds great potential for enabling liquid superlubricity in industrial applications in the future.

Acknowledgments

This work was financially supported by the National Natural Science Foundation of China (51905294, 51527901, and 51875303), China Postdoc Innovation Talent Support Program (BX20180168), and the China Postdoctoral Science Foundation (2019M650654).

References

[1] S.S. Perry, W.T. Tysoe, Frontiers of fundamental tribological research, Tribol. Lett. 19 (3) (2005) 151–161.

[2] S. Wen, P. Huang, Principles of Tribology, John Wiley & Sons, 2012.

[3] C. Matta, L. Joly-Pottuz, M.I. De Barros Bouchet, J.M. Martin, M. Kano, Q. Zhang, et al., Superlubricity and tribochemistry of polyhydric alcohols, Phys. Rev. B 78 (8) (2008) 085436.

[4] K. Holmberg, A. Erdemir, Influence of tribology on global energy consumption, costs and emissions, Friction 5 (3SI) (2017) 263–284.

[5] C. Tan, X. Cao, X. Wu, Q. He, J. Yang, X. Zhang, et al., Recent advances in ultrathin two-dimensional nanomaterials, Chem. Rev. 117 (9) (2017) 6225–6331.

[6] D. Dietzel, U.D. Schwarz, A. Schirmeisen, Nanotribological studies using nanoparticle manipulation: principles and application to structural lubricity, Friction 2 (2) (2014) 114–139.

[7] T. Kuila, S. Bose, A.K. Mishra, P. Khanra, N.H. Kim, J.H. Lee, Chemical functionalization of graphene and its applications, Prog. Mater. Sci. 57 (7) (2012) 1061–1105.

[8] D. Berman, A. Erdemir, A.V. Sumant, Approaches for achieving superlubricity in two-dimensional materials, ACS Nano 12 (3) (2018) 2122–2137.

[9] H. Li, J. Wu, Z. Yin, H. Zhang, Preparation and applications of mechanically exfoliated single-layer and multilayer MoS_2 and WSe_2 nanosheets, Acc. Chem. Res. 47 (4) (2014) 1067–1075.

[10] Y. Lin, T.V. Williams, J.W. Connell, Soluble, exfoliated hexagonal boron nitride nanosheets, J. Phys. Chem. Lett. 1 (1) (2009) 277–283.
[11] Y. Peng, Y. Li, Y. Ban, H. Jin, W. Jiao, X. Liu, et al., Metal-organic framework nanosheets as building blocks for molecular sieving membranes, Science 346 (6215) (2014) 1356–1359.
[12] J.W. Colson, A.R. Woll, A. Mukherjee, M.P. Levendorf, E.L. Spitler, V.B. Shields, et al., Oriented 2D covalent organic framework thin films on single-layer graphene, Science 332 (6026) (2011) 228–231.
[13] J. Zhang, Y. Chen, X. Wang, Two-dimensional covalent carbon nitride nanosheets: synthesis, functionalization, and applications, Energ. Environ. Sci. 8 (11) (2015) 3092–3108.
[14] C. Zhi, Y. Bando, C. Tang, H. Kuwahara, D. Golberg, Large-scale fabrication of boron nitride nanosheets and their utilization in polymeric composites with improved thermal and mechanical properties, Adv. Mater. 21 (28) (2009) 2889–2893.
[15] M. Dienwiebel, G.S. Verhoeven, N. Pradeep, J.W.M. Frenken, J.A. Heimberg, H.W. Zandbergen, Superlubricity of graphite, Phys. Rev. Lett. 92 (12) (2004) 126101.
[16] Z. Liu, J. Yang, F. Grey, J.Z. Liu, Y. Liu, Y. Wang, et al., Observation of microscale superlubricity in graphite, Phys. Rev. Lett. 108 (20) (2012) 205503.
[17] D. Berman, S.A. Deshmukh, S.K.R.S. Sankaranarayanan, A. Erdemir, A.V. Sumant, Macroscale superlubricity enabled by graphene nanoscroll formation, Science 348 (6239) (2015) 1118–1122.
[18] H. Li, J. Wang, S. Gao, Q. Chen, L. Peng, K. Liu, et al., Superlubricity between MoS_2 monolayers, Adv. Mater. 29 (27) (2017) 1701474.
[19] S. Liu, H. Wang, Q. Xu, T. Ma, G. Yu, C. Zhang, et al., Robust microscale superlubricity under high contact pressure enabled by graphene-coated microsphere, Nat. Commun. 8 (2017) 14029.
[20] Y. Liu, A. Song, Z. Xu, R. Zong, J. Zhang, W. Yang, et al., Interlayer friction and superlubricity in single-crystalline contact enabled by two-dimensional flake-wrapped atomic force microscope tips, ACS Nano 12 (8) (2018) 7638–7646.
[21] Q. Zheng, Z. Liu, Experimental advances in superlubricity, Friction 2 (2) (2014) 182–192.
[22] M. Hirano, Atomistics of superlubricity, Friction 2 (2) (2014) 95–105.
[23] E. Meyer, E. Gnecco, Superlubricity on the nanometer scale, Friction 2 (2) (2014) 106–113.
[24] J. Xu, K. Kato, Formation of tribochemical layer of ceramics sliding in water and its role for low friction, Wear 245 (1) (2000) 61–75.
[25] U. Raviv, S. Giasson, N. Kampf, J. Gohy, R. Jérôme, J. Klein, Lubrication by charged polymers, Nature 425 (6954) (2003) 163–165.
[26] J. Klein, Hydration lubrication, Friction 1 (1) (2013) 1–23.
[27] J. Li, C. Zhang, J. Luo, Superlubricity behavior with phosphoric acid-water network induced by rubbing, Langmuir 27 (15) (2011) 9413–9417.
[28] L. Sun, C. Zhang, J. Li, Y. Liu, J. Luo, Superlubricity of Si_3N_4 sliding against SiO_2 under linear contact conditions in phosphoric acid solutions, Sci. China Technol. Sci. 56 (7) (2013) 1678–1684.
[29] M. Deng, C. Zhang, J. Li, L. Ma, J. Luo, Hydrodynamic effect on the superlubricity of phosphoric acid between ceramic and sapphire, Friction 2 (2) (2014) 173–181.
[30] J. Li, C. Zhang, M. Deng, J. Luo, Investigations of the superlubricity of sapphire against ruby under phosphoric acid lubrication, Friction 2 (2) (2014) 164–172.

[31] J. Li, C. Zhang, L. Ma, Y. Liu, J. Luo, Superlubricity achieved with mixtures of acids and glycerol, Langmuir 29 (1) (2013) 271–275.

[32] J. Li, C. Zhang, J. Luo, Superlubricity achieved with mixtures of polyhydroxy alcohols and acids, Langmuir 29 (17) (2013) 5239–5245.

[33] X. Ge, J. Li, C. Zhang, Y. Liu, J. Luo, Superlubricity and antiwear properties of in situ-formed ionic liquids at ceramic interfaces induced by tribochemical reactions, ACS Appl. Mater. Interfaces 11 (6) (2019) 6568–6574.

[34] J. Li, C. Zhang, L. Sun, X. Lu, J. Luo, Tribochemistry and superlubricity induced by hydrogen ions, Langmuir 28 (45) (2012) 15816–15823.

[35] X. Ge, J. Li, H. Wang, C. Zhang, Y. Liu, J. Luo, Macroscale superlubricity under extreme pressure enabled by the combination of graphene-oxide nanosheets with ionic liquid, Carbon 151 (2019) 76–83.

[36] Z. Chen, X. Liu, Y. Liu, S. Gunsel, J. Luo, Ultrathin MoS_2 nanosheets with superior extreme pressure property as boundary lubricants, Sci. Rep. 5 (2015) 12869.

[37] Z. Chen, Y. Liu, S. Gunsel, J. Luo, Mechanism of antiwear property under high pressure of synthetic oil-soluble ultrathin MoS_2 sheets as lubricant additives, Langmuir 34 (4) (2018) 1635–1644.

[38] H. Wang, Y. Liu, W. Liu, R. Wang, J. Wen, H. Sheng, et al., Tribological behavior of NiAl-layered double hydroxide nanoplatelets as oil-based lubricant additives, ACS Appl. Mater. Interfaces 9 (36) (2017) 30891–30899.

[39] H. Wang, Y. Liu, Z. Chen, B. Wu, S. Xu, J. Luo, Layered double hydroxide nanoplatelets with excellent tribological properties under high contact pressure as water-based lubricant additives, Sci. Rep. 6 (1) (2016) 22748.

[40] H. Kinoshita, Y. Nishina, A.A. Alias, M. Fujii, Tribological properties of monolayer graphene oxide sheets as water-based lubricant additives, Carbon 66 (2014) 720–723.

[41] H. Song, N. Li, Frictional behavior of oxide graphene nanosheets as water-base lubricant additive, Appl. Phys. A 105 (4) (2011) 827–832.

[42] X. Ge, J. Li, R. Luo, C. Zhang, J. Luo, Macroscale superlubricity enabled by the synergy effect of graphene-oxide nanoflakes and ethanediol, ACS Appl. Mater. Interfaces 10 (47) (2018) 40863–40870.

[43] H. Wang, Y. Liu, J. Li, J. Luo, Investigation of superlubricity achieved by polyalkylene glycol aqueous solutions, Adv. Mater. Interfaces 3 (19) (2016) 1600531.

[44] H. Wang, Y. Liu, W. Liu, Y. Liu, K. Wang, J. Li, et al., Superlubricity of polyalkylene glycol aqueous solutions enabled by ultrathin layered double hydroxide nanosheets, ACS Appl. Mater. Interfaces 11 (22) (2019) 20249–20256.

Chapter 11

Superlubricity of carbon nitride coatings in inert gas environments

Koshi Adachi
Department of Mechanical Systems Engineering, Tohoku University, Sendai, Japan

11.1 Introduction

Carbon coatings such as diamond-like carbon (DLC) and carbon nitride (CN_x) have promising and attractive properties such as low friction and wear, which make them considerably appropriate for use in demanding mechanical systems. In the case of CN_x coatings, a superlow friction coefficient on the order of 0.001 is provided for sliding against both silicon nitride (Si_3N_4) and itself in an atmosphere of an inert gas such as nitrogen (N_2) [1–4]. The beneficial effect of an inert gas on reducing the friction of the system with CN_x coatings is significantly enhanced by selecting a counter material, atmospheric humidity, oxygen concentration, and especially running-in process [1].

The various aspects and comprehensive overviews of the superlow friction of systems with CN_x coatings are described in this chapter for further discussion of both the low-friction mechanism and the low-friction control.

11.2 Methods for coatings and experiments

11.2.1 CN_x coatings

The CN_x coatings introduced in this chapter were fabricated on a ball-and-a-disk substrate of Si_3N_4 by using an ion-beam-assisted deposition system, as shown in Fig. 11.1. The coatings were deposited on the substrates from a carbon target of 99.999% purity, along with the bombardment and mixing of carbon with the nitrogen ions that were simultaneously generated using an ion-beam gun.

Superlubricity. DOI: https://doi.org/10.1016/B978-0-444-64313-1.00011-9

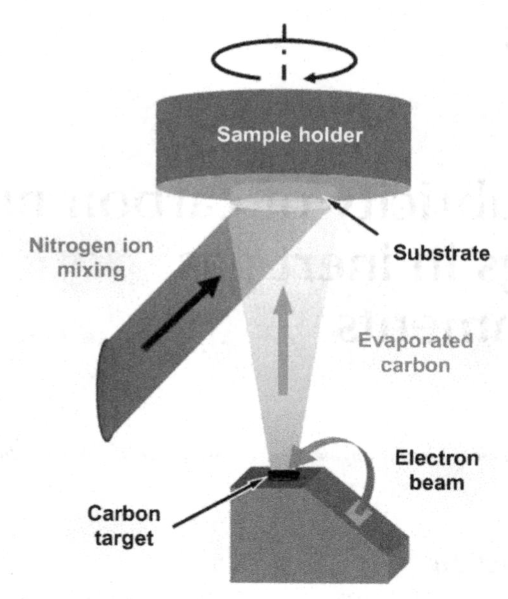

FIGURE 11.1 A schematic image of coating apparatus of an ion-beam-assisted deposition system with nitrogen ion gun for mixing, electron beam source for evaporation, carbon target, and rotary disk substrate for the coating.

The substrates were sequentially cleaned in an ultrasonic bath using acetone, ethanol, and deionized water for 20 min each before placing them into the deposition chamber. They were further sputter-cleaned via 5-min bombardment of nitrogen ions before starting deposition, to remove the native oxides and other adsorbed species from their surfaces.

The thickness of the coatings on both the ball and disk was approximately 400 nm. The atomic concentration of nitrogen in the CN_x coatings was approximately 10–12 at.%, as confirmed using X-ray photoelectron spectroscopy analysis. And a microstructure of the CN_x coatings was an amorphous. The surface roughness of the disk after coating was similar to that of the original disk (e.g., from 11.5 to 12.6 nm). The details of the coating method and fundamental properties of CN_x coatings are described in the first version of the book [1].

11.2.2 Experimental apparatus and environmental gases

A customized ball-on-disk friction apparatus with two types of gas-supply methods, namely gas-stream (see Fig. 11.2A) and gas-atmosphere (see Fig. 11.2B) methods, were used to perform the sliding test in various gas environments [2,3]. In the gas-stream method, specifically, the gas

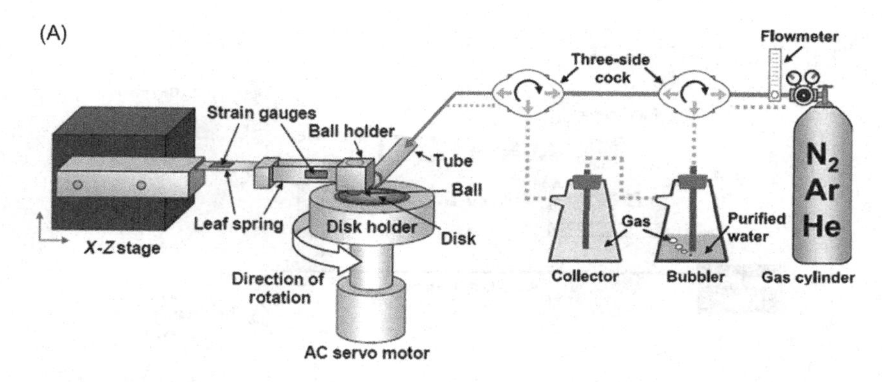

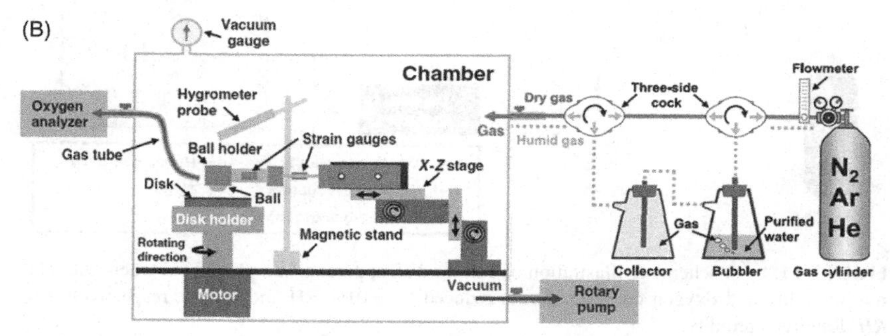

FIGURE 11.2 A schematic diagram of ball-on-disk friction apparatus with (A) gas-blowing system and (B) gas-atmosphere control system of mixture gas.

environment around the sliding interface was controlled by directly blowing the gas to the sliding interface in air using a tube with a diameter of 4 mm.

To conduct the friction tests in the pure N_2 gas atmosphere, the ball-on-disk friction apparatus was installed inside a glove box, in which the relative humidity and oxygen concentration were controlled within the ranges of 0.01%–40% RH and $1-1\times10^5$ ppm, respectively. A schematic image of the apparatus is shown in Fig. 11.3. Before performing the friction tests, the glove box was evacuated to have less than 5 kPa of atmospheric pressure by using a rotary pump, and it was purged using N_2 gas for one to three cycles to obtain a relative humidity and oxygen concentration of less than 3% RH and 10,000 ppm, respectively. In addition, the residual water and oxygen presented in the N_2 gas environment inside the glove box were removed using molecular sieves and a high-temperature copper catalyst, respectively. Subsequently, the relative humidity and oxygen concentration were adjusted as per requirements by using a nitrogen gas bubbling system and introducing oxygen gas as drawn in Fig. 11.3.

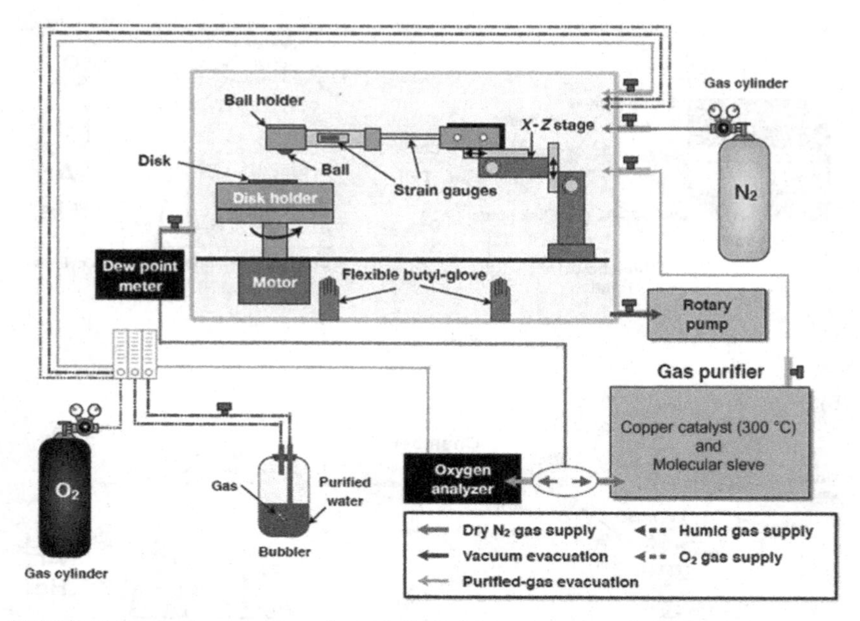

FIGURE 11.3 A schematic illustration of ball-on-disk apparatus in a glove box, where the relative humidity and oxygen concentration is reduced by 0.01% RH and 1 ppm, respectively [5]. *RH*, Relative humidity.

11.3 Superlubricity of CN_x coatings sliding against Si_3N_4 ball or CN_x-coated Si_3N_4 ball in inert gas environment

Fig. 11.4A and B show representative friction properties of Si_3N_4/CN_x and CN_x/CN_x at 250 rpm with a sliding velocity of 0.2–0.27 m/s under 400 mN corresponding to an initial mean contact pressure of 340–360 MPa for various gas streams such as argon (Ar), helium (He), N_2, carbon dioxide (CO_2), dry air, oxygen (O_2), and ambient air with 30% RH. The friction system with CN_x coatings exhibits high and unstable friction in reactive gas streams such as the streams of O_2, dry air, and ambient air. However it generates a low (friction coefficient: 0.007–0.05) and stable friction in inert gas environments, that is, Ar, He, and N_2, irrespectively of the ball materials used. The potential of inert gases of Ar, He, and N_2, for the effective reduction of friction is also confirmed via friction tests (see Fig. 11.2B) in a constant gas atmosphere, as shown in Fig. 11.5A and B.

In the case of high-friction conditions observed in Figs. 11.4 and 11.5, the friction slowly increases after the short running-in process and finally shows the steady state with a high fluctuation. However in inert gas environments, the friction gradually decreases during the running-in process, and reaches the steady state with a low fluctuation of low friction ($\mu < 0.05$).

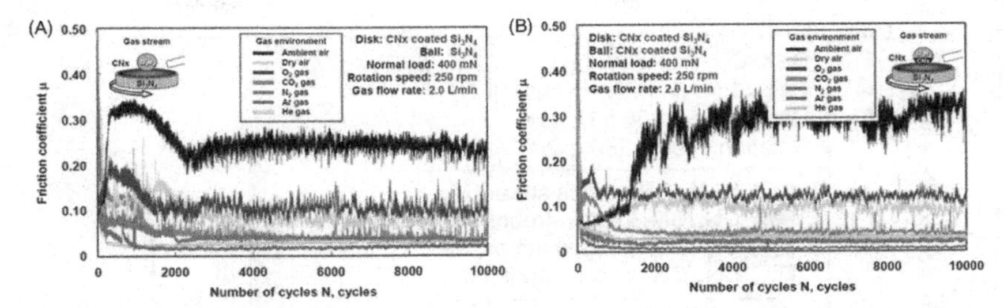

FIGURE 11.4 Friction properties of (A) Si_3N_4/CN_x and (B) CN_x/CN_x at 250 rpm under 400 mN in the various gas streams such as Ar, He, N_2, CO_2, dry air, O_2, and ambient air with 30% RH. *RH*, Relative humidity.

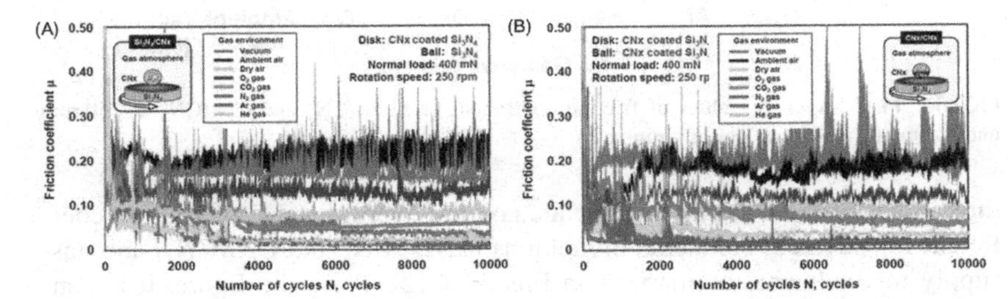

FIGURE 11.5 Friction properties of (A) Si_3N_4/CN_x and (B) CN_x/CN_x at 250 rpm under 400 mN in the various gas atmosphere such as Ar, He, N_2, CO_2, dry air, O_2, ambient air with 30% RH, and vacuum condition. *RH*, Relative humidity.

The average value of stable friction coefficient for two types of contact combination (Si_3N_4/CN_x and CN_x/CN_x) in various gas environments using two types of gas-supply methods (gas stream and gas atmosphere) is presented in Fig. 11.6. From the figure, it is evident that the friction of the system with CN_x coatings is closely related to the gas species around sliding interface irrespective of the mating material and gas-supply method. Importantly, friction coefficients of less than 0.05 are always achieved in the three inert gas environments.

Fig. 11.7A and B show friction properties and average values of friction coefficient at the steady state of CN_x/CN_x in the gas streams that contain both He and O_2 with different concentrations. By introducing O_2 to the He gas environment, the friction coefficient gradually increases from a low value less than 0.05 in the He gas environment to larger than 0.10 in the O_2 gas environment. Particularly, in the He gas environment, the friction coefficient slowly decreases and reaches to a value below 0.05 after running-in at the beginning of sliding. However in oxygen-containing He gas environments, no running in is observed and the friction increases from the beginning of sliding. The correlation between the oxygen concentrations in the

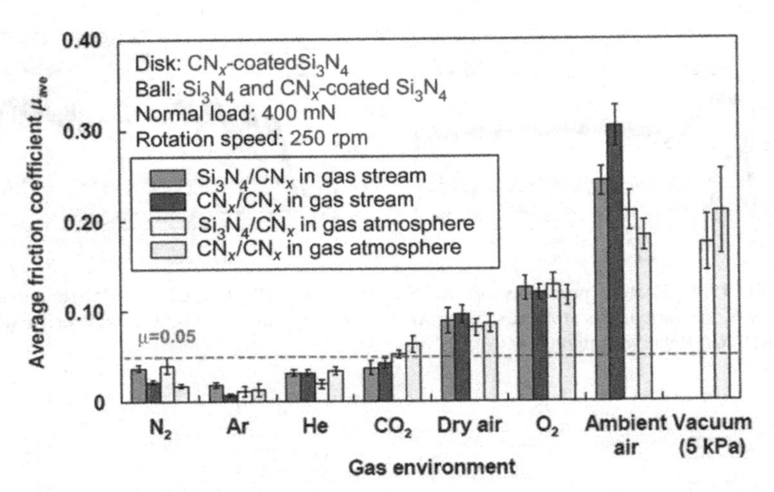

FIGURE 11.6 Average values of friction coefficient of Si_3N_4/CN_x and CN_x/CN_x at 250 rpm under 400 mN in various gas environments.

three types of inert gas environments, and the average value of friction coefficient for different contact combinations (Si_3N_4/CN_x and CN_x/CN_x) and gas-supply methods are summarized in Fig. 11.8 [2]. From the figure, it is confirmed that the friction coefficient sharply increases from the low value of 0.007–0.02 obtained in the three types of inert gas environments to approximately 0.10 upon increasing the oxygen concentration from 0 to 20 vol.%. Thereafter, it slowly increases to approximately 0.12 upon further increasing the oxygen concentration. From Fig. 11.8, it is evident that an inert gas environment with less oxygen concentration results in low friction of Si_3N_4/CN_x and CN_x/CN_x. In addition, the increase in the oxygen concentration might change the tribochemical reaction at the contact interface of CN_x and Si_3N_4. Therefore tribochemical products at the contact interface so-called tribofilm or tribolayer are introduced in the next section to understand low-friction mechanism and low-friction control.

11.4 Low-friction interface in tribological system with CN_x coatings

11.4.1 Contact interface for low friction in tribological system with CN_x coatings

The optical images of the wear scars on the ball in the friction conditions of Si_3N_4/CN_x and CN_x/CN_x for different gas streams are shown in Figs. 11.9 and 11.10. The CN_x coatings on the ball had worn out during the initial running-in process in almost all the experiments [1]. Therefore the contact

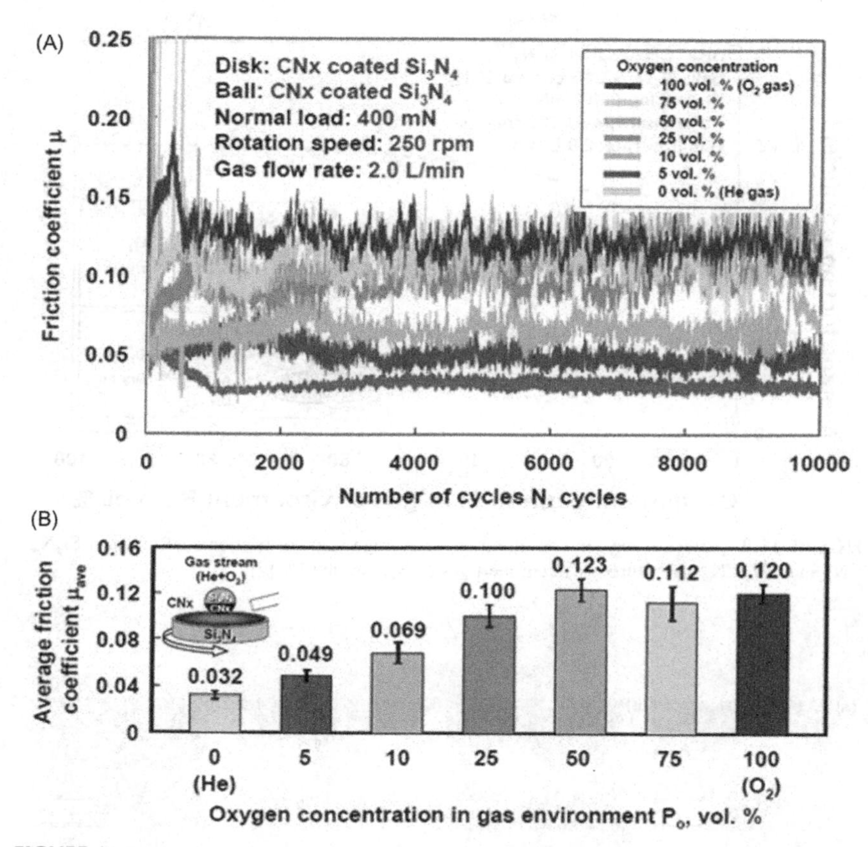

FIGURE 11.7 (A) Friction properties and (B) average values of friction coefficient at the steady state of the CN_x/CN_x in mixture gas streams of He and O_2 with different oxygen concentration.

area of the CN_x-coated Si_3N_4 ball was considered as the worn surface of the Si_3N_4 substrate, similar to that of the Si_3N_4/CN_x experiments.

The wear scars on the ball after performing the friction tests of Si_3N_4/CN_x and CN_x/CN_x in different gas environments with both the types of gas-supply methods are classified into the following three types, according to the condition and amount of visible tribofilm accumulated inside the contact area: no tribofilm (Type I), partial tribofilm (Type II), and almost full tribofilm (Type III), as shown in Fig. 11.11 [3]. The correlation between the average value of friction coefficient and wear scar types defined mentioned before is shown in Fig. 11.12 [3]. Here, low friction is defined as a friction coefficient of less than 0.05, which is easily realized in inert gas environments.

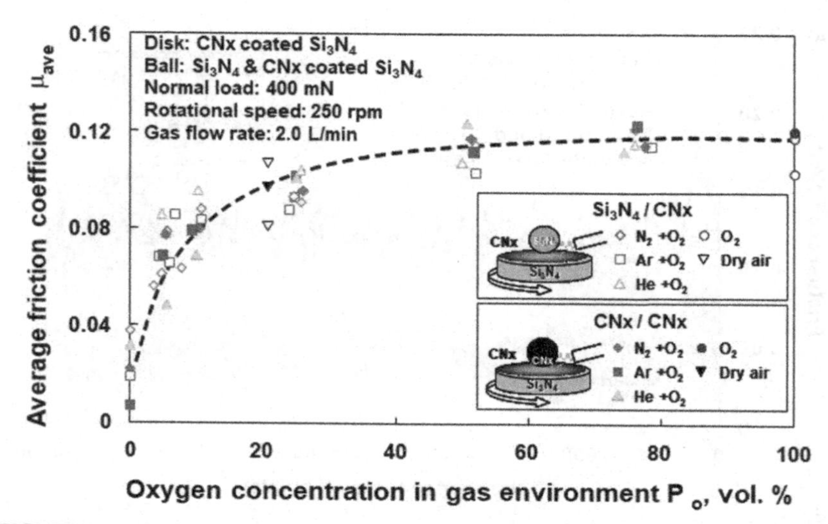

FIGURE 11.8 Effect of oxygen concentration on average value of friction coefficient of Si_3N_4/CN_x and CN_x/CN_x under three different inert gas of N_2, Ar, and He [2].

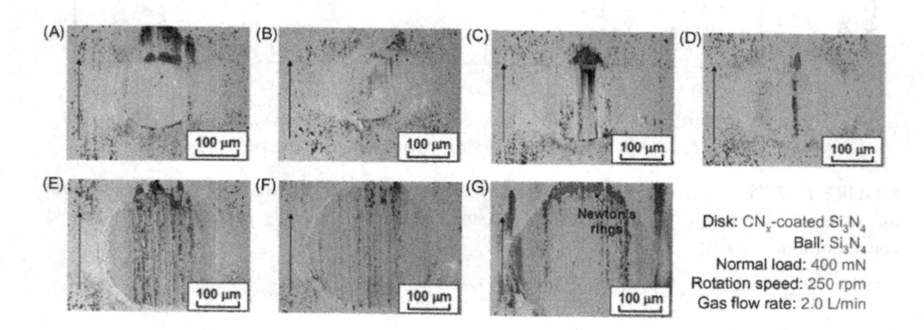

FIGURE 11.9 Optical images of the wear scars on the ball in friction condition of Si_3N_4/CN_x in different gas streams: (A) Ar gas, (B) He gas, (C) N_2 gas, (D) CO_2 gas, (E) dry air, (F) O_2 gas, (G) ambient air.

In the case existence of tribofilm on the wear scar of the ball as Type II and Type III, both low and high friction are generated. It is therefore difficult to explain the low friction on the basis of the existence of the tribofilm. The amount, composition, and structure of the tribofilm must be discussed for understanding the low-friction mechanism of tribological systems with CN_x coatings.

Fig. 11.13A and B [3] show the Raman analyses of the Type II wear scars of the ball, showing low friction ($\mu = 0.013$) generated in the Ar gas

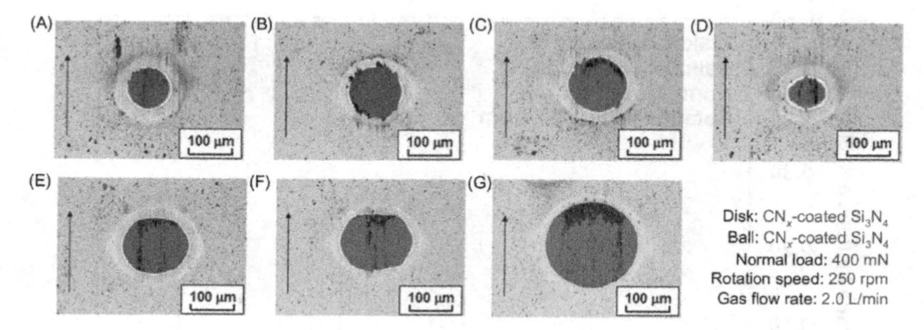

FIGURE 11.10 Optical images of the wear scars on the ball in friction condition of CN_x/CN_x in different gas streams: (A) Ar gas, (B) N_2 gas, (C) He gas, (D) CO_2 gas, (E) dry air, (F) O_2 gas, (G) ambient air.

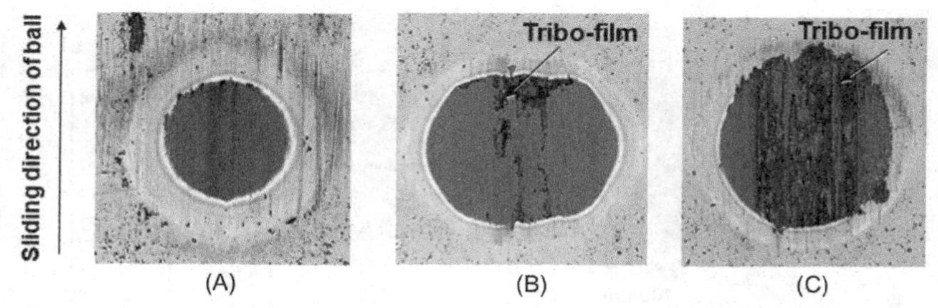

FIGURE 11.11 Three types of wear scars on the ball, which are observed in Si_3N_4/CN_x and CN_x/CN_x in different environments, according to the condition and amount of visible tribofilm accumulated inside the contact area [3]: (A) no tribofilm (Type I), (B) partial tribofilm (Type II), and (C) almost full tribofilm (Type III).

environment and relatively high friction ($\mu = 0.116$) in the O_2 gas environment. The Raman spectrum of the initial CN_x coatings exhibits a broad band with two shoulders, which correspond to the D and G bands with peaks at approximately 1350 and 1550 cm^{-1}, respectively [1,4]. The D and G bands of the spectrum of the tribofilm clearly observed in the case of low friction (see Fig. 11.13A) evolve from shoulders to two individual peaks. The shift of the G-band peak from 1550 to 1596 cm^{-1} indicates a possibility of graphitic structure formation so-called graphitization [6] on the top surface of the tribofilm. However the Raman spectrum of the tribofilm observed in the case of relatively high friction (see Fig. 11.13B) is similar to that of the polymer-like carbon [7,8]. Therefore it is considered that the main structure of the carbon in the tribofilm is polymer like. Notably, the existence of graphitization on the top surface of tribofilm has been reported and is always

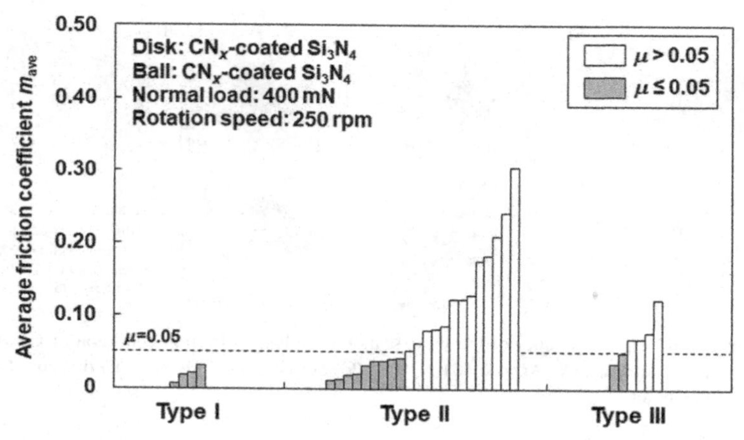

FIGURE 11.12 Correlation between the average value of friction coefficient and wear scar types shown in Fig. 11.11 [3].

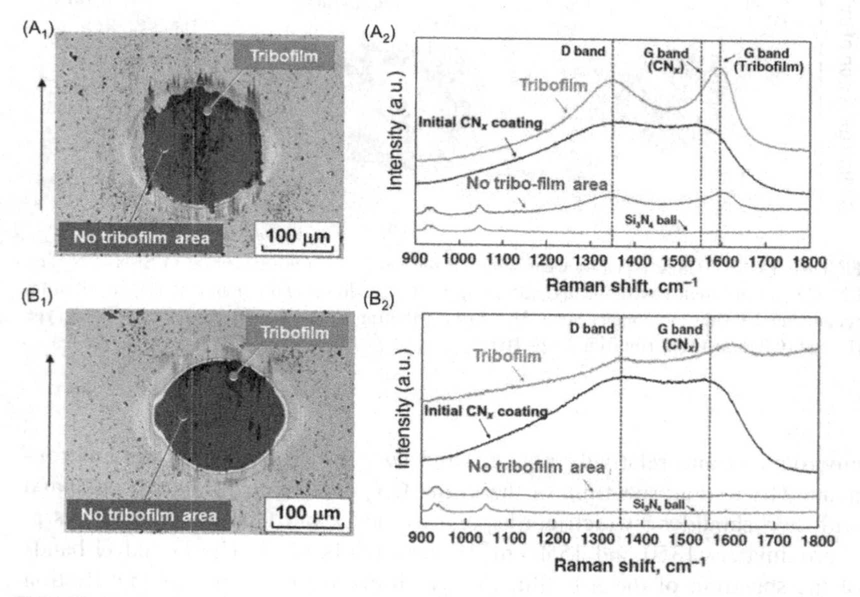

FIGURE 11.13 Raman analyses of the Type II wear scars of the ball, showing (A) low friction ($\mu = 0.013$) generated in the Ar gas environment and (B) relatively high friction ($\mu = 0.116$) in the O_2 gas environment [3].

claimed to be the key to achieve low friction of carbon-related solid lubrication coatings [1,9,10].

Furthermore, the Raman spectrum of the no tribofilm area in the wear scar for the case of high friction (see Fig. 11.13B) is identical to that of the

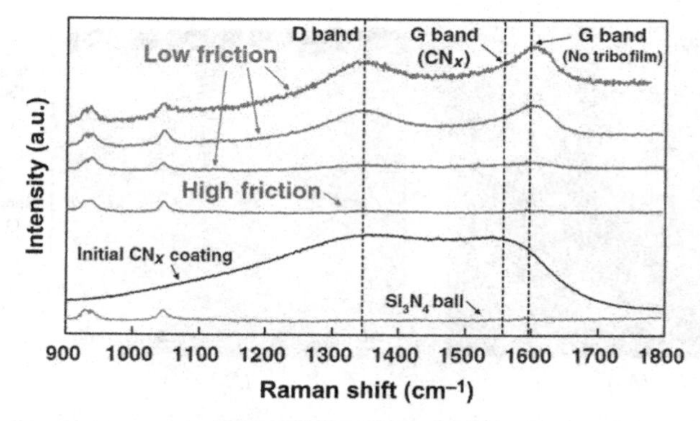

FIGURE 11.14 Raman spectra of the no tribofilm area on the wear scar of the ball after sliding of CN_x/CN_x in various gas conditions [3].

Si_3N_4 ball. However graphitization is observed at no tribofilm area on the Si_3N_4 substrate for the case of low-friction condition (see Fig. 11.13A). Fig. 11.14 [3] shows several Raman spectrum at no tribofilm area on the worn ball surfaces for both low- and high-friction conditions. It is evident from the figure that a carbon layer, which is invisible to the naked eyes, is formed on the worn Si_3N_4 substrate surface with no tribofilm when low friction is generated due to the friction of Si_3N_4/CN_x and CN_x/CN_x in an inert gas environment.

A significant finding is that stable and low-friction values of systems with CN_x coatings are always achieved in the case of Type I wear scar (see Fig. 11.12), where invisible carbon layer, other than the tribofilm is formed as the top surface of the wear scar. The formation of the aforementioned carbon layer (hereafter called tribolayer) on the wear scar is more effective than that of tribofilm, and it might be the key to obtain stable and low friction of system with CN_x coatings.

11.4.2 Characterization of tribofilm and tribolayer formed on wear scars of the ball

The cross-sectional transmission electron microscopy images of the tribofilm and tribolayer formed on the wear scar of CN_x/CN_x after performing the friction test in a He gas atmosphere are shown in Fig. 11.15A and B [3], respectively. The corresponding electron energy loss spectroscopy analyses of the tribofilm and tribolayer are also shown in Figs. 11.16 and 11.17 [3], respectively. From Fig. 11.15A_1 and A_2, it is evident that the tribofilm formed on the wear scar comprises two layers with thicknesses of approximately 30 and 100 nm, respectively. Specifically, the top layer, which comprises amorphous carbon (a-C), mainly consist of carbon sp^2 bonding, as shown in Fig. 11.16D, which corresponds to the existence of partial graphitization

FIGURE 11.15 Cross-sectional TEM images of (A) tribofilm and (B) tribolayer formed on the wear scar of CN_x/CN_x after performing the friction test in He gas environment [3]. *TEM*, Transmission electron microscopy.

confirmed by the Raman analysis. The second layer is confirmed to be a SiO_2 layer, as shown in Fig. 11.16E. The formation of the SiO_2 layer is attributed to the oxidation of the Si_3N_4 substrate of the ball. The SiO_2 layer is considered as the initial surface and/or as a result of tribochemical reaction between Si_3N_4 substrate and residual oxygen gas in air. It is inferred that the tribofilm in oxygen-containing gas environment mainly comprises of SiO_2 layer, without the formation of an a-C layer on the top surface [2].

However, from Fig. 11.15B_1 and B_2, the existence of a tribolayer, which further comprises two layers with thicknesses of approximately 10 and 20 nm, respectively, is confirmed. Specifically, the first top layer is confirmed to be made of a-C mainly comprising carbon sp^2 bonding, as evident from Fig. 11.17D. On the other hand, the second layer is a combination carbon (C)-silicon (Si)-oxygen (O) layer, as shown in Fig. 11.17D and E.

Accordingly, an invisible thin carbon-based tribolayer with a thickness of tens of nanometers is observed as the subsurface of the wear scar on the ball, where the conventional tribofilm is not observed by naked eyes as well as

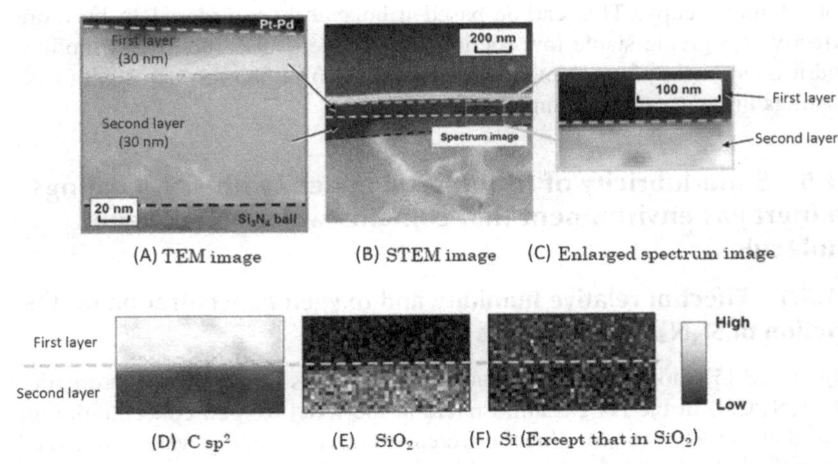

FIGURE 11.16 Tribofilm formed on the wear scar of CN_x/CN_x after performing the friction test in He gas environment. (A) TEM image, (B) STEM image, (C) enlarged STEM image from the rectangle-enclosed area of (B) and corresponding EELS maps of (D) carbon sp^2, (E) SiO_2, and (F) Si except that in SiO_2 [3]. *EELS*, Electron energy loss spectroscopy; *STEM*, scanning transmission electron microscopy; *TEM*, transmission electron microscopy.

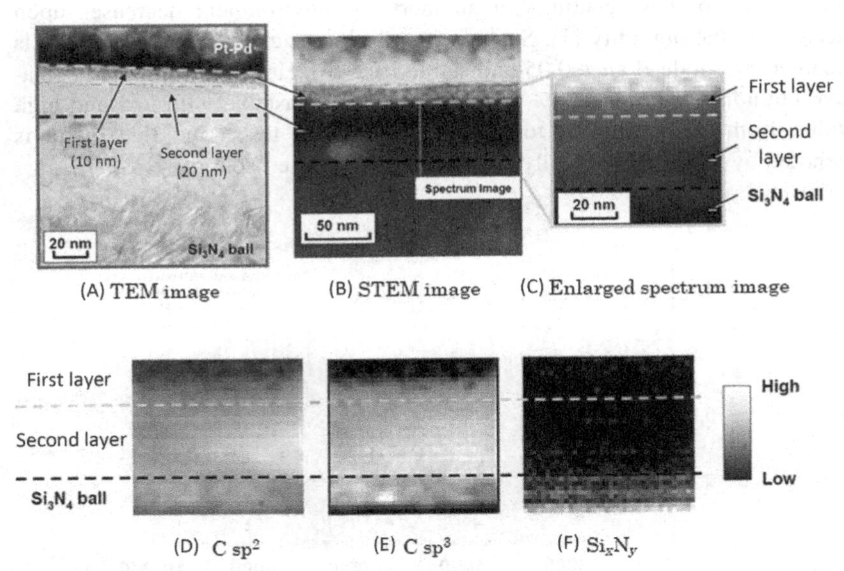

FIGURE 11.17 Tribolayer formed on the wear scar of CN_x/CN_x after performing the friction test in He gas environment. (A) TEM image, (B) STEM image, (C) enlarged STEM image from the rectangle-enclosed area of (B) and corresponding EELS maps of (D) carbon sp^2, (E) carbon sp^3 (in SiC or CN_x), and (F) Si_xN_y [3]. *EELS, Electron energy loss spectroscopy; STEM*, scanning transmission electron microscopy; *TEM*, transmission electron microscopy.

optical microscopy. This carbon-based tribolayer is considered to be more effective for giving stable low friction than the general carbon-rich tribofilm, and it is promising for achieving stable and low friction of systems with CN_x coatings in inert gas environments.

11.5 Superlubricity of tribological system with CN_x coatings in inert gas environment that contains water and oxygen molecules

11.5.1 Effect of relative humidity and oxygen concentration on the friction of Si_3N_4/CN_x in N_2 gas atmosphere

Fig. 11.18 [5] shows the representative three modes of the friction properties of Si_3N_4/CN_x in the N_2 gas atmosphere at the fixed oxygen concentration of less than 1100 ppm, for three different relative humidity values obtained using the ball-on-disk friction apparatus mounted inside the glove box shown in Fig. 11.3. The friction coefficient is approximately 0.1–0.3 from the beginning to the end of the friction test with 9.70% RH. This friction mode is denoted by Mode I ($0.05 < \mu$). For the case wherein the relative humidity is 0.20% RH, the friction coefficient dropped to 0.024 after running in, and keep friction low value on the order of 0.01. These results clearly show that the friction of CN_x coatings in an inert gas environment decreases upon decreasing the humidity [1]. Such constantly low-friction-generation mode is denoted by Mode II ($\mu < 0.05$ and stable friction). However for a lower relative humidity value of 0.04% RH, the friction is unstable with low and high value from the beginning to the end of friction tests, and this mode is denoted by Mode III (partially $\mu < 0.05$ and unstable friction).

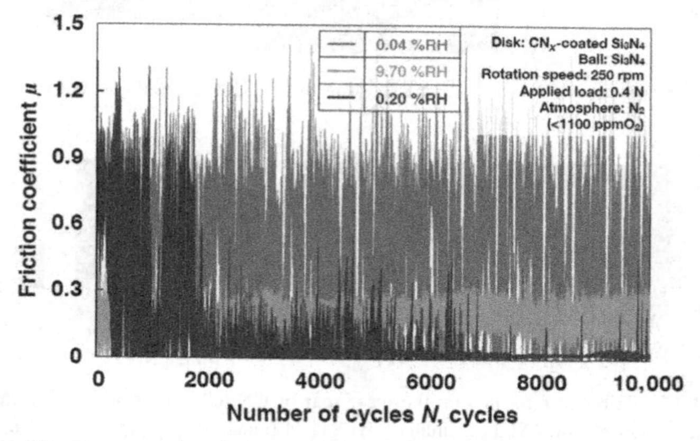

FIGURE 11.18 Representative three modes of friction properties of Si_3N_4/CN_x under N_2 gas atmosphere at a fixed oxygen concentration less than 1100 ppm for three different relative humidity values [5].

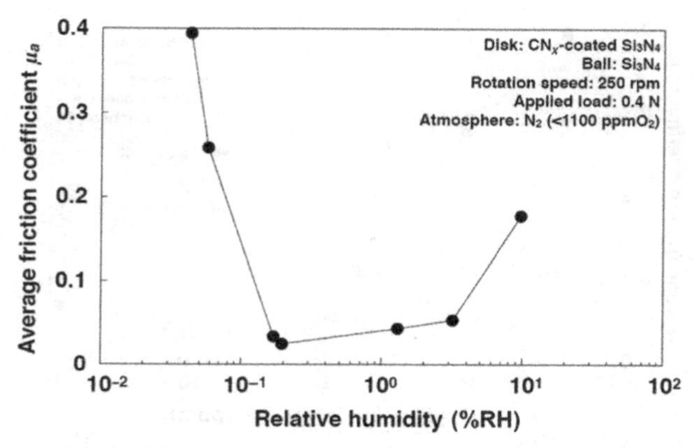

FIGURE 11.19 Effect of the relative humidity on average value of friction coefficient of Si_3N_4/CN_x in N_2 gas atmosphere with fixed oxygen concentration condition [5].

The effect of relative humidity on the average value of friction coefficient of Si_3N_4/CN_x in the N_2 gas atmosphere is summarized in Fig. 11.19 [5]. The low friction ($\mu < 0.05$, Mode II) of Si_3N_4/CN_x in the N_2 gas atmosphere is generated for the humidity range of 0.1%–1.0% RH. This clearly indicates the existence of an optimum relative humidity for obtaining low friction, and that a certain amount of water molecules is necessary to obtain low and stable friction.

The effect of oxygen concentration on the average value of the friction coefficient of the Si_3N_4/CN_x for three different ranges of relative humidity is shown in Fig. 11.20 [5]. Low friction ($\mu < 0.05$, Mode II) of Si_3N_4/CN_x in the N_2 gas atmosphere with 0.04%–0.06% RH is generated at the oxygen concentration of 5000 ppm. When the relative humidity is in the range of 0.1%–0.4% RH or 1.0%–4.0% RH, lower friction ($\mu < 0.05$, Mode II) of Si_3N_4/CN_x in the N_2 gas atmosphere is generated for wide range of oxygen concentration condition. There might be an optimum oxygen concentration that depends on the relative humidity, and a certain level of oxygen is critical to obtaining low ($\mu < 0.05$) and stable friction of Si_3N_4/CN_x in the N_2 gas atmosphere.

The friction modes of the Si_3N_4/CN_x as a function of relative humidity (0.01%–40% RH) and oxygen concentration ($1-1 \times 10^5$ ppm) are shown in Fig. 11.21 [5]. A certain amount of water molecules and oxygen concentration is required for Si_3N_4/CN_x to be in friction mode II with a stable friction coefficient of <0.05. The friction mode changed upon changing both the relative humidity and oxygen concentration, thereby suggesting that the interaction between water and oxygen molecules influences the friction of systems with CN_x coatings.

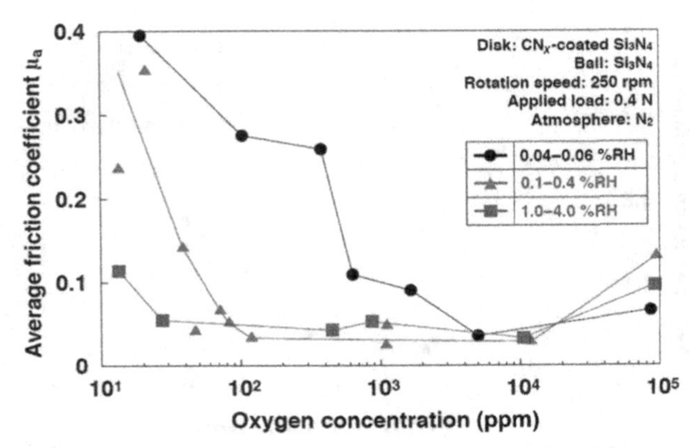

FIGURE 11.20 Effect of the oxygen concentration on average value of friction coefficient of Si_3N_4/CN_x in N_2 gas atmosphere with three different relative humidity values [5].

11.5.2 Tribochemical reaction of CN_x coatings with water molecules

The optical microscope images and Raman spectra of a representative wear scar of the Si_3N_4 ball in Mode I ($\mu = 0.18$; 10% RH), Mode II ($\mu = 0.02$; 0.3% RH), and Mode III ($\mu = 0.5$; 0.04% RH) are shown in Fig. 11.22A–C. The wear surface in Mode III (see Fig. 11.22C_1) is covered with the tribofilm formed by wear particles, and its Raman spectra are similar to those of the deposited CN_x coatings (see Fig. 11.22C_2). The tribofilm formed on the wear scar in Mode I appears to be made of polymer-like carbon (see Fig. 11.22A_2, Points 2 and 3), and no tribofilm area is identical to the Si_3N_4 surface (Fig. 11.22A_2, Point 1). In addition, oxidation products are also detected on the ball surface. However both tribofilm and no tribofilm areas in the wear scar in Mode II shown in Fig. 11.22B_2 indicate the existence of graphitized carbon, thereby resulting in low friction, as previously mentioned in Section 11.4. These results show that a certain amount of water molecules is effective for the graphitization of the tribofilm and tribolayer.

The worn surfaces of both the ball and disk after performing the friction test of CN_x/CN_x were analyzed by a time-of-flight secondary ion mass spectrometry (TOF-SIMS) to clarify both the chemical state of the wear scars and the significance of water molecules in the N_2 gas environment for friction reduction. Before performing the friction test, deuterated water [D_2O ($2H_2O$)] was introduced into the glove box shown in Fig. 11.3, and any water molecules (H_2O) in the N_2 gas environment were replaced with D_2O to distinguish the chemical adsorption of H_2O during the friction test from natural adsorption due to the atmosphere. Fig. 11.23A and B [5] shows the TOF-SIMS images of both the wear scar (Type I in Fig. 11.11) on the CN_x-coated Si_3N_4 disk and the wear scar on the CN_x-coated Si_3N_4 ball, when Mode II of

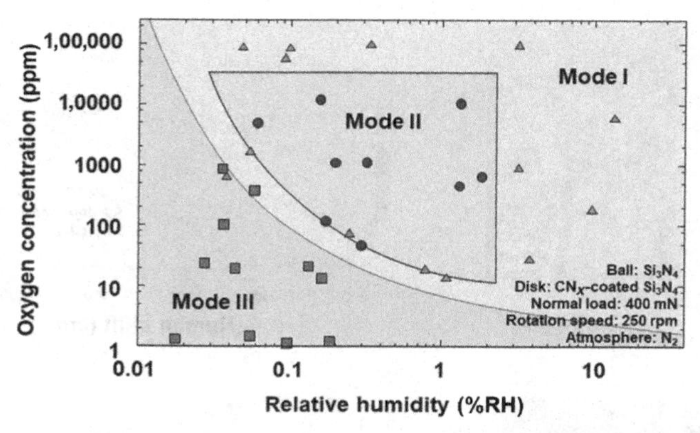

FIGURE 11.21 Distribution of friction modes for the Si_3N_4/CN_x in N_2 gas atmosphere as a function of relative humidity (0.01%–40% RH) and oxygen concentration ($1-1\times10^5$ ppm) [5].

friction property is exhibited in the N_2 gas environment ($\mu = 0.015$; O_2 concentration of 500 ppm; 5.4% RH). The TOF-SIMS images in Fig. 11.23A_1–A_3 show that $^2H^-$, O^2H^-, and CNO^- are adsorbed onto the wear scar of the CN_x coatings on the disk. The adsorption of $2H^-$ and O^2H^- is also clearly observed onto the wear scar on the CN_x-coated Si_3N_4 ball (see Fig. 11.23B_1 and B_2). These results suggest that the water molecules in the N_2 gas environment dissociated into H^- and OH^- because of the frictional energy applied at the contact interface, and both these ion species chemisorb onto the tribolayer of the graphitization formed on the wear scar.

11.5.3 Tribochemical reaction of CN_x coatings with oxygen molecules

The optical microscope images and Raman spectra of the representative wear scars of the Si_3N_4 ball in Modes II and III obtained using the oxygen concentrations of 1100 and 70 ppm, respectively, are shown in Fig. 11.24A and B [5]. Both the tribofilm and no tribofilm areas in the wear scar in Mode II, as shown in Fig. 11.24 (A_2, Points 1 and 2), indicate the existence of graphitized carbon, which result in low friction, as previously mentioned in Section 11.4. However the wear scar obtained using low oxygen concentration (Mode III) shows the carbon structure similar to that of deposited CN_x coating [see Fig. 11.24 (B_2, Point 3)]. The structure of carbon on the wear scar changed from that of the carbon in the deposited CN_x, in an environment with high oxygen concentration (1100 ppm) although carbon transfer occurred from the CN_x coating to Si_3N_4 ball irrespective of the oxygen concentration. These results show that a certain amount of oxygen concentration in the N_2 gas environment is effective for the graphitization of tribofilm and tribolayer.

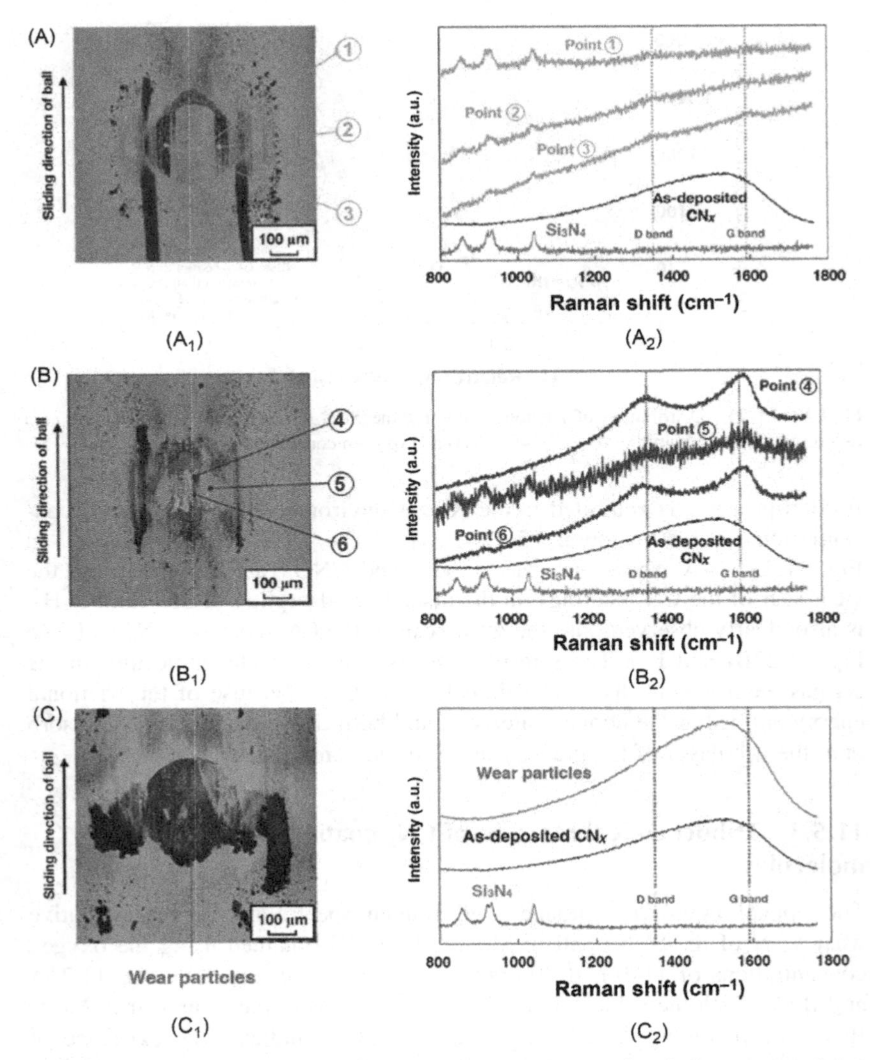

FIGURE 11.22 Optical microscope images and Raman spectra of typical wear scar of the Si_3N_4 ball in (A) Mode I ($\mu = 0.18$; 10% RH), (B) Mode II ($\mu = 0.02$; 0.3% RH), and (C) Mode III ($\mu = 0.5$; 0.04% RH) [5]. *RH*, Relative humidity.

Concerning the TOF-SIMS results shown in Fig. 11.23, which proved that oxygen in the N_2 gas environment chemisorbed onto the wear surface of the CN_x-coated Si_3N_4 disk (see Fig. 11.23A_3) and ball (Fig. 11.23B_3), both the adsorption of oxygen molecules onto the CN_x coatings and the tribochemical reaction result in the graphitization in the carbon on the wear scar of the Si_3N_4 ball and coatings.

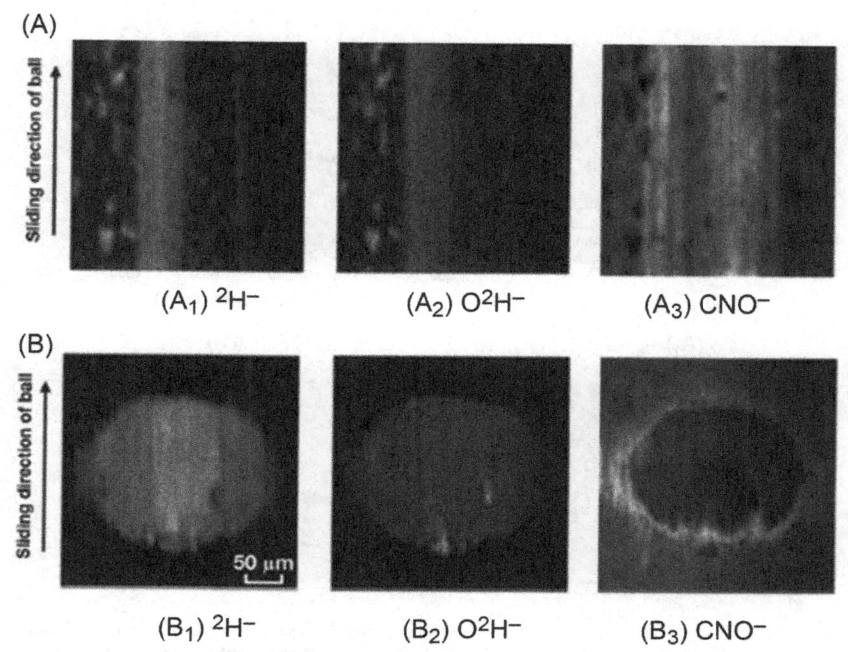

FIGURE 11.23 $^2H^-$, O^2H^-, and CNO^- TOF-SIMS images of the wear scar (Type I in Fig. 11.11) of (A) the CN_x-coated Si_3N_4 disk and (B) the wear scar on the CN_x-coated Si_3N_4 ball when Mode II of friction property is exhibited in N_2 gas environment ($\mu = 0.015$; O_2 concentration of 500 ppm; 5.4% RH) [5]. *TOF-SIMS*, Time-of-flight secondary ion mass spectrometry.

11.5.4 Roles of water and oxygen molecules in generating the low friction of tribological systems with CN_x coatings

Optimizing the amounts of water and oxygen molecules is the key to obtain stable and low friction of systems with CN_x coatings. The amounts of water and oxygen molecules chemisorbed during tribochemical reactions are significant for determining the friction of CN_x coatings.

For the optimum amount of water molecules in the N_2 gas environment (e.g., 0.1%−1.0% RH; Mode II), the water molecules are constantly chemisorbed onto the surface of CN_x coatings because of the friction energy, thereby resulting in low and stable friction. However when the water content of the N_2 gas environment was low (<0.1% RH, Mode III), the carbon that was transferred to the Si_3N_4 ball reacted directly with the carbon in the CN_x coating, thereby forming C−C bonds. This resulted in the wear of the CN_x coating and high friction. When the water content of the N_2 gas environment was high (>1.0% RH, Mode II), friction increased because of the fact that

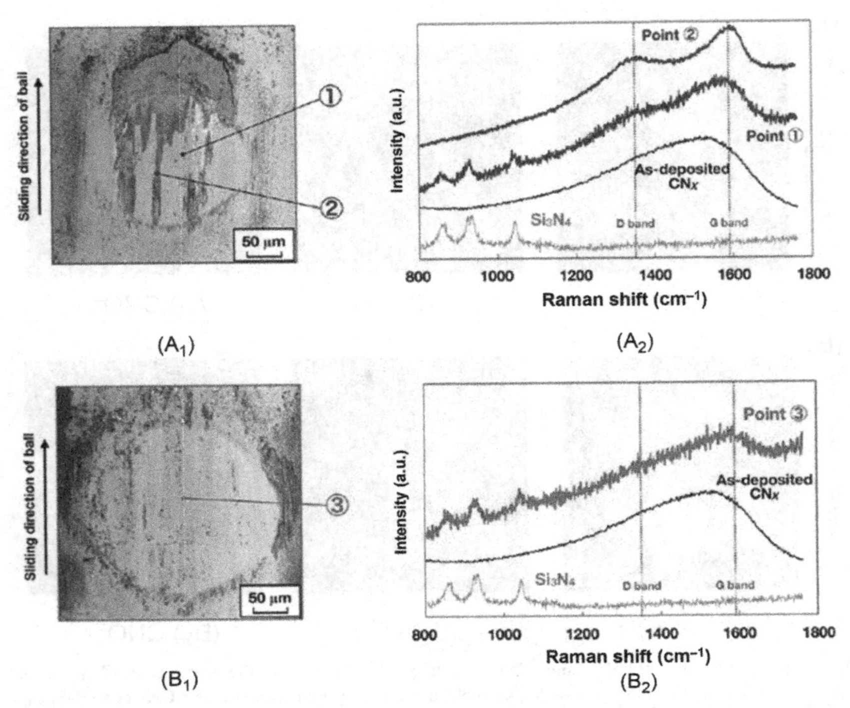

FIGURE 11.24 Optical microscope images and Raman spectra of typical wear scars of the Si_3N_4 ball in (A) Mode II and (B) Mode III obtained with oxygen concentration of 1100 and 70 ppm [5].

the structure of the transferred carbon was not maintained and because of the oxidation products formed on the surface of the Si_3N_4 ball [2].

Kubo et al. showed that both the hydrogen termination of DLC or CN_x coatings and the hydroxyl termination of DLC contributed to the generation of low friction, by using a tight-binding quantum chemical molecular dynamics simulation [11–14]. They also reported that nonhydrogen-terminated DLC coatings displayed considerably high friction because of the formation of C–C bonds [12]. In this study, both the hydrogen and hydroxyl derived from the water molecules terminated the carbon dangling bonds and suppressed the formation of C–C bonds [12,14]. This resulted in the formation of a low shearing interface where only weak van der Waals forces [15–17] existed between the CN_x coatings, thereby reducing the stable friction coefficient to less than 0.05. Therefore the tribochemical reaction of the CN_x coatings with the water molecules resulted in a stable and low (<0.05) friction coefficient.

Generally, the wear of CN_x and DLC coatings increases upon increasing the oxygen levels because of the reactions between the coatings and oxygen atoms [2,18]. Xu et al. reported that the tribochemical reactions of DLC coatings in dry air resulted in the oxidation of carbon, thereby forming CO_2 [19]. The structure of the worn CN_x changed from that of the deposited CN_x because of the desorption of nitrogen atoms inside the coating [20]. Accordingly, the surface-carbon desorption due to tribochemical reactions with oxygen atoms, resulting in CO_2 formation, can reorient the carbon networks on the surface and, consequently, result in structural changing and smoothing of the surfaces observed in the CN_x coatings.

11.6 Control of low-friction interface (tribolayer) by running-in

This section shows the possibility of running-in using a unique experimental method for low friction and friction control of the system with CN_x coatings. In the method, presliding is performed before introducing inert gas to the sliding interface to show the possibility of low friction in inert gas environments that contain water and oxygen molecules.

Using a ball-on-disk-type friction tester with a gas-supply unit, a two-step friction test, as shown in Fig. 11.25, is performed to show the possibility of running-in on the friction control [21]. The first step, "Step 1," is called presliding and is separated from the subsequent step, called "Step 2." The ball is replaced by a new one in between both the steps, consequently, the effect of presliding condition on the friction of CN_x/rubbed CN_x in the N_2 gas stream (Step 2) is investigated.

The representative effect of presliding on friction in the N_2 gas stream is shown in Fig. 11.26. Introducing 500 cycles of presliding to the CN_x-coated disk using the Si_3N_4 ball in ambient air (Step 1), the friction coefficient reduces from 0.07 to 0.004 in the subsequent CN_x/CN_x sliding in the N_2 gas stream (Step 2). The effect of the atmospheric species in Step 1 on the friction of CN_x/CN_x in Step 2 in N_2 gas stream is shown in Fig. 11.27. When the presliding of 1000 cycles under ambient air is introduced in Step 1, the lowest friction coefficient of 0.004 is generated in Step 2. When presliding is

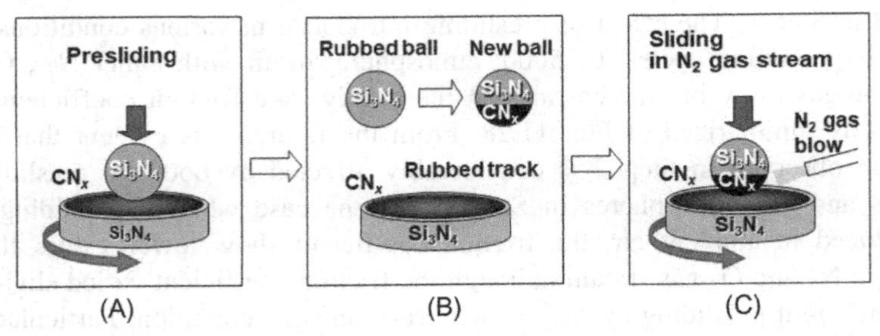

FIGURE 11.25 A schematic illustration of two-step friction test using ball-on-disk-type friction apparatus with a gas-supply unit: (A) Step 1, (B) exchange of ball, and (C) Step 2.

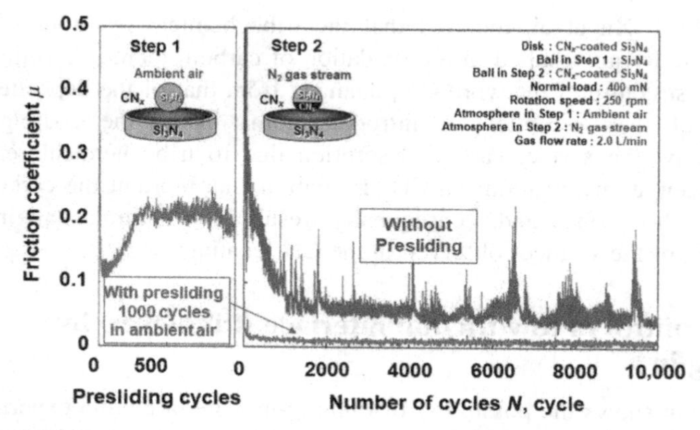

FIGURE 11.26 Drastic improvement of friction of CN_x/CN_x in N_2 gas stream (Step 2) by introducing of presliding in ambient air (Step 1).

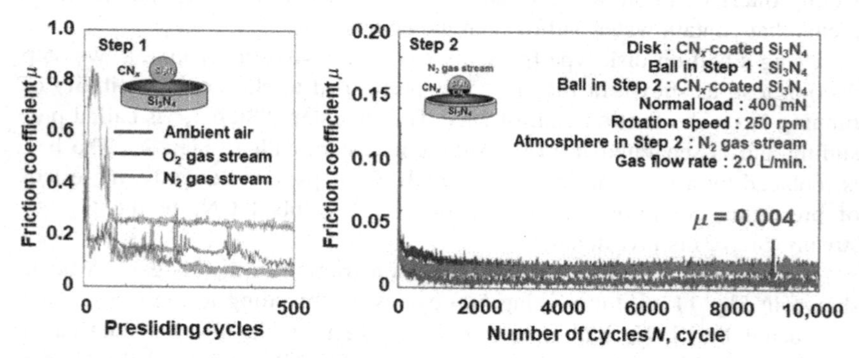

FIGURE 11.27 Effect of atmospheric gas species for Step 1 on friction of CN_x/CN_x in N_2 gas stream (Step 2).

introduced in the N_2 or O_2 gas stream in Step 1, the friction coefficient show higher values although the same numbers of presliding cycles were introduced in Step 1. The effect of presliding introduced in various conditions in Step 1 (presliding cycles: 0–5000; atmosphere: in air with either N_2-, O_2-, or none-gas blow in ambient air) on the steady-state friction coefficient in Step 2 is summarized in Fig. 11.28. From the figure, it is evident that the friction observed in Step 2 is considerably affected by both the presliding cycles and gas atmospheres in Step 1. For the case wherein presliding is introduced in ambient air, the friction coefficient show lower values than those in N_2 and O_2 gas stream although the friction coefficient varied slightly with different presliding cycles in each presliding gas condition. Particularly, the minimum friction coefficient of 0.004 was observed in Step 2 after optimum presliding condition in Step 1. It is confirmed that the two-step friction

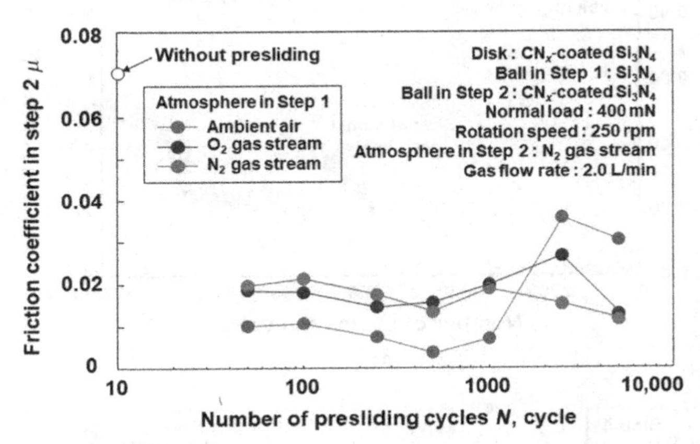

FIGURE 11.28 Effect of presliding in various conditions in Step 1 (presliding cycles: 0–5000, atmosphere: in air with either N_2^-, O_2^-, or none-gas blow in ambient air) on average value of friction coefficient at steady state in Step 2.

test is a simple and effective technique for achieving stable and low-friction coefficient of CN_x coatings in the N_2 gas environment. These results show that the atmospheric gases during running in have important rule for the formation of tribolayer, which determines the following friction properties.

The typical friction curves of CN_x/CN_x in the Ar gas environment with progressively increasing relative humidity both without and with presliding are shown in Fig. 11.29A and B, respectively.

The effect of the relative humidity in the inert gas environment on the friction of CN_x/CN_x without and with presliding is summarized in Fig. 11.30 [22]. These figures clearly show that the sensitivity of friction to water vapor for the sliding contact of CN_x/CN_x drastically decreased by introduction optimum presliding (running in), and the low-friction coefficient of 0.025 is achieved in the Ar and N_2 gas environments with relative humidity up to 37% RH. These results show the possibility of a stable a low friction of system with CN_x in an inert gas environment that contain wide range of amount of water and oxygen molecules.

11.7 Summary

It was confirmed that Si_3N_4/CN_x and CN_x/CN_x in an inert gas environment provided the friction coefficient less than 0.05, even without using any lubricants.

Furthermore, it was clarified that an invisible thin a-C layer of the thickness of several nanometers (so called as tribolayer) was formed because of the friction in the optimum condition of relative humidity and oxygen concentration, in the inert gas environment, and such surface resulted in a stable and low-friction coefficient ranging from 0.007 to 0.02. The key to

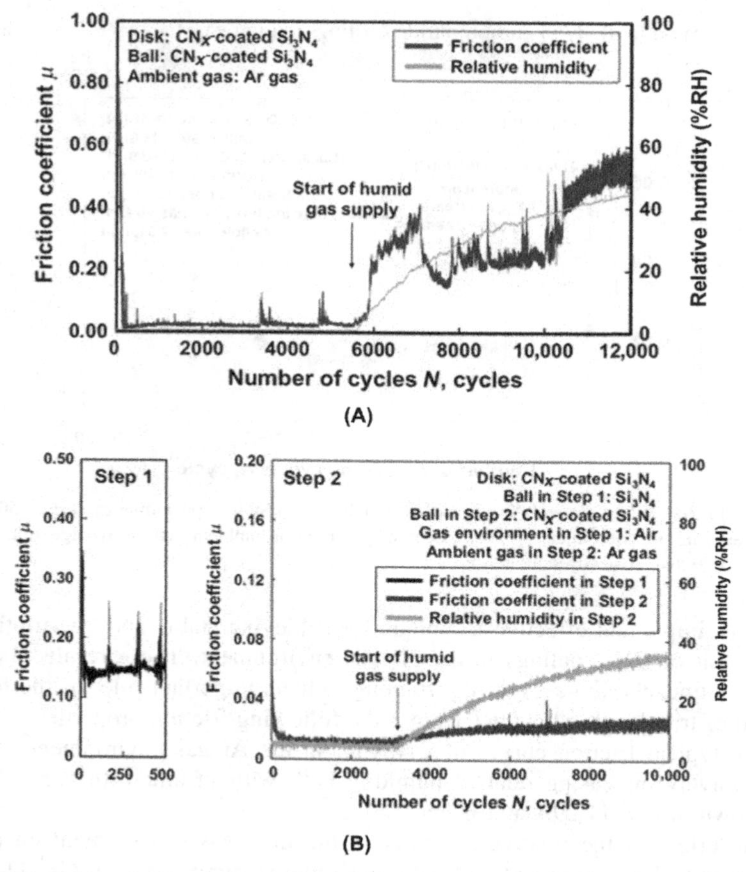

FIGURE 11.29 Representative friction curves of CN_x/CN_x in the argon gas atmosphere with progressively increasing relative humidity (A) without presliding and (B) after introducing presliding [22].

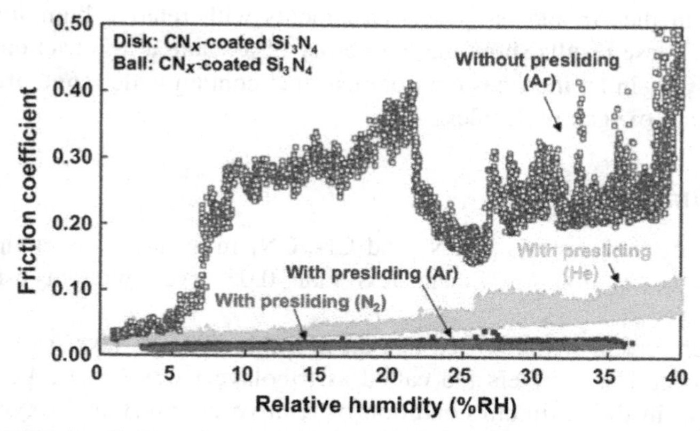

FIGURE 11.30 Drastic improvement of friction of CN_x/CN_x in three inert gas environment by introducing of presliding [22].

understanding the mechanism and control of low friction for friction systems with CN_x coatings is to understand the continuous formation mechanisms of such layers. Superlow friction systems in air without any lubricant are expected to be widely used in future.

References

[1] K. Kato, K. Adachi, Superlubricity of CN_x coatings in nitrogen gas atmosphere, in: A. Erdemir, J.-M. Martin (Eds.), Superlubricity, Elsevier, 2007, pp. 341–364.

[2] P. Wang, K. Adachi, Effect of oxygen concentration in inert gas environments on friction and wear of carbon nitride coatings, Tribol. Online 6 (6) (2011) 265–272.

[3] P. Wang, M. Hirose, Y. Suzuki, K. Adachi, Carbon tribo-layer for super-low friction of amorphous carbon nitride coatings in inert gas environments, Surf. Coat. Technol. 221 (2013) 163–172.

[4] K. Adachi, K. Kato, Tribology of carbon nitride coatings, in: C. Donnet, A. Erdemir (Eds.), Tribology of Diamond-Like Carbon Films, Springer, New York, 2008, pp. 339–361.

[5] N. Yamada, T. Watari, T. Takeno, K. Adachi, Role of water and oxygen molecules in the lubricity of carbon nitride coatings under a nitrogen atmosphere, Tribol. Online 11 (2) (2016) 308–319.

[6] Y. Liu, A. Erdemir, E.I.U. Meletis, A study of the wear mechanism of diamond-like carbon films, Surf. Coat. Technol. 82 (1996) 48–56.

[7] T. Watanabe, K. Yamamoto, Y. Koga, A. Tanaka, Formation of a-C thin films by plasma-based ion implantation, Sci. Technol. Adv. Mater. 2 (2001) 539–545.

[8] Y. Wang, Y. Ye, H. Li, J. Chen, H. Zhou, A magnetron sputtering technique to prepare a-C:H films: effect of substrate bias, Appl. Surf. Sci. 257 (2011) 1990–1995.

[9] N. Umehara, M. Tatsuno, K. Kato, Nitrogen lubricated sliding between CN_x-coatings and ceramic balls, Proceedings of the International Tribology Conference, Nagasaki, Japan, 2000, pp. 1007–1012.

[10] T. Tokoroyama, M. Kamiya, N. Umehara, C. Wang, D. Diao, Influence of UV irradiation in low frictional performance of CN_x coatings, Lubr. Sci. 24 (2012) 129–139.

[11] K. Hayashi, K. Tezuka, N. Ozawa, T. Shimazaki, K. Adachi, M. Kubo, Tribochemical reaction dynamics simulation of hydrogen on a diamond-like carbon surface based on tight-binding quantum chemical molecular dynamics, J. Phys. Chem. C. 115 (2011) 22981–22986.

[12] K. Hayashi, S. Sato, S. Bai, Y. Higuchi, N. Ozawa, T. Shimazaki, et al., Fate of methanol molecule sandwiched between hydrogen-terminated diamond-like carbon films by tribochemical reactions: tight-binding quantum chemical molecular dynamics study, Faraday Discuss. 156 (2012) 137–146.

[13] S. Sato, S. Bai, T. Ishikawa, Y. Higuchi, N. Ozawa, K. Adachi, et al., First-principles and tight-binding quantum chemical molecular dynamics studies for low friction mechanism of carbon nitride coatings, Proc. 5th World Tribol. Congr. 1 (2013) 678–679.

[14] M. Kubo, First-principles molecular dynamics and tight-binding quantum chemical molecular dynamics simulations on tribochemical reaction dynamics and low-friction mechanism of diamond-like carbon, J. Comput. Chem., Jpn. 12 (2013) A3–A13.

[15] A. Erdemir, The role of hydrogen in tribological properties of diamond-like carbon films, Surf. Coat. Technol. 146–147 (2001) 292–297.

[16] A. Erdemir, C. Donnet, Tribology of diamond-like carbon films: recent progress and future prospects, J. Phys. D: Appl. Phys. 39 (2006) R311–R327.
[17] J. Fontaine, C. Donnet, A. Grill, T. Le Mogne, Tribochemistry between hydrogen and diamond-like carbon films, Surf. Coat. Technol. 146–147 (2001) 286–291.
[18] J.C. Sánchez-López, A. Erdemir, C. Donnet, T.C. Rojas, Friction-induced structural transformations of a diamondlike carbon coatings under various atmospheres, Surf. Coat. Technol. 163–164 (2003) 444–450.
[19] X. Wu, T. Ohana, T. Nakamura, A. Tanaka, Gaseous tribochemical products of hydrogenated DLC film and stainless steel pair in air detected by mass spectrometry, Tribol. Lett. 57 (2015) 1–7.
[20] T. Tokoroyama, M. Goto, N. Umehara, T. Nakamura, F. Honda, Effect of nitrogen atoms desorption on the friction of the CN_x coating against Si3N4 ball in nitrogen gas, Tribol. Lett. 22 (2006) 215–220.
[21] P. Wang, M. Sugo, K. Adachi, Stable and super-low friction of amorphous carbon nitride coatings in nitrogen gas by using two-step ball-on-disk friction test, Lubr. Sci. 27 (2015) 137–149.
[22] P. Wang, K. Adachi, Low frictions of self-mated CN_x-coatings in dry and humid inert gas environments, Surf. Coat. Technol. 258 (2014) 1137–1144.

Chapter 12

Diamond-like carbon films and their superlubricity

Ali Erdemir
Texas A&M University, J. Mike Walker '66 Department of Mechanical Engineering, College Station, TX, United States

12.1 Introduction

Superlubricity, a new and exciting field of tribology, is broadly defined as a sliding regime in which friction or resistance to sliding essentially vanishes. The theoretical foundation of superlubricity has been introduced back in the early 1990s [1,2] but the more comprehensive experimental and computational studies occurred during last decade (see a collection of most relevant work in Ref. [3)]. Friction which works against superlubricity has always been our lives from the very beginning. As humankind has constantly been on the move throughout the history, they had to always deal with friction and develop effective ways to reduce its undesirable effects. As we live in a highly mechanical and mobile world at these days, friction has become more intertwined with our daily activities than ever before. We spent huge amount of energy to reduce its intensity in all industrial sectors. In fact, it is now estimated that nearly one-fifth of the total energy we produce globally in a year is spent to overcome friction between moving mechanical systems [4]. In most of the industrialized nations, it is estimated that 2%–5% of their gross national products are spent on friction-related energy losses [5]. Accordingly, if realized fully and employed in all types of moving mechanical systems, vanishing friction (or achieving superlubricity) can have a huge beneficial impact on energy security, environmental sustainability, and economic well-being of all humanity.

Thanks to much intensified research efforts in recent years, superlubricity has been shown feasible through the uses of many types of advanced materials, coatings, and lubricants [6–8]. In particular, certain solids (including 2D materials), liquids (acids, ionic liquids, alcohols, and glycols), and some gaseous media were all shown to bring friction coefficients well below 0.01 threshold which is considered as the upper limit of the superlubric sliding

Superlubricity. DOI: https://doi.org/10.1016/B978-0-444-64313-1.00012-0

regime. At present, it is hoped that all of these developments will in the near future lead to truly sustainable superlubricity at macroscales and enable the design of nearly frictionless mechanical systems consuming little or no energy and emitting far less greenhouse gases to our environments than before.

Realizing its societal, environmental, and economic impacts, number of research activities on vanishing friction has increased tremendously during the last two decades. In particular, advances in computational, analytical, and experimental capabilities have greatly expanded superlubricity research all over the world. At present, we have some very powerful experimental tools like atomic force microscopy and friction force microscopy, scanning tunneling microscopy, surface force apparatus, etc. combined with high-power computational methods (such as ab initio, molecular dynamics, quantum chemistry simulations, and artificial intelligence combined with machine learning algorithms) at our disposal to more precisely understand, predict, and control friction at the smallest possible length, force, and time scales. Furthermore, with the discovery of novel self-lubricating nanomaterials like graphene, nanoonions, nanotubes, etc. superlubricity research has gained much increased momentum [9–14].

Among those materials that afforded superlubricity; DLCs have always occupied a special place [15,16]. With their very unique chemical, structural, and mechanical properties, they have emerged as some of the best tribological coatings providing exceptional friction and wear properties under both lubricated and dry sliding conditions. Among the many types of DLCs, a new breed with very high hydrogen content was shown to provide superlubricity at macroscales under a wide range of test conditions [17]. In other studies, the pivotal role of hydrogen in DLC's superlubricity was further corroborated [18–21]. Superlubricity was also achieved in hydrogen-free DLC films in the presence of highly polar liquids like glycerol and/or oleic acid [22].

Over the years, DLC films have matured in so many ways and made their ways into many diverse and large-scale applications in several cross-cutting industries. Despite all these, there is a continuous research effort in both the fundamental and application-oriented aspects of these films. In fact, the volume of published work on DLC has expanded tremendously (i.e., over 5000 during the last decade alone); while those dealing with their superlubricity are more than 40 in the Web of Science. In this chapter, an overview of the recent advances in superlubricity of DLC films is presented. Special emphasis is placed on fundamental understanding of what makes or break surperlubricity in such films in relation to many intrinsic and extrinsic factors.

12.2 Carbon-based materials and their superlubricity

Carbon-based materials are used extensively by industry in so many ways and forms. They exist in a variety of crystalline and amorphous structures including diamond, diamond-like carbons (DLCs), graphite, graphene, glassy carbon, carbon nanotubes, carbon nanoonions, buckyballs, carbon nanofibers, etc. Fig. 12.1 presents the evolution and fundamental lubrication mechanisms of some of these carbon forms. As a class, almost all of them afford very low friction and wear to sliding surfaces [23–26]. In particular, graphene and its derivatives are found to exhibit very impressive friction and wear properties among all other carbon forms [9,11,12,14,27–30]. Furthermore, most of the polymers with low friction and wear properties like polytetrafluoroethylene, polyether-ether-ketone, and polyethylene are also carbon based and used extensively for tribological applications.

Mechanistically, the ultralow friction and wear properties of carbon-based materials are primarily attributed to their ability to attain a very favorable interfacial shear property during sliding [8,13,31,32]. In the case of graphite, graphene, and nanotubes, low shear is primarily due to the easy shear between incommensurate atomic sheets or layer, while in the case of diamond and DLCs, a favorable surface chemistry or chemical passivity are the key reasons for low friction. Graphene and other 2D materials provide structural superlubricity mainly because of an incommensurate state of contact during sliding [10,12,25,33]. Specifically, when atomic sheets of such materials are rotated out of full atomic registry, an instant transition from stick-slip to smooth sliding occurs and the friction literally vanishes [25,34].

Besides the graphene and graphite, superlubricity was also observed with carbon nanotubes. In a recent study, Zhang et al. showed that the inner shell of a centimeter-long double-walled carbon nanotube can be pulled out effortlessly despite the very long (1 cm) nature of the nanotube thus confirming that incommensurability mechanism can also apply to such 1D carbon materials [35].

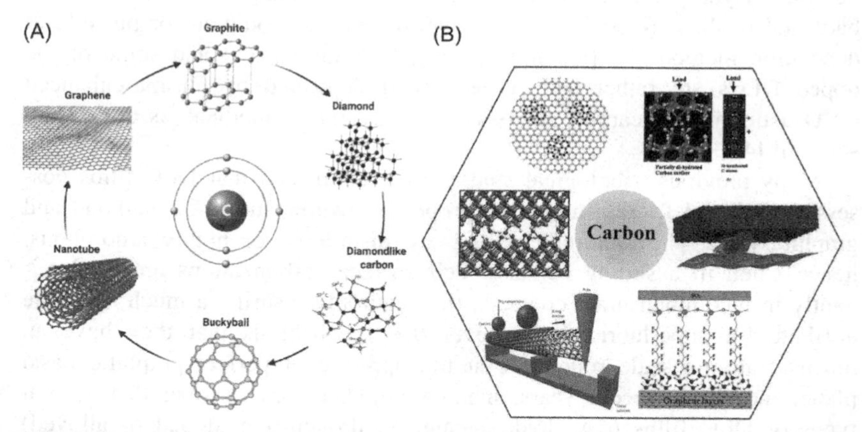

FIGURE 12.1 Illustration of (A) evolution and (B) proposed lubrication mechanisms of various carbon forms including graphene, DLC, carbon nanoscroll, carbon nanotube, and carbon nanostructures associated with liquid media. *(B) Reproduced with permissions from X. Chen, J. Li, Carbon, 158 (2020) 1.*

Concerted friction studies on bulk natural diamond and thin diamond coatings have also shown the possibility of achieving ultralow friction on such carbon-based materials. In particular, an ultrananocrystalline diamond film has proven to provide the best overall friction and wear properties [32,36–38]. In much earlier studies by Bowden and Young, the ultralow-friction property of diamond has been attributed to its ability to achieve a chemically inert surface condition insuring very little adhesion and hence friction [31]. Among the many gaseous species that have been investigated, hydrogen, oxygen, or water molecules were found to readily interact with the dangling σ bonds of diamond with the result that a dramatically reduced adhesion and hence friction (i.e., friction coefficient as low as 0.008) may be achieved in humid test conditions [38].

12.2.1 Development of diamond-like carbon films

The earliest reporting of DLCs goes back to the 1950s [16,39] but real advances in their design, synthesis, and large-scale industrial applications occurred during the last two decades [15,16,40]. Depending on the carbon sources and specific deposition processes employed, a variety of carbon-carbon bonding (i.e., sp^1, sp^2, and sp^3) is feasible and the specific fraction of such bonds plays an important role in the quality and tribological property of most DLC films [41]. In addition to their hydrogen-free forms, highly hydrogenated and doped DLCs (with N, B, F, Ti, Cr, W, etc.) can also be produced using physical vapor deposition (PVD) and chemical vapor deposition (CVD) methods with unique and tunable mechanical and tribological properties [42]. Hydrogen-free DLCs with mostly sp^3 bonding could be extremely hard and resilient (especially when deposited by cathodic-arc or pulse-laser deposition methods [43]); while the highly hydrogenated and some of the doped DLCs are rather soft [especially if deposited by plasma-enhanced CVD using a hydrocarbon gas (such as acetylene or methane as the carbon source)] [43].

Many previous tribological studies have confirmed that DLC films possess exceptional friction and wear properties rivaling those of diamond and graphite [15,23]. Structurally, DLCs are disordered or highly amorphous; hence when in a sliding contact, their surface carbon atoms are predominantly in incommensurate contacts; thus naturally insuring a much desirable attribute for superlubricity. However the drawback is that they have an intrinsic atomic scale roughness at the opposite of perfect graphitic basal plane. Indeed, in recent years, many researchers have shown that several types of DLC films (e.g., hydrogenated, hydrogen-free, doped or alloyed) can afford friction coefficients of less than 0.01 [15,17,18,20,21]. Fig. 12.2 summarizes the friction coefficient of such carbon films that were developed using various research groups over the years. Overall, these DLC films have

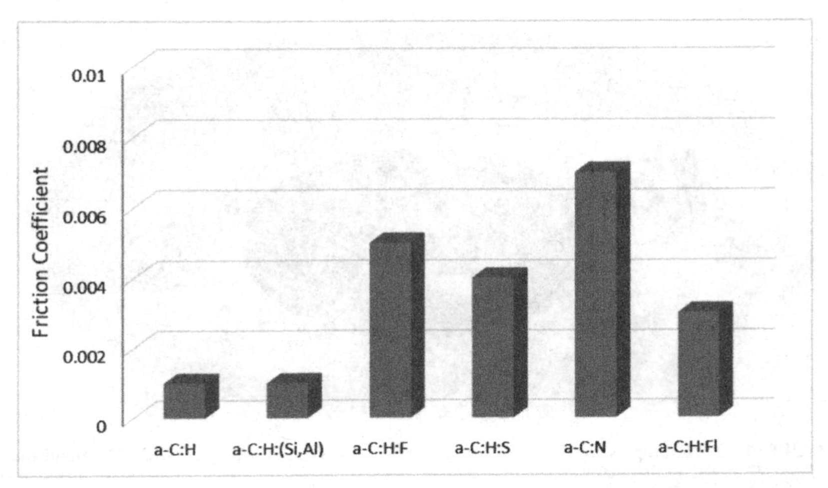

FIGURE 12.2 Superlow friction behavior of various DLC films under controlled test environments. *a-C:H:Fl*, Fullerene-like DLC; *DLC*, diamond-like carbon.

emerged as the leading tribological coatings in recent years and are now used in many industrial applications such as the ones shown in Fig. 12.3.

12.2.2 Unusual tribological properties of diamond-like carbon films

Systematic studies have shown that the unusual lubricity or superlow friction mechanism of DLC films is very different from those of other carbon forms aforementioned and strongly dependent on a range of chemical, structural, and environmental factors. Besides the environmental effects, structural, topographical, and chemical irregularities may also affect friction and wear behaviors of DLC films. As opposed to crystalline diamond and graphite which exhibit low friction and wear in moist air or in the presence of some polar molecules in their surrounding atmosphere [31,32,44,45]; some of the DLC films (i.e., highly hydrogenated DLCs) may provide the lowest friction and wear coefficients in vacuum or inert gas environments [17–20]. However just like diamond, hydrogen-free tetrahedral amorphous carbon DLCs also depend on moisture or water molecules in the surrounding air for low friction [23,36] as shown in Fig. 12.4. However hydrogenated DLCs tend to exhibit relatively high friction in moist test environments.

From a surface chemistry point of view, the existence of severe covalent-bond interactions in a DLC film always lead to strong adhesion (cold welding) and hence high friction [15,43]. This is particularly very common for sliding diamond and hydrogen-free DLC films whose surface carbon atoms can readily bond to one another via covalent σ bonds and the presence of

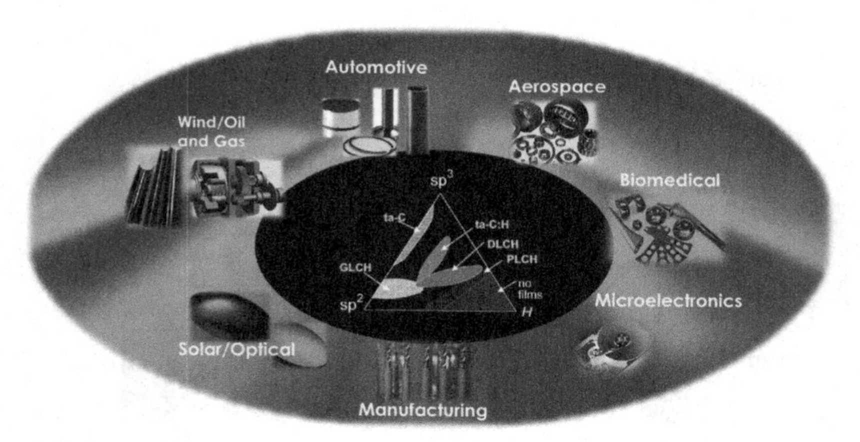

FIGURE 12.3 Examples of industrial applications of DLC coatings. *DLC*, Diamond-like carbon.

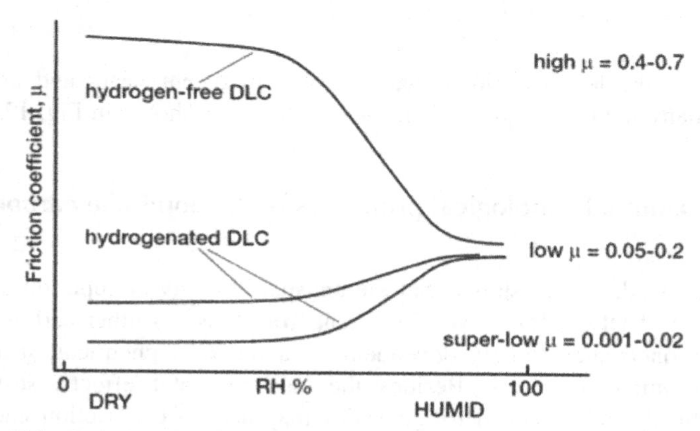

FIGURE 12.4 Variation of friction coefficient of hydrogenated and hydrogen-free DLC films with respect to relative humidity (RH). *Reproduced with permission from H. Ronkainen, K. Holmberg, Environmental and thermal effects on the tribological performance of DLC coatings, in: Tribology of Diamond-like Carbon Films, Springer, Boston, MA, 2008, pp. 155–200, Copyright 2008, Springer.*

such a strong bonding across the sliding interface can lead to very high friction [23,31,37]. When diamond and hydrogen-free DLCs are tested in dry nitrogen or high vacuum, friction coefficients of 0.5 to more than 1 are often observed [46–48]; again largely because of the high levels of covalent σ-bond interactions between sliding carbon atoms. Furthermore, depending on the situation or type of carbon film, relatively weaker van der Waals forces, π-π* interactions, capillary forces, and electrostatic attractions may also come into play and further contribute to high friction behavior.

When some polar or chemically active species (e.g., water molecules, oxygen, polar liquids like oleic acid, etc.) are present in the surrounding environment or at sliding interface, the surface carbon atoms of diamond and H-free DLC films may interact with them tribochemically and the resultant interface chemistry may in turn dominate their friction and wear properties [22,23,36,47]. Some of the highly polar species like water molecules, oxygen, oleic acid, or alcohols have high affinity to chemically adsorb on such diamond and DLC surfaces and then effectively passivate their dangling σ bonds with the result that friction coefficients coming down to 0.01 level or below. However previous research has also confirmed that oxygen and water molecules could be very detrimental to the friction and wear performance of hydrogenated DLC films. These films often provide very low friction (i.e., less than 0.01) in vacuum or inert gases as explained before, but when tested in humid air or oxygen, their friction coefficients increase to more than 0.2 [49].

12.2.3 Mechanisms of superlubricity in diamond-like carbon films

As obvious from the foregoing, tribology of DLC films is very much dominated by a range of intrinsic (film-specific) and extrinsic (environmental) factors. Hence, for achieving superlow friction and wear in these films, one has to control all of these factors in a logical manner. First and foremost, such film-specific factors as surface roughness, structural/chemical homogeneity, chemical composition in bulk and on the surface, uniform film thickness, etc. have to be at the most desirable or optimum levels in order to avoid any adverse contributions from them to friction. Surface roughness tends to be a major problem with H-free DLC films deposited by cathodic-arc or pulsed-laser deposition processes; as they cause nano/microscale droplets or protrusions on the growing films' surface. Fortunately, with the latest PVD and CVD methods, DLC films can now be deposited with a high degree of surface smoothness and uniform thickness down to a few nanometers. Their highly disordered or amorphous structures naturally avoid growth steps or textures which can otherwise contribute to nanoscale surface roughness and hence higher friction. Constant ionic and/or atomic bombardment during deposition process sputters off or etch high points or protrusions and thus results in a high degree of surface smoothness and thickness uniformity.

Besides the surface irregularities mentioned, chemical inertness of DLC is also very important. There are multiple ways to achieve a high degree of chemical inertness in DLC films. For example, introduction of large amounts of hydrogen during deposition process (see Fig. 12.5) can certainly provide a highly hydrogenated DLC both in bulk and on its surface; and thereby assures a very high degree of chemical inertness [17–19,21,43,50,51]. Note that in its protonic state, hydrogen is very small and can easily be incorporated into the DLC films at very high concentrations (e.g., as much as 50 at.

%) during deposition. In CVD processes, the carbon source for the synthesis of DLC is primarily made of CH_4 or C_2H_2 gases. During deposition process, hydrogen is dissociated from carbon and some of it is subsequently incorporated into the film structure or reacted with the film surface. To achieve a very high degree of hydrogenation, additional hydrogen (up to 90 vol.%) may also be blended with CH_4 or C_2H_2 during the deposition processes and obviously, these films will attain even much higher degree of hydrogenation with a typical hydrogen content of 38–50 at.% [52].

Fully hydrogen-terminated DLC surfaces will in general have very weak chemical interactions with the opposing surfaces during tribotesting. Specifically, there will be very little σ-bond interactions and this will certainly lead to low friction. Accordingly, in addition to bulk hydrogenation as mentioned before, if hydrogen made available to hydrogen-free or -poor DLCs during tribotesting in their surrounding atmosphere, they can also readily react with hydrogen at the sliding contact interfaces and thus achieve a high degree of in situ hydrogen termination and thereby achieve ultralow friction [6,17–20,53]. Specifically, when hydrogen available in the surrounding atmosphere, they can bond to those active or free σ bonds of the surface carbon atoms and thus help eliminate stiction and hence friction.

As described in Fig. 12.5, a high degree of hydrogenation in DLC films can have a dramatic effect on their frictional behavior. Specifically, hydrogen termination of sliding DLC surfaces can significantly lower friction during tribotesting. On the contrary, if little or no hydrogen exists in a DLC film, the friction coefficients tend to be very high due to extensive covalent

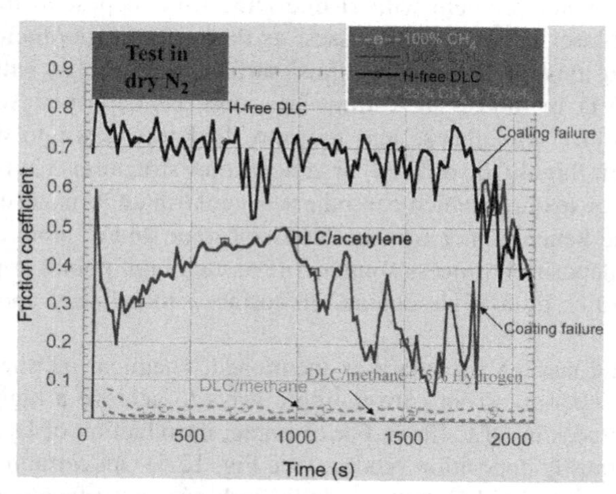

FIGURE 12.5 Friction coefficients of DLC coatings derived from various source gases in a plasma-enhanced CVD system. Note that the lowest friction (i.e., 0.003) is provided by the DLC grown in a 25% methane +75% hydrogen containing plasma (test conditions: load, 10 N; speed, 0.5 m/s; temperature, 22°C–23°C; test environment, dry N_2). *DLC*, Diamond-like carbon.

or σ-bond interactions. Typical friction coefficient of a hydrogen-free DLC film may range from 0.7 to more than 1 in inert or vacuum test environments, while a highly hydrogenated DLC film synthesized from a gas discharge plasma consisting of 25% methane and 75% hydrogen can provide friction coefficients of less than 0.01 as shown in Fig. 12.5 [51,52,54].

As mentioned, superlow friction coefficients can also be achieved on hydrogen-free or -poor DLC films by simply supplying hydrogen into the test chambers during tribological testing [18,20,55]. Specifically, as shown in Fig. 12.6, when hydrogen is supplied into the test chamber at increasing partial pressures, hydrogen-poor DLC films can also attain superlow friction coefficients for a long time.

Based on the results from a large body of previous studies, one can draw the following conclusions on the pivotal role of hydrogen on the superlubricity of DLC films. Hydrogen is known to have a very high chemical affinity to bond to carbon atoms with very strong covalent bonds. This will effectively passivate the dangling σ bonds of carbon atoms on the surface. Such

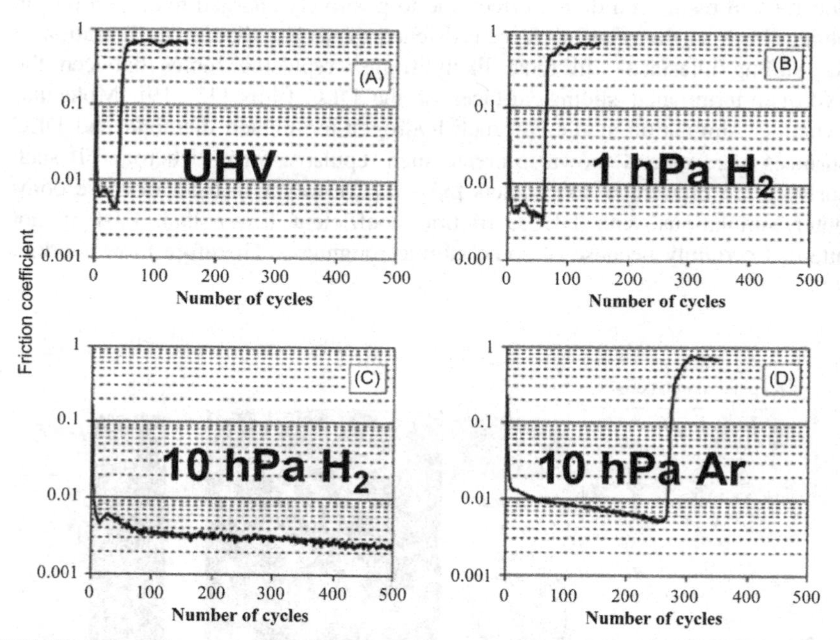

FIGURE 12.6 Effect of partial pressure of hydrogen and argon on friction behavior of DLC films: (A) in UHV, (B) 1 hPa partial pressure of hydrogen, (C) 10 hPa partial pressure of hydrogen, and (D) 10 hPa of argon. Low friction is attained when a sufficient amount of hydrogen is present in the vacuum chamber [32]. *DLC*, Diamond-like carbon; *UHV*, Ultra-high vacuum. *Reproduced with permission from A. Erdemir, O.L. Eryilmaz, Superlubricity in diamondlike carbon films, in: A. Erdemir, J.-M. Martin (Eds.), Superlubricity, Elsevier, 2007. pp. 155–200, Copyright 2007, Elsevier.*

passivation will prevent any type of adhesive interactions taking place across the sliding DLC interfaces and this will insure minimal stiction and hence ultralow friction. For instance, if we consider the highly hydrogenated DLC films, unoccupied σ bonds of their carbon atoms are effectively passivated by hydrogen; hence they are chemically very passive and do not contribute to stiction or adhesion. When and if needed, hydrogen can diffuse out of the bulk to the sliding surface and thus replenish or replace those hydrogen atoms that may have been lost or removed due to mechanical sliding action.

In summary, DLC films created in a highly hydrogenated gas discharge plasmas or tested in hydrogen-rich test environments can attain a high degree of hydrogenation on their sliding surfaces. Undoubtedly, such a situation will insure a highly inert or chemically insensitive surface providing superlow friction as illustrated in Fig. 12.7 [18–20]. Specifically, when surface carbon atoms react with hydrogen, their dangling σ bonds are neutralized or passivated. Consequently, the electrical charge density of hydrogen atoms is shifted in such a way that their positively charged nucleus with proton is now positioned near the outer surface. As is obvious, such a shift in electron charge density will result in a dipole effect due to positively charged hydrogen proton being closer to the outer surface. Existence of such a dipole configuration at the sliding interface will most likely trigger repulsive forces between the hydrogen-terminated sliding surfaces of the DLC films [17–19]. Molecular dynamics simulation studies of such hydrogen-terminated diamond and DLC surfaces were indeed shown to create such repulsive forces when and if such surfaces are brought in to contacts [55–58]. However in most of these computer simulations with DLCs, friction coefficient lower than 0.01 is not attained certainly because of some atomic roughness. Therefore more work is

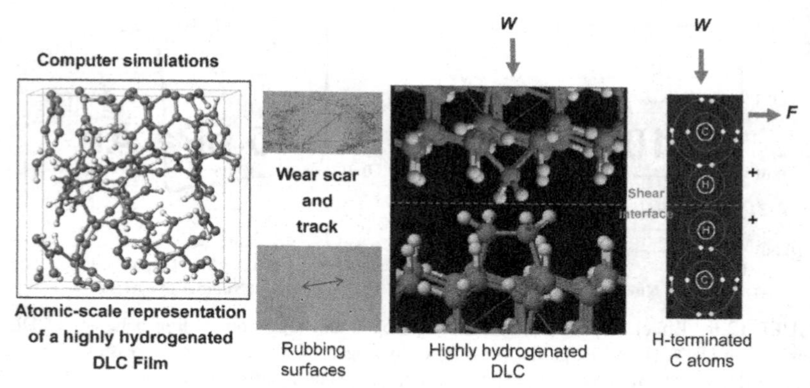

FIGURE 12.7 A mechanistic model for the superlow friction behavior of highly hydrogenated DLC films. *DLC*, Diamond-like carbon. *Reproduced with permission from A. Erdemir, O.L. Eryilmaz, Superlubricity in diamondlike carbon films, in: A. Erdemir, J.-M. Martin (Eds.), Superlubricity, Elsevier, 2007. pp. 155–200, Copyright 2007, Elsevier.*

necessary to deeply understand the low friction of H-terminated carbon surfaces.

12.3 Superlubricity in other diamond-like carbon films

Besides the hydrogenated and hydrogen-free DLC films discussed before, many other types of DLC coatings exist nowadays and these can also provide superlow friction and wear to sliding surfaces. For example, doped DLC films with some special elements have been synthesized in recent years for a wide range of industrial applications and some of them have been shown to provide friction coefficients of less than 0.01 [59]. Nitrogen-doped DLCs became very popular for magnetic hard-disk applications, while tungsten and silicon-doped DLCs were used for automotive applications mainly because of their much desirable physical, mechanical, and tribological properties. Depending on the type of dopant or alloying element, one can achieve a combination of superior toughness, hardness, thermal stability, and tribological performance. Furthermore, as presented earlier (see Fig. 12.2), some of them are able to exhibit superlubricity in sliding contacts and they will be briefly discussed further.

Carbon nitride (CN_x) films: The CN_x films have been around for a very long time and used successfully by magnetic recording industry for many years [60]. These films are largely made of carbon but can ideally contain 5–15 at.% nitrogen as a dopant. Just like hydrogen-free and hydrogenated DLCs, they possess an amorphous structure. Due to very high degree of sp^3 bonding, they are generally very hard and stiff and hence afford high wear resistance to sliding surfaces. For the magnetic recording media, they are always used with polar lubricating oils (like Z-dol, Pennzane, etc.) to further enhance their tribological performance [61].

There have been many experimental studies on the tribological properties of CN_x coatings. In particular, systematic studies by Kato, Adachi, and their coworkers have led to the development of a superlubricious CN_x film that can provide friction coefficients below 0.01 [62–64]. Superlubric sliding behavior their CN_x film has been the subject of many dedicated studies during the past two decades. Superlow friction coefficients of 0.005–0.007 were achieved after testing CN_x in streams or environments containing controlled amounts of water/oxygen molecules prior to testing in dry nitrogen or other inert gases. Specifically, it has been proposed that such CN_x films have a unique ability to attain a special surface termination state by reaction with such molecules during the initial running-in or conditioning periods and thus subsequently attaining a highly inert and nonadhesive surface state creating very little stiction and hence friction.

Silicon-doped DLCs: Silicon is a very popular alloying or doping element for DLC coatings and used particularly to enhance thermal stability, structural integrity, and surface reactivity for especially lubricated sliding conditions. Si-doped DLCs were also shown to exhibit superlubricity in high vacuum against steel ball in recent studies [65–67]. For the superlubric sliding behavior, the formation of a very desirable nanostructured Si-rich transfer layer enabling different types of carbon rehybridization and effective shielding against the deleterious influences of reactive environmental species during sliding.

Fluorine-doped DLCs: Fluorine can also be incorporated into DLC films during deposition process as a dopant. It has been claimed to lower surface energy and internal stress level while at the same time improving thermal stability and biocompatibility [68]. However due to its high electronegativity and hygroscopic nature, the amount of fluorine that can safely be introduced into DLC is somewhat limited. Nevertheless, the presence of fluorine in DLCs appears to be beneficial to the tribological performance at some optimum concentrations as demonstrated by Fontaine et al. [69]. In fact, they showed that a moderately fluorine-doped hydrogenated DLC films could provide as low a friction coefficient as 0.005 in ultrahigh vacuum. Such a superlow friction property was correlated with a high hardness and the significantly higher viscoplasticity of such films. In another theoretical study involving DLC surfaces terminated with fluorine atoms, superlow friction was primarily attributed to the much stronger repulsive Coulombic forces exerted by such F-terminated surfaces [70,71].

Sulfur-doped DLCs: Sulfur doping of DLC has not been very common but, it was postulated that such a doping may also result in ultralow friction even in ambient air with high humidity. Specifically, Freyman et al. achieved a superlow friction coefficient of 0.004 under ambient air conditions with a sulfur-doped DLC produced in a magnetron sputtering system using a gas discharge plasma made of $Ar/H_2/H_2S$ [72]. Mechanistically, the superlubricious behavior of such films was attributed to the much weaker bonding between the adsorbed water molecules on the surface and the dipole effects of sulfur and hydrogen. In a more recent study, scientists produced fluorine- and sulfur-doped DLC films and achieve ultralow friction down to 0.01 [73]. They noticed that there was a graphite-like transfer layer on the ball surface and potentially providing a low-shear interface due to –CF2 and –CF3 termination. Specifically, the electrostatic repulsion between the F-terminated transfer layer and the S- and F-terminated DLC surface was most likely responsible for the ultralow friction behavior.

Fullerene-like DLCs: Recently, researchers have accomplished short-range fullerene-like ordering in otherwise highly amorphous random network structure of DLC films. Friction and wear tests of such films have shown that through the formation of a hydrogen-rich tribochemical film on rubbing surfaces, friction coefficients of less than 0.01 are feasible [74,75]. These films were also shown to be superelastic and hence could tolerate significant amounts of elastic deformation to provide much larger contact areas than some of the other DLC film types. Postdeposition thermal annealing of such films was also shown to be very beneficial to the friction and wear performance of such films. Specifically, it was shown that by postdeposition annealing in dry nitrogen for 30 min at 250°C–500°C, the friction coefficients of such films could be brought down to 0.004–0.006 levels [76].

Superlubricity of DLC films against 2D and other materials: Recent studies have confirmed that highly hydrogenated DLC films have a unique ability to trigger superlow friction when sliding against graphene, MoS_2, and other 2D materials especially under inert test environments. In some of these studies, achieving superlubricity was preconditioned to the existence of nanodiamond particles where altogether, they formed nanoscrolls or carbon nanoonions at sliding interfaces leading to superlubricity [29,77,78].

Hydrogenated DLC films against stabilized zirconia balls were also shown to provide superlow friction coefficients well below 0.001 (in fact down to 0.0001 which is referred to as friction fadeaway) in hydrogen. Such a combination of test pairs was found to result in a polymer-like tribofilm especially under the influence of very high Hertzian contact pressures (i.e., 2.6 GPa). These researchers attributed such extremely low-friction coefficients to a very special catalytic effect that the zirconia balls can provide for more intense triboreactions between hydrogen molecules and the carbon atoms of DLC film and thus enabling much stronger repulsive forces [79,80].

12.4 Concluding remarks

Superlubricity research has come a long way and in particular gained a high momentum during the last decade or so. Thanks to many new scientists who entered the field of superlubricity, several new variants of DLCs and other materials (in particular 2D materials) have been offered to the superlubricity field and they have been the subject of intense worldwide research in recent years. These efforts were critically important for the advancement of superlubricity research worldwide. Fig. 12.8 summarizes the progress in the development of superlubricious materials and coatings over the years. Among all others, DLC films still continue to attract significant attentions from both the scientific and industrial communities. What is so intriguing is that under the right environments and/or test conditions, many of these films provide superlubricity at macroscopic scales. Large body of knowledge gained over the years on DLC is now used effectively in the design and synthesis of more robust DLC films providing significant energy-saving benefits in key industrial sectors including automotive. Highly hydrogenated DLC films with much desirable surface and structural chemistry are perhaps the oldest ones providing friction coefficients down to 0.001 level. Because of their amorphous structure, most DLC films conform to the incommensurability requirement for superlubricity. By and large, their superlow friction behavior is primarily related to their very passive or chemically inert surfaces causing very little or no adhesive bonding during sliding.

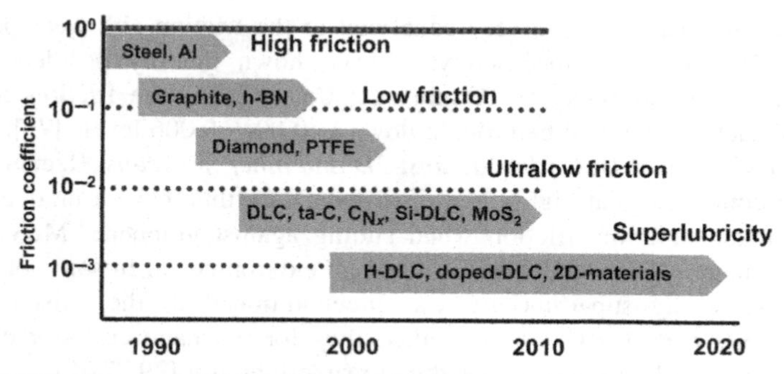

FIGURE 12.8 Progress in low-friction materials research over the years.

Besides the highly hydrogenated DLCs, there now exists a wide range of hydrogen-free and doped (or alloyed) DLC films as well. They were also shown to provide superlow friction coefficients (some were even shown to provide friction coefficients lower than 0.001, although accurate measurement could be questionable). With all these new and exciting developments, it appears that the DLC films are poised to enable superlubricity in a wide range of real industrial applications and thus save huge amounts of energy. In particular, hydrogen-free DLC films were confirmed to be very compatible with lubricated environments providing friction coefficients of 0.01 or below in the presence of some organic friction modifiers like glycerol, oleic acid, etc. under lubricated conditions. Overall, with recent advances in industrial-scale production of DLC films on various engine components, DLC films hold great promises for meeting the increasingly harsher operational conditions of future automotive applications associated with much reduced oil viscosities and antifriction and -wear additives. Further optimization of their friction and wear properties is currently underway by taking into consideration both the intrinsic (or film-specific) and extrinsic (or operational) factors. Intrinsically, optimization effort is directed toward the control film microstructure, chemistry, internal stress, adhesion, and surface finish. These efforts are also intensifying toward the creation of more favorable synergy between DLC film itself and its operating environments such as lubricating oils and additives. In particular, type and proportion of doping elements and the fraction of sp^2 versus sp^3 bonding appear to influence the extent and nature of all tribochemical interactions at sliding contact interface and these are all expected to ultimately determine the long-term tribological performance of all DLC films.

Acknowledgments

This review is based on the collaborative research efforts by many students and research colleagues from all over the world over the past three decades, and hence the author is truly indebted to all of them for their help and inspiration. Further, the financial support by the US Department of Energy for author's work over many years is gratefully acknowledged.

References

[1] J.B. Sokoloff, Phys. Rev. B 42 (1990) 760.
[2] M. Hirano, K. Shinjo, Wear 168 (1993) 121.
[3] A. Erdemir, J.-M. Martin (Eds.), Superlubricity, Elsevier, 2007.
[4] K. Holmberg, A. Erdemir, Friction 5 (3) (2017) 263.
[5] S.S. Perry, W.T. Tysoe, Trib. Lett. 19 (2005) 151.
[6] J.-M. Martin, A. Erdemir, Phys. Today 71 (2018) 40.
[7] X. Ge, J. Li, J. Luo, Front. Mech. Eng. 5 (2019) 2.
[8] D. Berman, et al., ACS Nano 12 (2018) 2122.
[9] W. Zhai, K. Zhou, Adv. Funct. Mater. 29 (2019) 1806395.
[10] M. Baykara, M.R. Vazirisereshk, A. Martini, Appl. Phys. Rev. 5 (2018) 041102.
[11] X. Chen, J. Li, Carbon 158 (2020) 1.
[12] O. Hod, et al., Nature 563 (2018) 485.
[13] S. Zhang, et al., Mater. Today 26 (2019) 67.
[14] Y. Song, et al., Small (2019)) 1903018.
[15] C. Donnet, A. Erdemir (Eds.), Tribology of Diamond-Like Carbon Films: Fundamentals and Applications, Springer Science & Business Media, 2008.
[16] J. Vetter, Surf. Coat. Technol. 257 (2014) 213.
[17] A. Erdemir, O.L. Eryilmaz, G. Fenske, J. Vac. Sci. Technol. A 18 (2000) 1987.
[18] A. Erdemir, O. Eryilmaz, Friction 2 (2014) 140.
[19] A. Erdemir, Surf. Coat. Technol. 146 (2001) 292.
[20] C. Donnet, et al., Tribol. Lett. 9 (2001) 137.
[21] A. Erdemir, O.L. Eryilmaz, Superlubricity in diamondlike carbon films, in: A. Erdemir, J.-M. Martin (Eds.), Superlubricity, Elsevier, 2007, pp. 155−−00.
[22] M. Kano, et al., Friction 2 (2014) 156.
[23] A. Erdemir, J.-M. Martin, Curr. Opin. Solid State Mater. Sci. 22 (2019) 243.
[24] N. Matsumoto, et al., Tribol. Mater. Surf. Interfaces 6 (2012) 116.
[25] M. Dienwiebel, et al., Phys. Rev. Lett. 92 (2004) 126101.
[26] K. Miura, D. Tsuda, N. Sasaki, J. Surf. Sci. Nanotechnol. 3 (2005) 21.
[27] D. Berman, A. Erdemir, A.V. Sumant, Carbon 54 (2013) 454.
[28] D. Berman, Adv. Funct. Mater. 24 (2014) 6640.
[29] D. Berman, et al., Science 348 (2015) 1118.
[30] I. Leven, et al., J. Phys. Chem. Lett. 4 (2013) 115.
[31] F.P. Bowden, J.E. Young, Proc. Roy. Soc 208 (1951) 444.
[32] A. Erdemir, et al., Surf. Coat. Technol. 50 (1991) 17.
[33] M. Hirano, Wear 254 (2003) 932.
[34] Y. Song, et al., Nat. Mater. 17 (2018) 894.
[35] R. Zhang, et al., Nat. Nanotechnol. 8 (2013) 912.
[36] A. Erdemir, et al., Surf. Coat. Technol. 121 (1999) 565.
[37] A.R. Konicek, et al., Phys. Rev. Lett 100 (2008) 235502.
[38] N. Kumar, et al., Tribol. Int. 44 (2011) 2042.
[39] H. Schmellenmeier, Experimentelle Technik der Physik 1 (1953) 49.

[40] S.C. Cha, A. Erdemir (Eds.), Coating Technology for Vehicle Applications, Springer, Switzerland, 2015.
[41] J. Robertson, Mat. Sci. Eng 37 (2002) 129.
[42] C. Donnet, Surf. Coat. Technol. 100 (9) (1998) 180.
[43] A. Erdemir, C. Donnet, J. Phys. D Appl. Phys. 39 (2006) R311.
[44] G.W. Rowe, Wear 3 (1060) 274.
[45] M.N. Gardos, B.L. Soriano, J. Mater. Res. 5 (1990) 2599.
[46] H. Ronkainen, K. Holmberg, Environmental and thermal effects on the tribological performance of DLC coatings, Tribology of Diamond-Like Carbon Films, Springer, Boston, MA, 2008, pp. 155–200.
[47] K. Miyoshi, et al., J. Appl. Phys. 74 (1993) 4446.
[48] J. Andersson, R.A. Erck, A. Erdemir, Surf. Coat. Technol. 163 (2003) 535.
[49] O.L. Eryilmaz, A. Erdemir, Wear 265 (2008) 244.
[50] C. Donnet, et al., Tribol. Let. 4 (1998) 259.
[51] A. Erdemir, et al., Surf. Coat. Technol 133–134 (2000) 448.
[52] A. Erdemir, et al., Diam. Rel. Mater. 9 (2000) 632.
[53] J. Fontaine, et al., Tribol. Int. 37 (2004) 869.
[54] A. Erdemir, Mat. Res. Soc. Symp. Proc 697 (2002) 391.
[55] Y. Qi, E. Konca, A.T. Alpas, Surf. Sci. 600 (2006) 2955.
[56] S. Dag, S. Ciraci, Phys. Rev. B 70 (2004) 241401.
[57] K. Hayashi, et al., J. Phys. Chem. C 115 (2011) 22981.
[58] J.D. Schall, G. Gao, J.A. Harrison, J. Phys. Chem. C 114 (2010) 5321.
[59] J.C. Sánchez-López, A. Fernández, Doping and alloying effects on DLC coatings, Tribology of Diamond-Like Carbon Films, Springer, Boston, MA, 2008, pp. 311–338.
[60] T.E. Karis, Nano-tribology of thin film magnetic recording media, Nanotribology, Springer, Boston, MA, 2003, pp. 291–325.
[61] P.H. Kasai, Tribol. Lett. 13 (2002) 155.
[62] K. Kato, N. Umehara, K. Adachi, Wear 254 (2003) 1062.
[63] K. Kato, K. Adachi, Superlubricity of CN_x-coatings in nitrogen gas atmosphere, Tribology of Diamondlike Carbon Films: Fundamentals and Applications, Springer, New York, 2008, pp. 342–364.
[64] N. Yamada, et al., Tribol. Online 11 (2016) 308.
[65] I. Sugimoto, S. Miyake, Appl. Phys. Lett. 56 (1990) 1868.
[66] X. Chen, et al., Nat. Commun. 8 (2017) 1.
[67] X. Chen, et al., Sci. Adv. 6 (2020) 1272.
[68] L. Zhang, et al., RSC Adv. 5 (2015) 9635.
[69] J. Fontaine, et al., Tribol. Lett. 17 (2004) 709.
[70] S. Bai, et al., J. Phys. Chem. C 116 (2012) 12559.
[71] S. Bai, et al., RSC Adv. 4 (2014) 33739.
[72] C.A. Freyman, Y. Chen, Y.W. Chung, Surf. Coat. Technol. 201 (2006) 164.
[73] F. Wang, L. Wang, Q. Xue, Carbon 96 (2016) 411.
[74] Z. Wang, et al., Tribol. Lett. 41 (2011) 607.
[75] Y. Wang, et al., Carbon 77 (2014) 518.
[76] Z. Wang, et al., Solid State Sci. 90 (2019) 29.
[77] D. Berman, et al., Nat. Commun. 9 (2018) 1.
[78] K.C. Mutyala, Carbon 146 (2019) 524.
[79] M. Nosaka, et al., Proc. Inst. Mech. Eng. J. Eng. Tribol 230 (2016) 1389.
[80] M. Nosaka, et al., Tribol. Online 12 (2017) 274.

Chapter 13

Superlubricity of glycerol is enhanced in the presence of graphene or graphite

Yun Long[1], Maria-Isabel De Barros Bouchet[1], Andjelka Ristic[2], Nicole Dörr[2] and Jean Michel Martin[1]

[1]*Université de Lyon, Ecole Centrale de Lyon, LTDS, CNRS, Ecully, France,* [2]*AC2T research GmbH, Wiener Neustadt, Austria*

13.1 Superlubricity of glycerol by self-sustained chemical polishing

The mechanism of superlubricity [friction coefficient (CoF) < 0.01] of steel sliding against tetrahedral amorphous carbon (ta-C)-coated steel in presence of glycerol has already been investigated in a previous work [1]. Briefly, polishing of the steel wear scar is responsible for settling a sub-nm-thick film into contact area while the contact pressure remains at a relatively high value (about 265 MPa). Hence, experiments take place in thin film elastohydrodynamic lubrication (EHL) regime. Thanks to low-pressure viscosity coefficient and viscosity of glycerol at 50°C, the formed thin film is easy to shear. However the mechanism of polishing is not straightforward. Moreover, sliding contact surfaces are covered with a lamellar material (here FeOOH) possibly together with additional graphene oxide layers. Both contact surfaces are terminated by hydroxyl groups, ensuring a low CoF in the boundary lubrication regime (about 0.004). As to the counterpart, ta-C coating is characterized by a certain amount of sp^2-hybridized carbon. The role of this graphitic carbon in the polishing mechanism needs further investigation. In the first part of this chapter, we propose an answer by using two different ta-C coatings. Alternatively, we used the penciling technique to deposit graphite/graphene directly on the surface.

In the second part, we turn to ceramic materials lubricated by glycerol where superlubricity could also be demonstrated. In this case, the graphitic material is found embedded in the material, here polycrystalline silicon carbide (SiC). Raman spectroscopy, infrared spectroscopy, X-ray photoelectron

Superlubricity. DOI: https://doi.org/10.1016/B978-0-444-64313-1.00013-2

spectroscopy (XPS), and Orbitrap mass spectrometry (MS) analyses are performed to study the origin of superlubricity of the SiC/Si_3N_4 friction pair lubricated by glycerol. Special attention is paid to tribochemical reactions between glycerol and Si-based materials in forming aromatic compounds.

13.2 Materials and tribological experiments

Pure glycerol (≥99.5%) was purchased from Sigma-Aldrich (St. Louis, the USA). AISI 52100 steel balls with a diameter of 12.7 mm and steel disc specimens with an initial roughness *Ra* of 2.2 nm were acquired from PCS Instruments (London, UK). About 0.3-μm-thick ta-C-coated steel disc [ta-C (1)] was purchased from the Onward Company (Nomi, Japan). The 2-μm-thick ta-C coating [ta-C(2)] was produced by Fraunhofer IWS (Dresden, Germany) using a physical vapor deposition process (arc-ion plating) using a graphite target for the deposition onto a polished bearing steel disc. Coating hardness and Young modulus were measured by microindentation (Fig. 13.1). Deducing from the correlation between Young modulus and sp^2 content of ta-C proposed by Mabuchi et al. [2]. The sp^2 content of ta-C(1) and ta-C(2) was estimated as 32% and 44%, respectively.

The ball-on-disc sliding experiments were performed using a reciprocating tribometer with a sinusoidal speed profile. Before the tribo-experiments, both balls and discs were ultrasonically cleaned in heptane for 30 min followed by acetone for 5 min. In the contact area between the ball and disc, 50 μL of glycerol were supplied using a syringe. In order to ensure that the tribo-experiments were performed in a boundary lubrication regime according to the initial conditions, the normal load was fixed at 3 N corresponding to a maximum Hertzian contact pressure (P_{max}) of 577 MPa and an average pressure of 382 MPa. The maximum sliding speed in our experiments was 3 mm s^{-1}. Three temperatures were studied, that is, 30°C, 50°C, and 80°C. The stroke length was fixed at 2 mm to ensure a large kinematic length compared with the diameter of the contacting area (approximately 0.1 mm). All

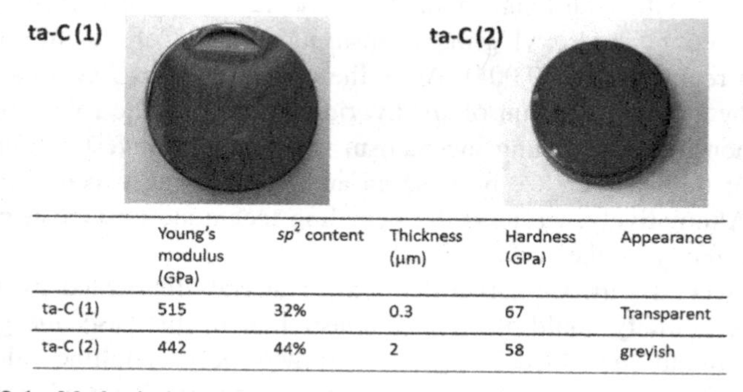

	Young's modulus (GPa)	sp^2 content	Thickness (μm)	Hardness (GPa)	Appearance
ta-C (1)	515	32%	0.3	67	Transparent
ta-C (2)	442	44%	2	58	greyish

FIGURE 13.1 Mechanical properties of ta-C-coated steel discs.

experiments were performed three times and demonstrated excellent repeatability. With the advantage of reciprocating movement, the accuracy of the measured CoF for this system was approximately 0.002. However, CoFs less than 0.004 are difficult to measure, and we took special care in the data processing (see further details about data processing in ref. [1]).

13.3 Friction and wear results of the steel/ta-C contact configuration

Starting with the friction properties of the two ta-C coatings in the steel/ta-C configuration lubricated by glycerol at 50°C, it is immediately seen that the lowest friction is rapidly obtained with the ta-C coating with the higher content in sp^2-hybridized carbon (Fig. 13.2). Regarding the wear scars on the ball, chemical polishing is visible in both cases but is significantly more pronounced for the ta-C coating containing the highest sp^2 content causing the smallest ball diameter scar. This may be due to a progressive conversion of sp^3 into sp^2-hybridized carbon under the friction conditions. Based on these data, we can conclude that there is a synergistic interaction between graphitic-like carbon and glycerol to reach superlubricity state in the experiment. It is remarkable that no wear debris is observed at the end of the experiment in used glycerol, even after centrifugation, suggesting dissolution of iron wear particles in glycerol after chemical attack.

Considering now the case of ta-C(1) coating in experiments carried out at different temperatures (Fig. 13.3), it is shown that chemical polishing is not obtained at 80°C and that superlow friction could not be established under the chosen conditions.

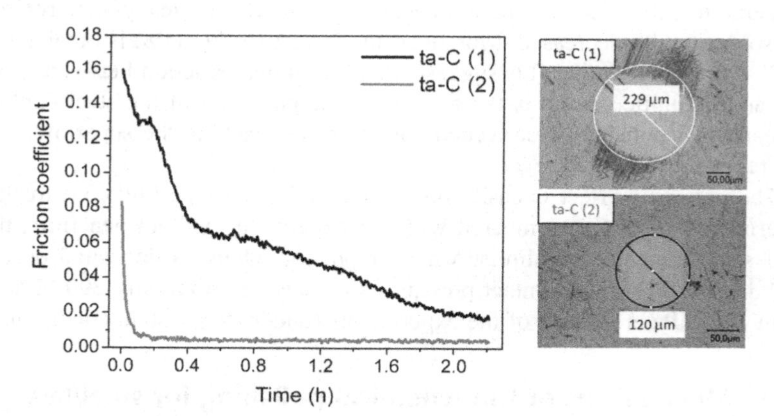

FIGURE 13.2 Evolution of friction coefficient of steel/ta-C(1) and steel/ta-C(2) friction pairs lubricated by glycerol at 50°C. Optical pictures taken by Keyence (VHX-6000) of the corresponding wear scars on the steel ball are shown. Only ta-C(2) containing a higher content of sp^2 carbon can provide superlubricity (friction coefficient below 0.01).

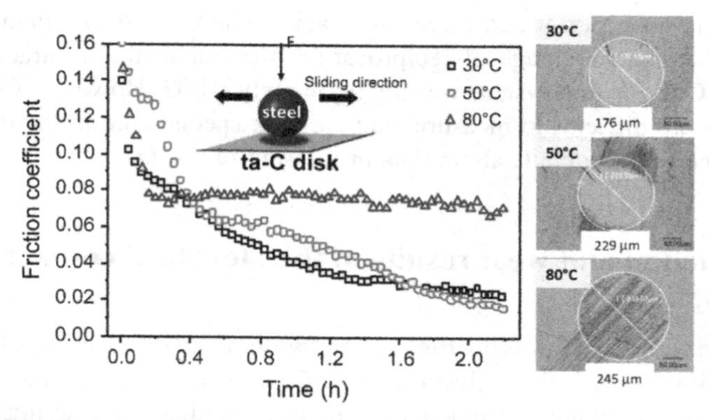

FIGURE 13.3 Evolution of friction coefficient of steel/ta-C(1) contact in glycerol at three temperatures. Optical pictures of the corresponding wear scars on the steel ball are shown. At 80°C, chemical polishing is not obtained and friction remains at a relatively high level.

As the presence of graphitic carbon seems to be a necessary condition to obtain superlow friction under glycerol lubrication, we used here the penciling technique already described by Zhang's group [3]. We selected four pencils HB, 2B, 4B, and 6B containing different graphite/graphene ratios (see Fig. 13.4). Pencil material was deposited manually on the ta-C(1) surface, whereby the graphitic layer was covering the whole area where the tribological experiments were carried out.

In Fig. 13.5, experiments with glycerol were performed exactly under the same conditions as those in Fig. 13.4, but at 50°C. Results clearly show that penciling affects the decrease of friction. At 30°C, this decrease is proportional to the graphite content in the pencil (see insert in Fig. 13.4). At both temperatures, the induction period necessary to reach the steady-state regime is also drastically decreased. Moreover, the lowest CoF (0.008) is obtained at 50°C after only 30 min. This suggests that a chemical reaction between glycerol and graphitic carbon is the reason for the polishing of the steel contact area and that polishing is a needed condition for reaching the superlow friction regime.

The polishing effect is easily observed in all cases (see Fig. 13.6 for the experiments at 50°C). Compared with the experiment without penciling, the wear scar diameters are similar when graphite/graphene is deposited on the ta-C disc. The average contact pressure is reduced from initially 390 MPa to about 130 MPa at the end of the experiments (under steady-state conditions).

13.4 Mechanisms of tribochemical polishing for steel/ta-C friction pairs

Fig. 13.7A illustrates the chemical polishing mechanism. A part of steel material has been progressively removed and the steel ball is truncated at the

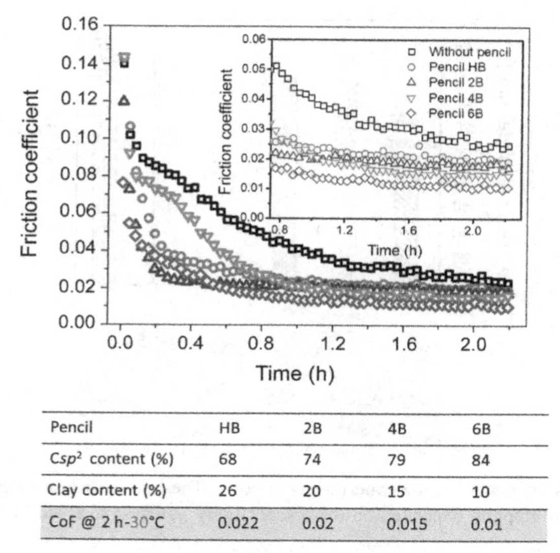

Pencil	HB	2B	4B	6B
Csp^2 content (%)	68	74	79	84
Clay content (%)	26	20	15	10
CoF @ 2 h-30°C	0.022	0.02	0.015	0.01

FIGURE 13.4 Evolution of friction coefficient of steel/ta-C(1) contact in glycerol at 30°C. Effect of penciling of the ta-C(1) disc prior to the friction experiment. The presence of graphite leads to friction decrease and superlubricity is easily obtained after using the 6B pencil.

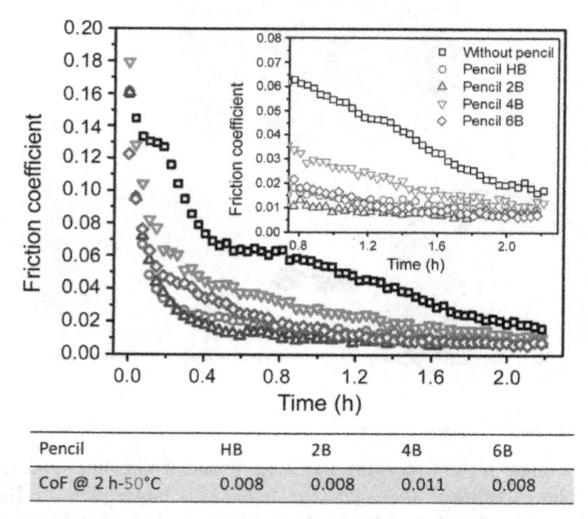

Pencil	HB	2B	4B	6B
CoF @ 2 h-50°C	0.008	0.008	0.011	0.008

FIGURE 13.5 Evolution of friction coefficient of steel/ta-C(1) contact in glycerol at 50°C. Effect of penciling of the ta-C(1) disc on the friction coefficient. The presence of graphite leads to friction decrease and superlubricity is obtained by using HB, 2B, and 6B pencils.

end (see schematic in Fig. 13.7B). The roughness inside the wear scar measured by atomic force microscopy indicates the existence of *Ra* values below 1 nm sub-nm *Ra* values. Compared with lubricant film thickness calculations [4], it is very likely shown that an ultrathin EHL film is established and

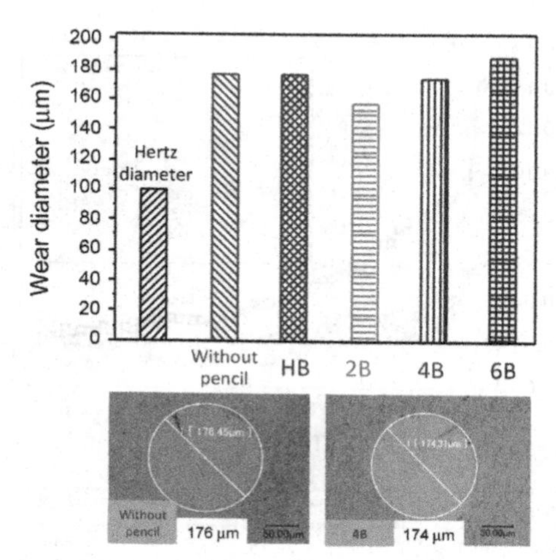

FIGURE 13.6 Wear results of the experiments at 50°C. The ball wear scar diameters are similar with or without graphite deposited on the discs. Highly polished steel surfaces are generated in all case.

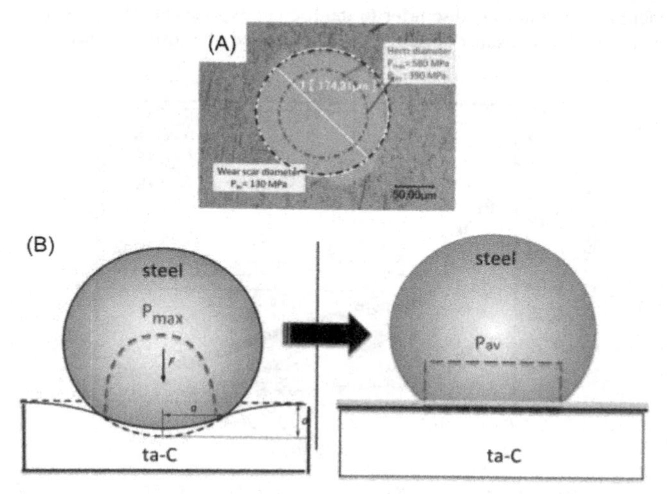

FIGURE 13.7 Schematic picture showing the mechanism of chemical polishing of the steel ball. At the end of the experiment, the wear scar diameter is larger than the Hertzian diameter (A) but the perfectly disc surface is highly polished and functionalized (B) on the right. The contact pressure does not collapse and remains at about 130 MPa, sufficient to form 1-nm-thick fluid film of glycerol to form a 1-nm thick fluid film of glycerol hydrodynamic film of glycerol.

further chemical polishing is therefore stopped (Fig. 13.7B). This model is supported by lambda ratio calculations (λ around 1 after polishing steel surface) [1]. EHL by glycerol in rolling-sliding conditions has been previously studied by Bjorling and Shi [5]. The superlow CoF is attributed to several factors including (1) the low pressure-viscosity coefficient of glycerol, (2) its high viscosity-temperature dependence, and (3) the role of thermally insulating diamond-like carbon material in favoring elevated temperature in the interface.

Special attention is paid to the polishing mechanism that has a key role in the establishment of superlubricity. The conversion of glycerol into acidic species has been widely reported [6]. The formation of ketones, aldehydes, and acids is favored in presence of metallic catalysts supported on graphite. Fig. 13.8 illustrates numerous oxidation products including acidic species that can be formed. In this case, it is thought that graphitic parts from the ta-C material concentrates at the surface and contains some iron nanoparticles formed by a transfer film from the steel in the induction period [1]. This ensemble could act as a catalyst in the contact to transform glycerol into acidic species. Therefore iron can be removed atomically by acidic attack, starting in zones of higher pressure, and this generates the smooth disc surface necessary for an effective thin film enabling EHL. Simultaneously, OH termination of the surface provides also a lower friction under the boundary lubrication regime. This mechanism looks similar to the lubrication of ceramics such as SiC and Si_3N_4 by water, which has been extensively studied in the past [8,9]. However, water is characterized by a significantly lower viscosity than glycerol. Thus higher sliding speeds are necessary to form a hydrodynamic water film that is supported by a drastic decrease of the contact pressure in the MPa range, associated with tremendous ball wear [10,11]. Some authors point out that silica gel can help to increase the viscosity of water. Nevertheless, the latter case is not acceptable for practical use. This is the advantage of using glycerol instead of water. Concluding, glycerol could be considered as a kind of "viscous water" in lubrication studies.

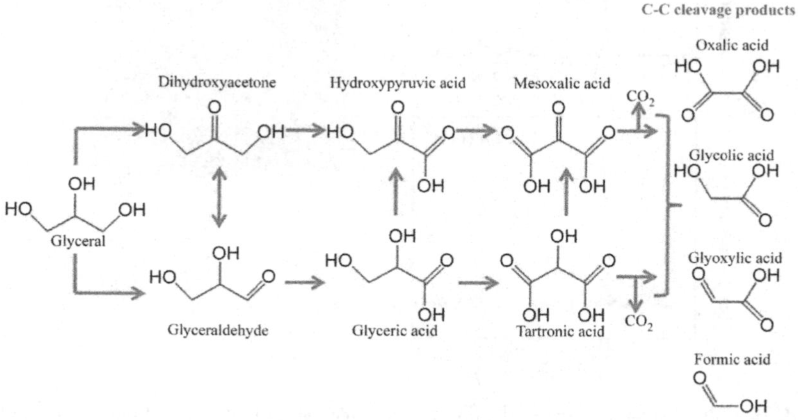

FIGURE 13.8 Degradation products of glycerol formed by oxidation [7].

13.5 SiC/Si_3N_4 contact configuration in glycerol

13.5.1 Materials and tribological conditions

In the second part, tribological performances of ceramics in glycerol were studied. SiC ball with a diameter of 11.12 mm was bought from Metalball (Crisolles, France) while Si_3N_4 disc was purchased from LianYunGang HighBorn Technology (LianYunGang, China). Both the ball and disk are manufactured by hot pressure sintering and are commercially available. The SiC ball was analyzed first by Raman spectroscopy, as shown in Fig. 13.9, where white areas and dark spots are observed by optical microscopy. It is obvious from the Raman spectra that the dark spots contain higher amount of graphitic material whereas white areas are mainly composed of silicon carbide. As for the Si_3N_4 disc surface, Raman analysis does not show the presence of any carbon-related peaks (result not shown here).

Concerning tribological test conditions, we used same tribometer, cleaning process of samples, stroke length, and sliding speed as in the case of steel/ta-C in glycerol. The normal force applied in these experiments is 1.9 N that corresponds to P_{max} and the average contact pressure of 750 and 494 MPa, respectively. Moreover, test temperatures were not only conducted at 30°C, 50°C, and 80°C but also at 120°C. Except the variation of temperature and P_{max}, all other tribological conditions are the same as in the first part of this chapter.

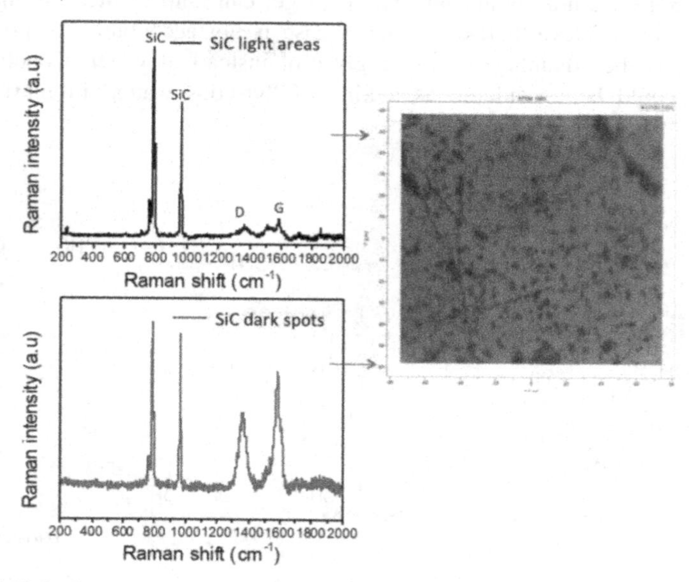

FIGURE 13.9 Raman analysis of the SiC ball. The dark spots in the optical image are rich in graphitic carbon.

13.5.2 Results and characterizations

The results of tribological experiments performed at different temperatures are presented in Fig. 13.10A. The findings are highly remarkable since superlubricity is observed at all temperatures after an induction period of about 0.5 h (corresponding to 1350 cycles). The detailed analysis of the friction cycles (Fig. 13.10B) clearly indicates the evolution of friction as a function of speed (or location along the stroke). After a duration of 2 h, the CoF is superlow and the friction force at the reversal points also becomes very low, where the friction is even not measurable with the equipment at hand. As for optical image of the SiC ball after friction, image (Fig. 13.10C) at 30°C has been chosen as a representative of 50°C and 80°C since wear diameter at those three temperatures are all around 140 μm. Taking Hertzian contact area diameter (70 μm) as reference, the apparent contact pressure drops from 494 MPa down to 124 MPa, which is reasonable value from the mechanical point of view. At 120°C, wear diameter on the SiC ball increases to 191 μm (Fig. 13.10E). Interferometer image (Fig. 13.10F) shows that the top of the SiC ball has been truncated. Moreover, the roughness of the SiC ball worn surface is measured as 11.5 nm (Fig. 13.10E). Considering the initial roughness of virgin SiC equal to 12.8 nm, no significant surface polishing has been observed unlike a previous work using glycerol as lubricant reported [1]. Since reciprocating tribometer has been used in friction

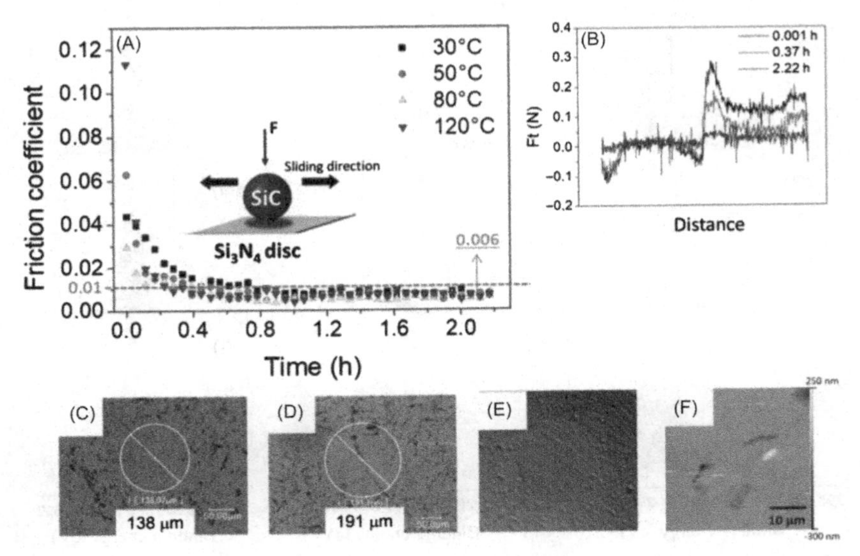

FIGURE 13.10 (A) Friction curves at different temperatures, (B) at 120°C, accurate friction measurements over one cycle at different durations, (C) optical image of SiC wear scar at 30°C, and (D)–(F) optical, interferometer, and AFM image of SiC ball after friction at 120°C. (*AFM*: Atomic force microscopy).

experiments, actual speed at one cycle varies from 0 to 3 mm s^{-1}. Even taking maximum speed (3 mm s^{-1}), EHL film thickness is calculated below 1 nm. For this value, the EHL contribution is highly questionable. Also the CoF is below 0.01, near the zero speed under the boundary conditions. Moreover, surface roughness values after friction are both above 10 nm, EHL film less than 1 nm would not be able to separate completely contacting surfaces. The contact of surfaces is dominated by contacts between asperities.

XPS analyses were performed on the SiC ball. As for virgin and outside wear scar, dominant contribution of C_{1s} peak is C = C bond since SiC samples is known to contain some graphitic materials. In comparison with virgin sample, the XPS analysis outside wear scar shows a decrease of SiC peak, indicating that less signal is detected from SiC substrate (Fig. 13.11A and B). Hence, adsorption layer of used glycerol is thicker than that in air. Moreover, C_{1s} signal augments compared with virgin sample (Table 13.1). This increase mainly comes from C = C content, that indicates that C = C bond maybe formed in used glycerol. C = C bond is also one of main peak in the wear scar (Fig. 13.11C). The other dominant peak is SiC, meaning that tribolayer is only

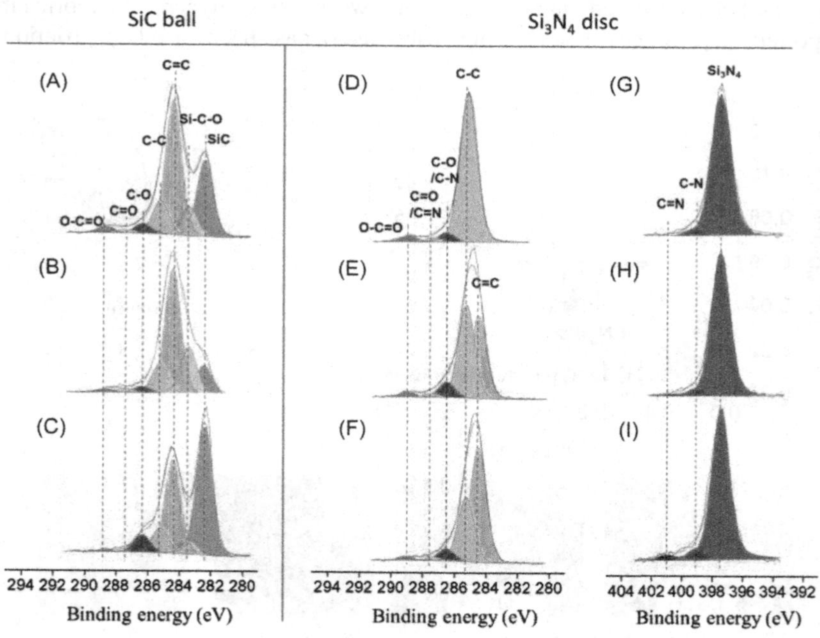

FIGURE 13.11 [(A)–(C)] XPS spectra of SiC sample's C_{1s} peaks: virgin, outside, and inside wear scar shown by sequence at take-off angle 45 degrees, [(D)–(F)] C_{1s} peaks of Si_3N_4 sample: virgin and inside wear scar with take-off angle 45 degrees and 25 degrees, respectively, [(G)–(I)] N_{1s} peaks of Si_3N_4 sample, the arrange sequence is same as (D)–(F).

TABLE 13.1 Element content of the different chemical states of SiC sample.

	C_{1s}	O_{1s}(%)	Si_{2p}(%)	N_{1s}(%)
Virgin	56.3	22.5	21.2	0
Outside	64.6	23.2	12.2	0
Inside	71.9	12.1	16.0	0

a few nm-thick. Inside the wear scar, more C–O contribution is also observed than outside. Furthermore, no nitrogen-containing species are transferred to the SiC ball (Table 13.1).

As for the XPS analyses of the Si_3N_4 disc, the biggest difference between the virgin sample and the wear scar of the sample is the presence of C = C (Fig. 13.11D–F). This either comes from transfer of graphitic materials of SiC or from tribochemical reaction betweeen ceramics and glycerol. Furthermore, there is an increase of C–O or C–N content. When information depth of angle-resolved XPS analyses approaches the very top surface (Fig. 13.11E and F), the major peak of C_{1s} peak shifts toward lower binding energy side which corresponds to the increase of C = C content. C = C bond becomes the main contribution of the first 4.4 nm of the upper surface. Regarding to N 1 s peak, C–N and C = N signals slightly show up with decrease of detection angle.

Infrared spectrometry was employed to analyze both virgin and used glycerol lubricant (Fig. 13.12). Similar spectra contributions of –OH, –CH–, and –C–O– are observed. However, only in used glycerol, a new peak is detected at 1650 cm^{-1}[12,13]. This peak is attributed to stretching modes of C = C and/or C = N bonds. However, because no C–N has been observed in the IR spectrum, this peak only comes from C = C stretching mode. This result also confirms XPS results showing an increase of C = C bond content inside wear track.

Complementary to surface analysis by XPS, high-resolution mass spectrometry was applied (Orbitrap MS) to glycerol collected from discs after the tribological experiments at 120°C. Electrospray ionization in the positive ion mode revealed aromatic structures with the involvement of nitrogen and oxygen, most probably included in the cycles, as exemplarily depicted in Fig. 13.13A and B. These compounds, $C_8H_{10}ON$ at *m/z* 136.0754 and $C_{10}H_{10}N$ at *m/z* 144.0805, were not detected in virgin glycerol, thus it is concluded that they were formed in the sliding SiC/Si_3N_4 contact.

The analysis of glycerol from tribological experiments via electrospray ionization in the negative ion mode provided mass spectra of further

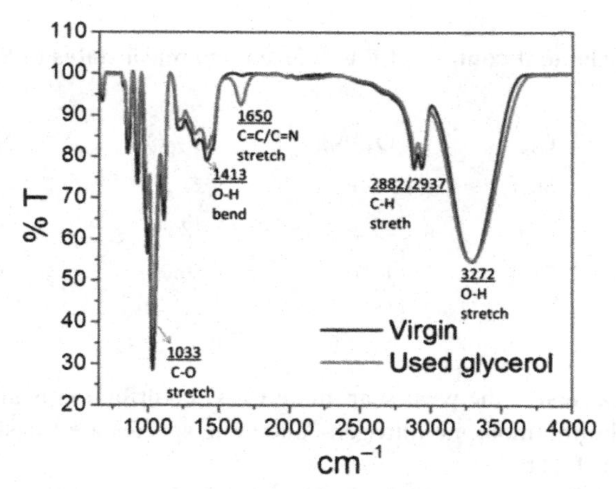

FIGURE 13.12 Infrared spectrums of virgin and used glycerol.

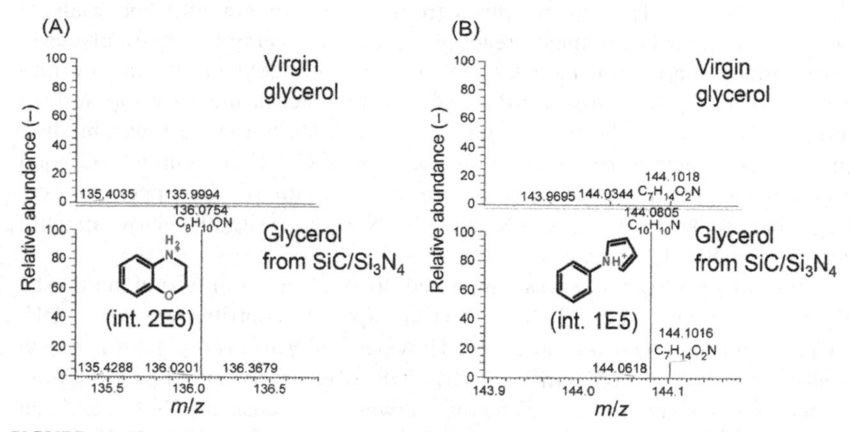

FIGURE 13.13 Detailed spectra in the positive ion mode obtained by Orbitrap MS. Virgin glycerol as reference (top). Intensities are given in brackets. A number of aromatic compounds were identified in glycerol collected after the tribological experiments at 120°C (bottom). (A) and (B) correspond to two different mass.

heterocyclic compounds such as $C_4H_4O_3N$ at *m/z* 1114.0199 in Fig. 13.14. Besides these tribochemical reaction products, also silicate-based reaction products such as CH_3O_6Si at *m/z* 138.9708 were detected. The structure of the latter is silicic acid and carbonic acid.

The chemical structures provided in Figs. 13.13 and 13.14 are based on the sum formulas obtained by high-resolution MS measurements. Structural confirmation usually provided by fragmentation of *m/z* of interest was not possible for all ions detected, especially those with aromatic moieties. Thus the chemical structures proposed are based on plausibility concerns.

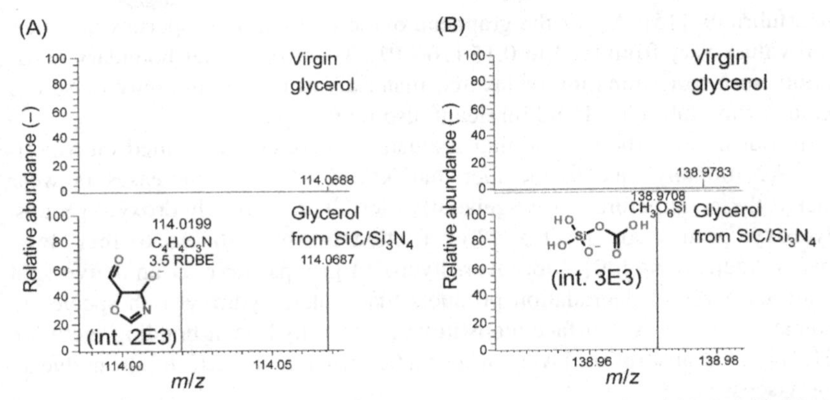

FIGURE 13.14 Detailed spectra in the negative ion mode obtained by Orbitrap MS. Virgin glycerol as reference (top). Intensities are given in brackets. A number of aromatic and silicate compounds were identified in glycerol collected after the tribological experiments at 120°C (bottom). (A) and (B) correspond to two different mass.

For the unambiguous assignment of the origin of carbon atoms in the tribochemical reaction products, either from SiC or glycerol, further investigations based on stable isotope labelling, in detail by ^{12}C and ^{13}C, are necessary.

13.6 Discussion

Superlubricity has been easily achieved from 30°C to 120°C for SiC sliding against Si_3N_4 in glycerol. This temperature change correlates with a glycerol liquid film thickness decreasing from 10 nm at 30°C to 0.5 nm at 120°C. Considering that no significant polishing effect has been observed, λ is always well below unity. Hence, contacts between ceramics surfaces come from their asperities with no hydrodynamic fluid film. In our case, contacts among asperities give a CoF equal to 0.006 due to only chemical terminations. Based on XPS analyses, both ceramic surfaces are terminated by a sp^2-carbon-rich layer that also has been detected in used glycerol. Orbital MS results suggest that structure may be aromatic.

Settling carbon-containing aromatic structures and confining them between contact asperities has proved to be extremely effective in reducing friction to the superlow regime. The generation of graphene oxide on ta-C surfaces by friction could lead to a CoF as low as 0.005 in mixed lubrication regime due to high content sp^2 passivating surfaces to inhibit interface C-C bond formation [14]. However once λ is inferior to 1, CoF increases up to 0.03. This is in good agreement with Xiangyu Ge's work where CoF between graphene-graphene is around 0.02 in macroscale and ethanediol liquid layer is indispensable to further separate contact surfaces and achieve superlubricity [15]. As for the graphene oxide lubrication properties in water, CoF values vary from 0.03 to 0.15 [16–19]. Therefore under boundary lubrication, surface termination of layered materials cannot be the only reason to achieve superlubricity. Fluid lubricant also plays a part.

In our case, carbon-containing aromatic structures are formed on asperities. Additionally, due to the fact that XPS C–O peak increases in wear track, these structures are generally terminated by hydroxyl groups. Hydroxyl groups could act as "low friction brushes" thanks to their low-energy interactions [20]. Moreover, hydroxyl groups serve as an active spot to attract glycerol degradation products like water, hydroxyl-rich species to aromatic structures' surface constituting so a hydrogen-bonding network [21,22]. This absorption layer on aromatic structure is easy to shear due to low viscosity.

13.7 Conclusions

This work reveals the underlying correlations between glycerol and graphite in lubrication to reach superlubricity regime. Both glycerol and graphite serve as carbon sources to form aromatic lamellar structures between contacting surfaces asperities. Glycerol permits a tribochemical polishing effect leading to very flattened surface asperities functionalized with higher load capacity. It also provides an environment rich in hydroxyl groups, which further contributes to friction reduction in the most severe situations. Thereby, these synergetic effects make possible to realize superlubricity experiments under boundary lubrication conditions.

References

[1] Y. Long, M.I.D.B. Bouchet, T. Lubrecht, T. Onodera, J.M. Martin, Sci. Rep. 9 (1) (2019) 6286.
[2] Y. Mabuchi, T. Higuchi, V. Weihnacht, Tribol. Int. 62 (2013) 130–140.
[3] B. Zhang, Y. Xue, K. Gao, L. Qiang, Y. Yu, Z. Gong, et al., Solid State Sci 75 (2018) 71–76.
[4] B.J. Hamrock, D. Dowson, J. Lubr. Technol 98 (1976) 223–228.
[5] M. Björling, Y. Shi, Tribol. Lett. 67 (1) (2019) 23.
[6] C. Crotti, E.J. Farnetti, Mol. Catal. A Chem 396 (2015) 353–359.
[7] J.C. Colmenares, R. Luque, Chemi. Soc. Rev 43 (3) (2014) 765–778.
[8] M. Chen, K. Kato, K. Adachi, Int. 35 (3) (2002) 129–135.
[9] H.C. Wong, N. Umehara, K. Kato, Tribol. Lett 5 (4) (1998) 303–308.
[10] M. Chen, K. Kato, K. Adachi, Wear 250 (1-12) (2001) 246–255.
[11] D.A. Rani, Y. Yoshizawa, H. Hyuga, K. Hirao, Y.J. Yamauchi, Eur. Ceram. Soc. 24 (10-11) (2004) 3279–3284.
[12] L. Zhong, K. Yun, Int. J. Nanomedicine 10 (Spec Iss) (2015) 79.
[13] R.L. Pecsok, L.D. Shield, *Modern Methods of Chemical Analysis*, J. Wiley and Sons, New-York, 1968.

[14] M.I.D.B. Bouchet, J.M. Martin, J. Avila, M. Kano, K. Yoshida, T. Tsuruda, et al., Sci. Rep. 7 (2017) 46394.

[15] X. Ge, J. Li, R. Luo, C. Zhang, J. Luo, ACS Appl. Mater. Interfaces 10 (47) (2018) 40863–40870.

[16] H. Kinoshita, Y. Nishina, A.A. Alias, M. Fujii, Carbon 66 (2014) 720–723.

[17] H. Xie, B. Jiang, J. Dai, C. Peng, C. Li, Q. Li, et al., Materials 11 (2) (2018) 206.

[18] H.J. Song, N. Li, Appl. Phys. A 105 (4) (2011) 827–832.

[19] H.J. Kim, D.E. Kim, Sci. Rep. 5 (2015) 17034.

[20] M. Kano, Y. Yasuda, Y. Okamoto, Y. Mabuchi, T. Hamada, T. Ueno, et al., Tribol. Lett. 18 (2) (2005) 245–251.

[21] J.M. Martin, M.I.D.B. Bouchet, C. Matta, Q. Zhang, W.A. Goddard III, S. Okuda, et al., Phys. Chem. C 114 (11) (2010) 5003–5011.

[22] W. Wang, G. Xie, J. Luo, ACS Appl. Mater. Interfaces 10 (49) (2018) 43203–43210.

Chapter 14

The role of lubricant and carbon surface in achieving ultra- and superlow friction

Volker Weihnacht and Stefan Makowski
Fraunhofer Institute for Material and Beam Technology (IWS), Dresden, Germany

14.1 Introduction

The fascinating phenomenon of superlow friction describes the presence of very low coefficients of friction in the order of $\mu = 0.01$ and below. Classical materials exhibiting superlubricity behavior are molybdenum and tungsten disulfide, substances from the class of dichalcogenides also known as solid lubricants. Extremely low friction of such materials can be explained by crystal-structural incommensurability. However, it later became apparent that the material systems with extremely low friction are very diverse and not restricted to crystalline structures, that combinations of different materials can also lead to superlubricity and that superlubricity can occur under all lubrication conditions, from dry friction to hydrodynamic lubrication [1,2].

Amorphous carbon (a-C) coatings, which are sometimes also referred to as diamond-like carbon (DLC), have an outstanding position among materials with superlubricity properties. These coatings, as defined in ISO 20523, consist of different ratios of diamond-like (sp^3-) to graphite-like (sp^2-) hybridized carbon atoms and comprise hydrogen-free amorphous carbon coatings (a-C), hydrogen-free tetrahedral amorphous carbon coatings (ta-C), and hydrogen-containing amorphous carbon coatings (a-C:H) with further addition of metals (a-C:H:Me) and nonmetals (a-C:H:X).

Superlubricity was found on a-C:H coatings with a high H content [3] and on a-C:H:Si coatings [4]. With these amorphous structures, very smooth, self-passivated surfaces can slide on each other, repel the counterbody and prevent covalent bonds with it. If the surface passivation is damaged, it can regenerate from elements contained in the bulk (e.g. H termination of a-C:H coatings) or ambient gas (e.g. N termination in CN_x coatings in N_2 atmosphere), so that very smooth passive surfaces can be formed by running-in.

Superlubricity. DOI: https://doi.org/10.1016/B978-0-444-64313-1.00014-4

The state of superlubricity is disturbed in the case of gas adsorption processes [5], for example, by polar gases such as water vapor and atmospheric gases.

While superlubricity on hydrogen-containing amorphous carbon occurs under vacuum and N_2-atmosphere conditions, superlubricity on hydrogen-free, dominantly sp^3-bonded ta-C is only observed in combination with certain fluids. However, superlubricity with ta-C coatings is of particular interest for technical applications, since ta-C coatings with their high hardness not only reduce friction, but also provide excellent wear protection under various conditions.

For the low-friction behavior of a-C:H coatings in vacuum and with gas lubrication there are plausible models, all based on (self-)passivation of surface dangling bonds by internal hydrogen or gas species from the atmosphere. In contrast, the comprehensive description of the ta-C superlubricity in combination with certain fluids is apparently a complex topic. Therefore, there are different explanation models for this.

In pioneering work in the early 2000s, Kano et al. investigated the interaction of ta-C coatings with the friction modifier additive glycerol monooleate (GMO), a simple fatty acid ester with glycerol, which is used on an industrial scale as an emulsifier and friction reducer. Even in low concentrations of 1% in a synthetic polyalpha olefin (PAO) base oil, unexpectedly low friction values in oscillation ($\mu = 0.03$) and rotation ($\mu < 0.001$) were achieved in tribometer tests in a ta-C/steel ball contact [6,7]. Surface analysis with X-ray photoelectron spectroscopy (XPS) and secondary ion mass spectrometry (SIMS) indicated hydroxylation of the surface carbon atoms. For a-C:H coatings this effect does not occur [6,8,9].

These studies took place in the environment of the automotive industry and so Nissan introduced ta-C-coated valve-train tappets into series production for large-volume gasoline engines shortly afterward. With a special ester-additive engine oil, significant friction advantages were achieved, but not as low as in the laboratory [10].

The exact mechanisms leading to the low friction were unknown, but the formation of self-assembled monolayers according to the Hardy model as on steel surfaces was implausible [11] and could not be confirmed by experiments as well [12].

Martin et al. took up the topic and also found superlubricity for glycerol and other polyalcohols [13–15]. From extensive experiments [11,15,16] with deuterated gases in vacuum, hydrogen peroxide activation, molecular dynamic (MD) simulation as well as analyses with electron energy loss spectroscopy (EELS), SIMS, and XPS the hydrogen bond network model was derived, which consists of two essential steps [17]: (1) partial decomposition of the hydrocarbon lubricants and formation of water in the lubricant mixture, (2) hydroxylation of the ta-C surface, on which now polar surface water molecules form a low-viscosity slip plane.

For these processes an activation is assumed, since these partly take place only above a certain temperature [11] or only after a certain running-in period [6]. More recent tests with glycerol-like ethylene glycol, in which the lubricant is replaced by new ones in the briefly interrupted test, achieve low coefficients of friction [12] without renewed running-in.

Recent analyses, using depth-resolved X-ray near-edge spectroscopy on a superlubricity tribosystem with oleic acid lubrication, shows graphene oxide with a thickness of only 1 nm on the surface of the ta-C coating. The partially oxidized and hydroxylated surface prevents covalent welding with the counterbody, the smooth surface promotes friction in the assumed elastohydrodynamic lubrication range. It is speculated that the double bond of oleic acid contributes to the formation of the aromatic graphene-oxide surface [9].

Such ordered, atomically thin surface structures were also formed on a-C:H:Si coatings running in a passive atmosphere [4] or can be produced on surfaces by means of dispersed graphene-oxide nanoflakes [18] and detected by transmission electron microscopy.

Both during low friction in humid air and during superlubricity with GMO or polyhydric alcohols, there is a concisely opposite behavior of ta-C and a-C:H coatings. One theory for this could be the process of Pandey reconstruction, which only takes place on hydrogen-free carbons and is inhibited by hydrogen. Under tribological load and presence of water, this process leads to the formation of planar, graphene-like structures on the surface, which are characterized by low roughness, stiffness, and stable passivation and have led to significantly reduced friction in simulations [19].

For some time now, the influence of rehybridization processes on carbon coatings near the surface has also been discussed, which are investigated by EELS, energy filtered transmission electron microscopy (EFTEM), and MD simulation.

After tribological loading, an amorphous layer with an increased proportion of sp^2-hybridized carbon atoms is found on the surface, which was converted from originally sp^3-hybridized carbon. The thickness of the TPT layer (tribo-induced phase transformation) is several tens of nanometers and was found in ta-C coatings [16], coarse diamond [20], and nanocrystalline diamond [21]. Other authors measure converted areas up to 200 nm thickness on top of ta-C and a-C:H coatings by spectroscopic reflectometry, which influence the friction properties [22]. A decrease in hardness of a ta-C surface after heavy wear with a soot-blended engine oil could also be detected by nanoindentation [23].

In some works, the questionable conclusion is drawn that the low frictional properties of the surface are due to the "graphitized" surface [24–26], implying graphite-like friction properties. However the process only involves an increase of sp^2-hybridized bonds in the amorphous state and at best the formation of partially structured regions (nanoclustering), which increases the D-peak in the Raman spectrum [27]. A planar structure with planes of the graphite sliding on each other is not generated. Therefore, the term rehybridization instead of graphitization is more appropriate for the described material change.

Since the rehybridization of a-C is not known as a mechanism of solid lubrication and therefore possibly only an accompanying phenomenon of tribological loading [28], it is questioned by other authors as a low friction mechanism [4,19,20,29].

Also with regard to the wear properties of ta-C under ultra- and superlow friction, there is a very differentiated behavior both in the experimental description and with regard to the tribological model ideas. In some of the early works of Kano et al. [6] and de Barros Bouchet et al. [13], the wear of ta-C coatings was not measurable. In later investigations, however it was shown that ta-C coatings can also be subject to relatively high wear while exhibiting ultralow friction. Due to the very low coefficient of friction, not abrasive, but tribochemical processes are held responsible as wear mechanisms [30–32].

In the preliminary work of the authors [31] in contacts lubricated with biodiesel, there was a sudden drop in friction from initial $\mu > 0.1$ to $\mu < 0.04$, both spontaneous and temperature activated. Furthermore, the contact pressure seemed to have a considerable influence on wear. Comparable results were achieved in contacts lubricated with ethylene glycol [12].

In some works, the role of the metallic counterbody is addressed [32–35]. For example, when lubricating with PAO [33] or rapeseed oil [35], higher wear occurs on the hard ta-C coating when running against a steel mating body than against a similar ta-C coating. In connection with high wear of ta-C with PAO lubrication, a carbon diffusion into the mating body is discussed which depends on its affinity and carbide formation. It is pronounced for a steel counterbody, but not present when coated with noncarbide-forming metals [32].

A similar wear-associated, temperature-activated mechanism of carbon diffusion has already been described for a PAO-lubricated a-C:H/steel contact [34].

Several authors report that a reduction in friction and an increase in wear is significantly more pronounced for ta-C than for a-C:H. Lubrication with a pure PAO base oil results in wear at ta-C which is 50 times higher than at a-C:H [30], although ta-C is much harder. The difference could be explained by the self-passivation of a-C:H, which quickly saturates the dangling bonds resulting from shear processes [3,36–38] and prevents the reactivity observed on ta-C coatings.

However, some studies also conclude that fatty acid-based lubricants have a positive effect on wear. For example, when oleic acid is added to vegetable oil for artificial aging in an a-C/steel system, wear is observed to decrease [24]. In a ta-C/steel system the wear decreases with the content of unsaturated C = C bonds in the vegetable oil [39].

In summary, the studies show a large discrepancy in the friction and wear results achieved on ta-C coatings. This is probably due to the fact that the mechanochemical behavior of ta-C surfaces is relatively complex in interaction with certain fluids and depends on various factors. An important fact is that the previous studies were carried out with very different loading and lubrication conditions. The type of ta-C coatings investigated also differed in the studies.

The present contribution therefore aims to carry out systematic experimental studies on the friction and wear behavior of ta-C coatings under potential superlubricity conditions and to study the influence of various parameters of the tribological system. The aim is to close experimental gaps and to clarify previously contradictory behavior. The following influences are to be investigated in particular:

- sp^3-bond fraction in ta-C coatings, comparison to a-C, a-C:H, and crystalline diamond
- loading condition/contact pressure
- chemical functions of lubricants

The work was carried out using coating machines, coating systems, and characterization methods that are already employed in industrial series production. Furthermore, the results take into account the authors' experience and comprehensive understanding of the whole value added chain of ta-C technology. Therefore the tribological investigations are of great relevance with regard to their fast transferability to technical applications.

In addition to the motivation to achieve extremely low coefficients of friction with the ta-C layer in order to achieve enormous CO_2 savings, this study also focuses on the aspect of green lubrication. So far, natural fatty acids and their glycerides in particular have shown the greatest potential for reducing friction on carbon coatings. In the most favorable case, the superlubricity of ta-C can also be combined with the use of vegetable oils.

14.2 Experimental details

14.2.1 Materials and coatings

The standard base material, used as an uncoated reference and as substrate for all coatings, was a hardened low-alloy chromium steel (100Cr6, EN 1.3505, SAE 52100), typically polished to $R_a < 20$ nm.

All a-C and ta-C coatings were deposited by the pulsed laser-arc technique with plasma filtering, using a commercial physical vapor deposition coating system, consisting of a deposition chamber with an attached laser-arc module [40,41] and a plasma-filtering unit. The module is a plasma source of highly ionized carbon species operating with a pulsed laser-induced vacuum arc process.

Prior to carbon coating, an Ar^+ ion etching was performed, using a hollow-cathode plasma source. After etching, a chromium adhesion interlayer with thickness of about 100 nm was deposited by magnetron sputtering. The samples were mounted in a twofold rotation arrangement corresponding to a real situation of an industrial coating of tribological components. The deposition rate during a-C and ta-C deposition by laser-arc was about $1\ \mu m\ h^{-1}$.

All a-C and ta-C coatings in this study have been smoothed prior to tribological testing to minimize the release of hard asperities and growth defects into tribological contact. The smoothing process used in this study was conventional lapping with diamond slurry, which is well suited for flat surfaces and small lot numbers, and commonly results in a mirror finish. Roughness was measured using the profile method (ISO 3274, ISO 4288), applying a diamond tip radius of $r_{tip} = 2\ \mu m$ and a filter cut off of $l_c = 80\ \mu m$. A roughness of $R_a < 20$ nm was found for all coatings.

Mechanical properties of the amorphous carbon coatings were adjusted by varying the carbon energy during deposition and hence the fraction of sp^3-bonded carbon, which directly correlates with Young's modulus E and hardness H. Using laser-induced surface acoustic wave spectroscopy, the coating Young's modulus was quantified [42] with a commercial LAwave measurement system. Additionally, hardness of coatings was measured with either an UNAT (ASMEC, Germany) or ZHN (Zwick Roell, Germany) nanomechanical hardness tester, with both systems being well comparable by sharing same routines for calibration, measurement, and evaluation. A Berkovich indenter was loaded to a maximum force of 100 mN, using the quasicontinuous stiffness method (QCSM). Considering the relation $E = 10 \bullet H$, valid for ta-C [43], both methods agree well for the mechanical properties. Furthermore, the sp^3 fraction was calculated from Young's modulus [44] obtained by LAwave method. Coating thickness was measured by crater-grinding method and confirmed by the LAwave measurement, which yields film thickness as well.

In this study, a-C coatings are identified by their Young's modulus given as a subscript index. They range from a-C_{140}(low stiffness, low hardness) to ta-C_{770} (high stiffness, superhardness), corresponding to an estimated 4% and 97% sp^3 fraction, respectively.

In addition to a-C and ta-C, also an a-C:H coating was investigated. The a-C:H was synthesized by a commercial supplier using a plasma-enhanced chemical vapor deposition process. Its hardness was about 25 GPa, which corresponds to a hydrogen content of approx. 20% [45]. After coating, the surface was sufficiently smooth and did not require additional smoothing.

Finally, also single crystalline diamond was tested. The diamond was grown in a microwave-powered plasma discharge to a thickness of more than 1 mm, using a H_2/CH_4 atmosphere [46,47]. This diamond plate was glued onto a 100Cr6 steel disc with a {100} surface facing upward and then lapped to $R_a < 10$ nm. An overview of all tested surfaces is given in Table 14.1.

As counterpart specimens for tribological tests polished balls ($d = 10$ mm) or pins with spherical face ($d = 200$ mm) were used, both made from hardened 100Cr6 steel. For some experiments pins were coated with ta-C_{430} or, alternatively, ta-C_{650}

14.2.2 Lubricants

The focus of this work is on lubricants that can be obtained from vegetable oils. They are mainly composed from fatty acids, which consist of a linear hydrocarbon molecule, often with a carbon chain length of 18 (C18) and, when present as a free fatty acid, a terminal acid group (COOH). The aliphatic chain of fatty acids often contains one or more unsaturated double bonds. Fatty acids do not normally occur in nature as free fatty acids, but in

TABLE 14.1 Properties of steel and carbon materials used for tribological experiments. Young's modulus was measured using LAwave technique, coating thickness was measured using crater-grinding method, sp^3 fraction was calculated from Young's modulus and hardness was measured using nanoindentation.

Surface material	Deposition process	Young's modulus (GPa)	Hardness (GPa)	sp^3 fraction (%)	Coating thickness (μm)
Steel 100Cr6	None	220	9	–	–
a-C_{140}	PVD	140	15	4	2.6
ta-C_{420}	PVD	420	45	52	2.8
ta-C_{430}	PVD	430	40	53	3.7
ta-C_{650}	PVD	650	70	83	3.1
ta-C_{770}	PVD	770	N/a	97	2.6
a-C:H	PECVD	240	25	–	2.8
Single crystalline diamond	CVD	1050[a]	~100[a]	100[a]	>1000

Notes: *PECVD*, plasma-enhanced chemical vapor deposition; *PVD*, physical vapor deposition.
[a] *Theoretical value.*

vegetable oil in the form of triglycerides, which can be technically used by decomposition of the molecule by hydrolysis and transesterification. The fatty acid-based molecules used and their characteristic functions are discussed below:

- The oleic acid as starting point for the systematic variation is an unsaturated C18-fatty acid with the carboxyl group (COOH) and the double bond (C = C) in characteristic cis-configuration, which is the major constituent in many vegetable oils.
- Elaidic acid is an isomer of oleic acid and has the double bond in trans-configuration.
- Stearic acid is also a C18 fatty acid, but has no unsaturated bonds.
- GMO is an ester (COO*R*) of oleic acid and glycerol. The latter thus has still two remaining hydroxyl groups (OH). GMO is produced by hydrolysis when two fatty acids are separated from a triglyceride. The substance is used on an industrial scale with lower purity as a friction reducer and food emulsifier.
- Oleic acid methyl ester is an ester of oleic acid and methanol and is the main component of biodiesel.
- The rapeseed oil investigated here is a refined cooking oil. It consists essentially of triglycerides, with major constituent of oleic acid [35].

Reference lubricants are investigated to classify the observed effects:

- Glycerol (propane-1,2,3-triol) is a trivalent alcohol (OH) with industrial use. It is one of the few lubricants for which, in addition to GMO and oleic acid, superlubricity has also been described and investigated in the literature.
- Octadecane is a straight alkane with a chain length of 18 carbon atoms. It has no other functional groups and therefore serves as a direct reference for all C18 fatty acids.
- A fully additivated engine oil based on mineral oil and a classification of 0W30 was included in the test series to reflect the state of the art.
- An overview of all tested lubricants is given in Table 14.2.

14.2.3 Tribological testing

For the experiments, a tribometer-type SRV®4 from *Optimol Instruments* with the standard oscillation setup was used, which corresponds to the translatory tester according to DIN 51834-1:2010-11.

In this tribometer, the upper test piece is subjected to a defined normal force via a spring preloaded by a servomotor and oscillated in a horizontal direction via a voice coil linear motor.

For the tribological test, a lower disc was paired with an upper ball or, alternatively, cylindrical pin with a spherical face. The test specimen and

TABLE 14.2 Overview of the lubricants used in the work, the purity and the number of molecular groups characteristic of fatty acids.

Name	Formula	Purity	C_n	–OH	–C=C–	–COOH–	–COOR
Oleic acid	$C_{17}H_{33}COOH$	99%	18	–	1 (cis)	1	–
Elaidic acid	$C_{17}H_{33}COOH$	–	18	–	1 (trans)	1	–
Stearic acid	$C_{17}H_{35}COOH$	98%	18	–	–	1	–
Glycerol monooleate (GMO)	$C_{17}H_{33}COOC_3H_5(OH)_2$	~40%	18	2	1 (cis)	–	1
Oleic acid methyl ester	$C_{17}H_{33}COOCH_3$	>99%	18	–	1 (cis)	–	1
Rapeseed oil	e.g. $(C_{17}H_{33}COO)_3C_3H_5$	–	For example, 3 × 18	traces	2–3 (cis)	traces	3
Octadecane	$C_{18}H_{38}$	–	18	–	–	–	–
Glycerol	$C_3H_5(OH)_3$	>98%	3	3	–	–	–
Engine Oil 0W30	–	–	–	–	–	–	–

holder are cleaned ultrasonically in high-purity benzene before the test. Approximately, 0.3 ml lubricant is applied to the contact. Before the test, the disc is brought to the test temperature by means of a heater in the sample table and equilibrated for 5 min. The tribological test usually lasts 60 min, unless the test is interrupted earlier. The control program records the friction coefficients at a sampling rate of 2 Hz.

In this work two different parameter sets were used with regard to the tribological loading.

Parameter set I was selected for broad screening tests to determine friction and wear. Characteristic features are high contact pressure and asymmetric pairing of a coated disc and an uncoated ball counterpart. The parameters are based on the standard test procedures for lubricating oil testing according to DIN 51834-2 or ASTM D6425, but with adaptations to the conditions of a thin coating.

Parameter set II was used for the selective investigation of the superlubricity phenomenon. In this case a flat-curved pin was used to achieve low contact pressures. This geometry also allowed for a long-term stable coating on the counterpart, and therefore allowed paring of self-mated ta-C-coated surfaces.

Both parameter sets are shown in detail in Table 14.3. If deviations were made for individual test series, these are explicitly mentioned later.

The following specific considerations apply to parameter set II:

- The cylindrical pin with a spherical face as counterbody enables ta-C-coated surfaces on both sides to be brought into defined contact and a rather low and constant average Hertzian pressure of approx. $p_{\mathrm{Hertz}} = 160$ MPa to be achieved.
- Due to the significantly larger contact diameter of more than 600 μm, there would be no sufficient rewetting with lubricant at an oscillating amplitude of 1 mm. The amplitude is therefore increased to 5 mm, to guarantee reliable rewetting.
- The oscillation frequency is reduced to 10 Hz so that an average sliding speed of 0.1 m s^{-1} is achieved for both parameter sets.

The measuring principle of the SRV® tribometer is to determine the mean value of the maximum friction forces in both directions from an oscillation cycle and write it to the buffer memory, when it is higher than the buffer value. At an oscillation frequency of 50 Hz, the maximum value of 25 consecutive oscillation maxima is used, given the sampling rate of 2 Hz selected in this paper. In contrast to many other friction data in the literature, based on average or effective values of an oscillation cycle, the friction data in this work are "peak friction values" due to the type of tribometer used.

The coefficient of friction is recorded during the whole tribological experiment. Depending on the curve characteristics, one of two different methods was used to extract a mean friction coefficient: (1) In the case that

TABLE 14.3 Overview of the test parameters used for the screening studies (I) and for selective studies of superlubricity (II).

Parameters	Set I	Set II
Oscillation frequency (Hz)	50	10
Stroke (mm)	1	5
Average sliding speed (m s^{-1})	0.1	0.1
Test time (min)	60	60
Temperature (°C)	80	80
Lower specimen	Disc, uncoated	Disc, uncoated/coated
Upper specimen	Ball (d = 10 mm), uncoated	Spherical pin (d = 200 mm), uncoated/coated
Normal load (N)	50	50
Initial average contact pressure (MPa)	≈1,150	≈160
Initial contact radius (μm)	120	320
Lubricant volume (mL)	≈0.3	≈0.3
Application of lubricant	Initial	Initial
Environment	Ambient air	Ambient air

the friction slowly reaches a saturation value, the average of the last 30 min is given. (2) If increased wear occurs on the coating, the friction decreases continuously until the coating is worn out. The friction will then increase again, forming a local minimum. In such a case, the experiment was interrupted before the end of the 60-min test period and the minimum coefficient of friction was specified.

The coating wear for *parameter set I* was determined using the wear volume on the coated disc measured by topographic white light interferometry. The wear coefficient k was calculated by normalizing the wear volume with normal force and sliding distance.

For *parameter set II*, the wear volume or wear coefficient was not determined. The reason for this is that in many cases no measurable wear occurred, so that the wear measurement could not be carried out systematically.

The wear on the balls could not be measured systematically and volumetrically. In order to calculate the actual wear volume, it is necessary to sufficiently record the ball geometry around the wear track. In the case of heavy wear, however, the area was so large that the surrounding ball surface was too steep to be recorded by means of white light interferometry. The wear was therefore quantified by measuring the wear track diameter using an optical microscope. The geometric surface pressure at the end of the tribological test could also be determined via the projected contact surface and normal force.

14.2.4 Raman spectroscopy

A Renishaw inVia Raman microscope with a 50 × objective and an excitation wave length of 514 nm was used. The laser power on the sample was estimated to be 0.8 mW. The spectra were fitted using a linear base line, a Lorentzian line for the D peak and a Breit-Wigner-Fano line for the G peak [27]. For numerical evaluation, the position of the G peak maximum and the peak height ratio of D and G peaks were calculated accordingly.

14.3 Results and discussion

14.3.1 Screening of friction and wear properties (parameter set I)

In the following investigations, tribometer parameter set I was used, a typical ball on disc configuration with small contact area and relatively high contact pressure allowing to measure both friction and wear coefficients.

First, the friction and wear behavior of a standard ta-C coating with medium hardness (ta-C_{550}) was investigated against a 100Cr6 steel ball in a screening test with the following lubricants: engine oil, glycerol, oleic acid, GMO, oleic acid methyl ester, rapeseed oil, and stearic acid. Fig. 14.1 shows the results for friction and wear of the disc.

All fatty acid-based lubricants, including GMO are shown in orange color. They all have extremely low friction, regardless if molecules are present as free fatty acids (oleic acid and stearic acid), as monoesters (GMO and oleic acid methyl ester) or as a natural triglyceride with a mixture of

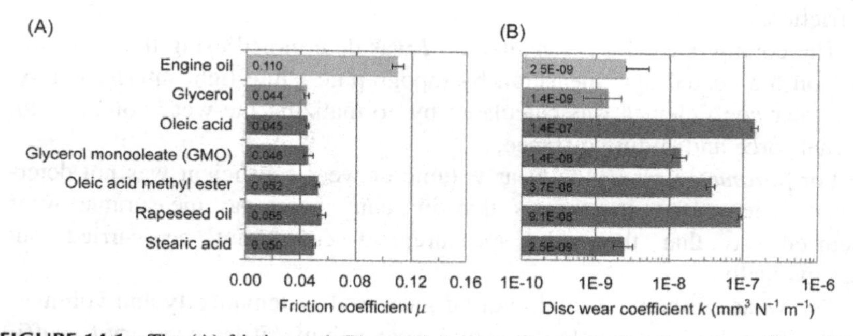

FIGURE 14.1 The (A) friction and (B) wear behavior of a standard ta-C coating with medium hardness (ta-C_{550}) was investigated against 100Cr6 steel in a screening with various lubricants. The color coding corresponds to engine oil, glycerol, and all fatty acid-based lubricants.

different fatty acids (rapeseed oil). Comparable low friction is also obtained for glycerol, while friction is much higher for engine oil.

The wear of the ta-C-coated disc does not correlate directly with friction and varies strongly. All lubricants based on fatty acids, with the exception of stearic acid, result in an increase in ta-C wear of 1–2 orders of magnitude compared to the lowest observed level of wear.

A direct comparison of stearic acid and oleic acid suggests that the double bond is crucial for high wear. The oleic acid is almost identical to stearic acid in terms of chain length and structure, but has a reactive double bond in the alkyl chain. As oleic acid is contained in GMO, oleic acid methyl ester, and rapeseed oil as main component, wear is increased with these lubricants as well.

While the low friction occurs with various organic, oxygen-containing molecular functions, for example, with ester, acid, and hydroxyl groups, the remarkably increased wear occurs in connection with unsaturated fatty acids. For free oleic acid the wear is highest, for GMO it is slightly lower.

The phenomenon of the sometimes extraordinarily high differences in wear was investigated more intensively by comparing glycerol-lubricated ta-C with oleic acid-lubricated ta-C, applying Raman spectroscopy inside and outside the wear track. Due to the extremely low wear coefficient with glycerol lubrication, the test duration was extended to 10 h in order to obtain a visible wear track. The test with oleic acid was already terminated after 15 min due to the heavy wear of the coating in order to leave a remaining ta-C layer for Raman investigation.

The results of Raman spectroscopy are shown qualitatively in Fig. 14.2. The ta-C coating lubricated with glycerol (Fig. 14.2A) has an identical Raman spectrum in and outside the wear track. The absence of a D peak indicates a completely amorphous structure, the symmetrical shape of the G

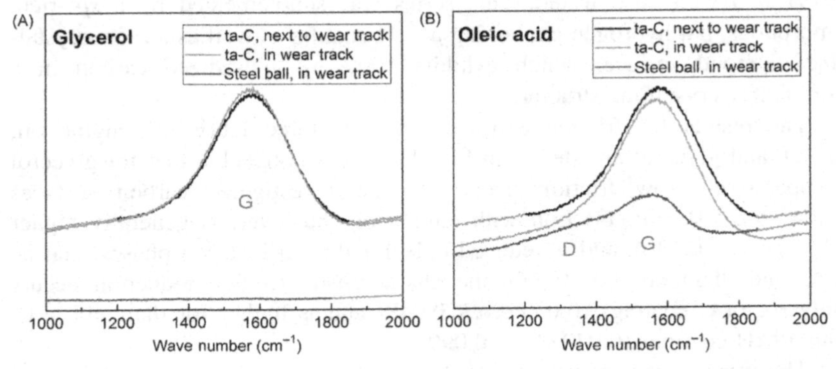

FIGURE 14.2 Raman spectra for the ta-C coating in the wear track, for the ta-C coating next to the wear track and on the steel ball in the wear track for the lubricants: (A) glycerol and (B) oleic acid.

peak is characteristic for a high sp^3 content. The tribological stress, therefore, did not lead to any change of the initial state on the coating. The Raman spectrum of the steel sphere does not contain any relevant signal, as expected for a metal.

On the ta-C coating lubricated with oleic acid (Fig. 14.2B), the signals are similar, but with values shifted slightly to smaller wave numbers within the wear track compared to outside values. This shift may possibly result from residual stress relaxation, but is more likely to result from rehybridization of sp^3 to sp^2-hybridized carbon. The comparatively small shift of the G peak, however, only corresponds to a small decrease in hardness, so that the coating can still be classified as "superhard." Since no D peak occurs, clustering of the amorphous phase can be excluded during the structural transformation.

However a Raman spectrum with a pronounced D peak at 1550 cm^{-1}, typical for a-C or a-C:H coatings, occurs at the associated tribocontact at the steel ball. The high signal strength indicates a sufficiently high excited material volume of presumably at least 100 nm. The existence of a D peak is evidence of ordered atomic structures, which have a larger proportion of sp^2 carbon than the hard original coating. The carbon-containing layer on the steel ball thus differs considerably in structure from the original coating. A detachment and transfer of whole ta-C particles, which would still have generated a ta-C signal, could therefore not have taken place. Instead, an a-C or a-C:H coating must have synthesized atom by atom.

In all spectra recorded on balls and discs, no iron oxides could be found in the typical range from 200 to 800 cm^{-1}. Therefore ball wear by oxidation does not seem to play a significant role.

After the study of lubricant chemistry, the role of the carbon coating for increased wear was investigated. In addition to the ta-C_{650} (83% sp^3) investigated so far, the variants a-C_{140} (4% sp^3), ta-C_{420} (52% sp^3), and ta-C_{770} (97% sp^3) were also tested. This series was supplemented by a sp^3-rich, amorphous, but hydrogen-containing a-C:H coating as well as a monocrystalline diamond surface, which exhibits 100% sp^3-hybridized carbon in a completely crystalline structure.

The results for friction and wear for the three lubricants engine oil, GMO, and glycerol are shown in Fig. 14.3. It is noticeable that for glycerol comparatively low friction occurs for all investigated carbon surfaces ($0.045 < \mu < 0.063$). Friction with engine oil, however, is generally higher ($0.080 < \mu < 0.140$), and is reduced only for the hard carbon phases, that is, ta-C and diamond. For GMO, the characteristic friction reduction occurs only for ta-C coatings ($0.039 < \mu < 0.048$) and is higher for diamond, a-C, and a-C:H coatings ($0.070 < \mu < 0.099$).

The wear on the investigated carbon surfaces is extremely small with all lubricants and, therefore, usually not measurable. The exceptions are ta-C coatings with GMO lubrication, where wear is significantly increased and is

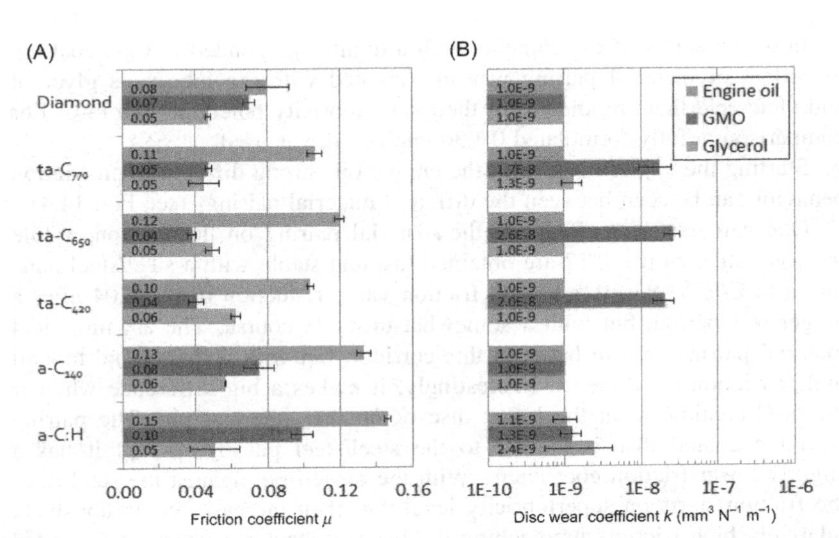

FIGURE 14.3 (A) Friction and (B) coating wear for diamond, a-C:H and various ta-C coatings, each with the lubricants engine oil, GMO, and glycerol.

at least 10 times higher than the measurement threshold. These observations confirm the previous observations with GMO and oleic acid (see Fig. 14.1) and the discussion that all oleic acid containing lubricants cause wear on ta-C due to the reactive double bond in the alkyl chain.

Further studies on ta-C friction and wear behavior lubricated by GMO were done by varying testing temperature [48]. In these experiments, it could be shown that a thermal activation is necessary to achieve an ultralow friction level in a ta-C/steel contact. In an interval of 60°C–100°C, the friction reduces to about 0.04. At the same time the wear increase by about 2 orders on magnitude.

Due to the strong influence of carbon bonding structure, lubricant chemistry, and temperature a tribochemical wear mechanism of ta-C coatings is assumed, where a detrimental interaction of carbon double bonds in the lubricant and radicals in the ta-C coating are triggered by elevated temperature and catalytic function of the steel counterbody. A more detailed discussion is given elsewhere [48].

14.3.2 Superlubricity (parameter set II)

The following experiments were carried out with tribometer parameter set II (see Table 14.3), creating a low and stable contact pressure situation ($p < 200$ Pa) over the course of the experiment due to the low curvature of the pin. This setup allows the study of low friction phenomena without causing any measurable wear of the ta-C coating.

In a first series of experiments with a highly sp^3-bonded ta-C_{650} coating, the effect of material pairing was investigated with the lubricants glycerol and oleic acid that are known for their superlubricity potential with ta-C. For comparison, a fully formulated 0W30 engine oil was used.

Starting the experiments with the engine oil, strong differences in friction behavior can be seen between the different material pairings (see Fig. 14.4).

One can see clear effects of the material pairing on the friction. While friction values of $\mu \approx 0.17$ are obtained fast and stable with a steel/steel pairing, a ta-C/ta-C pairing allows a friction value reduction below 0.04 after a longer test period, but with a somewhat unstable course. The asymmetrical material pairings are in between this corridor, but with a clear trend toward higher friction coefficients. Interestingly, it makes a big difference whether the ta-C coating is on the lower disc or on the spherical pin. The pairing with the coated disc is similar to the steel/steel pairing, except it has a slightly lower friction coefficient. With the coated pin against the steel disc, the friction starts on superlubricity level, but then increases continuously to relatively high friction approaching the level of steel pin versus ta-C-coated disc.

The reason of the observed friction behavior is assumed in the ability of the engine-oil additives to form triboreactive layers with iron surfaces that improve wear resistance but also increase friction due to increased topography. Such layers can easily be formed on steel surfaces under tribological load. On the steel pin, which is in permanent tribological contact, the tribolayer forms after short time, resulting in immediate high friction from the very beginning. The formation of the tribolayer on the uncoated steel disc, however, takes much longer, because the interacting surface area is much larger compared to the pin contact. Hence, the friction of the uncoated disc paired with the ta-C-coated pin slowly increases and approaches the friction level of the steel/steel pairing after a long time.

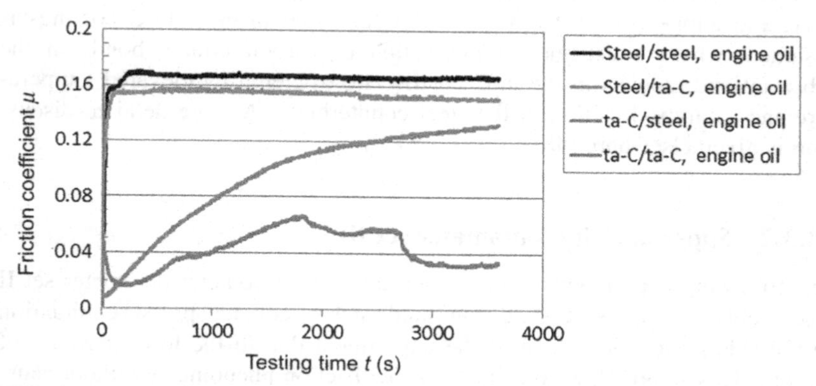

FIGURE 14.4 Friction versus time for different material pairing with engine oil lubrication.

In the next experiments, oleic acid lubrication was applied with the same pairings of steel and ta-C as described earlier. The measured friction curves can be seen in Fig. 14.5.

With oleic acid lubrication, all pairings except the steel/steel pairing show constant superlubricity with $\mu \approx 0.01$ after a more or less pronounced running-in time.

In direct comparison of the single side ta-C-coated experiment, it can be seen that superlubricity is virtually instantly achieved with the ta-C-coated pin on steel, but only after about 3 min of running-in for the steel pin on the coated disc. The reason is believed to be analog to engine oil, where in this case the reaction of oleic acid with the ta-C surface is necessary to achieve superlubricity. This process can occur with faster speed on the coated pin, where the coating is permanently in contact, opposed to the coated disc, which is not in permanent contact, thus requiring a longer running in.

The ta-C/ta-C pairing also achieves superlubricity, with a high initial friction and a long running-in of 3 min. A direct correlation of running-in time and wear track analysis of repeated experiments reveals that running-in is directly correlated to the amount of wear marks in contact area, as shown as in Fig. 14.6. This suggests that running-in in this case is actually a mechanical process due to two mating hard surfaces, leading to smooth and conforming surfaces. In other cases, where no wear occurs), superlubricity is instantly obtained.

This running-in process might occur for ta-C/steel contacts as well, but not as severe as the ta-C counterbody is much softer, leading to less friction during running-in.

The steel/steel contact does not show any superlubricity. Instead, after running-in friction increases, then decreases to saturate at $\mu = 0.05$ eventually. Other than glycerol, oleic acid cannot achieve superlubricity without a

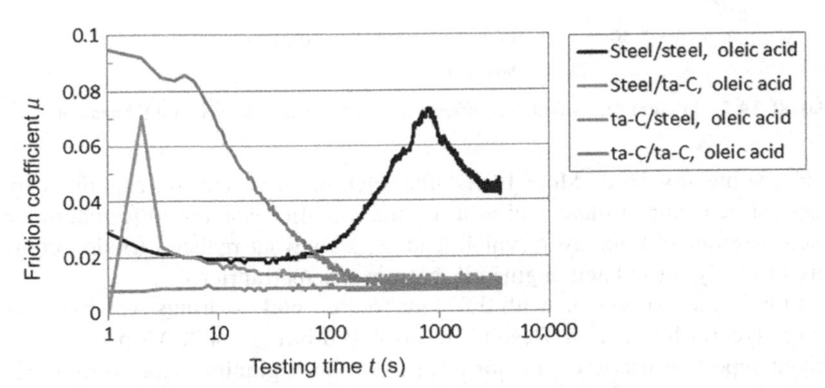

FIGURE 14.5 Friction versus time for different material pairing under oleic acid lubrication. Note that time scale is logarithmic to highlight the small differences in running-in behavior.

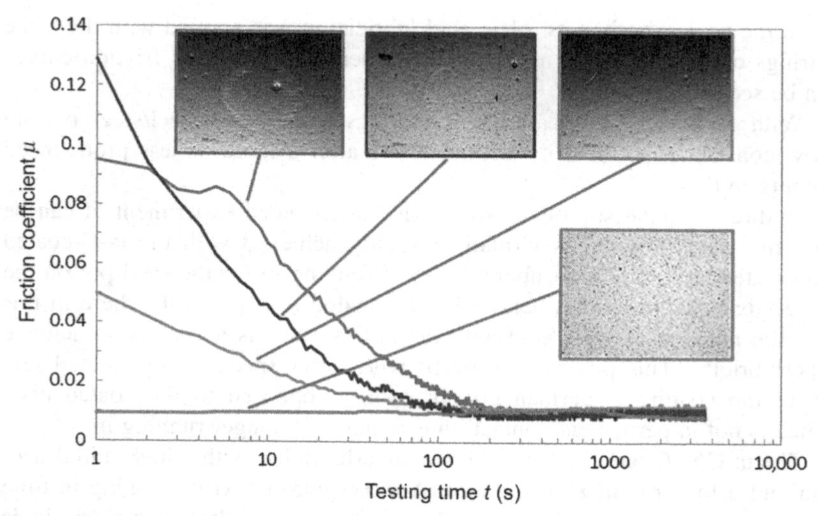

FIGURE 14.6 Running-in behavior of four ta-C/ta-C paired experiments with oleic acid lubrication. All tests result in superlubricity after about 3 min. The amount of wear marks observed on the ta-C-coated pin correlates well with running-in behavior.

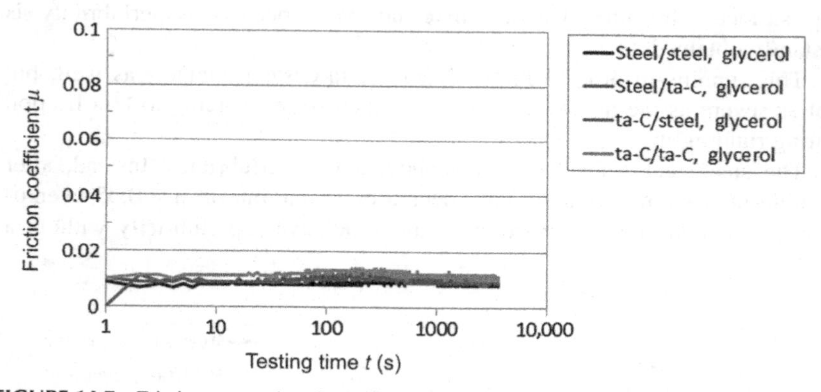

FIGURE 14.7 Friction versus time for different material pairing with glycerol lubrication.

ta-C coating involved. More likely, the friction curve can be explained by classical behavior of oleic acid as a friction modifier, leading to formation of a self-assembled monolayer, which leads to somewhat reduced friction compared to fully formulated engine oil, but not to superlubricity.

Finally the experiment with different friction body pairings was repeated using glycerol lubrication. Results are displayed in Fig. 14.7. All parings displayed superlow friction behavior right from the beginning with no running-in, even though ta-C and steel surfaces from the same batch as before were used. The extremely low friction under the chosen tribological condition seems to be a feature of the glycerol lubrication rather than the result of a tribochemical interaction to the steel or ta-C-coated surfaces.

This is in contrast to similar works by Long et al. [49] who achieved superlubricity for a glycerol lubricated ta-C/steel contact. For a steel/steel pair, the friction was $\mu = 0.02$ for a short time only and then increased, showing typical signs of adhesive failure. In their experiments, they used a higher contact pressure, which might have led to the finding that the steel surfaces were not able to stabilize a superlubricity state.

In the following, the influence of certain chemical functionalities of a lubricant series with the same basic chemical structure will be investigated in detail. Fluids from the C18 molecule series were chosen for this purpose. This series contains variations with regard to the presence and type of a double bond as well as with regard to the acid head. The lubricants were octadecane, stearic acid, elaidic acid, and oleic acid. Glycerol was also used for comparison.

The tribological investigations were carried out with self-mated ta-C coating using parameter set II (see Table 14.3). A ta-C coating with an average sp^3 content of 52% (ta-C_{430}) was selected for the tests.

The coefficient of friction for the different lubricants is shown in Fig. 14.8A, the corresponding structural formulas of the lubricants in Fig. 14.8B. The octadecane with its linear and nonfunctionalized C18 alkane chain has by far the highest friction with $\mu = 0.095$. Of all the lubricants examined here, the molecular structure has the greatest similarity with a synthetic base oil, for example, PAO.

The same C18 basic structure with a terminal acid group results in the saturated fatty acid stearic acid. With $\mu = 0.062$ the friction is already noticeably reduced compared to octadecane. If one adds a double bond in cis-configuration, characteristic of unsaturated fatty acids, to the stearic acid,

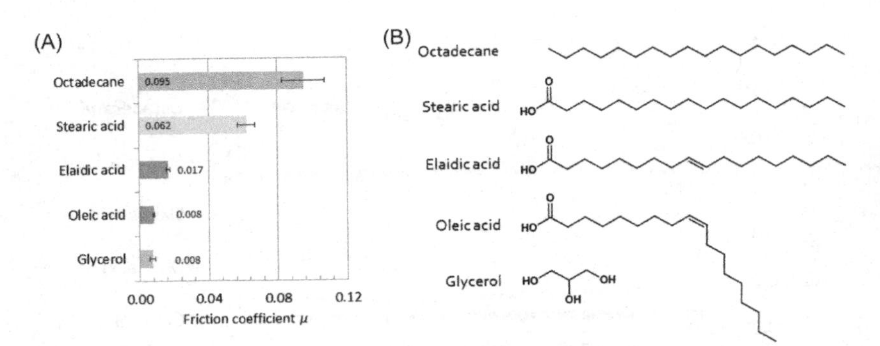

FIGURE 14.8 Superlubricity for self-mated ta-C-coated surfaces occurs only for some specific lubricants, such as glycerol and oleic acid. In contrast, the friction of stearic acid and octadecane is considerably higher. Shown are (A) the coefficients of friction and (B) the corresponding structural formulae of the lubricant molecules.

one obtains the commonly occurring oleic acid. The coefficient of friction drops significantly ($\mu = 0.008$) and is thus less than one-tenth of that of octadecane.

Another variant is elaidic acid, whose structure is almost identical to oleic acid, but the double bond is in linear trans-configuration, thus of less reactivity. The friction coefficient of elaidic acid, $\mu = 0.017$, lies between oleic acid and stearic acid. Glycerol, which consists of a short-chain propane with three hydroxyl groups, also achieves the state of superlubricity with $\mu = 0.008$ in this experimental setup.

Fig. 14.9 shows a characteristic trend of the coefficient of friction over time for each of the lubricants investigated. All lubricants except the octadecane quickly achieve a very stable state of friction, which is maintained throughout the entire experiment. The octadecane, on the other hand, exhibits a much greater fluctuation in the short term and the coefficient of friction does not reach a saturation value within the test period, but increases continuously.

The wear mark appearance on the disc and the pin can be well related to the friction behavior. The amounts of wear on both bodies are too small for a quantitative evaluation, but clear differences can be seen with the light microscope. Fig. 14.10 shows a characteristic pair of disc and pin with spherical face for each lubricant. For glycerol, oleic acid, and elaidic acid lubricants with very low friction, wear is extremely low and hardly visible. It essentially consists of a slight smoothing effect, particularly on the edges of the wear marks where lubricant film thickness is smaller [50], so that their contours become clear. For stearic acid, the wear is slightly greater and is manifested on almost the entire contact surface by smoothing the surface. The octadecane, on the other hand, shows strong abrasive wear on the contact surfaces, which is characterized by grooves on the entire contact surface.

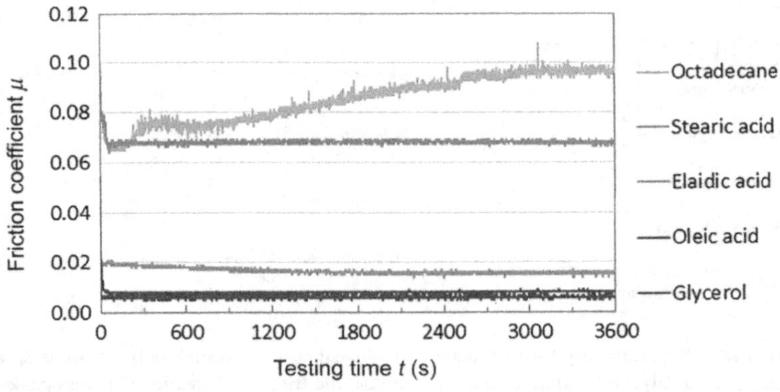

FIGURE 14.9 Experimental data from Fig. 14.8, one characteristic test each plotted as coefficient of friction as a function of testing time.

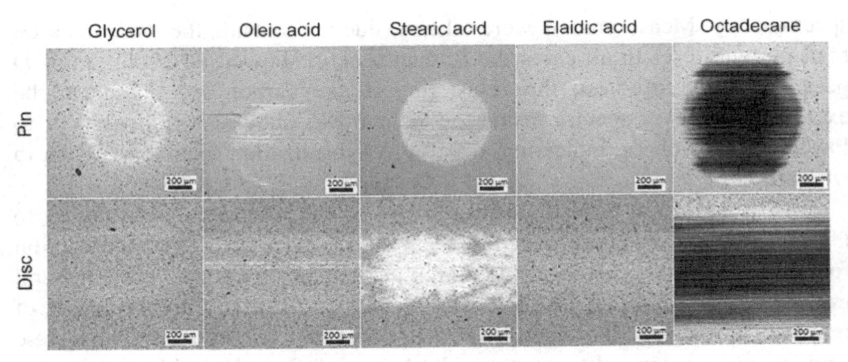

FIGURE 14.10 Wear marks of the ta-C-coated discs and the ta-C-coated pins after testing with different lubricants.

In summary, there is a clear correlation between the molecular structure of the lubricant, friction, and wear behavior. In the case of a completely nonpolar alkane chain (octadecane), the lubricant has no wear-protective effect. The surfaces are not sufficiently separated and come into direct contact, which leads to strong abrasion and an unstable coefficient of friction at a high level. If the alkane chain has an acid group, as is the case with stearic acid, this reduces friction and wear, as one would expect from a conventional friction modifier. The tribological system is in a stable and low-wear condition compared to the octadecane. If the lubricant molecules have at least two reactive centers, such as an acid group, hydroxyl group, or double bond, this leads to very low and stable friction and hardly measurable wear on ta-C coatings.

The mechanism behind the observed tribological behavior was studied by Kuwahara et al. [51] using quantum molecular statics and dynamics simulations. The chemical reactivity of an alkane, an alkene, an alkanoic acid as well as a trans- and a cis-alkenoic acid confined between two ta-C surfaces was studied, employing self-consistent-charge density-functional-based tight-binding. Additional studies were done by classical molecular dynamics. In these atomistic simulations, a fundamental model to explain superlow friction of ta-C was developed. It revealed that, due to the simultaneous presence of two reactive centers (carboxylic group and $C=C$ double bond), unsaturated fatty acids can concurrently chemisorb on both ta-C surfaces and bridge the tribogap. Sliding-induced mechanical strain triggers a cascade of molecular fragmentation reactions releasing passivating hydroxyl, keto, hydrogen, and olefinic groups.

In conclusion, this study unveiled that the synergy between two functional groups in unsaturated fatty acids promotes the mechanochemical formation of superlubricious layers on ta-C.

After the screening of C18 lubricants on self-mated ta-C coatings, the structural properties of carbon were studied in the wear tracks using Raman spectroscopy. Measurements were taken inside and outside the wear track on both pin and disc. In all cases the Raman spectra showed no evolution of D peaks after the tribotest, thus clustering of the carbon sp^2 phase can be excluded. However spectra inside the wear tracks showed a G peak shift to lower wave numbers, indicating a partial rehybridization from sp^3 carbon to sp^2 carbon [27].

In Fig. 14.11 the G peak position of inside the wear area with respect to the outside area is shown over the shear stress, calculated from the friction force divided by the actual contact area at the end of the test. As a general trend it can be seen that for higher shear stresses there is a higher degree of rehybridization. In case of superlubricity obtained with glycerol and oleic acid, there is no or only a very small change of G-peak position, while for high friction and high wear test as obtained with octadecane the shift in G peak is very strong. Furthermore, the G-peak shift on the pin is always higher, which can be attributed to the higher degree of interaction of the pin coating.

In recent works a graphene-oxide layer on top of the ta-C coating after achieving superlubricity has been evidenced by experiment [9] and simulation [51].

While Raman spectroscopy probes the volume of at least 1 μm into the surface it is possible to study subsurface volume rehybridization, but not the formation of a graphene-oxide layer on top of the ta-C coating.

Based on these recent finding, the structural changes that can occur during sliding of a ta-C coating must be considered as two different mechanisms: First, a general shear-induced mechanisms that occur in the subsurface volume, which can easily be evidenced by Raman spectroscopy, but does not necessarily occur for superlubricity due to low shear forces; and second,

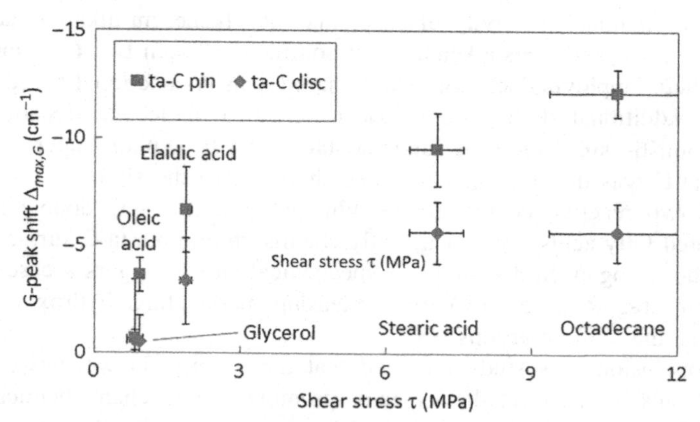

FIGURE 14.11 The G-peak shift, characteristic for a sp^3 to sp^2 transformation of the subsurface volume, is higher for higher shear stress; supporting the assumption of shear-induced rehybridization.

formation of a graphene-oxide layer, which is unique to superlubricity and require high sophisticated detection methods described elsewhere [9].

14.4 Conclusion and outlook

In the presented work, interaction of ta-C coatings with variations of *sp*^{3}-content, counterbody, and temperature have been systematically studied in the boundary lubrication regime with emphasis on fatty acid-based lubricants. Very manifold phenomena regarding wear and friction have been observed that exceed the conventional understanding of lubricant-surface interaction. Some of them are unique features of hydrogen-free a-C coatings.

The main findings are:

- Under unfavorable conditions a tribochemical wear process for superhard ta-C coatings can be triggered, resulting in wear rates exceeding abrasive wear by 3 orders of magnitude. Such conditions require high contact pressure, a steel counterbody, elevated temperature above a threshold, and a fatty acid-based lubricant, featuring an unsaturated carbon chains. Though the atomistic mechanism has not been understood in detail, a tribochemical wear process, involving a catalyzed radical process is assumed [48].
- Ultralow friction as low as $\mu = 0.04$ of single-side coated ta-C coatings can be obtained with fatty acid-based lubricants for moderate contact pressures of $p < 500$ GPa, whereas a-C:H coatings or uncoated steel surfaces have much higher friction under same conditions. In case of tribological model tests with ball-on-plate geometry, tribochemical wear might be triggered, resulting in friction reduction through significant change of contact pressure.
- Distinct superlubricity behavior of ta-C with stable friction coefficients of $\mu \approx 0.01$ or below was found with glycerol and oleic acid lubrication under moderate tribological loading conditions and low contact pressure $p < 200$ MPa. All configurations of ta-C/ta-C-, ta-C/steel-, and steel/ta-C-pairing resulted in a constant and long-term stable superlow friction situation. A short running-in behavior associated with minor macroscopic surface smoothing was observed for some of the experiments.
- Interestingly, the steel/steel-paring with glycerol lubrication displayed stable superlow friction as well, showing that superlubricity is a property of the lubricant solely and does not require participation of ta-C coating in general. However as the steel/steel system tends to show adhesive seize [49], a ta-C coating prevents such failure and will provide a nonadhesive surfaces for stable superlubricity.
- With oleic acid, on the other hand, superlubricity could be achieved only with at least one ta-C-coated partner. It was shown that both the acid function and the cis-configured double bond are essential features of the

oleic acid molecule to achieve superlubricity, while similar C18 molecules like elaidic acid, stearic acid, and octadecane did not achieve superlubricity. This strongly indicates a necessary reaction of the surface with the lubricant as previously assumed by other authors [9].
- The subsurface rehybridization of sp^3 carbon to sp^2 carbon after tribological loading was studied and showed a general correlation with shear stress in the contact. Raman spectroscopy indicated a higher G-peak shift for higher shear stresses, and no G-peak shift for superlubricity achieved with glycerol. This observed mechanism which takes place in the volume below the surface is most likely separate from formation of a graphene oxide on the topmost surface layer that is associated with superlubricity.

In application, the tribochemical wear mechanism is of less concern. In most cases of technical sliding contacts, the contact pressure seldom exceeds 500 MPa. Furthermore, the wear can be controlled by the choice of the counterbody material [48].

In contrast, the low friction properties of ta-C coatings, including superlubricity have been observed for medium ($p < 500$ MPa) and low contact pressures ($p < 200$ MPa) which are common contact situations in application.

Of high interest is also the beneficial interaction of fatty acid-based lubricants that are derived from vegetable oils, as they provide an environment-friendly alternative to mineral based oil and can work without sulfur- and phosphorous-containing additives.

In general, the application of ta-C coatings with lubricants from renewable sources to achieve superlubricity, sometimes termed *green superlubricity*, is a very realistic scenario. However it will require the full redesign of the tribological system to account for the necessary interaction of surface and lubricant, rather than only modifying the lubricant or adding a coating to an existing surface.

References

[1] M.Z. Baykara, M.R. Vazirisereshk, A. Martini, Emerging superlubricity: a review of the state of the art and perspectives on future research, Appl. Phys. Rev. 5 (4) (2018) 41102.

[2] X. Ge, J. Li, J. Luo, Macroscale superlubricity achieved with various liquid molecules: a review, Front. Mech. Eng. 5 (2019) 41102.

[3] A. Erdemir, Genesis of superlow friction and wear in diamondlike carbon films, Tribol. Int. 37 (11-12) (2004) 1005–1012.

[4] X. Chen, C. Zhang, T. Kato, X.-A. Yang, S. Wu, R. Wang, et al., Evolution of tribo-induced interfacial nanostructures governing superlubricity in a-C:H and a-C:H:Si films, Nat. Commun. 8 (1) (2017) 1675.

[5] J.A. Heimberg, K.J. Wahl, I.L. Singer, A. Erdemir, Superlow friction behavior of diamond-like carbon coatings: time and speed effects, Appl. Phys. Lett. 78 (17) (2001) 2449–2451.

[6] M. Kano, Y. Yasuda, Y. Okamoto, Y. Mabuchi, T. Hamada, T. Ueno, et al., Ultralow friction of DLC in presence of glycerol mono-oleate (GMO), Tribol. Lett. 18 (2) (2005) 245–251.

[7] Krell, L., Makowski, S., Weihnacht, V., Rausch, J., Luther, R., Schnagl, J., Superlubricity: Phänomenologische Untersuchung des tribologischen Verhaltens tetraedischer amorpher Kohlenstoffschichten mit organischen Fluiden, Reibung, Schmierung und Verschleiß, GfT, Aachen, 2012.

[8] M. Kano, J.-M. Martin, K. Yoshida, M.-I. De Barros Bouchet, Super-low friction of ta-C coating in presence of oleic acid, Friction 2 (2) (2014) 156–163.

[9] M.-I. De Barros Bouchet, J.-M. Martin, J. Avila, M. Kano, K. Yoshida, T. Tsuruda, et al., Diamond-like carbon coating under oleic acid lubrication: evidence for graphene oxide formation in superlow friction, Sci. Rep. 7 (2017) 46394.

[10] M. Kano, Super low friction of DLC applied to engine cam follower lubricated with ester-containing oil, Tribol. Int. 39 (12) (2006) 1682–1685.

[11] J.-M. Martin, M.-I. De Barros Bouchet, C. Matta, Q. Zhang, W.A. Goddard, S. Okuda, et al., Gas-phase lubrication of ta-C by glycerol and hydrogen peroxide. Experimental and computer modeling, J. Phys. Chem. C. 114 (11) (2010) 5003–5011.

[12] S. Bachmann, M. Schulze, L. Krell, R. Merz, M. Wahl, R. Stark, Ultra-low friction on tetrahedral amorphous diamond-like carbon (ta-C) lubricated with ethylene glycol, Lubricants 6 (3) (2018) 59.

[13] M.-I. De Barros Bouchet, C. Matta, T. Le-Mogne, J.-M. Martin, T. Sagawa, S. Okuda, et al., Improved mixed and boundary lubrication with glycerol-diamond technology, Tribol. Mater. Surf. Interfaces 1 (1) (2007) 28–32.

[14] C. Matta, L. Joly-Pottuz, M.-I. De Barros Bouchet, J.-M. Martin, M. Kano, Q. Zhang, et al., Superlubricity and tribochemistry of polyhydric alcohols, Phys. Rev. B 78 (8) (2008).

[15] C. Matta, M.-I. De Barros Bouchet, T. Le-Mogne, B. Vachet, J.-M. Martin, T. Sagawa, Tribochemistry of tetrahedral hydrogen-free amorphous carbon coatings in the presence of OH-containing lubricants, Lubr. Sci. 20 (2) (2008) 137–149.

[16] L. Joly-Pottuz, C. Matta, M.-I. De Barros Bouchet, B. Vacher, J.-M. Martin, T. Sagawa, Superlow friction of ta-C lubricated by glycerol: an electron energy loss spectroscopy study, J. Appl. Phys. 102 (6) (2007) 64912.

[17] Martin, J.-M., De Barros Bouchet, M.-I., Kano, M., The ultimate of green lubrication with hydrogen-free DLC coatings, in: Yoshimoto Shigeka (Ed.), Proceedings of International Tribology Conference 2015, Tokyo, 2015, 17aB-01.

[18] X. Ge, J. Li, R. Luo, C. Zhang, J. Luo, Macroscale superlubricity enabled by the synergy effect of graphene-oxide nanoflakes and ethanediol, ACS Appl. Mater. Interfaces 10 (47) (2018) 40863–40870.

[19] T. Kuwahara, G. Moras, M. Moseler, Friction Regimes of Water-Lubricated Diamond (111): Role of Interfacial Ether Groups and Tribo-Induced Aromatic Surface Reconstructions, Phys. Rev. Lett. 119 (9) (2017) 96101.

[20] X. Zhang, R. Schneider, E. Müller, M. Mee, S. Meier, P. Gumbsch, et al., Electron microscopic evidence for a tribologically induced phase transformation as the origin of wear in diamond, J. Appl. Phys. 115 (6) (2014) 63508.

[21] M.-I. De Barros Bouchet, C. Matta, B. Vacher, T. Le-Mogne, J.-M. Martin, J. Lautz, et al., Energy filtering transmission electron microscopy and atomistic simulations of tribo-induced hybridization change of nanocrystalline diamond coating, Carbon 87 (2015) 317–329.

[22] K. Ohara, N.A.B. Masripan, N. Umehara, H. Kousaka, T. Tokoroyama, S. Inami, et al., Evaluation of the transformed layer of DLC coatings after sliding in oil using spectroscopic reflectometry, Tribol. Int. 65 (2013) 270–277.

[23] Y. Mabuchi, T. Higuchi, D. Yoshimura, M. Murashima, H. Kousaka, N. Umehara, Influence of carbon black in engine oil on wear of H-free diamond-like carbon coatings, Tribol. Int. 73 (2014) 138–147.

[24] J. Tang, Q. Ding, G. Zhang, L. Hu, The influence of total acid number of ester oil in tribological behavior of DLC contacts, Tribol. Trans. 58 (5) (2015) 849–858.

[25] K.A.H. Al Mahmud, M.A. Kalam, H.H. Masjuki, H.M. Mobarak, N.M. Zulkifli, An updated overview of diamond-like carbon coating in tribology, Crit. Rev. Solid. State Mater. Sci. 40 (2) (2014) 90–118.

[26] L. Bai, N. Srikanth, E.A. Korznikova, J.A. Baimova, S.V. Dmitriev, K. Zhou, Wear and friction between smooth or rough diamond-like carbon films and diamond tips, Wear 372-373 (2017) 12–20.

[27] A.C. Ferrari, J. Robertson, Interpretation of Raman spectra of disordered and amorphous carbon, Phys. Rev. B 61 (20) (2000) 14095–14107.

[28] A.R. Konicek, D.S. Grierson, A.V. Sumant, T.A. Friedmann, J.P. Sullivan, P.U.P.A. Gilbert, et al., Influence of surface passivation on the friction and wear behavior of ultrananocrystalline diamond and tetrahedral amorphous carbon thin films, Phys. Rev. B 85 (15) (2012) 155448.

[29] X. Deng, H. Kousaka, T. Tokoroyama, N. Umehara, Tribological behavior of tetrahedral amorphous carbon (ta-C) coatings at elevated temperatures, Tribol. Int. 75 (2014) 98–103.

[30] C. Héau, C. Ould, P. Maurin-Perrier, Tribological behaviour analysis of hydrogenated and nonhydrogenated DLC lubricated by oils with and without additives, Lubr. Sci. 25 (4) (2013) 275–285.

[31] S. Makowski, V. Weihnacht, F. Schaller, A. Leson, Ultra-low friction of biodiesel lubricated ta-C coatings, Tribol. Int. 71 (2014) 120–124.

[32] S. Lafon-Placette, J. Fontaine, M.-I. De Barros Bouchet, C. Heau, Critical role of a metallic counterpart on the tribochemical wear of ta-C coatings in base oil, Wear 402-403 (2018) 91–99.

[33] H.A. Tasdemir, M. Wakayama, T. Tokoroyama, H. Kousaka, N. Umehara, Y. Mabuchi, et al., Ultra-low friction of tetrahedral amorphous diamond-like carbon (ta-C DLC) under boundary lubrication in poly alpha-olefin (PAO) with additives, Tribol. Int. 65 (2013) 286–294.

[34] K.A.M. Aboua, N. Umehara, H. Kousaka, X. Deng, H.A. Tasdemir, Y. Mabuchi, et al., Effect of carbon diffusion on friction and wear properties of diamond-like carbon in boundary base oil lubrication, Tribol. Int. 113 (2016) 389–398.

[35] A. Mannan, M.F.M. Sabri, M.A. Kalam, H.H. Masjuki, Tribological properties of steel/steel, steel/DLC and DLC/DLC contacts in the presence of biodegradable oil, J. Jpn. Petrol. Inst. 62 (1) (2019) 11–18.

[36] C.W. Chen, J. Robertson, Surface atomic properties of tetrahedral amorphous carbon, Diam. Relat. Mater. 15 (4-8) (2006) 936–938.

[37] J. Shi, Z. Gong, Y. Wang, K. Gao, J. Zhang, Friction and wear of hydrogenated and hydrogen-free diamond-like carbon films: relative humidity dependent character, Appl. Surf. Sci. 422 (2017) 147–154.

[38] J.D. Schall, G. Gao, J.A. Harrison, Effects of adhesion and transfer film formation on the tribology of self-mated DLC contacts †, J. Phys. Chem. C 114 (12) (2010) 5321–5330.

[39] K.A.H. Al Mahmud, M.A. Kalam, H.H. Masjuki, M.F.B. Abdollah, Tribological study of a tetrahedral diamond-like carbon coating under vegetable oil-based lubricated condition, Tribol. Trans. 58 (5) (2015) 907–913.

[40] H.-J. Scheibe, D. Drescher, B. Schultrich, M. Falz, G. Leonhardt, R. Wilberg, The laser-arc: a new industrial technology for effective deposition of hard amorphous carbon films, Surf. Coat. Technol. 85 (3) (1996) 209–214.

[41] Fraunhofer Institute for Material and Beam Technology, Annual Report 2012, 2013.

[42] D. Schneider, T. Schwarz, H.-J. Scheibe, M. Panzner, Non-destructive evaluation of diamond and diamond-like carbon films by laser induced surface acoustic waves, Thin Solid Films 295 (1-2) (1997) 107–116.

[43] D. Schneider, C.F. Meyer, H. Mai, B. Schöneich, H. Ziegele, H.-J. Scheibe, et al., Non-destructive characterization of mechanical and structural properties of amorphous diamond-like carbon films, Diam. Relat. Mater. 7 (7) (1998) 973–980.

[44] A.C. Ferrari, J. Robertson, M.G. Beghi, C.E. Bottani, R. Ferulano, R. Pastorelli, Elastic constants of tetrahedral amorphous carbon films by surface Brillouin scattering, Appl. Phys. Lett. 75 (13) (1999) 1893.

[45] H. Ito, K. Yamamoto, Mechanical and tribological properties of DLC films for sliding parts, Kobelco Technol. Rev. 35 (2017) 55–60.

[46] K.-P. Kuo, J. Asmussen, An experimental study of high pressure synthesis of diamond films using a microwave cavity plasma reactor, Diam. Relat. Mater. 6 (9) (1997) 1097–1105.

[47] J. Asmussen, T.A. Grotjohn, T. Schülke, M.F. Becker, M.K. Yaran, D.J. King, et al., Multiple substrate microwave plasma-assisted chemical vapor deposition single crystal diamond synthesis, Appl. Phys. Lett. 93 (3) (2008) 31502.

[48] S. Makowski, F. Schaller, V. Weihnacht, G. Englberger, M. Becker, Tribochemical induced wear and ultra-low friction of superhard ta-C coatings, Wear 392-393 (2017) 139–151.

[49] Y. Long, M.-I.D.B. Bouchet, T. Lubrecht, T. Onodera, J.M. Martin, Superlubricity of glycerol by self-sustained chemical polishing, Sci. Rep. 9 (1) (2019) 6286.

[50] G. Nijenbanning, C.H. Venner, H. Moes, Film thickness in elastohydrodynamically lubricated elliptic contacts, Wear 176 (2) (1994) 217–229.

[51] T. Kuwahara, P.A. Romero, S. Makowski, V. Weihnacht, G. Moras, M. Moseler, Mechano-chemical decomposition of organic friction modifiers with multiple reactive centres induces superlubricity of ta-C, Nat. Commun. 10 (1) (2019) 151.

Chapter 15

Friction of diamond-like carbon: Run-in behavior and environment effects on superlubricity

Zhe Chen and Seong H. Kim
Department of Chemical Engineering and Materials Research Institute, Pennsylvania State University, University Park, PA, United States

15.1 Introduction

Diamond-like carbon (DLC) films are composed of varying degrees of sp^2- and sp^3-type C–C bonds and hydrogen contents. The chemical and mechanical properties of DLC films vary substantially depending on their compositions. Nonhydrogenated DLC films can provide a hardness as high as bulk ceramic materials such as nitrides [1], and hydrogenated DLC (H-DLC) films can be tailored to possess nearly frictionless and nearly wearless properties [2]. The lowest coefficient of friction (COF) reported for H-DLC films in the literature is close to 0.001 and the wear rate is lower than 10^{-9} $mm^3 \cdot N^{-1} \cdot m^{-1}$ [3–5]. Thus H-DLC has a great potential of reducing parasitic energy loss and material degradation due to friction in various engineering systems.

However as reported in numerous research on the friction and wear properties of H-DLC films [4,6,7], such ultralow friction or superlubricity is observed only in vacuum, or inert gas atmosphere, or H_2 gas atmosphere. Even in these ideal environmental conditions, the superlubricious behavior is observed after an induction period (called a run-in period) during which the friction is initially high and then gradually decreases to an ultralow value [6]. In atmospheric environments containing O_2 and water vapor, H-DLC films lose their superlubricity; their COF are in the $0.1 \sim 0.2$ range; [8] in contrast, the COF of nonhydrogenated DLC is very high (as high as nearly 1) in vacuum and decreases to $0.1 \sim 0.2$ as the oxygen or water partial pressure increases. It is interesting and important to note that the COF of $0.1 \sim 0.2$ is within the range of typical COF of boundary lubrications in the presence of

Superlubricity. DOI: https://doi.org/10.1016/B978-0-444-64313-1.00015-6

organic vapors or the Stribeck curve plots of organic molecules [9,10]. The run-in behavior and environment dependence are two crucial parameters to consider for the superlubricity of H-DLC films. Understanding these issues will lead to more advanced and knowledge-based design and development of more efficient DLC coatings that provide superior lubrication and wear protection in ambient conditions; such coatings are of great interest for energy saving from friction and prevention of mechanical devices from material loss due to wear in various engineering systems operating in ambient air.

In this chapter, the structure and surface chemistry of DLC films are briefly reviewed first. Then, the hypotheses of the origin of the run-in behavior are discussed. The discussions on the transfer film, which is often found on the countersurface after the run-in behavior when the countersurface is not DLC, are also reviewed. At last, research on the environmental effects on the superlubricity of DLC films is summarized focusing on the origins of high friction in humid and O_2 atmosphere and promising solutions to enhance the environment tolerance of DLC films.

15.2 Structure and surface chemistry of DLC films

Because DLC has an amorphous structure, the carbon atoms in DLC have very broad distributions in bond length and angle, regardless of synthetic methods. Fig. 15.1 compares the radial and angular distributions of C–C bonds in DLC with those of crystalline allomorphs of carbon (diamond and graphite) [11,12]. Because the covalent bonds formed during the high-energy deposition process cannot be relaxed or rearranged without annealing at

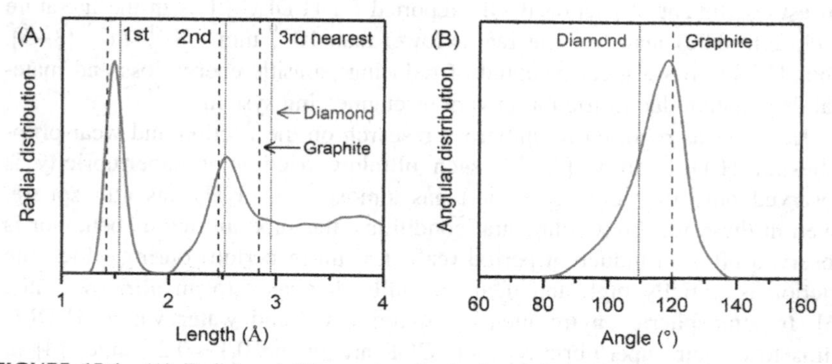

FIGURE 15.1 (A) Radial and (B) angular distributions of chemical bonds in DLC *(solid curve)*, diamond *(vertical dotted line)*, and graphite *(vertical dashed line)*. The data of DLC are obtained from MD simulation. The diamond and graphite data are based on the crystallographic structure. (*DLC*: Diamond-like carbon; *MD*: molecular dynamics). *Reproduced from X.-W. Li, M.-W. Joe, A.-Y. Wang, K.-R. Lee, Surf. Coat. Technol. 228 (2013) S190.*

extremely high temperature, the lengths and angles of many bonds in DLC deviate from the ideal structures of sp^2- and sp^3-hybridized carbon atoms. Then, it is conceivable that the C–C bonds with local geometries far from the minimum-energy structures of ideal sp^2 and sp^3 hybridization are weaker and more reactive than the bonds with parameters close to the ideal structures.

For the surfaces of DLC films, chemical analyses with X-ray photoelectron spectroscopy (XPS), X-ray absorption near-edge structure (XANES), and time-of-flight secondary ion mass spectrometry (TOF-SIMS) showed the presence of oxygenated species [13–15]. Based on the results of X-ray reflectivity analysis and XPS analyses after ion sputtering, the thickness of the oxide layer was found to be in the range of 2–3 nm [16].

Considering that physisorbed contaminants will readily desorb in ultrahigh vacuum (UHV) condition used for chemical composition analysis, the possibility that the detected oxygen-containing species come from the physisorption of organic contaminants is unlikely; even if they were detected, its contribution is relatively small. Moreover, the measured oxide layer thickness and the observed oxygen-to-carbon ratio is too high for one monolayer of small organic molecules chemisorbed on DLC surfaces. Therefore it is believed that the surface of the DLC films is oxidized.

Surface oxidation of solid materials upon exposure to air is a well-known phenomenon for metals and semiconductor materials such as aluminum, stainless steel, and silicon [17]. In previoue literature, it was assumed that the DLC surface would be chemically inert [18,19]. It is true that the DLC surface is relatively inert to chemical reactions compared to metals and semiconductor materials, but that does not mean that the DLC surface is completely resistant to oxidation in air. The data shown in the literature imply that the air-exposed DLC surface is indeed covered with an oxidized layer. The oxidation of the DLC surface in ambient air must be thermodynamically favorable so that the oxygenated species in the surface layer are in equilibrium with the O_2 and H_2O vapor in air. The thickness of the oxidized surface layer will depend on the transport of oxidizing species into the DLC film at given temperature and pressure [20,21].

This finding explains the lack of any dependence of *ex situ* chemical analysis of DLC films after friction tests in controlled environments [13,14]. Unless the DLC samples are deposited, friction tested, and spectroscopically analyzed *in situ* in a single chamber with UHV and atmospheric gas control capabilities, the samples are inevitably exposed to air between the deposition and the environment-controlled friction test and between the friction test and the chemical analysis. If the chemical and structural changes that are spontaneously occurring at the DLC surface upon exposure to air are not properly considered, the surface characterization data will be incorrectly interpreted, and this could lead to incorrect understanding of the DLC friction behaviors.

The quantitative composition of the oxide layer may reflect the reactivity of the DLC surface in different environmental conditions. However determination of specific oxidized species is often difficult in routine XPS analysis, because some species cannot be unambiguously distinguished from the binding energy shift. For example, both hydroxyl (C–OH) and ether (C–O–C) functional groups appear at the same nominal C 1s peak position (i.e., ~1.5 eV higher than the C 1s of saturated hydrocarbon). To resolve this problem, Yang et al. [22] used a chemical derivatization method to obtain the surface composition of hydroxyl, ether, carbonyl, and carboxyl groups. In the chemical derivatization method, trifluoroacetic anhydride was used to differentiate hydroxyl and ether groups, and trifluoroethyl hydrazine was used to enhance the detection of carbonyl groups. Moreover, with the use of the inelastic mean free path of X-ray photoelectrons and typical densities of the carbon materials, the average thickness of the oxide layer was estimated to be 2–3 nm based on the O/C ratio obtained from the XPS analysis [22].

15.3 Effect of run-in behavior on superlubricity

15.3.1 Removal of the oxide layer

Based on the chemical analysis of DLC surfaces, it was believed that the high friction during the run-in period is due to the presence of the oxidized surface layer, and ultralow friction can be achieved after the oxide layer is removed [6,16]. This statement was supported by the TOF-SIMS analysis, from which it was shown that the oxygen content inside the wear track of a nearly frictionless H-DLC film after friction test in dry N_2 is lower than that outside the wear track [14]. This analysis also indicates that, although the DLC surface will be oxidized when exposed to air, the degree of oxidation may change depending on the exposure period and condition.

To prove the hypothesis that the surface oxide layer is responsible for the run-in period and check how readily the DLC surface gets oxidized, Marino et al. [13] performed a control experiment. In this experiment, self-mated H-DLC-coated ball and plate slid against each other in dry Ar atmosphere. After a steady-state ultralow friction was achieved, the relative sliding motion was paused, and the ball was lifted off the plate. The ball and plate were then exposed to another gas or vapor, without being exposed to the ambient air, for varying periods of time. Then, the environmental atmosphere was switched back to dry Ar. Once the system was completely purged with dry Ar, the friction measurement was resumed at the same wear track.

As shown in Fig. 15.2, the environment conditions during the pause period included dry Ar, dry air (can be considered as dry O_2 with its pressure at about 20% of the saturation value), humid Ar with 40% relative humidity (RH), and a mixture of n-pentanol and Ar with a partial pressure of n-pentanol at 40% of its saturation pressure. When the environment is still Ar

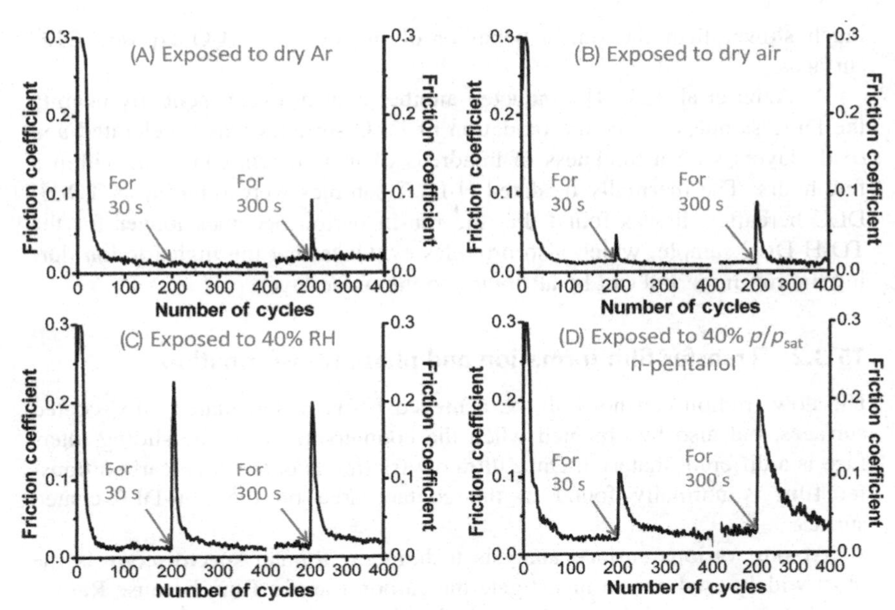

FIGURE 15.2 COF of self-mated H-DLC as a function of cycle in dry Ar ($O_2 < 8$ ppm; $H_2O < 10$ ppm). The sliding ball was posed and lifted from the substrate after the 200th sliding cycle and then exposed to different gases [(A) dry Ar, (B) dry air, (C) Ar with 40% RH, and (D) Ar with 40% p/p_{sat} n-pentanol] for a certain period (indicated in the graph). After the gas exposure, the sliding was resumed in dry Ar. (*COF*: Coefficient of friction; *H-DLC*: hydrogenated diamond-like carbon). *From M.J. Marino, E. Hsiao, Y. Chen, O.L. Eryilmaz, et al. Langmuir 27 (2011) 12702.*

during the paused period (Fig. 15.2A), the COF does not increase after 30 s pause; but after 300 s pause, it jumps to ~0.1 and then decreases quickly back to its original steady-state value. It should be noted that the ultrahigh purity Ar contains a trace amount of O_2 (<4 ppm) and H_2O (<5 ppm). Thus the data in Fig. 15.2A imply that even a trace amount of O_2 and H_2O in the gas can induce the surface oxidation of DLC if the exposure time is long enough. The surface oxidation occurs more readily (less than 30 s) in atmospheric pressure O_2 (Fig. 15.2B) and humid Ar (Fig. 15.2C) environments. The influence of the humidity-induced surface oxidation seems to be more drastic on the run-in behavior than that of the O_2-induced oxidation. The increase of the COF upon resuming the frictional sliding after the same period of gas exposure is much higher in humid Ar (40% RH) than dry O_2. The exposure to n-pentanol also results in the increase of COF upon resuming the sliding motion (Fig. 15.2D). This clearly indicates that the n-pentanol also induces surface oxidation (or changes the surface composition) of DLC. The surface reaction products formed upon exposure to n-pentanol seem to be different from those formed by O_2 or H_2O exposure; their removal rate is much slower than the run-in behavior of the O_2- and H_2O-exposed DLC surfaces.

Al-Azizi et al. [23,24] conducted another control experiment. By heating the DLC samples in air, the oxidation of DLC surfaces was accelerated and oxide layers with a thickness of hundreds of nanometers can form within a few hours. The thermally oxidized H-DLC samples were referred as TO-H-DLC hereafter. It was found that the run-in period becomes longer for the TO-H-DLC sample, which also provides evidence that the high friction during the run-in period can be attributed to the oxide layer.

15.3.2 Transfer film formation and phase transformation

Ultralow friction can not only be achieved between self-mated DLC-coated surfaces, but also be obtained when the countersurface of the sliding interface is a different material. Once ultralow friction is obtained, a carbon transfer film is normally found in the contact area on the non-DLC-coated surface.

Among various surface analysis techniques, Raman spectroscopy is the most widely used one to investigate the carbon transfer film, because Raman spectroscopy is sensitive to the sp^2/sp^3 ratio, clustering of sp^2 phase, bond-length disorder, hydrogen content, and density [25–27]. In the Raman spectra of DLC films, there are two characteristic peaks, known as G band at $\sim 1600\ cm^{-1}$ (E_{2g} mode) and D band at $\sim 1350\ cm^{-1}$ (A_{1g} mode) [28]. The G band originates from the stretching of all pairs of sp^2 atoms in both chains and rings. The breathing modes of sp^2 atoms in ring structure give rise to the D band. Formation of ring structures or local ordering within amorphous carbon (a-C) matrix enhances ring structure breathing modes or D-band intensity. The pristine H-DLC consists of a-C matrix and has weaker D band. The formation of the sp^2-rich phase upon heat treatment has been reported [29], and the growth of D band for the heated sample confirms formation of the sp^2-rich phase.

For the carbon transfer film as well as the wear track on DLC-coated surface, almost all Raman spectra reported in the literature showed a growing D band, indicating the formation of sp^2-rich phases [4,15,30,31]. Meanwhile, with the help of the techniques of focused ion beam and transmission electron microscopy (TEM), nanosized graphite flakes can be found in the carbon transfer film and the wear track [23,32].

As mentioned in Section 15.1, the most stable allomorph of carbon is graphite. Conversion of diamond to graphite (which involves conversion of all sp^3 carbons to sp^2 carbons) is extremely difficult and requires high-energy processes such as compression to ~ 100 GPa or bombardments with ions [33,34]. Conversion of a-C to graphite can be done only at extremely high temperatures [35]. The reported formation of graphitic flakes or domains at the sliding interfaces of DLC and a-C coatings under the contact pressure of a few GPa at ambient temperatures demonstrates that local graphitization of DLC at the shearing interface is much easier than compression or thermal reaction conditions. Such facile conversions must be attributed to the presence of local strains in chemical bonds in a-C networks (Fig. 15.1); in addition, mechanical shear can further decrease the activation barrier of such processes [36].

Based on the data from Raman spectroscopy and TEM mentioned earlier, some papers attributed the superlubricity of H-DLC after the run-in behavior to the formation of nanocrystalline graphite at the sliding region [37]. However nanocrystalline flakes were typically observed in trace amounts; it is difficult to imagine that the superlubricity of the entire sliding contact region originates from a few graphitic flakes. Instead, it is more reasonable to interpret these data as evidence for the occurrence of graphitization at the sliding interface (although difficult to distinguish ultrathin graphitic layers from the a-C phase especially in ex situ analysis); the observed nanocrystalline flakes must be a few aggregates that were big enough to be seen in ex situ TEM analysis.

One thing needs to be noted again is that, due to the reactivity of the amorphous carbon surface, its surface structure and composition may have changed before any ex situ analysis. To overcome the limitation of ex situ analysis, Manimunda et al. [23] carried out in situ tribo-Raman study of H-DLC films. Fig. 15.3A shows the schematics of the experiment setup. A sapphire hemisphere was used as the countersurface. The thickness of the transfer film was also collected based on the Newton's rings. Both pristine H-DLC and TO-H-DLC were investigated.

Fig. 15.3B plots the COF and transfer film thickness as a function of sliding cycle. It can be seen that, for both pristine H-DLC and TO-H-DLC, the COF and the transfer film become stable after a period of run in. During the steady region, the transfer film thickness is larger in the case of pristine H-DLC than TO-H-DLC, but ultralow friction is only achieved by the TO-H-DLC.

The in situ Raman spectra recorded during the friction tests are shown in Fig. 15.3C, and the G-band position as well as the intensity ratio between D band and G band of the Raman spectra as a function of sliding cycle are plotted in Fig. 15.3D and E. In the case of pristine H-DLC, the changes in G-band position and I_D/I_G ratio appear to correlate with the trend in the transfer film thickness. However such a correlation is not observed for the case of TO-H-DLC. Regardless of the transfer film thickness change, the common features of the in situ Raman spectra for both pristine and TO-H-DLC films are that the G-band position increases to $1590-1600\ cm^{-1}$ and the I_D/I_G ratio asymptotically approaches 0.75 ± 0.02 as the sliding cycle continues even though their initial values before sliding are different. The shifting of G band and I_D/I_G ratio indicated the sp^2-rich transfer film.

Previously, it was believed that graphitization of a thick transfer film is necessary to achieve ultralow friction in the DLC system [38,39]. However

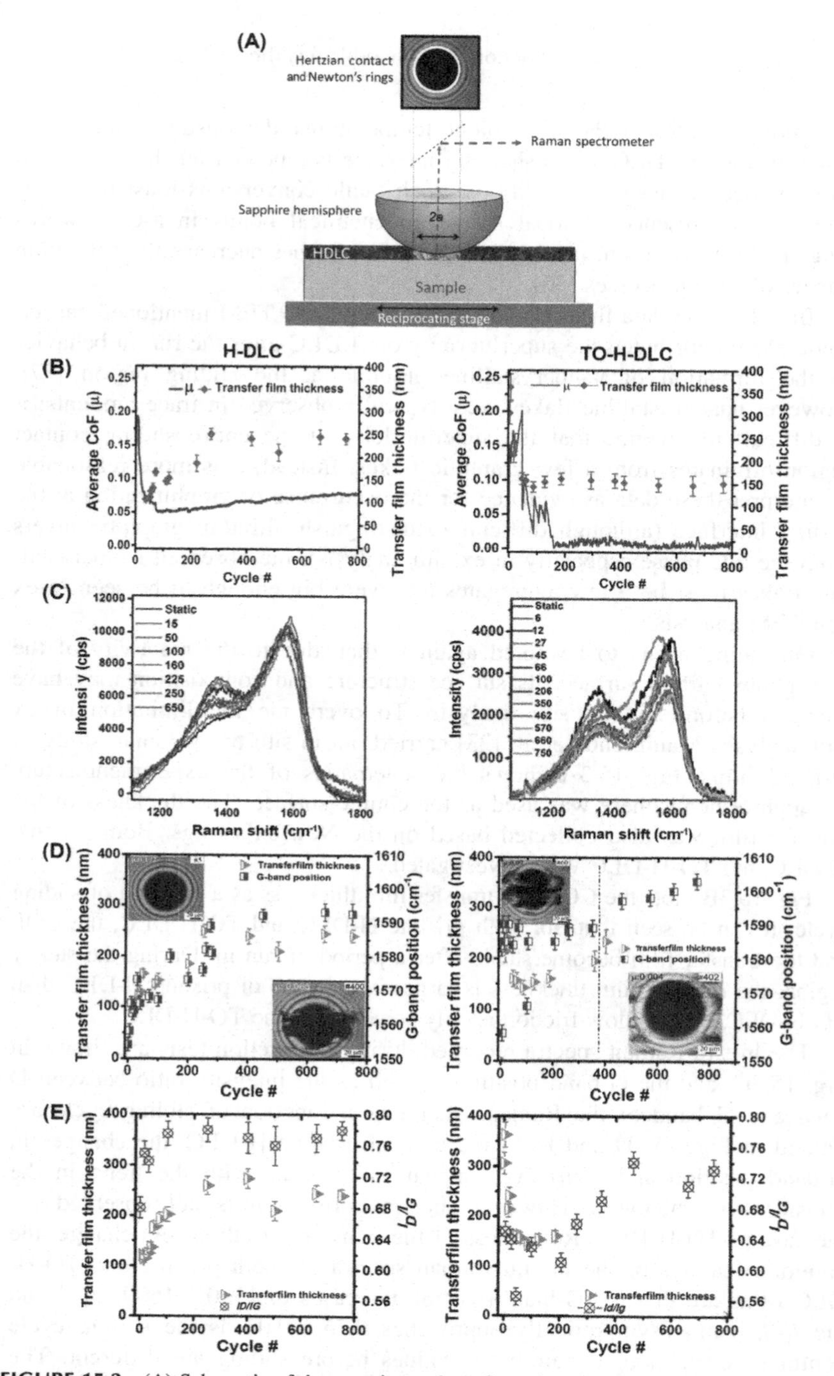

FIGURE 15.3 (A) Schematic of the experimental configuration. Ball on a flat tribometer coupled with a Raman microscope. (B) COF and transfer film thickness, (C) in situ Raman spectra, (D) G-band position, and (E) I_D/I_G ratio recorded for a sapphire ball sliding on pristine H-DLC and TO-H-DLC surfaces, respectively. *Reproduced from P. Manimunda, A. Al-Azizi, S.H. Kim, R.R. Chromik, ACS Appl. Mater. Interfaces 9 (2017) 16704.*

the in situ analysis suggests that the key determinant of the superlubricious state is not the transfer film thickness, but the type and uniformity of the transfer film. At a first glance, the shift of the G-band position toward 1600 cm^{-1} and the higher I_D/I_G ratio could be interpreted as the possibility of nanocrystalline graphite formation. However it was difficult to predict whether the transfer film has undergone the full transformation from the in situ Raman study alone, because the in situ Raman spectroscopy averages signals not just from the surface, but also from the deep subsurface region within the light penetration depth. So, the data cannot be definitive about the location of the graphitic domains within the probe depth.

Apart from the Raman spectroscopy analysis, Al-Azizi et al. [40] compared the friction properties of H-DLC films with and without topographic texturing of the substrate. Two kinds of nanotextures—nanodomes and nanowells—were investigated. Stainless steel balls were used as the countersurfaces. It was found that ultralow friction can always be obtained in dry N_2 regardless of the nanotexture, but the transfer film on the steel ball is much less when it is sliding against the H-DLC film with nanodomes. This experiment also supports the statement that the thickness of the transfer film is not a critical factor to ultralow friction.

15.4 Effect of environment on the superlubricity

Beside the run-in behavior before superlubricity, another and even more serious challenge is the environmental sensitivity of the DLC friction especially to H_2O vapor. The effect of humidity on H-DLC is quite different from that on nonhydrogenated DLC (often called a-C). In inert gas environments, H-DLC shows ultralow friction (COF $\leq$ 0.01) after a run-in period, while a-C shows high friction (COF $>$ 0.5). With increasing H_2O partial pressure in the tribo-test environments, the COF of these two surfaces become closer and converge to the same value ($\sim$0.1) [41].

For the H-DLC, some papers attributed the friction increase to the high shear force required to slide on adsorbed H_2O layer [41]. The higher shear force could be related to the increase in capillary force between H-DLC surface and the countersurface due to the adsorbed H_2O in the contact [42]. However other experimental evidence suggested that oxidation causes chemical as well as structural changes of the surface layer of H-DLC [14]. The hydrogen-terminated defect sites may be easily oxidized because the oxidized form is thermodynamically more stable [17].

To get further insights into the effects of environment, especially H_2O and O_2, on the friction of H-DLC, Alazizi et al. [8] measured the friction between a stainless steel (SS) ball and a H-DLC-coated surface under the environments of dry N_2, dry O_2, dry air, and humid N_2, respectively. The test results are displayed in Fig. 15.4.

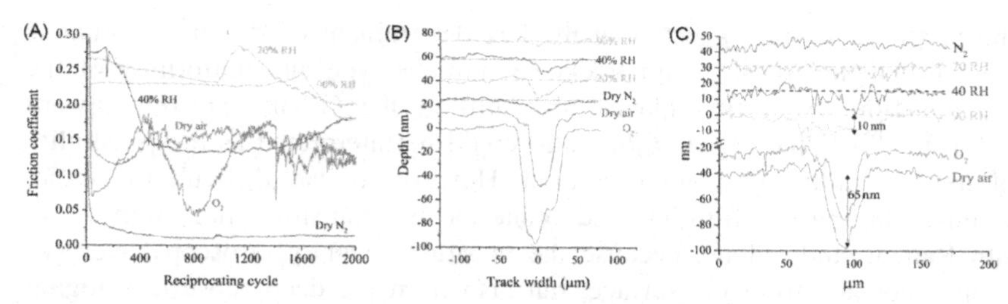

FIGURE 15.4 (A) COF of 440 C SS balls sliding on H-DLC-coated surfaces in dry N_2, dry O_2, dry air, and humid N_2, respectively. (B) Cross sections of the wear tracks on the H-DLC surfaces. (C) Cross sections of the wear scars on the SS balls. (*H-DLC*: Hydrogenated diamond-like carbon; *SS*: stainless steel). *Reproduced from A. Alazizi, A. Draskovics, G. Ramirez, A. Erdemir, et al., Langmuir 32 (2016) 1996.*

In dry N_2, typical H-DLC friction and behaviors are observed. After a short run-in period, the friction drops to an ultralow value (COF $\leq$ 0.02). The wear of the H-DLC surface measured after the friction test is very mild, and the SS ball wear is negligible.

In dry O_2 and dry air, there is also a run-in period initially, but after this period, COF is noisy and varied randomly within 0.05 and 0.2. Both the H-DLC surface and the SS ball have some wear. It is believed that continuous formation of oxide layers and their removal by the countersurface result in the wear of H-DLC [43]. This is evident by the higher wear rate in dry O_2 (O_2 pressure = 760 Torr) compared to dry air (O_2 pressure = 152 Torr), indicating that the amount of O_2 in the environment plays a critical role. Removal of the oxidized layers from the DLC surface produces wear debris that can have abrasive effects on the countersurface resulting in mutual wear of both surfaces [44]. The abrasive effect of wear particles cause the third-body contacts, resulting in noisy COF in dry O_2 and air environments.

In humid N_2, the H-DLC/SS friction did not show the run-in period; instead, the COF varies gradually and smoothly between 0.15 and 0.3. The wear of H-DLC is very mild. As Fig. 15.4B shows, at 90% RH, the wear of H-DLC is almost completely suppressed. However, while the H-DLC wear decreases as the RH increases, the SS ball wear increases slightly, and iron oxide was found in the wear debris.

Based on the infrared spectroscopic analysis of H-DLC films in humid environment, the H_2O adsorption isotherm on H-DLC can be obtained [40]. It was found that the thickness of adsorbed H_2O increases quickly to a monolayer of H_2O ($\sim$0.3 nm thick) at around 20% RH. The adsorbed H_2O layer grows slowly as RH increases from 20% to 90% RH. It reaches $\sim$0.5 nm at 40% RH and increases to a maximum thickness of $\sim$1 nm near saturation RH. The adsorbed H_2O layer can work as a lubricant preventing wear and giving a friction of $0.15-0.2$ as long as no mechanical failure takes place [9]. While H_2O vapor can induce oxidation, which can be a competing mechanism beside molecular lubrication, some wear can still be caused. As RH approaches the saturation point, multilayers of H_2O are adsorbed on the H-DLC surface. The presence of thick multilayers of H_2O can introduce another mechanism beside lubrication by physisorbed molecules. When the adsorbed H_2O layer becomes thick enough, they can act as an electrolyte, allowing an electrochemical reaction, or corrosion, between two dissimilar materials with different corrosion potentials [45]. This electrochemical reaction could explain the formation and accumulation of iron oxide particulates around the sliding contact region after friction tests in 90% RH and, to a lesser extent, 40% RH.

Fig. 15.5 shows schematically how the electrochemical reaction can take place within the contact area between steel ball and DLC surface. The electrochemical reaction occurring here is also called Galvanic corrosion. It occurs when two materials with different electrochemical potentials are in direct contact in the presence of an electrolyte; this leads to a preferential corrosion of the more anodic material. The galvanic corrosion potential of SS is more anodic than that of graphite. Based on the analysis mentioned earlier, both oxidation and interfacial shears can lead to the growth of graphitic carbons at the sliding interface. When a graphitic layer is present, then the SS countersurface could be electrochemically corroded by the sp^2-rich carbon in the presence of sufficient amounts of H_2O adsorbed from the gas phase.

It is of great interest to overcome this humidity sensitivity of H-DLC superlubricity, so that H-DLC can be employed more effectively in ambient air. Several papers reported that doping of Si into H-DLC can maintain the ultralow friction ($COF < 0.1$) behavior in humidity conditions [46,47]. Alazizi et al. [48] also showed that chemical vapor reactions of H-DLC surfaces with silane gases can yield a similar humidity tolerance effect. The mechanism for the Si-mediated humidity tolerance effect of H-DLC is not well understood. Initially, the low friction mechanism of the Si incorporating H-DLC films was attributed to the formation of silica-sol, which may function as a sort of liquid-lubricant [49]. However neither Si nor silica glasses exhibits lower friction than Si-DLC in humid conditions [50]. Then, it was reported that H_2O can dissociate directly at the silicon sites incorporated at the DLC surface leading to the formation of Si–OH groups, and speculated the formation of hydrophilic surfaces can enhance the formation of a H_2O layer working as a lubricating film [51]. However this speculation neglected the fact that there is also a layer of H_2O molecules on the oxidized H-DLC surface, and the adsorbed H_2O layer gives a COF of $0.1 \sim 0.2$.

15.5 Conclusions

Based on the information reviewed in this chapter, it can be summarized that the friction properties of the DLC film are largely dependent on its structure

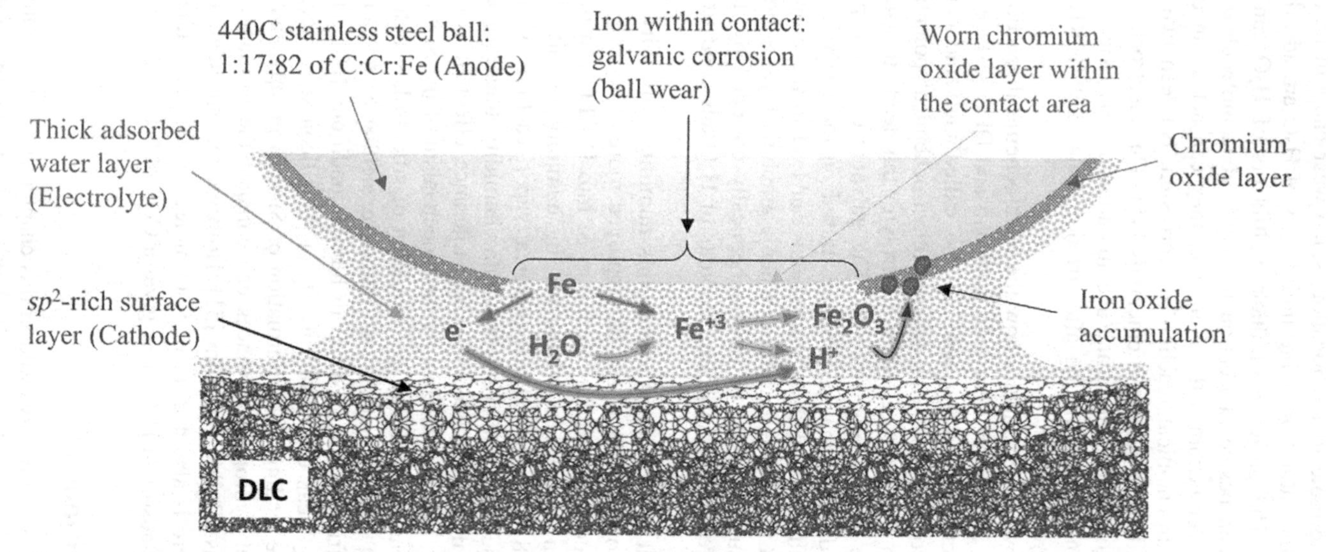

FIGURE 15.5 A schematic showing how an electrochemical reaction can take place within the contact area of steel/DLC in the presence of sp^2-rich transfer film and adsorbed water. (*DLC*: Diamond-like carbon). *From A. Alazizi, A. Draskovics, G. Ramirez, A. Erdemir, et al., Langmuir 32 (2016) 1996.*

and chemical reactivity. The amorphous nature of the DLC film facilitates the oxide layer formation upon exposure to H_2O and O_2. Before the superlubricity is achieved in inert environmental conditions, the oxide layer is removed during the run-in period. If the countersurface is not DLC, a transfer film is usually formed on the countersurface. Transformation of sp^3 to sp^2 is detected for the carbon atoms in the transfer film and inside the contact area. The friction seems to be dependent on the extent of the phase transformation. The H-DLC film loses its superlubricity in O_2- or H_2O-containing environment due to the facile formation of surface oxide layers. Adsorbed H_2O layer on the oxide layer can work as a lubricant, thus leading to less wear and smoother friction compared to the case of dry O_2. When the adsorbed H_2O layer is thick enough, it can work as an electrolyte, leading to Galvanic corrosion of the countersurface, if the countersurface is more anodic than the DLC surface. Si-incorporated H-DLC films show ultralow friction even in humid environments as long as RH is low, but the mechanism is still not well understood.

Acknowledgment

This review work was supported by the National Science Foundation (Grant Nos. CMMI-1727571 and CMMI-1912199).

References

[1] N. Savvides, T. Bell, Thin Solid Films 228 (1993) 289.
[2] A. Erdemir, Tribol. Int. 37 (2004) 1005.
[3] A. Erdemir, O.L. Eryilmaz, G. Fenske, J. Vac. Sci. Technol. A 18 (2000) 1987.
[4] A. Erdemir, O.L. Eryilmaz, I.B. Nilufer, G.R. Fenske, Surf. Coat. Technol. 133-134 (2000) 448.
[5] A. Erdemir, O.L. Eryilmaz, I.B. Nilufer, G.R. Fenske, Diamond Relat. Mater. 9 (2000) 632.
[6] C. Donnet, J. Fontaine, A. Grill, T. Le Mogne, Tribol. Lett. 9 (2001) 137.
[7] A. Erdemir, O. Eryilmaz, Friction 2 (2014) 140.
[8] A. Alazizi, A. Draskovics, G. Ramirez, A. Erdemir, et al., Langmuir 32 (2016) 1996.
[9] A.J. Barthel, S.H. Kim, Langmuir 30 (2014) 6469.
[10] J. Dawczyk, N. Morgan, J. Russo, H. Spikes, Tribol. Lett. 67 (2019) 34.
[11] X.-W. Li, M.-W. Joe, A.-Y. Wang, K.-R. Lee, Surf. Coat. Technol. 228 (2013) S190.
[12] G. Jungnickel, T. Köhler, T. Frauenheim, M. Haase, et al., Diamond Relat. Mater. 5 (1996) 175.
[13] M.J. Marino, E. Hsiao, Y. Chen, O.L. Eryilmaz, et al., Langmuir 27 (2011) 12702.
[14] O.L. Eryilmaz, A. Erdemir, Surf. Coat. Technol. 201 (2007) 7401.
[15] O.L. Eryilmaz, A. Erdemir, Wear 265 (2008) 244.
[16] N.J. Mehta, S. Roy, J.A. Johnson, J. Woodford, et al., Mater. Res. Soc. Symp. Proc. 843 (2004). T2.7.
[17] C.T. Campbell, Phys. Rev. Lett. 96 (2006) 066106.
[18] A. Erdemir, C.J. Donnet, J. Phys. D Appl. Phys. 39 (2006) R311.

[19] M. Kalin, I. Velkavrh, J. Vižintin, L. Ožbolt, Meccanica 43 (2008) 623.
[20] M. Morita, T. Ohmi, E. Hasegawa, M. Kawakami, et al., J. Appl. Phys. 68 (1990) 1272.
[21] J. Evertsson, F. Bertram, F. Zhang, L. Rullik, et al., Appl. Surf. Sci. 349 (2015) 826.
[22] M. Yang, M.J. Marino, V.J. Bojan, O.L. Eryilmaz, et al., Appl. Surf. Sci. 257 (2011) 7633.
[23] P. Manimunda, A. Al-Azizi, S.H. Kim, R.R. Chromik, ACS Appl. Mater. Interfaces 9 (2017) 16704.
[24] A.A. Al-Azizi, O. Eryilmaz, A. Erdemir, S.H. Kim, Langmuir 31 (2015) 1711.
[25] A.C. Ferrari and J. Robertson, Philos. Trans. R. Soc. A 362 (2004) 2477.
[26] C. Casiraghi, A.C. Ferrari, J. Robertson, Phys. Rev. B 72 (2005) 085401.
[27] C. Casiraghi, F. Piazza, A.C. Ferrari, D. Grambole, et al., Diamond Relat. Mater. 14 (2005) 1098.
[28] A.C. Ferrari, Diamond Relat. Mater 11 (2002) 1053.
[29] H. Li, T. Xu, C. Wang, J. Chen, et al., Thin Solid Films 515 (2006) 2153.
[30] J.C. Sánchez-López, A. Erdemir, C. Donnet, T.C. Rojas, Surf. Coat. Technol. 163-164 (2003) 444.
[31] H. Song, L. Ji, H. Li, X. Liu, et al., RSC Adv. 5 (2015) 8904.
[32] D.-S. Wang, S.-Y. Chang, Y.-C. Huang, J.-B. Wu, et al., Carbon 74 (2014) 302.
[33] A.A. Gippius, R.A. Khmelnitsky, V. Dravin, A.V. Khomich, Diamond Relat. Mater. 12 (2003) 538.
[34] J. Gaudin, N. Medvedev, J. Chalupský, T. Burian, et al., Phys. Rev. B 88 (2013) 060101.
[35] T.A. Friedmann, K.F. Mccarty, J.C. Barbour, M.P. Siegal, et al., Appl. Phys. Lett. 68 (1996) 1643.
[36] X. He, A.J. Barthel, S.H. Kim, Surf. Sci. 648 (2016) 352.
[37] Y. Liu, A. Erdemir, E.I. Meletis, Surf. Coat. Technol. 86-87 (1996) 564.
[38] Y. Liu and E.I. Meletis, *J. Mater. Sci.* 32 (1997) 3491.
[39] T.W. Scharf, I.L. Singer, Tribol. Lett. 14 (2003) 137.
[40] A.A. Al-Azizi, O. Eryilmaz, A. Erdemir, S.H. Kim, Carbon 73 (2014) 403.
[41] H.I. Kim, J.R. Lince, O.L. Eryilmaz, A. Erdemir, Tribol. Lett. 21 (2006) 51.
[42] J. Shi, Z. Gong, Y. Wang, K. Gao, et al., Appl. Surf. Sci. 422 (2017) 147.
[43] X. Wu, T. Ohana, T. Nakamura, A. Tanaka, Tribol. Lett. 57 (2015) 5.
[44] T.W. Scharf, I.L. Singer, Tribol. Trans. 45 (2002) 363.
[45] A.J. Barthel, M.D. Gregory, S.H. Kim, Tribol. Lett. 48 (2012) 305.
[46] X. Chen, T. Kato, M. Kawaguchi, M. Nosaka, et al., J. Phys. D Appl. Phys. 46 (2013) 255304.
[47] I. Tanaka, T. Ikeda, T. Nakano, H. Kousaka, et al., Jpn. J. Appl. Phys. (58) (2018) SAAC06.
[48] A. Alazizi, D. Smith, A. Erdemir, S.H. Kim, Tribol. Lett. 63 (2016) 43.
[49] M.-G. Kim, K.-R. Lee, K.Y. Eun, Surf. Coat. Technol. 112 (1999) 204.
[50] F.G. Sen, X. Meng-Burany, M.J. Lukitsch, Y. Qi, et al., Surf. Coat. Technol. 215 (2013) 340.
[51] N. Kato, H. Mori, N. Takahashi, Phys. Status Solidi C 5 (2008) 1117.

Chapter 16

Tribo-induced interfacial nanostructures stimulating superlubricity in amorphous carbon films

Xinchun Chen[1], Takahisa Kato[2], Chenhui Zhang[1] and Jianbin Luo[1]
[1]State Key Laboratory of Tribology, Tsinghua University, Beijing, P.R. China, [2]Surface Science and Tribology Laboratory, The University of Tokyo, Tokyo, Japan

16.1 Introduction

Diamond-like carbon (DLC), that is, amorphous carbon, is a metastable carbon material possessing a wide variety of bonding structures and mechanical properties [1]. Since the discovery of near-frictionless performances (friction coefficient down to 0.001) in 2000 [2,3], it has emerged as one of the most promising solid lubricants to realize macroscale superlubricity. Over the past decade, the unique lubricity of DLC and its structural diversity have endowed the research community with successive discoveries of superlubricious DLC varieties and relevant tribomaterials such as fullerene-like hydrogenated amorphous carbon (FL-C:H) [4], silicon-doped hydrogenated amorphous carbon (a-C:H):Si [5,6], polymer-like carbon (PLC) [7], DLC-coupled graphene nanoscroll [8], and glycerol-lubricated tetrahedral a-C with aromatic passivation layer [9]. The vanishing level of friction coefficient has also advanced from the thousandth to ten thousandth, and meanwhile the adapted environments have been expanded from the initial dry inert gaseous atmosphere (i.e., dry N_2) [3] and ultrahigh vacuum (UHV) [2] to hydrogen-mediated gas [7], liquid medium [10], ambient environment [6,11], or even cryogenic condition. For instance, Kato et al. found a friction fading-out scenario with the friction coefficient vanishing to 0.0001 for PLC films under heavy loads in hydrogen atmosphere [12–14]. Besides the initial a-C:H superlubricious systems, Luo's group achieved a robust superlubricity in self-mated a-C:H:Si films with optimized silicon incorporation [15]. Recently, they are capable of realizing superlow friction

Superlubricity. DOI: https://doi.org/10.1016/B978-0-444-64313-1.00016-8

about 1 of the 10,000 between hydrocarbon surfaces under an extremely high contact pressure (larger than 2 GPa) and a cryogenic environment with the temperature around 120K.

Until now the majority of studies in superlubricity in regard to DLC are concentrated on a-C:H with the emphasis on the critical role of hydrogen in achieving superlow friction [16–22]. As proposed by Erdemir [16], the most prevailing explanation for the superlow friction state observed in self-mated a-C:Hs is that most carbon dangling bonds are speculated to be passivated by hydrogen terminations and a nonreactive sliding interface was then created. From the atomic-scale viewpoint, the authors proposed that the redistribution of electrical charge density and the repulsive forces generated between two positively charged hydrogen protons along the sliding interface are the origins of suppression of adhesive interactions and hence friction [17]. In this case, the canceling out of friction is solely dependent on the interfacial electric charge characteristics, namely without formation of a tribolayer. Nevertheless, in most reported experimental results [23–28], the tribo-induced graphitization or phase transformation of a-C:Hs films was frequently observed, especially for the heterogeneous sliding interface when an uncoated counterface was used. During this process, interfacial material transfer (generally from the film side to the non-DLC side) occurs between the two surfaces [24,25]. Accompanying the process is the in situ formation of a graphitized tribolayer on the heterogeneous surface [29]. Meanwhile, the topmost tribo-affected region of a-C:H film surface is also subjected to structural change. The bonding structure of the as-formed tribolayer is significantly different from the original hydrocarbon matrix and usually possesses unprecedented atomic features, which exerts the decisive influence in the establishment of a superlubricity state and the following sustainability of this combating-friction status. Therefore in order to broaden the operating environments and achieve robust superlubricity of a-C:H films, it is quite essential to resolve the structural characteristics of the nanostructured tribolayer.

The present chapter is organized to provide a summarizing overview of behaviors of superlubricious tribolayers encountered in different a-C:Hs and rubbing conditions. Major efforts and specific emphasis are devoted to the interfacial nanostructures formed in the tribolayer and their relationship with the friction state of a-C:H films. In the first part, the common characterization techniques that enable the analysis of a contact area are compared with the focus on the advantage and the shortage of each technique. As a unique and powerful characterization method, scanning transmission electron microscope (STEM) and electron energy-loss microscopy (EELS) are highlighted for their imaging capability at atomic-scale resolution and the superiority to detect the local structural details of nanoscale tribolayers. The second part provides several representative case studies addressing the growth dynamics and bonding features of the nanostructured tribolayers formed on the newly constructed contact surface. Finally the concluding section is devoted to a brief summarization on the underlying lubrication mechanisms regarding to each sliding condition through the analysis of the as-formed tribolayers in carbon films.

16.2 Techniques to detect the sliding interface at the atomic scale

The intrinsic hinder to clarify the superlubricity mechanisms is that most contact activities are taking place in a buried sliding interface and it is always a great challenge to extract isolated signals for common characterization techniques such as Raman spectroscopy, X-ray photoelectron spectroscopy (XPS), and time-of-flight secondary ion mass spectrometry (TOF-SIMS). Raman spectroscopy is a powerful method to detect the ring-like or chain-like sp^2-C phases and qualitatively determine the bonding fraction of sp^2-C/sp^3-C [30], and the probing depth can approach to a few hundred nanometers, reflecting the structural signals from the bulk [31]. However it is quite hard to extract the exact signals of the topmost sliding interface. In addition, it is incapable to directly capture the bonding information of the light element of hydrogen. XPS is effective to record the chemical-bonding signals of most elements except H and He from the sample surfaces with a depth of a few nanometers, and it is feasible to semiquantitatively determine the evolution of sp^2-C/sp^3-C fraction for carbon-related bonds [32]. TOF-SIMS, possessing superior acquisition and spatial imaging resolutions, is more sensitive to the surface chemical function species with a near-surface detection depth of 1 nm. It is possible to detect all the elements in the periodic table of elements, including hydrogen [33,34]. Nevertheless, the as-acquired signals are delivering information of elemental composition and distribution rather than an analysis of chemical-bonding state. Meanwhile, all these characterization methods are incapable of direct imaging of the interested rubbing region, which impedes the visual inspection of the sliding interface. To overcome this, some researchers have attempted to resolve the tribo-induced structural transformation in carbon lubricants at the sliding interface using transmission electron microscope (TEM) and EELS [23,35–39]. Encouraging microscopic evidence regarding the bonding information of carbon-related triboproducts (i.e., tribolayer) after sliding has been achieved, even though most of these studies were not dealing with the research topic of superlubricious mechanisms [37–39]. In addition, a few in situ tribological tests based on nanomanipulator in TEM facility were designed to capture the real-time details of interfacial activities and phase transformation of carbon films [35]. This is the most desirable circumstance to reveal the underlying mechanisms upon sliding. However it should be pointed out that the contact length in this case is usually at the nanoscale, which is different from the macroscale range for the most reported superlubricity phenomena of carbon films. Thus it cannot guarantee the fact that the rubbing activities are occurring in a superlubricious state and it may lose the opportunity to reveal the real nature of superlow friction behaviors in a-C:Hs.

Taking these facts into account, a complementary characterization method is recently developed to probe the tribo-induced interfacial nanostructures using the most advanced STEM and EELS after slicing a nanometer-thick lamella from the contact area by a site-specific focused ion beam (FIB). Compared to traditional TEM, STEM is a more powerful microscope with its advanced *Z* contrast (*Z* denotes the atomic number) imaging capability at atomic-scale resolution in high-angle annular dark-field (HAADF) mode [40]. This high compositional sensitivity is particularly useful in distinguishing the elemental distribution across the tribolayer formed on heterogeneous surfaces. Along with element-analytical EELS, compositional and bonding information can be determined at high spatial resolution. It should be pointed out that only successful FIB preparation of specimens with sufficient quality for atomic-resolution imaging by STEM allows this objective to be realized. With the accumulation of experience regarding FIB sample preparation, it is now feasible to fabricate lamellar specimens with the possibility to keep the protective surface layers intact during the milling process [41], which allows the region of interests close to or at the sample surface to be analyzed by STEM and EELS. It is reported that such a combination of complementary methods allows the achievements of structural information of the sliding interfaces at the atomic scale, making it possible to understand the relevant mechanisms of superlubricity in a-C:Hs in a more in-depth way.

As reported in the literature, the methodological setup of characterization pathway regarding the superlubricious sliding interfaces of carbon films is summarized in Fig. 16.1. The superlubricity test is first carried out in an atmosphere-controlled tribometer by purging dry N_2 gas into the chamber (Fig. 16.1A). After that, the contact areas including the ball wear scar and wafer wear track are basically characterized by optical microscopy, 3D surface profilometer, Raman spectroscopy, scanning electron microscope (SEM), and so on (Fig. 16.1B). Then, nanometer-thick lamellar specimens in situ sliced from the region of interests on the contact surfaces are fabricated by the dual beam SEM-FIB system based on site- and material-specific lift-out technique (Fig. 16.1C). Afterward, the as-prepared lamellar specimens are first analyzed by traditional TEM facility for a rough inspection. Finally, the samples are analyzed by the state-of-the-art dual-aberration-corrected STEM system with a bright field (BF) imaging resolution of 0.14 nm and a HAADF resolution of 0.08 nm (Fig. 16.1D). For EELS spectra, the DualEELS function is used for simultaneous acquisition of the core-loss and low-loss (including the zero-loss peak) spectra. Generally, the recorded EELS spectra can be classified into zero-loss (0 eV), low-loss (0–50 eV), and core-loss ($>$50 eV) spectra. As a useful evaluation, the plasmon peak deriving from the excitation of all valence electrons can be employed to

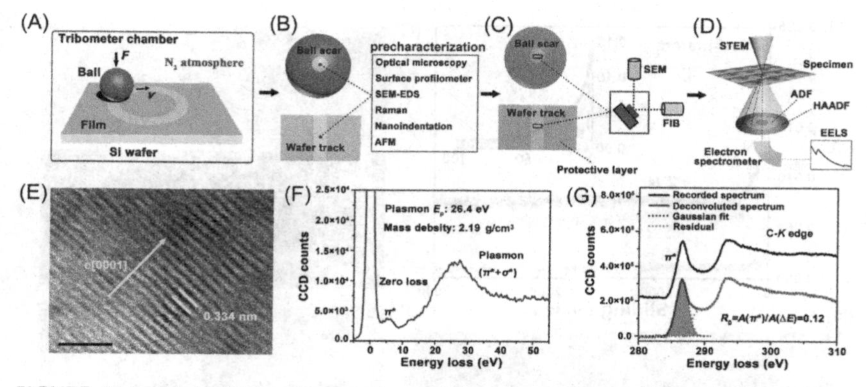

FIGURE 16.1 Methodology showing the experimental setup for probing the superlubricity mechanisms including: (A) tribological test, (B) basic characterization of the contact area, (C) FIB lift-out technique of lamellar specimen, (D) STEM and EELS atomic-resolution imaging of the sample, (E) HOPG as reference for EELS bonding calculation, (F) EELS plasmon peak for mass density calculation, and (G) sp^2-C/sp^3-C fraction calculation based on Gaussian fitting procedures regarding to the C-*K* core-edge [15]. *EELS*: Electron energy-loss microscopy; *FIB*: focused ion beam; *HOPG*: highly oriented pyrolytic graphite; *STEM*: scanning transmission electron microscope.

calculate the mass density of the targeted area (Fig. 16.1F) [42]. The bonding fractions of sp^2- and sp^3-related carbon phases can be calculated by fitting the C-*K* core-loss edges with several Gaussian peaks based on the π^* and σ^* transition states [43,44], in which the ratio of the integrated area of the individual peak to the total ($\pi^* + \sigma^*$) area in the energy window is used as the basis for the calculation (Fig. 16.1G). To do this, a 100% sp^2-bonded standard sample such as highly oriented pyrolytic graphite is used as a reference one (Fig. 16.1E).

16.3 Tribo-induced interfacial nanostructures governing superlubricity in different pathways

16.3.1 Stability and local ordering in self-mated hydrocarbon network assisting friction vanishment

As the superlubricity phenomenon was discovered in self-mated a-C:H films [2,3], the lubricity nature intrinsic to hydrocarbon contact asperities is the most fundamental basis to understand the antifriction mechanisms in a-C:H films. Among the tribotesting parameters, contact pressure is one of the most crucial factors that can highly affect the structural stability and the phase transformation of hydrocarbon network. As shown in Fig. 16.2A, the vanishing level of friction coefficient (in the range of 0.001–0.008) in self-mated a-C:H tribocouple highly depended on the applied normal load (contact pressure), namely a higher contact pressure resulting in a lower friction

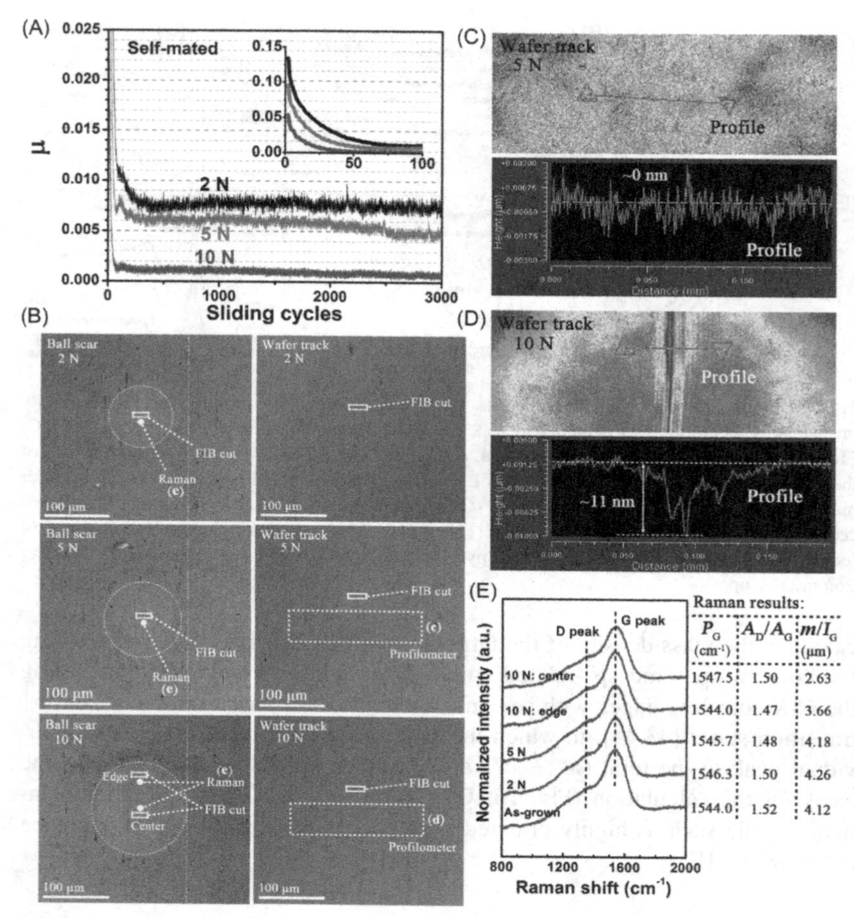

FIGURE 16.2 Superlubricity behaviors of self-mated a-C:H films and the corresponding characterization of the contact areas. (A) Friction tests at different applied normal loads in dry N_2, (B) ball wear scars and wafer wear tracks produced on the contact areas, (C and D) 3D white-light interference images showing the morphologies and cross-sectional profiles of the wear tracks, and (E) Raman spectra and the analysis results [15].

coefficient. The wear depth of the wear tracks (Fig. 16.2B) was below the detection limitation for applied normal loads up to 5 N (Fig. 16.2C), and a maximum of ~11 nm was produced for 10 N (Fig. 16.2D). As clearly indicated by Raman spectra (Fig. 16.2E), STEM image and EELS results (Fig. 16.3A), the film bonding structures of the contact zones were almost intact for loads up to 5 N (contact pressure of 0.93 GPa). However local clustering and orientation of sp^2-C phases along the outermost sliding interface was found when the applied normal load increased to 10 N (contact pressure of 1.17 GPa). The as-formed shear band possessed a higher fraction of sp^2-C

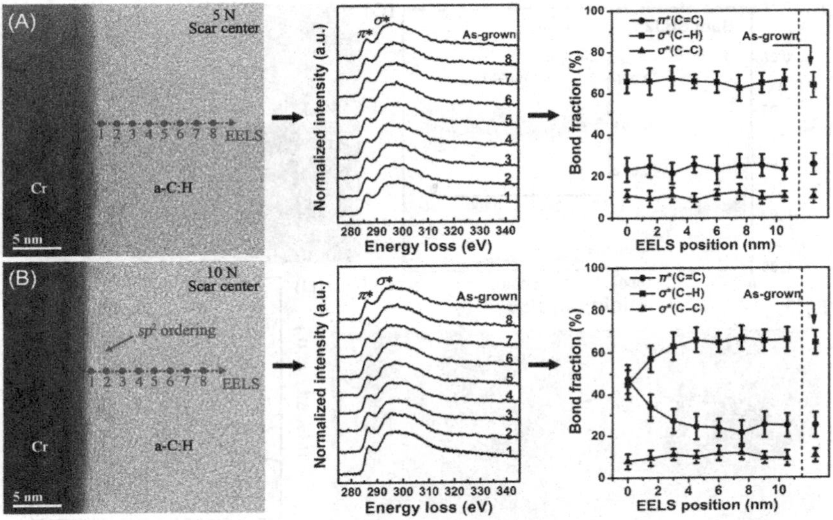

FIGURE 16.3 BF-STEM image, EELS C-*K* core-loss edges and bonding calculation results showing the nearly intact bonding structure and local clustering and ordering near the sliding interface at (A) 5 N and (B) 10 N, respectively [15]. *BF-STEM*: Bright field scanning transmission electron microscope; *EELS*: electron energy-loss microscopy.

phase (47.5%) as compared to that of the pristine film (25%). Meanwhile, a high density of sp^3-C phase in the form of C—H was maintained. Therefore the microstructure of the shear sublayer could be defined as a highly hydrogenated graphite-like carbon [45]. These findings clearly demonstrate that the self-mated hydrogen-rich hydrocarbon network is capable of withstanding a contact pressure up to 0.93 GPa for establishing a superlubricity state with few detectable phase transformations and the film material loss. Nevertheless, tribo-induced localization and ordering of hydrocarbon network along with the occurrence of nanoscale wear is initialized at a pressure up to 1.17 GPa. This is further confirmed by the significant reduction in the shear strength of the sliding interface at 10 N (0.62 MPa) in comparison to the cases of 2 N (2.16 MPa) and 5 N (2.47 MPa).

16.3.2 Nanostructured tribolayers in situ grown on heterogeneous surfaces

In comparison to the case of self-mated a-C:H films, the interfacial material behaviors are significantly changed when one of the rubbing surfaces is replaced by other materials such as metal or ceramic counterfaces. As shown in Fig. 16.4A, a superlow friction coefficient of 0.004 was achieved for the tribocouple of a-C:H versus bare steel after an obvious running-in period. It

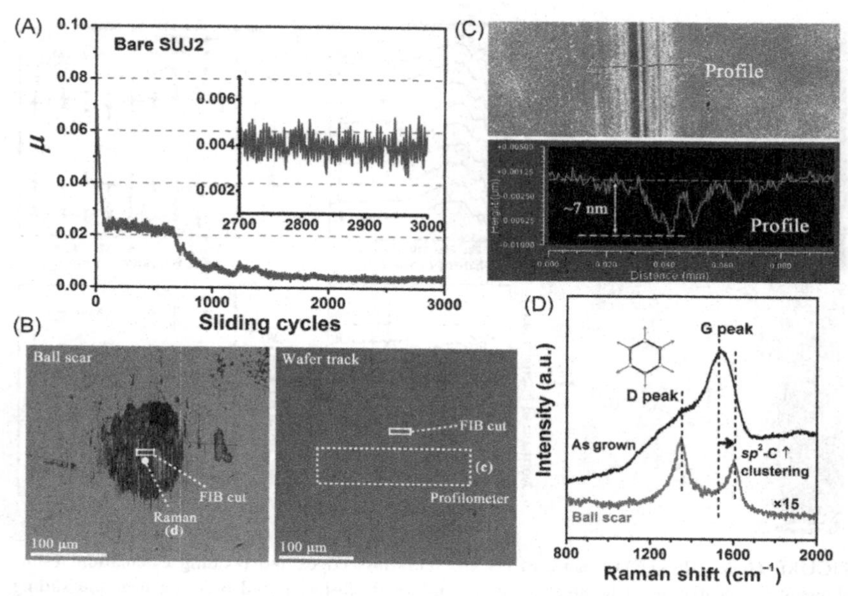

FIGURE 16.4 Superlubricity behavior of a-C:H film when sliding against bare steel ball and the corresponding characterization of the contact areas. (A) Friction test in dry N_2, (B) ball wear scar and wafer wear track produced on the contact area, (C) 3D white-light interference images showing the morphology and cross-sectional profile of the wear track, and (D) Raman spectra of the ball wear scar and pristine film for comparison [15].

is usually argued that this running-in stage is necessary to establish the superlubricity state for heterogeneous rubbing surfaces. After the test, a dark-brown wear scar was produced on the bare steel ball surface, indicating the existence of new triboproducts in the contact area (Fig. 16.4B). However the wear track on the film surface was still barely visible, for which the wear depth was measured to be only about 7 nm (Fig. 16.4C). The splitting of D and G peaks as distinguished in the Raman spectrum strongly implied the promoted formation of sp^2-C phases and local clustering of these aromatic nanostructures [30,45].

The STEM results in Fig. 16.5 confirmed the presence of a uniform carbon-rich tribolayer with a thickness of ~27 nm on the ball scar surface. This in situ formed tribolayer was rich in structural characteristics, especially the formation of local nanostructures. The whole tribolayer could be divided into three sublayers including a Fe-nanoparticle-dispersed sublayer near the sliding interface (Fig. 16.5B), a C-rich low-density sublayer in the middle (Fig. 16.5C) and a C-Fe-O heterogeneously crystalized sublayer close to the steel surface (Fig. 16.5D). This layered structure was substantially confirmed by EELS analysis. As seen from the EELS elemental maps of C, Fe, and O in Fig. 16.5E, the spatial distribution of interfacial atoms was in the form of

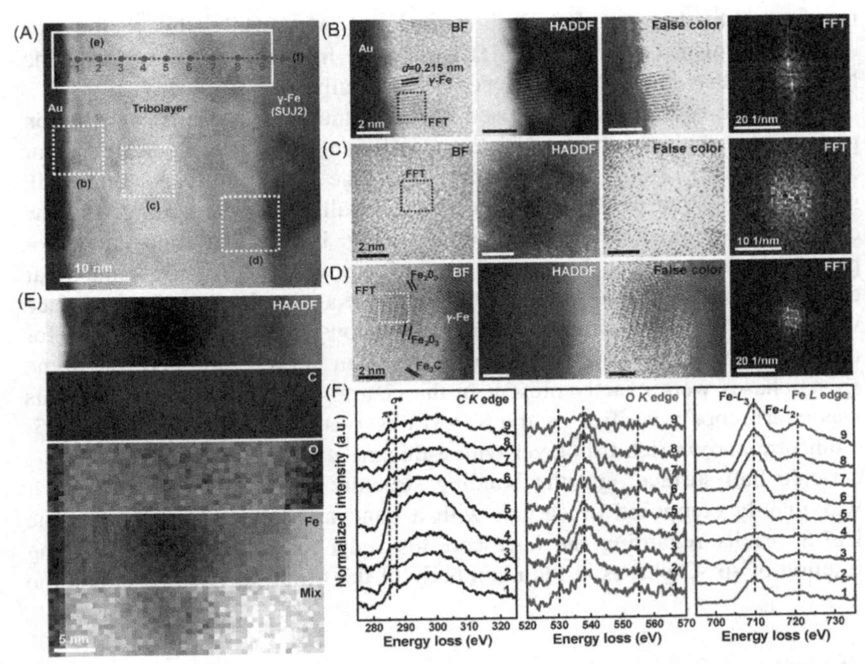

FIGURE 16.5 STEM and EELS characterization of the tribolayer formed on the bare steel ball surface. (A) BF-STEM showing a 27-nm-thick tribolayer in situ grown on the steel surface, (B–D) BF-, HAADF-, false-colored BF-STEM, and FFT images showing three nanostructured sublayers evolved in the tribolayer, (E) HAADF-STEM and EELS-SI mapping images indicating the elemental distribution across the tribolayer, and (F) evolution of C-*K*, O-*K*, and Fe-*L* core-edge spectra acquired across the tribolayer [15]. *BF-STEM*: Bright field scanning transmission electron microscope; *EELS*: electron energy-loss microscopy; *FFT*, fast fourier transform; *HAADF*: high-angle annular dark-field; *SI*: spectrum image; *STEM*: scanning transmission electron microscope.

multilayer and local aggregation. Fig. 16.5F presents the EELS spectrum image (SI)-line spectra of C-*K*, O-*K*, and Fe-*L* core edges recorded point by point across the tribolayer, revealing the evolution of bonding environment for each element. Obviously, carbon was mainly involved in forming sp^2-C phases in the middle region of the tribolayer as revealed by the sharpened π^* peaks in the C-*K* edges (EELS points 4 and 5). The bonding fraction of sp^2-C phases could reach to 65%–70%. Based on the Raman spectrum (Fig. 16.4D), STEM images and fast fourier transform result (Fig. 16.5C), the authors argued that these sp^2-C phases were mainly in the form of amorphous structure, but with nanostructures ordered locally. In addition, it is interesting to note that iron gathered at the outermost sliding interface in the form of nanoparticles, probably through shear-induced diffusion from the steel surface (EELS points 1–3). These nanoparticles were supposed to passivate the outermost shear band since they could remain in this antifriction interface until the end of the tribological test. Meanwhile, this region was rich in hydrogen [~58% sp^3(C-H) as derived from the C-*K* edge] and the hydrogen passivation effect produced a chemically inert surface.

When the ceramic ball was used as the counterpart, a similar antifriction performance with an ultralow friction coefficient of 0.004 was achieved for a-C:H film (Fig. 16.6A). One noticeable change is the shortened running-in stage as compared to the case of bare steel ball. After the sliding test, the contact surfaces became smooth and clean (Fig. 16.6B), and almost no tribo-products were found in the central area of the wear scar, while some wear debris was piled up around the scar edge. Once again, the wear track surface was almost wearless with the wear depth below 7 nm (Fig. 16.6C). As for the scar surface, however, the Raman spectrum strongly implied that some sp^2-C phases were exactly present in the central area of the wear scar. This was microscopely verified by the STEM observation. As shown in Fig. 16.7, a uniform carbon-rich tribolayer with a thickness of 5 nm was formed on the ceramic ball surface, and the bonding structure was mainly amorphous. It was mainly composed of carbon with a concentration of 65–80 at%. The EELS results regarding the C-*K* core-loss edge indicate that the bonding fraction of sp^2-C phases in the main body of the tribolayer was 80%, and the

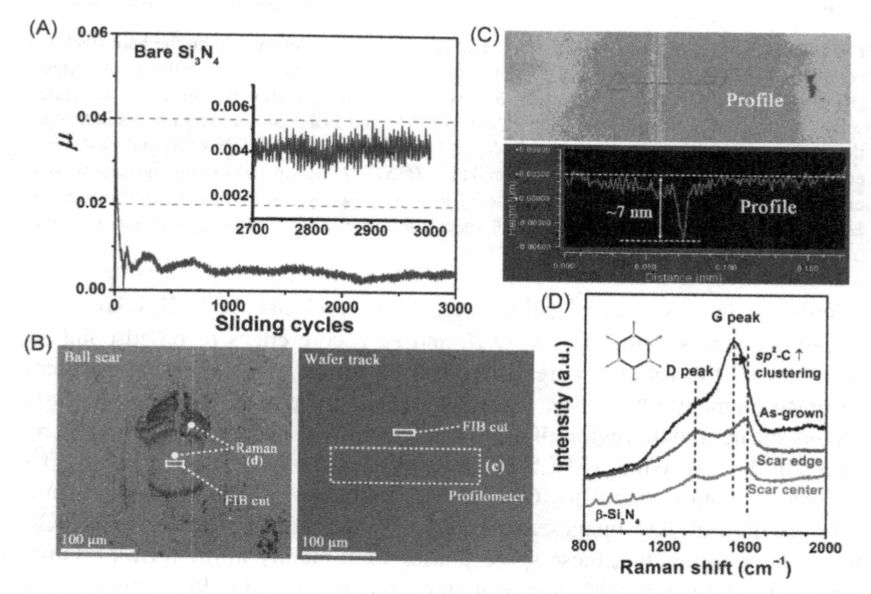

FIGURE 16.6 Superlubricity behavior of a-C:H film when sliding against bare Si_3N_4 ball and the corresponding characterization of the contact areas. (A) Friction test in dry N_2, (B) ball wear scar and wafer wear track produced on the contact area, (C) 3D white-light interference images showing the morphology and cross-sectional profile of the wear track, and (D) Raman spectra of the ball wear scar and pristine film for comparison [15].

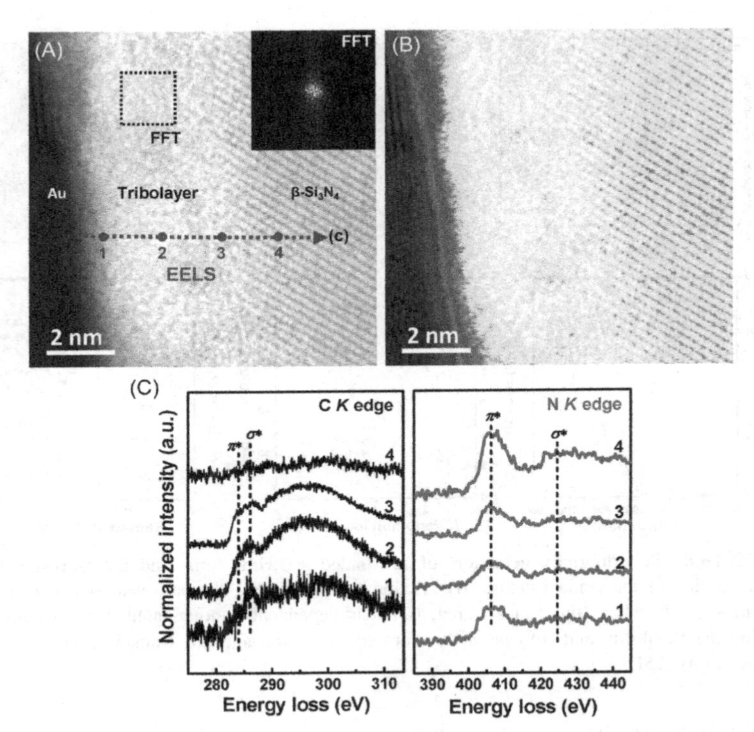

FIGURE 16.7 STEM and EELS characterization of the tribolayer formed on the bare Si_3N_4 ball surface. (A and B) BF-STEM and the corresponding false-colored BF-STEM images showing a 5-nm-thick tribolayer in situ grown on the steel surface, and (C) evolution of C-*K* and N-*K* core-edge spectra acquired across the tribolayer [15]. *BF-STEM*: Bright field scanning transmission electron microscope; *EELS*: electron energy-loss microscopy; *STEM*: scanning transmission electron microscope.

near-surface region was rich in sp^3(C–H) bonds with the bonding fraction up to 40%. In addition, the presence of trace elements such as N and Si and the recorded N-*K* edges imply that the ceramic surface was involved in tribochemical reactions to some extent in spite of its chemical inertness and high hardness. All these findings demonstrate that a nanoscale tribolayer with delicate nanostructures formed is capable of establishing a robust superlubricity state on the heterogeneous sliding surface.

16.3.3 Tribo-induced polymerization assisting shear softening of the sliding interface

Numerous studies have indicated that metal or nonmetal doping can significantly affect the bonding structures, mechanical and tribological properties of amorphous carbon films. As compared to the pure a-C:H film, the

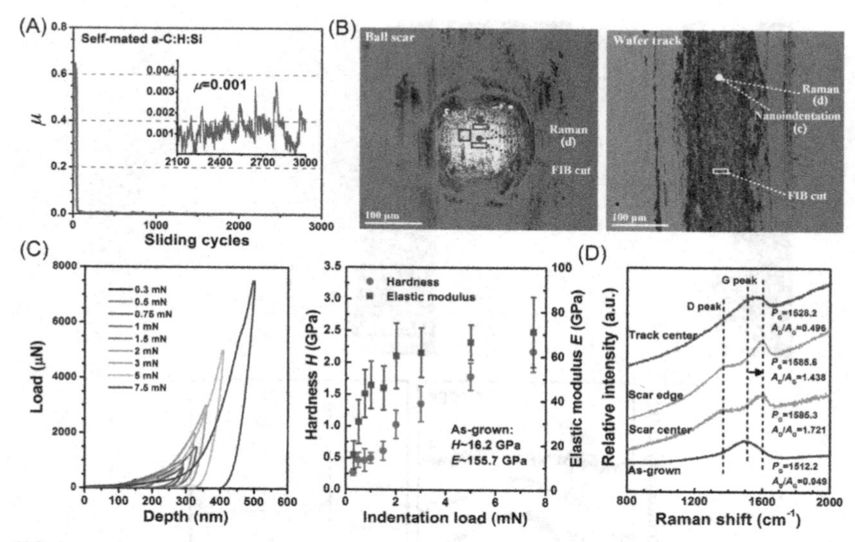

FIGURE 16.8 Superlubricity behaviors of self-mated a-C:H:Si films and the corresponding characterization of the contact areas. (A) Friction test in dry N_2, (B) ball wear scar and wafer wear track produced on the contact area, (C) load-dependent nanoindentation measurements revealing the flexibility and softness of the formed tribolayer, and (D) Raman spectra and the analysis results [15].

incorporation of silicon into the hydrocarbon matrix provides another pathway to realize superlubricity. Fig. 16.8A shows the antifriction performance of this film structure. It can be seen from the friction curve that a stable superlow friction coefficient down to 0.001 was readily achieved for the self-mated a-C:H:Si films without a noticeable running-in stage. During this process, all the film materials were transferred to the counterfacing a-C:H:Si film surface, and a thick tribolayer was in situ formed in the wear track (Fig. 16.8B). This tribolayer was quite soft and the as-measured hardness was about 0.25 GPa (Fig. 16.8C), resembling that of a hydrocarbon polymer. Raman results also reveal the presence of plentiful sp^2-C phases in the contact areas. The STEM and EELS characterization further clarifies the reconstruction of the sliding interface by confirming the presence of nanostructured tribolayers on the contact surfaces. As shown in Fig. 16.9A–E, a highly crystalized tribolayer with lattice structures ordered along the sliding direction was formed in the central region of the wear scar. On the scar edge (Fig. 16.9F), a thicker tribolayer of ~20 nm was also in situ formed, containing well-evolved multilayer structure with three sublayers as shown in Fig. 16.9G–I. Specifically, for the sublayer close to the sliding interface, a well-crystalized Fe-rich shear band with their shear layers oriented parallel to the sliding direction was formed (Fig. 16.9G). These highly ordered shear bands were expected to act as passivation layers for producing a chemically

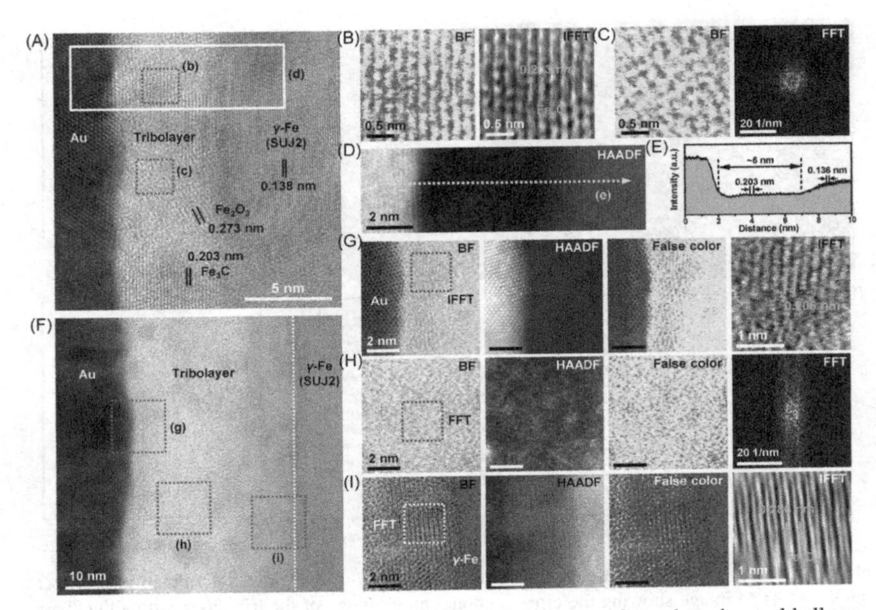

FIGURE 16.9 STEM and EELS characterization of the tribolayer formed on the steel ball surface for the self-mated a-C:H:Si films after the superlubricity test. (A) BF-STEM image showing a nanostructured tribolayer with a thickness of 5 nm in the scar center, (B and C) BF-STEM and IFFT (or FFT) images showing the highly crystalized shear band and amorphous local structure in the tribolayer, respectively, and (D and E) HAADF-STEM and intensity profile showing the nanostructure of the tribolayer. (F) BF-STEM image showing a nanostructured tribolayer with a thickness of 20 nm in situ grown on the scar edge, (G–I) BF-, HAADF-, false-color displayed BF-STEM, and local IFFT (or FFT) images showing the local nanostructures from three different regions in the tribolayer [15]. *BF-STEM*: Bright field scanning transmission electron microscope; *EELS*: electron energy-loss microscopy; *FFT*, fast fourier transform; *HAADF*: high-angle annular dark field.

inert interface. For the counterfacing a-C:H:Si film on the surface of the Si wafer, the cross-sectional morphology by TEM observation (Fig. 16.10A) confirms the wavy interfacial profiles of the sliding interface, namely the rough surface characteristics of the tribolayer. An outermost region of 30 nm in the wear track was found to undergo significant structural transformation (Fig. 16.10B and C). As further revealed by EELS calculations, this tribo-affected region possessed a higher fraction of sp^2-C phases (up to $\sim$57%), meanwhile containing a high density of σ^*-hybridized bonds such as C–H, C–Si, and C–C bonds. In view of the low hardness, the bonding structure of the tribolayer was hence more like a polymeric hydrocarbon compound. Clearly, the presence of Si atoms in the lubricating hydrocarbon matrix was expected to bring forth a new pathway to achieve the superlubricity state. This was experimentally confirmed by the fact that a silicon-rich nucleation sublayer was found in the bottom region of the tribolayer (Fig. 16.11),

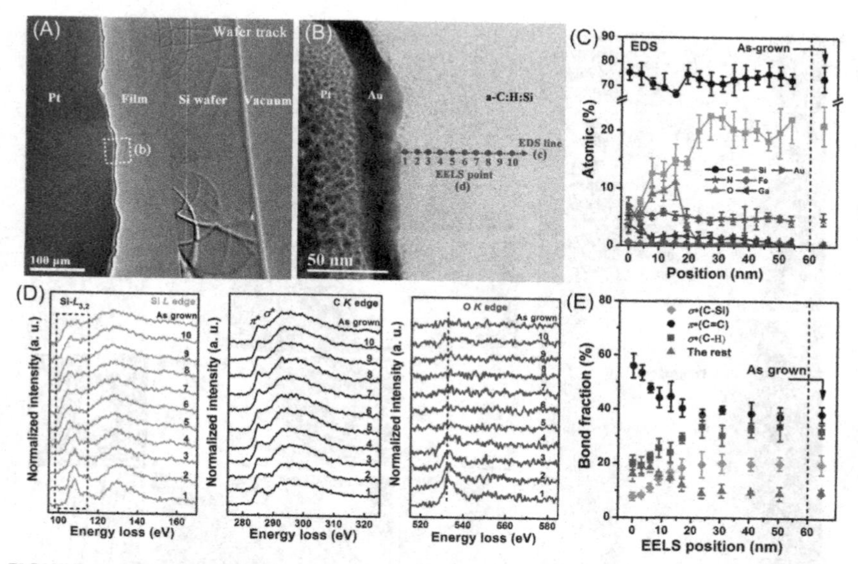

FIGURE 16.10 TEM, STEM, and EELS characterization of the tribolayer formed in the wear track. (A) TEM image showing the cross-sectional morphology of the tribolayer with a thickness of 5 nm in the scar center, (B) BF-STEM image showing the enlarged detail of the tribolayer, (C) EDS-elemental distribution of the tribolayer from the sliding interface toward the underlying intact a-C:H:Si film, and (D) evolution of Si-*L*, C-*K*, and O-*K* EELS core-loss edges and the calculated C-bonds fractions recorded across the tribolayer [15]. *BF-STEM*: Bright field scanning transmission electron microscope; *EDS*, energy dispersive spectrum, *EELS*: electron energy-loss microscopy; *STEM*: scanning transmission electron microscope; *TEM*: transmission electron microscope.

namely the initial starting point for the tribolayer growth. It seems that this Si-nucleation layer behaved as a catalytic unit to trigger the following growth of a polymeric tribolayer upon the sliding contact as silicon is well known for its role in promoting a porous nanocomposite structure in amorphous Si_xC_y:H matrix [46].

16.3.4 Friction fading-out phenomenon at heavy loads in hydrogen environment

In comparison to the case of dry N_2, hydrogen, or dilute hydrogen gas is a source of atomic hydrogen to tribochemically passivate the sliding interface. Under the shear stress and flash temperature exerted at the contact asperities, hydrogen molecules are expected to be decomposed into atomic hydrogen and take part in the tribochemical reactions with the carbon film surfaces. This process favors the production of numerous hydrocarbon termination groups on the contact surface and promotes the hydrogenation of the sliding interface. A recent discovery regarding this condition further emphasizes the

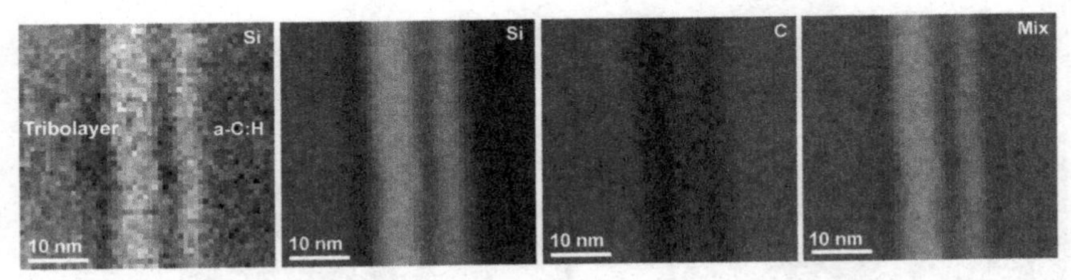

FIGURE 16.11 EELS-SI mapping image showing a Si-nucleation sublayer formed in the bottom region of the tribolayer and the corresponding extracted Si-*L*, C-*K*, and mix map [15]. *EELS-SI*: Electron energy-loss microscopy spectrum image.

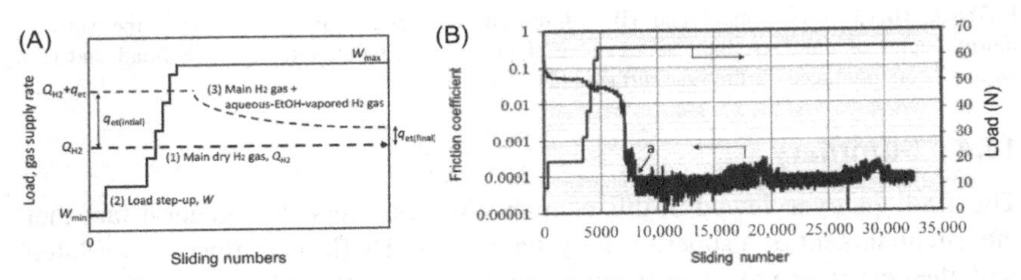

FIGURE 16.12 Friction fading-out phenomenon at heavy loads in hydrogen environment: (A) schematic diagram showing the test procedures and (B) a stable and extremely low friction coefficient of 0.0001 achieved under a heavy contact pressure of 2.6 GPa in aqueous-ethanol-vapored H_2 atmosphere [13].

possibility to lower the friction coefficient down to the level of 0.0001 when a PLC film was slid against ZrO_2 ball in H_2 atmosphere [7,12]. Nosaka et al. designed a straightforward manner to tailor the running-in stage by optimizing the supply rate of main H_2 gas, feeding time and rate of aqueous ethanol vapor and the load step-up mode [13], as shown in Fig. 16.12A. The friction coefficient gradually decreased with the applied normal load stepped up from 9.8 to 63.7 N, and reached to a robust near-frictionless state with the friction coefficient to 1 of 10,000 at a heavy contact pressure of 2.6 GPa (Fig. 16.12B).

The authors argued that the ZrO_2 ball behaves as a catalytic to promote the dissociative reactions of H_2 gas when the H_2 molecules are coming near to the Zr and O atoms of ZrO_2. The dissociative H atoms adsorb on the catalytic agent and tend to be involved in hydrogenation reactions with carbon atoms possessing double or triple bonds, that is, transforming C_2H_2 into C_2H_4 [12]. As shown in Fig. 16.13, a blister-rich and polymer-like tribofilm was formed on the ZrO_2 contact surface. In addition to the repulsive forces derived from paired interfacial H atoms, the lubricating effect from some produced tribochemical substances and the possibility of elastohydrostatic gas bearing lubrication [14] are also proposed by the authors to explain this range of superlow friction.

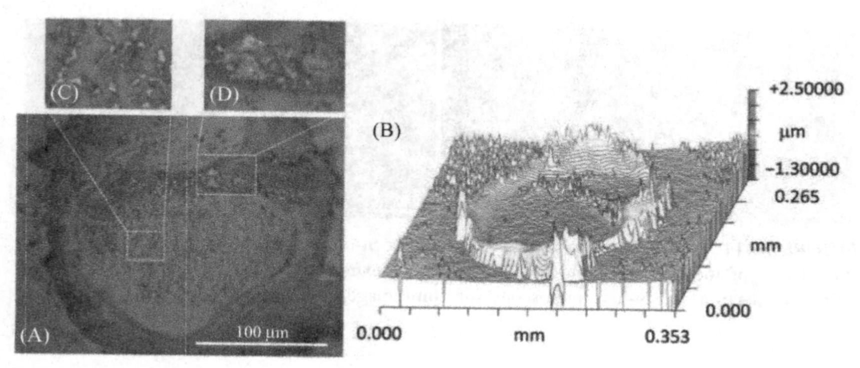

FIGURE 16.13 (A) Optical and (B) white-light interference images showing the surface morphologies of the ZrO_2 ball wear scar after the friction fading-out test. (C) Small and (D) large blisters produced on the scar surface [12].

16.4 Summary

The findings overviewed in this chapter obviously reveals a general fact that the establishment of a superlubricity state in a-C:Hs films is always correlated with the occurrence of an antifriction nanostructure at the sliding interface. It is capable for a-C:Hs to achieve superlow friction in a wide range of tribological conditions, generally via in situ growth of a nanostructured carbon-based tribolayer in the contact area. Under an appropriate tribotesting condition, a newly grown superlubricious tribolayer with specific composition was expected to be formed in the running-in stage rather than in the superlubricious state. The as-formed tribolayer can possess highly oriented and ordered surface nanostructures near the sliding interface, which assists the canceling out of the frictional forces during the sliding and hence the realization of superlubricity.

In summary, the achievements as mentioned earlier have clearly illustrated the exceptional roles of the tribo-induced interfacial nanostructures in superlubricity. Relying on the bonding diversity of carbon, more and more carbon-based superlubricious lubricants with unique structural features are emerging. The controllable manners in the arrangement of interfacial atoms, the prevailing of interrelated elements in the nanostructured sliding interfaces, and the possibility to achieve superlubricity in various rubbing conditions endow these types of carbon-based lubricants with broadening applications in engineering fields. Moreover, the as-achieved understandings and fundamentals can be extended to other surface and interface-related materials with the potential to acquire robust properties through designing and tailoring specific nanostructures.

Acknowledgment

This work was financially supported by National Natural Science Foundation of China (Grant Nos. 51527901 and 51975314).

Competing interests

The authors declare no competing financial interests.

References

[1] J. Robertson, Diamond-like amorphous carbon, Mater. Sci. Eng. R Rep. 37 (4) (2002) 129–281.

[2] C. Donnet, J. Fontaine, A. Grill, T. Le Mogne, The role of hydrogen on the friction mechanism of diamond-like carbon films, Tribol. Lett. 9 (3) (2001) 137–142.

[3] A. Erdemir, O.L. Eryilmaz, G. Fenske, Synthesis of diamondlike carbon films with superlow friction and wear properties, J. Vac. Sci. Technol. A 18 (4) (2000) 1987–1992.

[4] C. Wang, S. Yang, Q. Wang, Z. Wang, J. Zhang, Super-low friction and super-elastic hydrogenated carbon films originated from a unique fullerene-like nanostructure, Nanotechnology 19 (22) (2008) 225709.

[5] X. Chen, T. Kato, Growth mechanism and composition of ultrasmooth a-C:H:Si films grown from energetic ions for superlubricity, J. Appl. Phys. 115 (4) (2014) 044908.

[6] X. Chen, T. Kato, M. Nosaka, Origin of superlubricity in a-C:H:Si films: a relation to film bonding structure and environmental molecular characteristic, ACS Appl. Mater. Interfaces 6 (16) (2014) 13389–13405.

[7] M. Nosaka, A. Mifune, M. Kawaguchi, T. Shiiba, T. Kato, Friction fade-out at polymer-like carbon films slid by ZrO_2 pins under hydrogen environment, Proc. Inst. Mech. Engineers, Part. J: J. Eng. Tribol. 229 (8) (2015) 1030–1038.

[8] D. Berman, S.A. Deshmukh, S.K.R.S. Sankaranarayanan, A. Erdemir, A.V. Sumant, Macroscale superlubricity enabled by graphene nanoscroll formation, Science 348 (6239) (2015) 1118–1122.

[9] T. Kuwahara, P.A. Romero, S. Makowski, V. Weihnacht, G. Moras, M. Moseler, Mechano-chemical decomposition of organic friction modifiers with multiple reactive centres induces superlubricity of ta-C, Nat. Commun. 10 (1) (2019) 151.

[10] C. Matta, L. Joly-Pottuz, M.I. De Barros Bouchet, J.M. Martin, M. Kano, Q. Zhang, et al., Superlubricity and tribochemistry of polyhydric alcohols, Phys. Rev. B 78 (8) (2008) 085436.

[11] P. Huang, W. Qi, X. Yin, J. Choi, X. Chen, J. Tian, et al., Ultra-low friction of a-C:H films enabled by lubrication of nanodiamond and graphene in ambient air, Carbon 154 (2019) 203–210.

[12] M. Nosaka, R. Kusaba, Y. Morisaki, M. Kawaguchi, T. Kato, Stability of friction fade-out at polymer-like carbon films slid by ZrO_2 pins under alcohol-vapored hydrogen gas environment, Proc. Inst. Mech. Engineers, Part. J: J. Eng. Tribol. 230 (11) (2016) 1389–1397.

[13] M. Nosaka, Y. Morisaki, T. Fujiwara, H. Tokai, M. Kawaguchi, T. Kato, The run-in process for stable friction fade-out and tribofilm analyses by SEM and nano-indenter, Tribol. Online 12 (5) (2017) 274–280.

[14] T. Kato, H. Matsuoka, M. Kawaguchi, M. Nosaka, Possibility of elasto-hydrostatic evolved-gas bearing as one of the mechanisms of superlubricity, Proc. Inst. Mech. Engineers, Part. J: J. Eng. Tribol. 233 (4) (2017) 532–540.

[15] X. Chen, C. Zhang, T. Kato, X.A. Yang, S. Wu, R. Wang, et al., Evolution of tribo-induced interfacial nanostructures governing superlubricity in a-C:H and a-C:H:Si films, Nat. Commun. 8 (1) (2017) 1675.

[16] A. Erdemir, The role of hydrogen in tribological properties of diamond-like carbon films, Surf. Coat. Technol. 146-147 (2001) 292–297.

[17] A. Erdemir, C. Donnet, Tribology of diamond-like carbon films: recent progress and future prospects, J. Phys. D Appl. Phys. 39 (18) (2006) R311–R327.

[18] G.T. Gao, P.T. Mikulski, J.A. Harrison, Molecular-scale tribology of amorphous carbon coatings: effects of film thickness, adhesion, and long-range interactions, J. Am. Chem. Soc. 124 (24) (2002) 7202–7209.

[19] J. Fontaine, T. Le Mogne, J.L. Loubet, M. Belin, Achieving superlow friction with hydrogenated amorphous carbon: some key requirements, Thin Solid Films 482 (1) (2005) 99–108.

[20] H. Guo, Y. Qi, X. Li, Predicting the hydrogen pressure to achieve ultralow friction at diamond and diamondlike carbon surfaces from first principles, Appl. Phys. Lett. 92 (24) (2008) 241921.

[21] G. Zilibotti, M.C. Righi, M. Ferrario, Ab initio study on the surface chemistry and nanotribological properties of passivated diamond surfaces, Phys. Rev. B 79 (7) (2009) 075420.

[22] S. Bai, T. Onodera, R. Nagumo, R. Miura, A. Suzuki, H. Tsuboi, et al., Friction reduction mechanism of hydrogen- and fluorine-terminated diamond-like carbon films investigated by molecular dynamics and quantum chemical calculation, J. Phys. Chem. C. 116 (23) (2012) 12559–12565.

[23] Y.A.N. Liu, E.I. Meletis, Evidence of graphitization of diamond-like carbon films during sliding wear, J. Mater. Sci. 32 (13) (1997) 3491–3495.

[24] T.W. Scharf, I.L. Singer, Quantification of the thickness of carbon transfer films using Raman tribometry, Tribol. Lett. 14 (2) (2003) 137–145.

[25] J.C. Sánchez-López, A. Erdemir, C. Donnet, T.C. Rojas, Friction-induced structural transformations of diamondlike carbon coatings under various atmospheres, Surf. Coat. Technol. 163-164 (2003) 444–450.

[26] N.S. Tambe, B. Bhushan, Nanoscale friction-induced phase transformation of diamond-like carbon, Scr. Mater. 52 (8) (2005) 751–755.

[27] P. Stoyanov, P.A. Romero, R. Merz, M. Kopnarski, M. Stricker, P. Stemmer, et al., Nanoscale sliding friction phenomena at the interface of diamond-like carbon and tungsten, Acta Mater. 67 (2014) 39–408.

[28] Y. Wang, J. Xu, J. Zhang, Q. Chen, Y. Ootani, Y. Higuchi, et al., Tribochemical reactions and graphitization of diamond-like carbon against alumina give volcano-type temperature dependence of friction coefficients: a tight-binding quantum chemical molecular dynamics simulation, Carbon 133 (2018) 350–357.

[29] Y. Wang, K. Gao, B. Zhang, Q. Wang, J. Zhang, Structure effects of sp^2-rich carbon films under super-low friction contact, Carbon 137 (2018) 49–56.

[30] A.C. Ferrari, J. Robertson, Interpretation of Raman spectra of disordered and amorphous carbon, Phys. Rev. B 61 (20) (2000) 14095–14107.

[31] R. Arenal, A.C.Y. Liu, Clustering of aromatic rings in near-frictionless hydrogenated amorphous carbon films probed using multiwavelength Raman spectroscopy, Appl. Phys. Lett. 91 (21) (2007) 211903.

[32] J.A. Johnson, D. Holland, J.B. Woodford, A. Zinovev, I.A. Gee, O.L. Eryilmaz, et al., Top-surface characterization of a near frictionless carbon film, Diam. Relat. Mater. 16 (2) (2007) 209–215.

[33] O.L. Eryilmaz, A. Erdemir, On the hydrogen lubrication mechanism(s) of DLC films: An imaging TOF-SIMS study, Surf. Coat. Technol. 203 (5) (2008) 750–755.

[34] O.L. Eryilmaz, A. Erdemir, TOF-SIMS and XPS characterization of diamond-like carbon films after tests in inert and oxidizing environments, Wear 265 (1) (2008) 244–254.

[35] A.P. Merkle, A. Erdemir, O.L. Eryilmaz, J.A. Johnson, L.D. Marks, In situ TEM studies of tribo-induced bonding modifications in near-frictionless carbon films, Carbon 48 (3) (2010) 587–591.

[36] A. M'Ndange-Pfupfu, O. Eryilmaz, A. Erdemir, L.D. Marks, Quantification of sliding-induced phase transformation in N3FC diamond-like carbon films, Diam. Relat. Mater. 20 (8) (2011) 1143–1148.

[37] X. Zhang, R. Schneider, E. Müller, M. Mee, S. Meier, P. Gumbsch, et al., Electron microscopic evidence for a tribologically induced phase transformation as the origin of wear in diamond, J. Appl. Phys. 115 (6) (2014) 063508.

[38] D.S. Wang, S.Y. Chang, Y.C. Huang, J.B. Wu, H.J. Lai, M.S. Leu, Nanoscopic observations of stress-induced formation of graphitic nanocrystallites at amorphous carbon surfaces, Carbon 74 (2014) 302–311.

[39] M.I. De Barros Bouchet, C. Matta, B. Vacher, T. Le-Mogne, J.M. Martin, J. von Lautz, et al., Energy filtering transmission electron microscopy and atomistic simulations of tribo-induced hybridization change of nanocrystalline diamond coating, Carbon 87 (2015) 317–329.

[40] D.A. Muller, Structure and bonding at the atomic scale by scanning transmission electron microscopy, Nat. Mater. 8 (2009) 263–270.

[41] M. Schaffer, B. Schaffer, Q. Ramasse, Sample preparation for atomic-resolution STEM at low voltages by FIB, Ultramicroscopy 114 (2012) 62–71.

[42] G. Soto, J.A. Díaz, W. de la Cruz, Copper nitride films produced by reactive pulsed laser deposition, Mater. Lett. 57 (26) (2003) 4130–4133.

[43] S.D. Berger, D.R. McKenzie, P.J. Martin, EELS analysis of vacuum arc-deposited diamond-like films, Philos. Mag. Lett. 57 (6) (1988) 285–290.

[44] A.C. Ferrari, A. Libassi, B.K. Tanner, V. Stolojan, J. Yuan, L.M. Brown, et al., Density, sp^3 fraction, and cross-sectional structure of amorphous carbon films determined by x-ray reflectivity and electron energy-loss spectroscopy, Phys. Rev. B 62 (16) (2000) 11089–11103.

[45] C. Casiraghi, A.C. Ferrari, J. Robertson, Raman spectroscopy of hydrogenated amorphous carbons, Phys. Rev. B 72 (8) (2005) 085401.

[46] D.L. Williamson, A.H. Mahan, B.P. Nelson, R.S. Crandall, Microvoids in amorphous $Si_{1-x}C_x$:H alloys studied by small-angle x-ray scattering, Appl. Phys. Lett. 55 (8) (1989) 783–785.

Chapter 17

Superlubricity in carbon nanostructural films: from mechanisms to modulating strategies

Junyan Zhang[1,2] **and Yongfu Wang**[1,2]
[1]State Key Laboratory of Solid Lubrication, Lanzhou Institute of Chemical Physics, Chinese Academy of Sciences, Lanzhou, P.R. China, [2]Center of Materials Science and Optoelectronics Engineering, University of Chinese Academy of Sciences, Beijing, P.R. China

17.1 Introduction

Hydrogenated diamond-like carbon (DLC) films are often characterized by the measurements of sp^2:sp^3 carbon bonding ratio and hydrogen content. At higher sp^3 content, there is a specific type of a-C designated as highly tetrahedral hydrogenated amorphous carbon or "ta-C:H." The sp^3 content is less, the sp^2 fraction increases, and the films are more graphitic. If the films contain many C–H bonds, they are more "polymeric." These types of DLC can be seen on a ternary phase diagram [1].

Generally, DLC films are "soft coatings," in terms of classification of coatings with respect to hardness and coefficient of friction (COF) [2]. Similar to the solid lubricants such as polymers, soft metals (e.g., indium, silver, and gold), or lamellar solids (e.g., graphite and MoS_2), DLC films are easily sheared. Nevertheless, they exhibit higher hardness, than most metals and/or alloys. Thus DLC films not only afford very high wear resistance but also impressive friction coefficients (generally in the range of 0.05–0.2). In some cases, the friction coefficient is reduced to less than 0.01, and the phenomenon is called "superlubricity." Their excellent ability of friction-reducing and wear resistance attract ever-growing attention from scientific and industrial communities.

The superlubricity is a regime in which friction or resistance to sliding nearly vanishes between two sliding solid surfaces. It is proposed by Hirano and Sokoloff [3] in the early 1990s when they predicted that the frictional

Superlubricity. DOI: https://doi.org/10.1016/B978-0-444-64313-1.00017-X

forces should vanish between two sliding surfaces in an incommensurate contact. After this discovery, two main types of solid superlubricity have been defined: (1) structural superlubricity, realized in the cases such as double-walled carbon nanotubes [4], graphene and graphite [5–8]; (2) disordered solid interface-based superlubricity, demonstrated by DLC films [9–11]. In this chapter, we focus on the disordered solid interface-based superlubricity. Researchers have developed a series of hydrogenated and nitrogen-containing DLC films that exhibit superlubricity, and have shown that superlubricity is attributed to their atomically smooth and chemically inert surfaces from the high levels of hydrogenation (≥40 at.% hydrogen). However when special carbon nanostructures are introduced into hydrogenated DLC films, the mechanisms unlike the root causes above are proposed and discussed. For example, hydrogenated fullerene-like (FL) carbon (FL-C:H) films have the H content of less than 25 at.%, but attain superlubricity [11,12].

In the preface of this book, the review about the structural superlubricity and their lubrication mechanisms helps to understand the mechanisms for superlubricity in nanostructural carbon films, and is essential for understanding how it may relate to, or differ from, the superlubricity behavior of the films. In this chapter, we discuss the classification of nanostructural carbon films, devote much of our attention to the root causes of the films and review their modulating strategies that attain superlubricity.

17.2 Nanostructural carbon films

Carbon has several allotropic forms capable of creating a variety of nanocarbons with nanoscale size shapes, such as graphene, fullerenes, nanoonions, etc. These sp^2-hybridized nanocarbons show great promise in a wide variety of applications in electronic, biological, tribological applications, and so on. Driven by the advent of nanocarbons, most of the attempts from tribological researchers have resulted in sp^2-dominated carbon films, and researchers never stop their forward progress on the way.

17.2.1 Graphene-embedded carbon films

Graphene sheet consists of sp^2-bonded carbon atoms arranged in hexagon lattice. Because of its outstanding properties, various pathways were proposed to prepare single-layer or multilayer graphene. However few researches referred to the access to growing graphene sheets in plasma environments. This is because the traditional plasma-assisted film deposition always employs ions etching on the film. Recently, graphene-embedded carbon films have been prepared by electron cyclotron resonance plasma [13], or unbalanced magnetron sputtering technique with the assistance of an external-field effect [14]. The graphene nanostructure helps to significantly improve the tribological properties of the amorphous carbon film, which exhibit a superlow friction (~0.005) in a vacuum environment [14]. Their lubrication mechanisms are similar to the structural superlubricity for graphene, and thus are not further discussed again.

17.2.2 Graphite-like carbon films

At present, in many literatures, FL and graphite-like (GL) structures are not strictly defined and distinguished. In many cases, FL structures are well defined, but graphite like are usually obscured. Always, a structure with high sp^2 content can be called graphite like. For example, the structure characterizations of the introduced films are affected by deposition parameters [such as deposition temperature, ion energy, ion density, gas source type (Ar, C_2H_2, CH_4, H_2, etc.), gas flow ratio, substrate bias voltage, etc.]. When the parameters are changed, the microstructure characterizations of the film (such as size, orientation, stacking, curvature, and cross-linking of graphene layers) are different, and the macroproperties are also distinguishing [15–19]. As shown in Fig. 17.1A and B, the microstructures of sp^2-rich carbon films are tailored by varying the flow rate of hydrogen gas, that is, the orientation, stacking, and curvature of graphene sheets changes along with the H_2 flow rate [20]. The films grown at the H_2 flow rate of 5 SCCM

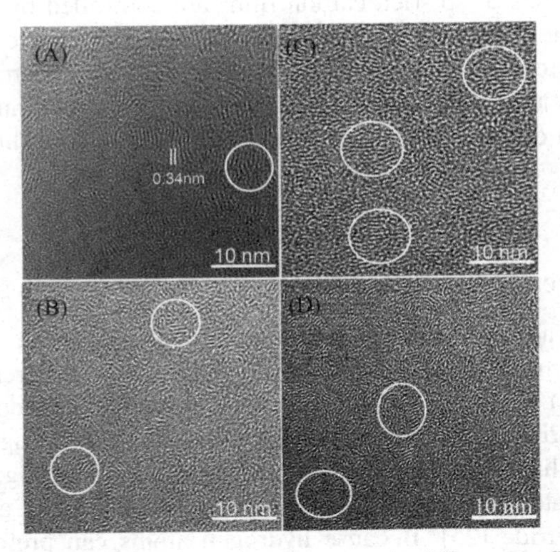

FIGURE 17.1 HRTEM micrographs of FL-C:H films under different gas flow rate of H_2. (A) 0 SCCM, (B) 2.5 SCCM, (C) 5 SCCM, and (D) 10 SCCM. *HRTEM*, High-resolution transmission electron microscopy; *FL-C:H*, hydrogenated fullerene-like carbon. *Adapted from Y. Wang, J. Guo, K. Gao, B. Zhang, A. Liang, et al., Understanding the ultra-low friction behavior of hydrogenated fullerene-like carbon films grown with different flow rates of hydrogen gas, Carbon 77 (2014) 518–524, with permission from Elsevier.*

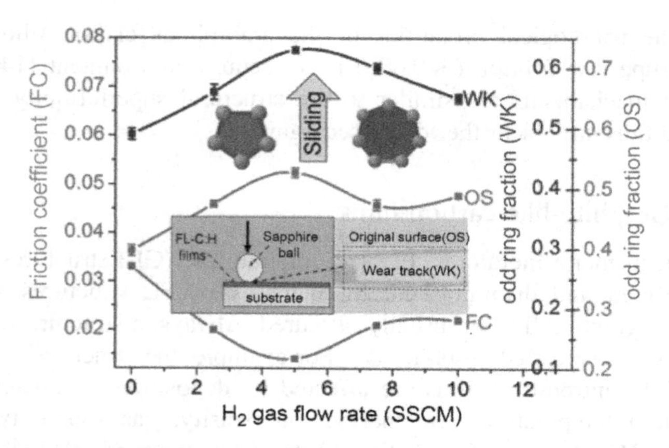

FIGURE 17.2 Friction coefficient (FC) and odd ring fraction of hydrogenated fullerene-like carbon (FL-C:H) films as a function of gas flow rate of H_2. All the central wear tracks (WK) had higher odd (pentagonal and heptagonal) carbon ring fraction than the originally deposited surfaces (OS). *Adapted from Y. Wang, J. Guo, J. Zhang, Y. Qin, Ultralow friction regime from the in-situ production of a richer fullerene-like nanostructured carbon in sliding contact. RSC Adv. 5 (2015) 106476 [21], with permission from Royal Society of Chemistry.*

exhibit lowest friction and wear in humid air (Fig. 17.2) [20]. The hardness and elastic modulus of sp^2-rich carbon films are controlled by the number of cross-linking sites between graphene sheets, and the elastic recovery rate mainly depends on the sheet size (Fig. 17.3A–C) [16]. Therefore it is very necessary to clearly understand and distinguish the structural characterizations of sp^2-rich carbon films, in order to clarify the relationship between the characterizations and behaviors of the superlubricity film.

17.2.3 Fullerene-like carbon films

Driven by the advent of the elusive superhard crystalline phase β-C_3N_4 (harder than diamond), SjÖstrÖm et al. [22] discovered fullerene-like carbon nitride (FL-CN_x) films in 1995. The FL-CN_x films are a predominantly sp^2-hybridized material, which consists of the substitution of nitrogen for carbon, bent, and cross-linked graphitic basal planes, as shown in Fig. 17.4. Such a structure was later reported for pure carbon films [23,24] and hexagonal boron carbon nitride [25]. Because hydrogen atoms can preferentially etch the sp^2 phase during the deposition process, FL-C:H films are reported until 2008 [12,26]. The film presents outstanding mechanical properties due to the presence of the 3D sp^2-based FL structure. This is unlike traditional DLC films, which is attributed to the presence of sp^3 hybrids with tetrahedral coordination.

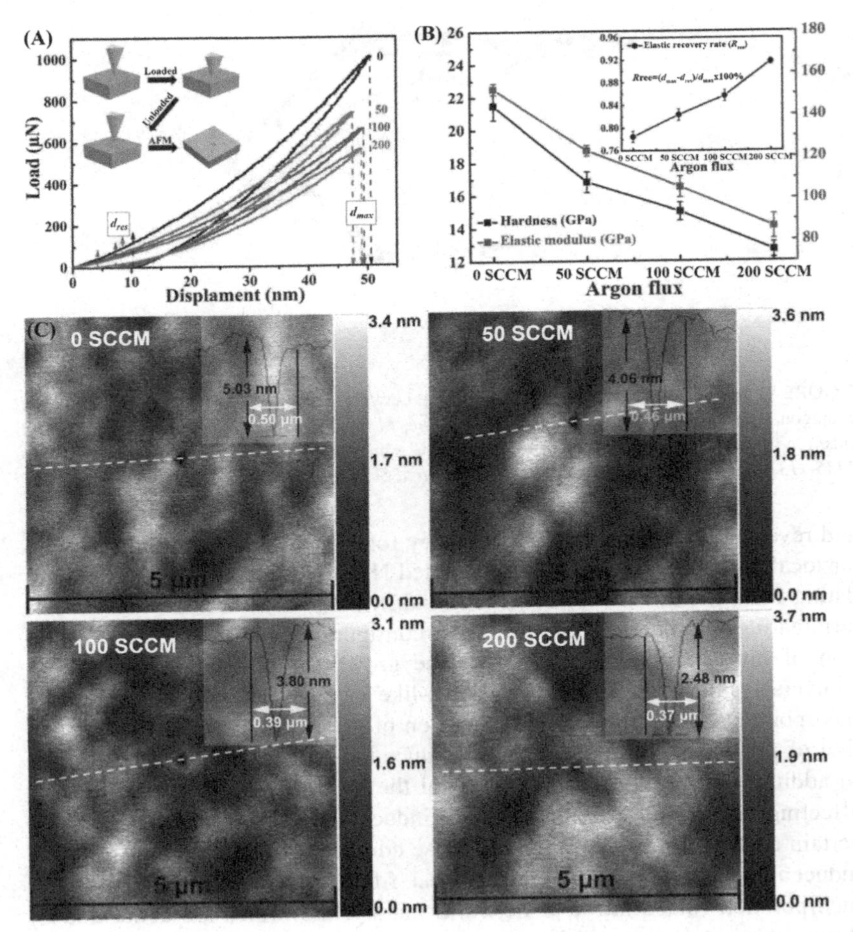

FIGURE 17.3 (A) Load–displacement curves of FL-C:H films prepared at 0, 50, 100, and 200 SCCM argon flux; the left upper inset shows the corresponding schematic representations of the proposed nanoindentation test. (B) The hardness, elastic modulus, and elastic recovery rate (Rrec) calculated from the load–displacement curves as a function of argon flux. (C) AFM images of the indented areas in the FL-C:H films prepared at different argon flux: 0, 50, 100, and 200 SCCM. The right upper insets show the corresponding line profiles extracted from the dashed lines in the indented areas. *FL-C:H*, Hydrogenated fullerene-like carbon. *Adapted from Z. Cao, W. Zhao, Q. Liu, A. Liang, J. Zhang, Super-elasticity and ultralow friction of hydrogenated fullerene-like carbon films: associated with the size of graphene sheets. Adv. Mater. Interfaces 5 (6) (2018) 1701303, with permission from Wiley.*

17.2.3.1 Fullerene-like carbon nitride films

Because the formation and control of FL structure in CN_x is much easier, for example, occurring at lower substrate temperatures, the mechanisms inducing curvature and the nature of the cross-linking sites are extensively discussed

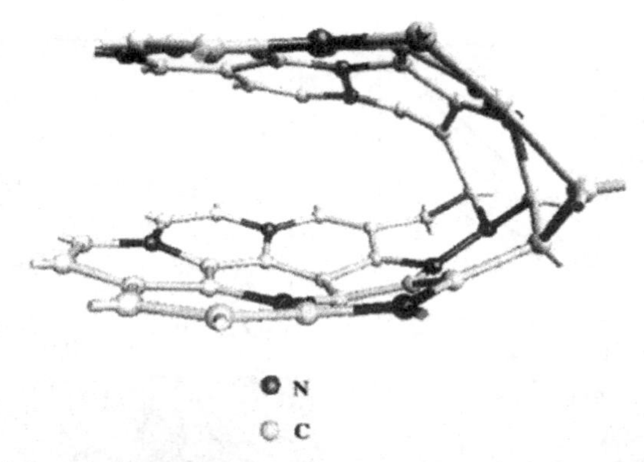

FIGURE 17.4 Schematic picture of cross-linking between graphite-based planes caused by the pentagon. *Adapted from H. SjÖstrÖm, S. StafstrÖm, M. Boman, J.-E. Sundgren, Superhard and elastic carbon nitride thin films having fullerene like microstructure. Phys. Rev. Lett. 75 (1995) 1336–1339, with permission from American Physical Society.*

and revealed. N incorporation plays a key role for the formation of a particular local atomic arrangement. Incorporated N has the capability of accommodating local atomic structures such as substituting C in aromatic clusters or terminating pyridine-like or nitrile-like configurations [27,28]. The substitution of nitrogen for carbon hinders the growth of aromatic clusters [29]. Conversely, the termination of pyridine-like or nitrile-like configurations by incorporated N can promote the extension of graphene sheets and the formation of pentagons that is believed to induce the curvature in the sheets [30]. In addition, incorporated N may control the localization of the electrons by affecting the cluster size and order, or induces doping by incorporating into certain configurations. The effect of these competing local bonding structures induced by incorporated N on the final film properties depend on the N incorporation routes into a C network, mainly, the routes are controlled by the synthesis technique and treatment conditions [28]. Apparently, the final film properties can be tailored by controlling the curvature, size, and connectivity of graphene sheets. In other words, they can determined by different proportions of chain and aromatic structures and different degrees of extension, bending, and cross-linking of aromatic clusters. The ability of N to FL structures makes these materials promising coatings for tribological applications. The alleged nitrogen-induced cross linkage between graphitic sheets contributes considerably to the film strength by preventing interplanar slip [31]. The result leads to an extremely resilient material where the contact stresses are dissipated over large volumes. Thus FL-CN_x films have better tribological properties than amorphous CN_x films. However friction coefficients of FL-CN_x films in ambient air remain high values ranging from 0.2 to 0.4, and the coefficients depend on the nitrogen content and growth temperature [31]. Moreover, the large spread in wear rate values was observed,

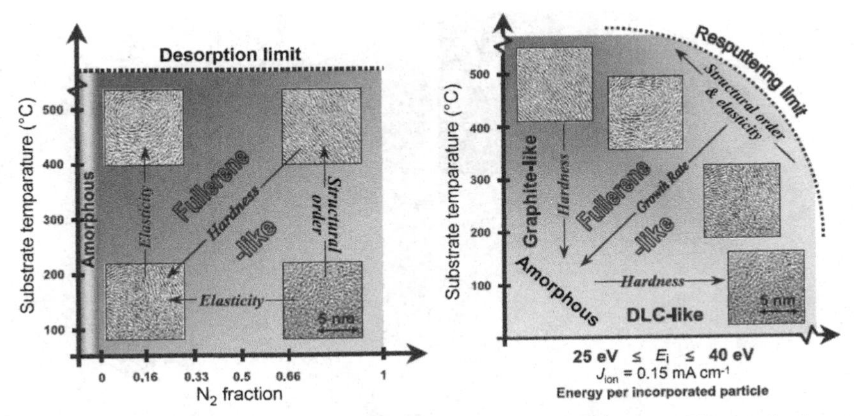

FIGURE 17.5 The effect of substrate temperature, N_2 fraction, and degree of ion bombardment on the structures and properties for magnetron-sputtered FL-CN_x coatings. *FL-CN*, Fullerene-like carbon nitride. *Adapted from J. Neidhardt, L. Hultman, E. Broitman, T.W. Scharf, I.L. Singer, Structural, mechanical and tribological behavior of fullerene-like and amorphous carbon nitride coatings, Diam. Relat. Mater. 13 (10) (2004) 1882–1888, with permission from Elsevier.*

depending on the change in microstructure of the FL-CN_x films which is determined by the N_2 fraction and growth temperature, as shown in Fig. 17.5. That is, the tribological behaviors of FL-CN_x films can be tuned by tailoring their microstructure controlled by the synthesis technique and treatment conditions [31].

17.2.3.2 FL-C:H films

Despite the successful synthesis of FL-C and CN_x materials, FL structures were prepared in hydrogen-containing deposition environments until 2008. This is because hydrogen atoms can preferentially etch the sp^2 phase during the deposition process and destroy FL structures to a certain extent [1]. As compared with C and CN_x materials, the deposition parameters yielding FL structure is very narrow, with the structure being difficult to be controlled. The FL features in FL-C:H films are not so clear compared to those in FL-C and FL-CN_x materials, for example, the closed rings of planes [32]. Moreover, the basal planes in the FL-C:H films are more frequently curved and cross-linked than FL-C and FL-CN_x materials. Thus FL-C:H have worse mechanical properties than FL-C and FL-CN_x materials. But the friction coefficients of FL-C:H films remain excellently low values, below 0.01 in ambient air [12].

Designing other nanostructures in FL-C:H films can provide an opportunity to significantly enhance material properties earlier. Generally, the FL features in FL-C:H films have three diffraction rings at 1.15, 2, and 3.5 Å [12]. When C60 and fullerene crystalline nanoparticles were introduced,

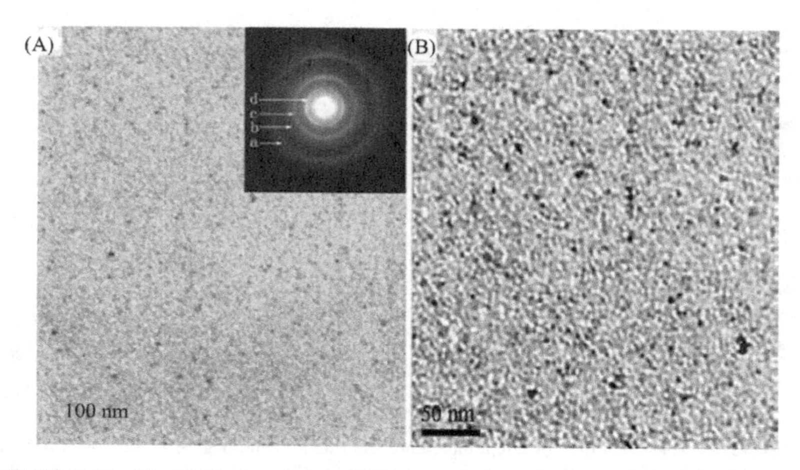

FIGURE 17.6 (A) TEM micrograph and (B) the corresponding selected area electron diffraction pattern of the hydrogenated carbon films prepared by dc-pulse plasma CVD. *CVD*, Chemical vapor deposition; *TEM*, transmission electron microscopy. *Adapted from Q. Wang, C. Wang, Z. Wang, J. Zhang, D. He, Fullerene nanostructure-induced excellent mechanical properties in hydrogenated amorphous carbon. Appl. Phys. Lett. 91 (2007) 141902, with permission from American Institute of Physics.*

an additional diffraction ring at 2.25 Å can be observed, as shown in Fig. 17.6 [33]. Later, diamond particles with the size of about 100 nm were embedded in a FL-C matrix film at elevated deposition substrate temperature (300ºC), and at the same time, the (111) lattice fringes of diamond with a separation of 2.1 Å can be seen [34]. Recently, similar to those in FL-CN_x films, the FL features have several nanometers ordered domains consisting of less frequently cross-inked graphitic planes, with the presence of one new ring at ~1.8 Å [35]. All the nanostructural films not only exhibit good mechanical properties, that is, high hardness (> 17 GPa) and elastic recovery (~85%), but also attain ultralow friction coefficients (as low as 0.009) in the open air (Fig. 17.7).

Unlike the role of N in the FL-CN_x films, H plays a key role in the formation of FL structures. At the beginning of the preparation of FL-C:H films, the films are mainly extracted from methane and/or hydrogen gases in a discontinuous discharged plasma environment by employing a high pulse bias, and pulse duty cycle is thus assumed to be the key factor in the formation of FL-C structures in terms of "annealing" effects [32,36]. Recently, FL-C:H films can be formed in a continuous discharged plasma environment [20]. The FL-C:H films were not fully thermodynamic relaxed, which was essential for the formation of a well-aligned FL structure as observed in FL-C and FL-CN_x materials [22–24]. FL structures are formed in equilibrium growth environment and on the depositing surface. Conversely, for FL-C:H films, the structures are growing in far away from equilibrium growth

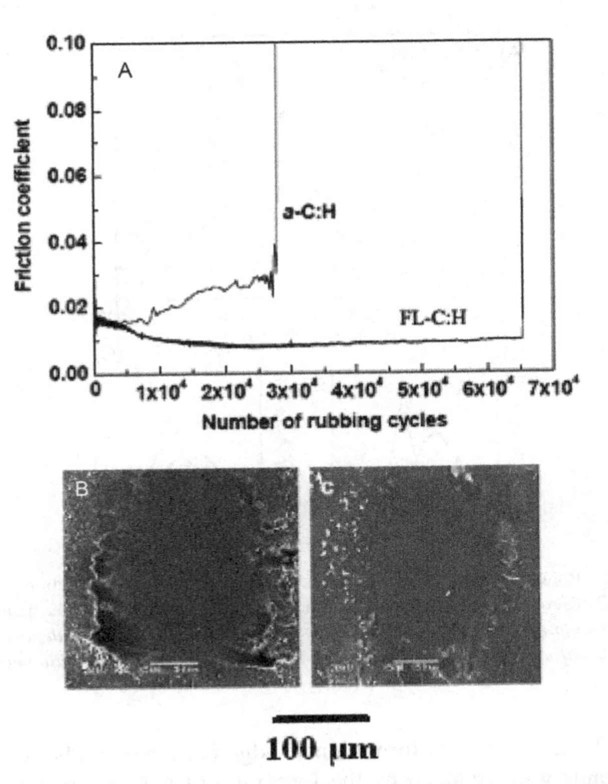

FIGURE 17.7 Tribological tests on the FL-C:H in air with 20% relative humidity at ambient temperature: (A) the friction coefficient as a function of time for a-C:H (top curve) and FL-C:H (bottom curve); (B) the worn surface of an Si_3N_4 ball against FL-C:H; and (C) a-C:H. *a-C:H*, Hydrogenated amorphous carbon; *FL-C:H*, hydrogenated fullerene-like carbon. *Adapted from C. Wang, S. Yang, Q. Wang, Z. Wang, J. Zhang, Super-low friction and super-elastic hydrogenated carbon films originated from a unique fullerene-like nanostructure, Nanotechnology 19 (2008) 225709, with permission from IOP Publishing Ltd.*

environment in a very short pulse period at low temperature, and they are formed under the depositing surface because of higher compressive stress induced by CH_n^+ and H^+ [37]. We carried out the transformation experiments of GL to FL structures by placing high-temperature steel substrates in the depositing environment which can form FL-C:H films (Fig. 17.8) [38]. We also investigated the changes of bond mixtures, H content, aromatic clusters, and internal stress at the transformation process (Fig. 17.9). In H-containing environments, the size of aromatic clusters and accordingly graphene planes and the formation of edge dangling bonds was a key step for the FL structure formation. More detailed, H^+ bombardments leaded to the spitting of large graphene planes (at GL stage) into more and smaller

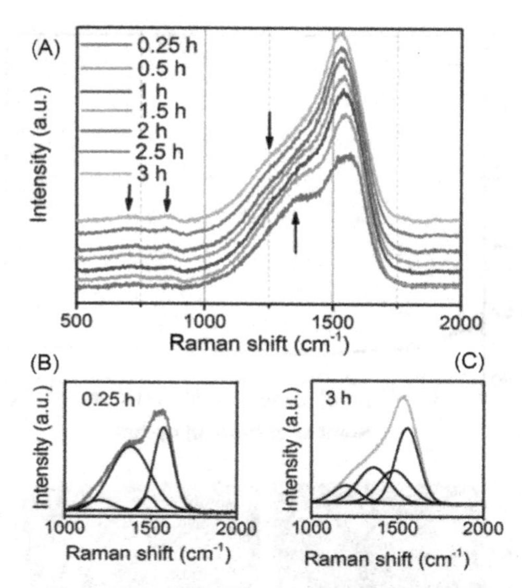

FIGURE 17.8 Raman spectra (A) of the films with the temperature reduction and deposition time; (B and C) deconvolutions of the films' Raman spectra. *Adapted from Y. Wang, K. Gao, J. Zhang, Observation of structure transition as a function of temperature in depositing hydrogenated sp2-rich carbon films. Appl. Surf. Sci. 439 (2018) 1152–1157, with permission from Elsevier.*

planes (at FL stage) and the formation of edge dangling bonds; some of these dangling bonds were reduced by the formation of pentagons and subsequent curving of the smaller planes, which were an indicator of FL structures [38].

17.3 Superlubricity mechanism

DLC films can achieve superlow friction, but the superlubricity sliding regimes are only feasible under special conditions with effectively controlled factors or parameters (surface roughness, chemical and tribochemical interactions, transfer films, thermal effects, etc.). Erdemir et al. [39] indicated that adhesive or chemical interactions play more important roles than smooth surfaces and environmental species. How to reduce adhesive or chemical interactions between two sliding surfaces can be discussed here.

17.3.1 Surface passivation

Highly hydrogenated DLC films can achieve superlow friction in high vacuum with certain types, hydrogen. Hydrogen can bond to the carbon atoms that make up the bulk DLC films and satisfy the unoccupied or free σ bonds to make the surfaces become chemical passivated seriously, when the H

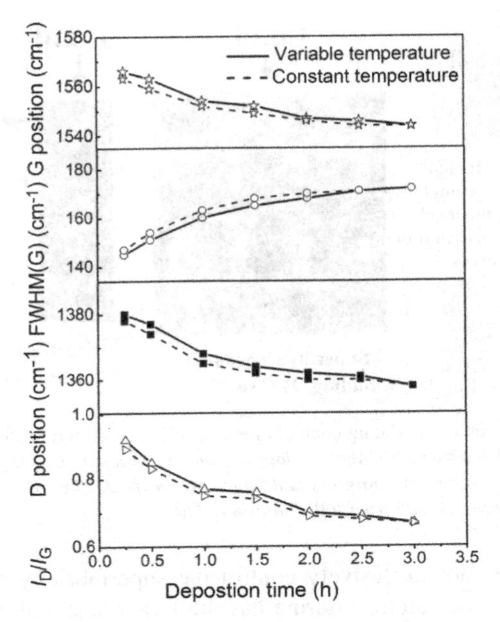

FIGURE 17.9 The change curves (solid lines) of the indicators, D and G peaks' positions, FWHM(G), and I_D/I_G from the Raman spectra (Fig. 17.2A) of the films with the uncontrolled temperature reduction and the deposition time. The dashed lines show corresponding change curves of the indicators for the samples obtained at controlled or constant temperature. *FWHM*, Full width at half maximum. *Adapted from Y. Wang, K. Gao, J. Zhang, Observation of structure transition as a function of temperature in depositing hydrogenated* sp^2*-rich carbon films. Appl. Surf. Sci. 439 (2018) 1152–1157, with permission from Elsevier.*

content is above 40 at.%. The totally passivated surface becomes chemically inert, causing very little adhesive interactions between two sliding surfaces and yielding the superlow friction. More clearly, bonded H can exist at the carbon surface because of stronger C–H bonds than C–C bonds, and is difficult to remove; free H within the films as a reservoir, can replenish or replace the hydrogen atoms that may be lost or removed from the surface due to the shear sliding. More importantly, H can induce a dipole configuration to give rise to repulsion (rather than attraction) at the sliding interface. The surface passivation from H can occur on the unreconstructed (100) surfaces of diamond structures under special conditions. The superlow friction model is presented in Fig. 17.10.

Passivation by other species such as oxygen, water molecules, and hydroxyl ions, always work during sliding, because the carbon atoms are already bonded to hydrogen during deposition [40]. The other species passivate free σ bonds when the hydrogen atoms may be lost or removed from the surface due to the shear sliding.

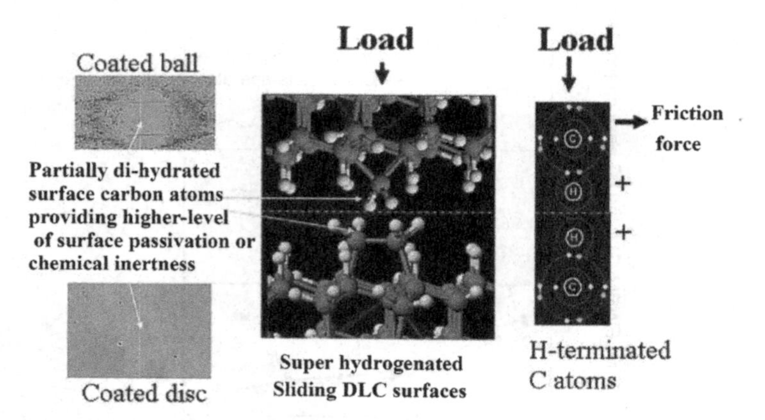

FIGURE 17.10 Illustration of sliding contact interface of hydrogen-terminated a-C:H surfaces. *a-C:H*, Hydrogenated amorphous carbon. *Adapted from A. Erdemir, C. Donnet, Tribology of diamond-like carbon films: recent progress and future prospects, J. Phys. D Appl. Phys. 39 (18) (2006) 311–327, with permission from IOP Publishing Ltd.*

Hydrogen does not exclusively control the superlubricity of DLC films. Compared to hydrogen atom, fluorine has the larger size and electronegativity, and the films with only 5 at.% of hydrogen (18 at.% of fluorine is incorporated) can achieve superlow friction [41]. Silicon incorporation into a-C:H films may also be responsible for friction values as low as 0.007, and the reason is due to hydrocarbon-transferred films formed and oriented along the sliding direction [42,43].

Nitrogen incorporation in DLC can induce superlow friction in the nitrogen gas. For example, López et al. [44] reported that a-C:N films exhibited friction values as low as 0.007 in the nitrogen gas whatever the nature of the counterface material. Nitrogen-surface interactions might have been the reason for the phenomenon of CN_x films [45].

The sulfur incorporated DLC films with 30 at.% hydrogen and 5 at.% sulfur can achieve superlow friction of 0.004 in ambient humid air. The reason is weaker binding between C–S–H surface and adsorbed water molecules which causes smaller dipole moment of S–H bonds than C–H bonds [46].

Overall, reducing adhesive or chemical interactions between two sliding surfaces by incorporating other elements is the key for the realization of superlow friction.

17.3.2 Friction-induced spherical nanoparticles

In many cases, sp^2-rich carbon films are also called as FL-C or GL-C films. FL-C is a carbon-phase structure characterized by highly curved graphite basal planes and frequently cross-linkings between adjacent basal planes [22,47]. The basal planes in the FL-C are interlocked with covalent bonds, in which bond length is shorter than the van der Waal bonds in graphite and turbostratic carbon. FL-C structure is well defined, in contrast, GL-C is usually obscured because it is featured by carbon films with high sp^2 content and sp^2 clustering [48]. In other words, FL-C is also called GL-C. Therefore the prevalent definitions of sp^2-rich carbon clusters weaken useful reference for the development and application of superlubricity carbon films.

Some works [28,38,49,50] have tried to redefine the films according to the structure characteristics of sp^2-rich clusters. In the pure carbon films [50], sheets of (flat) sixfold ring clusters can stack in high density to form nanocrystalline planar graphitic configurations, and curve in low density with the incorporation of five- and seven-membered rings, to achieve FL-C similar to fullerenes. In the carbon nitride [28,49] and hydrogenated carbon films [38], the atomic structure transforms with the increasing substrate temperature from three-dimensional bent and cross-linked graphite basal planes to planar graphitic configurations with crystallinity on the nanoscale, with a progressive promotion in size and ordering of sixfold ring clusters. Thus GL-C and FL-C have differences in size, orientation, stacking, curvature, and cross-linking of graphene layers.

In our group, we have refined the two structural films: FL-C is featured by highly curved and frequently cross-linked graphite base planes with shorter size ($\sim$1 nm), and GL-C is large stacks (3–5 nm) of graphite basal planes with less cross-linking between adjacent base planes (Fig. 17.11A and B) [38]. With the transformation of GL-C to FL-C, the D shoulder peak at $\sim 1330\ cm^{-1}$ vanishes, the position of G peak downshifts from $\sim$1530 to $\sim 1500\ cm^{-1}$, and a new peak at $\sim 1200\ cm^{-1}$ emerges (Fig. 17.11C). The $\sim 1200\ cm^{-1}$ peak is strongly linked with pentagons or heptagons in the curved graphene sheets, as observed and discussed in the cases such as fullerene-like carbon [51], carbon onions [52], etc. Owing to the $\sim 1200\ cm^{-1}$ peak, symmetric multipeak fits are proposed to replace simple two symmetric-line fits with the characteristic D and G peaks, which involve four peaks at $\sim$1200, $\sim$1360, $\sim$1470, and $\sim 1560\ cm^{-1}$. The reasonability of the fits is successively investigated by Schwan et al. [53], Siegal et al. [50], Rao et al. [54], and Zhang et al. [35,51]. In addition, we have realized the deposition of the two films on steel balls by the combination of plasma-enhanced chemical vapor deposition (PECVD) and plasma nitriding preprocess [37,55]. The frictional behaviors and structure evolutions of self-mated FL-C and GL-C films have been discussed in N_2 gas from low to superlow friction with the increasing normal load and sliding cycle [11].

The initial structures of sp^2-rich carbon films can predetermine friction-induced structure transformation pathways, frictional produces, and friction reduction mechanisms at their sliding interfaces [11]. The FL-C films were capable of yielding superlubricity only at the normal load or contact pressures higher than the GL-C (Fig. 17.11D), and the friction coefficients of FL-C were a little higher and unstable (Fig. 17.11E). This was because the

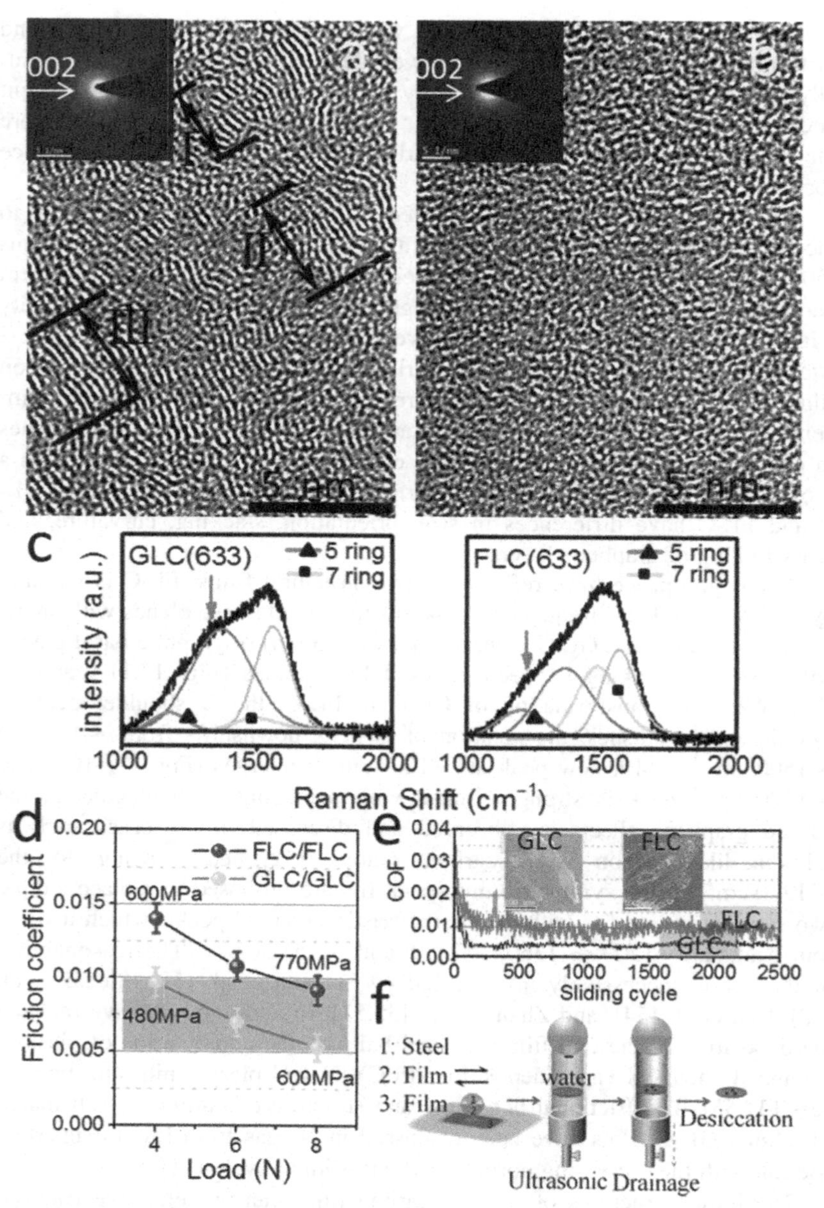

FIGURE 17.11 TEM images (A and B), Raman spectra at 633 nm (C), frictional behaviors (D and E) under different loads and sliding cycles of GL-C and FL-C. (F) Schematic diagram of gathering wear debris. The sets in (E) show the corresponding wear scars. (A color version of this figure can be viewed online.) *FL-C*, Fullerene-like carbon; *GL-C*: graphite-like carbon; *TEM*: transmission electron microscopy. *Adapted from Y. Wang, K. Gao, B. Zhang, Q. Wang, J. Zhang, Structure effects of sp2-rich carbon films under super-low friction contact, Carbon 137 (2018) 49–56, with permission from Elsevier.*

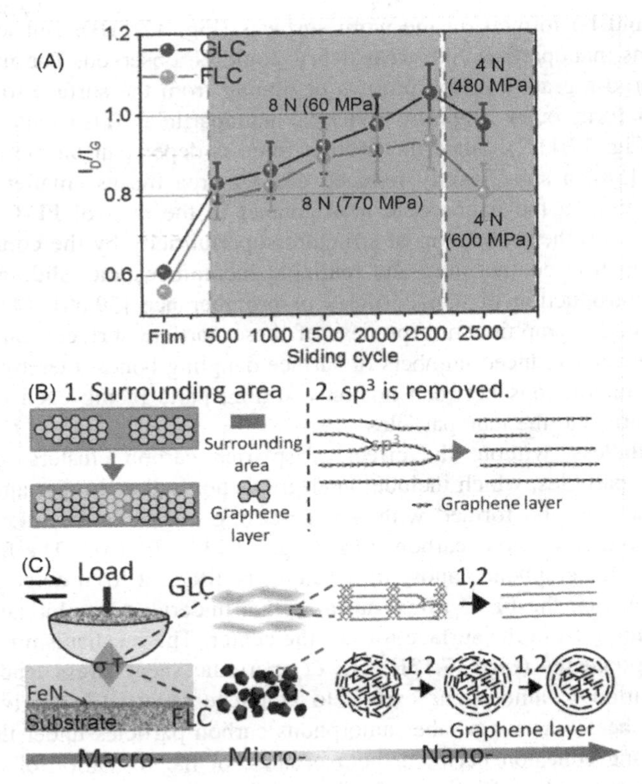

FIGURE 17.12 (A) The I_D/I_G changes at different sliding cycles and loads, (B) friction-induced rehybridization sites, and (C) evolution processes from macro- to nanoscale of GL-C and FL-C groups. FL-C, Fullerene-like carbon; *GL-C*, graphite-like carbon. *Adapted from Y. Wang, K. Gao, B. Zhang, Q. Wang, J. Zhang, Structure effects of sp²-rich carbon films under super-low friction contact, Carbon 137 (2018) 49–56, with permission from Elsevier.*

large graphene layer stacks in the GL-C films had certain capability of slippage between layers, and whereas the "distorted graphene" structure of the FL-C films endowed the films better resistance to destruction under friction [11]. As shown in Fig. 17.12A, the FL-C films had lower rehybridization degree at both low and high loads than that the GL-C films. This indicated that the "distorted graphene" structure could prevent the layer realignment and resisted long-range reorganization into nanocrystalline planar graphitic configurations [56,57]. We collected the debris from the worn surfaces on the coated balls by a smart device in Fig. 17.11F, which was detected under transmission electron microscopy. Influenced by the difference, the GL-C was transformed into larger more ordered graphene, in terms of adaptively sliding-induced rehybridizations and reorganization along the sliding direction (Fig. 17.11B and C). As a response in the changes, flake-like wear debris could be formed on the worn surfaces (Fig. 17.11B). But as for the FL-C films, nanoparticle-like wear debris could be observed. The nanoparticles undergo a graphitization process beginning from the surface toward the center, to form outer graphitic spherical nanoparticles (also called carbon onions) (Fig. 17.11C). Since the friction force is dependent on the true contact area [58], a significantly reduced contact area means smaller friction. The reduction in the nanoscopic area contact in the case of FL-C films is consistent with the realization of structure superlubricity by the combination of reducing true contact area and realizing incommensurate sliding contact with the introduction of nanoparticles or protuberance [59,60]. Meanwhile, the preferential graphitization process of these particle surfaces can lead to the significantly reduced numbers of surface dangling bonds, thereby bestowing weak interactions between not only sliding parts facing each other but also the parts and the nanoparticles.

Nevertheless, without the effect of sp^2-rich carbon clusters, graphene nanoscroll particles, which include outer graphene shell and inner amorphous carbon core, can be formed with the increasing cycles at the interfaces of hydrogenated amorphous carbon films (Fig. 17.13A–D) [61]. The formation mechanism for graphene nanoscroll particles is different from outer graphitic nanoparticles on the FL-C films' surfaces that undergo a graphitization process beginning from the surface toward the center. The mechanism is divided into two processes (Fig. 17.13E–H): (1) periodic shear stress leads to the transformation of amorphous carbon to graphene sheets; (2) newly formed graphene sheets can wrap other amorphous carbon particles under the effect of the strong adhesion between them because of the formation of dangling bonds. Corresponding friction reduction mechanism is (1) formed graphene sheets creates slippage between layers and (2) a large number of graphene nanoscroll particles triggers the incommensurate contact and rolling mechanism.

In a word, the structures of carbon films predetermine their transforming routes, resulting produces and friction reduction mechanisms under macrofriction. That is, superlubricity is yielded by layer slippage from larger more ordered graphene on the GL-C films' surfaces, or significantly reduced nanoscopic contact area from outer graphitic spherical nanoparticles on the surfaces of FL-C films.

17.4 Modulating strategies: the introduction of spherical nanoparticles

Similar to the structural superlubricity, the carbon films with the introduction of highly graphitized sp^2 carbon clusters such as nanocrystalline graphene [14], can lose superlubricity at higher humidity. This is because the incommensurate structures with superclear interfaces are contaminated by H_2O, O_2, etc. from operating environments. Since machines are generally used in

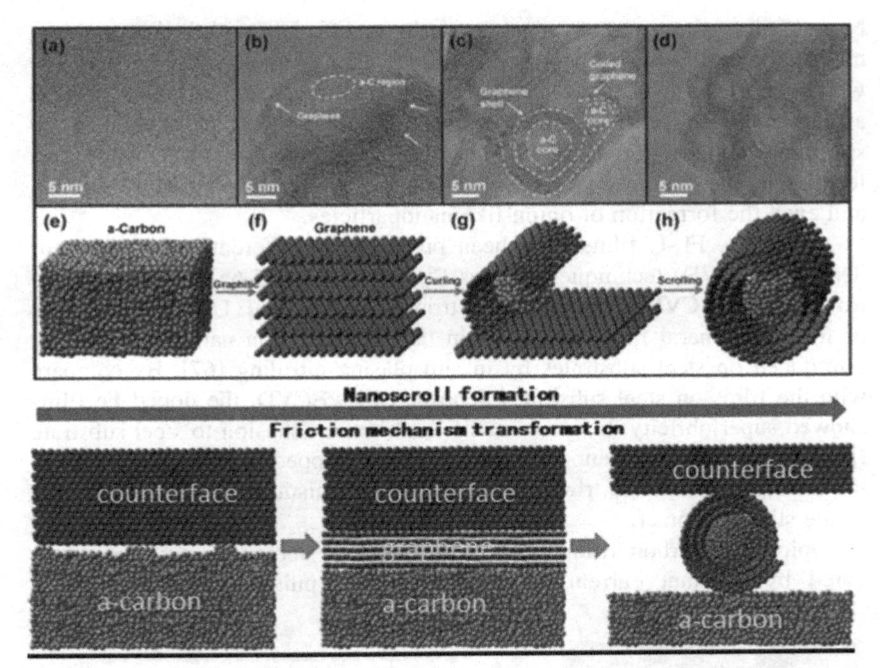

FIGURE 17.13 Schematic diagram of friction-induced evolution processes of hydrogenated amorphous carbon films without the effect of sp^2-rich carbon clusters. (A) HRTEM images of original films. (B–D) tribofilm debris of 200, 1000, and 7500 friction cycles, at 10 N, 10 Hz. (E–H) Schematic diagram of nucleation and growth model for graphene nanoscroll particles. The bottom sketch map show the transformation of friction mechanism of friction interface. (A color version of this figure can be viewed online.) *HRTEM*, High-resolution transmission electron microscopy. *Adapted from Z. Gong, J. Shi, B. Zhang, J. Zhang, Graphene nano scrolls responding to superlow friction of amorphous carbon, Carbon, 116 (2017) 310–317, with permission from Elsevier.*

open air, the superlubricity in N_2 gas and humid environments due to significantly reduced nanoscopic contact area from outer graphitic spherical nanoparticles on the surfaces of FL-C films is very desirable. Inspired by that, tuning film and interface structures to achieve superlubricity in humid environments are proposed, and recent related scientific results are reviewed here.

Nowadays, general FL-C is characterized by three Raman peaks at 400, 700, and 1200 cm^{-1} and three diffraction rings at 1.15, 2, and 3.5 Å in the selected area electron diffraction patterns. By optimizing the deposition parameters of forming general FL-C structures, a well-structured FL-C was prepared with the additional diffraction ring at ~1.8 Å. The presence and intensity of ~3.5 and ~1.75 Å (if it is present) rings, as an indicator of the short/medium-range graphitic order, could assess F-LC structure characteristics [62]. The films had ultralow friction and wear in the humid air and

N_2 gas. Since the trend of reducing friction coefficient with increasing normal load have been always observed on the DLC film surfaces [63,64], the well-structured FL-C films had the potential of realizing superlow friction at higher contact pressures in N_2 gas and humid air [35]. Compared to the starting films, friction facilitated the transformation into FL-C structures, leading to the presence of stably tribolayers with richer FL-C structures, and even the formation of onion-like nanoparticles.

Generally, FL-C films have been prepared by different chemical vapor deposition (CVD) techniques such as PECVD [12,20,35,65], electron cyclotron resonance CVD [26], and magnetron sputtering [66]. Under the premise of forming general FL-C structures in the PECVD, iron nanoparticles were introduced on steel substrates by in situ plasma nitriding [67]. By compare with the films on steel substrates deposited by PECVD, the doped Fe films showed superlubricity, longer wear life and better adhesion to steel substrate (Fig. 17.14). The iron nanoparticles would be wrapped by graphene sheets at the sliding interfaces to trigger the rolling mechanism and attain incommensurate sliding contact.

Onion-like carbon films (with a thickness of about 517 nm) were prepared by constant current high-frequency dual-pulsed PECVD [68]. As

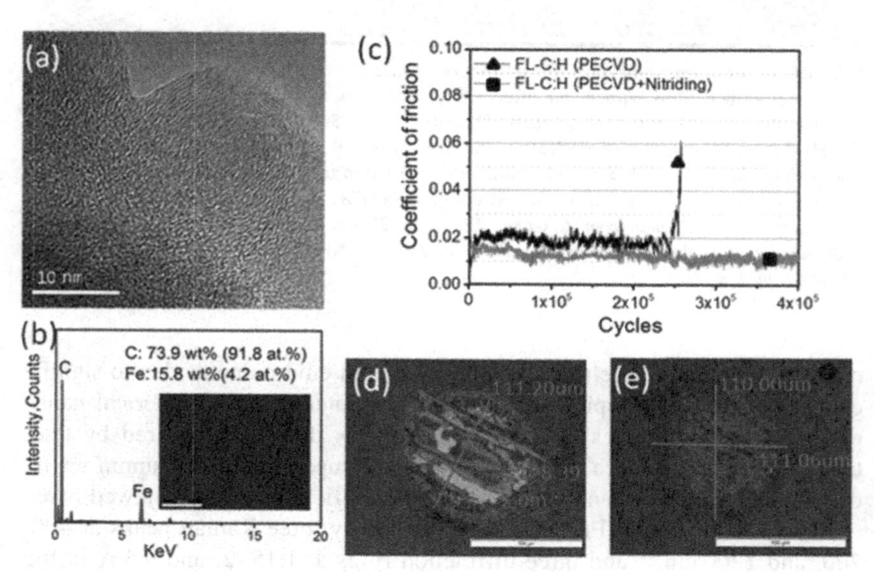

FIGURE 17.14 (A) HRTEM of FL-C:H in PECVD; (B) EDX of FL-C:H in PECVD + plasma nitriding; friction behaviors (C); and wear scars (D and E) of FL-C:H films on PECVD and PECVD + plasma nitriding (color online). *EDX*, Energy-dispersive X-ray spectroscopy; *FL-C:H*, hydrogenated fullerene-like carbon; *HRTEM*, high-resolution transmission electron microscopy; *PECVD*, plasma-enhanced chemical vapor deposition. *Adapted from Y. Wang, Z. Yue, Y. Wang, J. Zhang, K. Gao, Synthesis of fullerene-like hydrogenated carbon films containing iron nanoparticles, Mater. Lett. 219 (2018) 51–54, with permission from Elsevier.*

shown in Fig. 17.7C and D, a few spherical carbon clusters could be observed. The clusters had disordered structures in their centric shells, more or less, and their outlines were clearer. The concentric layer structure was similar to the structure of carbon onions. The films not only showed super-high elastic recovery of 92%, but also obtained superlow friction coefficient and wear rate (about 6.41×10^{-18} m^3/N/m) in humid air. The onion-like carbon clusters could move out from the films and resided between two sliding surfaces, where rolling friction and incommensurate contact played important role because the detached clusters played as "molecular bearing" in the friction process, resulting in both superlow friction and wear rate.

More interesting, when nanodiamond particles were embedded in amorphous carbon network, they could be converted into onion carbons at the loads greater than 20 N ($\approx$2 GPa contact pressure), and achieved the superlubricity for velocity ranging from 1 to 15 cm/s [69]. This was because the nanodiamond particles were transformed into amorphous carbon at $<$1 GPa, graphitic carbon at 1−2 GPa, and onion carbon at $>$2 GPa under macrofriction (Fig. 17.15).

These results further indicate that macroscale superlubricity can be achieved by introducing some beneficial structures, which results in the in situ formation of graphene-wrapped nanoparticles at the interface to significantly reduce true contact area and realize incommensurate sliding contact. So it is expected that tuning film and interface structures for hydrogenated carbon film would hold considerable prospects of achieving excellent frictional behaviors.

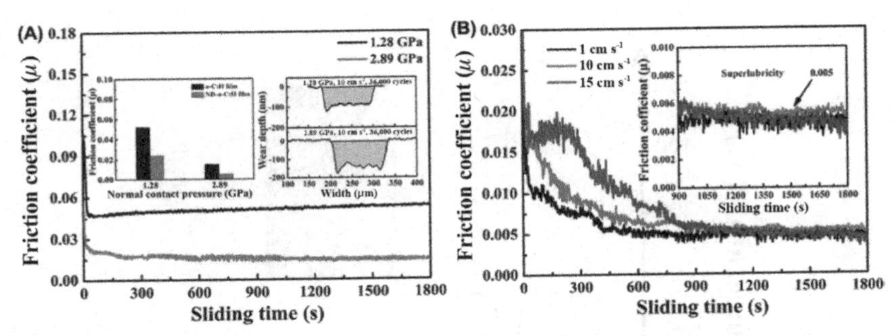

FIGURE 17.15 (A) Friction coefficient curves of a-C:H film at contact pressure of 1.28 and 2.89 GPa; the insets show the friction coefficient at stable sliding period and the cross-sectional profiles of wear tracks. (B) Experimental demonstration of the superlubricity regime for nanodiamond particles embedded a-C:H film under 1, 10, and 15 cm/s linear velocity (air environment, room temperature, 2.89 GPa contact pressure). *a-C:H*, Hydrogenated amorphous carbon. *Adapted from Z. Cao, W. Zhao, A. Liang, J. Zhang, A general engineering applicable superlubricity: hydrogenated amorphous carbon film containing nano diamond particles, Adv. Mater. Interfaces 4 (14) (2017) 1601224, with permission from Wiley.*

17.5 Conclusion

The preparations of the carbon nanostructural films can permit to deposit the suitable film to be appropriate for different applications, especially, a particular type of applications, such as in miniaturized devices and large-scale automobile industry. Nevertheless, the carbon nanostructural films, for FL-C:H films, have a very narrow synthesis window (Fig. 17.16) [38], compared with the FL-CN_x films [28,70] that have high COFs in humid air. Therefore research in the near future should focus on improving the superlubricity properties and preparation technologies of solid lubricants.

Our studies on the friction-induced interfacial formation of outer graphitic spherical nanoparticles confirm that the well-known friction reduction mechanisms, the transfer films or graphitization theory are too broad, and need to be refined and further explored. Moreover, the understanding between the realization of macroscale superlubricity and the mechanism of outer graphitic nanoparticles assembly on the surface of hydrogenated carbon film is still relatively poor. The spherical nanoparticles are ideal produces, actually, the outer graphitic nanoparticles have diverse shapes (such as quasi-spherical, elliptical, and deformed onions), large size ranges, and different rehybridization degrees in real macrocontact [11]. The variation brings unpredictable numbers of surface dangling bonds, and thus the friction force is uncontrollable. All these factors result in the evident fluctuation of the COFs therewith [11]. Thus how to tune film structure to obtain uniform interfacial structure and achieve more stably superlubricity behaviors needs to be further explored.

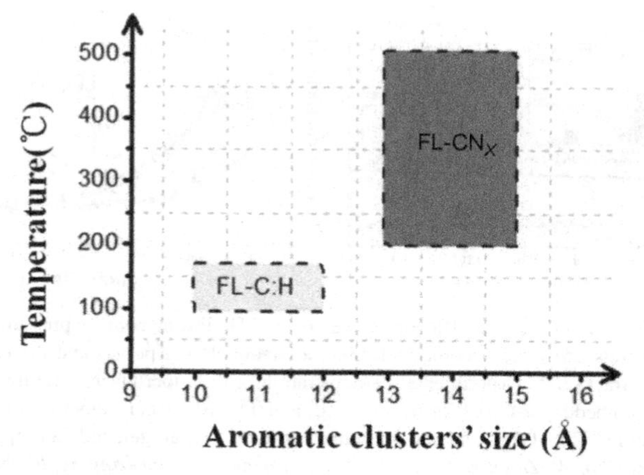

FIGURE 17.16 The formation window of FL-CN_x and FL-C:H films [28,38]. *FL-C:H*, Hydrogenated fullerene-like carbon; *FL-CN*, fullerene-like carbon nitride.

Acknowledgments

All authors commented on the manuscript and discussed the results. This work is supported by the National Natural Science Foundation of China (51661135022 and 51905517) and Youth Innovation Promotion Association CAS (2017459).

References

[1] J. Robertson, diamond-like amorphous carbon, Mater. Sci. Eng. R Rep. 37 (2002) 129–281.

[2] K. Holmberg, A. Matthews, Coatings Tribology: Properties, Techniques and Applications in Surface Engineering, Elsevier Science B.V., Amsterdam, 1994.

[3] M. Hirano, K. Shinjo, Atomistic locking and friction, Phys. Rev. B 41 (1990) 11837.

[4] R. Zhang, Z. Ning, Y. Zhang, Q. Zheng, Q. Chen, et al., Superlubricity in centimetres-long double-walled carbon nanotubes under ambient conditions, Nat. Nanotechnol. 8 (2013) 912–916.

[5] M. Dienwiebel, G.S. Verhoeven, N. Pradeep, J.W. Frenken, J.A. Heimberg, et al., Superlubricity of graphite, Phys. Rev. Lett. 92 (12) (2004) 126101.

[6] Z. Liu, J. Yang, F. Grey, J. Liu, Y. Liu, et al., Observation of microscale superlubricity in graphite, Phys. Rev. Lett. 108 (2012) 205503.

[7] E. Koren, E. Lörtscher, C. Rawlings, A.W. Knoll, U. Duerig, Adhesion and friction in mesoscopic graphite contacts, Science 348 (6235) (2015) 679–683.

[8] X. Feng, S. Kwon, J. Park, M. Salmeron, Superlubric sliding of graphene nanoflakes on graphene, ACS Nano 7 (2013) 1718–1724.

[9] J.A. Heimberg, K.J. Wahl, I.L. Singer, A. Erdemir, Superlow friction behavior of diamond-like carbon coatings: time and speed effects, Appl. Phys. Lett. 78 (17) (2001) 2449.

[10] A. Erdemir, Genesis of superlow friction and wear in diamond like carbon films, Tribol. Int. 37 (2004) 1005.

[11] Y. Wang, K. Gao, B. Zhang, Q. Wang, J. Zhang, Structure effects of sp^2-rich carbon films under super-low friction contact, Carbon 137 (2018) 49–56.

[12] C. Wang, S. Yang, Q. Wang, Z. Wang, J. Zhang, Super-low friction and super-elastic hydrogenated carbon films originated from a unique fullerene-like nanostructure, Nanotechnology 19 (2008) 225709.

[13] C. Wang, D. Diao, X. Fan, C. Chen, Graphene sheets embedded carbon film prepared by electron irradiation in electron cyclotron resonance plasma, Appl. Phys. Lett. 100 (2012) 231909.

[14] H. Song, L. Ji, H. Li, X. Liu, W. Wang, et al., External-field-induced growth effect of an a-C:H Film for manipulating its medium-range nanostructures and properties, ACS Appl. Mater. Interfaces 8 (10) (2016) 6639–6645.

[15] J. Shi, Y. Wang, Z. Gong, B. Zhang, C. Wang, et al., Nanocrystalline graphite formed at fullerene-like carbon film frictional interface, Adv. Mater. Interfaces 4 (8) (2017) 1601113.

[16] Z. Cao, W. Zhao, Q. Liu, A. Liang, J. Zhang, Super-elasticity and ultralow friction of hydrogenated fullerene-like carbon films: associated with the size of graphene sheets, Adv. Mater. Interfaces 5 (6) (2018) 1701303.

[17] Y. Wang, K. Gao, Q. Wang, J. Zhang, The correlation between nano-hardness and elasticity and fullerene-like clusters in hydrogenated amorphous carbon films, Chem. Phys. Lett. 692 (2018) 258–263.

[18] Y. Wang, X. Ling, Y. Wang, J. Zhang, Probing the effect of doped F and N on the structures and properties of fullerene-like hydrogenated carbon films, Diam. Relat. Mater. 79 (2017) 32–37.

[19] L. Qiang, B. Zhang, K. Gao, Z. Gong, J. Zhang, Hydrophobic, mechanical, and tribological properties of fluorine incorporated hydrogenated fullerene-like carbon films, Friction 1 (4) (2013) 350–358.

[20] Y. Wang, J. Guo, K. Gao, B. Zhang, A. Liang, et al., Understanding the ultra-low friction behavior of hydrogenated fullerene-like carbon films grown with different flow rates of hydrogen gas, Carbon 77 (2014) 518–524.

[21] Y. Wang, J. Guo, J. Zhang, Y. Qin, Ultralow friction regime from the in-situ production of a richer fullerene-like nanostructured carbon in sliding contact, RSC Adv. 5 (2015) 106476.

[22] H. SjÖstrÖm, S. StafstrÖm, M. Boman, J.-E. Sundgren, Superhard and elastic carbon nitride thin films having fullerene like microstructure, Phys. Rev. Lett. 75 (1995) 1336–1339.

[23] I. Alexandrou, H.-J. Scheibe, C.J. Kiely, A.J. Papworth, G.A.J. Amaratunga, et al., Carbon films with an sp^2 network structure, Phys. Rev. B 60 (1999) 10903–10907.

[24] G.A.J. Amaratunga, M. Chhowalla, C.J. Kiely, I. Alexandrou, R. Aharonov, et al., Hard elastic carbon thin films from linking of carbon nanoparticles, Nature 383 (1996) 321–323.

[25] N. Hellgren, T. Berlind, G.K. Gueorguiev, M.P. Johansson, S. Stafström, et al., Fullerene-like BCN thin films: a computational and experimental study, Mater. Sci. Eng. B 113 (3) (2004) 242–247.

[26] J.G. Buijnsters, M. Camero, R. Gago, A.R. Landa-Canovas, C. Gómez-Aleixandre, I. Jiménez, Direct spectroscopic evidence of self-formed C60 inclusions in fullerene-like hydrogenated carbon films, Appl. Phys. Lett. 92 (2008) 141920.

[27] A.C. Ferrari, S.E. Rodil, J. Robertson, Interpretation of infrared and Raman spectra of amorphous carbon nitrides, Phys. Rev. B 67 (2003) 155306.

[28] G. Abrasonis, R. Gago, M. Vinnichenko, U. Kreissig, A. Kolitsch, et al., Sixfold ring clustering in sp^2-dominated carbon and carbon nitride thin films: a Raman spectroscopy study, Phys. Rev. B 73 (2006) 125427.

[29] R. Gago, J. Neidhardt, M. Vinnichenko, U. Kreissig, Z.S. Czigány, et al., Synthesis of carbon nitride thin films by low-energy ion beam assisted evaporation: on the mechanisms for fullerene-like microstructure formation, Thin Solid Films 483 (2005) 89.

[30] R. Gago, I. Jiménez, J. Neidhardt, B. Abendroth, I. Caretti, et al., Correlation between bonding structure and microstructure in fullerene like carbon nitride thin films, Phys. Rev. B 71 (2005) 125414–125419.

[31] J. Neidhardt, L. Hultman, E. Broitman, T.W. Scharf, I.L. Singer, Structural, mechanical and tribological behavior of fullerene-like and amorphous carbon nitride coatings, Diam. Relat. Mater. 13 (10) (2004) 1882–1888.

[32] Z. Wang, J. Zhang, Deposition of hard elastic hydrogenated fullerene like carbon films, J. Appl. Phys. 109 (10) (2011) 103303.

[33] Q. Wang, C. Wang, Z. Wang, J. Zhang, D. He, Fullerene nanostructure-induced excellent mechanical properties in hydrogenated amorphous carbon, Appl. Phys. Lett. 91 (2007) 141902.

[34] C. Wang, Q. Wang, Z. Wang, S. Yang, J. Zhang, Nanocrystalline diamond embedded in hydrogenated fullerene like carbon films, J. Appl. Phys. 103 (5) (2008) 056110.

[35] Y. Wang, K. Gao, J. Zhang, Structure, mechanical, and frictional properties of hydrogenated fullerene-like amorphous carbon film prepared by direct current plasma enhanced chemical vapor deposition, J. Appl. Phys. 120 (4) (2016) 045303.

[36] L. Ji, H. Li, F. Zhao, W. Quan, J. Chen, et al., Effects of pulse bias duty cycle on fullerene like nanostructure and mechanical properties of hydrogenated carbon films prepared by plasma enhanced chemical vapor deposition method, J. Appl. Phys. 105 (10) (2009) 106113.

[37] Y. Wang, Y. Wang, X. Zhang, J. Shi, K. Gao, et al., Hydrogenated amorphous carbon films on steel balls and Si substrates: nanostructural evolutions and their trigging tribological behaviors, Appl. Surf. Sci. 420 (2017) 586–593.

[38] Y. Wang, K. Gao, J. Zhang, Observation of structure transition as a function of temperature in depositing hydrogenated sp^2-rich carbon films, Appl. Surf. Sci. 439 (2018) 1152–1157.

[39] A. Erdemir, C. Donnet, Tribology of diamond-like carbon films: recent progress and future prospects, J. Phys. D Appl. Phys. 39 (18) (2006) 311–327.

[40] A. Konicek, D. Grierson, P. Gilbert, W. Sawyer, A. Sumant, R. Carpick, Origin of ultralow friction and wear in ultrananocrystalline diamond, Phys. Rev. Lett. 100 (23) (2008) 235502.

[41] J. Fontaine, J.L. Loubet, T.L. Mogne, A. Grill, Superlow friction of diamond-like carbon films: a relation to viscoplastic properties, Tribol. Lett. 17 (2004) 709–714.

[42] I. Sugimoto, S. Miyake, Oriented hydrocarbons transferred from a high performance lubricative amorphous C: H: Si film during sliding in a vacuum, Appl. Phys. Lett. 56 (19) (1990) 1868.

[43] X. Chen, T. Kato, M. Nosaka, Origin of superlubricity in a-C:H:Si films: a relation to film bonding structure and environmental molecular characteristic, ACS Appl. Mater. Interfaces 6 (2014) 13389.

[44] J.C. Sánchez-López, M. Belin, C. Donnet, C. Quiros, E. Elizalde, Friction mechanisms of amorphous carbon nitride films under variable environments: a triboscopic study, Surf. Coat. Technol. 160 (2002) 138.

[45] C. Wang, B. Li, X. Ling, J. Zhang, Superlubricity of hydrogenated carbon films in a nitrogen gas environment: adsorption and electronic interactions at the sliding interface, RSC Adv. 7 (2017) 3025.

[46] C.A. Freyman, Y. Chen, Y.W. Chung, Synthesis of carbon films with ultra-low friction in dry and humid air, Surf. Coat. Technol. 201 (2006) 164.

[47] R. Gago, G. Abrasonis, A. Mücklich, W. Möller, Z. Czigány, et al., Fullerene like arrangements in carbon nitride thin films grown by direct ion beam sputtering, Appl. Phys. Lett. 87 (7) (2005) 071901.

[48] C. Casiraghi, F. Piazza, A.C. Ferrari, D. Grambole, J. Robertson, Bonding in hydrogenated diamond-like carbon by Raman spectroscopy, Diam. Relat. Mater. 14 (3-7) (2005) 1098–1102.

[49] D. Roy, M. Chhowalla, N. Hellgren, T.W. Clyne, G.A.J. Amaratunga, Probing carbon nanoparticles in CN_x thin films using Raman spectroscopy, Phys. Rev. B 70 (3) (2004) 035406.

[50] M.P. Siegal, D.R. Tallant, L.J. Martinez-Miranda, J.C. Barbour, R.L. Simpson, et al., Nanostructural characterization of amorphous diamond like carbon films, Phys. Rev. B 61 (2000) 10451–10462.

[51] C. Wang, S. Yang, H. Li, J. Zhang, Elastic properties of a-C:N:H films, J. Appl. Phys. 101 (1) (2007) 013501.

[52] D. Roy, M. Chhowalla, H. Wang, N. Sano, I. Alexandrou, et al., Characterisation of carbon nano-onions using Raman spectroscopy, Chem. Phys. Lett. 373 (1-2) (2003) 52–56.

[53] J. Schwan, S. Ulrich, V. Batori, H. Ehrhardt, S. Silva, Raman spectroscopy on amorphous carbon films, J. Appl. Phys. 80 (1) (1996) 440–447.

[54] J. Rao, K. Lawson, J. Nicholls, The characterisation of e-beam evaporated and magnetron sputtered carbon films fabricated for atomic oxygen sensors, Surf. Coat. Technol. 197 (2) (2005) 154–160.

[55] Y. Wang, X. Ling, Y. Wang, J. Zhao, J. Zhang, The tribological behaviors between fullerene-like hydrogenated carbon films produced on Si substrates, steel and Si_3N_4 balls, Tribol. Int. 115 (2017) 518–524.

[56] C.L. Burket, R. Rajagopalan, H.C. Foley, Overcoming the barrier to graphitization in a polymer-derived nanoporous carbon, Carbon 46 (3) (2008) 501–510.

[57] J.S. McDonald-Wharry, M. Manley-Harris, K.L. Pickering, Reviewing, combining, and updating the models for the nanostructure of non-graphitizing carbons produced from oxygen-containing precursors, Energy Fuels 30 (10) (2016) 7811–7826.

[58] J.A. Greenwood, J.B.P. Williamson, Contact of nominally flat surfaces, Proc. R. Soc. Lond. Ser. A Math. Phys. Sci. 295 (1966) 300.

[59] D. Berman, S. Deshmukh, S. Sankaranarayanan, A. Erdemir, A. Sumant, Macroscale superlubricity enabled by graphene nanoscroll formation, Science 348 (6239) (2015) 1118–1122.

[60] S.W. Liu, H.P. Wang, Q. Xu, T.B. Ma, G. Yu, et al., Robust microscale superlubricity under high contact pressure enabled by graphene-coated microsphere, Nat. Commun. 8 (2017) 14029.

[61] Z. Gong, J. Shi, B. Zhang, J. Zhang, Graphene nano scrolls responding to superlow friction of amorphous carbon, Carbon. 116 (2017) 310–317.

[62] J. Neidhardt, L. Hultman, Z.S. Czigány, Correlated high resolution transmission electron microscopy and X-ray photoelectron spectroscopy studies of structured CN_x ($0 < x < 0.25$) thin solid films, Carbon 42 (12–13) (2004) 2729–2734.

[63] W. Zhang, A. Tanaka, Tribological properties of DLC films deposited under various conditions using a plasma-enhanced CVD, Tribol. Int. 37 (11-12) (2004) 975–982.

[64] W. Zhuang, X. Fan, W. Li, H. Li, L. Zhang, J. Peng, et al., Comparing space adaptability of diamond-like carbon and molybdenum disulfide films toward synergistic lubrication, Carbon 134 (2018) 163–173.

[65] J. Wang, Z. Cao, F. Pan, F. Wang, A. Liang, et al., Tuning of the microstructure, mechanical and tribological properties of a-C:H films by bias voltage of high frequency unipolar pulse, Appl. Surf. Sci. 356 (2015) 695–700.

[66] P. Wang, X. Wang, W. Liu, J. Zhang, Growth and structure of hydrogenated carbon films containing fullerene-like structure, J. Phys. D Appl. Phys. 41 (8) (2008) 085401.

[67] Y. Wang, Z. Yue, Y. Wang, J. Zhang, K. Gao, Synthesis of fullerene-like hydrogenated carbon films containing iron nanoparticles, Mater. Lett. 219 (2018) 51–54.

[68] Z. Gong, C. Bai, L. Qiang, K. Gao, J. Zhang, et al., Onion-like carbon films endow macro-scale superlubricity, Diam. Relat. Mater. 87 (2018) 172–176.

[69] Z. Cao, W. Zhao, A. Liang, J. Zhang, A general engineering applicable superlubricity: hydrogenated amorphous carbon film containing nano diamond particles, Adv. Mater. Interfaces 4 (14) (2017) 1601224.

[70] N. Hellgren, M.P. Johansson, E. Broitman, P. Sandstrom, L. Hultman, et al., Role of nitrogen in the formation of hard and elastic CN_x thin films by reactive magnetron sputtering, Phys. Rev. B 59 (1999) 5162.

Chapter 18

Superlubricity of water-based lubricants

Jinjin Li[1], Xiangyu Ge[2] and Jianbin Luo[1]
[1]State Key Laboratory of Tribology, Tsinghua University, Beijing, P.R. China, [2]School of Mechanical Engineering, Beijing Institute of Technology, Beijing, P.R. China

18.1 Introduction

The term of superlubricity was proposed at the beginning of the 1990s to describe a phenomenon that the friction force between two sliding surfaces nearly vanishes [1]. However in an engineering domain, when the sliding coefficient of friction (COF) is less than 0.01, the state of friction is referred to as superlubricity [2]. Because the friction force is very small relative to applied load in superlubricity state, the energy dissipation in sliding process and the wear of the solid surfaces would become very small. Considering that the annual cost of energy losses from friction and wear is estimated to be about 5% of their gross domestic products in highly industrialized nations [3], the investigations of superlubricity and its underlying mechanisms are very important for saving energy in the traditional mechanical and nanomechanical systems.

The liquid superlubricity phenomenon was first discovered by Tisza [4] with helium II at the temperature of 2.17K in 1938, which was found under no pressure conditions and it was called as superfluidity. However achieving liquid superlubricity under a contact pressure is still a challenge. In 1987, water was found to provide superlubricity at the Si_3N_4/Si_3N_4 interface under low contact pressures with a COF of even less than 0.002 [5]. During the past 20 years, and particularly in the last 10 years, superlubricity field has expanded very fast and attracted a large amount of attention from researchers in many fields [6–10]. Luo et al. in 2001 gave detailed discussions and proposed several research orientations on liquid superlubricity [11], including the relationship between the degree of molecular ordering and superlubricity, the mechanisms of superlubricity, the basic conditions for the transition from nonsuperlubricity to superlubricity state, and the physical and chemical characteristics that lubricant molecules should possess in order to achieve

Superlubricity. DOI: https://doi.org/10.1016/B978-0-444-64313-1.00018-1

superlubricity. At present, liquid superlubricity can be achieved with many lubricating systems, for example, ceramic material lubricated by water [12,13], polymer brushes with water [14,15], glycerol solution with polyhydric alcohol or boric acid [16,17], mucilage from plants [18–20], and room-temperature ionic liquids (RTILs) [21,22]. The superlubricity mechanisms of them are usually linked to the properties of confined liquid or thin-film lubrication, such as forming brush, hydration layer, hydrogen-bond network, and tribochemical layer, etc.

In 2011, our group found that the superlubricity was achieved with the lubrication of mixture of glycerol and boric acid after a wearing-in period [17]. After that, our group established a novel system of liquid superlubricity between a Si_3N_4 ball and glass plate lubricated by phosphoric acid solution [23,24]. According to the corresponding mechanisms, the superlubricity system of mixture of polyhydroxy alcohol and acid solution was established by further studies [25,26]. In addition, RTILs are well known as ecofriendly lubricants and lubricating additives. Because RTILs can be designed by combining a variety of cations and anions, researchers have developed RTILs possessing excellent superlubricity property. Further, a mixture of RTIL and graphene oxide (GO) nanosheets was found to achieve superlow wear, and thereby achieving superlubricity under high pressure [22]. At the nanoscale, the liquid superlubricity was achieved by various surfactant micelles, which are possible to be widely used in the nanomachine lubrication [27–30]. Therefore in the present chapter, the mechanisms of liquid superlubricity of water-based lubricants designed by our group are introduced in detail, hoping to offer perspectives and inspirations for future research on liquid superlubricity.

18.2 Phosphoric acid

Our group found that the superlubricity between a Si_3N_4 ball and glass plate can be obtained with phosphoric acid solution. As shown in Fig. 18.1, as the Si_3N_4 ball sliding against glass disk with the lubrication of H_3PO_4 solution (pH = 1.5), the COF decreased from 0.45 to 0.004 after a wearing-in period. It is found that the wearing-in period is necessary for obtaining superlubricity [31]. Li et al. [6] divided the wearing-in period into two stages according to the evolution of COF with time. In each stage, different mechanisms dominated the reduction of friction.

In the first stage, the COF is reduced from 0.45 to 0.05 rapidly and such phenomena could be found in other kinds of acid solutions as the pH value of solutions is less than 2 [32]. As shown in Table 18.1, the COFs of acid solutions could finally reach the range of 0.03–0.05. However in cases of NaCl solution with pH value of 7 and NaOH solution with pH value of 14, their COFs is kept as high as 0.3–0.5. Because the viscosity of all the liquids above is close to that of water, it is hard to form the hydrodynamic films

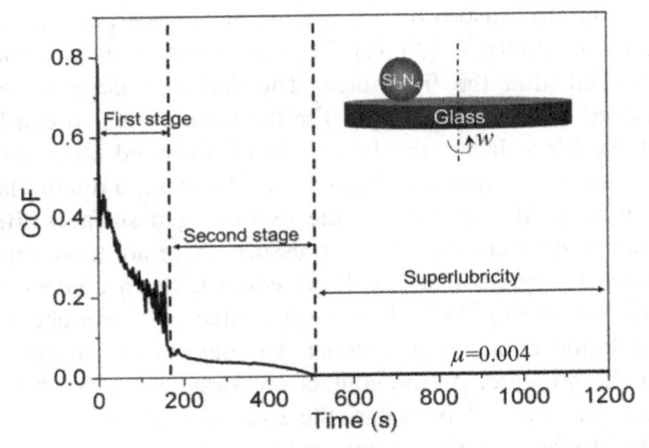

FIGURE 18.1 Evolution of COF with time lubricated by H_3PO_4 solution (pH = 1.5) under the load of 3 N (corresponding to a maximum contact pressure of 700 MPa) with the sliding velocity of 0.075 m/s at the room temperature. *COF*, Coefficient of friction. *Adapted from X. Ge, J. Li, J. Luo, Macroscale superlubricity achieved with various liquid molecules: a review. Front. Mech. Eng. 5 (2019) 2. Copyright 2019, Frontiers Media SA.*

TABLE 18.1 COF after the wearing-in period lubricated by various liquids. T_{rin} means the wearing-in period.

Lubricant	Formula	pH	T_{rin}	COF
Sulfuric acid	H_2SO_4	1.0	150	0.04
Hydrochloric acid	HCl	1.0	180	0.03
Oxalic acid	$H_2C_2O_4$	1.0	190	0.05
Sulfamic acid	H_3NO_3S	1.0	160	0.05
Sodium chloride	NaCl	7.0	–	0.49
Sodium hydroxide	NaOH	14	–	0.35

COF, Coefficient of friction.
Source: Adapted from J. Li, C. Zhang, L. Sun, X. Lu, J. Luo, Tribochemistry and superlubricity induced by hydrogen ions, Langmuir 28 (45) (2012) 15816–15823. Copyright 2012, American Chemical Society.

under such experiment condition. It is therefore concluded that the hydrogen ions play the key role in the friction reduction in the first stage. Because there are ≡SiOH bonds on the surfaces of glass and Si_3N_4, the hydrogen ions in H_3PO_4 solution can be adsorbed on solid surfaces by surface protonation reaction in the first stage. In the meantime, the wear would occur on the solid surfaces during the first stage, leading to the original surface being polished off. In such a case, there would appear many dangling bonds

(≡SiO$^-$) that can also adsorb hydrogen ions by surface protonation reaction (≡SiO$^-$ + 2H$^+$ → ≡SiOH$_2$$^+$) [33,34]. Thus the solid surfaces would become positively charged after the first stage. The surface charge is balanced by equally dissolved counterions ($H_2PO_4^-$) in the solution which can lead to the formation of the Stern layer (the first layer of adsorbed counterions on the solid surface) and the diffuse double layer. In this case, a double-layer repulsive force can be produced between the two charged surfaces. Because the repulsive force is less than the contact pressure, there are three kinds of contact in the contact region; that are direct contact, Stern contact, and liquid contact, as shown in Fig. 18.2. The Stern contact can produce a hydration force and the liquid contact can produce a double electrical layer repulsive force to bear the pressure. As the contact pressure reduces with the increase in the contact area caused by wear, the hydration force, double electrical layer repulsive force, and the contact pressure are in an equilibrium state after the first stage, and thereby the low friction state prevails ($\mu = 0.05$).

Although hydrogen ions are closely linked to friction reduction in the first stage, it does not mean that the hydrogen ions can themselves enable superlubricity. In fact, more than six kinds of acid solutions have been tested, and only H_3PO_4 solution could attain superlubricity under the working condition after the second wearing-in stage, meaning the second stage is unique for H_3PO_4 solution. It is found that the H_3PO_4 solution is liquid in the first

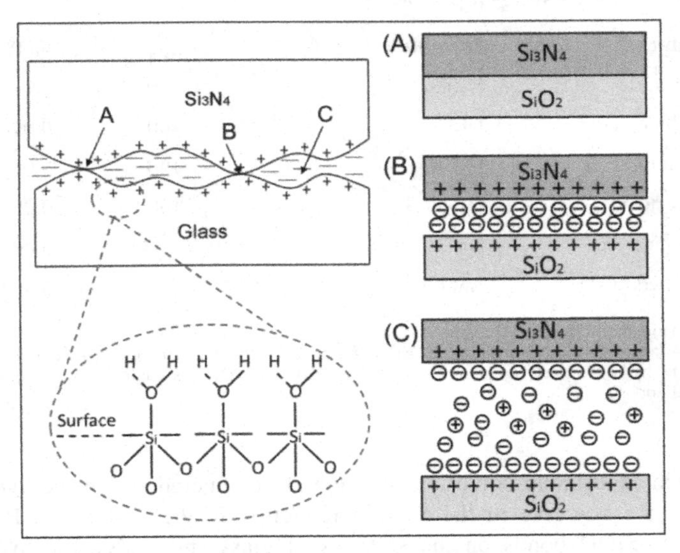

FIGURE 18.2 The schematic illustration of lubrication model. The picture in the dash line circle is the illustration of surface protonation reaction. (A), (B), and (C) are the further illustrations of the three kinds of contacts. *Adapted from J. Li, C. Zhang, L. Sun, X. Lu, J. Luo, Tribochemistry and superlubricity induced by hydrogen ions, Langmuir 28 (45) (2012) 15816–15823. Copyright 2012, American Chemical Society.*

stage and almost the whole second stage. At the end of second stage, no liquid can be observed on the surfaces and a solid-like film was formed on the surfaces. It is because free water molecules in solution evaporated completely at the end of the second stage, leading to only hydrated water molecules being left in the solid-like film. It means that the second stage is related with the water evaporation process. The results of the tests carried out with different temperatures and different humidities proves such relationship. At a lower temperature or a higher humidity, a longer wearing-in period is needed due to the water being evaporating more slowly, as shown in Fig. 18.3A and B.

Raman spectroscopy studies of the solid-like film on the surface indicate that the film formed after wearing-in period is composed by the hydrogen-bond network between H_3PO_4 and H_2O molecules [35]. If the hydrogen-bond network was broken by adding a small amount of deionized water (2 μL) in the track, it is found that COF would become high suddenly. This indicates that the superlubricity is closely linked to the hydrogen-bond network. In the first stage, the hydrogen ions in H_3PO_4 solution were adsorbed on the two solid surfaces to form the Stern layer. Therefore after the second stage, the hydrogen-bond network between H_3PO_4 and H_2O molecules was adsorbed on the Stern layer, which plays an important role on bearing load. Besides bearing load, the hydrogen-bond network has another important function that is holding water molecules in the contact region. It is inferred that there is a thin hydrated water layer adsorbed on the surface of hydrogen-bond network due to strong hydrogen-bond effect of H_3PO_4 molecules, just like a water layer adsorbed on ice surface [36]. When two surfaces slide with each other, the shearing occurs in the layer of hydrated water due to the excellent fluidity of hydration layer under high pressure [37]. The superlubricity model of phosphoric acid is proposed in Fig. 18.4, which is attributed to

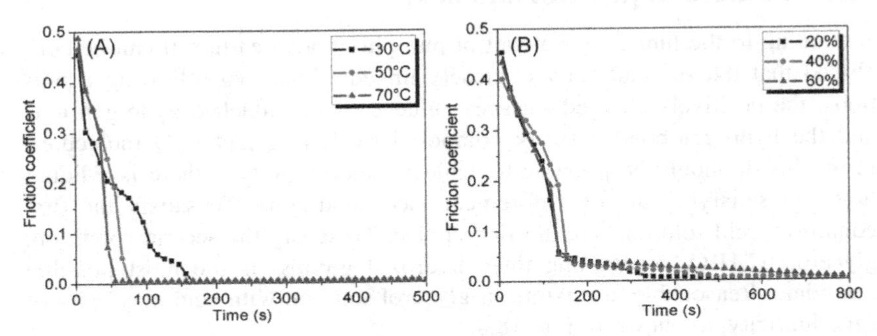

FIGURE 18.3 (A) Evolution of COF with time lubricated by H_3PO_4 (pH = 1.5) with different temperatures. (B) Evolution of COF with time lubricated by H_3PO_4 (pH = 1.5) with different humidities. COF, Coefficient of friction. *Adapted from X. Ge, J. Li, J. Luo, Macroscale superlubricity achieved with various liquid molecules: a review, Front. Mech. Eng. 5 (2019) 2. Copyright 2019, Frontiers Media SA.*

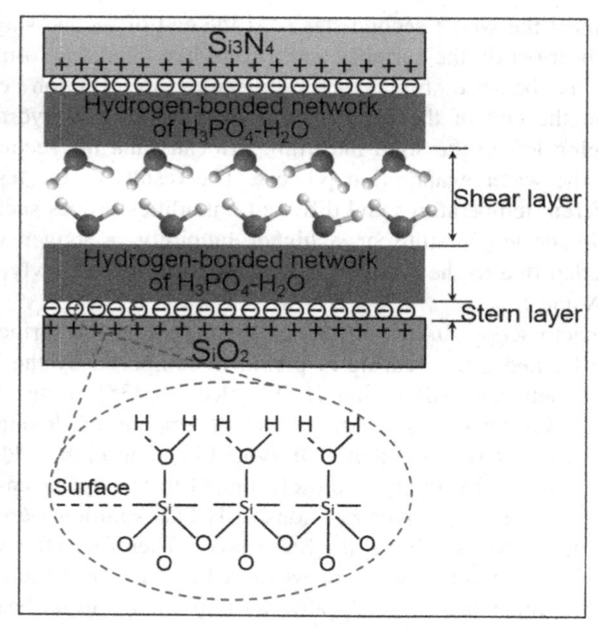

FIGURE 18.4 Illustration of superlubricity model. The *small* atom represents hydrogen and the *big* atom represents oxygen, respectively. *Adapted from J. Li, C. Zhang, L. Ma, Y. Liu, J. Luo, Superlubricity achieved with mixtures of acids and glycerol, Langmuir 29 (1) (2013) 271–275. Copyright 2012, American Chemical Society.*

the formation of a hydrated water layer between the hydrogen-bond networks on the Stern layer.

18.3 Mixture of glycerol and acid

According to the lubrication model of phosphoric acid earlier, it can be concluded that the superlubricity is closely linked to the two following conditions: the positively charged surfaces induced by the attached hydrogen ions and the hydrogen-bond network composed by H_3PO_4 and H_2O molecules. From this, it should be possible to achieve superlubricity if there is a liquid lubricant satisfying the two aforementioned conditions. To satisfy the first condition, acid solutions are the first choice. To satisfy the second condition, glycerol ($C_3H_8O_3$, containing three hydroxyl groups) is the most possible candidate. Reasonably, a mixture of glycerol solution with acid may achieve superlubricity, as shown in Fig. 18.5.

Ma et al. [17] found that the superlubricity at Si_3N_4/glass interface could be achieved with the lubrication of mixture of glycerol and boric acid. The superlubricity mechanism is attributed to the strongly adsorbed diglycerin borate layer on the sliding surface and the hydration effect, which make

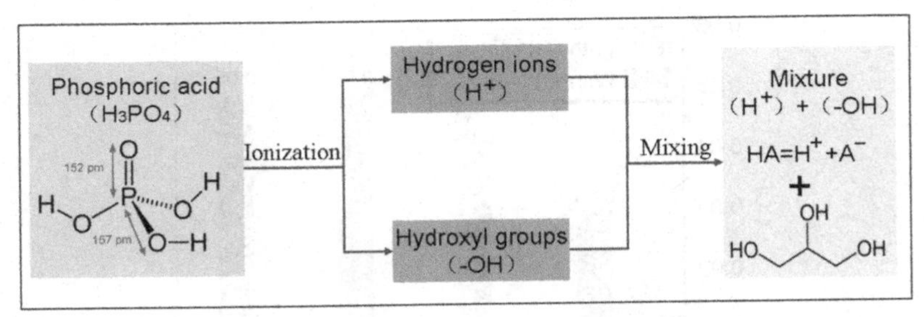

FIGURE 18.5 Illustration of promoting phosphoric acid to mixture of glycerol and acid.

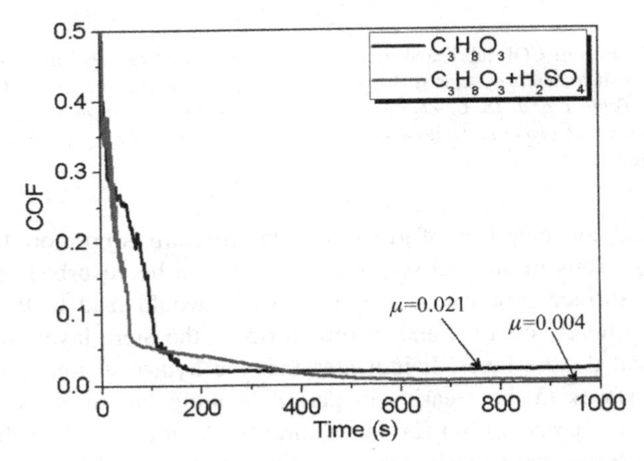

FIGURE 18.6 Evolution of COF with time lubricated by $C_3H_8O_3$ (20%, v/v) and mixture of $C_3H_8O_3$ (20%, v/v) and H_2SO_4 (pH = 1), respectively. COF, Coefficient of friction. *Adapted from J. Li, C. Zhang, L. Ma, Y. Liu, J. Luo, Superlubricity achieved with mixtures of acids and glycerol, Langmuir 29 (1) (2013) 271–275. Copyright 2013, American Chemical Society.*

water molecules act as a lubricant in the contact region. In addition, such method has been proved by conducting the tests with the mixture of $C_3H_8O_3$ (20%, v/v) solution and H_2SO_4 (pH = 1) solution with a volume ratio of 10:1. The phenomenon is quite similar as that lubricated with H_3PO_4 solution. After the wearing-in period, superlubricity with a COF of 0.004 is achieved, as shown in Fig. 18.6. However if the glass plate and Si_3N_4 ball are lubricated by $C_3H_8O_3$ (20%, v/v) only, superlubricity cannot be achieved. Moreover, the mixture of glycerol solution with different acid solution, including HCl, $C_3H_6O_3$, $H_2C_2O_4$, and H_3NO_3S, all can achieve superlubricity, as shown in Fig. 18.7. It indicates that superlubricity is linked to the attached hydrogen ions.

The superlubricity closely depends on the pH value of the mixed solution and the initial concentration of glycerol. As shown in Fig. 18.8, the superlubricity can be obtained only when the pH value of the mixture is less than 2

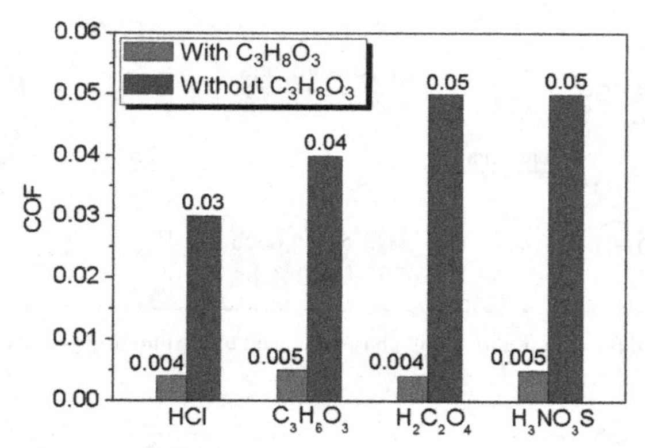

FIGURE 18.7 Lowest COF lubricated by mixtures of glycerol (20%, v/v) and acids (pH = 1) and lowest COF lubricated by acids (pH = 1) only. COF, Coefficient of friction. *The data are obtained from Table 1 of J. Li, C. Zhang, L. Ma, Y. Liu, J. Luo, Superlubricity achieved with mixtures of acids and glycerol, Langmuir 29 (1) (2013) 271–275. Copyright 2013, American Chemical Society.*

and the initial concentration of glycerol in the mixture is no more than 40%. The hydrogen ions in an acid solution ($pH \leq 2$) can be adsorbed on ≡SiOH surfaces by surface protonation reaction, which would lead to the surfaces becoming positively charged and thereby forming the Stern layer and the diffuse electrical double layer. It indicates that the hydrogen ions attached on the solid surfaces in the wearing-in period, making the surfaces positively charged, are the precondition for superlubricity. As for glycerol solution, the viscosity increases with the increase of concentration, which may influence the film thickness between two sliding surfaces. Based on the Hamrock–Dowson theory, the film thickness of glycerol solutions with different concentrations could be estimated and shown in Fig. 18.8B. It is seen that when the concentration of glycerol is lower than 40%, the film thickness is less than 2 nm. Comparing with the combined roughness of surface ($R_a = 5$ nm), it indicates that there is no elastohydrodynamic film formed between the two solid surfaces when the concentration is lower than 40%. Therefore it can be concluded that less elastohydrodynamic film formed at the beginning of test is also the precondition for superlubricity.

Combining the relationship between superlubricity and hydrogen ions and concentration of glycerol earlier, it can be concluded that the superlubricity mechanism of mixture of $C_3H_8O_3$ and H_2SO_4 is in accordance with that of H_3PO_4, which is attributed to the formation of a hydrated water layer between the hydrogen-bond networks of $C_3H_8O_3$-H_2O on the Stern layer induced by hydrogen ions. Therefore if the concentration of hydrogen ions is too low ($pH > 2$), the ionization of hydrogen ions from surfaces ($\equiv SiOH \rightarrow \equiv SiO^- + H^+$) would tend to dominate, which would lead to the surfaces becoming negatively charged. In this case, there are no adsorbed hydrogen ions to form the positively charged sites

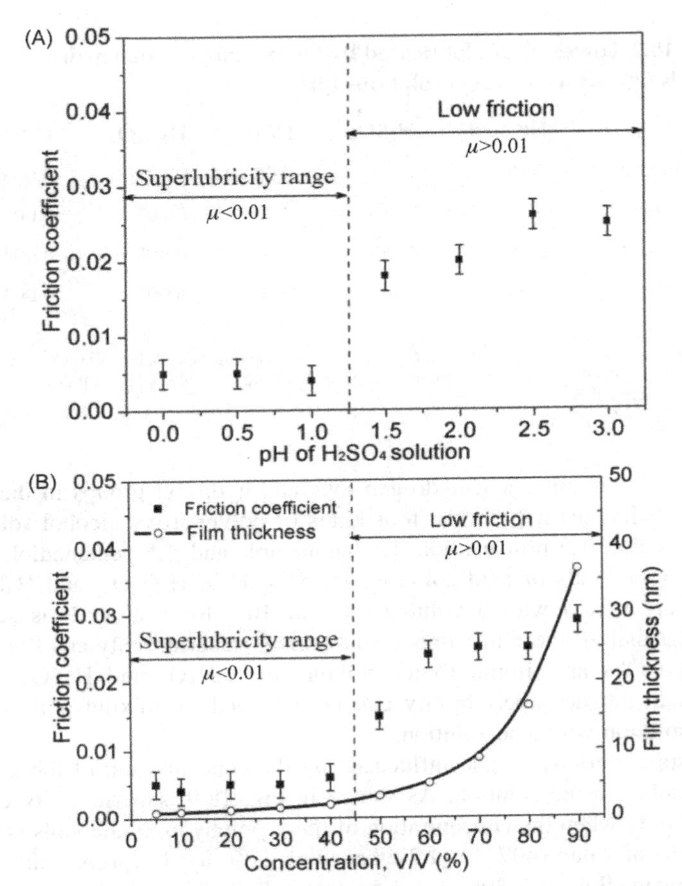

FIGURE 18.8 (A) Lowest COF lubricated by mixture of $C_3H_8O_3$ (20%, v/v) and H_2SO_4 with different pH values; (B) lowest COF lubricated by mixture of H_2SO_4 (pH = 1) and $C_3H_8O_3$ with different concentrations and film thickness predicted by the H-D theory. COF, Coefficient of friction; *H-D*: Hamrock–Dowson. *Adapted from J. Li, C. Zhang, L. Ma, Y. Liu, J. Luo, Superlubricity achieved with mixtures of acids and glycerol, Langmuir 29 (1) (2013) 271–275. Copyright 2013, American Chemical Society.*

on solid surfaces. In addition to this, if the concentration of glycerol is too high, the elastohydrodynamic film can be established between two solid surfaces at the beginning of test. In this case, the two solid surfaces would be separated by the mixture at the beginning of test, and thereby the rubbing action cannot occur in the wearing-in period. It can be seen that the aforementioned two cases cannot satisfy the two conditions for superlubricity, which is the reason why the superlubricity cannot be achieved.

Based on the superlubricity mechanism of the mixture of $C_3H_8O_3$ and H_2SO_4, it can be inferred that superlubricity is possible to be achieved as

TABLE 18.2 Lowest COFs lubricated by the mixtures of polyhydroxy alcohols (20%, v/v) and acid solutions (pH = 1).

	No acid	H_2SO_4	HCl	$H_2C_2O_4$	H_3NO_3S
1,2-Ethanediol	0.05	0.004	0.004	0.004	0.003
1,3-Propanediol	0.07	0.004	0.003	0.003	0.003
1,4-Butanediol	0.06	0.004	0.005	0.005	0.004
1,5-Pentanediol	0.06	0.004	0.006	0.004	0.003

COF, Coefficient of friction.
Source: The data are obtained from J. Li, C. Zhang, J. Luo, Superlubricity achieved with mixtures of polyhydroxy alcohols and acids, Langmuir 29 (17) (2013) 5239–5245. Copyright 2013, American Chemical Society.

long as there are sufficient hydrogen ions and hydroxyl groups in the solution. To testify this hypothesis, four kinds of polyhydroxy alcohol solutions (1,2-ethanediol, 1,3-propanediol, 1,4-butanediol, and 1,5-pentanediol, 20%, v/v) and four kinds of acid solutions (H_2SO_4, HCl, $H_2C_2O_4$, and H_3NO_3S, pH = 1) are mixed with a volume ratio of 10:1 for testing. It is seen in Table 18.2 that all these mixtures could achieve superlubricity and the corresponding COFs are similar to the mixture of $C_3H_8O_3$ and H_2SO_4. These results indicate the superlubricity can be achieved by mixing polyhydroxy alcohol solution with acid solution.

The superlubricity is also influenced by the concentration of the polyhydroxy alcohol in the solution. As shown in Fig. 18.9, superlubricity can be achieved only when the concentration of these polyhydroxy alcohols is lower than a critical value (40% for 1,2-ethanediol, 50% for 1,3-propanediol, 40% for 1,4-butanediol, and 30% for 1,5-pentanediol). This result is similar to that of glycerol; the superlubricity can be obtained only when polyhydroxy alcohol concentration is low, so that no elastohydrodynamic film is formed between the two contact surfaces due to the low viscosity at the beginning of test. Therefore it can be concluded that the superlubricity mechanism of the mixture of polyhydroxy alcohol and acid is similar to that of the mixture of glycerol and acid. The hydrogen-bond network between these polyhydroxy alcohol and water molecules can also be formed on the Stern layer (induced by hydrogen ions) after the wearing-in period, which is helpful to achieve superlubricity.

Although acid-based lubricants can achieve macroscale liquid superlubricity, the strong acidity of them limits their applications in practical mechanical systems. Recently, a new type of acid-based lubricant, which is a mixture of boric acid and polyethylene glycol aqueous solution ($PEG_{(aq)}$), was synthesized and found to provide superlubricity (COF ≈ 0.004) in neutral conditions (pH ≈ 6.4) at Si_3N_4/SiO_2 interfaces, as depicted in Fig. 18.10 [38].

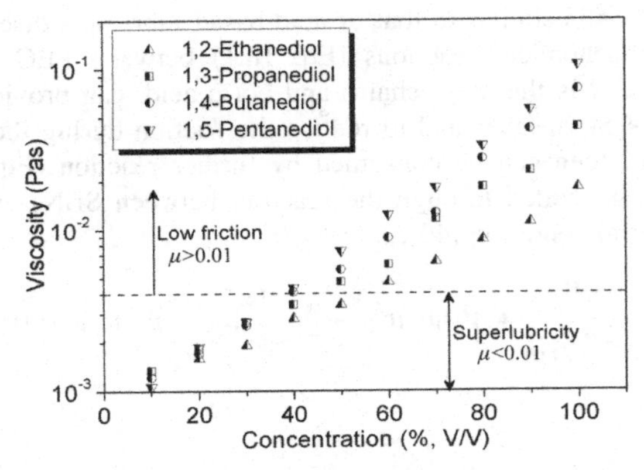

FIGURE 18.9 Relationship between lowest COF and concentration of polyhydroxy alcohols. COF, Coefficient of friction. *Adapted from J. Li, C. Zhang, J. Luo, Superlubricity achieved with mixtures of polyhydroxy alcohols and acids, Langmuir 29 (17) (2013) 5239–5245. Copyright 2013, American Chemical Society.*

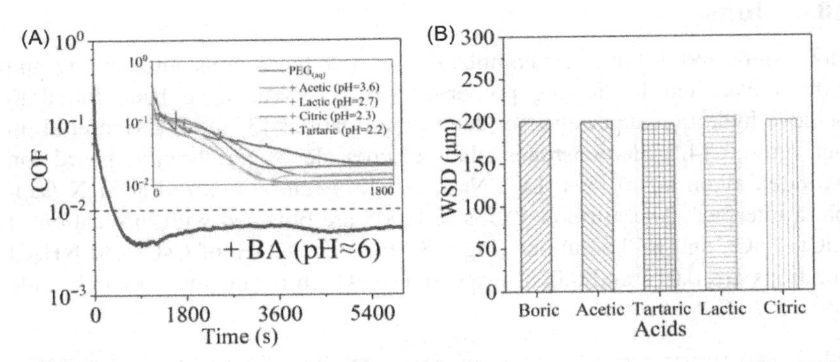

FIGURE 18.10 (A) COFs and (B) WSDs of the mixtures of $PEG_{(aq)}$ and acids. COF, Coefficient of friction; PEG(aq): polyethylene glycol aqueous solution. *Adapted from X. Ge, J. Li, C. Zhang, J. Luo, Liquid superlubricity of polyethylene glycol aqueous solution achieved with boric acid additive, Langmuir 34 (12) (2018) 3578–3587. Copyright 2018, American Chemical Society.*

$PEG_{(aq)}$, which has proved to be a good lubricant [39], was used as base solution. It is observed that all the five acids as additives in $PEG_{(aq)}$ could reduce friction despite different pH values of them. However only boric acid as additive could achieve superlubricity in neutral conditions, whereas the other acids exhibited stronger acidity and higher friction (nonsuperlubricity). In addition, the wear scar diameters (WSDs) as well as the contact pressure with the lubrication of these acid-based lubricants are nearly the same, indicating under the same contact pressure, boric acid as additive has better lubricity than other acids. The mechanism of this superlubricity achieved in neutral conditions is similar to that of acid-based lubricants discussed previously. The tribochemical reactions (Eq. 18.1) between PEG (denoted as HO*R*OH, where *R* is the alkyl chain) and boric acid, can provide sufficient H^+ to form the Stern layer and to reduce the friction during the wearing-in period. The H^+ then can be consumed by further reaction (Eq. 18.2) with NH_3, which is generated through the reaction between Si_3N_4 and water, to maintain a neutral condition [40].

$$B(OH)_3 + HOROH \longrightarrow [R(O)_2B(O)_2R]^- H^+ + H_2O \quad (18.1)$$

$$[R(O)_2B(O)_2R]^- H^+ \xrightarrow{NH_3} [R(O)_2B(O)_2R]^- NH_4^+ \quad (18.2)$$

18.4 Ions

Both salts and RTILs are completely formed by cations and anions and exhibit excellent lubricating properties [37,41]. Salts have been found to achieve hydration superlubricity due to their cations [37,42,43]. Han et al. in our group [42] demonstrates that macroscale superlubricity based on hydrated alkali metal ions (Li^+, Na^+, and K^+) can be realized at Si_3N_4/sapphire interface. The ultralow COFs of 0.005 are obtained with lubrication of LiCl, NaCl, and KCl solutions (Fig. 18.11A). The COFs of CsCl and NH_4Cl solutions are 0.014 and 0.025, respectively, which means these two chloride

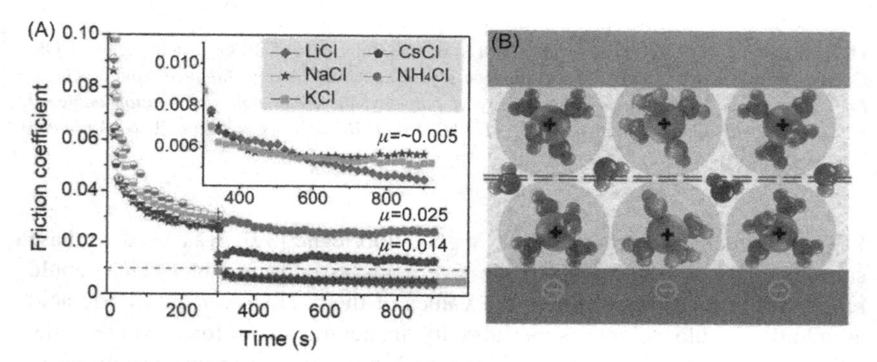

FIGURE 18.11 (A) COFs of chloride salt solutions between Si_3N_4 and sapphire surfaces after wearing-in with H_3PO_4 solutions (pH = 1.5). (B) The hydration lubrication mechanism schematic. COF, Coefficient of friction. *Adapted from T. Han, C. Zhang, J. Luo, Macroscale superlubricity enabled by hydrated alkali metal ions, Langmuir 34 (38) (2018) 11281–11291. Copyright 2018, American Chemical Society.*

salt solutions could not achieve superlubricity. Apparently, the only difference among these five different salt solutions is the cations. The Li^+, Na^+, and K^+ can form hydrated ions by combining three or more water molecules. However each Cs^+ and $NH_4{}^+$ can only combine one or less water molecule, which results in their relatively weak hydration strength. In addition, based on the hydrated radii, hydration number, and hydration energy of these ions, it can be concluded that the hydration intensity decreases in the sequence $Li^+ > Na^+ > K^+ > Cs^+ > NH_4{}^+$. Therefore it can be concluded that the superlubricity achieved by hydrated alkali metal ions is in high accordance with the hydration strength of these hydrated ions. The hydrated alkali metal ions in the salt solutions have good capacities including not only sustaining a large normal load through the hydration repulsion of the hydrated ions but also a fluid response to shear due to the low viscosity of the monovalent salt solutions. Therefore these good capacities could contribute to the achievement for the macroscale superlubricity of hydrated alkali metal ions significantly (Fig. 18.11B).

To reveal the specific contribution of silica layer formed through tribochemical reactions to hydration superlubricity. Han et al. [42] resolved the amorphous structure on the atomic scale and determined the thickness of the tribo-induced silica layer (Fig. 18.12A). Then the mechanical properties of the 6-nm-thick silica layer generated on a Si_3N_4 ball was revealed, which has a smaller elastic modulus of 75 GPa. Through friction experiments and zeta-potential analyses, they report on two main effects of the silica layer on achieving superlubricity. First, the silica layer can reduce friction resistance between ceramic surfaces. Second, the silica layer increases the number of hydroxyl groups on the surface of Si_3N_4, thereby increasing the surface negative potential and hydrophilicity, which ensures that more hydrated cations can be adsorbed on the solid surfaces (Fig. 18.12C and D). The superlubricity could be obtained as long as the pH was no less than 5.5. The COF increased gradually with the decrease in pH for pH values below 5.5. Both surfaces were negatively charged when pH > IEP (the isoelectric point) and then could adsorb hydrated cations (Fig. 18.12B), which resulted in hydration repulsive force supporting the van der Waals attractive force and the normal load, leading to the ultralow COF of 0.005, but become positive at pH < IEP. Therefore only when the surface pairs are negatively charged will the KCl solution exhibit superlubricity. Consequently, these insights largely account for the observations that the superlubricity of hydrated ions can be obtained not only between two mica surfaces but also for ceramic surface pairs with lower surface charge density, higher elastic modulus, and larger surface roughness. Moreover, these results strongly suggest a synergistic mechanism behind macroscale hydration superlubricity and reveal fundamentals guiding a new direction for applications of water-based lubricants.

In surface science and engineering, RTILs are known as excellent neat lubricants and lubricant additives [44,45], which play indispensable roles in

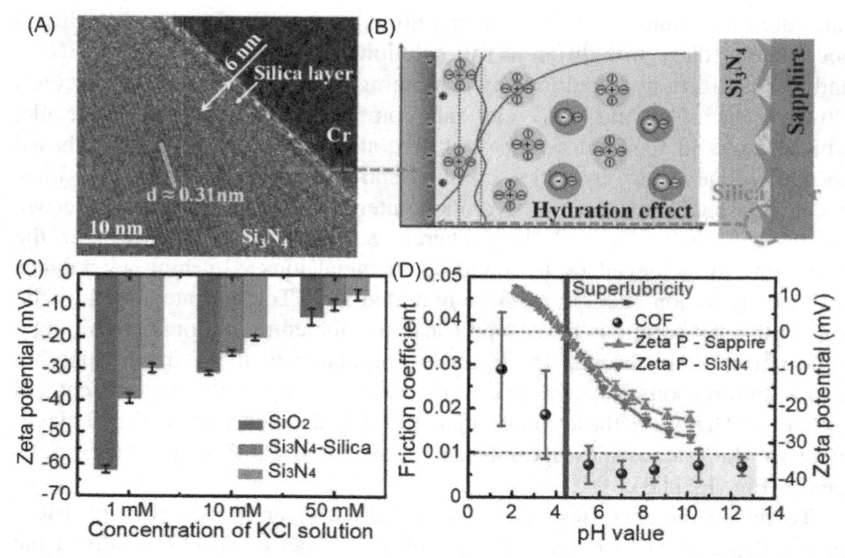

FIGURE 18.12 (A) HRTEM image showing the cross section of the worn region on the Si_3N_4 ball. (B) Schematics of hydration effect and the tribo-induced silica layer. (C) Comparison of zeta potential for three different plates in neutral KCl solutions (pH = 5.3 ± 0.2). (D) COF of KCl solution as well as the zeta potential of sapphire and Si_3N_4 as a function of pH value in 50 mM KCl solution. COF, Coefficient of friction; *HRTEM*, high-resolution transmission electron microscopy. *Adapted from T. Han, C. Zhang, X. Chen, J. Li, W. Wang, J. Luo, Contribution of a tribo-induced silica layer to macroscale superlubricity of hydrated ions, J. Phys. Chem. C. 123 (33) (2019) 20270–20277 [43]. Copyright 2019, American Chemical Society.*

industry because they can enhance the lifetime of mechanical systems via reducing friction and wear and improving energy efficiency [46–51]. Our group explored the robust superlubricity of the aqueous solutions of four kinds of RTILs [52], including 1-ethyl-3-methylimidazolium trifluoromethanesulfonate ([EMIM]TFS), 1-ethyl-3-methylimidazolium tetrafluoroborate ([EMIM]BF_4), 1-ethyl-3-methylimidazolium dicyanamide ([EMIM]DCA), and 1-ethyl-3-methylimidazolium thiocyanate ([EMIM]SCN).

As shown in Fig. 18.13A, pure water's COF stayed larger than 0.45 throughout the test, while all the four RTILs exhibited steady COFs of 0.04–0.07. The four kinds of 40% $RTILs_{(aq)}$ (RTILs aqueous solution with RTIL concentration of 40%) were also tested under the same conditions, as shown in Fig. 18.13B. During the wearing-in period, the COFs with the lubrication of $[EMIM]BF_{4(aq)}$, $[EMIM]DCA_{(aq)}$, and $[EMIM]SCN_{(aq)}$ quickly reduced to a very low level (0.015–0.02) and then remained stable. The superlubricity could not be achieved by these $RTILs_{(aq)}$ because their final COFs were always above 0.01. However the COF with the lubrication of $[EMIM]TFS_{(aq)}$ gradually reduced to 0.002 after 300 s of wearing-in process, and the COF remained stable with only a slight increase (0.002–0.004) for

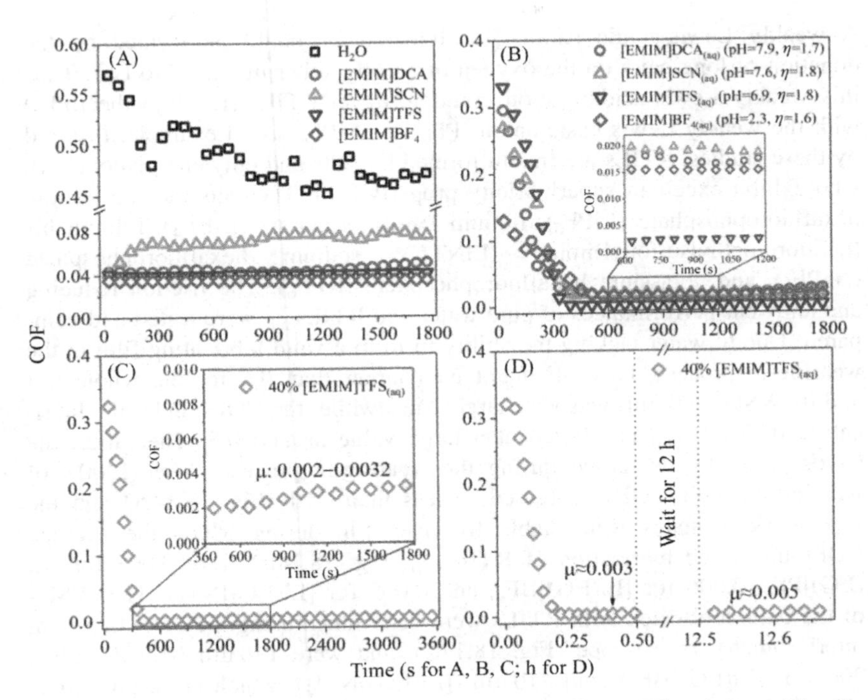

FIGURE 18.13 COFs when lubricated (A and B) with various RTIL solutions; (C) with $[EMIM]TFS_{(aq)}$; and (D) with $[EMIM]TFS_{(aq)}$ with 12 h of suspension. COF, Coefficient of friction; *[EMIM]TSF(aq)*, 1-ethyl-3-methylimidazolium trifluoromethanesulfonate aqueous solution; *RTIL*: room-temperature ionic liquid. *Adapted from X. Ge, J. Li, C. Zhang, Z. Wang, J. Luo, Superlubricity of 1-ethyl-3-methylimidazolium trifluoromethanesulfonate ionic liquid induced by tribochemical reactions, Langmuir 34 (18) (2018) 5245–5252. Copyright 2018, American Chemical Society.*

more than 1 h, as shown in Fig. 18.13C. Because these RTILs have the same cations, it can be inferred that the superlubricity of $[EMIM]TFS_{(aq)}$ is mainly owing to its anions, which may lead to the formation of a firm and effective lubricating film via a possible tribochemical reaction between its anions and the solid surfaces during the wearing-in period. Additionally, the superlubricity was still realized after the experiment was halted for 12 h in the absence of shear and pressure, and the COFs before and after suspension were nearly the same, as shown in Fig. 18.13D, indicative of the robustness of the lubricating film in its natural state.

This result proves that RTILs can achieve superlubricity at the macroscale, and notably, RTILs can be designed by combining various cations and anions, giving rise to numerous potential RTILs with the superlubricity property. It is well known that RTILs can be synthesized with lithium salts via in situ formation in oligoethers, such as ethers, tetraglyme, and PEG [53–55].

A weakly Lewis-acidic cation $[Li(oligoether)]^+$ could be formed by the donation of lone pairs on the oxygen of an oligoether molecule to Li^+. Thus this weakly Lewis-acidic cation could form an RTIL ($[Li(oligoether)]PF_6$) with the weakly Lewis-basic anion (PF_6^-) of $LiPF_6$, as an example. Inspired by these studies, RTILs are in situ formed by salts and ethylene glycol (EG), who exhibit excellent superlubricity property [21]. The salts include lithium hexafluorophosphate ($LiPF_6$), lithium tetrafluoroborate ($LiBF_4$), lithium bis (trifluoromethylsulfonyl)imide ($LiNTf_2$), sodium hexafluorophosphate ($NaPF_6$), and potassium hexafluorophosphate (KPF_6). The friction reducing and antiwear performances of pure water and $RTILs_{(aq)}$ were tested and compared. Due to water lacking the ability to form a valid lubricating film in the wearing-in process, its COF maintained larger than 0.6 for the whole test and its WSD (700 μm) was very large. Meanwhile, the COFs under the lubrication of $RTILs_{(aq)}$ all started at a large value around 0.3. Thereafter, the COFs gradually decreased during the wearing-in process. After 1200 s of wearing-in, their COFs decreased to less than 0.01 (Fig. 18.14A) and the superlubricity states were stable for over 1 h, during which the average COFs under the lubrication of $RTILs_{(aq)}$ (Fig. 18.14B) were 0.003 for $[Li(EG)]PF_6$, 0.006 for $[Li(EG)]BF_4$, and 0.008 for $[Li(EG)]NTf_2$. The WSDs of the balls lubricated with RTILs were determined using the 3D white light interferometry microscope (Fig. 18.14C), and were 170 μm ($[Li(EG)]PF_6$), 300 μm ($[Li(EG)]BF_4$), and 410 μm ($[Li(EG)]NTf_2$), which corresponded to the contact pressures of 132, 42, and 23 MPa, respectively. Given the similar viscosities of the RTILs (Table 18.3), generally, the COFs should be small if the contact pressure is low. However the highest contact pressure corresponds to the smallest COF in the present case, thereby indicating the unique lubricity properties of $[Li(EG)]PF_6$. Due to these RTILs possessing the same cation, it is inferred that the PF_6^- anion plays the dominant role in the greatest superlubricity and antiwear properties among these anions.

Meanwhile, to probe the effect of cations on friction and wear, the anion PF_6^- was fixed and three different kinds of monovalent metal cations (Li^+, Na^+, and K^+) with different hydration abilities were selected. The lubricities of $[Li(EG)]PF_{6(aq)}$, $[Na(EG)]PF_{6(aq)}$, and $[K(EG)]PF_{6(aq)}$ were tested and compared, as shown in Fig. 18.14D. The COFs during the superlubricity period were similar and followed the sequence of $COF(Li^+) < COF(Na^+) < COF(K^+)$ (Fig. 18.14E). WSDs of the balls at the end of the superlubricity test are shown in Fig. 18.14F. They were 175 μm for Li^+, 425 μm for Na^+, and 480 μm for K^+, which corresponded to the contact pressures of 125, 21, and 17 MPa, respectively. Similar to the case of the anions, the highest contact pressure corresponds to the smallest COF, which indicates that the Li^+ cation possesses the greatest antiwear property among these cations. Integrating the results of anions and cations from the lubricity experiments, it is summarized that $[Li(EG)]PF_6$ exhibits the most effective lubricity and antiwear properties.

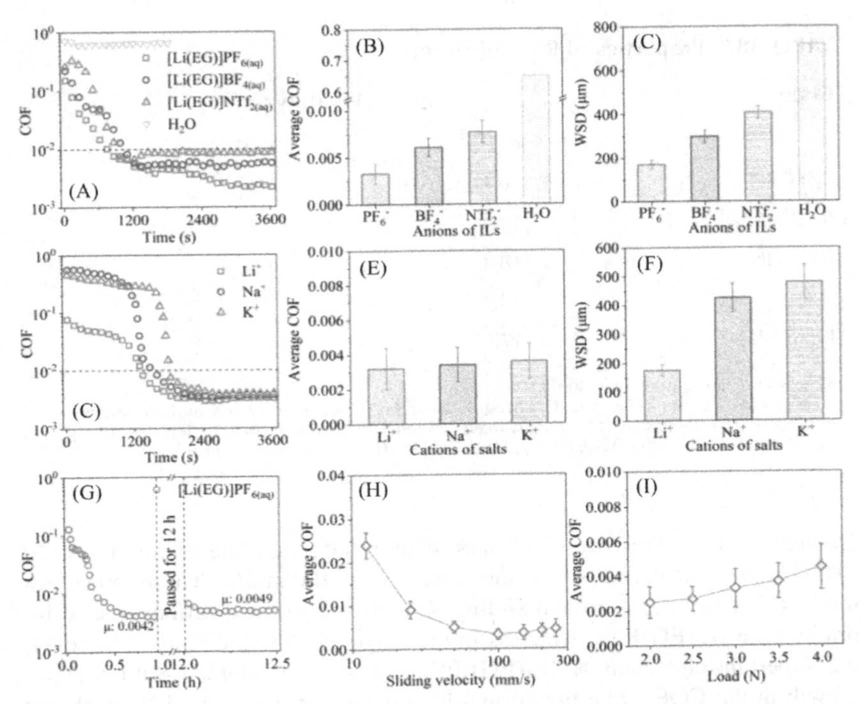

FIGURE 18.14 Friction experiments were conducted (A – C) with various anions and the fixed $Li(EG)^+$ cation, (D – F) with various cations and the fixed PF_6^- anion. For (G – I), experiments were conducted with $[Li(EG)]PF_{6(aq)}$. (G) COFs under the lubrication of $[Li(EG)]PF_{6(aq)}$ with 12 h of suspension; (H) relationship between COFs and sliding velocities; and (I) relationship between COFs and normal load. *COF*, Coefficient of friction. *Adapted from X. Ge, J. Li, C. Zhang, Y. Liu, J. Luo, Superlubricity and antiwear properties of in situ-formed ionic liquids at ceramic interfaces induced by tribochemical reactions, ACS Appl. Mater. Interfaces 11 (6) (2019) 6568–6574. Copyright 2019, American Chemical Society.*

The robustness of the superlubricity state achieved with $[Li(EG)]PF_6$ was studied. After the superlubricity state was achieved with $[Li(EG)]PF_{6(aq)}$, the experiment was paused for 12 h in the absence of any change. Then, on the same wear track, the experiment was resumed. As depicted in Fig. 18.14G, the superlubricity state could be restored and maintained at a steady value of 0.0049, which was indicative of the robustness of the superlubricity state achieved with $[Li(EG)]PF_6$. In addition, the superlubricity property of $[Li(EG)]PF_6$ at different sliding velocities was determined, as depicted in Fig. 18.14H. The COF under the lubrication of $[Li(EG)]PF_{6(aq)}$ could only decrease to around 0.024 when the sliding velocity was 12.5 mm/s, indicative of the failure of superlubricity. Meanwhile, its COF was at the boundary of the superlubricity regime (COF of approximately 0.01) when the sliding velocity was 25 mm/s. When the sliding velocities ranged from 50 to

TABLE 18.3 Properties of RTIL lubricants.

Code	Viscosity (mPa s)	
	RTIL	$RTIL_{(aq)}$
$[Li(EG)]PF_6$	20.1	1.3
$[Li(EG)]BF_4$	19.4	1.4
$[Li(EG)]NTf_2$	18.3	1.5
$[Na(EG)]PF_6$	17.2	1.3
$[K(EG)]PF_6$	17.4	1.3

RTIL, Room-temperature ionic liquid.
Source: Adapted from X. Ge, J. Li, C. Zhang, Y. Liu, J. Luo, Superlubricity and antiwear properties of in situ-formed ionic liquids at ceramic interfaces induced by tribochemical reactions, ACS Appl. Mater. Interfaces 11 (6) (2019) 6568–6574. Copyright 2019, American Chemical Society.

250 mm/s, the COFs were all less than 0.005, and the minimum COF (0.0033) was acquired when the speed was 100 mm/s. These outcomes implies that there is a critical sliding velocity for the realization of superlubricity with $[Li(EG)]PF_6$, which is in the range of 50 – 250 mm/s. Moreover, the superlubricity state of $[Li(EG)]PF_{6(aq)}$ could be retained with a slight growth in the COF while the normal load increased from 2 to 4 N, as shown in Fig. 18.14I.

The analysis indicates that there are three critical factors for the realization of superlubricity by in situ-formed RTILs. One factor is the formation of the composite tribochemical layer, which is the production of the tribochemical reactions between the anions of RTILs and surfaces during the wearing-in period. Moreover, PF_6^-, which is transformed into phosphates, could provide the greatest superlubricity and antiwear performance among the selected anions. Another factor is the hydration layer formed by the adsorption of the monovalent metal cations, which is surrounded by water molecules due to their hydration ability. It has been found that the hydration layer can help lower friction, and the stronger hydration ability of a monovalent metal cation leads to better antiwear performance. The third factor is the fluid film formed by the lubricant, whose low viscosity and shearing resistance further lower the friction and wear. Because different cations and anions of RTILs were studied, this work enriches the range of alternative RTILs for macroscale superlubricity. Among them, $[Li(EG)]PF_6$ possesses the greatest superlubricity and antiwear properties. The mixture of $[Li(EG)]PF_6$ and GO nanosheets [56–58] can also help achieve robust superlubricity at the macroscale [22]. Further, the average contact pressure of the friction pairs during the superlubricity period reaches 600 MPa, much higher than the current upper limitation of contact pressure (300 MPa) [59]. Even though the exploration on RTILs with superlubricity property is still in the initial state and more researches would be performed in the near future, the superlubricity and antiwear properties of RTILs at Si_3N_4 interfaces (commonly used in industry) make it a potential promising lubricating material in engineering applications.

18.5 Surfactant micelles

At the nanoscale, the liquid superlubricity also can be achieved by various surfactant micelles, which are widely used in the micro-/nanomachine lubrication. The hexadecyltrimethylammonium bromide ($C_{16}TAB$) solution, with a concentration equal to three times of the critical micelle concentration (CMC) (CMC = 0.9 mM), was studied. As shown in Fig. 18.15A, when the load exceeded 183 nN (corresponding to an average contact pressure of 3.6 MPa), the friction force increased hundreds of times, and hence, this load can be defined as the critical load (F_{crit}). The lubrication state was divided into two regimes according to F_{crit}. The first regime is the superlubricity regime where the load is less than F_{crit}, and the corresponding COF was $\mu \approx 0.0007$, attained by the slope of the linear fitting of the friction force points in the superlubricity regime. The second regime is the high friction regime where the load is greater than F_{crit}, and the corresponding COF was $\mu \approx 0.3$, which is the same as the COF between two bare silica surfaces across water. These aforementioned results show that the superlubricity of surfactant solutions can be achieved only when the load is very small. At the same time, the relationship between superlubricity and sliding speed was shown in Fig. 18.15B. It was observed that the friction behaviors at different sliding speeds were identical to that in Fig. 18.15A. The F_{crit} and COF in the ultralow friction regime were both independent of the sliding speed, indicating that the lubrication state between two silica surfaces with $C_{16}TAB$ solution is boundary lubrication as superlubricity appears.

The friction behaviors, critical loads (F_{crit}), and maximum normal forces (F_{max}) for $C_{16}TAB$ solution with seven different concentrations were measured and compared, as shown in Fig. 18.16A. The friction behaviors for these different concentrations were found to be similar to that shown in Fig. 18.15A; that is, the friction force remained at the superlow value until the load exceeded the critical load limit, which indicates that all the concentrations can lead to the achievement of superlubricity when the load is very small. However the critical loads were different with the variation of concentrations. When the concentration was less than 1 CMC, both F_{crit} and F_{max} increased with the increase of concentration, which indicates that the load-bearing capacity of superlubricity increases with concentration increasing in this region. When the concentration was greater than 1 CMC, both F_{crit} and F_{max} remained constant ($F_{max} = 214$ nN and $F_{crit} = 183$ nN), which indicates that the load-bearing capacity of superlubricity is independent of the

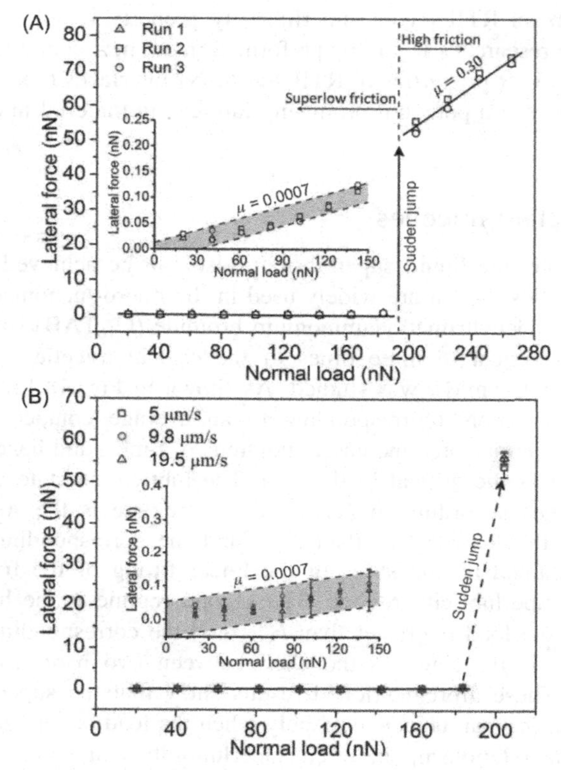

FIGURE 18.15 Friction force versus normal load across the $C_{16}TAB$ solution (3 CMC) (A) with the sliding velocity of 5 μm/s; (B) measured under various sliding velocities. *$C_{16}TAB$,* Hexadecyltrimethylammonium bromide. *Adapted from J. Li, C. Zhang, P. Cheng, X. Chen, W. Wang, J. Luo, AFM studies on liquid superlubricity between silica surfaces achieved with surfactant micelles, Langmuir 32 (22) (2016) 5593–5599. Copyright 2016, American Chemical Society.*

concentration in this region. Moreover, it is clear to observe that F_{crit} is always less than F_{max}. According to the aforementioned results, it is confirmed that F_{crit} is determined by F_{max}. In other words, the superlubricity can only be achieved when the two solid surfaces are in the repulsive region (the adsorbed micelle layer does not collapse).

According to these results and analyses, a lubrication model of the $C_{16}TAB$ micelles was proposed, as presented in Fig. 18.16B. Owing to the electrostatic interaction, the surfactant molecules are adsorbed on the two silica surfaces in the form of micellar layers (>1 CMC) or bilayer structure (<1 CMC). After that, the positively charged headgroups on the micellar layer or bilayer structure can firmly absorb water molecules to form the hydration layer. When the two solid surfaces slide against each other, the

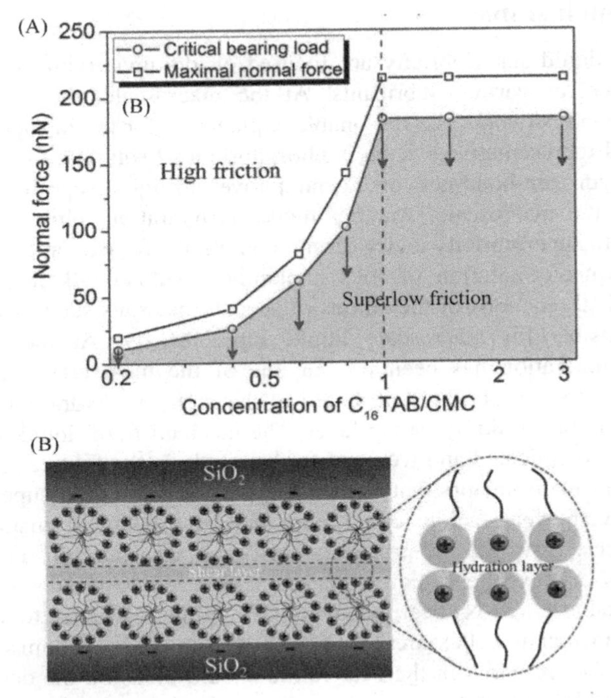

FIGURE 18.16 (A) Maximum normal forces and critical bearing loads with seven different concentrations. (B) Lubrication model of $C_{16}TAB$ solution in the superlubricity regime. *Adapted from J. Li, C. Zhang, P. Cheng, X. Chen, W. Wang, J. Luo, AFM studies on liquid superlubricity between silica surfaces achieved with surfactant micelles, Langmuir 32 (22) (2016) 5593–5599. Copyright 2016, American Chemical Society.*

hydration repulsion produced by the hydration layer can bear the applied load and the shear occurs in the hydration layer, where the shear strength is extremely low due to its excellent fluidity [60]. As a result, the superlubricity can be achieved when the micelle layers are not ruptured. However if the load exceeds the maximum normal force, the micellar layers would be destroyed, and thus, superlubricity phenomenon would disappear because the uniform hydration layer is destroyed. From these results, it can be inferred that the superlubricity of surfactant solutions can be achieved as long as the following three conditions are satisfied. First, the surfactant molecules can be adsorbed on the solid surfaces to form the micelles or bilayers as a result of the electrostatic interaction. Second, there exist positively charged headgroups on the surfactant molecules to attach water molecules and produce the hydration repulsion. Finally, the applied load cannot go beyond the maximum normal force produced by the hydration layer. Therefore most cationic surfactants can be selected to achieve superlubricity between negatively charged surfaces, such as mica and silica, at the nanoscale.

18.6 Conclusions

Studies on liquid superlubricity are focused on the underlying mechanisms, which differ for various lubricants. At the macroscale, the formation of hydrogen-bond networks is a reasonable explanation for liquid superlubricity. One typical representative system is phosphoric acid solution. Of course, the model of hydrogen-bond networks cannot cover all liquid superlubricity phenomena at the macroscale. Another model is hydration lubrication, which could lead to superlubricity under certain conditions, for example, lubrication with the aqueous solution of salts containing hydrated alkali metal ions. Tribochemical reactions of the anions of special materials such as RTILs are also responsible for macroscale liquid superlubricity. At the nanoscale, hydration lubrication has been proven one of the most effective ways to achieve liquid superlubricity, attributed to the repulsive hydration force and the formation of a fluid hydration layer. The mechanism of liquid superlubricity is very complicated and worth of further study in the future.

There are many factors that limit the application of liquid superlubricity, such as severe wear, acidic conditions, and complexity of manufacturing ideal surfaces. However the prospects for liquid superlubricity in industrial applications are excellent. For instance, acid-based lubricants and RTILs may be suitable for Si_3N_4 bearings, who have strong resistance to acidic corrosion. Surfactant micelles are possible to be applied to nanomachine systems. Therefore research in the near future should focus on the development of new liquid lubricants to offer lubricating performance with extremely low friction and wear, as well as be associated with practical operating conditions in mechanical systems.

Acknowledgments

This work was financially supported by National Natural Science Foundation of China (Grant Nos. 51775295 and 51527901), Foundation from State Key Laboratory of Tribology (Grant No. SKLT2019C01), and China Postdoctoral Science Foundation (Grant No. 2019M650652).

Competing interests

The authors declare no competing financial interests.

References

[1] K. Shinjo, M. Hirano, Dynamics of friction-superlubric state, Surf. Sci. 283 (1–3) (1993) 473–478.

[2] A. Erdemir, J. Martin, Superlubricity, first ed., Elsevier, New York, 2007.

[3] S. Perry, W. Tysoe, Frontiers of fundamental tribological research, Tribol. Lett. 19 (3) (2005) 151–161.

[4] L. Tisza, On the thermal supraconductibility of liquid helium II and the Bose-Einstein statistics, C. R, Hebd. Seances Acad. Sci. D 207 (1938) 1035–1037.
[5] H. Tomizawa, T. Fischer, Friction and wear of silicon-nitride and silicon-carbide in water: hydrodynamic lubrication at low sliding speed obtained by tribochemical wear, ASLE Trans. 30 (1987) 41–46.
[6] X. Ge, J. Li, J. Luo, Macroscale superlubricity achieved with various liquid molecules: a review, Front. Mech. Eng. 5 (2019) 2.
[7] J. Li, J. Luo, Advancements in superlubricity, Sci. China Technol. Sci 56 (12) (2013) 1–11.
[8] Y. Hu, T. Ma, H. Wang, Energy dissipation in atomic-scale friction, Friction 1 (1) (2013) 24–40.
[9] J. Martin, C. Donnet, T. Lemogne, T. Epicier, Superlubricity of molybdenum-disulfide, Phys. Rev. B 48 (14) (1993) 10583–10586.
[10] A. Erdemir, O. Eryilmaz, Achieving superlubricity in DLC films by controlling bulk, surface, and tribochemistry, Friction 2 (2) (2014) 140–155.
[11] J. Luo, X. Lu, S. Wen, Developments and unsolved problems in nano-lubrication, Prog. Nat. Sci. 11 (3) (2011) 173–183.
[12] J. Xu, K. Kato, Formation of tribochemical layer of ceramics sliding in water and its role for low friction, Wear 245 (1–2) (2000) 61–75.
[13] C. Xiao, J. Li, L. Chen, C. Zhang, N. Zhou, T. Qing, et al., Water-based superlubricity in vacuum, Friction 7 (2) (2019) 192–198.
[14] U. Raviv, S. Giasson, N. Kampf, J. Gohy, R. Jerome, J. Klein, Lubrication by charged polymers, Nature 425 (6954) (2003) 163–165.
[15] M. Muller, X. Yan, S. Lee, S. Perry, N. Spencer, Lubrication properties of a brushlike copolymer as a function of the amount of solvent absorbed within the brush, Macromolecules 38 (13) (2005) 5706–5713.
[16] C. Matta, L. Joly-Pottuz, M. Bouchet, J. Martin, Superlubricity and tribochemistry of polyhydric alcohols, Phys. Rev. B 78 (8) (2008) 085436.
[17] Z. Ma, C. Zhang, J. Luo, X. Lu, S. Wen, Superlubricity of a mixed aqueous solution, Chin. Phys. Lett. 28 (5) (2011) 056201.
[18] S. Arad, L. Rapoport, A. Moshkovich, D. van Moppes, M. Karpasas, R. Golan, et al., Superior biolubricant from a species of red microalga, Langmuir 22 (17) (2006) 7313–7317.
[19] J. Li, Y. Liu, J. Luo, P. Liu, C. Zhang, Excellent lubricating behavior of *Brasenia schreberi* mucilage, Langmuir 28 (20) (2012) 7797–7802.
[20] Z. Jin, D. Duncan, Bio-friction, Friction 1 (2) (2013) 100–113.
[21] X. Ge, J. Li, C. Zhang, Y. Liu, J. Luo, Superlubricity and antiwear properties of in situ-formed ionic liquids at ceramic interfaces induced by tribochemical reactions, ACS Appl. Mater. Interfaces 11 (6) (2019) 6568–6574.
[22] X. Ge, J. Li, H. Wang, C. Zhang, Y. Liu, J. Luo, Macroscale superlubricity under extreme pressure enabled by the combination of graphene-oxide nanosheets with ionic liquid, Carbon 151 (2019) 76–83.
[23] J. Li, C. Zhang, J. Luo, Superlubricity behavior with phosphoric acid-water network induced by rubbing, Langmuir 27 (15) (2011) 9413–9417.
[24] L. Sun, C. Zhang, J. Li, Y. Liu, J. Luo, Superlubricity of Si_3N_4 sliding against SiO_2 under linear contact conditions in phosphoric acid solutions, Sci. China Technol. Sci 56 (7) (2013) 1678–1684.

[25] J. Li, C. Zhang, L. Ma, Y. Liu, J. Luo, Superlubricity achieved with mixtures of acids and glycerol, Langmuir 29 (1) (2013) 271–275.

[26] J. Li, C. Zhang, J. Luo, Superlubricity achieved with mixtures of polyhydroxy alcohols and acids, Langmuir 29 (17) (2013) 5239–5245.

[27] J. Li, C. Zhang, P. Cheng, X. Chen, W. Wang, J. Luo, AFM studies on liquid superlubricity between silica surfaces achieved with surfactant micelles, Langmuir 32 (22) (2016) 5593–5599.

[28] J. Li, Z. Dou, Y. Liu, J. Luo, J. Xiao, Improvement of load bearing capacity of nanoscale superlow friction by synthesized fluorinated surfactant micelles, ACS Appl. Nano Mater. 1 (2) (2018) 953–959.

[29] J. Li, J. Luo, Nonlinear frictional energy dissipation between silica-adsorbed surfactant micelles, J. Phys. Chem. Lett. 8 (10) (2017) 2258–2262.

[30] J. Li, J. Luo, Normal and frictional force hysteresis between self-assembled fluorosurfactant micelle arrays at the nanoscale, Adv. Mater. Inter. 5 (4) (2018) 1700802.

[31] J. Li, C. Zhang, J. Luo, Friction process of superlubricity, in: Proceedings of the 2012 ASME/STLE International Joint Tribology Conference, Denver, America, 2012.

[32] J. Li, C. Zhang, L. Sun, X. Lu, J. Luo, Tribochemistry and superlubricity induced by hydrogen ions, Langmuir 28 (45) (2012) 15816–15823.

[33] S. Sjoberg, Silica in aqueous environments, J. Non-Cryst Solids 196 (1996) 51–57.

[34] N. Sahai, Is silica really an anomalous oxide? Surface acidity and aqueous hydrolysis revisited, Env. Sci. Technol. 36 (3) (2002) 445–452.

[35] J. Li, L. Ma, S. Zhang, C. Zhang, Y. Liu, J. Luo, Investigations on the mechanism of superlubricity achieved with phosphoric acid solution by direct observation, J. Appl. Phys. 114 (11) (2013) 114901.

[36] J. Dash, H. Fu, J. Wettlaufer, The premelting of ice and its environmental consequences, Rep. Prog. Phys. 58 (1) (1995) 115–167.

[37] J. Klein, Hydration lubrication, Friction 1 (1) (2013) 1–23.

[38] X. Ge, J. Li, C. Zhang, J. Luo, Liquid superlubricity of polyethylene glycol aqueous solution achieved with boric acid additive, Langmuir 34 (12) (2018) 3578–3587.

[39] X. Ge, T. Halmans, J. Li, J. Luo, Molecular behaviors in thin film lubrication—Part three: superlubricity attained by polar and nonpolar molecules, Friction 7 (6) (2019) 625–636.

[40] L. Shteinberg, Effect of boric acid concentration on the catalysis of the reaction of 4-nitrobenzoic acid with ammonia, Russ. J. Appl. Chem. 82 (4) (2009) 613–617.

[41] Y. Zhou, J. Qu, Ionic liquids as lubricant additives: a review, ACS Appl. Mater. Interfaces 9 (4) (2017) 3209–3222.

[42] T. Han, C. Zhang, J. Luo, Macroscale superlubricity enabled by hydrated alkali metal ions, Langmuir 34 (38) (2018) 11281–11291.

[43] T. Han, C. Zhang, X. Chen, J. Li, W. Wang, J. Luo, Contribution of a tribo-induced silica layer to macroscale superlubricity of hydrated ions, J. Phys. Chem. C. 123 (33) (2019) 20270–20277.

[44] Z. Song, Y. Liang, M. Fan, F. Zhou, W. Liu, Lithium-based ionic liquids as novel lubricant additives for multiply alkylated cyclopentanes (MACs), Friction 1 (3) (2013) 222–231.

[45] C. Ye, W. Liu, Y. Chen, L. Yu, Room-temperature ionic liquids: a novel versatile lubricant, Chem. Commun. 21 (2011) 2244–2245.

[46] K. Holmberg, A. Erdemir, Influence of tribology on global energy consumption, costs and emissions, Friction 5 (3) (2017) 263–284.

[47] A. Somers, B. Khemchandani, P. Howlett, J. Sun, D. MacFarlane, M. Forsyth, Ionic liquids as antiwear additives in base oils: influence of structure on miscibility and antiwear performance for steel on aluminum, ACS Appl. Mater. Interfaces 5 (22) (2013) 11544–11553.

[48] T. Amann, F. Gatti, N. Oberle, A. Kailer, J. Ruhe, Galvanically induced potentials to enable minimal tribochemical wear of stainless steel lubricated with sodium chloride and ionic liquid aqueous solution, Friction 6 (2) (2018) 230–242.

[49] G. Huang, Q. Yu, Z. Ma, M. Cai, F. Zhou, W. Liu, Oil-soluble ionic liquids as antiwear and extreme pressure additives in poly-alpha-olefin for steel/steel contacts, Friction 7 (1) (2019) 18–31.

[50] W. Ma, Z. Gong, K. Gao, L. Qiang, J. Zhang, S. Yu, Superlubricity achieved by carbon quantum dots in ionic liquid, Mater. Lett. 195 (2017) 220–223.

[51] H. Li, R. Wood, M. Rutland, R. Atkin, An ionic liquid lubricant enables superlubricity to be “switched on” in situ using an electrical potential, Chem. Commun. 50 (33) (2014) 4368–4370.

[52] X. Ge, J. Li, C. Zhang, Z. Wang, J. Luo, Superlubricity of 1-ethyl-3-methylimidazolium trifluoromethanesulfonate ionic liquid induced by tribochemical reactions, Langmuir 34 (18) (2018) 5245–5252.

[53] M. Fan, Z. Song, Y. Liang, F. Zhou, W. Liu, In situ formed ionic liquids in synthetic esters for significantly improved lubrication, ACS Appl. Mater. Interfaces 4 (12) (2012) 6683–6689.

[54] X. Ge, Y. Xia, Z. Shu, Conductive and tribological properties of lithium-based ionic liquids as grease base oil, Tribol. Trans. 58 (4) (2015) 686–690.

[55] T. Tamura, T. Hachida, K. Yoshida, N. Tachikawa, K. Dokko, M. Watanabe, New glyme-cyclic imide lithium salt complexes as thermally stable electrolytes for lithium batteries, J. Power Sources 195 (18) (2010) 6095–6100.

[56] L. Wu, Z. Xie, L. Gu, B. Song, L. Wang, Investigation of the tribological behavior of graphene oxide nanoplates as lubricant additives for ceramic/steel contact, Tribol. Int. 128 (2018) 113–120.

[57] F. Zhao, L. Zhang, G. Li, Y. Guo, M. Qi, G. Zhang, Significantly enhancing tribological performance of epoxy by filling with ionic liquid functionalized graphene oxide, Carbon 136 (2018) 309–319.

[58] X. Ge, J. Li, R. Luo, C. Zhang, J. Luo, Macroscale superlubricity enabled by synergy effect of graphene-oxide nanoflakes and ethanediol, ACS Appl. Mater. Interfaces 10 (47) (2018) 40863–40870.

[59] J. Li, C. Zhang, M. Deng, J. Luo, Investigation of the difference in liquid superlubricity between water- and oil-based lubricants, RSC Adv. 5 (78) (2015) 63827–63833.

[60] U. Raviv, J. Klein, Fluidity of bound hydration layers, Science 297 (5586) (2002) 1540–1543.

Chapter 19

Superlubricity of lamellar fluids

Juliette Cayer-Barrioz[1], **Alexia Crespo**[1], **Hélène Fay**[1,2,3],
Nazario Morgado[1] **and Denis Mazuyer**[1]
[1]Laboratoire de Tribologie et Dynamique des Systèmes, CNRS UMR5513, Ecole centrale de Lyon, Ecully, France, [2]Centre de Recherche Paul Pascal, CNRS UMR5031, Université de Bordeaux, Pessac, France, [3]Now at Solvay, LOF, CNRS UMR5258, Université de Bordeaux, Pessac, France

19.1 State of the art

Organized lamellar systems exist in many industrial water-based lubricants, such as wire drawing processes, because they promote low friction between sliding surfaces in contact [1] even under high contact pressures. A prime example was the previous work of Refs. [2–5] showing that the lubricity was strongly enhanced after the formation of lamellar crystallites resulting from the metallic ionic interaction between the lubricant and brass surfaces. The authors [2,3] demonstrated the correlation between low-friction values with the presence of highly anisotropic lamellar crystallites in the contact [1]. The latter offer a solid-like crystalline order in the direction perpendicular to the lamellae, combined to a liquid-like order due to preferred shear direction in the plane parallel to the lamellae morphology [6].

Lamellar phases usually consist of regular stacks of planar surfactant bilayers, separated by solvent. At equilibrium, the ideal bilayers are parallel, infinitely long and defectless [1]. Their stability results from the balance between attractive and repulsive interactions between neighboring membranes [6]. Behaving as a solid system in one direction and liquid in the others, lamellar phases exhibit properties similar to that of smectic A phases, such as elasticity. Their deformation may modify the bilayer thickness and/or the bilayer periodicity, inducing a variation of the local concentration of surfactant. There exist two main defect types in lamellar phases: [6] defects of texture and defects of structure. The former are visible using polarized optical microscopy. The latter can be detected using cryotransmission electron microscopy, freeze-fracture, neutron scattering, etc. Various types of structure defects were identified in the literature: [7] bridges connecting surfactant lamellae, pores connecting the solvent area, and passages connecting

Superlubricity. DOI: https://doi.org/10.1016/B978-0-444-64313-1.00019-3

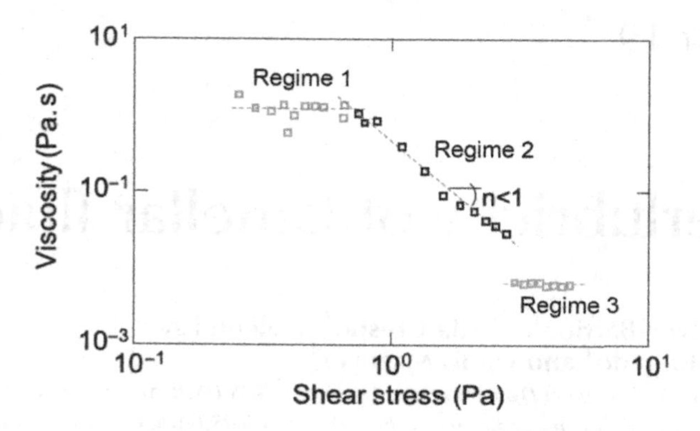

FIGURE 19.1 Viscosity versus shear stress for a volume fraction of membranes of 0.7 for a system composed of SDS/pentanol/dodecane/water. Different regimes (1), (2), and (3) were observed. They were interpreted as the signature of different organizations of the initial lamellar phase. This transition was controlled by a critical velocity mainly dependent on the solvent viscosity and the elasticity of the bilayers. Beyond this velocity, dislocations cannot follow the flow and local dislocations are created inducing a membrane undulation and vesicles appearance. In this second regime, the radius of the vesicles decreases when the shear rate increases, resulting in a macroscopic shear-thinning rheological behavior. At higher shear rates ($>10-1000\ s^{-1}$), lamellae orient parallel to the flow with less defects than in regime 1: the viscosity of the plateau in regime 3 is lower than that obtained in regime 1 [12]. SDS, sodium dodecyl sulfate. *Adapted from D. Roux, F. Nallet, O. Diat, Rheology of lyotropic lamellar phases, Europhys. Lett. 24 (1993) 53–58.*

the two media. It was also shown that the orientation of the phases and the existence of defects were strongly dependent on the applied shear [8–11]. Rheological characterizations were performed measuring the shear stress as a function of the shear rate. For instance, Roux et al. [10] evidenced three regimes, as illustrated in Fig. 19.1. At low shear rates due to low shear rates, typically less than $1\ s^{-1}$, the viscosity was high and the lamellae were oriented parallel to the shear direction. Many dislocation-like defects were observed. At intermediate shear rates, undulation instability generated by parallelism misalignment, lead to a transition from lamellae with defects to multilamellae vesicles of radius equal to few tenth of microns.

Under pressure, when confined in a contact between two sliding surfaces, the lamella flow becomes even more complex. Assuming that the lamellae structure was maintained, the interface can be schematized as shown in Fig. 19.2.

Once subject to shear, the interface accommodates sliding velocity and shear can localize within the lamella (Fig. 19.3A) and/or interlamellae (Fig. 19.3B).

Intralamella shear was the model system of boundary friction studied by Hardy [13]. It is usually supposed that slip was confined to a plane, probably midway between the solid surfaces to explain friction reduction with this

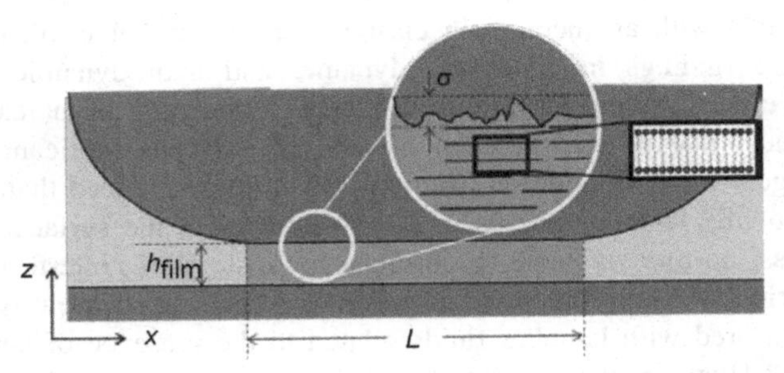

FIGURE 19.2 Schematic of a sliding contact under pressure, typically 10–500 MPa, assuming that the lamellae structure was maintained with relevant parameters such as h_{film} the film thickness and L the contact diameter. The confinement ratio was defined as h_{film}/L and is equal to 10^{-3}. The zoom shows the lamellar structure of the fluid.

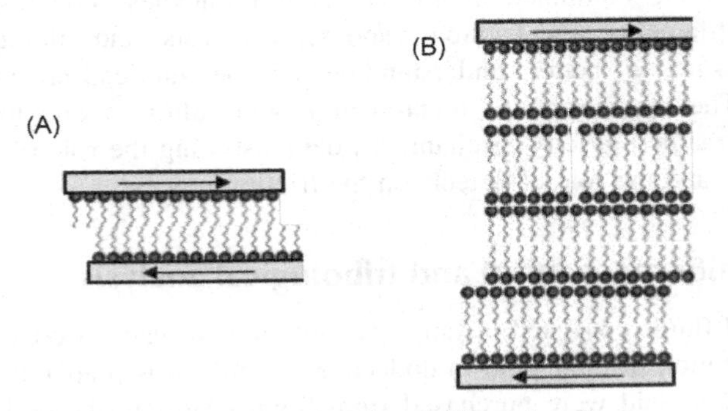

FIGURE 19.3 Shear localization (A) intralamella and (B) interlamellae.

type of bilayer [13]. This conclusion was confirmed for nanometric thin films of various fluids including thermotropic nematics [14], lamellar phases [15], and surfactants [16–21] under low contact pressures. Such interfaces display different friction responses depending on the experimental conditions. For instance, the sliding could transit from a steady kinetic state to an inverted stick-slip regime (the friction spikes point down rather than up) [18]. For transient conditions, such as stop and go experiments or sliding velocity change, friction static peaks could also be observed, depending on the history of sliding [19–21]. Velocity-weakening and strengthening behavior could also be observed [21]. These complex responses were function of the molecular architecture and organization on the surfaces, of the surface roughness, etc.

A prime example of investigation of interlamellae shear mechanisms is the work of Oswald [22,23] in which a lubrication theory of smectic A phases was proposed: lubrication mechanisms were controlled by the formation and motion of stacking defects during shear. The increase in stiffness was associated with an increase in elastic bearing. Fischer et al. [24] confirmed these findings in elastohydrodynamic and hydrodynamic regimes. Lockwood et al. [25] and Friberg et al. [26] also observed an increase in the film thickness due to the presence of a highly viscous lubricant fluid in the convergent zone. In addition, friction was strongly reduced thanks to the capability of the lamellar phases to adsorb parallel to the surfaces [27,28]. Nevertheless, no one was able to correlate microstructure, interfacial rheology, and friction. Therefore several questions remain open: how is the low friction measured with lamellar fluids related to the presence of lamellae in the contact? How do intra- or interlamellar slipping accommodate shearing velocity? Can we identify the superlubricity mechanisms involved?

To address these questions and investigate more deeply the intra- and interlamellae shear mechanisms, an in situ mechanical and tribological analysis was performed at both scale with model fatty acid–based lamellar fluids. From the combined evolution of film thickness composed of the bilayer/multilamellae and friction response, with time and sliding velocity, under pressure, a better understanding of the lubrication mechanisms emerged. The application of theoretical models allowed one to perfectly describe the superlubricity mechanisms, demonstrating the role of the lamellar structure and the role of defects on the friction process.

19.2 In situ mechanical and tribological analysis

The studied fluids consisted in fatty acid solutions in various concentrations. A dilute solution of oleic acid in dodecane at 2 mM was prepared: the dodecane and oleic acid were purchased from Sigma-Aldrich, the dodecane was first dehydrated by adding zeolites for several days before being filtrated [21–29], and the oleic acid had a purity greater than 99.0%. This low concentration solution had the same viscosity as dodecane, 1.5 mPa s at 23°C and presents a Newtonian rheological behavior at ambient pressure [29]. A more concentrated complex solution of fatty acids was prepared, composed of a mixture of oleic and linoleic acids, both 18-carbon chains fatty acids bearing, respectively 1 and 2 unsaturation. The fatty acid mixture was provided by Rhodia and then mixed under stirring at 80°C in a water/ethylene diamine (EDA) solution: $X_{\mathrm{FattyAcid}}$=44.8 wt.%, X_{water}=30 wt.%, and X_{EDA}=25.2 wt.% [1]. A detailed analysis of the microstructure [30] of the solution showed that the fatty acids formed bilayers of thickness 2.2 nm, shielding their alkyl chains from contact with water while exposing their polar carboxylic head to the homogeneous water/EDA solution separating the bilayers [1]. The interlamellar distance was equal to 4.9 nm. The rheological characterization of this lamellar fluid was performed and Fig. 19.4 presents the evolution of the viscosity as a function of the shear rate at room temperature and ambient pressure [1]. The data clearly evidenced shear thinning at intermediate shear rates. In addition, associated microstructure

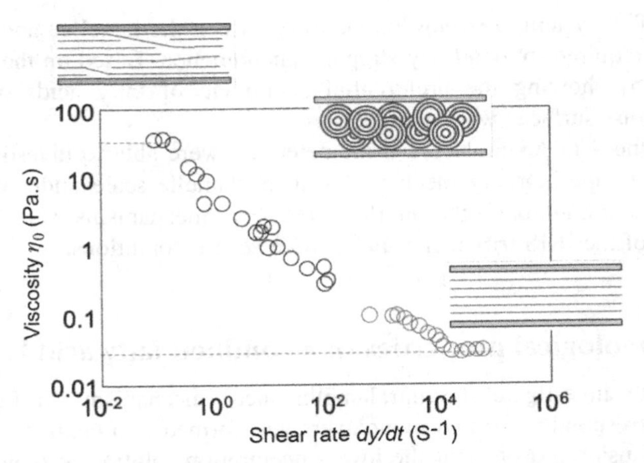

FIGURE 19.4 Rheology of the lamellar fatty acid fluid at room temperature and ambient pressure. A highly viscous plateau was obtained at low shear rate and a low-viscosity plateau at high shear rate. For intermediate shear rates, the viscosity decreases as the shear rate increases obeying a power law: $\eta_0 \alpha (d\gamma/dt)^{-0.8}$. A schematic of microstructure was proposed for each regime of organization [10].

schematics were proposed, based on the pioneering work of Diat et al. [10,31]. The highly viscous plateau, measured at low shear rates, was consistent with an organization of the lamellar phase in locally ordered domains with an orientation perpendicular to the shear gradient, and defects such as dislocations, appear. As shear rate increases, an undulation instability occurred [1] and monodisperse micrometric compact multilayer vesicles self-organizes. The radius of the vesicles decreases as the shear rate increases, leading to a decrease in viscosity [31–34]. At larger shear rates, lamellae were formed again, aligned along the flow direction without any defects, resulting in a low-viscosity Newtonian plateau.

The lubricating capabilities of these fatty acid solutions were investigated using two complementary friction measurement setups, developed in laboratoire de tribologie et dynamique des systèmes, the ATLAS molecular tribometer [35,36], and the IRIS tribometer [1,37,38]. These setups allowed us to analyze in situ the formation of the confined lamellar phases as well as their rheological behavior under pressure and their friction response, keeping the confinement ratio constant, h_{film}/contact size $\approx 10^{-3}$, over 10 decades of sliding velocity, from 10^{-10} to 1 m/s, with a contact pressure ranging from 10 to 500 MPa. The procedure followed in each case was described in detail in Refs. [1,29]. The sphere radius was, respectively, 2.00 $\pm$ 0.05 and 12.50 $\pm$ 0.05 mm for the ATLAS and IRIS experiments. In the case of the bilayer system analysis, smooth metallic surfaces were chosen. Such surfaces promote vertical physisorption of the fatty acids, thanks to weak interactions between the polar carboxylic group and metallic oxides. In the case of the multilamellae system, a strong interaction between the lamellae and the surfaces was required to avoid any slippage phenomenon. Based on the work of Refs. [2–5] showing the preferential anchorage of fatty acids on brass, polished brass surfaces were chosen.

Using the ATLAS molecular tribometer, we were able to investigate and identify the superlubricity mechanisms at the lamella scale under moderate contact pressure although the interlamellae shear mechanisms were analyzed by means of the IRIS tribometer under more severe conditions.

19.3 Rheological properties of a confined fatty acid lamella

In order to investigate the intralamella shear mechanisms, a fatty acid lamella corresponding of two monolayers was formed in a contact. This was carried out using a droplet of the low concentration solution of oleic acid in dodecane [21], deposited between the two cobalt-covered surfaces in the ATLAS tribometer. The evolution of the normal force during a quasistatic approach at 0.2 nm/s was plotted versus the separation distance in Fig. 19.5 after 3 and 5 h of adsorption.

It can be seen that:

- No force was detected for large distances.
- A small attractive force was measured at distances smaller than 10 nm and it could be adjusted by the theoretical van der Waals force for an interface cobalt/oleic acid/cobalt [39] with a Hamaker constant, $A_H = 1.4 \times 10^{-19}$ J, according to Eq. (19.1):

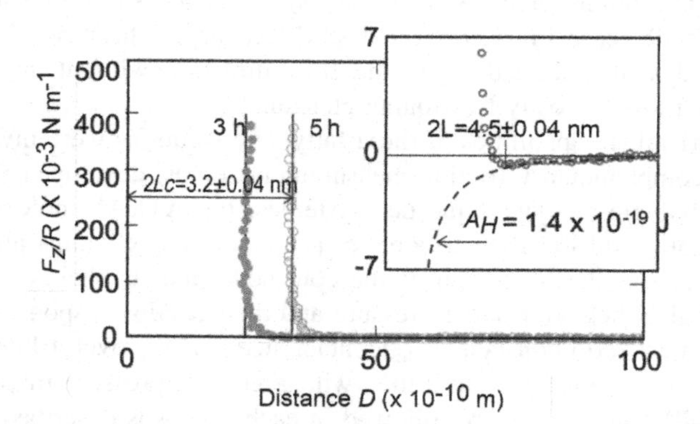

FIGURE 19.5 Evolution of the normal force, F_Z, normalized by the sphere radius, R, versus the separation distance, D, between the sphere and the plane for oleic acid solution. The adsorption kinetics was clearly shown with a thicker and more stable elastic wall after 5 h compared to 3 h (3.2 nm vs 2.1 nm). The inset presents the onset of normal force variation, F_Z/R, as well as the theoretical van der Waals force in discontinuous line for the attractive part with a Hamaker constant, $A_H = 1.4 \times 10^{-19}$ J. [21]

$$F_{\text{vdW}} = -\frac{A_H R}{6D^2} \quad (19.1)$$

- When surfaces come closer, a repulsive force appears from a distance, $2L = 4.5 \pm 0.04$ nm, corresponding to twice the length of the oleic acid molecules. The gradient of this repulsive force results from the combination of steric effects, due to the different configurations that could be taken by the molecules at the surfaces vicinity [29] and to the elastic deformation of the interface (depending on the mechanical properties of the solids and the confined layers).
- When the surfaces get even closer, the normal force increases at almost constant distance D. This thickness corresponds to the bilayer thickness, $2L_c$. Inward and outward force measurements collapse, showing no hysteresis and indicating an elastic wall. The thickness of this elastic wall was stable after 5 h.

The tangential stiffness, K_x, of the confined bilayer of oleic acid, increases from a distance corresponding to $2L_c$, to reach 11,500 N/m. From this value, the shear elastic modulus, G of the bilayer, can be calculated [29,35,40], assuming that:

- The bilayer is elastic and much less stiff than the substrate ($G_{\text{substrate}} = 51$ GPa).
- The oscillation amplitude of the dynamic measurement is small enough to avoid interfacial sliding [35].

We get a value of $G = 1.6 \pm 0.6$ MPa.

19.4 Intralamellar shear mechanisms

Friction experiments were then performed at constant normal force of 0.7 mN, corresponding to a contact pressure of the order of 30 MPa and sliding velocities ranging from 0.5 to 50 nm/s. The effect of a sudden change in sliding velocity on the friction force is illustrated in Fig. 19.6: the sliding velocity was increased from 0.7 to 1.5 nm/s, the friction force instantaneously increased before it slowly relaxed until it reached its steady-state value.

This behavior was observed for various changes in velocity in the range 0.5–50 nm/s. In addition, the interface recovers its initial state after a succession of identical steps of increasing/decreasing sliding velocities. The sliding distance beyond which the level of friction is stable, D_0, is equal to 4.8 ± 1.4 nm, regardless of the change in velocity. This friction evolution was previously observed in the literature for different confined systems [20,35,41]. The accommodation distance, D_0, was interpreted by Dieterich [42,43] as the memory distance beyond which the interface has forgotten the

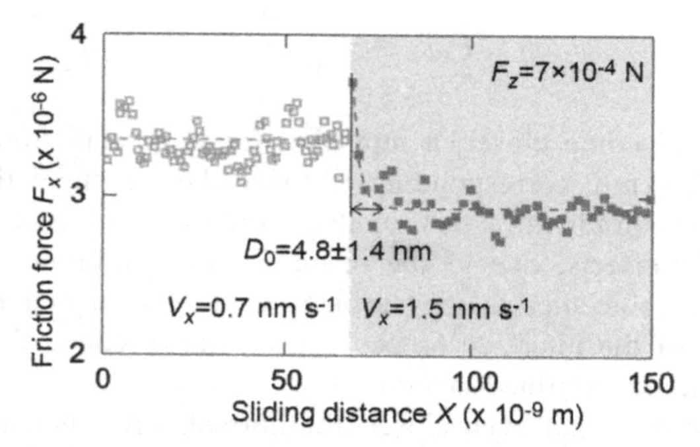

FIGURE 19.6 Typical evolution of the friction force during a sudden increase in sliding velocity under a constant normal force of 0.7 mN as a function of the sliding distance. A transient increase was observed before the force relaxed to reach a lower steady-state value. An accommodation distance, D_0, characteristic from this relaxation process, was defined at 4.8 nm, independently of the velocity.

history of the previous sliding kinematics, corresponding to the size of nanojunctions that are formed during sliding [20]. This explains why these transient effects are completely reversible.

The dependence of the friction force on the sliding velocity was plotted in Fig. 19.7. The friction force remains small over three decades of sliding velocities: it slightly decreases as the sliding velocity increases until 10 nm/s when the friction force becomes constant. In parallel, the film thickness decreases from 0.2 nm when the sliding velocity reaches 10 nm/s.

Fig. 19.8 presents the average tangential stiffness, K_x, measured during each velocity step, as a function of the sliding velocity. The stiffness first slightly decreases as the velocity increases until 10 nm/s. At large velocities, a sharp decrease in stiffness was measured. The viscous damping remained small. It was considered negligible compared to the tangential stiffness until a sliding velocity of 15 nm/s.

This correlated evolution of the friction force and the tangential stiffness with a sliding velocity up to 10 nm/s indicates that the interpenetration zone in which shear occurs, undergoes a series of molecular reorganization and dynamic phenomena of pinning/depinning, consistently with [20,35].

19.5 Formation kinetics and rheological properties of multilamellae boundary film

To get insights in the interlamellae shear mechanisms, a thick film composed of nanometric flat bilayers of fatty acids, composed of a mixture of oleic and linoleic acids, was first formed between two surfaces, one made of polished brass and the other of silica. The latter was transparent in order to visualize the film formation mechanisms, the evolution of the film thickness

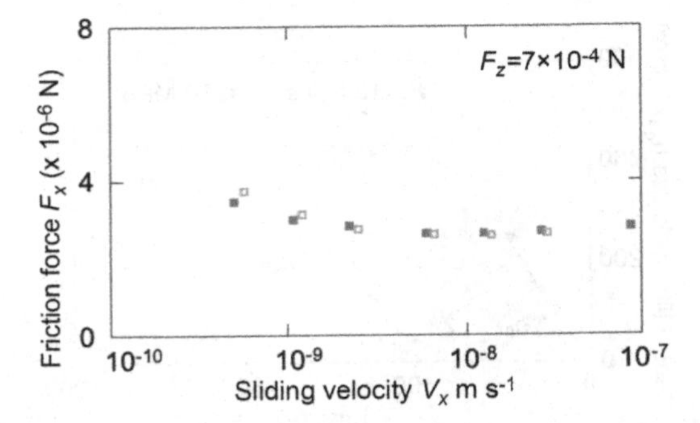

FIGURE 19.7 Variation of the steady-state friction force for sliding velocities varying from 0.5 to 50 nm/s under a constant normal force of 0.7 mN. Full symbols correspond to successive steps of increased velocity, respectively empty symbols for successive steps of decreased velocity. [21]

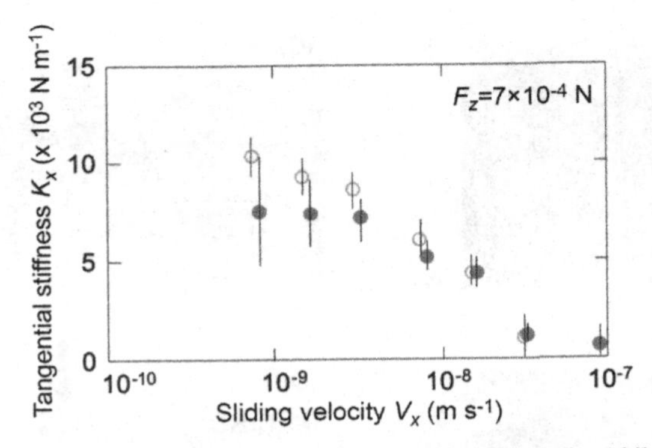

FIGURE 19.8 Variation of the average tangential stiffness of a confined bilayer of oleic acid for sliding velocities varying from 0.5 to 50 nm/s under 0.7 mN. Full symbols correspond to successive steps of increased velocity, respectively empty symbols for successive steps of decreased velocity. [21]

distribution with time [1,37,38] by means of the IRIS tribometer. The evolution of the central film thickness with time is presented in Fig. 19.9 for two different entrainment velocities under pure rolling conditions.

Surprisingly, the film thickness depends on time: [1–3,6] it first linearly increases with a slope, $(dh/dt)_{t=0}$, corresponding to the average value of film thickness increase per second (or per revolution) before it stabilizes at a plateau value, h_l, that depends on the mean entrainment velocity. The faster, the thicker: h_l (0.1 m/s) = 250 nm and h_l (0.2 m/s) = 430 nm. Contact interferograms (Fig. 19.10) at 320 s (A) and 800 s (B) confirm that the film build-up mechanisms differ from the classical eastohydrodynamic lubrication (EHL)

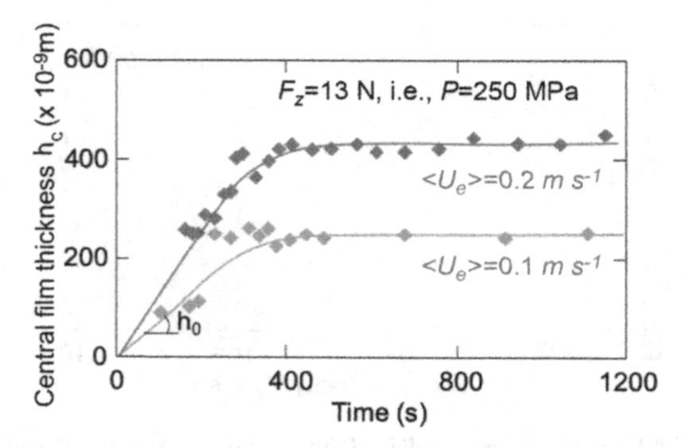

FIGURE 19.9 Evolution of the central film thickness with time for two entrainment velocities, $<U_e> = 0.1$ and 0.2 m/s under a constant normal force of 13 N, corresponding to a mean contact pressure of 250 MPa. After an initial linear increase of film thickness with time, the central film thickness reaches a plateau depending of the entrainment velocity [1–3,6].

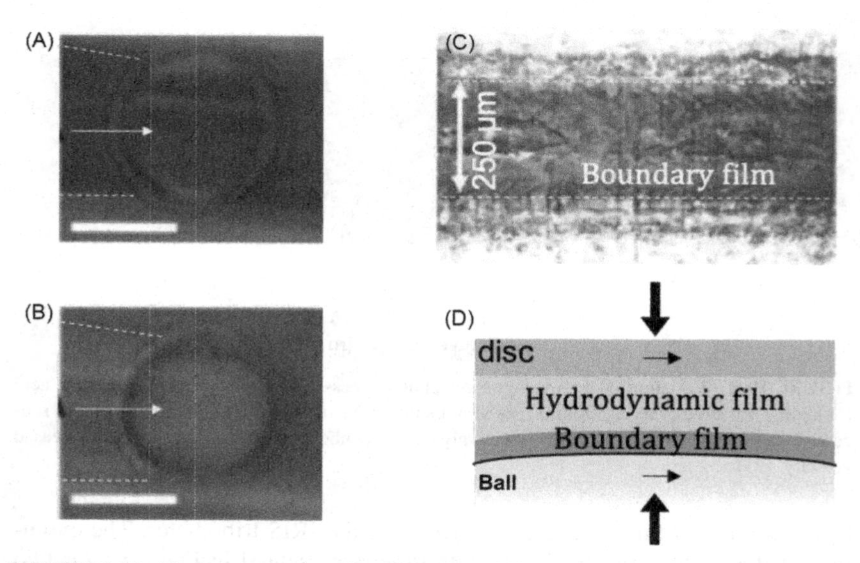

FIGURE 19.10 Typical contact interferograms (A) at 320 s and (B) 800 s under 13 N and pure rolling conditions at $<U_e> = 0.2$ m/s. The fluid is entrained from left to right as indicated by the white arrow. The contact diameter is 257×10^{-6} m, obeying Hertz theory. After the experiment, an observation of the ball surface (C) shows the presence of a boundary film adsorbed onto the surface in the contact zone. A schematic interpretation of the film build-up is presented in (D). *Adapted from H. Fay, J. Cayer-Barrioz, D. Mazuyer, O. Mondain-Monval, V. Ponsinet, S. Meeker, Lubrication mechanisms of lamellar fatty acid fluids, Tribol. Lett. 46 (2012) 285–297.*

theory. In addition to the time dependence, the expected "horse-shoe"-shape constriction was not clearly formed and is characteristic from a starved contact.

A microscopic observation of the surfaces after pure rolling experiments demonstrates the presence of a boundary film adsorbed onto the ball surface only (Fig. 19.10C). This residual film, about 200 nm thick, is adherent and covers the contact zone. The film build-up mechanisms were discussed in detail in Ref. [1]. With a higher solution pH than the point of zero charge, the surfaces are negatively charged and the fatty acids are mainly in the carboxylate form $R{-}COO^-$, as confirmed by IR spectroscopy [30]. However to maintain electroneutrality, the presence of protonated form of EDA is likely. It is known to form complexes with both copper and zinc [1], which could interact at the interface between the surface and the fatty acid lamellae. In addition, in the convergent zone of the contact, shear rates are high and beyond the transition onions/lamellas in the rheological-microstructure diagram (Fig. 19.4), resulting in a mechanism of piling up of the bilayers in the contact, facilitating the formation and growth of the boundary lamellar film. The schematic presented in Fig. 19.10D summarizes the lubricated interface formation: a boundary film adsorbed on the ball surface, due to lubricant/surface interactions, supramolecular organization, and contact pressure, with an additional hydrodynamic film, due to the entrainment conditions.

Another key information was provided by the observation of the interferograms (Fig. 19.10A and B): the contact was not fully flooded as shown by the presence of a lubricant/air meniscus in the inlet zone. This observation was consistent with the profile of film thickness distribution with the absence of "horse-shoe"-shape constriction in the high-pressure zone and the time dependence of the film thickness. This also indicates that the boundary layers formed on the surface were highly viscous. They were unable to fully flood the inlet zone, leading to starvation [1]. In this framework, according to Bouré et al. [3] and Chevalier [44], the film formation kinetics can be modeled. At the onset of the entrainment, the film thickness was assumed to increase at a constant rate, $(dh/dt)_{t=0}$. The rate of increase per second varies with the entrainment conditions: the faster, the higher. The film growth and stabilization kinetics can then be described using:

$$h_c(n) = \frac{h_0(n)h_{cff}}{\left[h_0(n)^\beta + h_{cff}^\beta\right]^{\frac{1}{\beta}}} \tag{19.2}$$

where $h_0(n)$ is the inlet film thickness at revolution n, h_{cff} the fully flooded central film thickness, and β a dimensionless parameter introduced by Chevalier [44] to characterize the starvation-induced thickness reduction. β slightly depends on the lubricant viscosity, the solids elasticity, and the working conditions. The application of this model to the experimental data can be seen in Fig. 19.9 in continuous line. It provides values of β, equal to

4.5 and 4, respectively for $<U_e> = 0.1$ and 0.2 m/s, meaning that the level of starvation increases when the velocity increases, as expected and numerically predicted [44]. The steady-state fully flooded film thickness was also given by the model: h_{cff}(0.1 m/s) = 680 nm and h_{cff}(0.2 m/s) = 1630 nm. These thick films are consistent with the highly viscous boundary-layer hypothesis.

Using Moes–Venner's theory for isoviscous fluids [45] leads to estimate the rheological properties, η_0 and α, respectively the viscosity at ambient pressure and zero shear rate and the piezoviscosity, of the boundary film. This analysis provides a viscosity, η_0=4.1 Pa s and a piezoviscosity, $\alpha = 2.10^{-9}$ Pa^{-1}. These, respectively high and low, values were also consistent with the occurrence of starvation and the aqueous nature of the lamellar lubricant.

19.6 From intralamella to interlamellae shear mechanisms

Introducing a relative sliding velocity between the two surfaces separated by a lamellar boundary film and simultaneously measuring the friction response and visualizing the film distribution in the high-pressure zone [1,37], allow one to analyze the interlamellae shear mechanisms. Fig. 19.11 presents the friction response after the boundary film formation. The film thickness remained constant during the friction experiment, at about 350 nm. This means that either the film was removed and reformed continuously, regardless of the sliding velocity, due to the constant entrainment velocity, or well withstood shear [1]. Friction increases more or less linearly as the sliding velocity increases up to 0.15 m/s. For higher values of sliding velocity, the friction increases more strongly and the values were slightly more dispersed. Nevertheless, the friction values remain of the order of magnitude of 0.01.

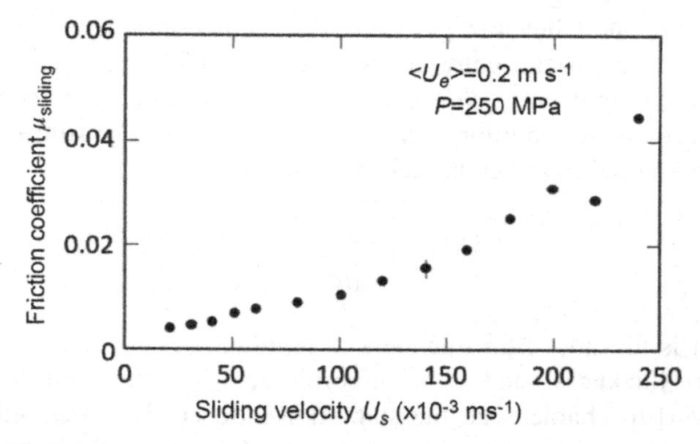

FIGURE 19.11 Friction evolution of a sheared lamellar film under a constant contact pressure of 250 MPa. The entrainment velocity was kept constant during the friction test, at 0.2 m/s and the sliding velocity was increased by steps. Each dot corresponds to an average value.

Using the value of the central film thickness, the apparent shear rate could be calculated:

$$\frac{d\gamma}{dt} = \frac{U_S}{h_c} \tag{19.3}$$

and the shear stress was:

$$\tau = \mu_{sliding} \frac{F_Z}{\pi a_H^2} \tag{19.4}$$

where $\mu_{sliding}$ is the friction coefficient and a_H is the Hertzian contact radius. Fig. 19.12 presents the associated rheogram, assuming that the frictional properties mainly depend on the rheological properties of the confined lamellar interface [46–48], since there is always a thin film that separates the surfaces.

Tentative approaches to describe and model the friction behavior of lamellar fluids were made, based on the assumption that the flow was homogeneous within ideally organized lamellae parallel to the flow direction [1]. Many descriptions were considered. The first macroscopic one accounted for the contact as a low-piezoviscous Newtonian fluid using the values of viscosity, η_0=4.1 Pa s and a piezoviscosity, $\alpha = 2.10^{-9}\,\mathrm{Pa}^{-1}$, obtained from the film establishment analysis. This leads to a predicted linear shear stress-shear rate dependence with a slope of 7–8 Pa s [1]. Even though the right order of magnitude was calculated, the linear dependence was not quite well

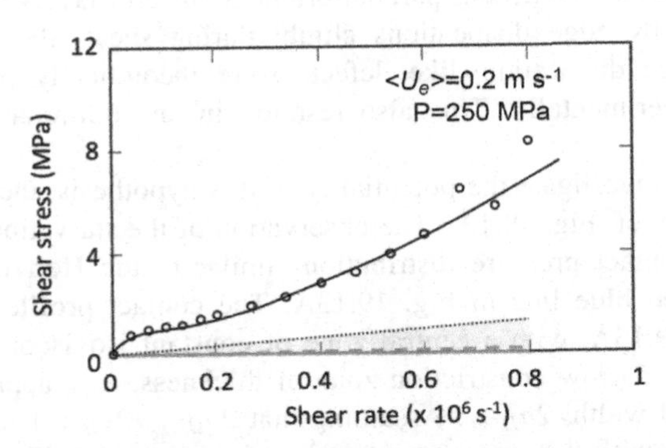

FIGURE 19.12 Rheological analysis of a sheared lamellar film under a constant contact pressure of 250 MPa. The film thickness was measured during the whole experiment. The shear stress was calculated measuring the friction coefficient and the contact area. The shear rate was calculated from the knowledge of the imposed sliding velocity and the measured central film thickness. Symbols correspond to average experimental points. Lines correspond to theoretical models: the continuous line corresponds to the case $\phi_{FA,P} = 0$, dashed line to the case $\phi_{FA,P} = 1$ and $\alpha_{FA} = 8.10^{-9}\mathrm{Pa}^{-1}$, and the dotted line to the case $\phi_{FA,P} = 1$ and $\alpha_{FA} = 15.10^{-9}\mathrm{Pa}^{-1}$ and finally the black continuous line corresponds to the 1D modeling of the friction of isoviscous lamellar fluid in a EHL contact. [1]

representative of shear stress/shear rate evolution presented in Fig. 19.12, especially at low shear rates. In addition, this view was inconsistent with the shear-thinning behavior expected for the high shear rates at which the interface is sheared all over the contact. Another description of the shear mechanisms at the microscopic level was made [1], taking into account the lamellar structure of the fluid. This approach assumes a heterogeneous shear flow in Newtonian fatty acid layer separated by aqueous Newtonian layers of water. Each Newtonian lamella was characterized by a viscosity under pressure, $\eta_{FA,P}$ and thickness δ. The aqueous Newtonian water layer was characterized by a thickness $(d-\delta)$, and a viscosity $\eta_{\text{water},P}$. The volume fraction, $\phi_{FA,P}$, corresponds to the volume fraction inside the film at local pressure. The application of such model is illustrated in Fig. 19.12 up to shear rates of $0.8 \times 10^6\ s^{-1}$ with orange-dashed line, pale blue continuous line, and dark blue dotted line, respectively for different volume fractions (covering the whole possible range 0–1) of fatty acid lamellae inside the film under pressure. It predicts very low values of shear stress compared to the experimentally measured ones. This naïve model of well-organized parallel lamellae was not representative of the real shear mechanisms occurring in the confined film.

These failed attempts to model the film shear stress dependence indicate that local effects combined to the lamellar structure need to be accounted for: from an analogy with liquid crystal shear mechanisms [23], the hypothesis of dislocation formation and motion under shear was then investigated. Indeed [23], showed that the application of a sliding velocity to the slightly inclined-top surface confining parallel-oriented smectic layers-induced plastic relaxation by edge dislocations climb: during shear, the creation and motion of edge dislocations-like defects were theoretically predicted and confirmed experimentally. This also resulted in an additional elastic load bearing.

In order to investigate the potentiality of this hypothesis, the contact was modeled in 1D (cf. Fig. 19.13). The observation of the starvation occurrence results in a contact pressure distribution similar to the Hertzian one indicated in dashed blue line in Fig. 19.13A. The contact profile can also be seen in Fig. 19.13A, with a central zone of constant isoviscous film thickness, h_c, and a narrow constriction zone of thickness, $h_{\min}$ approximated to $3h_c/4$ [49], and width, $2a_H$-x_c. Assuming that $2a_{H-x_c}/2a_H \ll 1$, as confirmed experimentally with the starved contact interferograms (cf. Fig. 19.10A and B), the contact can be approximated as in Fig. 19.13B. The two systems are equivalent in terms of hydrodynamic bearing. According to Ertel's model [50], the equivalent pressure profile is also indicated in dashed blue line in Fig. 19.13B showing a plateau of constant pressure, P_{moy}, over the contact zone with a symmetrical high-pressure gradient, over $l \ll 2a_H$, in the inlet and the outlet zones. This small value of l is also typical of starved contacts. The aqueous lamellar fluid was considered as isoviscous, consistently with

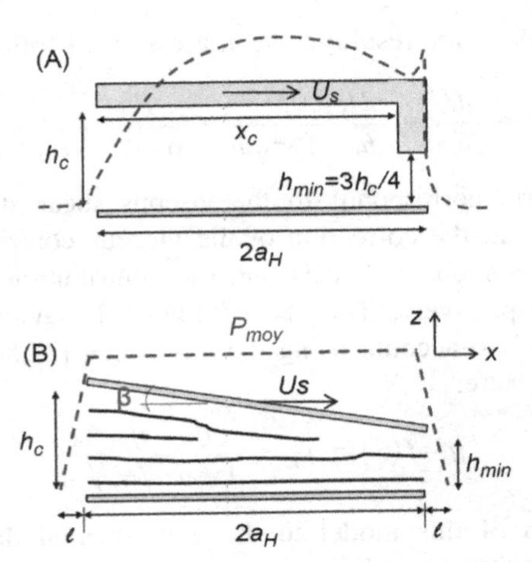

FIGURE 19.13 1D modeling of the EHL contact lubricated with an isoviscous lamellar fluid (A) schematic of the contact, (B) equivalent contact in case of starvation provided that $2a_{H-x_c}/2a_H \ll 1$ and $k = l/2a_H << 1$ the degree of starvation. The pressure profiles are indicated in both cases in dashed blue lines.

experimental observations and film thickness establishment conclusions. In addition, only the steady-state lubrication process was investigated. Once in the contact, an elementary element of fluid was subject to a compressive force, due to the diminution of the film thickness, creating a force opposed to the squeeze. The inclination angle, β, is small and equal to:

$$\beta = \frac{h_c - h_{\min}}{2a_H} = \frac{h_c}{8a_H} \tag{19.5}$$

The transient Reynolds equation is written in 1D:

$$\frac{\partial}{\partial x}\left[h^3.\frac{\partial P}{\partial x}\right] = 6\eta U_s \frac{\partial h}{\partial x} + 12\eta \frac{\partial h}{\partial t} \tag{19.6}$$

and the thickness variation at the position x is [23]:

$$\frac{\partial h}{\partial t} = -\beta(U_s - \nu) \tag{19.7}$$

where $\nu = m\sigma$, the dislocation velocity, σ the uniaxial stress associated to this motion and m the dislocation mobility.

Combining Eqs. (19.6) and (19.7) and integrating over the contact length, $2a_H$, the elastic bearing force induced by the presence and motion of the dislocations can be calculated. This contribution was then added in the calculation of the shear stress.

This model leads to the resulting shear stress, τ, as follows:

$$\tau = \frac{\eta U_s}{h_c} + \frac{\eta U_s}{h_c} \cdot \frac{h_c^2}{16 a_H m} + \frac{F_z}{64\eta k m}\left(\frac{h_c}{a_H}\right)^3 \quad (19.8)$$

where the first term corresponds to the viscous shear stress and the two others terms represent the correction of the viscous contribution due to the presence and propagation of defects and the contribution of the additional elastic bearing, respectively. The shear-thinning behavior of the lamellar fluid can be taken into account using the Cross law [37,51] to describe the viscosity under pressure:

$$\eta = f(\dot{\gamma}) = \eta_\infty + \frac{\eta_0 - \eta_\infty}{1 + \left(\dot{\gamma}/\dot{\gamma}_c\right)^n} \quad (19.9)$$

The application of this model to the experimental data, illustrated in Fig. 19.12 in continuous black line, gives access

- to the rheological behavior under pressure of the lamellar fluid with: $\eta_0 = 1.5$ Pa s, representative of the unknown lubricating phase in the contact under pressure, such as lamellae, defect concentration, etc., $\eta_\infty = 0.002$ Pa.s consistent with the value of an aqueous fluid at high shear rate, $\dot{\gamma}_c = 5.5 \times 10^4\ \mathrm{s}^{-1}$ of the same order of magnitude as that found for others lubricants [37,51];
- to the dislocation mobility, $m = 4.1 \times 10^{-12}\ \mathrm{m^3/N/s}$, lower than usual values for dislocation mobility of smectic phases [22]. Nevertheless, this lower value can be explained by the relatively high contact pressure at 250 MPa.

This model describes very well the friction response of the lamellar fluid accounting for the presence of defects, considered as dislocations induced by the confinement of the lamellar phases, at the local scale.

19.7 Superlubricity mechanisms with lamellar fluids

Two examples of superlubricity with lamellar fluids were illustrated in this work for intralamella shear and interlamellae shear.

In the case of intralamella shear, we showed that a bilayer of fatty acids was formed between the two surfaces in contact and its thickness under confinement and moderate contact pressure was smaller than twice the length of each molecule, indicating the formation of an interpenetration zone of about 1 nm. The bilayer was elastic with a shear elastic modulus of 1.6 ± 0.6 MPa. A series of steady-state and transient sliding experiments result in the determination of the size of nanojunctions, of about 4.8 nm of diameter, within the interpenetration zone, corresponding to the accommodation distance beyond which the sliding history was forgotten. The evolution of the tangential stiffness with the sliding velocity as well as the sliding dependence of the friction confirmed that local dynamics resulted from the kinetics of formation and rupture of adhesive bonds, associated with the memory distance.

For a more complex stacking of lamellae in the contact, we described the agglomeration process of these lamellae onto the surfaces leading to an adherent thick boundary film. The low viscous friction was attributed to the low piezoviscosity of the fluid. Unsuccessful attempts to explain the friction mechanisms at the microscopic level, taking into account the presence of lamellae, were first made before a more detailed theoretical model was proposed, based on both the shear-thinning rheological behavior of the lamellar fluid and the existence of edge dislocations-like defects in the lamellar structure that modify the viscous flow and promote an additional elastic bearing. This model well describes the physics at stake during the shear of lamellar fluids.

Acknowledgments

This work was supported by the Agence Nationale de la Recherche via the project Confluence ANR-13-JS09-0016-01, by the LabEx Manutech-Sise (ANR-10-LABX-0075) within the program "Investissements d'Avenir" (ANR-11-IDEX-0007), via the project DysCo, by CNRS and Solvay. Special thanks are due to Prof. O. Mondain-Monval and Dr. V. Ponsinet from CRPP and Dr. S. Meeker from Solvay.

References

[1] H. Fay, J. Cayer-Barrioz, D. Mazuyer, O. Mondain-Monval, V. Ponsinet, S. Meeker, Lubrication mechanisms of lamellar fatty acid fluids, Tribol. Lett 46 (2012) 285–297.

[2] P. Bouré, Lubrification et Usure du Contact Fil/Outil en Tréfilage Humide (Ph.D. thesis), Ecole Centrale de Lyon, 1999.

[3] P. Bouré, D. Mazuyer, J.M. Georges, A.A. Lubrecht, G. Lorentz, Formation of boundary layers with water-based lubricant in a concentrated elastohydrodynamic contact, Trans. ASME 124 (2002) 91–102.

[4] S. Hollinger, Comportement d'un Lubrifiant Aqueux dans un Contact à Très Hautes Pressions. Application au Tréfilage de Fils d'Acier Laitonnés (Ph.D. thesis), Ecole Centrale de Lyon, 1999.

[5] S. Hollinger, J.M. Georges, D. Mazuyer, G. Lorentz, O. Aguerre, Du Nguyen, High pressure lubrication with lamellar structures in aqueous lubricant, Tribol. Lett. 9 (3-4) (2000) 143–151.

[6] H. Fay, Films Lubrifiants Supramoléculaires Organisés: de la Microstructure aux Propriétés Tribologiques (Ph.D. thesis), Université de Bordeaux, 2011.

[7] W. Helfrich, Amphiphilic mesophases made of defects, in: R. Balian, M. Kléman, J.P. Poirier (Eds.), Physics of Defects, Amsterdam, The Netherlands, North Holland, 1980.

[8] C.R. Safinya, E.B. Sirota, R.F. Bruinsma, C. Jeppesen, R.J. Plano, L.J. Wenzel, Structure of membrane surfactant and liquid crystalline smectic lamellar phases under flow, Science, N. Ser. 261 (5121) (1993) 588–591.

[9] G.K. Auernhammer, H.R. Brand, H. Pleiner, The undulation instability in layered systems under shear flow—a simple model, Rheol. Acta 39 (1999) 215–222.
[10] D. Roux, F. Nallet, O. Diat, Rheology of lyotropic lamellar phases, Europhys. Lett. 24 (1993) 53–58.
[11] C. Meyer, S. Asnacios, C. Bourgaux, M. Kléman, Rheology of lyotropic and thermotropic lamellar phases, Rheol. Acta 39 (2000) 223–233.
[12] O. Diat, D. Roux, F. Nallet, Effect of shear on a lyotropic lamellar phase, J. Phys. II 3 (1993) 1427–1452.
[13] W.B. Hardy, I. Doubleday, Boundary lubrication—the paraffin series, Proc. R. Soc. Lond. Ser. A 100 (1922) 550–574.
[14] M. Ruths, S. Steinberg, J.N. Israelachvili, The effects of confinement and shear on the properties of thin films of thermotropic liquid crystals, Langmuir 12 (1996) 6637–6650.
[15] B. Cross, J. Crassous, Rheological properties of a highly confined film of lyotropic lamellar phase, Eur. Phys. J. E14 (2004) 249–257.
[16] K. Boschkova, B. Kronberg, M. Rutland, T. Imae, Study of thin surfactant films under shear using the tribological surface force apparatus, Tribol. Int. 34 (2001) 815–822.
[17] U. Raviv, J. Klein, Fluidity of bound hydration layers, Science 297 (2002) 1540–1543.
[18] C. Drummond, J. Israelachvili, P. Richetti, Friction between two weakly adhering boundary lubricated surfaces in water, Phys. Rev. E 67 (2003) 066110.
[19] J.M. Georges, A. Tonck, D. Mazuyer, Interfacial friction of wetted monolayers, Wear 175 (1994) 59–62.
[20] D. Mazuyer, J. Cayer-Barrioz, A. Tonck, F. Jarnias, Friction dynamics of confined weakly adhering boundary layers, Langmuir 24 (8) (2008) 3857–3866.
[21] A. Crespo, Compréhension de la Tribologie en Films Limites: de l'Organisation Moléculaire à la Réponse en Friction (Ph.D. thesis), Ecole Centrale de Lyon, 2017.
[22] P. Oswald, Rhéophysique ou Comment Coule la Matière, first ed., Belin (2005).
[23] P. Oswald, M. Kléman, Lubrication theory of smectic A phases, J. Phys. Lett. 43 (12) (1982) 411–415.
[24] T.E. Fisher, S. Bhattacharia, R. Sahler, J. Lauer, Y.J. Ahn, Lubrication by a smectic liquid crystal, Tribol. Trans. 31 (1988) 442–448.
[25] F.E. Lockwood, M.T. Benchaita, S.E. Friberg, Study of lyotropic liquid crystals in viscometric flow and elastohydrodynamic contact, ASLE Trans. 30 (4) (1987) 539–548.
[26] S.E. Friberg, A.J. Ward, F.E. Lockwood, Lyotropic liquid-crystals in lubrication, in: G. Biresaw (Ed.), Tribology and the Liquid-Crystalline State, ACS Symposium Series, vol., 441, American Chemical Society, Washington, DC, 1990.
[27] H.S. Lee, S.H. Winoto, W.O. Winer, M. Chiu, S.E. Friberg, Film thickness and frictional behavior of some liquid crystals in concentrated point contacts, in: G. Biresaw (Ed.), Tribology and the Liquid-Crystalline State, ACS Symposium Series, vol., 441, American Chemical Society, Washington, DC, 1990.
[28] K. Boschkova, J. Elvesjö, B. Kronberg, Frictional properties of lyotropic liquid crystalline mesophases at surfaces, Colloids Surf. A: Physicochem. Eng. Asp. 166 (2000) 67–77.
[29] A. Crespo, N. Morgado, D. Mazuyer, J. Cayer-Barrioz, Effect of unsaturation on the adsorption and the mechanical properties of fatty acid layers, Langmuir 34 (2018) 4560–4567.
[30] H. Fay, S. Meeker, J. Cayer-Barrioz, D. Mazuyer, I. Ly, F. Nallet, et al., Polymorphism of natural fatty acid liquid crystalline phases, Langmuir 28 (2012) 272–282.
[31] O. Diat, D. Roux, F. Nallet, Layering effect in a sheared lyotropic lamellar phase, Phys. Rev. E 51 (4) (1995) 3296–3299.

[32] R. Weigel, J. Laüger, W. Richtering, P. Lindner, Anisotropic small angle light and neutron scattering from a lyotropic lamellar phase under shear, J. Phys. II 6 (4) (1996) 529–542.

[33] J. Zipfel, P. Lindner, M. Tsianou, P. Alexandridis, W. Richtering, Shear induced structures in lamellar phases of amphiphilic block copolymers, Langmuir 15 (1999) 2599–2602.

[34] J. Zipfel, J. Berghausen, G. Schmidt, P. Lindner, P. Alexandridis, W. Richtering, Influence of shear on solvated amphiphilic block copolymers with lamellar morphology, Macromolecules 35 (2002) 4064–4074.

[35] D. Mazuyer, A. Tonck, J. Cayer-Barrioz, Friction control at the molecular level: from superlubricity to stick-slip, in: A. Erdemir, J.M. Martin (Eds.), Superlubricity, Elsevier B. V, Oxford, UK, 2005.

[36] A. Crespo, D. Mazuyer, N. Morgado, A. Tonck, J.M. Georges, J. Cayer-Barrioz, Methodology to characterize rheology, surface forces and friction of confined liquids at the molecular scale using the ATLAS apparatus, Tribol. Lett. 65 (2017) 138.

[37] M. Diew, A. Ernesto, J. Cayer-Barrioz, D. Mazuyer, Stribeck and traction curves under moderate contact pressure: from friction to interfacial rheology, Tribol. Lett. 57 (2015) 8.

[38] A. Ernesto, D. Mazuyer, J. Cayer-Barrioz, From full-film lubrication to boundary regime in transient kinematics, Tribol. Lett. 59 (2015) 23.

[39] J.N. Israelachvili, Intermolecular and Surface Forces, third ed., Elsevier, 2011.

[40] E. Gacoin, C. Fretigny, A. Chateauminois, A. Perriot, E. Barthel, Measurement of the mechanical properties of thin films mechanically confined within contacts, Tribol. Lett. 21 (2006) 245–252.

[41] C. Drummond, J. Israelachvili, Dynamic behavior of confined branched hydrocarbon lubricant fluids under shear, Macromolecules 33 (13) (2000) 4910–4920.

[42] J.H. Dieterich, Modeling of rock friction: 1. Experimental results and constitutive equations, J. Geophys. Res. Solid. Earth 84 (B5) (1979) 2161–2168.

[43] J.H. Dieterich, B.D. Kilgore, Direct observation of frictional contacts: new insights for state-dependent properties, Pure Appl. Geophys. 143 (1) (1994) 283–302.

[44] F. Chevalier, Modélisation des Conditions d'Alimentation dans les Contacts EHD Ponctuels (Ph.D. thesis), INSA de Lyon (1996).

[45] G.W. Stachowiak, A.W. Batchelor, Engineering Tribology, Tribology Series, vol. 24, Elsevier, 1993.

[46] E. Bou-Chakra, J. Cayer-Barrioz, D. Mazuyer, F. Jarnias, A. Bouffet, A non-Newtonian model based on Ree-Eyring theory and surface effect to predict friction in elastohydrodynamic lubrication, Tribol. Int. 43 (2010) 1674–1682.

[47] H. Spikes, Z. Jie, History, origins and prediction of elastohydrodynamic friction, Tribol. Lett. 56 (1) (2014) 1.

[48] V. Jadhao, M.O. Robbins, Rheological properties of liquids under conditions of elastohydrodynamic lubrication, Tribol. Lett. 67 (2019) 66.

[49] D. Dowson, G.R. Higginson, Elastohydrodynamic Lubrication, Pergamon, 1977.

[50] A. Mohrenstein-Ertel, Die Berechnungtierhydrodynamischen Schmierunggekrummter oberflaschen unter hoher Belastung und Relativbewegung, Fortschr. Bet. VDIZ., Reihe 1, No 115, Dsseldorf, ISBN3-18-141501-4, 1984.

[51] M.M. Cross, Polymer rheology: influence of molecular weight and polydispersity, J. Appl. Polym. Sci. 13 (1969) 765–774.

Chapter 20

Superlubricity with nonaqueous liquid

Chenhui Zhang[1], Ke Li[2] and Jianbin Luo[1]
[1]State Key Laboratory of Tribology, Tsinghua University, Beijing, P.R. China, [2]Intelligent Transport Systems Research Center, Wuhan University of Technology, Wuhan, P.R. China

20.1 Introduction: the challenge for oil-based superlubricity

Nonaqueous liquids, mainly consisting of oil-based liquids, are usually stable with changing environment, making them ideal lubricants for diverse practical applications. But it is difficult to achieve superlubricity with such nonaqueous liquids. The main aim of lubricants is to bear normal load to separate rubbing solid surfaces and avoid asperity-asperity contacts. Aqueous solution confined within two solid surfaces can provide strong surface repulsion such as electric double layer force and hydration force [1–3], while the repulsion force under nonaqueous liquid mainly derives from hydrodynamic effect in general. It requires high viscosity of nonaqueous liquid to enhance the hydrodynamic effect, which, however, increases the friction in full film lubrication according to Newton's law of viscosity.

$$\tau = \eta \frac{du}{dz}$$

In view of the Stribeck curve, the minimum friction coefficient exists in the transition stage from boundary lubrication to hydrodynamic lubrication, where the lubrication film is very thin. In addition, the pressure-viscosity coefficient of nonaqueous liquid is usually of large value. Therefore the viscosity will be higher and make it difficult to reduce friction coefficient to an ultralow level.

Formation of elasto-hydrodynamic lubrication (EHL) film is the dominant mechanism for the friction behavior of nonaqueous lubricants, for which we can make an approximate prediction [4]. According to the Hamrock–Dowson theory [5–7], the film thickness at the center of contact region in EHL

Superlubricity. DOI: https://doi.org/10.1016/B978-0-444-64313-1.00020-X

between ball and disk can be estimated, as described by the following equation.

$$h_c = 2.69\frac{\alpha^{0.53}\eta_0{}^{0.67}u^{0.67}R^{0.464}}{E^{0.073}w^{0.067}}\left(1 - 0.61e^{-0.73k}\right)$$

where α is the pressure-viscosity coefficient of the lubricant, η_0 is the bulk viscosity of the lubricant, w is the applied normal load, u is the average linear speed of the substrate and ball, k is ellipticity parameter and is set as 1.03 for circular contact to express the side-leakage factor [5]. R is the equivalent radius of the ball that can be described by the Hertz contact theory, as shown in the following equation.

$$R = \frac{2Eb^3}{3w}$$

where b is the radius of contact region and E is the effective elastic modulus of the friction pair defined by

$$E = \left[\frac{1}{2}\left(\frac{1-\nu_1^2}{E_1} + \frac{1-\nu_2^2}{E_2}\right)\right]^{-1}$$

where E_i and ν_i are the elastic modulus and Poisson's ratio of different solids. The Hamrock–Dowson equation of central film thickness can be simply regarded as average film thickness and used to predict friction coefficient of EHL. The friction coefficient can be approximately described by the following expression.

$$\mu_f = \frac{\tau A}{w}$$

where $A = \pi b^2$ is the area of contact region and τ is the shear stress of the confined fluid. Barus law [8] is applied to describe the variation of lubricant viscosity with changing pressure, giving the shear stress expression as follows:

$$\tau = \eta_0 e^{\alpha p}\frac{u}{h_c}$$

where p is the average contact pressure, following the equation as follows:

$$p = \frac{w}{A}$$

The aforementioned equations can be synthesized to establish the friction coefficient prediction of EHL, which is given by the following equation.

$$\mu_f = 1.397\frac{e^{\alpha p}\eta_0{}^{0.33}u^{0.33}}{E^{0.391}\alpha^{0.53}w^{0.165}p^{0.304}}$$

The exponent term of αp indicates the significant impact of pressure-viscosity coefficient and contact pressure on the friction coefficient of EHL.

Besides, the surface roughness is also considered statistically. It is assumed that the height of surface profile with RMS roughness σ follows Gaussian distribution. And the probability distribution function of surface profile is as follows:

$$P(h) = \frac{1}{\sqrt{2\pi}\sigma} e^{-\frac{h^2}{2\sigma^2}}$$

Thus the percentage of asperity contact load can be calculated by

$$\frac{w_c}{w} = \int_{h_c}^{\infty} P(h)dh$$

The friction coefficient of asperity contact region μ_c is estimated as 0.15. Therefore the total friction coefficient is determined by the following equation.

$$\mu = \mu_f\left(1 - \frac{w_c}{w}\right) + \mu_c \frac{w_c}{w}$$

In order to make a simple prediction for tribological behavior of nonaqueous lubricants, it is assumed that friction pair is made of sapphire and the effective elastic modulus is 442.24 GPa. Comprehensive RMS roughness of two surfaces is 5 nm. Bulk viscosity of lubricant is set as 30 mPa s to fit the value of a typical base oil—polyalpha olefin (PAO)4. Applied normal load is 3 N and sliding velocity is 0.1 m/s. All parameters except α and p are given. Friction coefficients are calculated when $0.8\ \text{GPa}^{-1} \leq \alpha \leq 30\ \text{GPa}^{-1}$ and $1\ \text{MPa} \leq p \leq 500\ \text{MPa}$. Superlubricity region is highlighted where friction coefficient is lower than 0.01, as shown in Fig. 20.1.

These calculation results confirm that liquid superlubricity is closely linked to the pressure-viscosity coefficient and contact pressure. Under high contact pressure, superlubricity can be achieved only when pressure-viscosity coefficient is small. However the pressure-viscosity coefficient of nonaqueous liquids is usually high and makes it difficult to realize ultralow friction under high contact pressure. Let us take PAO4 as example to investigate the friction coefficients at a wide range of contact pressure and sliding velocities. The pressure-viscosity coefficient is set as $18\ \text{GPa}^{-1}$ and sliding velocity ranges from 1 to 200 mm/s. Calculation results are shown in Fig. 20.2.

As depicted in the figure, superlubricity cannot be achieved under contact pressure higher than 270 MPa. At very low sliding velocity, hydrodynamic effect is so weak that asperity contact exists and friction coefficient is of large value even under low contact pressure. At higher sliding velocity, the critical contact pressure to achieve superlubricity decreases with the increase of velocity. Shear stress of nonaqueous fluid is high because of high viscosity (compared with water and aqueous solutions) and the growing sliding velocity. Even though friction coefficient lower than 0.01 can be achieved with nonaqueous liquid under some certain circumstances, it remains a

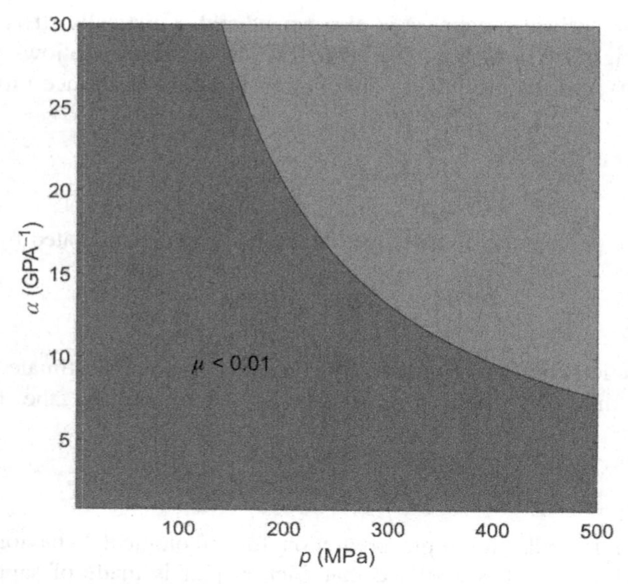

FIGURE 20.1 Superlubricity region from hydrodynamic effect.

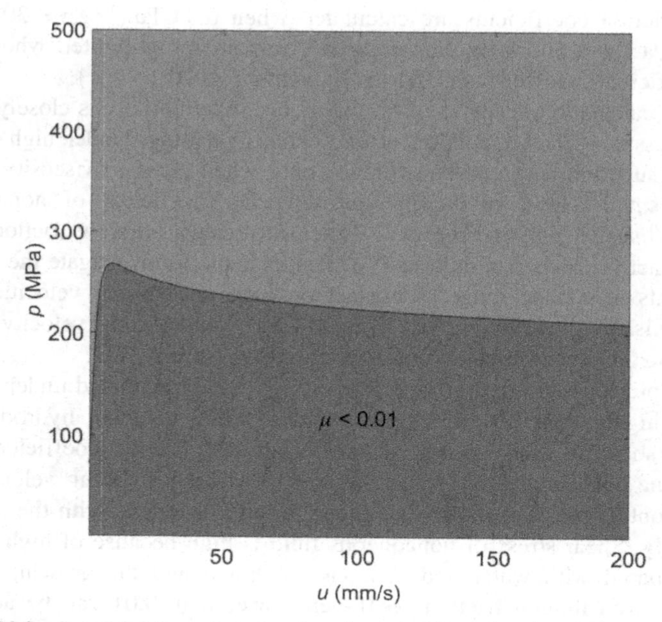

FIGURE 20.2 Superlubricity region when $1\text{mm/s} \le u \le 200\text{mm/s}$ and $1\text{MPa} \le p \le 500\text{MPa}$.

challenge to extend the superlubricity region to a wide range of velocity and pressure.

20.2 Theoretical analysis of confined thin oil films

According to the Stribeck curve, the minimal coefficient of friction (COF) appears during the transition from boundary lubrication to fluid lubrication regime, where the oil fluid is strongly confined into a very thin film as thin as several nanometers. In such thin film, the influence of solid surface on the molecules of lubricant cannot be neglected and the properties of such thin film are different from bulk liquid. In this section, we introduce the properties of confined thin oil film and their influence on attaining superlubricity.

20.2.1 Dynamic rheological model of thin-film lubrication

In 1996, Luo et al. [9] proposed a new lubrication state called "thin-film lubrication (TFL)" to describe the transition stage from EHL to boundary lubrication. The lubrication phenomena in TFL are different from those in EHL, in which the film thickness is strongly related to the speed, and from that in boundary lubrication in which the film thickness is mainly determined by molecular dimension and chemical characteristics of the lubricant.

The TFL model proposed by Luo et al. [10] has three different layers (Fig. 20.3). Specifically, layer close to the surfaces is the adsorbed layer which is depended on the physical and chemical interaction between the lubricating molecules and the solid surfaces. The layer furthest from the solid surface or in the central between the two solid surfaces is the fluid layer. It is formed by hydrodynamic effects and its thickness mainly depends on the velocity and the viscosity of liquids. The ordered layer, which is between the absorbed layer and the fluid layer, comes from the inducing effect of adsorbed molecules on the orientation of nearby liquid molecules. The stronger the polar end of the liquid molecules, the greater the thickness of the ordered layer. The variation of the thickness ratio of these three kinds of layers will dominate the transition of lubrication regimes between the two solid surfaces. If the solid surfaces are far apart from each other, the fluid

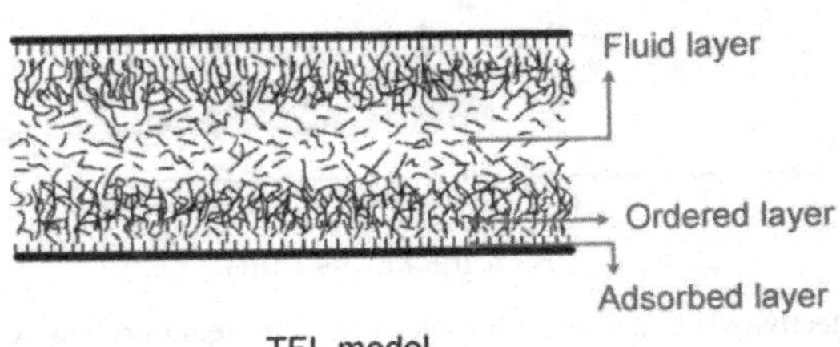

FIGURE 20.3 Thin-film lubrication model [10].

layer will become thick enough and the tribopair will act under hydrodynamic effects in terms of speed, pressure, lubricant viscosity, etc., and remain highly fluid in the EHL regime. If the two solid surfaces are so close to each other, that is, the lubricant film becomes very thin or there are only two or three molecular layers, it is the standard Hardy's definition of boundary lubrication, where the molecules are well firmly adsorbed on the solid surfaces, resulting in a solid layer. Thus a boundary lubrication regime will appear and it will be independent of the lubricant viscosity, speed, and so on. If the gap between two solid surfaces is a few of nanometers to tens of nanometers, the presence of a surface force and the orientation inducing effect of adsorbed molecules will induce the formation of the ordered layer, in which the order of the molecular arrangement gradually becomes weaker as the distance from the solid surface increases. The properties of this lubricating layer determine the performance of the entire tribopair. The film thickness and tribological properties in the TFL regime are different from those in both the EHL and boundary lubrication regimes, and they are affected by a number of factors, such as the hydrodynamic effect, the molecular size, and the molecular polarization.

The lubrication in the TFL regime has been regarded as the transition between the EHL and boundary lubrication which will exhibits a clearly observable higher effective viscosity than that of free molecules in experiments as shown in Fig. 20.4 [9,11]. However if the shear can take place just between the two ordered layers, a minimum COF and a low-COF region could be gotten due to its lower energy dissipation. Huang et al. [12] attributed the transition area on the Stribeck curve for smooth surfaces to a thin-

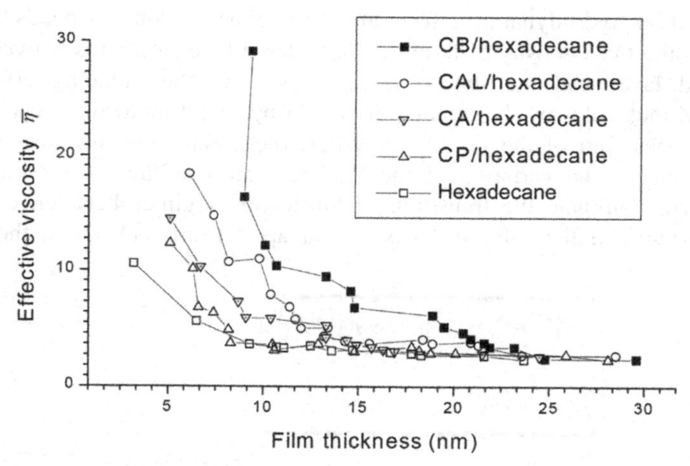

FIGURE 20.4 Effective viscosity with film thickness with liquid crystals [cholesteryl acetate (CA), cholesteryl pelargonate (CP), cholesteryl benzoate (CB), and cholesteryl acrylate (CAL)] concentration of 2 wt.% in hexadecane under the pressure of 0.174 GPa [11].

film lubrication state. It can be seen from Fig. 20.5 [13] that the Stribeck curve with lower surface roughness has a significantly lower minimum COF, a wider low-COF region, and sharper transition slopes than those of rough surfaces under the same normal load and sliding velocity. For a smooth surface in this transition regime, the film thickness typically ranges from 10 to 1000 nm. A lubricant's rheological and interfacial properties change when the confinement interspace is reduced to micro- or nanoscale.

Zhang et al. [13] proposed a dynamic rheology of the lubricant from the hydrodynamic lubrication regime to the TFL regime which is based on the flowing Maxwell constitutive equation. This rheology model includes the increased relaxation time and the yield stress of the confined lubricant thin film, as well as their dependences on the lubricant film thickness. Two proposed parameters in the dynamic rheological model, namely negative slipping length b and the characteristic relaxation time λ_0. The former is an indicator of the interfacial effect of the boundary layer, while the latter is an indicator of the viscoelasticity or solid-like phase of the confined thin-film layer. The negative slipping length b, which can be considered as the thickness of the boundary layer and ordered layer in TFL model, can be a quantitative indicator to describe the effect of the lubricant-substrate interaction on the lubrication properties. A lower b value indicates that the lubricant can be thoroughly adsorbed or spread onto the substrate to form a thin layer. The relaxation time of a liquid or a lubricant is the ratio of the viscosity to the elastic modulus of the liquid.

The negative slip length b plays a significant role in the COF curves of the thin-film lubrication regime (Fig. 20.6). As the negative slip length gradually decreases to 25 nm, the entire Stribeck curve shifts left on the velocity axis, the minimum COF decreases from approximately 0.05 to 0.015, and the low-COF regime widens. When b is less than 25 nm, the lubricant is closely adsorbed onto the substrate surface, resulting in a rather stable low-COF

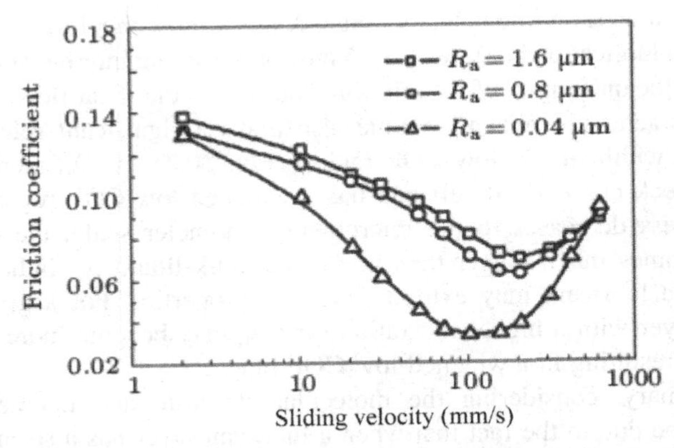

FIGURE 20.5 Typical experimental Stribeck curves for tribopairs of different roughness [13].

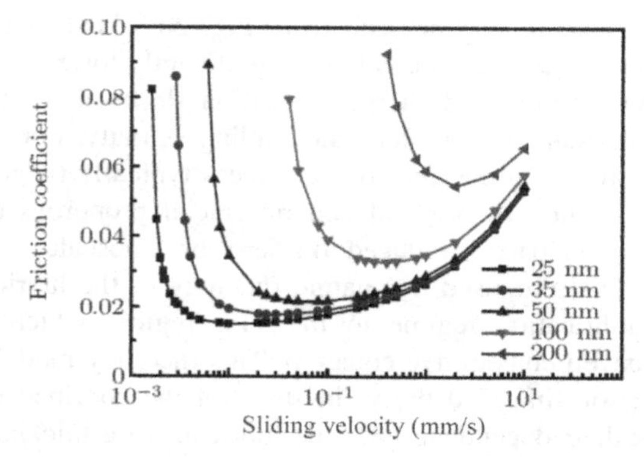

FIGURE 20.6 Stribeck curves calculated for different negative slip lengths b [13].

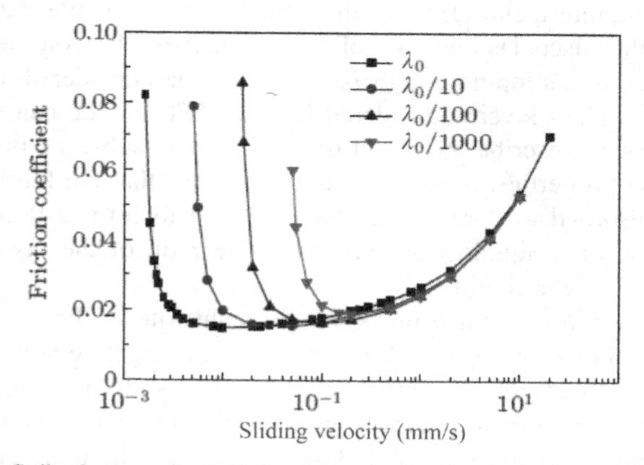

FIGURE 20.7 Stribeck curves for different characteristic relaxation time λ_0 [13].

regime over a wide sliding velocity range. A lubricant with a low b is preferred for thin-film lubrication because a low b indicates a strong interfacial interaction to maintain the uniformity of the thin film and sustain the shear flow.

The characteristic relaxation time also plays a significant role in determining the width of the low-COF region (Fig. 20.7). As λ_0 increases, the entire Stribeck curve shifts left and has a widened low-COF region. As the film thickness decreases to the micrometer/nanometer scale, the relaxation time λ becomes much longer than that of the bulk liquid λ_0, indicating that the confined lubricant may exhibit solid-like properties. For a given b, the thin-film layer with a higher relaxation time λ_0 may become more solid-like or ordered, resulting in a widened low-COF range.

In summary, considering the molecular structure and interaction, this result may be due to the fact that when a lubricant layer has a strong interfacial interaction with the substrate and also has a strong intermolecular interaction, a TFL with much lower friction than ordinary EHL will be promoted and maintained over a rather wide velocity range.

20.2.2 Molecular dynamics simulation of confined fluids

The central idea of TFL theory is the solid surface force- and adsorbed molecules orientation effect induced molecular ordering and packing. If a surface could be prepared so uniform that the fluid-surface potential of interaction was everywhere the same, this should be observable as very low friction (LF) when the solid surfaces are separated by fluid. This intuitive expectation [14] has been confirmed by numerous molecular dynamics (MD) simulations. However other experiments cast doubt on the generality of these conclusions. Experiments using the surface forces apparatus report that when nonpolar fluids are confined between atomically smooth mica sheets to a thickness of $<5-10$ molecular dimensions, the effective shear viscosity increases to the point that the frictional response turns solid like [15–17]. It was at first supposed that the fluids might be commensurate with the mica surface lattice, thus pinning near-surface fluid molecules [18]. This cannot be reconciled with the fact that confinement-induced solidification has been reported using a large family of confined fluids of different chemical structures and therefore different length scales.

In order to solve this paradox, Zhu et al. [19] used the method of Frantz and Salmeron [20] to cleave mica, investigated alkane fluids in a surface forces apparatus and confirm several predictions of MD simulation. Fig. 20.8 shows the force-distance profile for squalane, a branched alkane with a C24 backbone and six symmetrically placed methyl groups. As the surfaces were squeezed together, squalane drained smoothly until oscillatory forces of alternating attraction and repulsion were first detected at thickness ≈ 4 nm. This

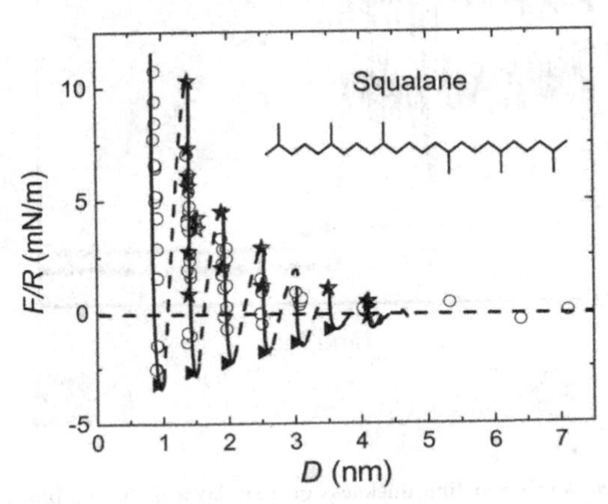

FIGURE 20.8 Static force-distance profile of squalane at room temperature of 25°C [19].

reflects the tendency of squalane to form parallel layers to the solids; application of pressure caused it to drain in discrete steps corresponding to squeezing out of successive layers. The liquid could ultimately be squeezed to ≈ 0.9 nm, twice the thickness of the chain backbone. There is agreement even in the details, the magnitudes of the force maxima exceed the magnitudes of the force minima and the magnitudes of the force maxima grow more strongly with decreasing separation than the magnitudes of the force minima.

Based on this layered structure, Jabbarzadeh et al. [21], to analyze the interplay between layering and friction, studied nonequilibrium MD (NEMD) simulations on a thin film of dodecane confined between model mica surfaces (Fig. 20.9). After 57 ns of shearing, a spontaneous and irreversible transition to a new low stress state was observed. This transition to low shear stress was accompanied with a sudden decrease of the film thickness to 2.8 nm. The disordered film structure after application of shear at a high-friction (HF) regime is converted to a molecular orientation in the layer after transitioning to a LF regime. This transition from the HF state to the LF state of the film was accompanied by a nematic-like alignment among the molecules as shown in the configuration "snapshot" in Fig. 20.8. The alignment in the LF state in this case involved five of the six layers and was parallel with the shear flow direction.

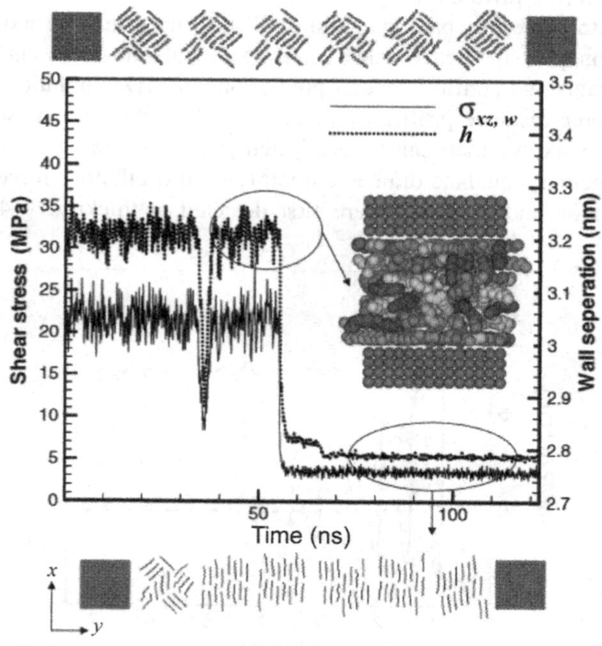

FIGURE 20.9 Shear stress and film thickness of a six-layer dodecane film that is undergoing shear at shear rate 10^{11} s^{-1} [21].

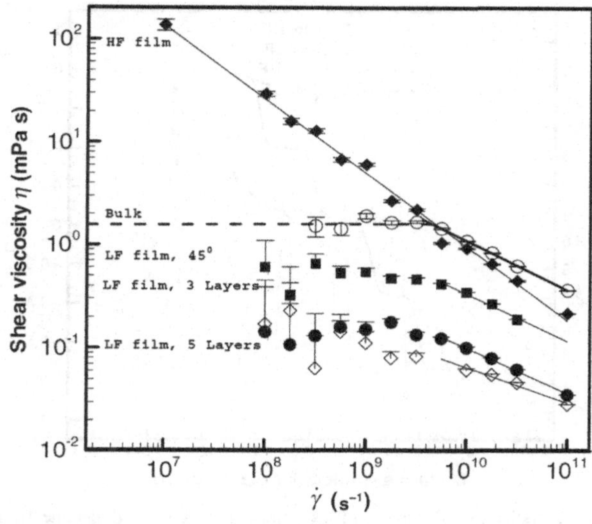

FIGURE 20.10 The shear viscosity is plotted against the shear rate for a bulk and confined dodecane films [21].

Fig. 20.10 presents the shear viscosity as a function of strain rate. To characterize the various LF states, they started with configurations generated at high shear rates and then examined their rheology over a range of lower shear rates. The viscosity of the LF state is found to decrease with an increasing degree of molecular alignment along the flow direction.

In Fig. 20.11, the average velocity and density profile through the HF and LF films at a strain rate of $10^{11}\ s^{-1}$ were compared. Neither state exhibits any sign of significant wall slip. The aligned LF state involves layer slippage throughout the dodecane film while the HF state exhibits a uniform shear gradient. While for the disordered HF film the velocity is almost linear at the middle of the film, for the LF film the velocity profile is stepwise, indicating the slip happening between the layers of the film.

In conclusion, application of a shear stress tends to align the molecules, giving a reduction in the apparent viscosity which can lead to the surface- and friction-induced molecular ordering and packing. An ultralow friction can be induced by the ultralow effective viscosity under a strong shear, which suggests that superlubricity obtained using oils is possible under certain conditions.

20.3 Some experimental achievements of oil-based superlubricity

20.3.1 1,3-Diketone oil on steel surfaces (thin-film lubrication)

Section 20.2 suggested that an ordered thin oil film may lead to superlubricity in the TFL regime under certain conditions of surface forces,

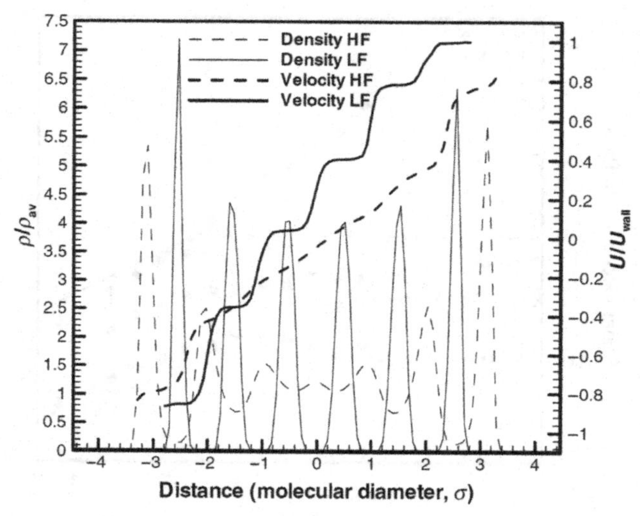

FIGURE 20.11 Velocity and density profiles across the six-layer dodecane film at high and low friction regimes at a shear rate of 10^{11} s^{-1} [21].

intermolecular forces, and shear stresses. However such oil-based superlubricity is mainly investigated in simulation work, and has never been experimentally obtained until the research on 1,3-diketone oil.

Since 2010 Amann et al. [22–24] and Li et al. [25–28] initiated the study of a synthetic oil 1,3-diketone, which then attracted the research attention of Zhang et al. [29,30]. 1,3-Diketone could reach a macroscopic ultralow COF (0.005) on steel surfaces under ambient atmosphere [22]. This research was the first reported oil-based superlubricity achieved through experimental methods. The typical studied 1,3-diketone is 1-(4-*e*thyl*p*henyl) *b*utane-1,3-*d*ione (EPBD), which has 2 and 6 carbons on the left and right side (Fig. 20.12). Friction tests were performed using a reciprocating cylinder-on-disc sliding geometry (SRV-III). Under a normal load of 50 N with a high frequency of 50 Hz in a stroke of 1 mm, EPBD reached an ultralow COF of 0.005 after a short running-in period on steel surfaces (100 Cr6) at 90°C, which was much lower than other reference oils [23]. After the 20 h tribotest, the color of EPBD changed from initially yellow to red (Fig. 20.13A). FTIR and UV-vis spectra proved the chemical reaction of EPBD into an iron complex during the friction test (Fig. 20.13B) [25].

Wear analysis using stylus profilometry (Fig. 20.14) [25] shows that in the beginning of the test, due to the fast reaction between EPBD and steel surfaces, both the worn area and the content of iron complex increased strongly in the first several hours. After 10 h or so, however, this increase leveled off strongly and kept nearly constant till 100 h with a large decrease of contact pressure from 130 to 6 MPa. These results indicate that once the

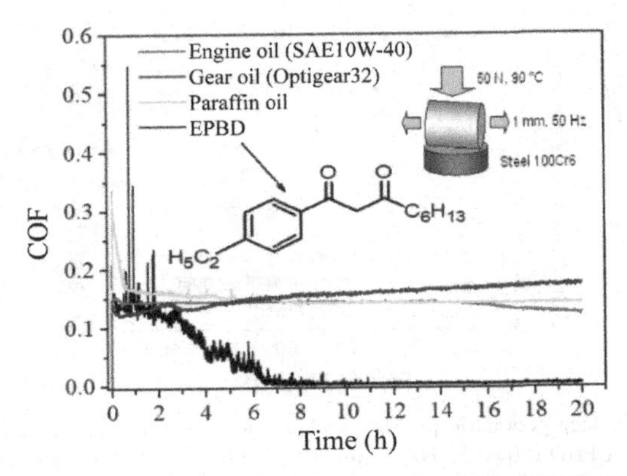

FIGURE 20.12 Superlubricity of 1,3-diketone on steel surfaces (50 N, 50 Hz, 1 mm, 90°C) [23].

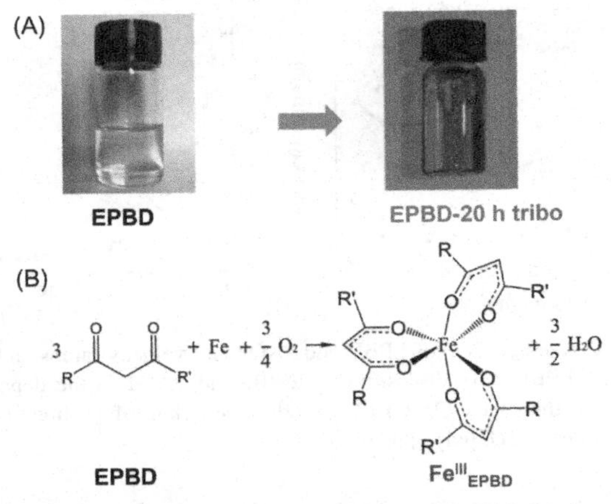

FIGURE 20.13 (A) Observed color change of EPBD after the 20 h friction test. (B) Chemical reaction between EPBD and iron [25]. *EPBD*, 1-(4-*e*thyl *p*henyl) *b*utane-1,3-*d*ione.

contact pressure is low enough to avoid solid-solid contacts, the tribochemical reaction between EPBD and iron asperity stops automatically without the presence of asperity flash temperature, and the wear is self-limited.

To study its Stribeck behavior and determine the lubrication regime, rotary friction tests were also performed [27]. After a running-in stage of 3 h 1,3-diketone 1-(4-*e*thyl *p*henyl) *b*utane-1,3-*d*ione (EPBD) reached superlubricity and generated a worn surface with a diameter of 568 μm (i.e., a contact pressure ≈ 20 MPa) on the steel ball. Then by washing away the lubricant oil but without position change of ball/disc, the Stribeck curves were obtained through a 7 velocity ramps procedure (Fig. 20.15A) [27]. EPBD showed

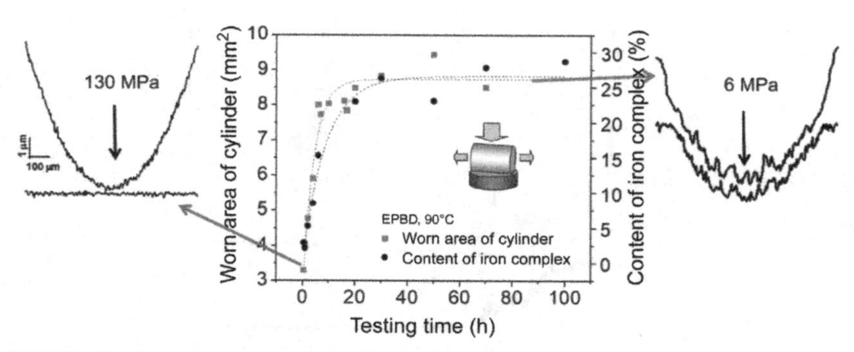

FIGURE 20.14 Surface geometric profiles, and contact pressures of steel surfaces before and after friction test of EPBD (50 N, 50 Hz, 1 mm, 90°C). Worn area of cylinder and the content of iron complex generated during the test are also depicted [25]. *EPBD*, 1-(4-*e*thyl *p*henyl) *b*utane-1,3-*d*ione.

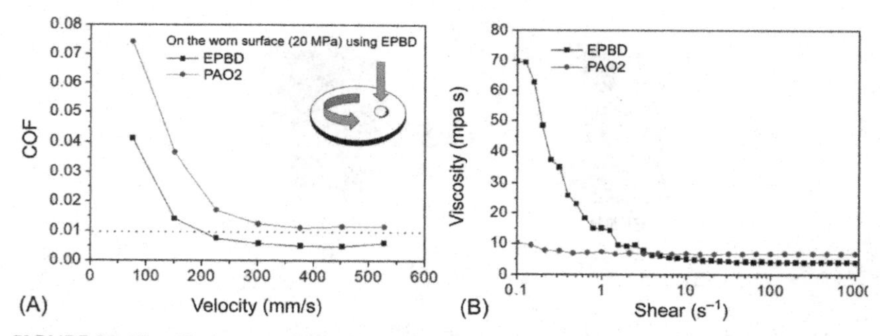

FIGURE 20.15 (A) Average COF of EPBD and PAO2 at 7 velocity ramps on the worn surface after running-in with EPBD. (B) Viscosities of EPBD and PAO2 in the dependence of shear rates measured by the rheometer (25°C) [27]. *COF*, Coefficient of friction; *EPBD*, 1-(4-*e*thyl *p*henyl) *b*utane-1,3-*d*ione; *PAO*, polyalpha olefin.

lower COF than that of PAO2 at the entire measuring range of velocities, which could be explained by the shear thinning of EPBD (Fig. 20.15B). As diketone lubricant has strong intermolecular forces due to its similar ellipsoid molecular structure as liquid crystals [26] (Fig. 20.16A), its shear-thinning behavior is most probably caused by a molecular alignment under the shear. This minimization of friction based on autonomous viscosity control at various velocities is very similar with the so-called "smart lubrication" of liquid crystal lubricant 5CB [31] (4-pentyl-4′-cyanobiphenyl, Fig. 20.16B).

Besides the strong intermolecular forces, the "smart lubrication" of 5CB also needs an external electrical field or surfactants to generate the molecular alignment. Although no electrical field or surfactants are applied in the case of 1,3-diketone, its chemisorption on the steel surface has been revealed by both XPS and ToF-SIMS [29], which also contributes to its superlubricity [27].

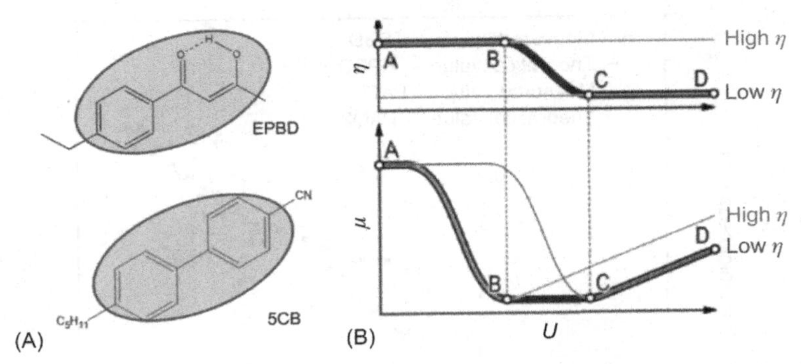

FIGURE 20.16 (A) Similar structure of EPBD with 5CB. (B) "Smart lubrication" of 5CB [31]. *EPBD*, 1-(4-Ethyl phenyl) butane-1,3-dione.

This experimental result is in good agreement with Zhang's work for TFL [13] (Figs. 20.4 and 20.5), which has proved that when a lubricant exhibits both strong interfacial interactions with the substrate and strong intermolecular interactions, the entire Stribeck curve shifts downward and the low-COF regime widens.

It is known that the distinguished difference between TFL and EHL is that the dependence of film thickness on rolling speed of TFL does not follow the Hamrock–Dowson equation which can well predict the situation in the EHL regime [9]. Hence the theoretical thickness using Hamrock–Dowson equation and the experimental thickness using the technique of relative optical interference intensity (ROII, NGY-6) [9] were compared in Fig. 20.17 [27]. It shows that the film thickness of EPBD became thicker than the theoretical value when the velocity was lower than 20 mm/s, which was induced by the strong intermolecular interaction when the fluid was strongly confined. This result supplies further evidence for the TFL regime of the superlubricity of 1,3-diketone.

In summary, the overall superlubricity mechanism of 1,3-diketone is illustrated in Fig. 20.18 [26]: (1) inspired by the asperity flash temperature, the tribochemical reaction between 1,3-diketone and iron eliminate the solid contacts between steel surfaces; (2) 1,3-diketone generates a strong adsorbed layer on steel surfaces via chemical bonds, which benefits the formation of potential molecular orientations; (3) 1,3-diketone has a rod-like molecular structure similar with liquid crystals, which supplies some strong intermolecular interactions to form the molecular alignment and reduce the shear stress; and (4) based on the combined effects of aforementioned factors, the superlubricity of 1,3-diketone is achieved in the TFL regime.

20.3.2 Superlubricity achieved with silicone oil and PAO oil

Some experiments have confirmed that hydrodynamic effect can help to achieve superlubricity [32–34]. The ultralow friction coefficient about 0.004

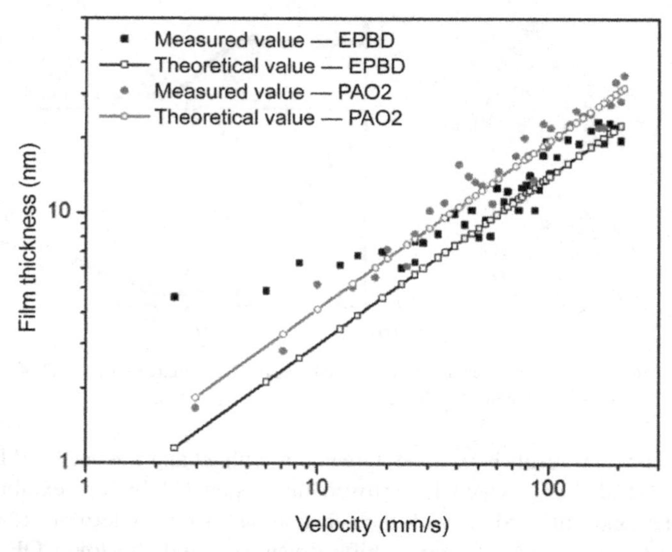

FIGURE 20.17 Lubricant film thicknesses of EPBD and PAO2 as the function of rolling speed [29]. *EPBD*, 1-(4-*e*thyl *p*henyl) *b*utane-1,3-*d*ione; *PAO*, polyalpha olefin.

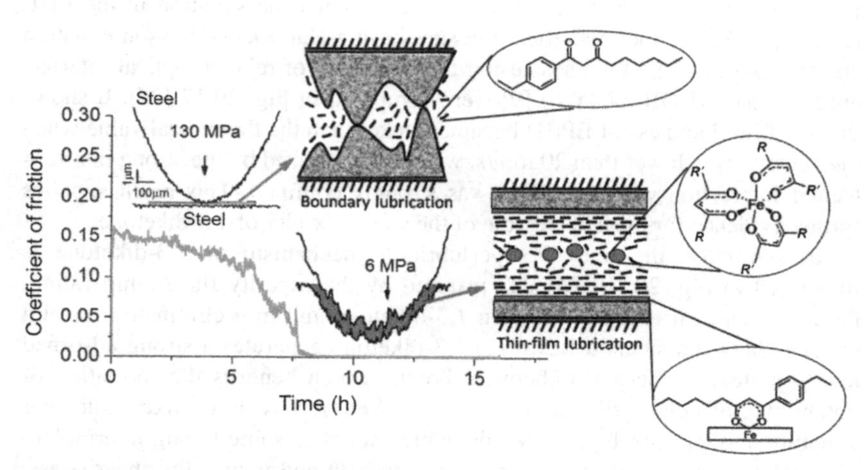

FIGURE 20.18 Artists impression of the lubrication mechanism of 1,3-diketones on steel surfaces [26].

was obtained between Si_3N_4/glass surfaces with the lubrication of silicone oil after running-in period with acid solution (Fig. 20.19A) [35]. The friction coefficient of the 100 mPa s silicone oil was reduced from 0.3 to 0.13 after a running-in period of 500 s, and then remained stable. However if the original

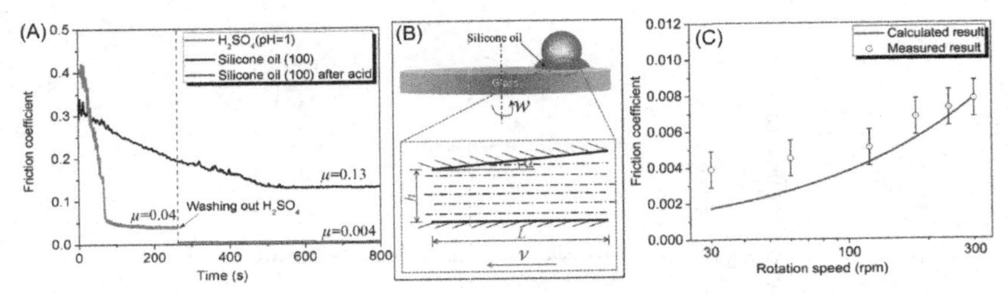

FIGURE 20.19 (A) Friction coefficient with the lubrication of the 100 mPa s silicone oil, H_2SO_4 (pH = 1) and the 100 mPa s silicone oil after running in with H_2SO_4 (pH = 1). (B) Lubrication model of silicone oil between the Si_3N_4/glass surfaces after running in with H_2SO_4 solution. (C) Calculated friction coefficient and measured friction coefficient of the 350 mPa s silicone oil between the Si_3N_4/glass surfaces [35].

surfaces of the Si_3N_4 and glass were lubricated with H_2SO_4 (pH = 1) for 250 s first, then the friction coefficient of the silicone oil was suddenly reduced to about 0.004 without a running-in period, so the silicone oil entered the superlubricity regime. Besides the ultralow friction, coefficient of 0.004 could be maintained for at least for 5 h, which indicates that the superlubricity state of the silicone oil was very stable. In order to analyze the lubrication state during superlubricity, the friction coefficient of silicone oil was calculated based on the hydrodynamic lubrication model (Fig. 20.19B) according to the Reynolds equation [34,36]. Comparing the calculated and measured results (Fig. 20.19C), the calculated friction coefficient was lower than the measured friction coefficient when the rotation speed was low, because the asperity contact has a great influence on the calculated result, but the effect of surface roughness was not considered in the calculation model. Therefore the silicone oil forms a hydrodynamic film above a certain speed to achieve superlubricity. According to this result, more experiments were conducted, such as polyether, oleic acid, ionic liquid, and liquid crystals (5CB). And their friction coefficients were all less than 0.01, which indicates that superlubricity can be obtained as long as the speed and the viscosity of these lubricants match up well [35]. These works confirm that liquid superlubricity can be obtained by hydrodynamic lubrication under some special conditions [37].

Since oil-based superlubricity can be achieved by hydrodynamic effect, the PAO as a commonly used base oil in industry was chosen to study [38]. The oil-based superlubricity cannot be obtained due to the high viscous-pressure coefficient under high contact pressure, so quartz lens (with the curvature radius of 60 mm) and glass disc were chosen as the tribopair. The contact pressure was 39.5 MPa under the applied normal load of 2 N. Fig. 20.20 shows PAO COFs with different viscosities. The COF decreased with the viscosity decreasing. The ultralow friction coefficient of 0.007 was obtained under the lubrication of PAO4. Therefore it is feasible to realize

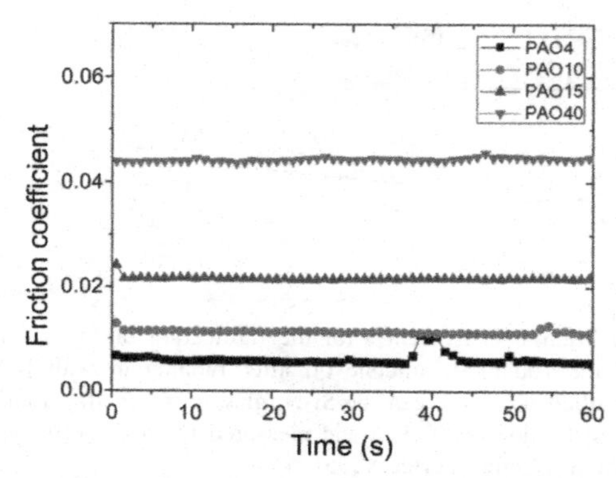

FIGURE 20.20 Friction coefficients of PAO oils between quartz lens (with the curvature radius of 60 mm) and glass disc. The applied load was 2 N and the velocity was 95 mm/s. *PAO*, Polyalpha olefin.

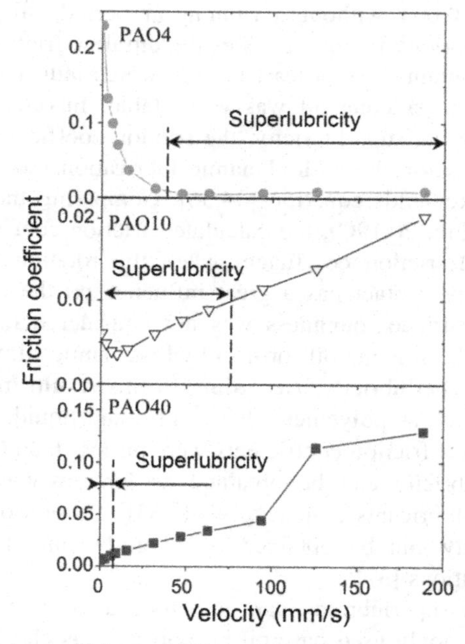

FIGURE 20.21 Friction coefficients of different PAO oils as a function of velocity. The friction pairs were quartz lens (with the curvature radius of 60 mm) and glass disc. *PAO*, Polyalpha olefin.

oil-based superlubricity through hydrodynamic effect. As shown in Fig. 20.21, the COF of PAO4 (34.1 mPa s) decreased with the velocity increasing, but for PAO10 (149 mPa s) with middle viscosity, the COF first decreased and then increased with the velocity increasing, and for PAO40

(1190 mPa s) with high viscosity, the COF increased with the velocity increasing. There always existed superlubricity regime at different velocities for different PAO oils. Therefore oil-based superlubricity can be achieved under proper conditions (including appropriate load, viscosity, viscosity-pressure coefficient, and velocity).

20.3.3 Superlubricity achieved with castor oil on Nitinol alloy surfaces

Zeng et al. [39] reported ultralow friction of castor oil on Nitinol alloy surfaces. Castor oil has three alkyl ester groups of long-chain fatty acids, which has an ability to form a thin film of intermolecular layer which promotes boundary lubrication hence helps to reduce friction between the interfaces. The high polarity of the entire base oil allows strong interactions with the lubricated solid surfaces.

Fig. 20.22A shows the COF of Nitinol 60 alloy pin sliding against GCr15 steel disk versus sliding time under castor oil lubrication. The COF decreases slightly to 0.008 and keeps stable within the testing time. From the SEM image of the worn area (Fig. 20.22B), the surface of Nitinol 60 alloy is scratched by the wear debris of the rotating steel disk. XPS analysis confirms the formation of a thin tribofilm and reveals that the original molecules have undergone chemical degradations. In the tribofilm, a major part of nickel is metallic and the rest is nickel oxide and/or hydroxide.

In the light of XPS analyses obtained here, the oxidation of castor oil is due to the presence of ester group, double bonds, and hydroxyl group in castor oil molecules and this offers high level of reactivity with oxygen, especially when it is placed in friction contact with humid air under HF heating and pressure simultaneously. Fig. 20.23 is the schematic representation of general radical reactions and the evolutionary process of castor oi in detail.

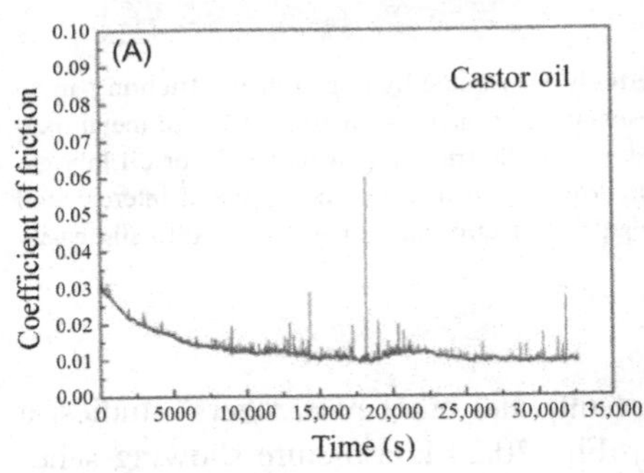

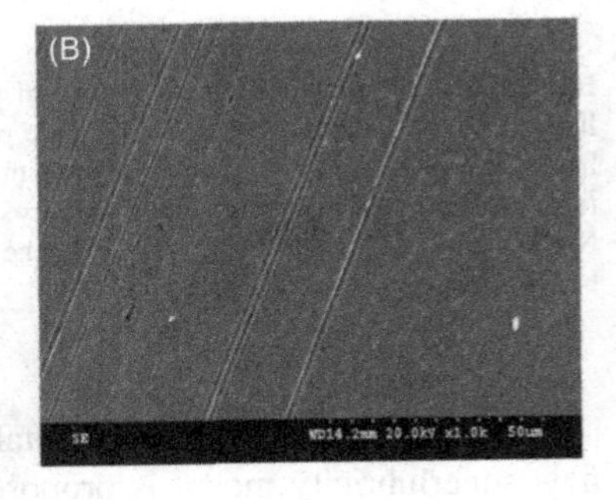

FIGURE 20.22 (A) COF and (B) SEM image of Nitinol 60 under castor oil lubrication [39]. *COF*, Coefficient of friction; *SEM*, scanning electron microscopy.

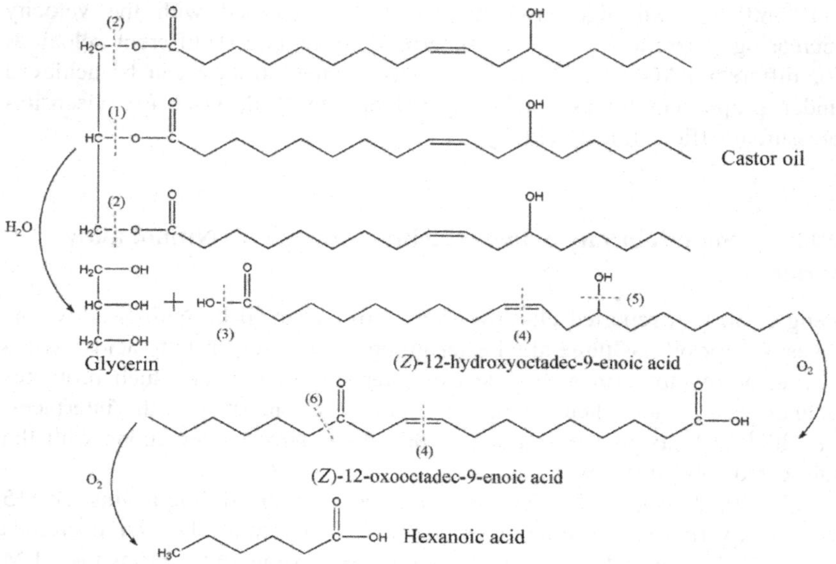

Castor oil

Glycerin

(*Z*)-12-hydroxyoctadec-9-enoic acid

(*Z*)-12-oxooctadec-9-enoic acid

Hexanoic acid

FIGURE 20.23 General radical reaction pathways for castor oil tribochemistry [40].

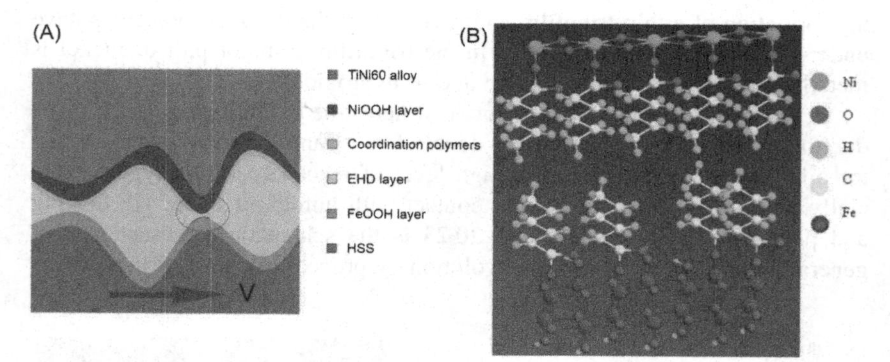

FIGURE 20.24 Schematic illustrations of partially dihydrated hydrogen of the friction pair in lubrication model and molecular schematic presentation of the coordination layer and metal oxyhydroxide. (A) The lubrication diagrammatic sketch of the friction pair under castor oil lubrication; (B) the hexanoic-acid-bonded metal oxyhydroxide and fragment of a general intercalation NiOOH and FeOOH structure displaying the repulsive electrostatic forces that lead to slid easily between the contacting asperities [41].

According to these experimental results and other researchers' studies, a new superlubricity model is proposed. Fig. 20.24 is a picture showing schematically the intercalation of hexanoic acid between nickel oxyhydroxide and iron oxyhydroxide. The dipole–dipole effects that form an interfacial

Coulomb repulsion force also make a contribution to superlow friction. Among all tribochemistry reaction products, there is also glycerol, which can form OH-terminated surfaces under HF heating and pressure at the contact surface of the friction pair. Alternatively, superlubricity of Nitinol 60 alloy is also attributed to easy glide on triboformed OH-terminated surface.

20.3.4 Superlubricity achieved with oleic acid on DLC films

Due to its wide band gap, semiconductive properties, high mechanical hardness, chemical inertness, and optical transparency, diamond-like carbon (DLC) films have a widespread application as protective coatings in areas as diverse as optical windows, magnetic storage disks, car engine parts, biomedical coatings, and microelectromechanical systems [41]. Recently, Bouchet et al. reported that superlubricity can be realized for engineering applications in bearing steel coated with ultrasmooth tetrahedral amorphous carbon (ta-C) under oleic acid lubrication [42].

Fig. 20.25 compares the friction results for the four possible DLC configurations under oleic acid lubrication at ambient temperature (RH of 45%); the test results were obtained by decreasing the sliding speed step by step from 100 to 0.01 mm/s. Under severer boundary lubrication conditions, amazing friction results were obtained for the ta-C-coated surfaces. Superlubricity has never been attained so far, but COF inferior to 0.04 is abnormally low for the boundary regime. At all speeds, the friction coefficients under oleic acid lubrication for the ta-C surfaces were much lower than the friction coefficients for the hydrogenated amorphous carbon (a-C:H)

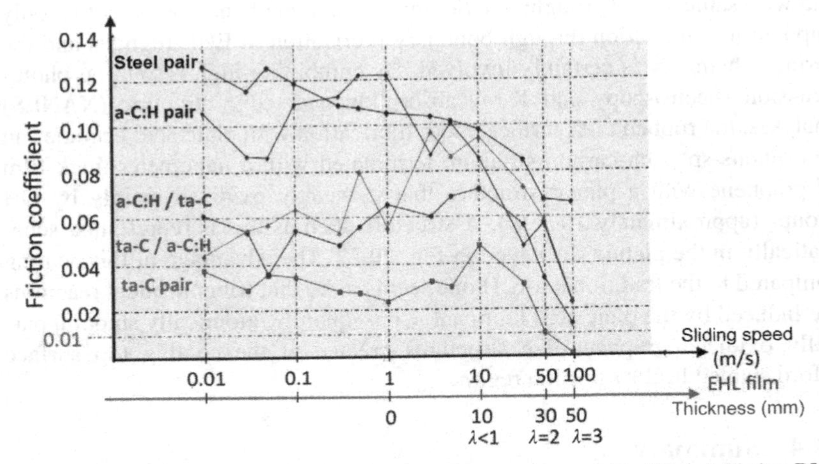

FIGURE 20.25 Friction results of decreasing sliding speed tests with different friction DLC surfaces lubricated by pure oleic acid at ambient temperature and RH 45% [42]. *DLC*, Diamond-like carbon; *RH*, relative humidity.

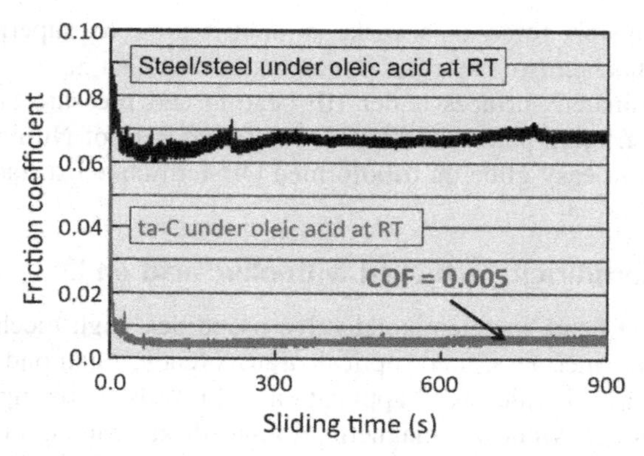

FIGURE 20.26 Friction results at constant sliding speed tests with ta-C pairs and oleic acid at ambient temperature featuring ML conditions with ultralow friction at 66% RH [42]. *ML*, mixed lubrication.

surfaces as well as for the mixed ta-C/a-C:H and a-C:H/ta-C combinations. So, ta-C surfaces was intensively investigated in the following section.

An example of superlubricity under the mixed lubrication regime and constant sliding speed is shown in Fig. 20.26 for a ta-C friction pair at a constant sliding speed of 50 mm/s, a mean contact pressure of 100 MPa and a RH of 66%. The friction coefficient starts at 0.1 and drops drastically below 0.01 after a test lasting a few tens of seconds long. Afterward, the regime remained at this very low-COF value for at least 900 s.

However such a remarkable ultralow friction coefficient is absolutely not observed for the traditional steel/steel pair under same lubrication conditions and with same surface roughness. So, this remarkable behavior cannot be only imputed to a transition through boundary lubrication to EHL regimes and the surface chemistry is certainly involved. By combining high-resolution photoemission spectroscopy and X-ray absorption near edge structure (XANES) analyses, the rubbed ta-C surface under lubrication with oleic acid becomes an amorphous sp^2 rich carbon structure terminated with a nanometer-thick film of graphene with a planar structure that is weakly oxidized mainly by OH groups (approximately 10 at.%). A structure such as this is represented schematically in the picture displayed in Fig. 20.27. The advantage of this coating compared to the traditional a-C:H one seems to be that tribochemical reactions are induced by the oleic acid lubricant. Consequently, atomically smooth, partially oxidized graphene-like structures created at the coating top surface afford stable ultralow friction regime.

20.4 Summary

This chapter has summarized oil-based superlubricity in both theoretical and experimental aspects. It can be seen that the separation of solid contacts is

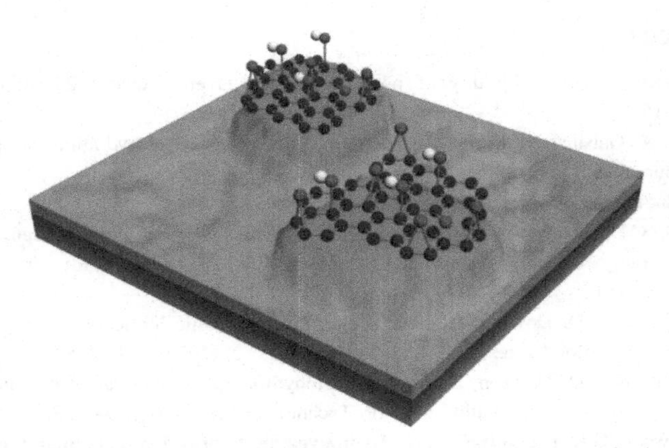

FIGURE 20.27 Schematic picture of the ta-C surface after the friction test in the presence of oleic acid (according to PES and XANES analyses). The protuberant parts correspond to the real contact area between the two antagonists. The flat part is noncontacting area [42]. *PES*, photoemission spectroscopy ; *XANES*, X-ray absorption near edge structure.

the precondition for the superlubricity. Under an appropriate operating condition (i.e., normal load and sliding velocity), the tribochemical reaction between the lubricant and the substrate (e.g., 1,3-diketone oil on steel surfaces and water on Si_3N_4 surfaces) can eliminate the direct asperity contacts. As the oil film is confined between two solid surfaces into a very thin-film, strong oil–solid interactions and oil intermolecular interactions play an important role and facilitate the superlubricity. Furthermore, a hard and smooth surface like DLC and an effective nanoadditive like graphene can improve the superlubricity system as well.

Considering potential industrial applications, the stability and the applicability of oil-based superlubricity need further investigation. It is expected to achieve a very stable macroscopic superlubricity at ambient atmosphere over a wide range of operating conditions. The diversity of friction pair materials especially metallic substrates (e.g., steel or alloy surfaces) should be paid more attention. Besides the LF, the comprehensive performance of the oil (e.g., the oxidation resistance, the detergency, the viscosity-temperature performance, and so on) in other aspects should be also investigated to develop novel formula oils with necessary relevant additives.

Acknowledgment

This work was financially supported by National Natural Science Foundation of China (Grant Nos. 51527901, 51925506, and 51975437).

Competing interests

The authors declare no competing financial interests.

References

[1] U. Raviv, J. Klein, Fluidity of bound hydration layers, Science 297 (5586) (2002) 1540–1543.

[2] L. Ma, A. Gaisinskaya-Kipnis, N. Kampf, J. Klein, Origins of hydration lubrication, Nat. Commun. 6 (2015) 6060.

[3] J. Klein, Hydration lubrication, Friction 1 (1) (2013) 1–23.

[4] J.J. Li, C.H. Zhang, M.M. Deng, J.B. Luo, Investigation of the difference in liquid superlubricity between water- and oil-based lubricants, RSC Adv. 5 (78) (2015) 63827–63833.

[5] B.J. Hamrock, D. Dowson, Isothermal elastohydrodynamic lubrication of point contacts: part III—fully flooded results, J. Lubr. Technol. 99 (2) (1977) 264–275.

[6] B.J. Hamrock, D. Dowson, Isothermal elastohydrodynamic lubrication of point contacts: part 1—theoretical formulation, J. Lubr. Technol. 98 (2) (1976) 223–228.

[7] D. Dowson, G.R. Higginson, Elasto-Hydrodynamic Lubrication, Pergamon Press, Oxford, 1977.

[8] C. Barus, Isothermals, isopiestics and isometrics relative to viscosity, Am. J. Sci. 45 (266) (1893) 87–96.

[9] J.B. Luo, S.Z. Wen, P. Huang, Thin film lubrication. 1. Study on the transition between EHL and thin film lubrication using a relative optical interference intensity technique, Wear 194 (1-2) (1996) 107–115.

[10] L.R. Ma, J.B. Luo, Thin film lubrication in the past 20 years, Friction 4 (4) (2016) 280–302.

[11] M.W. Shen, J.B. Luo, S.Z. Wen, J.B. Yao, Nano-tribological properties and mechanisms of the liquid crystal as an additive, Chin. Sci. Bull. 46 (14) (2001) 1227–1232.

[12] P. Huang, J.B. Luo, S.Z. Wen, The relation and transition between EHL and thin film lubrication, Tribology 19 (1) (1999) 72–77.

[13] X.J. Zhang, Y. Huang, Y. Guo, Y. Tian, Y.G. Meng, A dynamic rheological model for thin-film lubrication, Chin. Phys. B 22 (1) (2013) 016202.

[14] O. Vinogradova, Slippage of water over hydrophobic surfaces, Int. J. Miner. Process. 56 (1) (1999) 31–60.

[15] A.L. Demirel, S. Granick, Origins of solidification when a simple molecular fluid is confined between two plates, J. Chem. Phys. 115 (3) (2001) 1498–1512.

[16] J. Klein, E. Kumacheva, Simple liquids confined to molecularly thin layers. I. Confinement-induced liquid-to-solid phase transitions, J. Chem. Phys. 108 (16) (1998) 6996–7009.

[17] B. Bhushan, J.N. Israelachvili, U. Landman, Nanotribology: friction, wear and lubrication at the atomic scale, Nature 374 (6523) (1995) 607–616.

[18] J. Gao, W.D. Luedtke, U. Landman, Layering transitions and dynamics of confined liquid films, Phys. Rev. Lett. 79 (4) (1997) 705–708.

[19] Y. Zhu, S. Granick, Superlubricity: a paradox about confined fluids resolved, Phys. Rev. Lett. 93 (9) (2004) 096101.

[20] P. Frantz, M. Salmeron, Preparation of mica surfaces for enhanced resolution and cleanliness in the surface forces apparatus, Tribol. Lett. 5 (2-3) (1998) 151–153.

[21] A. Jabbarzadeh, P. Harrowell, R.I. Tanner, Very low friction state of a dodecane film confined between mica surfaces, Phys. Rev. Lett. 94 (12) (2005) 126103.

[22] T. Amann, A. Kailer, Ultralow friction of mesogenic fluid mixtures in tribological reciprocating systems, Tribol. Lett. 37 (2) (2010) 343–352.

[23] T. Amann, A. Kailer, Analysis of the ultralow friction behavior of a mesogenic fluid in a reciprocating contact, Wear 271 (9–10) (2011) 1701–1706.

[24] T. Amann, A. Kailer, Relationship between ultralow friction of mesogenic-like fluids and their lateral chain length, Tribol. Lett. 41 (1) (2011) 121–129.

[25] K. Li, T. Amann, M. Walter, M. Moseler, A. Kailer, J. Rühe, Ultralow friction induced by tribochemical reactions: a novel mechanism of lubrication on steel surfaces, Langmuir 29 (17) (2013) 5207–5213.

[26] K. Li, T. Amann, M. List, M. Walter, M. Moseler, A. Kailer, et al., Ultralow friction of steel surfaces using a 1,3-diketone lubricant in the thin film lubrication regime, Langmuir 31 (40) (2015) 11033–11039.

[27] K. Li, S.M. Zhang, D. Liu, T. Amann, C.H. Zhang, C. Yuan, et al., Superlubricity of 1,3-diketone based on autonomous viscosity control at various velocities, Tribol. Int. 126 (2018) 127–132.

[28] D. Liu, K. Li, S. Zhang, T. Amann, C. Zhang, X. Yan, Anti-spreading behavior of 1,3-diketone lubricating oil on steel surfaces, Tribol. Int. 121 (2018) 108–113.

[29] S.M. Zhang, C. Zhang, X. Chen, K. Li, J. Jiang, C. Yuan, et al., XPS and ToF-SIMS analysis of the tribochemical absorbed films on steel surfaces lubricated with diketone, Tribol. Int. 130 (2019) 184–190.

[30] S.M. Zhang, C.H. Zhang, K. Li, J.B. Luo, Investigation of ultra-low friction on steel surfaces with diketone lubricants, RSC Adv. 8 (17) (2018) 9402–9408.

[31] C. Tadokoro, T. Nihira, K. Nakano, Minimization of friction at various speeds using autonomous viscosity control of nematic liquid crystal, Tribol. Lett. 56 (2) (2014) 239–247.

[32] X.Y. Ge, T. Halmans, J.J. Li, J.B. Luo, Molecular behaviors in thin film lubrication—part three: superlubricity attained by polar and nonpolar molecules, Friction 7 (2019) 625–636.

[33] M.M. Deng, J.J. Li, C.H. Zhang, J. Ren, N.N. Zhou, J.B. Luo, Investigation of running-in process in water-based lubrication aimed at achieving super-low friction, Tribol. Int. 102 (2016) 257–264.

[34] M.M. Deng, C.H. Zhang, J.J. Li, L.R. Ma, J.B. Luo, Hydrodynamic effect on the superlubricity of phosphoric acid between ceramic and sapphire, Friction 2 (2) (2014) 173–181.

[35] J.J. Li, C.H. Zhang, M.M. Deng, J.B. Luo, Superlubricity of silicone oil achieved between two surfaces by running-in with acid solution, RSC Adv. 5 (39) (2015) 30861–30868.

[36] S.Z. Wen, P. Huang, Principles of Tribology, Tsinghua University Press, 2012.

[37] C.H. Zhang, Y.C. Zhao, M. Bjorling, Y. Wang, J.B. Luo, B. Prakash, EHL properties of polyalkylene glycols and their aqueous solutions, Tribol. Lett. 45 (3) (2012) 379–385.

[38] Q. Zeng, F. Yu, G. Dong, Superlubricity behaviors of Si_3N_4/DLC films under PAO oil with nano boron nitride additive lubrication, Surf. Interface Anal. 45 (8) (2013) 1283–1290.

[39] Q. Zeng, G. Dong, Influence of load and sliding speed on super-low friction of Nitinol 60 alloy under castor oil lubrication, Tribol. Lett. 52 (1) (2013) 47–55.

[40] Q. Zeng, G. Dong, J.M. Martin, Green superlubricity of Nitinol 60 alloy against steel in presence of castor oil, Sci. Rep. 6 (2016) 29992.

[41] J. Robertson, Diamond-like amorphous carbon, Mater. Sci. Eng. R. Rep. 37 (4) (2002) 129–281.

[42] M. Bouchet, J.M. Martin, J. Avila, M. Kano, K. Yoshida, T. Tsuruda, et al., Diamond-like carbon coating under oleic acid lubrication: evidence for graphene oxide formation in superlow friction, Sci. Rep. 7 (2017) 46394.

Chapter 21

Approaching superlubricity under liquid conditions and boundary lubrication—superlubricity of biomaterials

Yuhong Liu
State Key Laboratory of Tribology, Tsinghua University, Beijing, P.R. China

21.1 Introduction

Friction and wear are very common in our daily life, which can bring poor surface quality as well as an enormous loss of material and energy. Scientists and engineers have been looking for effective methods to decrease the friction from the remote past. Oil lubricants have been used as the great lubricants for decades. However with the shortage of resource and increasingly prominent environment problems, the negative effects of oil lubricants have not been ignored. New environmentally friendly lubricants are required to alleviate or avoid environmental pollution, which involves more negligible friction.

It has been long known that the normal activity of life is closely related to friction and wear in vivo. Excessive wear would have a serious influence on the health and properties of organs. It is found that biological mucus always plays a dominant role in the lubrication process and helps achieve ultralow friction coefficient in all kinds of organs and tissues. For example, synovial fluid could maintain a superlow friction coefficient lower than 0.005 under high pressures. In the last decades, Biotribology has attracted overwhelming attention from numbers of domestic and foreign researchers since it was first proposed in 1973 and has been one of the fastest-growing areas and most important researches in tribology [1,2]. Owing to the low viscosity-pressure index and great fluidity, biolubrication has great potential in achieving ultralow friction [3,4]. Quantities of studies on the lubricating properties of natural and artificial biological materials, such as the mucus, synovial fluid, polymer brushes, and hydrogels, have been carried out,

Superlubricity. DOI: https://doi.org/10.1016/B978-0-444-64313-1.00021-1

ranging from the macrocharacteristics to the micromechanism. For example, biological mucus is an adhesive substance, containing a variety of components, secreted by both animals and plants, which plays an essential role in biological water-based lubricating systems [5,6]. It is found that several kinds of mucus not only are essential lubricants in the organism, but also exhibit excellent lubricating behaviors in vitro, which indicates a promising application in biomedical devices and industrial applications. Moreover, various polymers and hydrogels with good mechanical properties are synthesized as excellent biolubricants, which promote the development of bionic researches. Especially, the lubricating mechanisms of these biological materials are more than elastohydrodynamic effects. The superior lubricating behaviors are more closely related to the structure of the macromolecules, the interactions between polymers and water molecules, and the tribochemical reactions during the tribological process.

21.2 Natural biological materials

21.2.1 Biological mucus

In nature, it is found that there are abundant phenomena closely related to water-based lubrication. Increasing numbers of natural materials are proved to exhibit superior lubricating behaviors. Herein, the research progress on lubricating properties of natural biological materials is briefly reviewed, including plant mucilage, animal mucus, and synovial fluid.

21.2.1.1 Plant mucilage

It has been long known that viscous mucilage can be secreted from various organs of plants, such as stems, leaves, and seeds, which plays a role in regulating growth, reproduction, and defense functions [7]. The extracted plant mucilage has also been found to have high medicinal value, industrial value as well as commercial value, and widely used in biomedicine, food processing, and industrial manufacturing [8]. In recent years, biological mucus has been investigated and developed as new environmentally friendly lubricants. Owing to the formation of hydrogen bond structure and strong adsorption of the polymer backbone, many kinds of plant mucilage are found to exhibit superlubricity behaviors in vitro. Polysaccharide extracted from red microalgae, *Brasenia schreberi* (BS) and basil seed gel (BSG) are three kinds of biological mucus to achieve macroscopic superlubricity.

In 2006, Arad et al. [9,10] conducted tribological researches on the polysaccharide extracted from red microalgae and found that the viscosity of the mucilage could keep stable in the temperature range from 20°C to 70°C. They evaluated the macrotribological properties of this plant mucilage by the tribometer with a ball-on-flat ceramic pair. As shown in Fig. 21.1, the friction coefficient of the polysaccharide mucilage could reach an ultralow value

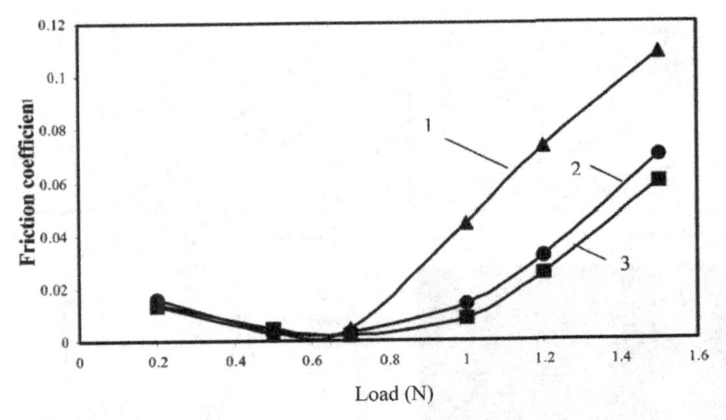

FIGURE 21.1 Effect of the load on the friction coefficient with a ball-on-flat device under severe contact conditions. The pairs were lubricated with hyaluronic acid (1%) −1, polysaccharide solutions (1%) − 2, and (2%) − 3. The test was carried out over 500 cycles. These experiments were carried out at a sliding velocity of 0.2 mm/s [9].

as low as 0.003, which entered the superlubricity region. However it was found that the superlubricity of this mucilage could only be maintained under a special range of low contact pressures. Since under high loads and low sliding velocities the main friction mechanism is boundary lubrication, they believed that boundary lubrication is the dominant mechanism under low sliding velocity and high pressure with the polysaccharide lubricants. From the atomic force microscopy (AFM) images of the worn region after lubrication, they found that molecular-level adsorption layers were formed on the solid surfaces, which improved the load-carrying capacity and the frictional properties. They inferred that the ultralow friction coefficient was probably attributed to the hydration layer on the absorbed polysaccharide macromolecules in the contact region.

In 2012, Li et al. [11] found a new plant-based lubricant named BS mucilage with superior lubricating behavior and confirmed the fundamental role of the polymer structure and hydration. They obtained a superlow friction coefficient with quartz friction pairs by a self-designed apparatus. Fig. 21.2A displayed the schematic illustration of the measurement system, and the yellow region represented the friction lock region. The circle inset was force analysis at the moment when the sample just entered the friction lock region with a critical angle of θ_m, where G was the gravity of the BS sample, N_1, N_2 were the normal forces given by two glass plates separately, and F_1, F_2 were the friction forces between the sample and glass plates. Fig. 21.2B showed the tribological results of the BS sample. They found that an ultralow friction coefficient with a minimal value of 0.005 was achieved stably. Through the observation on the microstructure of the freezed-dry BS sample

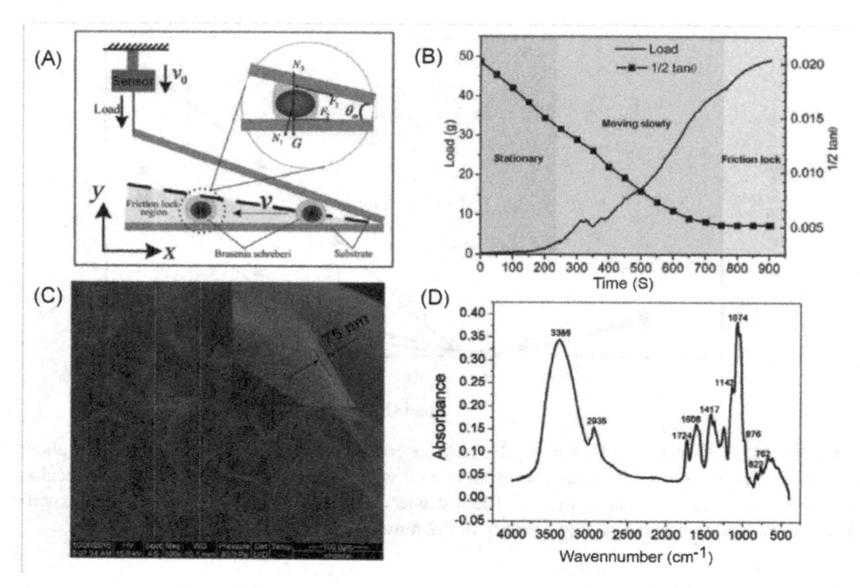

FIGURE 21.2 (A) Schematic illustration of the measurement system. (B) Friction measurement result based on the above measurement system. (C) SEM image of the mucilage after treating by vacuum-freeze-drying method. (D) FTIR spectra of the mucilage after treating by vacuum-freeze-drying method [11].

by a scanning electron microscope (SEM) under low vacuum circumstance, they found that there were many nanosheets in the mucilage, as shown in Fig. 21.2C. These nanosheets were solid and their thickness was about 75 nm. It was also found that there were some interspaces between these nanosheets in the dry mucilage, which was favorable to hold a large amount of water in the mucilage. The SEM image indicated that the mucilage had a special network structure, which was probably linked to the excellent lubrication property. In addition, the chemical composition of these nanosheets investigated by Fourier transform infrared (FTIR) was shown in Fig. 21.2D and large amounts of hydrophilic groups were found in the polysaccharide molecules, where water molecules can be adsorbed firmly by the hydrogen bond to form a hydrated layer. Thus the superior lubricating behaviors of the BS samples were mainly attributed to the strong hydration effect between nanosheets in the mucilage.

Liu et al. [12] conducted a further tribological measurement of the BS mucilage by universal mechanical tester (UMT) with a ring-on-disk rotary mode. They found that superlubricity could also be achieved with a range of sliding speed. They inferred that the superlubricity behavior was closely related to the adsorption layer of the polysaccharide and the hydration layer and proposed a probable lubrication model, as shown in Fig. 21.3. When the

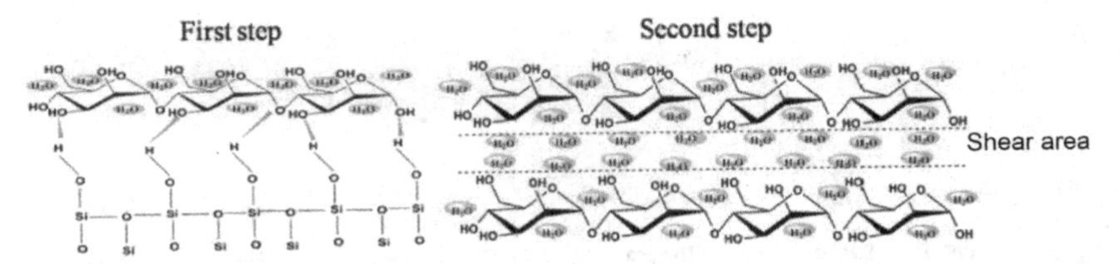

FIGURE 21.3 Schematic illustration of the lubricating processes that lead to liquid superlubricity state [12].

sample moved between the quartz surfaces, the outer polysaccharide molecules would adsorb firmly on the surfaces. The water molecules bound to the nanosheets would form a hydration layer between the solid surfaces. Owing to the strong hydrogen bonding, these water molecules were difficult to be extruded from the contact region even under high pressure. Shearing occurred between these hydrated nanosheets and superlubricity was achieved because of the good fluidity of the hydrated layer. These findings pointed out a way to find new effective biolubricants.

In 2016, Zhang et al. [8] found that BSG was a new superior biolubricant with a friction coefficient of around 0.005. Through the observation on the chemical characteristics of BSG samples, they found that the microstructure of BSG samples was similar to that of BS mucilage and also attributed the superlubricity to the adsorption of polysaccharide with strong hydration effect.

Generally, several kinds of plant mucilage are found to not only exhibit enhanced lubricating behaviors in the organism, but also be able to achieve stable macroscopic superlubricity in vitro. However only a few kinds of plant mucilage are demonstrated to have great lubricating properties. More microscopic studies need to be conducted to explore the nature of the lubricating mechanism.

21.2.1.2 *Animal mucus*

It has been known that mucosal cells of the animals could secret viscous and well-lubricated mucus to protect all kinds of tissues. Animal mucus also acts as a dominant role in various lubricating behaviors in vivo, such as the lubrication of the saliva in the mouth, the tear on the eye surface, and the mucus in the alimentary canal. To dated reports, animal mucus is mainly composed of water and some biological macromolecules, such as phospholipid, glycoprotein, and polysaccharide [5,13].

In recent years, mucin, a complex of polysaccharide and protein, has been found to play a dominate role in the lubricating behaviors in vivo. Mucin is essentially a kind of glycosylated glycoprotein with the molecular weight ranging from 0.5 to 20 MDa, which mainly consists of the core

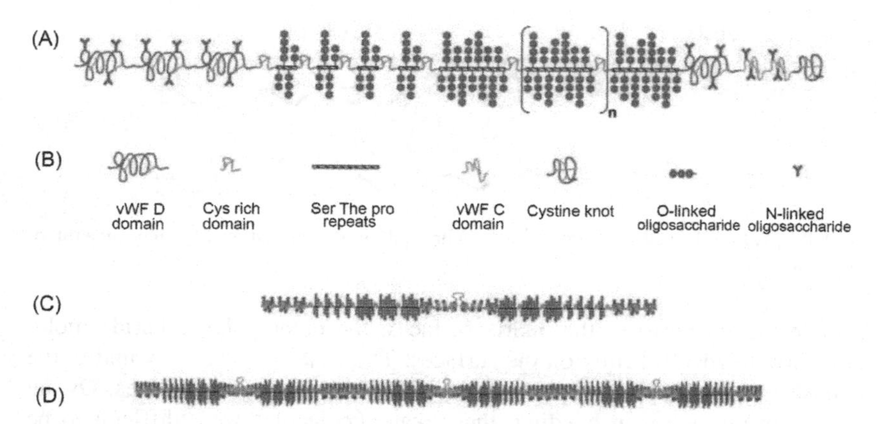

FIGURE 21.4 (a) A schematic diagram of the pig gastric mucin monomer consisting of glycosylated regions. (b) The symbols of the different domains in the sketch (a). (c) a dimer formed by two monomeric subunits linked via disulfide bonds in the non-glycosylated regions. (d) dimers that are further disulfide linked to form higher multimers [13].

protein chain and glycosylated oligosaccharide branches, as shown in Fig. 21.4. Owing to the unique structure of the bottle-brush branches and the highly glycosylated composition, mucin is widely considered as a potential water-based lubricant. The glycosylated branches have a large number of sialic acid groups, which lead to negative charges in the middle glycosylation region and positive charges in the nonglycosylation region at the end. This unique structure possesses particular properties similar to amphoteric polymer electrolyte, which is conducive to the formation of electrostatic interaction and the adsorption on the surfaces. Besides, there are quantities of –OH groups on the oligosaccharide branches, which can form strong hydrogen bindings with water molecules. Owing to these particular features, mucin is easy to have hydrogen-binding interactions, electrostatic interactions and hydrophilic–hydrophobic interactions with other molecules or functional surfaces. For example, mucin may be grafted to the sliding surfaces and improve the lubricating behaviors as polymer brushes [6,14,15].

Lee et al. [16] reported the lubricating properties of the gastric mucin from porcine (PGM) with poly(dimethylsiloxane) (PDMS) friction pairs on the tribometer. As shown in Fig. 21.5, they found that the friction coefficient approached superlubricity when the pH of PGM solution was around 2. They believed that the hydrophobic end of the PGM molecules could be adsorbed on the PDMS surfaces firmly through the electrostatic interactions, which was the reason for the low friction. The increase of pH would weaken the interactions and the absorbed PGM molecules on the solid surfaces would be destroyed in the initial few frictional cycles, which led to the increase of friction and wear. Then Nikogeorgos et al. [17] introduced chitosan with

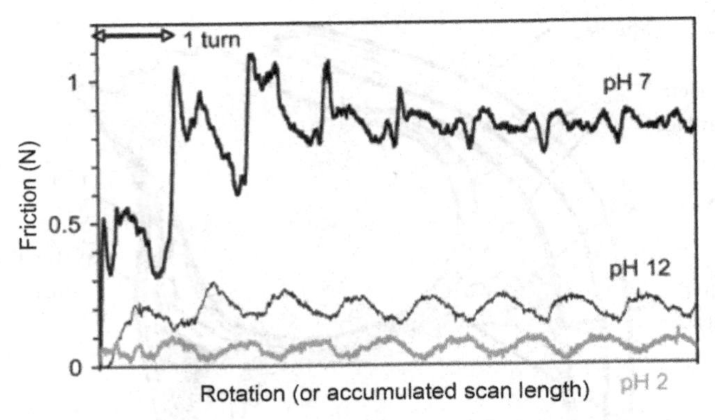

FIGURE 21.5 The tribological evaluation of PGM in solutions with different pHs [16].

abundant amino groups and hydroxyl groups to PGM solutions and demonstrated improved lubricating properties. They found that chitosan could connect the individual PGM molecules and reduce the entanglements between polymer chains. At present, increasing numbers of researches on the lubricating properties of mucin are conducted, which are mainly concentrated on the improvement of lubrication and corresponding micromechanism. However it has still been a long way to connect the macroscopic results and microscopic mechanism to explore the nature of the tribological process.

21.2.2 Synovial fluids

Natural synovial joint is one of the most important locomotive organs in the organism, which effectively helps decrease the friction and suffer the load, as shown in Fig. 21.6. Studies have shown that the enhanced mechanical properties and lubricating properties of synovial joints are mainly attributed to the synovial fluid in the articular cavity, which is secreted by the synovial membrane of the bursae and tendon sheaths. An ultralow friction coefficient as low as 0.001 is supposed to be stably obtained in joint lubrication even if the average pressure increases to 20 MPa [18–20]. To dated reports, increasing numbers of researchers believe that joint lubrication is the result of the synergy of several natural biolubricants and biological macromolecules in the synovial fluid, such as hyaluronan (HA), phospholipids, and glycoproteins. Specially, owing to the particular molecular structure and chemical properties, phospholipids are considered to be the main lubricating molecules in joint lubrication [21–30].

21.2.2.1 The lubrication of phospholipid

Phospholipids are typical amphiphilic molecules with hydrophilic head groups and highly hydrophobic tails, which are widely distributed in the

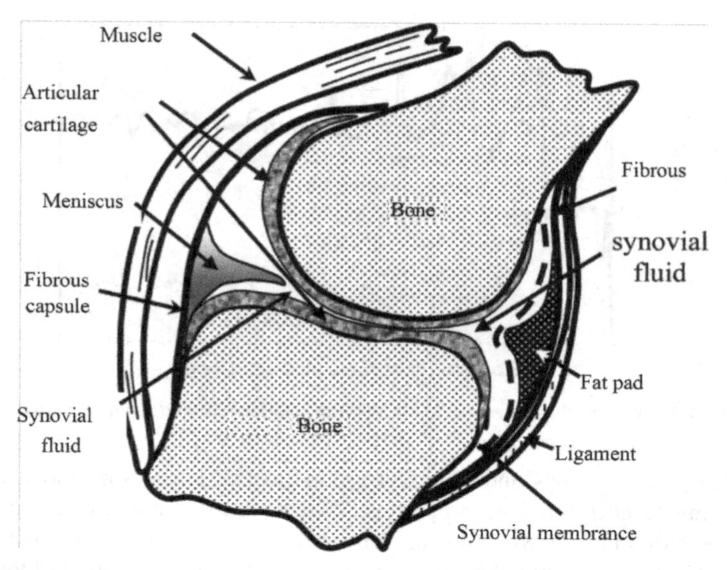

FIGURE 21.6 Diagrammatic representation of a synovial joint. [1].

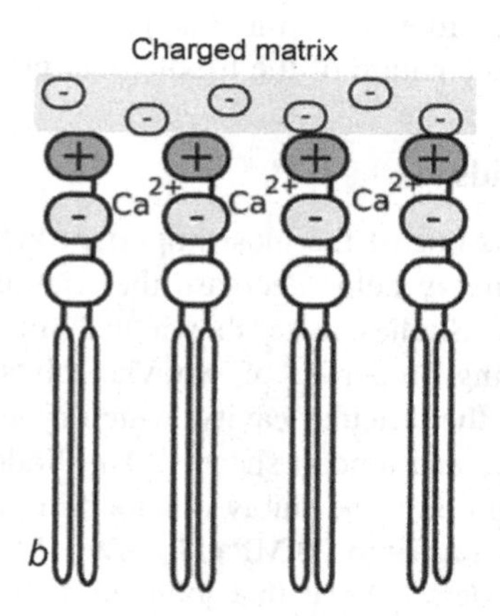

FIGURE 21.7 The adsorption of phospholipids on negatively charged surfaces [24].

various tissues and cells, as shown in Fig. 21.7. These molecules are easily adsorbed on negatively charged substrates such as HA or glycoprotein because of the positively charged head groups [24,31–34]. Generally, phospholipids have two typical self-assembled structures in water: vesicles and bilayers, which depend on the chain length, the chain volume, and the optimal area of the head group [35–38]. Recently, phospholipids have been proven to be good boundary lubricants owing to the reduced interfacial energy, strong adsorption and cohesion, and the high hydrophobicity of the acyl chain, which can provide good boundary lubrication [39–42].

Initially, superlubricity was achieved in few experiments under certain conditions at the macroscales. Trunfio-Sfarghiu et al. [43] conducted a frictional test by a homemade tribometer and found that the friction coefficient could be reduced to an ultralow value as low as 0.002 ± 0.0008 between the glass and hydroxy ethyl methacrylate surfaces covered with 1, 2-dipalmitoyl-sn-glycero-3-phosphocholine (DPPC) bilayers. Afterward, with the development of microscopic instruments, such as surface force balance (SFB) and AFM, more researchers have concentrated on exploring the deep lubrication mechanism of phospholipids at the nano- or microscales.

Goldberg et al. [44] measured the normal force and shear force between mica surfaces, which were bearing hydrogenated soy phosphatidylcholine (HSPC) small unilamellar vesicles (SUVs) at the physiologically high salt concentration (0.15 M $NaNO_3$). Owing to the topography observed by cryo-scanning electron microscopy, they found that each mica surface was covered by a closely packed HSPC-SUV layer, as shown in Fig. 21.8A. According to the measurement of the normal force, a clear attractive interaction between the liposome layers was found and a steric repulsive force occurred with further compression. In addition, the results of the shear forces indicated that a superlow friction coefficient of 0.0006–0.008 was achieved while the contact pressure might rise to 6 MPa (Fig. 21.8B). The excellent lubricating behavior was attributed to the hydration lubrication mechanism caused by the hydrated phosphatidylcholine head group layers at the outer surface of liposomes (Fig. 21.8C).

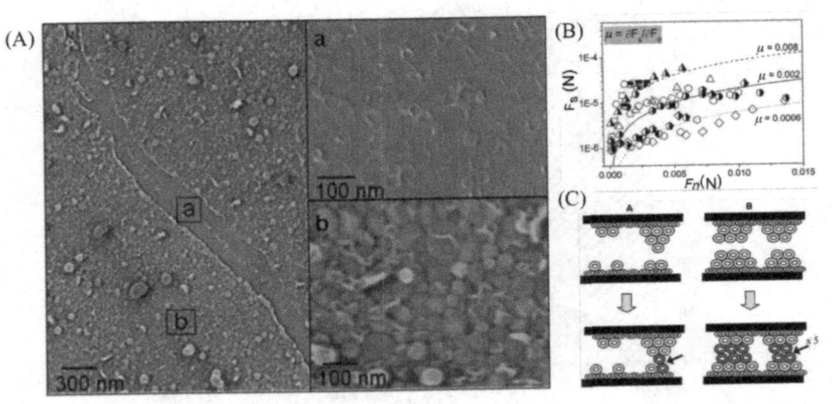

FIGURE 21.8 (A) Cryo-SEM imaging of HSPC-SUV on mica surface in 0.15 M $NaNO_3$ solution. (B) Friction force versus normal load between mica surfaces bearing HSPC. (C) Schematic illustration of possible models [44].

In their following works [45], they raised the load capacity to 12 MPa ($\mu \approx 2 \times 10^{-5} - 10^{-4}$) under pure water with HSPC preadsorbed onto mica surfaces. Such low values of friction coefficient at these high pressures have not been attained with any boundary lubricants before. Moreover, the relationship between the shear force and the sliding velocity was investigated and no significant correlativity was found, which indicated that the superlubricity was in the regime of boundary lubrication. Furthermore, since glucosamine sulfate (GAS) is widely used as a treatment for osteoarthritis, Gaisinskaya-Kipnis et al. [46] had systematic studies on it. The measurements were conducted on the mica surfaces coated with HSPC vesicles in deionized (DI) water and salt solution (0.15 M $NaNO_3$). It was found that the topography of HSPC liposomes was only slightly affected by GAS, as shown in Fig. 21.9A. When immersed both in DI water and physiological-level salt solutions, HSPC-SUVs with encapsulated GAS exhibited excellent

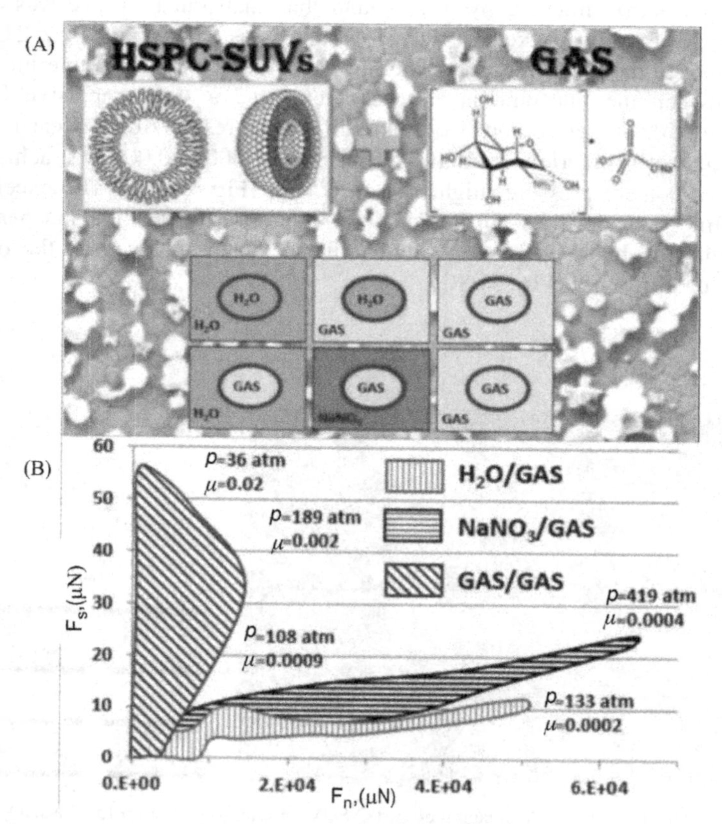

FIGURE 21.9 (A) Schematic diagram between HSPC-SUV and GAS. (b) Friction forces versus normal load between HSPC-SUV covered mica surface [46].

lubricating properties even at high compressions. The superlow friction was attributed to several parameters based on the hydration lubrication mechanism (Fig. 21.9B).

Moreover, the bearing capacity as well as topography (by AFM) and the normal load, as well as shear forces (by SFB), were investigated between two opposing surfaces bearing 1,2-distearoyl-sn-glycero-3-phosphocholine (DSPC) bilayers and vesicles across DI water [47]. It was found that the bilayers were easily punctured by the AFM tip and showed substantial hysteresis between approach and retraction curves, whereas liposomes were not punctured and showed purely elastic behavior. A friction coefficient between 0.002 and 0.008 could be achieved by bilayers up to pressures of more than 5 MPa, as shown in Fig. 21.10. However, the bilayers on the surfaces were less robust and tended to detach from the surface under shearing, leading to high friction for subsequent approaches at the same contact position. In contrast, liposomes showed ultralow friction coefficients around 10^{-4} under pressures up to 18 MPa and exhibited reversible and reproducible behaviors under shearing and compression. These superior properties were believed to be attributed to the increased mechanical stability of the self-closed, closely packed liposomes. To obtain softer biocompatible surfaces, Sorkin et al. [48,49] preadsorbed a soft biomimetic polymer layer (alginate-chitosan) onto the mica surface, (Fig. 21.11A). It was found that the adsorption of HSPC liposomes on the soft surfaces was more stable, achieving a friction coefficient around 10^{-4} under the high pressure around 35 MPa.

Recently, Liu et al. [50] found an efficient reduction of friction coefficient and wear on titanium alloy (Ti_6Al_4V) and ultrahigh molecular weight polyethylene surfaces under various velocities and loads when using DPPC liposomes as the lubricant at the macroscale. The existence of adsorbed liposome layers on both surfaces could reduce asperities contact and show great bearing capacity. Afterward, the tribological properties of liposomes on Ti_6Al_4V/polymer surface were studied by AFM at the nanoscale. The superlubricity with a friction coefficient around 0.007 was achieved under the maximal pressure of 15 MPa, as shown in Fig. 21.12, consisting with the lubrication condition of natural joints.

21.2.2.2 Effects on the friction of phospholipids

In addition to searching novel kinds of phospholipids to achieve superlubricity, researchers are devoted to investigating influence factors on friction and explore the deep lubricating mechanism. Currently, acyl chain length, temperature, vesicle size, synergy with other macromolecules, and some other effects are found to be essential influence effects on the tribological behaviors of phospholipids.

Acyl chain length is found to play a role in the superlubricity of phospholipids. Sorkin et al. [51] measured the friction force between two mica

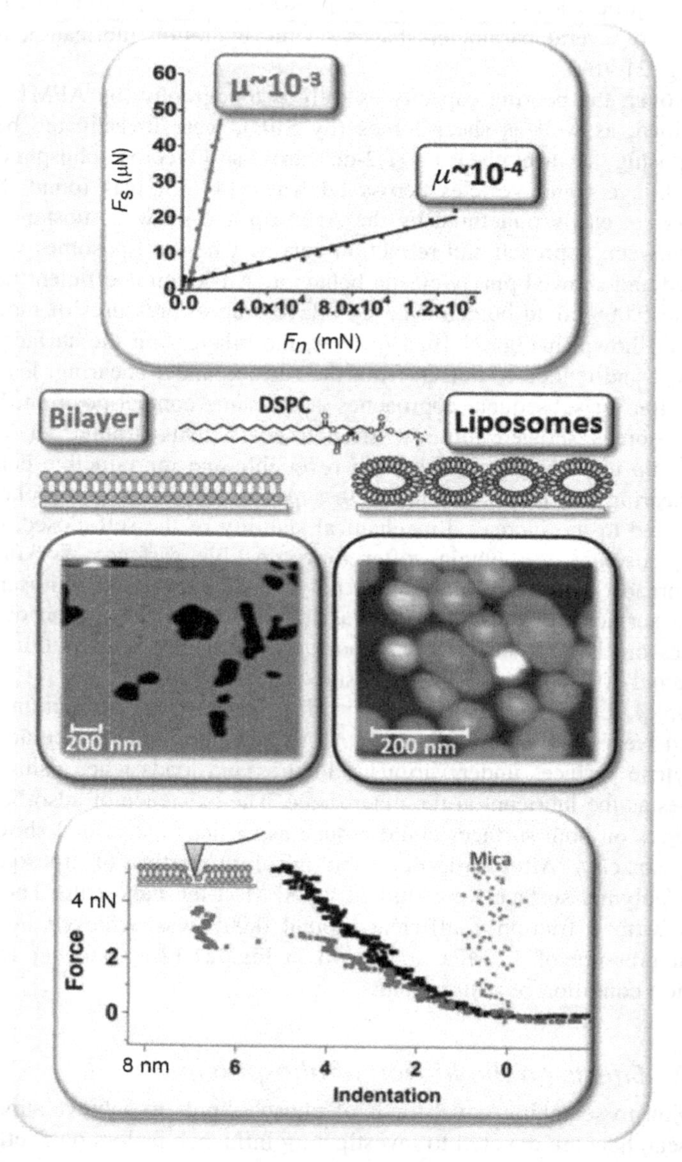

FIGURE 21.10 Comparison of lubrication and mechanical properties between DSPC vesicles and bilayers [47].

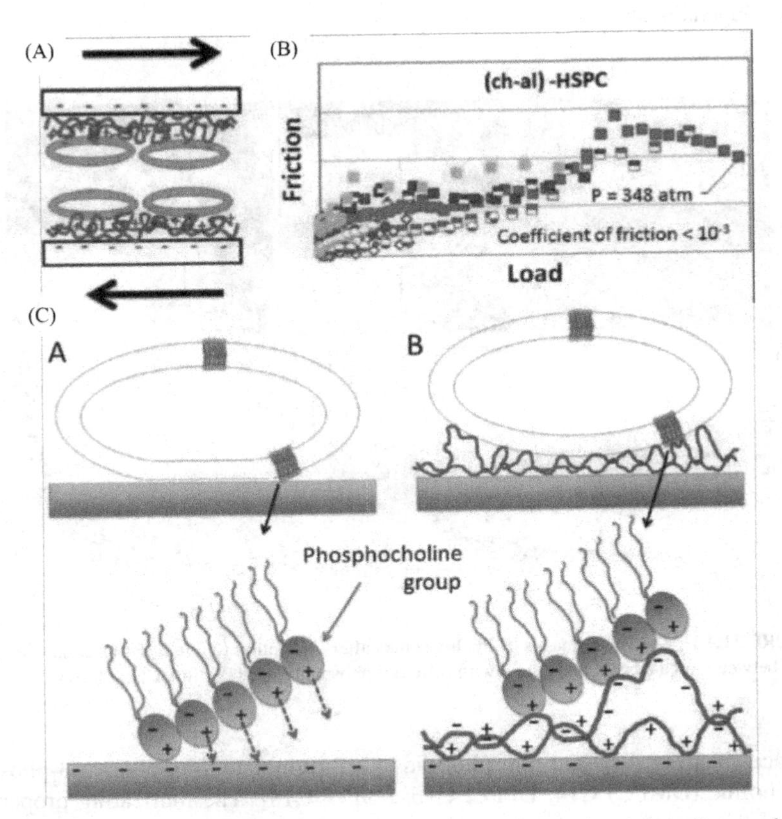

FIGURE 21.11 (A) Schematic diagram of DSPC vesicles on the mica surface after preadsorption of biomimetic polymer layer (alginate-chitosan). (B) Friction force versus normal load. (C) A tentative scenario to illustrate schematically the nature of the adsorption of PC liposomes [48].

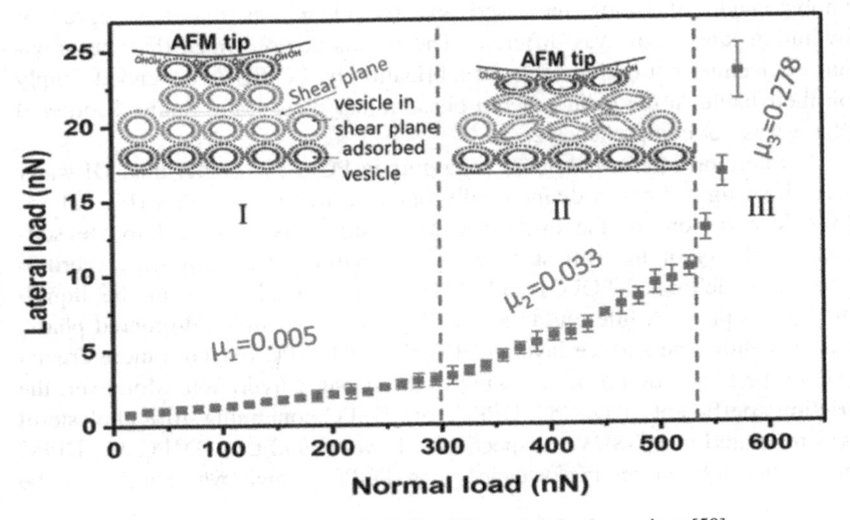

FIGURE 21.12 Friction coefficient of DPPC in different lubrication regions [50].

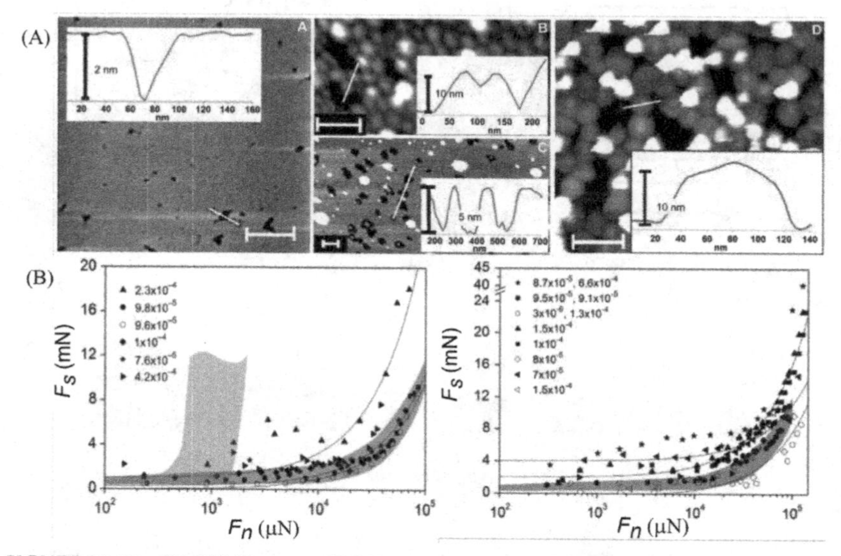

FIGURE 21.13 (A) AFM images of PC liposomes after adsorption to mica. (B) Friction coefficient between mica surfaces adsorbed with different phospholipid SUVs in DI water [51].

surfaces bearing different phospholipids [1, 2-dimyristoyl-sn-glycero-3-phosphocholine (DMPC) C14; DPPC, C16; DSPC, C18]. The lubricating properties of the vesicles significantly increased as the acyl chain length increased, which was closely related to the integrity of the surface structure, as shown in Fig. 21.13. The particular phenomenon might be attributed to two probable reasons. One was that the PC lipids had better robustness as the hydrophobic acyl tail length increased and the other was that the degree of hydration lubrication was different. The results demonstrated that the previous conjecture that highly efficient lubrication by PC-SUVs depended simply on their being in the solid-ordered phase rather than in the liquid-disordered phase was not comprehensive.

Furthermore, the tribological behaviors in PC-SUVs rather than DI water were investigated and a diametrically opposite trend was found (Fig. 21.14) [52]. The reason for the difference of friction was attributed to the self-healing ability of the soft surface layer. 1-palmitoyl-2-oleoyl-sn-glycero-3-phosphatidylcholine (POPC) and DMPC were found to be in the liquid-disordered phase, while DPPC and DSPC were both in solid-ordered phase. Consequently, the surface layers of POPC and DMPC played a more crucial role in the lubrication process owing to the stronger hydration. Moreover, the friction coefficient of DMPC, DPPC, and DSPC containing 40% cholesterol was measured in PC-SUVs, respectively. It was found that DMPC and DPPC had better lubrication performance than DSPC, which was similar to the

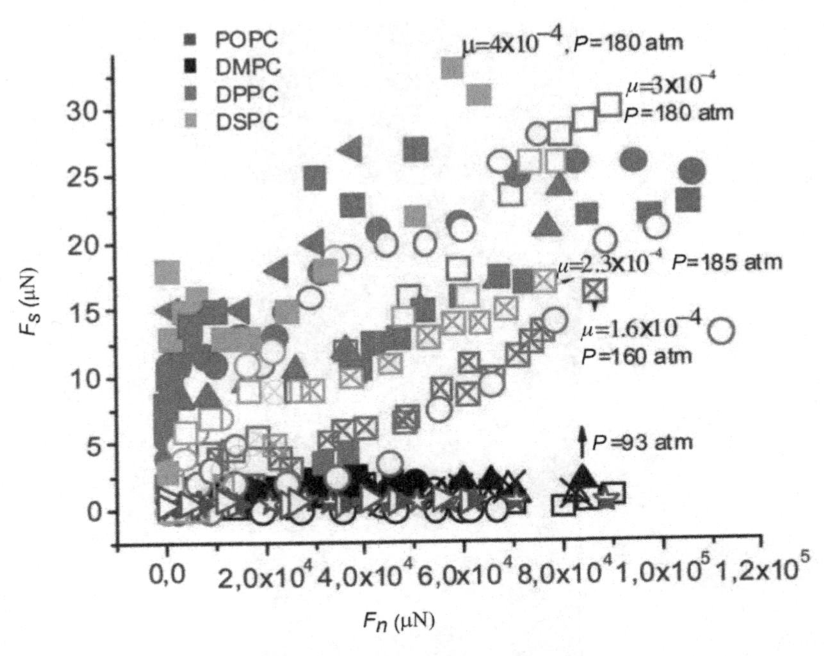

FIGURE 21.14 Friction coefficient between mica surfaces in PC-SUVs dispersion [52].

lubrication performance without cholesterol. It was indicated that superior lubricating behaviors mainly depended on the enhanced hydration lubrication mechanism [49].

Temperature is found to be another essential factor in influencing the friction of phospholipid due to the phase transition [35,36,53–56]. Some reports also showed that the phase transition temperature of the supported bilayer was different from that in the free state (not adsorbed on the surface). Tokumasu et al. [55] and Xie et al. [56] explored the phase transition of DMPC adsorbed on the mica surface. They found that DMPC in free state changed very rapidly from solid-ordered state to liquid-disordered state with a transition interval of less than 1°C at the transition temperature of 23.7°C. However when DMPC was adsorbed on mica, the transition interval was widened (8°C) and the transition temperature was moved to approximately 28°C, as shown in Fig. 21.15.

Furthermore, Wang et al. [57] performed detailed researches on the relationships between the increasing temperature and the friction force as well as the load-bearing capacity of DPPC bilayer adsorbed on the surface of the silicon wafer. The results suggested that the friction force would be maintained at low state throughout the whole temperature range, while the load-bearing capacity would improve with increasing temperature. The authors believed

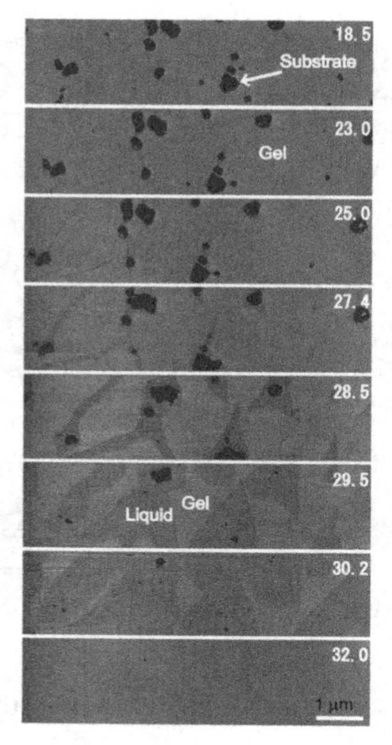

FIGURE 21.15 Topography of DMPC on the mica surface at different temperatures [55].

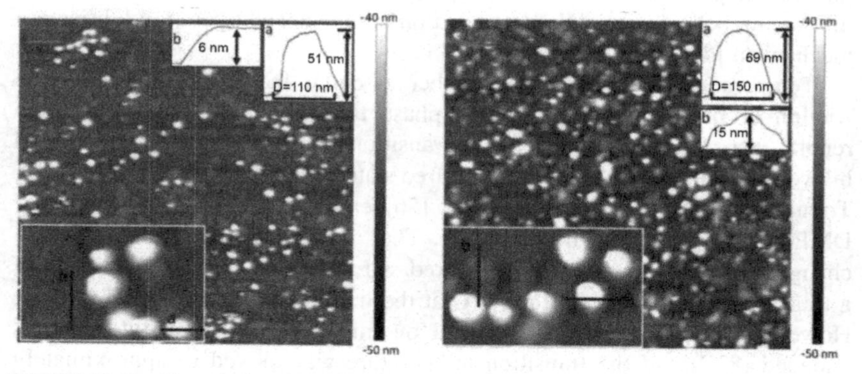

FIGURE 21.16 Topography of DPPC on titanium alloy surface at 60°C and 25°C [58].

that this particular phenomenon was attributed to the reason that DPPC bilayer had lower hardness and higher self-healing ability in the liquid-disordered phase than that in the gel state. Liu et al. [58] studied the mechanical properties of DPPC vesicles on titanium alloy surfaces at different temperatures (25°C and 60°C), as shown in Fig. 21.16. It was found that compared to the solid phase vesicles, the liquid phase vesicles were more metastable and more easily deformed to become bilayers. Besides, it has reversibility under multiple loading conditions for the solid-ordered phase, while the strength is lower in the liquid-disordered phase. Moreover, the defects will appear once the deformation is too large to recover because of the big pulling force.

In addition to the aforementioned effects, many other factors were found to have a significant influence on the tribological behaviors of phospholipids, such as vesicle size. Liu et al. [58] studied the effect of vesicle size on lubricating characteristics under macroscale conditions. It was found that the vesicles with smaller size had higher uniformity and could form a flat surface on the titanium alloy. The adsorption layer makes the energy of the interlaminar vesicles to overcome the deformation during the shearing process smaller, leading to a small friction coefficient, as shown in Fig. 21.17. As for the friction force with different liposome concentrations, the results showed that the thickness of the adsorption layer increased with the increasing concentration, which led to the effective reduction of the contact between the rough peaks. However, when the concentration increased to a certain extent, the shear resistance became large with some loose vesicles on top of the adsorption layer, resulting in the increased friction coefficient.

Regarding the synergy of HA and phospholipids, quantities of studies have confirmed that the lubricating properties of HA together with phospholipids were much better than that of HA itself [25,29,59–62]. Wang et al. [60,62] found that HA was difficult to be immobilized on the surface, while it could spontaneously adsorb on the phospholipid layer. Therefore they studied the synergy of HA and phospholipids on the silicon surface by quartz crystal microbalance with dissipation and AFM. As shown in Fig. 21.18, an adsorption structure model was proposed, in which the phospholipid bilayer was first adsorbed on the substrate, and then HA and phospholipid vesicles

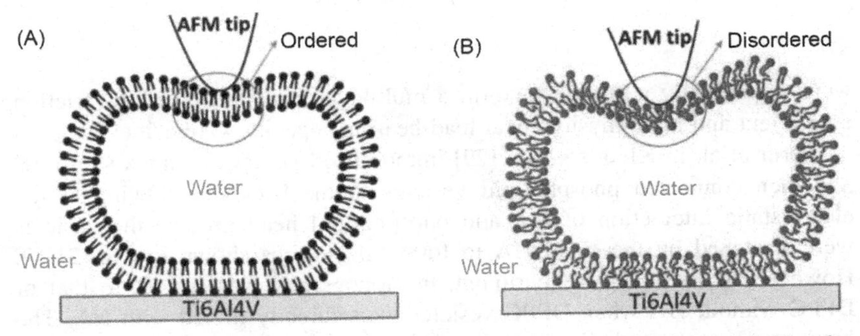

FIGURE 21.17 Schematic illustrations of vesicle deformation by AFM tip when lipid molecules are in (A) the gel phase and (B) the liquid phase [58].

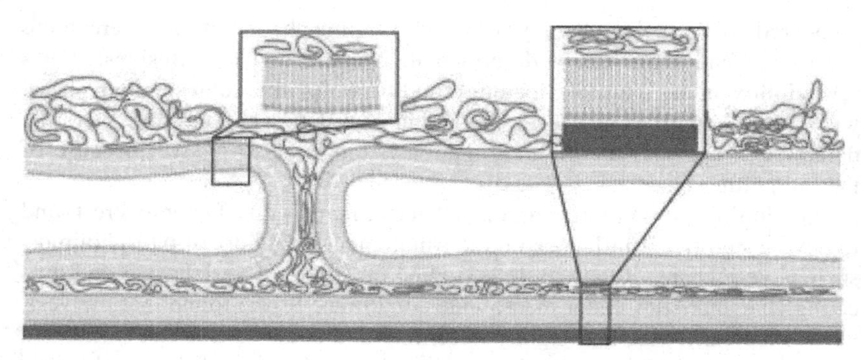

FIGURE 21.18 The adsorption schematic between HA and phospholipid [60].

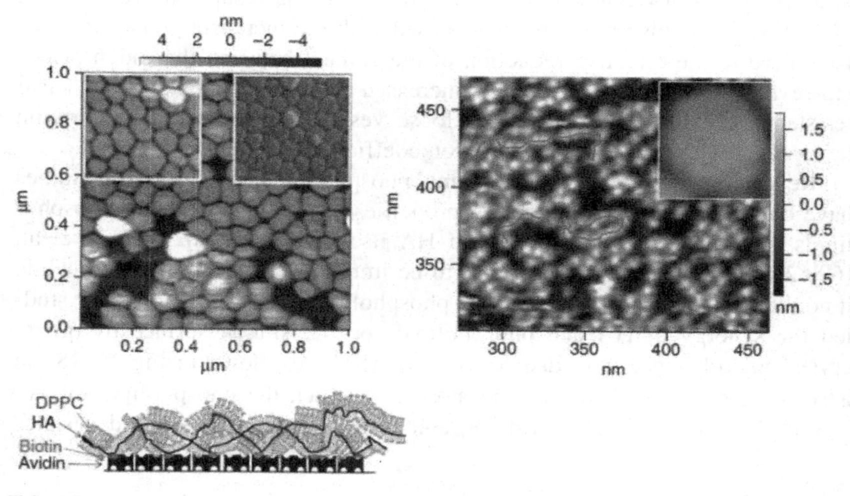

FIGURE 21.19 The topography of HA and phospholipids [29].

were sequentially adsorbed to form a multilayer structure. The low friction coefficient and high physiological load-bearing capacity were achieved.

Seror et al. in Klein's group [29] immobilized HA on the mica substrate, and then connected phospholipid vesicles to the backbone. Owing to the electrostatic interaction of HA and phospholipid head groups, the vesicles were stretched by the fixed HA to form bilayers, as shown in Fig. 21.19. However, in the control experiment, the topography was similar to that of DPPC without HA when DPPC vesicles were directly mixed with HA. The boundary layer revealed a friction coefficient around 0.001 under pressures up to 10 MPa, which is similar to that of natural joints. Moreover, HA was adsorbed on the mica surface and then phospholipid liposomes with different acyl chain lengths were added [63]. Then the friction force was measured by SFB after the mica surface was rinsed with pure water. The results revealed a low friction coefficient around 0.001 of HA-HSPC under physiological pressure conditions (15 MPa), while HA-DMPC and HA-POPC could only be effectively lubricated under very low pressure (1−2 MPa). The reduction of friction coefficient was attributed to the strong hydration effect by the highly hydrated PC head group. HA-HSPC had better lubricating behaviors and enhanced load-bearing capacity than that of HA-DMPC and HA-POPC under high normal load. Furthermore, the synergy of hyaluronic acid, lubricin, and PC phospholipids were also studied. The results suggested that HA, which would adsorb on the cartilage surfaces via the lubricin, had strong interactions with the liposomes, leading to the breakup of liposomes and the formation of the bilayer. Lately, at the macroscale, Liu et al. [64] found that the synergistic interaction of liposomes and biocompatible negative-charged polymer could efficiently improve the lubrication properties of liposomes.

21.3 Artificial biological materials

In addition to searching for new kinds of natural biological materials to explore the deep lubricating mechanism, quantities of studies are conducted to promote the development of artificial biological materials. Synthesis of artificial materials with superior lubricating properties and mechanical properties is the key incentive for bionic researches and industrial applications. Herein, the research progress on the lubricating properties of artificial biological materials is briefly reviewed, including polymer brushes and hydrogels.

21.3.1 Polymer brush

Polymer brush is one specific aggregation form of polymer chains in the interfaces, which is usually a monolayer structure with high grafting density of long-chain polymers. Generally, the distance between the molecular chains of the grafted polymer should be less than twice the z-average root-mean-square radius of gyration. Under the mutual influence of the steric resistance and the repulsive force between the chains, polymer chains are able to be perpendicular to the surface, which are extending outward and preventing large entanglements. In recent years, polymer brushes are found to be great boundary lubricants in liquid environments. Some kinds of linear polymer brushes and bottle-brushed polymer brush are demonstrated to achieve ultralow friction coefficient under high pressures, which is approaching the working condition in vivo.

21.3.1.1 Linear polymer brush

In 1966, Tabor et al. [65] began to study the tribological properties of polymers. They found that water film with the thickness of several diameters of a molecule could be formed on the hydrophilic surface and the nylon polymer surface. Ultralow friction coefficient could be achieved between these surfaces. Then lots of researchers [66–68] conducted a large number of studies on the tribological behaviors of polymer brushes since the 1990s. In early studies, polar polymers with high molecular weight were grafted on the mica surface via the polar groups on the chain, while the other head of the chain were floating in the toluene to form the "molecular brushes." Since the long-chain could be well extended in the solvent, strong osmotic pressure could be generated between the polymer brushes. It was demonstrated that the force field was in the range of the repulsive force field owing to the comprehensive effect of the reaction of the double layer repulsive force and the dispersive force between the polymers. As shown in Fig. 21.20, the repulsive force decreased rapidly as the distance of two mica surfaces increased. Consequently, during the tribological process, the mica surfaces could be separated by the polymer brushes due to the strong repulsion force, leading to a superlow friction coefficient of 0.001 [68].

Because water molecules have stronger polarity and fluidity than that of organic solvents, increasing numbers of researches select water as the solvent, which ensures that the polymer brushes could be stretched better to exhibit enhanced lubricating properties. Klein found that charged polymers (such as polyelectrolytes) could achieve a lower friction coefficient than that of other polymers when conducted tribological tests in an aqueous environment [69–71]. They grafted poly[2-(methacryloyloxy) ethyl phosphorylcholine]

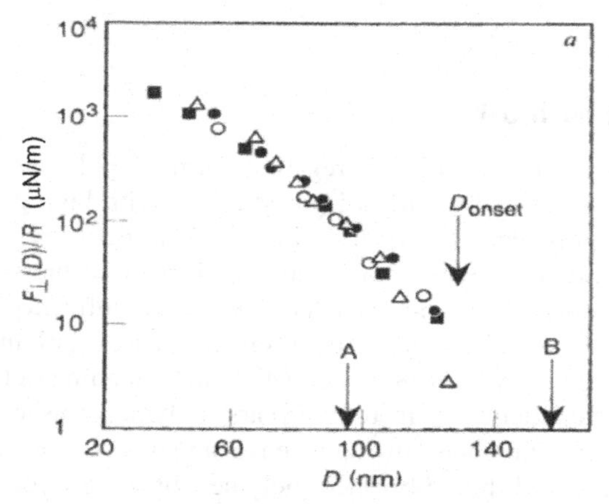

FIGURE 21.20 Relationship between repulsive force and surface distance formed by polymer brushes confined on the mica surface [66].

(PMPC) brushes on the mica surface, and measured the friction force by SFB [71,72,73]. It was found that the surface of mica with the modify of PMPC molecules was smooth, as shown in Fig. 21.21. And the lubrication performance of the modifying surface was excellent, which was attributed to the strong hydration effect of the phosphorylcholine group on the polymer chains.

Then, Tairy et al. [74] used the atom-transfer radical polymerization (ATRP) method with lower oxygen content to synthesize denser and longer-lasting PMPC brushes. They found that such PMPC brushes exhibited improved interfacial properties in both pure water and 0.2 M $NaNO_3$ solution. A friction coefficient with a value less than 10^{-4} was obtained under pressure over 150 MPa, as shown in Fig. 21.22.

Yu et al. [75] modified poly(glycidyl methacrylate) (PGMA) on the mica surface and then grafted PMPC brushes by the ATRP method, as shown in Fig. 21.23. The results showed that PGMA-PMPC brush was extremely stable and the thickness of the polymer film was reduced by only 1% after 4 weeks when immersing in saline solution and artificial seawater. Besides, the lubricating property of PGMA-PMPC was excellent with a friction coefficient of 0.0001–0.001 measured by AFM.

The scientists believed that there were a large number of oppositely charged ions moving around the charged brushes. The potentials generated by these ions suppressing the interpenetration between the polymer brushes are stronger than that of neutral polymer brushes. In this way, there were only very weak interpenetration effects between the charged brushes. Thus a narrow slip region was created between the two solid surfaces, in which the polymer fragments extending into the liquid were short and hard to get

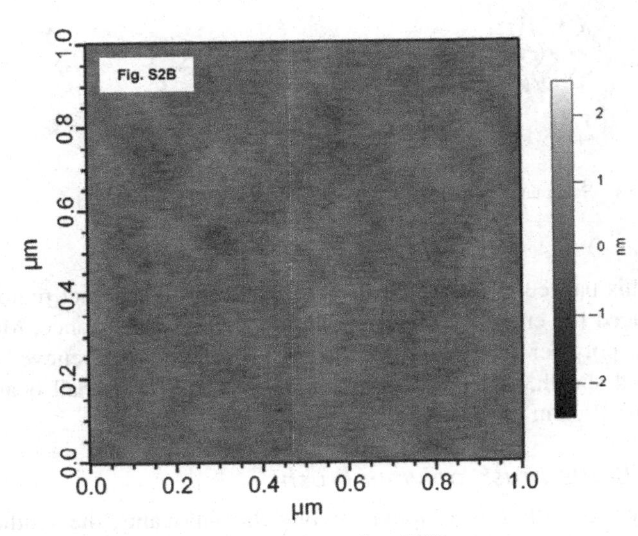

FIGURE 21.21 AFM image of PMPC on mica surface [73].

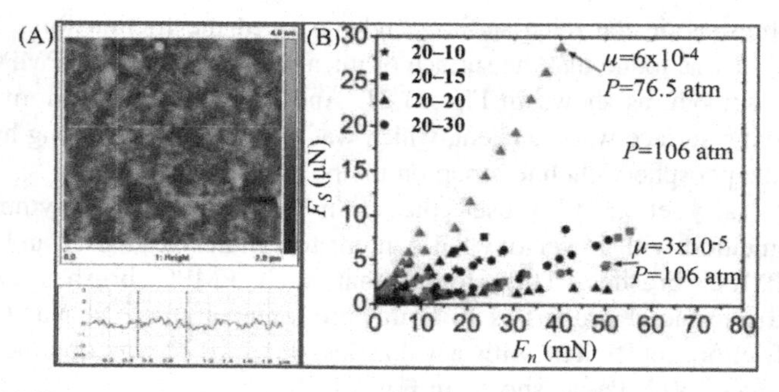

FIGURE 21.22 (A) AFM imaging of PMPC brushes. (b) Friction force of PMPC brushes obtained by ATRP methods in pure water and $NaNO_3$ solution [74].

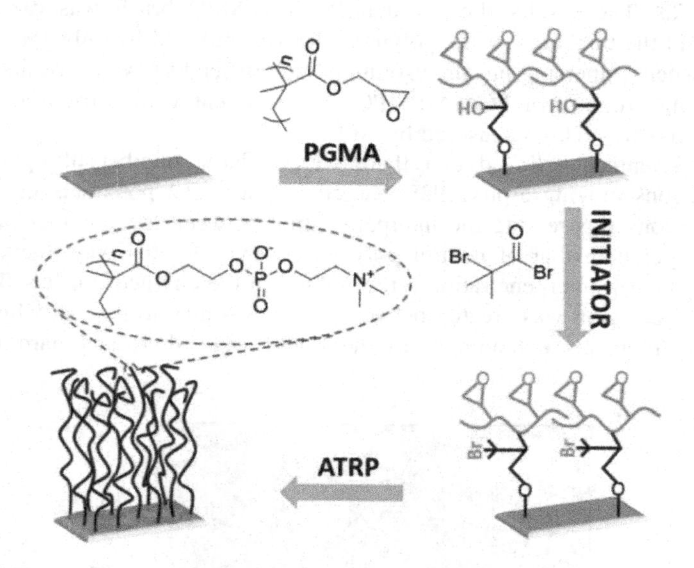

FIGURE 21.23 Schematic diagram of PGMA-PMPC by ATRP method [75].

tangled. This particular molecular-level structure in the contact region effectively reduced the energy dissipation and the frictional resistance. Moreover, the ionized polymer surrounded by the hydrated layer was believed to form the hydrated sheath, which was functioning as a molecular ball bearing and resulting in efficient lubrication [69].

21.3.1.2 Bottle-brush polymer brush

Inspired by the bottle-brush biomacromolecular lubricants, the synthesis and study of the lubricating properties of the bottle-brush polymers have attracted wide attention from researchers. Spencer et al. [76–80] grafted polyethylene glycol (PEG) onto polylysine, which adsorbed on the surface of charged silica by electrostatic interaction, as shown in Fig. 21.24. Thus the long chain of PEG could be extended into the aqueous solution to form the polymer brush, thereby greatly reducing the friction coefficient to a value around 0.0001 on the rolling test condition.

Israelachvili et al. [81] simulated the brush-like structure of the natural glycoprotein lubricant (lubricin) and prepared two bottle-shaped polymer brushes (quaternized 2-(dimethylaminoethyl) methacrylate (qDMAEMA)-*g*-PMPC and poly(methyl methacrylate) (PMMA)-*g*-PMPC). They reported a friction coefficient around 0.001 compared with that of 0.038 produced by lubricin, as shown in Fig. 21.25.

21.3.2 Hydrogel

Hydrogels are a kind of soft materials with a three-dimensional network structure formed by cross-linking hydrophilic polymers, as shown in Fig. 21.26. These hydrophilic polymer gels can be swollen in water without dissolving. The water molecules will be bound to the polymer and wrapped in the network. Owing to the low fluidity, hydrogels always have stable shapes and exhibit both solid-like properties, which can generate deformation under certain stress and then recover without the external force, and liquid-like properties, which can realize the fluid transport and solute diffusion in the polymer structure [82].

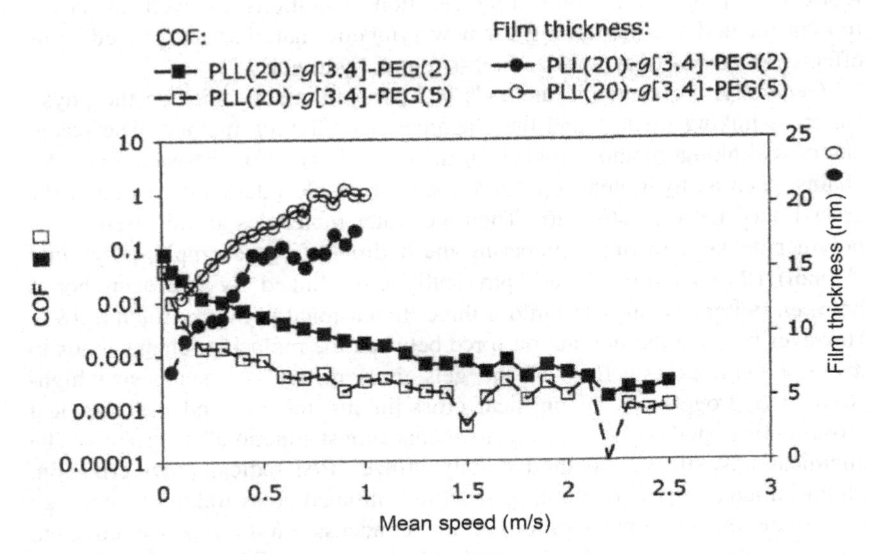

FIGURE 21.24 Friction coefficient versus speed [79].

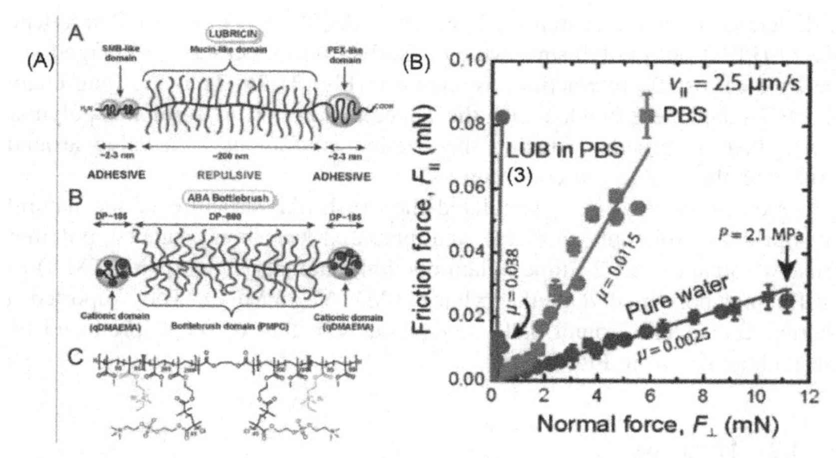

FIGURE 21.25 (A) Schematic representations of the protein lubricin and the mimicking bottle-brush polymer. (B) Friction force of synthetic polymer brushes (qDMAEMA-*g*-PMPC and PMMA-*g*-PMPC) [81].

21.3.2.1 High-strength hydrogels

Due to the particular properties and excellent biocompatibility, hydrogels are considered as valuable materials for bionic manufacturing. However, the mechanical strength of most hydrogels cannot reach the requirement of the joints, which limits the application of hydrogels. Numerous researches are carried out to synthesize hydrogels with enhanced mechanical properties and lubricating properties. Controlling reaction conditions as well as cross-linking methods and searching for new synthetic materials are proved to be effective methods to synthesize high-strength hydrogels [83].

Generally, the synthetic methods of hydrogels mainly include the physical cross-linking method and the chemical cross-linking method. The physical cross-linking method always utilizes the interaction between polymer chains, such as hydrogen bond, ion bond, or dipole interaction, to form the cross-linked network structure. Then the water molecules are diffused to the polymer network to finish preparing the hydrogels. For example, poly(vinyl alcohol) (PVA) hydrogels are physically cross-linked by hydrogen bonds between polymer chains to build a three-dimensional network structure [84]. However because the interaction force between the molecular chains tends to decrease with the swelling of the gels, it is difficult to synthesize high-strength hydrogels by the physical cross-linking method and the chemical cross-linking method is preferred to prepare most functional hydrogels. The chemical cross-linking method mainly utilizes free radical polymerization, photo-initiated polymerization, or radiation-initiated cross-linking technology to initiate the copolymerization or the condensation to produce covalent bonds to form cross-linking networks. For example, Gong et al. [84] first

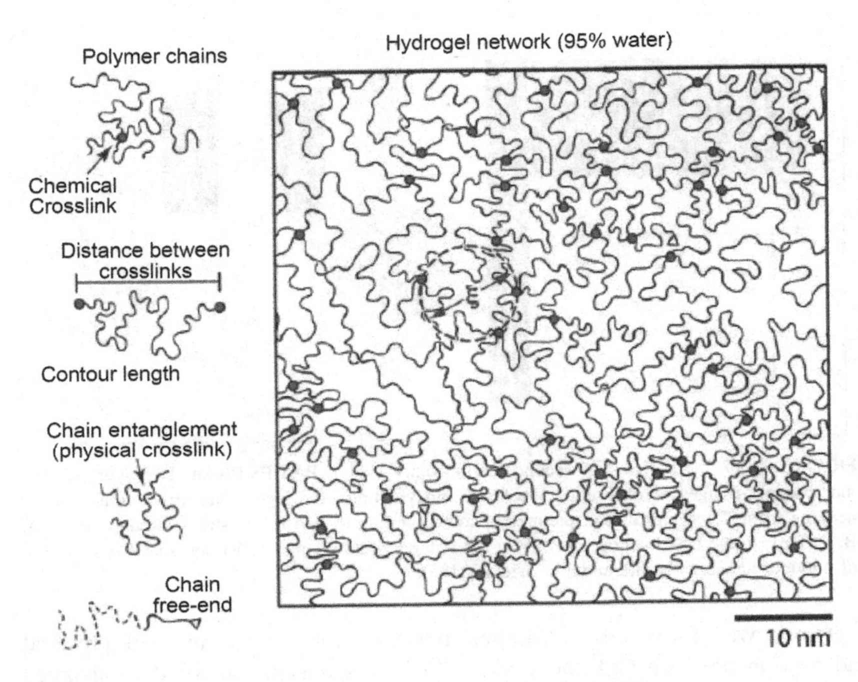

FIGURE 21.26 A schematic drawing of the structure of hydrogels [82].

synthesized poly (2-acrylamido-2-methylpropane sulfonic acid)/polyacrylamide (PAMPS/PAAm) double-network hydrogels through two-step free radical polymerization. They took PAMPS as the basic network to improve the rigidity and PAAm as the second network to improve the ductility. They found that the fracture energy was increased up to 1000 J/m^2, which was 0.1 J/m^2 for PAMPS hydrogels and 10 J/m^2 for Pam hydrogels, as shown in Fig. 21.27. They attributed this significant improvement to the unique network structure and strong interactions between polymers.

Moreover, introducing a few nanoparticles or ionogenic groups into the polymer network is widely used to improve the mechanical strength of hydrogels in recent years. The additives are able to enhance the effect of physical cross-linking through forming hydrogen bonds, intermolecular forces, and coordination bonds. Liu et al. [86] prepared polyacrylamide (PAM)/graphene oxide (GO) nanocomposite hydrogels (PGH) with GO nanosheets as cross-linkers were synthesized via in situ free-radical polymerization of acrylamide in an aqueous suspension of GO, as shown in Fig. 21.28. Compared to conventional PAM hydrogels (PBH), the tensile strength of PGH is about 4.5 times higher than that of PBH, and the elongation at break is over 3000%, nearly one order higher than that of PBH when introducing a small amount of GO. They found that an organic/inorganic

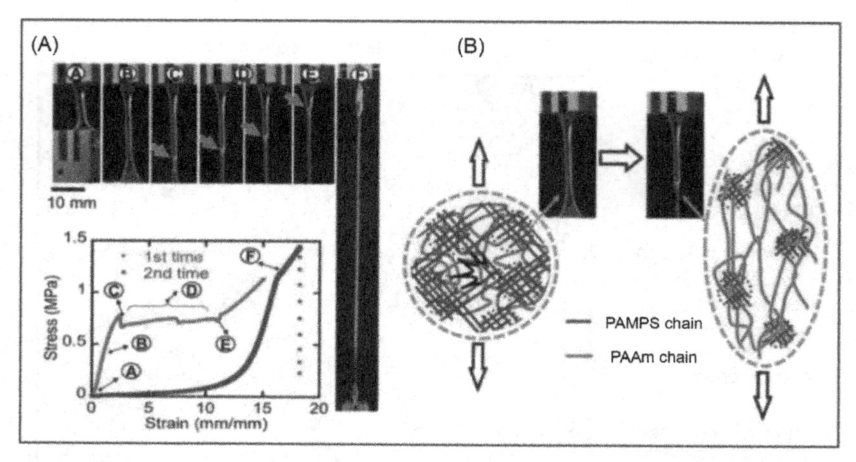

FIGURE 21.27 (A) First and second loading curves of a PAMPS/PAAm DN hydrogel and macroscopic images of elongation, necking, and yielding of the gel. The arrow indicates the necking front. Each alphabet in pictures is corresponding to that in the stress–strain curves (A, B, C, D, E in the first loading curve; F in the second loading curve). (b) Schematic illustrations of polymer networks before and after necking [85].

network was formed by hydrogen bonding, ionic bonding, and physical adsorption between GO and PAM, which was reasonable for the enhanced mechanical properties. Sun et al. [87] found that if the alginate hydrogels were immersed in the solution containing Ca^{2+}, Ca^{2+} would form coordination bonds with the negatively charged groups on the polymer chains, which could reinforce the strength of the hydrogel network. The synthesis process and structure of hydrogels reinforced with Ca^{2+} was shown in Fig. 21.29.

21.3.2.2 Lubricating properties of hydrogels

In recent years, researchers have carried out a series of studies on the tribological properties of high-strength hydrogels. Because the unique properties distinguished from the solid and liquid, all kinds of hydrogels exhibit various frictional behaviors. Besides environmental factors like temperature, load, and speed, the chemical structure, hydrophilic–hydrophobic interactions, charge effects, the morphology of frictional pairs, and preparations of templates all affect the lubricating properties of the hydrogels. It is found that the chemical structure of hydrogels is the most important factor to affect their lubricating performance. Gong et al. [82] compared the frictional properties of PVA hydrogels, gelatin hydrogels, and PAMPS hydrogels, and reported different friction-load curves. They found that the friction force was nearly unchanged with the increase of the applied load for the gelatin hydrogels, while the friction force would increase gradually with the increasing applied load for the nonionic PVA hydrogels and would increase

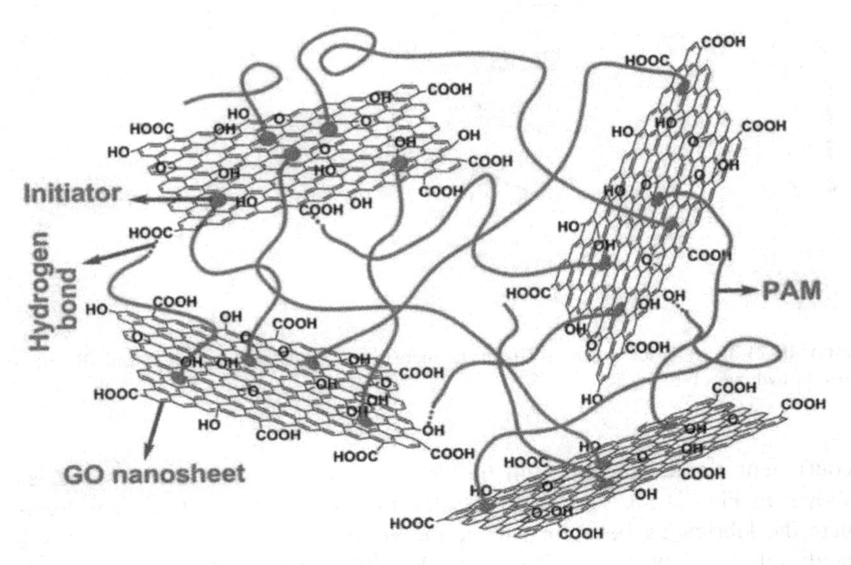

FIGURE 21.28 Hydrogels reinforced with GO [86].

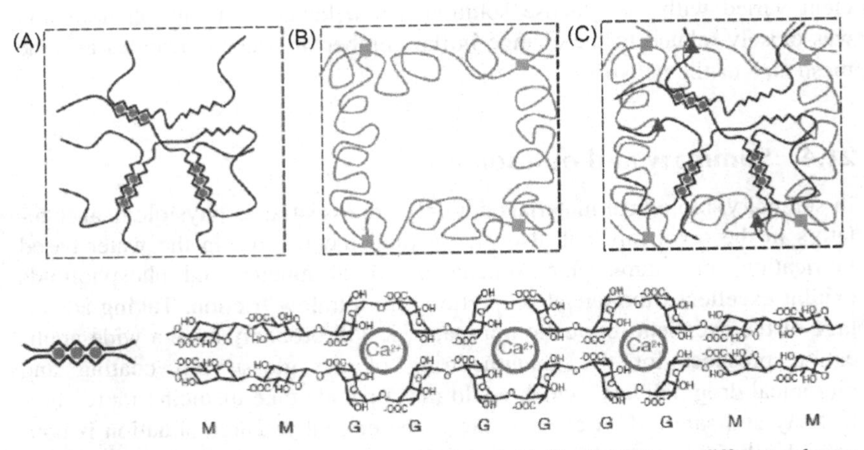

FIGURE 21.29 Schematics of three types of hydrogel. (a) The G blocks on different polymer chains form ionic crosslinks through Ca2 + in an alginate gel. (b) The polymer chains form covalent crosslinks through N,N-methylenebisacrylamide in a polyacrylamide gel. (c) The two types of polymer network are intertwined in an alginate–polyacrylamide hybrid gel [87].

significantly for the PAMPS hydrogels. They inferred that these differences between hydrogel were closely related to the structure of the polymer network, such as the mesh size. Saywer et al. [88,89] have taken systematic studies on the influence of the chemical structure and found that the friction

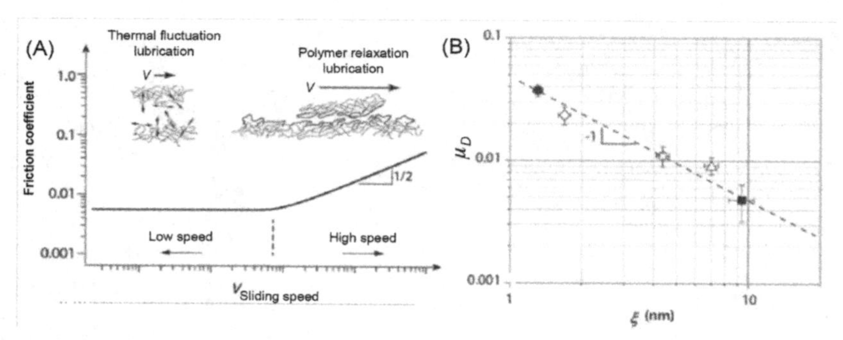

FIGURE 21.30 Friction coefficient with the increasing (A) sliding speed [88] and (B) mesh size of hydrogels [88].

coefficient would increase with the decrease of mesh size at low speed, as shown in Fig. 21.30. They believed that the thermal fluctuation may dominate the lubricating behaviors, while the non-Newtonian shearing effect was negligible at low speed. Zhao et al. [90] synthesized a new stimuli-responsive double-network hydrogel by introducing the glucose-sensitive group phenylboronic acid to the network. They found that the friction coefficient varied with the glucose solution and believed that this phenomenon was closely related to the change in the number of water molecules and the mesh size of the network.

21.4 Summary and outlook

In summary, biological materials not only are essential as physiological regulators in the organism, but also play a fundamental role in the water-based lubrication. In nature, plant mucilage, animal mucus, and phospholipids exhibit excellent tribological properties with ultralow friction. Taking advantage of the excellent lubrication property, it will probably have a wide application prospect for the manufacturing of the glossy pill coating and biological drug delivery, which could make people take medicine more comfortably and safely. However, as far as the current research situation is concerned, the exploration of the lubricating characteristics of the plant mucilage is still in a preliminary stage. More biolubricants with superior lubricating properties are expected to be discovered and evaluated. The deep lubricating mechanism is still of great significance to be explored.

To date reports, great progress has been made in the researches of artificial biological materials. Based on the structure of natural biolubricants, various polymer brushes are synthesized as efficient liquid superlubricants to get a microscopic ultralow friction coefficient. The osmotic pressure between the brushes can effectively suffer the normal load and the extended flexible chains help to reduce the shearing force between the solid surfaces. However, macroscale superlubricity is still hard to be achieved due to the high load and complex surface topography, which limits the industrial applications. Likewise, most hydrogels are too soft to suffer the applied load. More researches and experiments ought to be carried out to enhance the mechanical properties of artificial biological materials and finally achieve stable macrosuperlubiricty.

References

[1] D. Dowson, Bio-tribology, Faraday Discuss. 156 (2012).

[2] Z.R. Zhou, Z.M. Jin, Biotribology: recent progresses and future perspectives, Biosurface Biotribol. 1 (1) (2015) 3–24.

[3] J. Israelachvili, H. Wennerström, Role of hydration and water structure in biological and colloidal interactions, Nature 379 (6562) (1996) 219–225.

[4] L. Ma, C. Zhang, S. Liu, Progress in experimental study of aqueous lubrication, Chin. Sci. Bull. 57 (17) (2012) 2062–2069.

[5] S.K. Lai, Y.Y. Wang, D. Wirtz, J. Hanes, Micro- and macrorheology of mucus, Adv. Drug. Delivery Rev. 61 (2) (2009) 86–100.

[6] J.M. Coles, D.P. Chang, S. Zauscher, Molecular mechanisms of aqueous boundary lubrication by mucinous glycoproteins, Curr. Opin. Colloid Interface Sci. 15 (6) (2010) 406–416.

[7] Y. Dan, Biological functions of antioxidants in plant transformation, Vitro Cell. Develop. Biol. Plant. 44 (3) (2008) 149–161.

[8] L. Zhang, Y. Liu, Z. Chen, P. Liu, Behavior and mechanism of ultralow friction of basil seed gel, Colloids Surf. A Physicochem. Eng. Asp. 489 (2016) 454–460.

[9] S. Arad, L. Rapoport, A. Moshkovich, D. van Moppes, M. Karpasas, R. Golan, et al., Superior biolubricant from a species of red microalga, Langmuir 22 (17) (2006) 7313–7317.

[10] D. Gourdon, Q. Lin, E. Oroudjev, H. Hansma, Y. Golan, S. Arad, et al., Adhesion and stable low friction provided by a subnanometer-thick monolayer of a natural polysaccharide, Langmuir 24 (4) (2008) 1534–1540.

[11] J.J. Li, Y.H. Liu, J.B. Luo, P. Liu, C.H. Zhang, Excellent lubricating behavior of *Brasenia schreberi* mucilage, Langmuir 28 (20) (2012) 7797–7802.

[12] P. Liu, Y. Liu, Y. Yang, Z. Chen, J. Li, J. Luo, Mechanism of biological liquid superlubricity of *Brasenia schreberi* mucilage, Langmuir 30 (13) (2014) 3811–3816.

[13] R. Bansil, B.S. Turner, Mucin structure, aggregation, physiological functions and biomedical applications, Curr. Opin. Colloid Interface Sci. 11 (2) (2006) 164–170.

[14] G.J. Strous, J. Dekker, Mucin-type glycoproteins, Crit. Rev. Biochem. Mol. Biol. 27 (1-2) (1992) 57–92.

[15] R. Bansil, E. Stanley, J.T. Lamont, Mucin biophysics, Annu. Rev. Physiol. 57 (1) (1995) 635–657.

[16] S. Lee, M. Müller, K. Rezwan, N.D. Spencer, Porcine gastric mucin (PGM) at the water/poly (dimethylsiloxane) (PDMS) interface: influence of pH and ionic strength on its conformation, adsorption, and aqueous lubrication properties, Langmuir 21 (18) (2005) 8344–8353.

[17] N. Nikogeorgos, P. Efler, A.B. Kayitmazer, S. Lee, "Bio-glues" to enhance slipperiness of mucins: improved lubricity and wear resistance of porcine gastric mucin (PGM) layers assisted by mucoadhesion with chitosan, Soft Matter 11 (3) (2015) 489–498.

[18] H. Forster, J. Fisher, The influence of loading time and lubricant on the friction of articular cartilage, Proc. Inst. Mech. Eng. H. 210 (2) (1996) 109–119.

[19] W.A. Hodge, R.S. Fijan, K.L. Carlson, R.G. Burgess, W.H. Harris, R.W. Mann, Contact pressures in the human hip joint measured in vivo, Proc. Natl Acad. Sci. U.S.A. 83 (9) (1986) 2879–2883.

[20] W.A. Hodge, K.L. Carlson, R.S. Fijan, R.G. Burgess, P.O. Riley, W.H. Harris, et al., Contact pressures from an instrumented hip endoprosthesis, J. Bone Jt. Surg. Am. 71 (9) (1989) 1378–1386.

[21] J. Klein, Molecular mechanisms of synovial joint lubrication, Proc. Inst. Mech. Eng. Part J 220 (8) (2006) 691–710.

[22] T. Murakami, Y. Sawae, K. Nakashima, S. Yarimitsu, T. Sato, Micro- and nanoscopic biotribological behaviours in natural synovial joints and artificial joints, Proc. Inst. Mech. Eng. Part J 221 (3) (2007) 237–245.

[23] M. Kung, J. Markantonis, S. Nelson, P. Campbell, The synovial lining and synovial fluid properties after joint arthroplasty, Lubricants 3 (2) (2015) 394–412.

[24] M. Daniel, Boundary cartilage lubrication: review of current concepts, Wien. Med. Wochenschr. 164 (5–6) (2014) 88–94.

[25] S. Jahn, J. Klein, Hydration lubrication: the macromolecular domain, Macromolecules 48 (15) (2015) 5059–5075.

[26] S. Jahn, J. Seror, J. Klein, Lubrication of articular cartilage, Annu. Rev. Biomed. Eng. 18 (2016) 235–258.

[27] J. Yu, X. Banquy, G.W. Greene, D.D. Lowrey, J.N. Israelachvili, The boundary lubrication of chemically grafted and cross-linked hyaluronic acid in phosphate buffered saline and lipid solutions measured by the surface forces apparatus, Langmuir 28 (4) (2012) 2244–2250.

[28] S.E. Majd, R. Kuijer, A. Kowitsch, T. Groth, T.A. Schmidt, P.K. Sharma, Both hyaluronan and collagen type II keep proteoglycan 4 (lubricin) at the cartilage surface in a condition that provides low friction during boundary lubrication, Langmuir 30 (48) (2014) 14566–14572.

[29] J. Seror, L. Zhu, R. Goldberg, A.J. Day, J. Klein, Supramolecular synergy in the boundary lubrication of synovial joints, Nat. Commun. 6 (2015) 6497.

[30] A. Dedinaite, P.M. Claesson, Synergies in lubrication, Phys. Chem. Chem Phys 19 (35) (2017) 23677–23689.

[31] V. Saikko, T. Ahlroos, Phospholipids as boundary lubricants in wear tests of prosthetic joint materials, Wear 207 (1–2) (1997) 86–91.

[32] B.A. Hills, Boundary lubrication in vivo, Proc. Inst. Mech. Eng. H. 214 (1) (2000) 83–94.

[33] M. Hetzer, S. Heinz, S. Grage, T.M. Bayerl, Asymmetric molecular friction in supported phospholipid bilayers revealed by NMR measurements of lipid diffusion, Langmuir 14 (5) (1998) 982–984.

[34] B.A. Hills, B.D. Butler, Surfactants identified in synovial-fluid and their ability to act as boundary lubricants, Ann. Rheum. Dis. 43 (4) (1984) 641–648.

[35] R. Koynova, M. Caffrey, Phases and phase transitions of the phosphatidylcholines, Biochim. Biophys. Acta 1376 (1) (1998) 91–145.

[36] A. Charrier, F. Thibaudau, Main phase transitions in supported lipid single-bilayer, Biophys. J. 89 (2) (2005) 1094–1101.

[37] S. Tristram-Nagle, J.F. Nagle, Lipid bilayers: thermodynamics, structure, fluctuations, and interactions, Chem. Phys. Lipids 127 (1) (2004) 3–14.

[38] T. Oguchi, K. Sakai, H. Sakai, M. Abe, AFM surface morphology and friction force studies of microscale domain structures of binary phospholipids, Colloids Surf. B Biointerfaces 79 (1) (2010) 205–209.

[39] B.A. Hills, R.W. Crawford, Normal and prosthetic synovial joints are lubricated by surface-active phospholipid, J. Arthroplasty 18 (4) (2003) 499–505.

[40] M.-C. Corneci, F. Dekkiche, A.-M. Trunfio-Sfarghiu, M.-H. Meurisse, Y. Berthier, J.-P. Rieu, Tribological properties of fluid phase phospholipid bilayers, Tribol. Int. 44 (12) (2011) 1959–1968.

[41] A. Mavraki, P.M. Cann, Friction and lubricant film thickness measurements on simulated synovial fluids, Proc. Inst. Mech. Eng. Part J 223 (3) (2009) 325–335.

[42] R. Crockett, Boundary lubrication in natural articular joints, Tribol. Lett. 35 (2) (2009) 77–84.

[43] A.M. Trunfio-Sfarghiu, Y. Berthier, M.H. Meurisse, J.P. Rieu, Role of nanomechanical properties in the tribological performance of phospholipid biomimetic surfaces, Langmuir 24 (16) (2008) 8765–8771.

[44] R. Goldberg, A. Schroeder, Y. Barenholz, J. Klein, Interactions between adsorbed hydrogenated soy phosphatidylcholine (HSPC) vesicles at physiologically high pressures and salt concentrations, Biophys. J. 100 (10) (2011) 2403–2411.

[45] R. Goldberg, A. Schroeder, G. Silbert, K. Turjeman, Y. Barenholz, J. Klein, Boundary lubricants with exceptionally low friction coefficients based on 2D close-packed phosphatidylcholine liposomes, Adv. Mater. 23 (31) (2011) 3517–3521.

[46] A. Gaisinskaya-Kipnis, S. Jahn, R. Goldberg, J. Klein, Effect of glucosamine sulfate on surface interactions and lubrication by hydrogenated soy phosphatidylcholine (HSPC) liposomes, Biomacromolecules 15 (11) (2014) 4178–4186.

[47] R. Sorkin, Y. Dror, N. Kampf, J. Klein, Mechanical stability and lubrication by phosphatidylcholine boundary layers in the vesicular and in the extended lamellar phases, Langmuir 30 (17) (2014) 5005–5014.

[48] A. Gaisinskaya-Kipnis, J. Klein, Normal and frictional interactions between liposome-bearing biomacromolecular bilayers, Biomacromolecules 17 (8) (2016) 2591–2602.

[49] R. Sorkin, N. Kampf, J. Klein, Effect of cholesterol on the stability and lubrication efficiency of phosphatidylcholine surface layers, Langmuir 33 (30) (2017) 7459–7467.

[50] Y. Duan, Y. Liu, J. Li, S. Feng, S. Wen, AFM study on superlubricity between Ti6Al4V/polymer surfaces achieved with liposomes, Biomacromolecules 20 (4) (2019) 1522–1529.

[51] R. Sorkin, N. Kampf, Y. Dror, E. Shimoni, J. Klein, Origins of extreme boundary lubrication by phosphatidylcholine liposomes, Biomaterials 34 (22) (2013) 5465–5475.

[52] R. Sorkin, N. Kampf, L. Zhu, J. Klein, Hydration lubrication and shear-induced self-healing of lipid bilayer boundary lubricants in phosphatidylcholine dispersions, Soft Matter 12 (10) (2016) 2773–2784.

[53] S.V. Bennun, M. Longo, R. Faller, Phase and mixing behavior in two-component lipid bilayers: a molecular dynamics study in DLPC/DSPC mixtures, J. Phys. Chem. B 111 (32) (2007) 9504–9512.

[54] M.P. Goertz, B.L. Stottrup, J.E. Houston, X.Y. Zhu, Nanomechanical contrasts of gel and fluid phase supported lipid bilayers, J. Phys. Chem. B 113 (27) (2009) 9335–9339.

[55] F. Tokumasu, A.J. Jin, G.W. Feigenson, J.A. Dvorak, Atomic force microscopy of nanometric liposome adsorption and nanoscopic membrane domain formation, Ultramicroscopy 97 (1–4) (2003) 217–227.

[56] A.F. Xie, R. Yamada, A.A. Gewirth, S. Granick, Materials science of the gel to fluid phase transition in a supported phospholipid bilayer, Phys. Rev. Lett. 89 (24) (2002) 246103.

[57] M. Wang, T. Zander, X. Liu, C. Liu, A. Raj, W.D.C. Florian, et al., The effect of temperature on supported dipalmitoylphosphatidylcholine (DPPC) bilayers: structure and lubrication performance, J. Colloid Interface Sci. 445 (2015) 84–92.

[58] Y. Duan, Y. Liu, J. Li, H. Wang, S. Wen, Investigation on the nanomechanics of liposome adsorption on titanium alloys: temperature and loading effects, Polym. (Basel) 10 (4) (2018).

[59] R.W. Forsey, J. Fisher, J. Thompson, M.H. Stone, C. Bell, E. Ingham, The effect of hyaluronic acid and phospholipid based lubricants on friction within a human cartilage damage model, Biomaterials 27 (26) (2006) 4581–4590.

[60] M. Wang, C. Liu, E. Thormann, A. Dedinaite, Hyaluronan and phospholipid association in biolubrication, Biomacromolecules 14 (12) (2013) 4198–4206.

[61] A. Raj, M. Wang, T. Zander, D.C.F. Wieland, X. Liu, J. An, et al., Lubrication synergy: mixture of hyaluronan and dipalmitoylphosphatidylcholine (DPPC) vesicles, J. Colloid Interface Sci. 488 (2017) 225–233.

[62] C. Liu, M. Wang, J. An, E. Thormann, A. Dėdinaitė, Hyaluronan and phospholipids in boundary lubrication, Soft Matter 8 (40) (2012).

[63] L.Y. Zhu, J. Seror, A.J. Day, N. Kampf, J. Klein, Ultra-low friction between boundary layers of hyaluronan-phosphatidylcholine complexes, Acta Biomaterialia 59 (2017) 283–292.

[64] Z. Wang, J. Li, X. Ge, Y. Liu, J. Luo, D.G. Chetwynd, et al., Investigation of the lubrication properties and synergistic interaction of biocompatible liposome-polymer complexes applicable to artificial joints, Colloids Surf. B Biointerfaces 178 (2019) 469–478.

[65] S.C. Cohen, D. Tabor, Friction and lubrication of polymers, Proceedings of the Royal Society of London Series a—Mathematical and Physical Sciences 291 (1425) (1966) 186–207.

[66] J. Klein, D. Perahia, S. Warburg, Forces between polymer-bearing surfaces undergoing shear, Nature 352 (6331) (1991) 143–145.

[67] J. Klein, Y. Kamiyama, H. Yoshizawa, N.J. Israelachvili, G. Fredrickson, P. Pincus, et al., Lubrication forces between surfaces bearing polymer brushes, Macromolecules 26 (21) (1993) 5552–5560.

[68] J. Klein, E. Kumacheva, D. Mahalu, D. Perahia, L.J. Fetters, Reduction of frictional forces between solid surfaces bearing polymer brushes, Nature 370 (6491) (1994) 634–636.

[69] J. Klein, U. Raviv, S. Perkin, N. Kampf, L. Chai, S. Giasson, Fluidity of water and of hydrated ions confined between solid surfaces to molecularly thin films, J. Physics: Condensed Matter 16 (45) (2004) S5437–S5448.

[70] U. Raviv, S. Giasson, N. Kampf, J.-F. Gohy, R. Jérôme, J. Klein, Lubrication by charged polymers, Nature 425 (6954) (2003) 163–165.

[71] M. Chen, W.H. Briscoe, S.P. Armes, J. Klein, Lubrication at physiological pressures by polyzwitterionic brushes, Science 323 (5922) (2009) 1698–1701.

[72] M. Chen, W.H. Briscoe, S.P. Armes, H. Cohen, J. Klein, Polyzwitterionic brushes: extreme lubrication by design, Eur. Polym. J. 47 (4) (2011) 511–523.

[73] W.H. Briscoe, M. Chen, I.E. Dunlop, J. Klein, J. Penfold, R.M. Jacobs, Applying grazing incidence X-ray reflectometry (XRR) to characterising nanofilms on mica, J. Colloid. Interface Sci. 306 (2) (2007) 459–463.

[74] O. Tairy, N. Kampf, M.J. Driver, S.P. Armes, J. Klein, Dense, highly hydrated polymer brushes via modified atom-transfer-radical-polymerization: structure, surface interactions, and frictional dissipation, Macromolecules 48 (1) (2014) 140–151.

[75] Y. Yu, G.J. Vancso, S. de Beer, Substantially enhanced stability against degrafting of zwitterionic PMPC brushes by utilizing PGMA-linked initiators, Eur. Polym. J. 89 (2017) 221–229.

[76] M.T. Müller, X. Yan, S. Lee, S.S. Perry, N.D. Spencer, Preferential solvation and its effect on the lubrication properties of a surface-bound, brushlike copolymer, Macromolecules 38 (9) (2005) 3861–3866.

[77] M.T. Müller, X. Yan, S. Lee, S.S. Perry, N.D. Spencer, Lubrication properties of a brush-like copolymer as a function of the amount of solvent absorbed within the brush, Macromolecules 38 (13) (2005) 5706–5713.

[78] X. Yan, S.S. Perry, N.D. Spencer, S. Pasche, S.M. De Paul, M. Textor, et al., Reduction of friction at oxide interfaces upon polymer adsorption from aqueous solutions, Langmuir 20 (2) (2004) 423–428.

[79] M. Müller, S. Lee, H.A. Spikes, N.D. Spencer, The influence of molecular architecture on the macroscopic lubrication properties of the brush-like co-polyelectrolyte poly(L-lysine)-*g*-poly(ethylene glycol) (PLL-*g*-PEG) adsorbed on oxide surfaces, Tribol. Lett. 15 (4) (2003) 395–405.

[80] G.L. Kenausis, J. Vörös, D.L. Elbert, N. Huang, R. Hofer, L. Ruiz-Taylor, et al., Poly(L-lysine)-*g*-poly(ethylene glycol) layers on metal oxide surfaces: attachment mechanism and effects of polymer architecture on resistance to protein adsorption, J. Phys. Chem. B 104 (14) (2000) 3298–3309.

[81] X. Banquy, J. Burdynska, D.W. Lee, K. Matyjaszewski, J. Israelachvili, Bioinspired bottle-brush polymer exhibits low friction and Amontons-like behavior, J. Am. Chem. Soc. 136 (17) (2014) 6199–6202.

[82] J.P. Gong, Friction and lubrication of hydrogels—its richness and complexity, Soft Matter 2 (7) (2006) 544–552.

[83] H. Furukawa, K. Horie, R. Nozaki, M. Okada, Swelling-induced modulation of static and dynamic fluctuations in polyacrylamide gels observed by scanning microscopic light scattering, Phys. Rev. E 68 (3) (2003) 031406.

[84] T. Takigawa, H. Kashihara, K. Urayama, T. Masuda, Structure and mechanical properties of poly(vinyl alcohol) gels swollen by various solvents, Polymer 33 (11) (1992) 2334–2339.

[85] T. Nonoyama, J.P. Gong, Double-network hydrogel and its potential biomedical application: a review, Proc. Inst. Mech. Eng. Part H 229 (12) (2015) 853–863.

[86] R. Liu, S. Liang, X.Z. Tang, D. Yan, X. Li, Z.Z. Yu, Tough and highly stretchable graphene oxide/polyacrylamide nanocomposite hydrogels, J. Mater. Chem. 22 (28) (2012) 14160–14167.

[87] J.Y. Sun, X. Zhao, W.R.K. Illeperuma, O. Chaudhuri, K.H. Oh, D.J. Mooney, et al., Highly stretchable and tough hydrogels, Nature 489 (7414) (2012) 133–136.

[88] J.M. Urueña, A.A. Pitenis, R.M. Nixon, K.D. Schulze, T.E. Angelini, G.W. Sawyer, Mesh size control of polymer fluctuation lubrication in Gemini hydrogels, Biotribology 1–2 (2015) 24–29.

[89] A.A. Pitenis, J. Manuel Urueña, A.C. Cooper, T.E. Angelini, G.W. Sawyer, Superlubricity in Gemini hydrogels, J. Tribol. 138 (4) (2016).

[90] J. Zhao, P. Liu, Y. Liu, Adjustable tribological behavior of glucose-sensitive hydrogels, Langmuir 34 (25) (2018) 7479–7487.

Chapter 22

Superlubricity of black phosphorus as lubricant additive

Guoxin Xie, Wei Wang, Xiaoyong Ren, Shuai Wu, Hanjuan Gong and Jianbin Luo
State Key Laboratory of Tribology, Department of Mechanical Engineering, Tsinghua University, Beijing, China

22.1 Introduction

Superlubricity has been regarded as a lubrication state where the sliding friction coefficient is smaller than 0.01 [1–3]. It has been observed in aqueous solutions confined in between molecular polymeric brush layers [4,5] or ceramic materials [6–8]. The load-supporting capacity in liquid superlubricity gradually increased in the past decade, but in most cases, the average contact pressures (CPs) in the liquid superlubricity state were rarely over ~300 MPa. In spite of these efforts, it is still undesirably low for practical applications where high-contact-pressure operating conditions normally exist. Hence, it is of high significance to develop novel lubricant systems to achieve superlubricity at elevated CPs.

Two-dimensional (2D) nanomaterials, for example, graphene, molybdenum disulfide (MoS_2), and hexagonal boron nitride (*h*-BN), have attracted great scientific interest, owing to their remarkable electrical, thermal, optical and mechanical properties [9–13]. As a new kind of 2D material, black phosphorus (BP) has recently received considerable attention [14,15], and it has been intensely investigated for the potential applications in electronics and optoelectronics. Apart from these progresses, it is also of great scientific interests to explore whether it can be potentially used as effective lubricants. A relevant systematic research work including the synthesis and characterization of BP nanomaterials, as well as their tribological properties at different scales has been conducted [16–19]. It has been found that the degradation of BP flakes favored the enhancement of the lubrication properties, and a super-slippery degraded BP flake/hydrophilic silicon dioxide (SiO_2) surface with an ultralow interfacial shear strength (~0.029 MPa) was found. When aqueous suspensions with BP nanosheets were used as the lubricants between the frictional

Superlubricity. DOI: https://doi.org/10.1016/B978-0-444-64313-1.00022-3

pairs of Si_3N_4/sapphire, the robust liquid superlubricity state was achieved under a CP larger than 600 MPa. The results demonstrated that BP had shown very promising prospects as future high-performance lubricants.

22.2 Black phosphorus

Phosphorus (P) is the group V of the periodic table, and its allotropes include white phosphorus (WP), red phosphorus (RP), BP, etc. BP is nontoxic and the most thermodynamically stable at room temperature [20]. Being similar to bulk graphite, MoS_2, *h*-BN, bulk BP has a layered structure where individual layers are stacked together by weak van der Waals interactions. Every BP individual layer consists of two atomic layers and two kinds of P—P bonds, and each phosphorus atom has three bonding nearby atoms through sp^3 hybridized orbitals and one lone pair, forming a puckered honeycomb lattice [21]. The bond length of the nearest P atoms in the same atomic layer is 0.2224 nm, and the bond length between the P atoms of different atomic layers is 0.2244 nm [22]. The bond angles of the hexagonal structure are 102.09 and 96.34 degrees from the top view of BP, and the interlayer distance is 5.3 Å [23,24]. Moreover, two inequivalent directions at the edge of BP are present, that is, zigzag (along the puckered direction) and armchair (perpendicular to the puckered direction), giving rise to many anisotropic characteristics of BP.

22.3 Preparation of black phosphorus

In 1914, Bridgman firstly successfully converted WP to BP at 200°C under a pressure of 1.2 GPa [25], and various methods have been developed for the preparation of BP. This method was greatly optimized in 1980s, and the

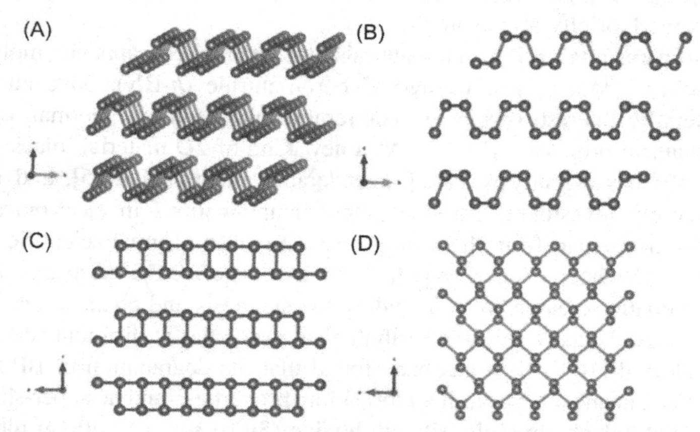

FIGURE 22.1 (A) Schematic diagram of the crystalline structure of BP flakes, (B) the armchair structure, (C) the zigzag structure, and (D) the top view of BP flakes. *Adapted with permission from W. Wang, G. Xie, J. Luo, Black phosphorus as a new lubricant, Friction 6 (2018) 116–142.*

quality and the size of BP single crystals were significantly improved [26]. The preparation of BP was also achieved by using the high-energy ball-milling (HEBM) methods [27], and the violent collisions of steel or ceramic balls in a milling vessel could provide high pressure and temperature for the transformation from RP to BP. Fig. 22.2A shows the XRD results of the RP powder and the inset photo is the image of RP. There are two broad peaks at $2\theta = 30$ and 55 degrees, indicating the amorphous structure of RP [28]. The TEM electron diffraction pattern in Fig. 22.2C shows a diffuse ring of RP, suggesting that the RP is amorphous. After BM for 2 h (Fig. 22.2B), the obvious diffraction peaks at $2\theta = 25$, 35, and 56 degrees indicate the phase transformation of RP to orthorhombic BP.

In the case of the preparation of phosphorene or few-layer BP, the top-down procedure is applicable because the weak van der Waals force dominates the interlayer interaction of BP. Hence, the method of micromechanical cleavage followed by being transferred onto a SiO_2/Si substrate has been

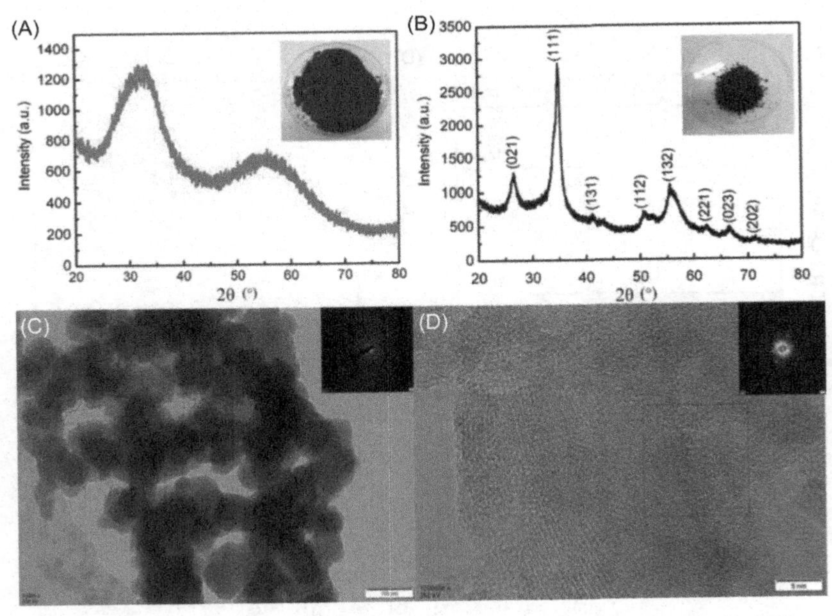

FIGURE 22.2 (A) The XRD pattern of RP material and the inset is a photograph of RP powder, (B) the XRD pattern of BP nanoparticles obtained for HEBM 2 h. The inset is a photograph of BP nanopowder. (C) The TEM images of RP. (D) The HRTEM images of BP. The milling vial was filled with argon gas and set in ball miller. The milling process was performed at different times. The revolution speed was set to 800 rpm. *(A and B) Reprinted (adapted) with permission from W. Wang, G. Xie, J. Luo, Superlubricity of BP as lubricant additive, ACS Appl. Mater. Interfaces 10(49) (2018) 43203–43210. Copyright (2018) American Chemical Society; (C and D) Adapted with permission from W. Wang, G. Xie, J. Luo, Black phosphorus as a new lubricant, Friction 6 (2018) 116–142.*

predominantly employed [29,30] for obtaining phosphorene or few-layer BP, being similar to other 2D materials. The Raman spectra of the few-layer BP flakes obtained with micromechanical cleavage is shown in Fig. 22.3A. The peaks at 359.9, 437.7, 468.0, and 519.2 cm^{-1} correspond to the $A_g{}^1$, $B_{2g}{}^1$, $A_g{}^2$ phonon modes and silicon substrate [31], indicating that the BP flakes remained crystalline after cleavage. After integrating the $A_g{}^1$ and Si peaks, the intensity ratios of the two peaks depend linearly with the layer number, as shown in Fig. 22.3B, offering the possibility of a method for determining the layer number of the flakes with an accuracy of two to three layers [24]. The XPS spectroscopy of the BP flakes after cleavage only contains two peaks, that is, P1 and P1, corresponding to the phosphorus atoms bonded to phosphorus atoms [32]. In Fig. 22.3C, new peaks emerge at 131.26, 132.39, and 134.19 eV after exposing in ambient condition for 24 h. The new peaks imply that apparent oxidation of the BP flake, and the oxides are mostly phosphorus pentoxide, and an intermediate stable component p-phosphorus tetraoxide (p-P_2O_4) [33]. Nevertheless, mechanical cleavage is relatively

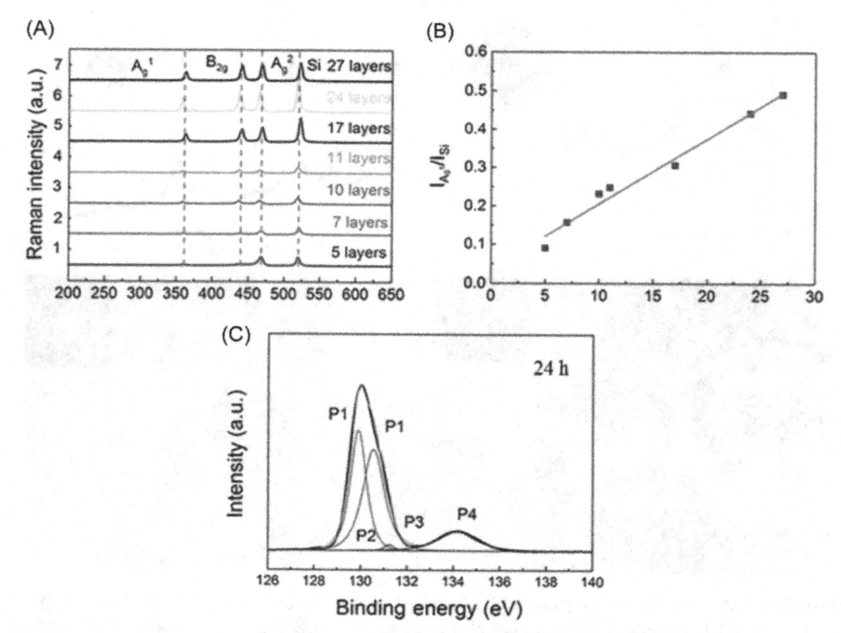

FIGURE 22.3 The basic properties of BP obtained with the micromechanical cleavage method. (A) Raman spectra of different BP flakes the layers ranging from 5 to 27, measured with a 514 nm excitation laser. (B) Linear dependence of the intensity ratio between the Ag1 and Si peaks on the layer numbers. (C)The XPS spectra of the BP flakes exposed to ambient conditions for 24 h. *Adapted with permission from Z. Cui, G. Xie, F. He, W. Wang, D. Guo, W. Wang, Atomic-scale friction of black phosphorus: effect of thickness and anisotropic behavior, Adv. Mater. Interfaces 4(23) (2017) 1700998. Copyright (2017) John Wiley and Sons.*

difficult to get monolayer phosphorene, and the usage of Ar plasma for thinning few-layer BP to monolayer phosphorene was demonstrated [34].

22.4 Black phosphorus: degradation favors lubrication

It has been considered that the degradation of BP with oxygen and moisture is one of the obstacles for its application in ambient conditions. Exfoliated BP sheets could degrade rapidly within a few hours and resulted in both compositional and physical changes [35]. Nevertheless, the high chemical reactivity and modifiable surface of BP sheets could also be beneficial in some cases. The environmental instability of BP has been proposed to be used in sensors [36] and neuroprotective nanodrug in biomedical applications [37]. Here, it is demonstrated that the formation of chemical groups (P—OH bonds) after the adsorption of water molecules on the degraded BP flake surface could be significantly in favor of the lubrication behavior besides its layered structure, by examining the nanofriction behavior of degraded BP sheets against an AFM tip [18]. Thereby, the mechanism of BP functioning as the potential lubricant would be clarified at a fundamental level.

Fig. 22.4A shows the exfoliated BP flakes were exposed in ambient conditions (25°C, 15% RH) with continuous light illumination for 40 h. The growth and coalescence of small blobs could be observed with the increase of exposure time. Finally, the bump-like structures would disappear at a longer exposure time owing to the possible water evaporation. The P2p and O1s XPS spectra of a newly exfoliated BP flake, the flake exposed for 2 h, 40 h, and 7 days are illustrated in Fig. 22.4B and C. The $P2p_{3/2}$ and $P2p_{1/2}$ peaks at 129.2 and 130.0 eV correspond to the characteristic peaks of the crystalline BP. The binding energies of the $P2p_{3/2}$ and $P2p_{1/2}$ peaks increase to 129.5 and 130.4 eV after degradation, suggesting that oxidized phosphorus (P_xO_y) was produced on the flake surface [20,38].

Fig. 22.5A shows the AFM height image of a BP flake (about 52 nm thick) on a SiO_2/Si substrate exposed to ambient conditions for 2 h. The AFM height image of the microdomains on the flake surface is shown in Fig. 22.5B1, and the morphology of one microbump is given in Fig. 22.5C1. As shown in Fig. 22.5C3, the bump height is 4.5 nm and the diameter is about 245 nm. The friction force in the bump area is about 50% of that of the flat area. Hence, the chemical reactions in the bump area should be emphasized. In Fig. 22.5D1, the dark spots on the flake surface are similar to the microbumps characterized by AFM in Fig. 22.5B1 and C1. As shown in Fig. 22.5D2, the STEM-EDS spectrum in Fig. 22.5D2 indicates the presence of phosphorus and oxygen in this area. The distribution maps of phosphorus element in Fig. 22.5D3 and the oxygen element in Fig. 22.5D4 in the area marked by the rectangle in Fig. 22.5D1 confirm the oxygen-enrichment of the dark spot.

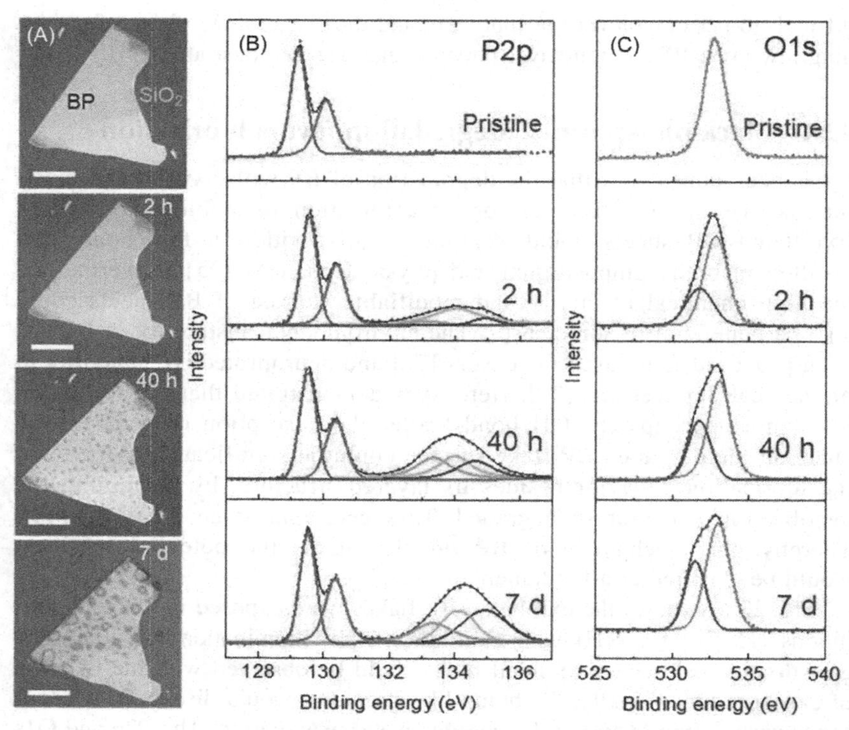

FIGURE 22.4 (A) The optical images of pristine BP flakes (~95 nm) on SiO_2/Si substrates and exposed in ambient conditions for 2 h, 40 h, and 7 days, respectively (scale bar: 20 μm); (B) P2p and (C) O1s XPS spectra of pristine BP flakes and exposed in ambient conditions for 2 h, 40 h, and 7 days, respectively. *Adapted with permission from S. Wu, F. He, G.X. Xie, Z.L. Bian, J.B. Luo, S.Z. Wen, Black phosphorus: degradation favors lubrication, Nano Lett 18 (2018) 5618–5627. Copyright (2018) American Chemical Society.*

The friction reduction in the degraded area was hypothesized to be related with the oxidized phosphorus. The AFM friction test was conducted under a low humidity (5% RH). The AFM height image and the corresponding friction signal obtained immediately after drying treatments are shown in Fig. 22.6A1 and A2. The friction force on the oxide layer is higher (83.4 nN) than that on the flat area (71.6 nN), suggesting that the loss of combined water molecules of the oxide layer could give rise to the friction increase. The AFM image and the friction signal of the BP surface immersed in a water droplet are shown in Fig. 22.6B1 and B2. As indicated in Fig. 22.6A3 and B3, the protrusion rapidly reduced in height because of the dissolution of the phosphorus oxide layer. Fig. 22.6B2 and C2 show the friction signals on the degraded BP surfaces immersed in water droplets for 15 and 25 min. The area covered with the oxide layer immersed in the water droplet shows the lower friction force than the flat area again. The results indirectly suggest

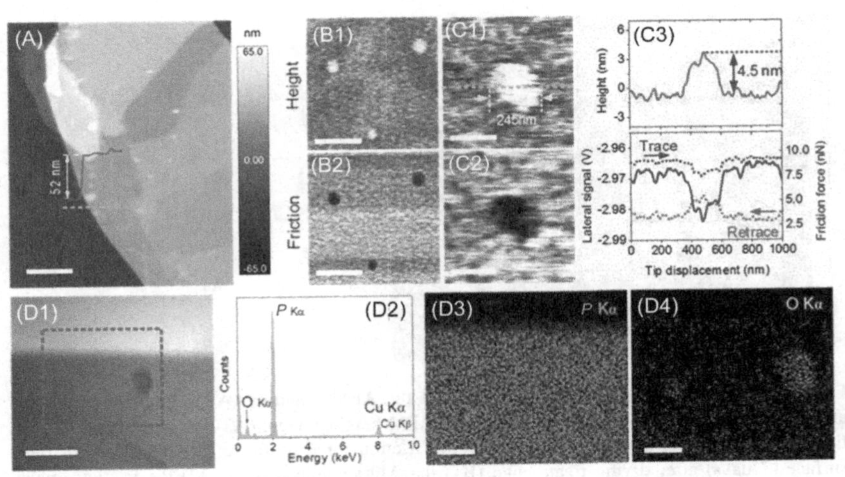

FIGURE 22.5 (A) The AFM height image of BP flake exposed for 2 h (scale bar: 10 μm); (B1) the AFM height and (B2) corresponding friction image of the BP flake surface with microbumps (scale bar: 1 μm); (C1) the AFM height and (C2) the corresponding friction image of a microbump on the BP surface (scale bar: 200 nm); (C3) the height profile, in situ lateral trace/retrace signal and the calculated friction force profile of the marked line in (C1). The AFM friction results were obtained with an applied normal force of 1.9 nN. (D1) The STEM bright-field image of BP flake degraded for 2 h; (D2) the STEM-EDS spectrum of local region in (D1) (scale bar: 500 nm); (D3) the P map and (D4) the O map of the area marked by the rectangle in (D1) (scale bar: 200 nm). *Reprinted with permission from S. Wu, F. He, G.X. Xie, Z.L. Bian, J.B. Luo, S.Z. Wen, Black phosphorus: degradation favors lubrication, Nano Lett, 18 (2018) 5618–5627. Copyright (2018) American Chemical Society.*

the important role of naturally absorbed water molecules in the friction reduction on the degraded BP surface. Therefore it can be said that the presence of water molecules is very essential for the lubrication behavior of the degraded BP flake.

22.5 Super-slippery degraded BP/silicon dioxide (SiO_2) interface

In the previous section, friction reduction was observed on the degraded BP surface owing to the naturally absorbed water molecules onto the oxidized region. It was found that water molecules could also accumulate at the interface between a hydrophilic surface (e.g., SiO_2 surface) and a degraded BP flake, imparting a super-slippery interface [39]. The ultralow interfacial shear strength (ISS) of the BP/SiO_2 interface was close to that of incommensurate rigid crystalline contacts in the superlubric regime.

The measurement principle of the ISS is schematically shown in Fig. 22.7A. Basically, an AFM tip was used to push the flake to move forward at a constant velocity and a constant tip-substrate distance to realize

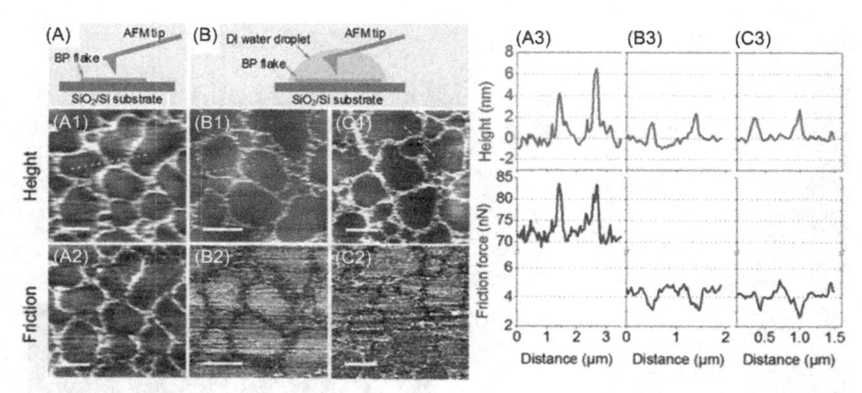

FIGURE 22.6 The schematic representation of the AFM scanning (A) in ambient conditions and (B) immersed in a water droplet; (A1) the AFM height image, (A2) the corresponding friction signal, and (A3) the height and the in situ friction force profile of the degraded BP flake surface (7 days) after drying treatments; (B1) the AFM height image, (B2) the friction image, and (B3) the height and the in situ friction force profile of the degraded BP flake surface (7 days) immersed in the water droplet for 15 min; (C1) the AFM height image, (C2) the friction signal, and (C3) the height and the in situ friction force profile of the degraded BP flake surface immersed in the water droplet for 25 min. The applied normal force was 1.9 nN (scale bar: 1 μm). *Reprinted with permission from S. Wu, F. He, G.X. Xie, Z.L. Bian, J.B. Luo, S.Z. Wen, Black phosphorus: degradation favors lubrication, Nano Lett. 18 (2018) 5618–5627. Copyright (2018) American Chemical Society.*

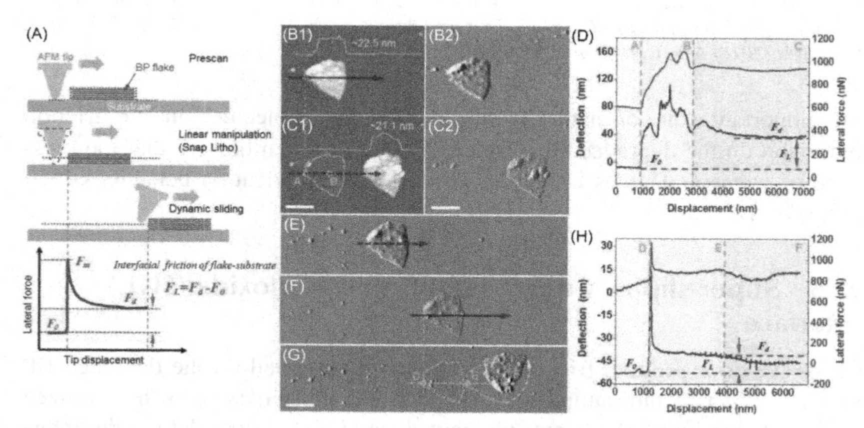

FIGURE 22.7 (A) Schematic of the lateral manipulation experiment of a BP flake on the SiO_2 surface. (B1, C1) AFM height images and (B2, C2) deflection error images (scale bars: 2 μm) of the BP flake before and after the lateral motion on the SiO_2 surface. (D) The corresponding lateral force-displacement plot and the deflection signal. (E–G) AFM deflection error image (scale bar: 2 μm) of the moved BP flake after ambient degradation for another 48 h during the lateral motion. (H) The corresponding lateral force-displacement plot and the deflection signal. *Reprinted with permission from S. Wu, F. He, G.X. Xie, Z.L. Bian, Y.L. Ren, X.Y. Liu, H.J. Yang, D. Guo, L. Zhang, S.Z. Wen, J.B. Luo, Super-slippery black phosphorus/silicon dioxide interface, ACS Appl. Mater. Interface 12(6) (2020) 7717–7726. Copyright (2020) American Chemical Society.*

the lateral manipulation of the BP flakes. The friction force at the flake/substrate interface is expressed as $F_L = F_d - F_0$, where F_d is the lateral force signal during the sliding process at the flake/substrate interface, and F_0 is the initial cantilever torsional signal prior to the tip's contact with the flake. If a uniform shear stress at the interfacial contact is considered, the interfacial shear strength τ is described as, $F_L = \tau \times A$, where A is the contact area [40].

In the case of a freshly exfoliated (within 2 h) BP flake on the SiO_2 surface, merely the abrasion or wrinkle at the edge of the flake was observed, and the flake did not move because of the interface adhesion. After the BP flake degraded in ambient conditions (25°C ± 5°C, 15% ± 5% RH) for 48 h, it could slide linearly on the SiO_2 surface upon the tip manipulation. As shown in Fig. 22.7B1 and B2, the lateral manipulation of the degraded BP flake was applied after a prescan of the tip along the path indicated by the arrow. In Fig. 22.7C1 and C2, the new positions of the flake after motion can be observed. From the lateral signal shown in Fig. 22.7D, the tip began to contact the flake from Point A to B. The flake was most likely static in this period, and the measured lateral force signal (i.e., tip deflection change) could be ascribed to the static friction of the flake-substrate interface. After the maximum static friction force was overcome, a sudden reduction in the lateral force was noted, suggesting that a transition from static friction to dynamic friction occurred. From Point B to C, the sliding at the flake-substrate interface predominated, and the tip-flake interface was relatively static. The subtraction of the measured lateral force (F_d) from Point B to C and the initial lateral force (F_0) on the tip prior to contacting the flake, gave rise to the friction force at the flake-substrate interface during the flake motion. In this case, $F_L = F_d - F_0 = 269.1 \pm 18.3$ nN, and an ISS of the BP/SiO_2 interface could be calculated as $\tau = F_L/A = 0.045 \pm 0.003$ MPa, where A was ~6 μm^2, as estimated from Fig. 22.7B1. The sample was exposed in ambient conditions for another 48 h, and the height of the BP flake increased slightly owing to further degradation. The interfacial sliding at the flake-substrate interface with the AFM tip manipulation was also observed (Fig. 22.7E–G), and the lateral force signal of the flake is shown in Fig. 22.7H. In this instance, the interfacial friction force and the ISS value of the linear motion from D to E were estimated as 176.78 ± 25.8 nN and 0.029 ± 0.004 MPa, respectively.

The ISS value of a monolayer graphene on the SiO_2 substrate was around 1.64 MPa on the basis of a different measuring method from ours [41], and hence the lateral manipulation by an AFM tip was also used for testing the graphene flake /SiO_2 and MoS_2 flake/SiO_2 interfaces. However neither the graphene flake/SiO_2 interface nor the MoS_2 flake/SiO_2 could be moved by the tip manipulation after several attempts, possibly owing to the large flake-substrate adhesion or interfacial interlocking. The measured ultralow ISS value (0.029 ~ 0.045 MPa) of the degraded BP flake/SiO_2 interface was close to those of the bilayer graphene interlayer (ISS 0.06 ~ 0.02) [41] and the

MoS_2/MoS_2 contact (ISS 0.12 ~ 0.038) [42] in the superlubric regime. As for the latter cases, the structural superlubric behavior of these interfaces could fail even after a small amount of foreign molecules was introduced into the contacts [43]. In contrast, the ultralow shear stress of the degraded BP/SiO_2 interface was likely to originate from that the diffusion of water molecules into the interface and the formation of a water layer. In order to underpin the speculation, the chemical compositions of the residuals on the SiO_2 surface after the flake motion were analyzed by using Fourier transform infrared nanospectroscopy (nano-FTIR) and solid-state nuclear magnetic resonance (NMR) experiments.

According to the AFM phase images in Fig. 22.8A and B, the residual on the SiO_2 surface is similar to the degraded BP surface. As shown in Fig. 22.8C, the nano-FTIR spectra of the residual and the degraded BP surface are very close, and the characteristic peaks of asymmetric stretching vibration ν_{as} (P–OH) at 924 cm^{-1}, bending vibration δ(P–OH) at ~940 cm^{-1}, unsolvated δ(P–OH) at 1050 cm^{-1}, and H_2O coupled δ(P–OH) at 1080 cm^{-1} can be identified [44]. Moreover, the peaks for the vibration modes of P–O bonds can be seen, for example, ν_{as}(P–O–P) at 935 cm^{-1}, ν(P–O) at 950 cm^{-1} and 970 cm^{-1}, $\nu_s(PO_4^{3-})$ at around 964 cm^{-1} and ν_{as} (P=O) at 990 cm^{-1} [45]. In contrast, weak peaks of possible P–O bonds can be identified on the fresh BP surface in Fig. 22.8C. Hence, it could be inferred that the compositions of the interfacial layer and the degradation product on the BP surface were almost the same, containing P–OH, P–O, and H_2O molecules.

As shown in Fig. 22.8D, the peaks at ~1.9 ppm and ~5.2 ppm on the single-pulse ^1H NMR spectra could be ascribed to the protons of the Si–OH groups and the free water molecules [46], respectively. The formation of Si–OH groups on the SiO_2 surface could be enhanced by the degradation of BP for the BP/SiO_2 system, as confirmed by the increased relative intensity of the ~1.9 ppm peak. Phosphate moiety HPO_4^{2-} groups could be inferred from the peaks from 8.0 to 16.0 ppm, –OH protons with acid character could the inferred from the peak at 11.0 ~ 15.0 ppm, and kinetically trapped –OH groups could be inferred from the peak at 7.0 ~ 9.0 ppm [47].

$\{^1H\text{-}^{31}P\}^1H$ double CP experiments were conducted to select and enhance the signals from specific rare ^{31}P nucleus, and spectra were recorded by using different contact times (0.1, 0.3, 0.5, 1, and 2 ms) to reveal the relative proximities between the protons and the ^{31}P nucleus on the degraded BP flake surface [48]. At $t_{CP}2 = 0.1$ ms, two resonances at ~0.9 ppm and ~5.0 ppm corresponded to the isolated -OH of the silanol group and water molecules, respectively. At $t_{CP}2 = 0.3$ ms, the nonacidic silanol groups at 2.3 ppm could be identified [46]. At $t_{CP}2 = 0.5$ ms, a peak at 5.8 ppm emerged, which might originate from the water molecules with nearby PO_4^{3-} groups through hydrogen bonding [47]. As the contact time increased, the resonances at 8.0 and 13.0 ppm, which were related with the phosphate

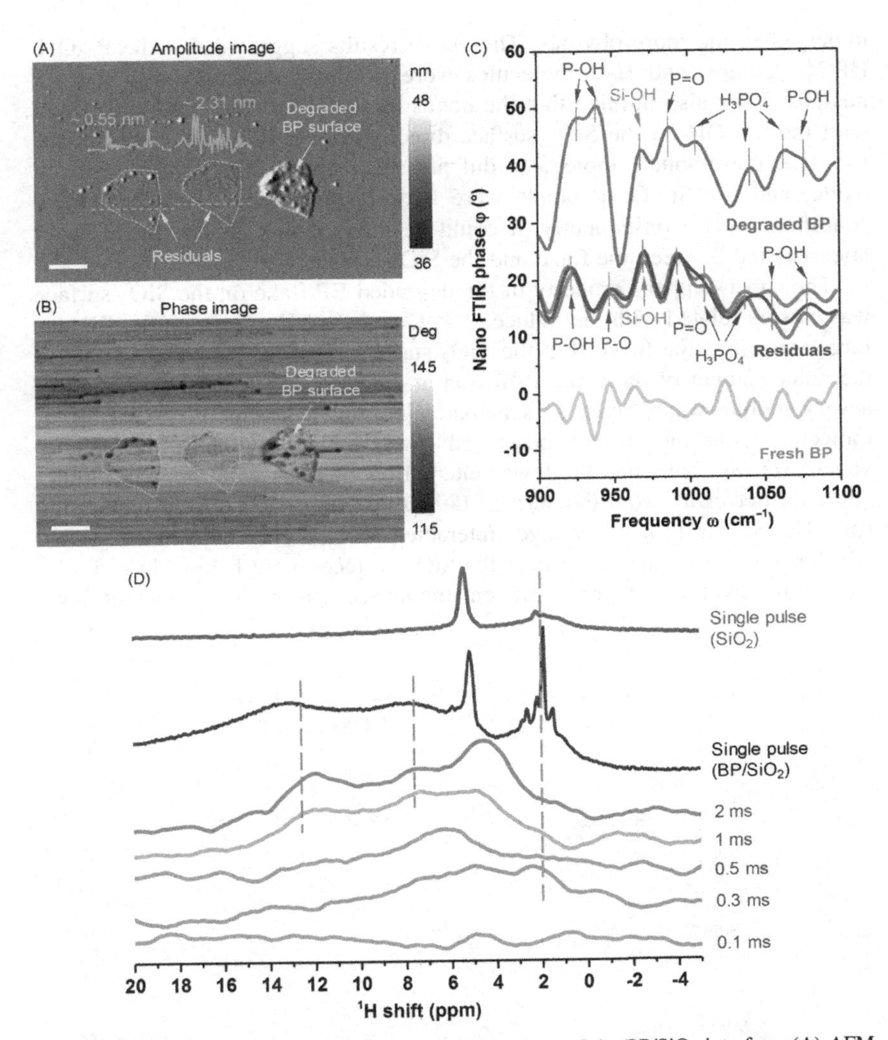

FIGURE 22.8 Characterizations of the chemical structures of the BP/SiO_2 interface. (A) AFM amplitude image and (B) phase image (scale bar: 2 μm) after the flake motion of the degraded BP flake; (C) normalized nano-FTIR phase (φ) versus frequency of the degraded BP surface, the residuals on the substrate, and the fresh BP flake (exfoliated within 30 min); (D) $\{^1H\text{-}^{31}P\}^1H$ double CP spectra of the degraded BP/SiO_2 sample; the first contact time ($t_{CP}1$) was 5 ms and variable second contact times ($t_{CP}2$) were labeled in the figure. *Reprinted with permission from S. Wu, F. He, G.X. Xie, Z.L. Bian, Y.L. Ren, X.Y. Liu, et al., Super-slippery black phosphorus/silicon dioxide interface, ACS Appl. Mater. Interface 12(6) (2020) 7717–7726. Copyright (2020) American Chemical Society.*

moiety, became more obvious. The above results suggested that the P–OH, HPO_4^{2-} groups, and H_2O molecules were in close proximity with the ^{31}P nucleus. They also implied that the bonds on the BP surface did not interact with the Si–OH on the SiO_2 surface directly. It was in agreement with the fact that phosphonate molecules did not react chemically with the silicon oxide, and the Si–O–P bonds were hydrolysis-vulnerable under ambient conditions [44]. Consequently, it could be inferred that an interfacial water layer existed between the flake and the SiO_2 surface.

The super-slippery property of the degraded BP flake on the SiO_2 surface was closely related with the induced P=O and P–OH groups on the BP surface, as well as the Si–OH at the SiO_2 surface. These groups contributed to the enhancement of the water diffusion at the interface and the formation of a very mobile water layer, as schematically shown in Fig. 22.9. The water molecules near the surface possessed fewer hydrogen bonds, the smaller water exchange rate, and the lower interaction energies with the water molecules further away from the surface [49], impacting itself a very lubricating role. Hence, the thin water layer interacted more firmly with the degraded BP flake, and the water layer near the SiO_2 surface might behave in a similar way to the hydration layer to prevent the surfaces from direct contact. As a

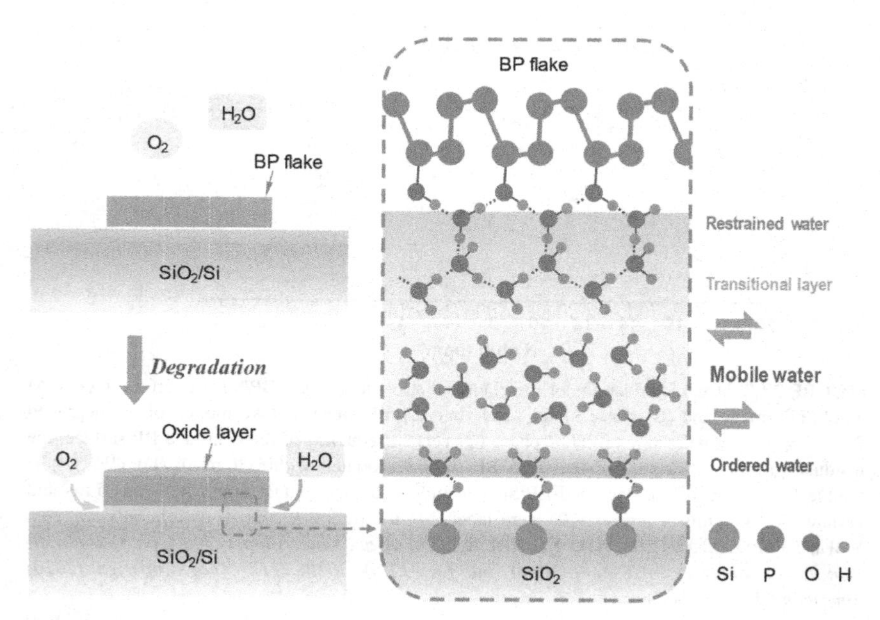

FIGURE 22.9 Schematic of the water layer at the degraded BP/SiO_2 interface. *Reprinted with permission from S. Wu, F. He, G.X. Xie, Z.L. Bian, Y.L. Ren, X.Y. Liu, et al., Super-slippery black phosphorus/silicon dioxide interface, ACS Appl. Mater. Interface 12(6) (2020) 7717–7726. Copyright (2020) American Chemical Society.*

result, the water layer at the degraded BP flake/SiO_2 interface could provide extremely low friction force.

22.6 Superlubricity of BP nanosheets as lubricant additive in aqueous solutions

In the previous parts, it was demonstrated that the presence of P–OH and P=O functional groups on the BP surface were conducive to the enhanced lubrication behavior of BP nanomaterials. In this part, the work on the active controls over the functional groups on the BP nanosheet surfaces to improve the lubrication performance was introduced. It has been found that aqueous lubricants with BP nanosheets modified by sodium hydroxide (NaOH) could reduce the friction dramatically, and superlubricity with a lowest coefficient of friction (COF) of 0.0006 was achieved [17]. Stable ultralow COF could be observed in a wide range of CPs, sliding velocities, and additive concentrations for a relatively long period. The schematic diagram of the preparation process of BP–OH powder is shown in Fig. 22.10. To begin with, BP powder was prepared by a HEBM technique using RP powder at room temperature and normal pressure. Subsequently, the mixture of the BP powder and the NaOH powder was milled to obtain the BP powder modified by NaOH. As shown in Fig. 22.11A, the XPS analyses of the BP–OH nanosheets show three main elements (P, O, and Na) on the full spectrum. In comparison with the spectra of the BP powder and the BP–OH powder, there is a new Na 1 s peak at 1073.2 eV for the BP–OH powder. Detailed analyses of the O 1 s, P 2p, and Na 1 s peaks in Fig. 22.11D–F show that there are two components for P–O and Na–O (metal oxide) at 530.3 and 535.2 eV. In combination with the FTIR result in Fig. 22.11B, it is indicative that the hydroxyl alkylation of BP was achieved in the presence of alkali. Moreover, together with Fig. 22.11F, it can be inferred that new metal phosphate (Na_xP) was formed after the ball milling of the mixture of the BP

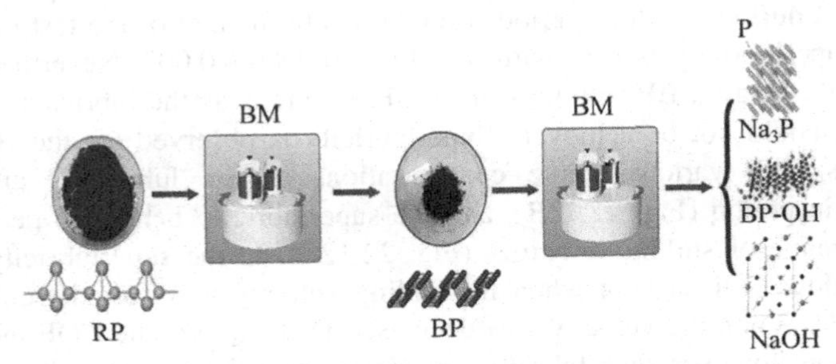

FIGURE 22.10 Schematic diagram of the preparation process of the BP and BP–OH powders. *Adapted with permission from W. Wang, G. Xie, J. Luo. Superlubricity of black phosphorus as lubricant additive, ACS Appl. Mater. Interfaces 10(49) (2018) 43203–43210. Copyright (2018) American Chemical Society.*

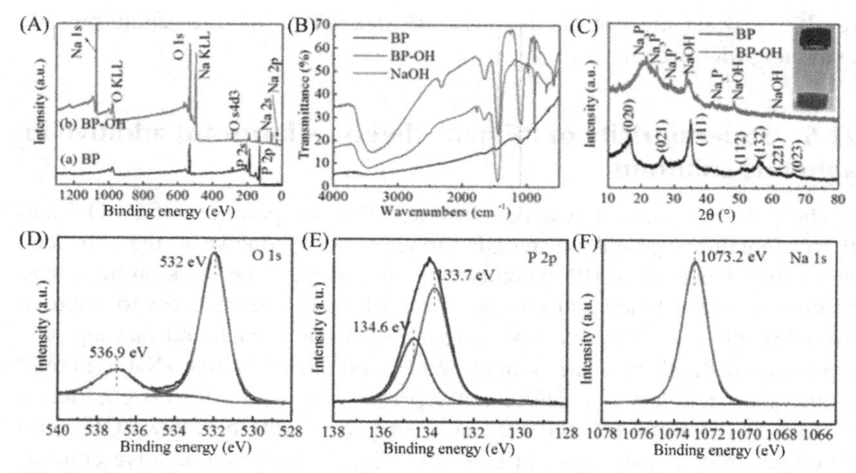

FIGURE 22.11 (A) The XPS full spectra of the BP powders before and after NaOH modification; (B) the FTIR spectra of BP, BP–OH and NaOH powders; (C) the XRD patterns of BP after BM for 48 h and BM with NaOH for 48 h, and the inset to (H) is the optical photo of the BP–OH solution; (D)–(F) are O 1 s, P 2p, and Na1s peaks in the XPS spectra of BP powders modified by NaOH, respectively. *Adapted with permission from W. Wang, G. Xie, J. Luo J, Superlubricity of black phosphorus as lubricant additive, ACS Appl. Mater. Interfaces 10(49) (2018) 43203–43210. Copyright (2018) American Chemical Society.*

powder and the NaOH powder. It can be seen from Fig. 22.11C that the XRD patterns of the BP powders after NaOH modification consist of three new sharp diffraction peaks at $2\theta = 22.5$, 25, and 30 degrees, further confirming that the new metal phosphate was Na_3P.

In Fig. 22.12A, the friction process with 5 wt.% BP–OH solution as the lubricant are divided into three regimes: (1) the period from 0 to 1000 s is the running-in process, and the COF reduces dramatically from 0.35 to 0.08; (2) the second period from 1000 to 3600 s is the transition process, and the COF decreases gradually from 0.08 to smaller than 0.01 (the superlubricity state); (3) during the third period from 3400 s to the end of the test (6500 s), stable superlubricity is kept with a COF of 0.0006 ~ 0.003. Nevertheless, in the cases of water, BP solutions or NaOH solutions as the lubricants, superlubricity could not be achieved. Superlubricity is observed for the BP–OH solutions with various solute concentrations as the lubricants after the running-in period (Fig. 22.12B), and the superlubricity behaviors persist for a wide range of sliding velocities (Fig. 22.12C). In the superlubricity state, the ultralow COF is kept when the sliding velocity was varied from 56 to 282 mm/s. When the velocity was decreased to 0.3 mm/s, the COF increases to 0.4, and however superlubricity reappears immediately after the velocity was increased back to 38 mm/s, which is the critical velocity for obtaining superlubricity. Obviously, the switch between superlubricity and high friction with the velocity change was reversible. In addition, if the maximum CP was

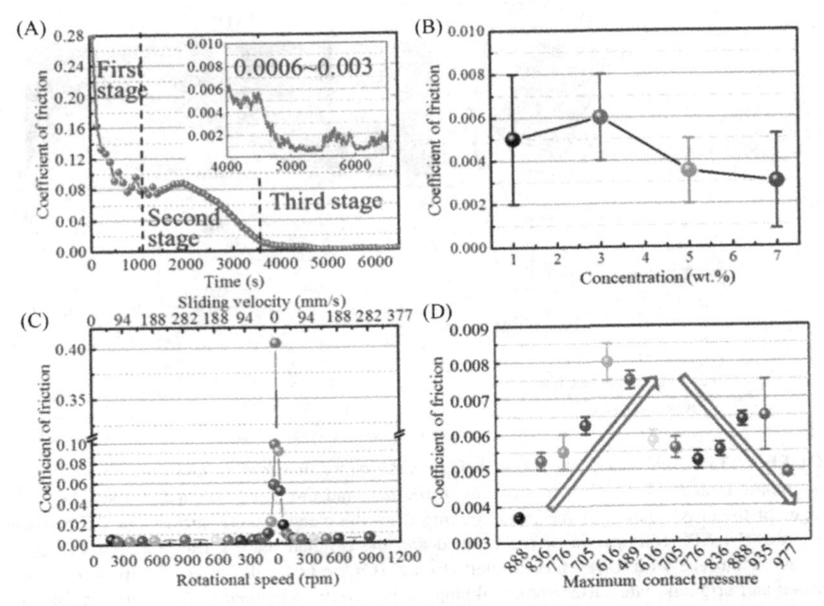

FIGURE 22.12 (A) The change of the COF over time with the BP–OH solution (solute concentration: 5 wt.%) as the lubricant. The inset is the magnified part after the running-in process; (B) the COFs of the BP–OH solutions as the lubricants; (C) the change of the COF of the 5 wt.% BP–OH solution as the lubricant with sliding velocity in the superlubricity state; (D) the COFs of the 5 wt.% BP–OH solution as the lubricant at different Hertzian CPs in the superlubricity state. *Adapted with permission from W. Wang, G. Xie, J. Luo J, Superlubricity of black phosphorus as lubricant additive, ACS Appl. Mater. Interfaces 10(49) (2018) 43203–43210. Copyright (2018) American Chemical Society.*

varied, the superlubricity behavior could be kept. It can be seen in Fig. 22.12D that the COF changes between 0.0035 and 0.008 after entering the superlubricity state under different Hertzian CPs.

In order to reveal the underlying mechanism, the cleaned surface features of the tribopairs after friction tests were analyzed. In Fig. 22.13A and B, there is an elliptic planar surface on the top part of the ball. To verify the occurrence of tribochemical reactions during the friction process, layers were cut from the worn surfaces of the ball and the plate after thoroughly rinsing, and the cross-section images of the layers were investigated by TEM observation. An obvious boundary between the substrate and the protective coating is indicated in Fig. 22.13C and D. From the TEM element analysis result, no phosphorous is present both for the ball and plate samples, implying that tribochemical reactions were not the dominant factors contributing to the superlubricity behavior. The original SiO_2 surface after rubbing and the SiO_2 surface without rubbing in contact with a BP–OH solution droplet were analyzed. It can be seen from Fig. 22.13E and F that there is no

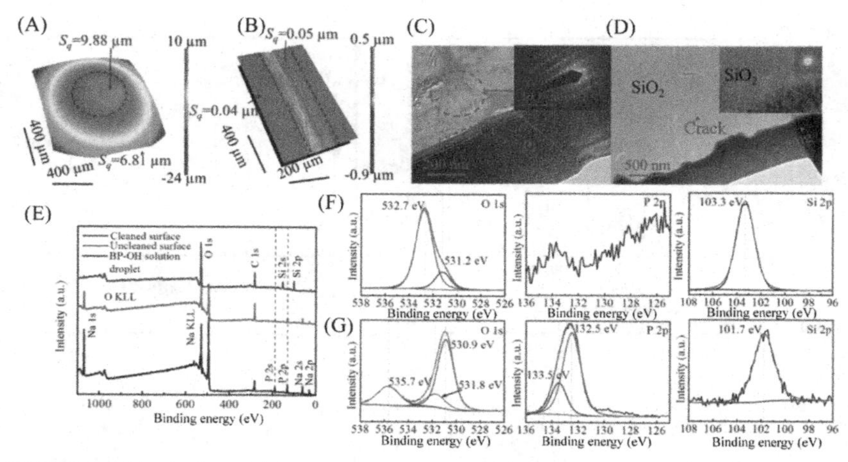

FIGURE 22.13 Analyses of the worn region on the tribopair surfaces. (A) and (B) are the 3D topographic images of the cleaned surfaces of the ball and the plate; (C) and (D) are the TEM images of the cross-section of the layers cutting from the cleaned worn surfaces of the ball and the plate; (E) XPS full spectra of the cleaned and the original plate surfaces after rubbing, as well as the surface with a BP–OH droplet; (F) and (G) are O 1 s, P 2p, and Si 2p peaks of the cleaned and original plate surface after rubbing, respectively. *Reprinted with permission from W. Wang, G. Xie, J. Luo, Superlubricity of black phosphorus as lubricant additive, ACS Appl. Mater. Interfaces 10(49) (2018) 43203–43210. Copyright (2018) American Chemical Society.*

phosphorous element on the cleaned SiO_2 surface after rubbing, and however the elements (O, P, and Na) on the uncleaned SiO_2 surface are the same as those in the BP–OH solution. The XPS spectra in Fig. 22.13G show that the Si 2p peak at 101.7 eV of the uncleaned SiO_2 surface after rubbing corresponds to Si–O (polymerized siloxane), the O 1 s peaks at 531.8, 530.9, and 535.7 eV correspond to Si–O, phosphorus oxide and hydroxyl group, and the P 2p peaks at 132.5 and 133.5 eV correspond to P–O(P^{5+}) and sodium phosphate [50,51]. The above results indicate that silica-gel and separate phosphorus compounds formed on the original SiO_2 surface after rubbing.

In the case of the ceramic tribopairs lubricated by aqueous solutions, silica-gel was proposed to form on the rubbing surfaces owing to the hydrolysis reaction ($\equiv$Si–O–Si$\equiv$ + H_2O → $\equiv$Si–OH...OH–Si$\equiv$) generated by the thermal activation in the highly stressed rubbing region [52]. The SEM result of the original Si3N4 ball surface after rubbing in Fig. 22.14A indicates the representative features of the gel formed by rubbing. The element distribution in the gel layer on the original Si_3N_4 ball surface after rubbing, as indicated in Fig. 22.14B, was analyzed by TEM observation, and the result is shown in Fig. 22.14C. In Region I, the phosphorus element was on the substrate surface, and it did not diffuse into the substrate, confirming the residuals on the tribopair surface were BP nanosheets. In Region II, there were oxygen, phosphorus and silicon elements in the silica-gel layer, and the

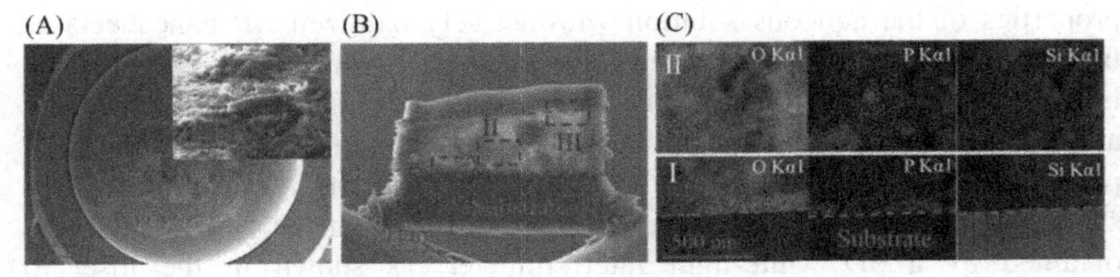

FIGURE 22.14 (A) The SEM result of the original Si_3N_4 ball surface after rubbing, and the inset is magnified photo; (B) the cross-section images of the layers cutting from the worn region of the original Si_3N_4 ball surface after rubbing; (C) the oxygen, phosphorus, and silicon element distributions in Region I and II obtained by scanning TEM analyses. *Adapted with permission from W. Wang, G. Xie, J. Luo, Superlubricity of black phosphorus as lubricant additive, ACS Appl. Mater. Interfaces 10(49) (2018) 43203–43210. Copyright (2018) American Chemical Society.*

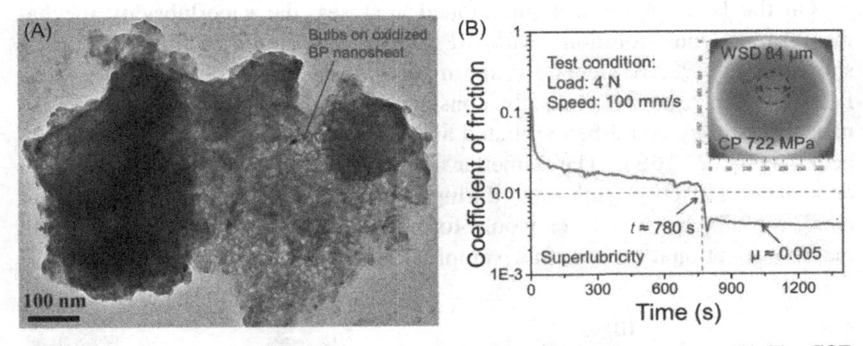

FIGURE 22.15 (A) TEM morphology of the actively oxidized BP nanosheet. (B) The COF variation curve over time with the lubrication of the aqueous solution with oxidized BP nanosheets as the additive (load: 4 N and sliding velocity: 100 mm/s). The inset is the 3D topographic image of the Si_3N_4 ball surface after the frictional tests. The worn scar diameter (WSD = 84 μm) and contact pressure (CP = 722 MPa).

phosphorus element distributed evenly. In Region III, similar results can be seen. It can be inferred that the BP–OH nanosheets in the silica-gel predominated the superlubricity behavior with the BP–OH solution.

In our recent work, it has been found that the modification of BP nanosheets by oxidizing agents, for example, diluted hydrogen peroxide solution, could be of great help to reduce friction, and hence actively oxidized BP nanosheets were tried to be used as the lubricant additive. The morphological result in Fig. 22.15A suggests that the actively oxidized BP nanosheet is a porous structure, which should be induced by the removal of water molecules owing to the presence of water molecules in the solution. Hence, it could be inferred that the formation of the porous structure was ascribed to the dissolution of part of the phosphorus oxide in water, as well as water evaporation in vacuum during the test of the morphology. The lubrication properties of the aqueous solution with actively oxidized BP nanosheets as the additive were investigated, and the result is shown in Fig. 22.15B. The tribopair was the combination of a Si_3N_4 ball and a sapphire disk, and the applied load was 4 N and the sliding velocity 100 mm/s. After the running-in period, the COF decreased from 0.02 to 0.005, and the low COF was kept stable afterward. The morphology of the worn scar on the ball surface was obtained by a 3D white-light interferometer, as shown in the inset to Fig. 22.15B, and the diameter of the worn scar (WSD) was around 84 μm. The corresponding CP during the superlubricity period (the ratio of the applied load and the area of the worn scar on the ball surface) was 722 MPa. To the best of the authors' knowledge, it was the largest value ever reported in the studies on liquid superlubricity. The actively oxidized BP nanosheets displayed excellent friction-reduction and antiwear performances, which is worthy of further study, and the relevant investigation is still under way.

On the basis of the aforementioned analyses, the superlubricity mechanism of aqueous solutions with BP nanosheets could be illuminated, as shown in Fig. 22.16. Generally, the maximum friction occurred at the interface between asperity peaks. In consideration of the layer structure of BP nanosheets, they could be exfoliated and absorbed on the surfaces of asperity peaks (Fig. 22.16B). The lamellar slip of BP nanosheets endowed the decreases of friction and wear during the running-in period. On the other hand, a retained water layer would form on the BP nanosheet surfaces with massive functional groups, for example, P–OH and P=O head groups. The

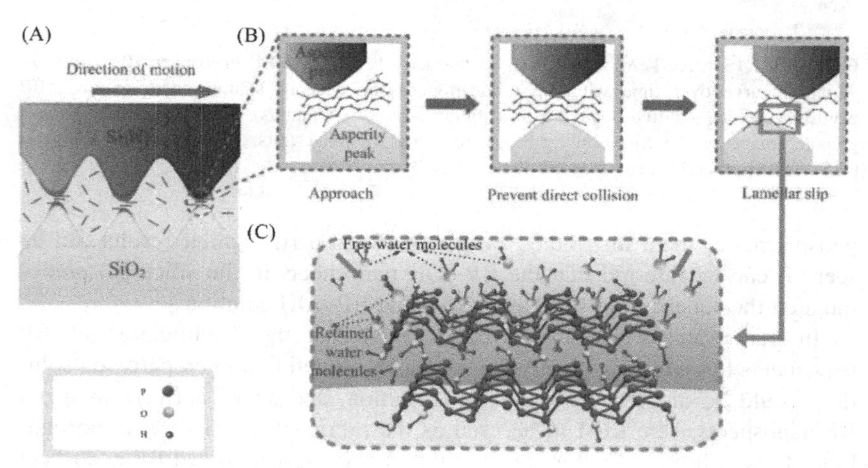

FIGURE 22.16 (A) Schematic of the sliding between rough surfaces lubricated by aqueous solutions with BP nanosheets. (B) The BP nanosheets between asperities prevented the direct contact between asperities, and lamellar sliding of the BP nanosheets would form between the sliding surfaces. (C) A retained water layer would form near the BP nanosheet surface with massive P=O and P–OH head groups, which was very conducive to the realization of robust liquid superlubricity.

dynamically stable water layer on the nanosheet surface could provide very low shear strength (Fig. 22.16C). The strong retaining ability of water molecules near the BP nanosheet surface with functional groups contributed greatly to the robust liquid superlubricity behavior.

22.7 Summary

In this chapter, the robust superlubricity in the aqueous solution with BP nanosheets as the additive has been observed, because of the low shear resistance of the water layer near BP nanosheets. It has been demonstrated that the degradation of BP flakes under ambient conditions could significantly promote the lubrication behavior. A super-slippery degraded BP flake/hydrophilic SiO_2 surface with an ultralow ISS ($\sim$0.029 MPa) was found. The combination of phosphorus oxides with P=O and P–O–P head groups with water molecules would give rise to the production of phosphorus oxyacids containing massive P–OH bonds. These functional groups could promote the formation of the water layer near the BP nanosheet surfaces, and the water layer was very critical for influencing the friction-reduction and antiwear behaviors. These results have important implications for the development of future high-performance aqueous lubricants to save energy via friction reduction, and prevent pollution from oil-based lubricant leakage.

Acknowledgments

This work was financially supported by National Natural Science Foundation of China (grant numbers 51527901 and 51822505), and the Beijing Natural Science Foundation of China (grant no. 3182010).

Competing interests

The authors declare no competing financial interests.

References

[1] M. Hirano, Superlubricity: a state of vanishing friction, Wear 254 (10) (2003) 932–940.

[2] J.M. Martin, A. Erdemir, Superlubricity: friction's vanishing act, Phys. Today 71 (4) (2018) 40–46.

[3] M.Z. Baykara, M.R. Vazirisereshk, A. Martini, Emerging superlubricity: a review of the state of the art and perspectives on future research, Appl. Phys. Rev. 5 (4) (2018) 041102.

[4] M. Chen, K. Kato, K. Adachi, Friction and wear of self-mated SiC and Si_3N_4 sliding in water, Wear 250 (1–12) (2001) 246–255.

[5] F. Zhou, X. Wang, K. Kato, Z. Dai, Friction and wear property of a-CNx coatings sliding against Si_3N_4 balls in water, Wear 263 (7–12) (2007) 1253–1258.

[6] M. Chen, W.H. Briscoe, S.P. Armes, J. Klein, Lubrication at physiological pressures by polyzwitterionic brushes, Science 323 (5922) (2009) 1698–1701.

[7] W.H. Briscoe, S. Titmuss, F. Tiberg, R.K. Thomas, D.J. McGillivray, J. Klein, Boundary lubrication under water, Nature 444 (7116) (2006) 191.

[8] X. Ge, J. Li, H. Wang, C. Zhang, Y. Liu, J. Luo, Macroscale superlubricity under extreme pressure enabled by the combination of graphene-oxide nanosheets with ionic liquid, Carbon 151 (2019) 76–83.

[9] K.S. Novoselov, D. Jiang, F. Schedin, T.J. Booth, V.V. Khotkevich, S.V. Morozov, et al., Two-dimensional atomic crystals, Proc. Natl. Acad. Sci. U. S. A. 102 (30) (2005) 10451–10453.

[10] T. Filleter, R. Bennewitz, Structural and frictional properties of graphene films on SiC (0001) studied by atomic force microscopy, Phys. Rev. B 81 (15) (2010) 155412.

[11] M. Dienwiebel, G.S. Verhoeven, N. Pradeep, J.W.M. Frenken, J.A. Heimberg, H.W. Zandbergen, Superlubricity of graphite, Phys. Rev. Lett. 92 (12) (2004) 126101.

[12] T. Chu, H. Ilatikhameneh, G. Klimeck, R. Rahman, Z. Chen, Electrically tunable bandgaps in bilayer MoS_2, Nano Lett. 15 (12) (2015) 8000–8007.

[13] X. Ma, W. Lu, B. Chen, D. Zhong, L. Huang, L. Dong, et al., Performance change of few layer black phosphorus transistors in ambient, AIP Adv. 5 (10) (2015) 107112.

[14] X. Ling, H. Wang, S. Huang, F. Xia, M.S. Dresselhaus, The renaissance of black phosphorus, Proc. Natl. Acad. Sci. U. S. A. 112 (15) (2015) 4523–4530.

[15] J. Tao, W. Shen, S. Wu, L. Liu, Z. Feng, C. Wang, et al., Mechanical and electrical anisotropy of few-layer black phosphorus, ACS Nano 9 (11) (2015) 11362–11370.

[16] W. Wang, G. Xie, J. Luo, Black phosphorus as a new lubricant, Friction 6 (2018) 116–142.

[17] W. Wang, G. Xie, J. Luo, Superlubricity of black phosphorus as lubricant additive, ACS Appl. Mater. Interfaces 10 (49) (2018) 43203–43210.

[18] S. Wu, F. He, G.X. Xie, Z.L. Bian, J.B. Luo, S.Z. Wen, Black phosphorus: degradation favors lubrication, Nano Lett. 18 (2018) 5618–5627.

[19] Y. Lv, W. Wang, G.X. Xie, J.B. Luo, Self-lubricating PTFE-based composites with black phosphorus nanosheets, Tribol. Lett. 66 (2018) 61.

[20] H. Liu, Y. Du, Y. Deng, D.Y. Peide, Semiconducting black phosphorus: synthesis, transport properties and electronic applications, Chem. Soc. Rev. 44 (9) (2015) 2732–2743.

[21] A. Castellanos-Gomez, Black phosphorus: narrow gap, wide applications, J. Phys. Chem. Lett. 6 (21) (2015) 4280–4291.

[22] Y. Du, C. Ouyang, S. Shi, M. Lei, Ab initio studies on atomic and electronic structures of black phosphorus, J. Appl. Phys. 107 (9) (2010) 093718.

[23] Z. Wang, P.X.L. Feng, Design of black phosphorus 2D nanomechanical resonators by exploiting the intrinsic mechanical anisotropy, 2D Mater. 2 (2) (2015) 021001.

[24] J.R. Brent, N. Savjani, E.A. Lewis, S.J. Haigh, D.J. Lewis, P. O'Brien, Production of few-layer phosphorene by liquid exfoliation of black phosphorus, Chem. Commun. 50 (87) (2014) 13338–13341.

[25] P. Bridgman, Two new modifications of phosphorus, J. Am. Chem. Soc. 36 (7) (1914) 1344–1363.

[26] S. Endo, Y. Akahama, S.-I. Terada, S.-I. Narita, Growth of large single crystals of black phosphorus under high pressure, Jpn J. Appl. Phys. 21 (Part 2, No. 8) (1982) L482–L484.

[27] C.M. Park, H.J. Sohn, Black phosphorus and its composite for lithium rechargeable batteries, Adv. Mater. 19 (18) (2007) 2465–2468.

[28] J. Sun, H.W. Lee, M. Pasta, H. Yuan, G. Zheng, Y. Sun, et al., A phosphorene–graphene hybrid material as a high-capacity anode for sodium-ion batteries, Nat. Nanotechnol. 10 (2015) 980.

[29] H. Liu, A.T. Neal, Z. Zhu, Z. Luo, X. Xu, D. Tománek, et al., Phosphorene: an unexplored 2D semiconductor with a high hole mobility, ACS Nano 8 (4) (2014) 4033–4041.

[30] Z. Cui, G. Xie, F. He, W. Wang, D. Guo, W. Wang, Atomic-scale friction of black phosphorus: effect of thickness and anisotropic behavior, Adv. Mater. Interfaces 4 (23) (2017) 1700998.

[31] Y. Qi, J. Park, B. Hendriksen, D. Ogletree, M. Salmeron, Electronic contribution to friction on GaAs: an atomic force microscope study, Phys. Rev. B 77 (18) (2008) 184105.

[32] A. Castellanos-Gomez, L. Vicarelli, E. Prada, J. Island, K. Narasimha-Acharya, S. Blanter, et al., Isolation and characterization of few-layer black phosphorus, 2D Mater. 1 (2) (2014) 025001.

[33] M.T. Edmonds, A. Tadich, A. Carvalho, A. Ziletti, M. O'Donnell, S. Koenig, et al., Creating a stable oxide at the surface of black phosphorus, ACS Appl. Mater. Interfaces 7 (27) (2015) 14557–14562.

[34] W. Lu, H. Nan, J. Hong, Y. Chen, C. Zhu, Z. Liang, et al., Plasma-assisted fabrication of monolayer phosphorene and its Raman characterization, Nano Res. 7 (6) (2014) 853–859.

[35] Y. Huang, J. Qiao, K. He, S. Bliznakov, E. Sutter, X. Chen, et al., Interaction of black phosphorus with oxygen and water, Chem. Mater. 28 (22) (2016) 8330–8339.

[36] A.N. Abbas, B. Liu, L. Chen, Y. Ma, S. Cong, N. Aroonyadet, et al., Black phosphorus gas sensors, ACS Nano 9 (5) (2015) 5618–5624.

[37] W. Chen, J. Ouyang, X. Yi, Y. Xu, C. Niu, W. Zhang, et al., Black phosphorus nanosheets as a neuroprotective nanomedicine for neurodegenerative disorder therapy, Adv. Mater. 30 (3) (2018) 1703458.

[38] S. Walia, Y. Sabri, T. Ahmed, M. Field, R. Ramanathan, A. Arash, et al., Defining the role of humidity in the ambient degradation of few-layer black phosphorus, 2D Mater. 4 (1) (2017) 015025.

[39] S. Wu, F. He, G.X. Xie, Z.L. Bian, Y.L. Ren, X.Y. Liu, et al., Super-slippery degraded black phosphorus/silicon dioxide interface, ACS Appl. Mater. Interface 12 (6) (2020) 7717.

[40] R.W. Carpick, E.E. Flater, K. Sridharan, D.F. Ogletree, M. Salmeron, Atomic-scale friction and its connection to fracture mechanics, JOM 56 (10) (2004) 48.

[41] G. Wang, Z. Dai, Y. Wang, P. Tan, L. Liu, Z.P. Xu, et al., Measuring interlayer shear stress in bilayer grapheme, Phys. Rev. Lett. 119 (3) (2017) 036101.

[42] H. Li, J. Wang, S. Gao, Q. Chen, L. Peng, K. Liu, et al., Superlubricity between MoS_2 monolayers, Adv. Mater. 29 (27) (2017) 1701474.

[43] G. He, M.H. Müser, M.O. Robbins, Adsorbed layers and the origin of static friction, Science 284 (5420) (1999) 1650.

[44] P. Thissen, T. Peixoto, R.C. Longo, W. Peng, W.G. Schmidt, K. Cho, et al., Activation of surface hydroxyl groups by modification of h-terminated Si(111) surfaces, J. Am. Chem. Soc. 134 (21) (2012) 8869.

[45] P. Stoch, A. Stoch, M. Ciecinska, I. Krakowiak, M. Sitarz, Structure of phosphate and iron-phosphate glasses by DFT calculations and FTIR/Raman spectroscopy, J. Non-Cryst. Solids 450 (2016) 48.

[46] L. Gill, A. Beste, B. Chen, M. Li, A. Mann, S. Overbury, et al., Fast MAS 1H NMR study of water adsorption and dissociation on the (100) surface of ceria nanocubes: a fully hydroxylated, hydrophobic ceria surface, J. Phys. Chem. C 121 (13) (2017) 7450.

[47] R. Mathew, C. Turdean-Ionescu, Y. Yu, B. Stevensson, I. Izquierdobarba, A. Garcia, et al., Proton environments in biomimetic calcium phosphates formed from mesoporous

bioactive $CaO-SiO_2-P_2O_5$ glasses in vitro: insights from solid-state NMR, J. Phys. Chem. C 121 (24) (2017) 13223.
[48] N. Folliet, C. Roiland, S. Bégu, A. Aubert, T. Mineva, A. Goursot, et al., Investigation of the interface in silica-encapsulated liposomes by combining solid state NMR and first principles calculations, J. Am. Chem. Soc. 133 (42) (2011) 16815.
[49] C. Wang, B. Wen, Y. Tu, R. Wan, H. Fang, Friction reduction at a superhydrophilic surface: role of ordered water, J. Phys. Chem. C 119 (21) (2015) 11679.
[50] T.V. Larina, L.S. Dovlitova, V.V. Kaichev, V.V. Malakhov, T.S. Glazneva, E.A. Paukshtis, et al., Influence of the surface layer of hydrated silicon on the stabilization of Co^{2+} cations in Zr-Si fiberglass materials according to XPS, UV-Vis DRS, and differential dissolution phase analysis, RSC Adv. 5 (97) (2015) 79898–79905.
[51] F.J. Sheu, C.P. Cho, Y.T. Liao, C.T. Yu, Ag_3PO_4-TiO_2-graphene oxide ternary composites with efficient photodegradation, hydrogen evolution, and antibacterial properties, Catalysts 8 (2) (2018) 57.
[52] Y. Nakamura, J. Muto, H. Nagahama, I. Shimizu, T. Miura, I. Arakawa, Amorphization of quartz by friction: implication to silica-gel lubrication of fault surfaces, Geophys. Res. Lett. 39 (21) (2012).

Chapter 23

Studying superlubricity with the oscillating relaxation tribometer

Michel Belin and Jean Michel Martin
Université de Lyon, Ecole Centrale de Lyon, LTDS, CNRS, Ecully, France

23.1 Introduction and context

Reducing friction is a general goal in tribology for many applications. Recently, major advances have been made in this field [1], leading to low and superlow friction tribosystems ($\mu < 0.01$). This requires the accurate quantification of low friction levels. Ultralow friction studies have been performed recently at the nanometer scale, using scanning probe techniques. However considering mesoscale contacts, the situation is more questionable. One possibility is to increase the precision of the friction force measurements of traditional tribometers. However these developments are limited because of technical difficulties. This chapter presents an alternative technique and apparatus that we have developed to overcome this limitation: the oscillating relaxation tribometer (ORT).

23.2 The oscillating relaxation method

23.2.1 From the literature

Previous works have been published that describe in detail the evolution of an elastic damped system with different types of damping [2–4]. The authors clearly showed the interest of considering the phenomena that govern the decrease of both position and speed, when finally returning to equilibrium. Up to now, this principle has not been used to characterize a tribological interface.

23.2.2 Principle

The basic principle of the "oscillating relaxation method" is to measure the damping instead of the friction force [5]. A sliding contact is included in a

Superlubricity. DOI: https://doi.org/10.1016/B978-0-444-64313-1.00023-5

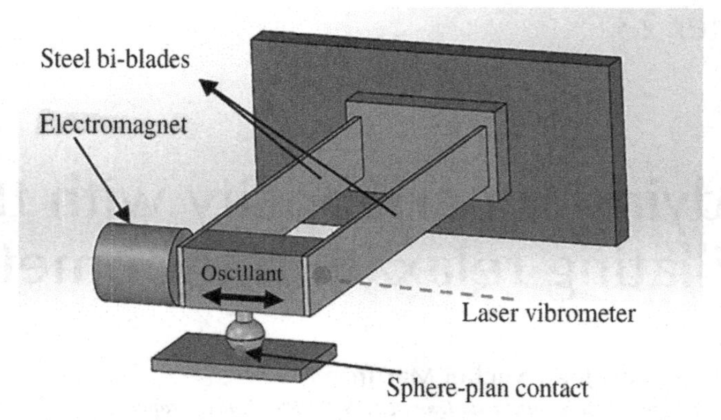

FIGURE 23.1 Schematic arrangement of the 1-DoF relaxation oscillating tribometer.

simple one-degree of freedom (1-DoF) mechanical oscillator, as shown in Fig. 23.1. After a normal force is applied, sliding is allowed in the plane of contact. First, we deform the elastic subsystem by an initial deflection, in a position out-of-equilibrium. Then at t_0, we release the mobile part and record in real time the position and velocity of the mobile part when it returns to mechanical equilibrium, which occurs when the motion stops. When processing the data, a suitable mechanical model is applied to the experimental data to determine the dissipative force generated by the sliding interface and its dependence on sliding velocity, as shown in Fig. 23.2. The two objectives of this technique are to:

- increase the sensitivity of the friction measurement;
- obtain the "friction law" using a robust and rapid method; and
- avoid wear during the experiment, which is very short, typically a few seconds.

23.3 Experimental

23.3.1 The tribological contact

We first consider a simple sphere-on-plane contact configuration. One part is fixed to the 1-DoF mechanical oscillator. The counterpart is considered as fixed to the frame of the instrument. An initial deformation of the subsystem is introduced via a small deflection from its equilibrium position. Therefore a given and limited elastic energy is stored in the mechanical system. At a given point, the moving part is released. The stored elastic energy is converted into kinetic energy and damping occurs during this conversion until the moving part stops completely. The time evolution of both the position and velocity of the moving part are recorded using a high precision laser velocimeter (POLYTEC PDV-100 or POLYTEC OFV-5000). This purely optical method provides high

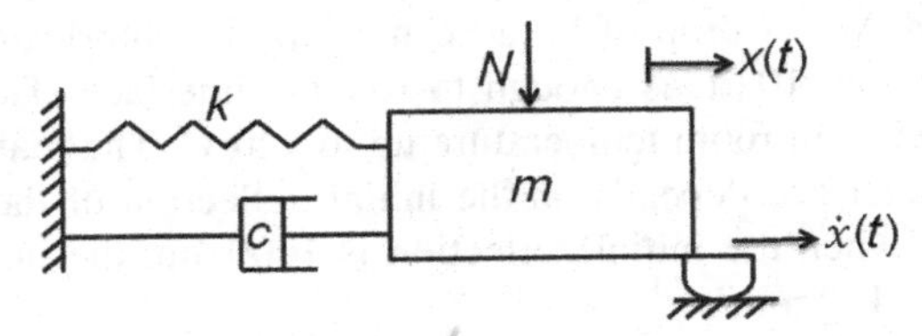

FIGURE 23.2 Schematic of the 1-DoF mechanical oscillator, including the sliding contact to study.

precision measurement and has strictly no influence on the motion of the oscillator. The measurements are digitally converted using a DAQ board (NATIONAL INSTRUMENTS DAQPAD-6221, 16-bit resolution).

23.3.2 The one-degree of freedom oscillator

Due to its simple structure, the mechanical oscillator naturally oscillates in a plane at its resonance frequency (Fig. 23.1). The resonance frequency F_0 is governed by the moving mass and the spring stiffness. In this instrument, the value of F_0 is 25 Hz.

The damping of such a system depends on three main factors:

- the mechanical assembly of blades, mainly designed with screws;
- the internal damping of the material of the two metallic strips, here stainless steel; and
- the pumping of air when the oscillator is moving in ambient atmosphere.

These different factors add damping to the motion. The nature and the value of damping are identified by a simple test in which the damped oscillations with an opened contact are measured. We showed previously that the damping of the free oscillator is purely "viscous-type" [6]. The quality factor of this mechanical oscillator is close to 300. The range of sliding velocity can be increased by choosing different characteristics of the oscillator: metallic strips, stiffness, and mass.

23.3.3 The contact conditions

In the apparatus developed, the control of the normal force N on the contact area is achieved by the elastic deformation of a steel strip, so that changing N has no influence on the resonance frequency of the oscillator. The normal force N can be adjusted between 25 and 1000 mN. Depending on both the diameter and the material of the sphere, the normal pressure can be adjusted to several tens of MPa and up to 1 GPa. Usually, the sphere used is an AISI 52100 bearing steel, and the counterface is a flat coupon. Both surfaces are mirror polished by smooth polishing with a 0.5 μm diamond suspension (Ra $<$ 30 nm). The contact can be operated in open air or under a gas flow

(N_2, dry air, and Ar). A drop of liquid can be used to lubricate the interface by a fluid meniscus: 10 μL is enough to wet the interface. The temperature can be controlled from room temperature up to 120°C. The maximum sliding velocity at the interface depends on the initial deflection of the elastic oscillator. Typically, when the initial deflection is 1000 μm, the initial amplitude of the velocity is 160 mm/s.

23.3.4 Additional information from the interface

The electrical contact resistance (ECR) can be measured when testing a tribosystem in which the two facing materials are conductive. This measurement provides additional information from the sliding contact that can be particularly useful for elucidating the lubrication regime and its possible evolution during an oscillating test when the system returns to mechanical equilibrium [6].

23.4 Data processing

23.4.1 Identification of the friction laws

In previous studies [5,6], the authors reported how the relaxation tribometer technique was able to discriminate different contributions to friction: a velocity-dependent ("viscous-type") one and a velocity-independent one ("Coulomb-type"). Another result showed that the evolution of friction depends on the velocity amplitude of the free oscillating response. On this basis, Majdoub et al. [7] proposed to determine the friction coefficient from the free-damped response of the sliding 1-DoF oscillator. In the present study, we introduce a very simple way to do this, by assuming a constant friction coefficient at each oscillation. Thus we consider the equation of motion written as follows, Eq. (23.1):

$$m\ddot{x} + kx = \pm \mu N \tag{23.1}$$

where x is the displacement, $\dot{x}$ the sliding velocity, $\ddot{x}$ the corresponding acceleration, m the moving mass and k the spring stiffness. N is the applied normal load and μ the kinetic friction coefficient to be determined. From this decaying amplitude of the oscillations, we can extract the friction coefficient using a classical method. As described in detail in a previous work [8], the analysis of this response of a damped 1-DoF system is well known and shows that the decaying variation between three successive peaks of maximum velocity $\dot{x}$ is related $\Delta\dot{x}_{3\text{peaks}}$ to the dissipation force and then the global friction coefficient by the following relation, Eq. (23.2):

$$\mu(V_{\max}) = \Delta\dot{x}_{3\text{peaks}}\frac{\sqrt{km}}{8N}, \quad N \neq 0 \tag{23.2}$$

23.5 Several applications

23.5.1 Studying low friction of polymer brushes

We explored the friction behavior of surfaces covered by polymer brush layers swelled in ionic liquid (ILPBs) [9]. In this study, different normal loads were applied on the contact. Briefly, we show that the relaxation response of such an interface in mild contact conditions ($P_{max} = 160-400$ MPa) is mainly "viscous-type," with a typical exponential decay of the velocity amplitude with time, as shown in Fig. 23.3A and B. The corresponding friction laws show a strictly linear decrease with maximum velocity, as shown in Fig. 23.4. The minimum value of the superlow friction coefficient calculated was below 0.01 and could be measured accurately. In this case, the lubrication regime can be interpreted as full-film EHL confined lubrication. Conversely, when the contact conditions become more severe at a contact pressure of 680 MPa, for example, the decay of the velocity amplitude takes on a more complex shape, indicating that the damping is no longer "viscous-type," Fig. 23.3C. The corresponding friction law is drastically changed and appears as a characteristic "Stribeck-like" evolution, suggesting that the lubrication regime changes from full-film to boundary lubrication when the maximum velocity drops to zero (Fig. 23.4). It should be recalled here that the whole Stribeck curve can be obtained directly by processing only one single experiment in no wear conditions (Fig. 23.4).

23.5.2 Characterizing surface films and additives

The characterization of surface films is a major application of this method due to: (1) its tunable contact conditions, (2) the noninvasive character of the test (the cumulated sliding length in an oscillating test can range from several mm to tens of mm), and (3) the no-wear sliding condition. We investigated different studies on the effect of polar additives on lubrication of diamond-like carbon coatings [10,11]. By way of example, we present the free-response of a contact between steel and amorphous hydrogenated carbon film (a-C:H), lubricated by pure oleic acid, for two values of normal force: 0.40 and 0.90 N (Fig. 23.6A and B). The velocity envelope presents a complex shape, suggesting that the friction value has two contributions: velocity dependent and solid-like. Fig. 23.7 shows the evolution of the solid-like contribution μ_0 as a function of the maximum contact pressure for different friction pairs. It is shown here that a ta-C/ta-C pair can be better lubricated by oleic acid [10].

23.5.3 Highlighting the lubrication regime

This method can be applied to a fluid-lubricated contact. By way of example, we reported the experimental results obtained on a steel-on-steel contact lubricated by pure glycerol [6]. In this simple experiment, we showed that

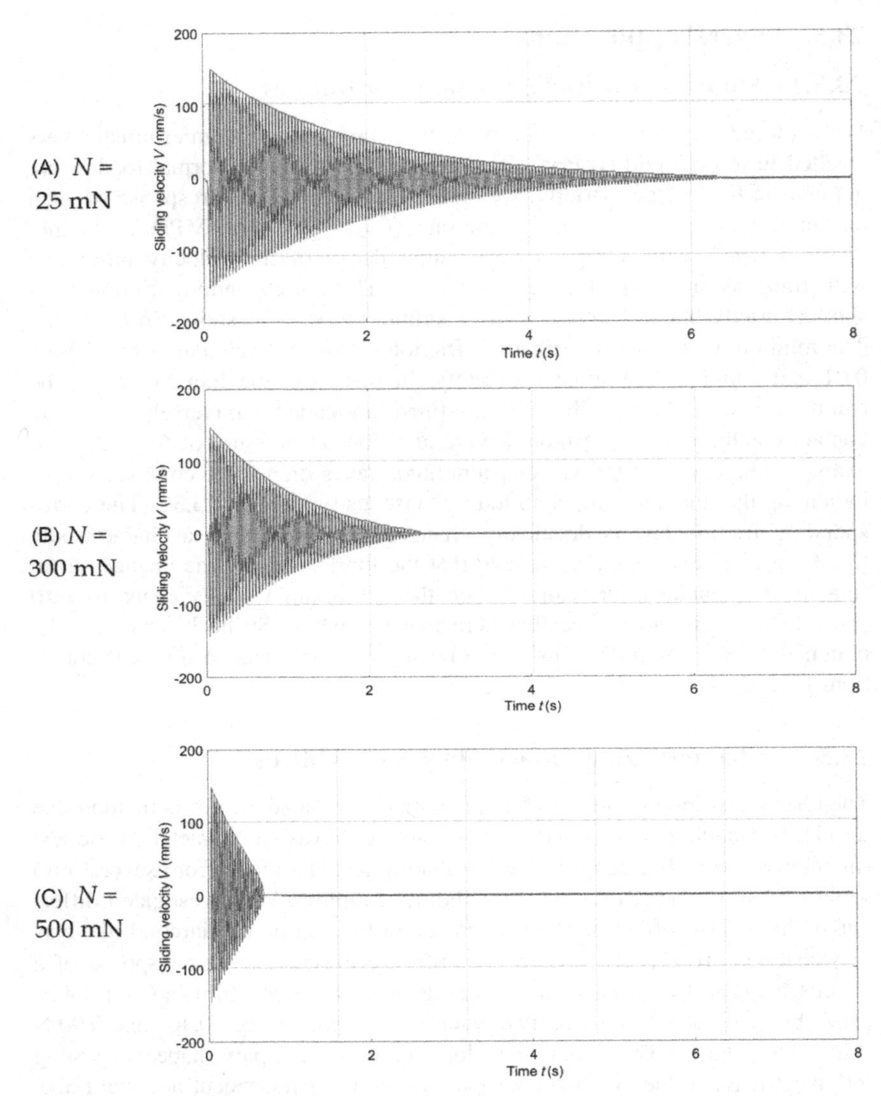

FIGURE 23.3 The typical evolution of the velocity of the oscillating sliding sphere, with time $V = f(t)$, and the envelope $V\max = f(t)$, and the corresponding friction law diagram $\mu = f(V\max)$, for three normal loads N: (A) 25 mN, (B) 300 mN, and (C) 500 mN, for thin ILPB grafted on steel, that was tested by a steel sphere, at 25°C. *After M. Belin, H. Arafune, T. Kamijo, J. Perret-Liaudet, T. Morinaga, S. Honma, et al., Low friction, lubricity, and durability of polymer brush coatings, characterized using the relaxation tribometer technique, Lubricants 6 (2018) 52.*

the friction response has two contributions, one is velocity-dependent, and the other is solid-friction type. ECR measurements were used to good effect, detecting intimate metallic contact interaction. It confirmed that the

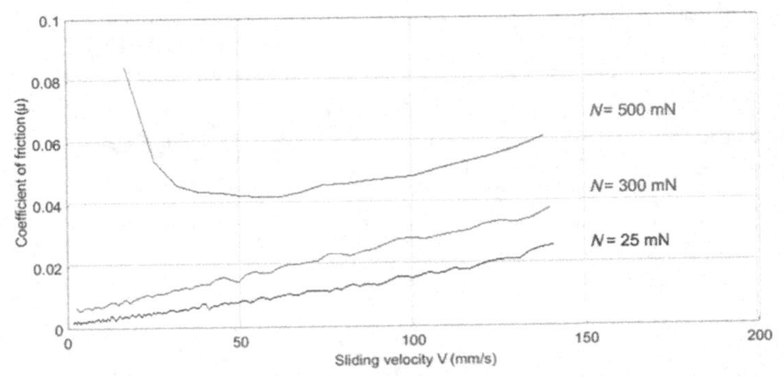

FIGURE 23.4 The friction laws $\mu = f(V\max)$ that were obtained on the thin ILBP that was grafted onto steel coupons. Friction was low and essentially viscous-type for the tests at normal force $N = 25$ mN and 300 mN. In the case of 500-mN-applied load, we found a combination of viscous and Coulomb-type contributions to friction, and a significant change in the lubrication regime. *After M. Belin, H. Arafune, T. Kamijo, J. Perret-Liaudet, T. Morinaga, S. Honma, et al., Low friction, lubricity, and durability of polymer brush coatings, characterized using the relaxation tribometer technique, Lubricants 6 (2018) 52.*

lubrication regime varies with the amplitude of sliding velocity, from a full-film regime at high velocity toward a boundary lubrication regime when the system returns to equilibrium as shown in Fig. 23.5 [6]. A previous paper showed how information can be extracted from the time response of the interface [12]. Other results especially dedicated to an EHD lubricated contact were published [13].

23.6 Perspectives

23.6.1 Exploring the nanometer length scale

The instrument described earlier is used to consider contacts at the mesoscale with a typical diameter of the apparent contact zone varying from tens to hundreds of micrometers and initial displacement amplitudes ranging between 100 and 1000 μm. We assume that such a method could be applied to nanometer-scale contacts in the AFM, especially those used in scanning probe techniques. In this case, the amplitude of motion will be considerably reduced, but the analysis of the damping phenomena could be transferred directly. This would open the way to measurements of superlubricity tribosystems at a much lower scale.

23.6.2 Capabilities of the relaxation technique

Considering the damping of a system to characterize friction at the interface can be extended to other geometries. For example, a recent paper showed how this technique can be extrapolated with other geometries by introducing the

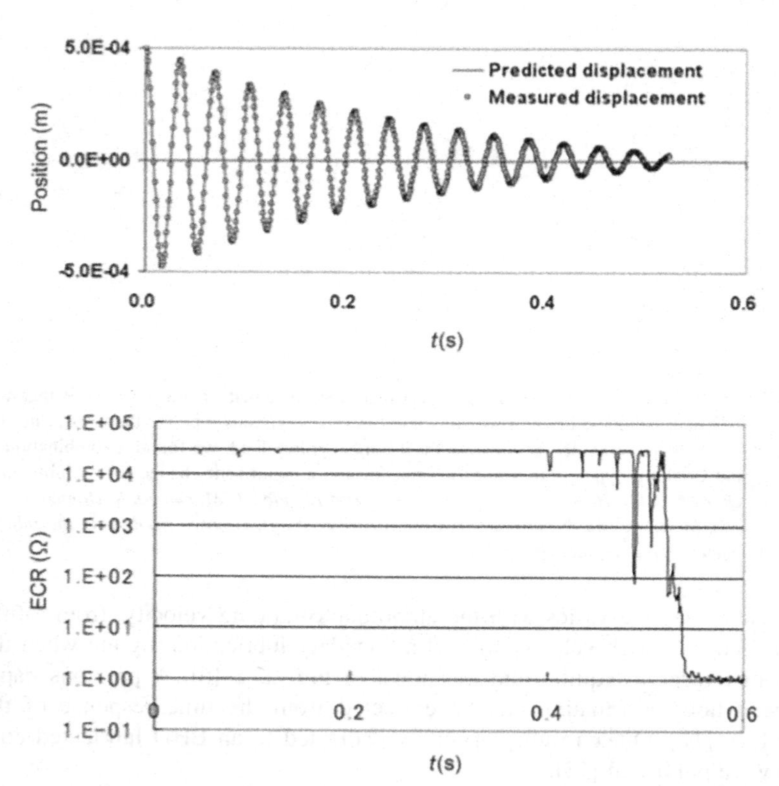

FIGURE 23.5 Time response of a sphere-on-plane contact lubricated by pure glycerol, $F_n = 50$ mN, temp = 24°C, AISI 52100 steel against itself. (A) Evolution of position: both experimental data for position (full line) and computed response (dotted line) are plotted. The model is based on a dual component behavior: zeta = 0.0186 and mu = 0.037. (B) Evolution of electrical contact resistance ECR, showing the full fluid film lubrication regime, changing to a boundary regime, when metallic contact occurs. *After M. Belin, M. Kakizawa, E. Rigaud, J.M. Martin, Dual characterization of boundary friction thanks to the harmonic tribometer: identification of viscous and solid friction contributions, J. Phys. Conf. Ser. 258 (2010) 012008.*

torsional relaxation test. The authors concluded that the relaxation principle has a vast but still insufficiently exploited potential to characterize friction [14].

23.6.3 Toward time-resolved characterization of an interface

In the results presented here, the characteristic time for an oscillating test ranged from 1 to 10 s. In the case of an interface undergoing physical evolution with time, it is possible to explore the relaxation response of the interface through consecutive multiple repeated oscillating tests. The first attempt was performed on a surface modified by self-assembled monolayers (SAM) layers, produced by alkylphosphonic acid interacting with steel surfaces [15].

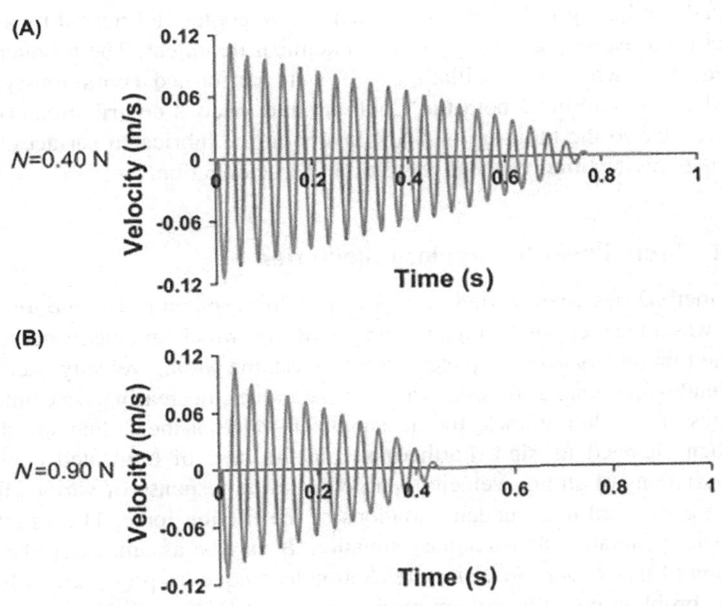

FIGURE 23.6 Velocity free-response for steel/a-C:H contacts lubricated with oleic acid at (Fig. 23.6A) $N = 0.40$ N, (Fig. 23.6B) $N = 0,90$ N. *After F. Majdoub, M. Belin, J.-M. Martin, J. Perret-Liaudet, M. Kano, K. Yoshida, Exploring low friction of lubricated DLC coatings in no-wear conditions with a new relaxation tribometer, Tribol. Int. 65 (2013) 278–285.*

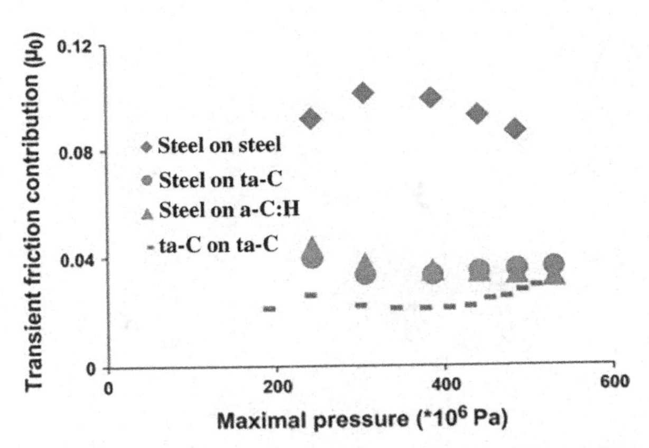

FIGURE 23.7 Transient friction contribution, μ_0, as a function of the maximal contact pressure. We can notice the drastic lubrication effect of the oleic acid on the friction behavior of the tribosystem. *After F. Majdoub, M. Belin, J.-M. Martin, J. Perret-Liaudet, M. Kano, K. Yoshida, Exploring low friction of lubricated DLC coatings in no-wear conditions with a new relaxation tribometer, Tribol. Int. 65 (2013) 278–285.*

Repeated oscillating tests were performed on the contact lubricated by a base product (here isopropanol) prior to any chemical treatment. The product was injected at t_1, while the oscillation tests were performed continuously. We showed the evolution of both the Coulomb and viscous contributions of the interface, due to the build-up of SAM layers on the lubricated surfaces. This opens the way to time-resolved relaxation characterization.

23.6.4 From linear to circular trajectories

In the method described earlier, we used a 1-DoF oscillator. The motion generated was linear reciprocating, the amplitude of which gradually decreased until the motion stopped completely. In this case, the sliding velocity oscillated in a pseudo-sine manner between the extreme values, decreasing with time. At the edges of the sliding track, the instantaneous value of the sliding speed was null, then changed its sign. Furthermore, in the case of lubricated contacts, these variations of sliding velocity can generate the response of various lubrication regimes and thus sudden variations of the friction force. The variations of velocity generate a nonstationary situation. It may be assumed that this is a limitation of this 1-DoF oscillatory relaxation technique, as presented earlier.

This problem could be solved by designing a 2-DoF oscillating relaxation tribometer based on two independent 1-DoF mechanical oscillators. These oscillators are strictly identical in mass, stiffness, resonance frequency and

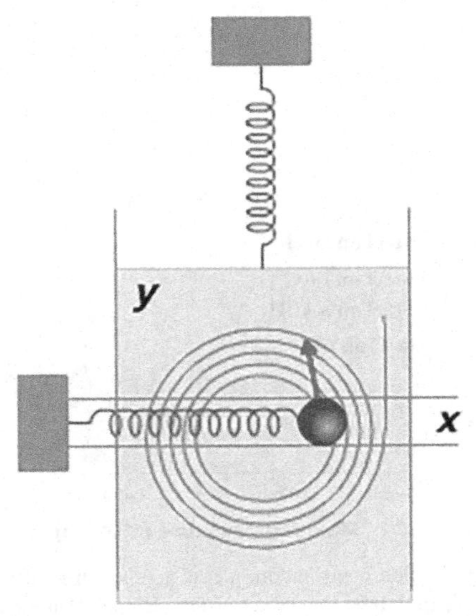

FIGURE 23.8 Schematic top-view arrangement of the 2-DoF oscillating relaxation tribometer. The combination of two identical 1-DoF oscillators, released in phase quadrature and damped by the sliding interface, generates a spiral trajectory.

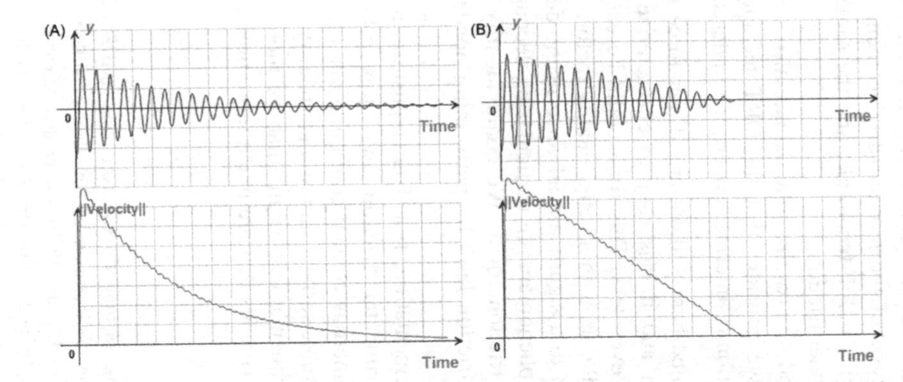

FIGURE 23.9 The projection of the motion on both x and y axes is oscillating in a pseudo-sine way, but the modulus of the velocity is monotonic decreasing with time, down to null value. (A) Case of pure viscous damping and (B) case of pure solid-type damping. We consider that in both cases, the velocity decay is close to a monotonic one, leading to stationary sliding velocity conditions for the interface.

internal damping. The principle consists in placing the plane on a similar bimetallic strip but perpendicular to the bimetallic strip of the slider (Fig. 23.8). By initially moving the two mechanical oscillators from their equilibrium position and releasing them with a delay of a quarter of their oscillation period (in quadrature), a damped circular displacement is generated. Therefore the trajectory is a spiral track with continuously decreasing speed without any stop situation as shown in Fig. 23.9. The fluctuations of speed with time will be sufficiently low considering that the sliding velocity is quasi-stationary. The analysis of the motion (displacement and speed) in each axis is the same as in the 1-DoF technique, and the same data processing can be applied to the model friction.

To conclude, based on both the results obtained and also several perspective and upcoming works, we consider that the ORT technique has great potential for the characterization of interfaces operating under superlubricity:

1. The friction was determined here without any direct force measurement, with unequaled precision and it was easy to characterize tribosystems with a CoF of 10^{-4} and even less. This limit could be easily overcome by using mechanical oscillators with a better quality factor.
2. this original approach led to a significant advance in the description and understanding of friction phenomena, in particular by decoupling the various components of the friction force. We could identify the different components of complex friction laws by performing nondestructive and very brief experiments.
3. By combining easy implementation, fast data acquisition and processing and direct access to experimental "friction laws," this technique allowed the characterization of numerous contact configurations (according to the temperature, contact pressure, surface nature and state, type of lubricant, etc.) by performing experimental campaigns that are more rapid in comparison to traditional experiments that are much more time consuming.

Acknowledgments

The authors would like to thank *Institut Carnot* I@L and iMUST Labex (project MUSCAT-2D) for partially funding the instruments, and also the following contributors: Drs. Joël Perret-Liaudet, Emmanuel Rigaud, Pierre-Emmanuel Mazeran, Fida Madjoub, and Amal Hirouech. We are especially grateful to Mrs. Genevieve Tulloue, for the graphical simulation of the 2-DoF damped mechanism and to Mr. Paul Colin and Nicolas Scremin, for their active contribution to the development of the RELAX-2D instrument.

References

[1] A. Erdemir, J.M. Martin (Eds.), Superlubricity, Elsevier B. V, Oxford, 2005.

[2] J.W.S. Rayleigh, The Theory of Sound, Dover Publications, New York, 1945. First published in 1877.

[3] J.W. Liang, B.F. Feeny, Identifying Coulomb and viscous friction from free-vibration decrements, Nonlinear Dynam 16 (1998) 337–347.
[4] B.F. Feeny, W. Liang, A decrement method for the simultaneous estimation of Coulomb and viscous friction, J. Sound Vib. 195 (1996) 149–154.
[5] E. Rigaud, J. Perret-Liaudet, M. Belin, L. Joly-Pottuz, J.-M. Martin, An original dynamic tribotest to discriminate friction and viscous damping, Tribol. Int. 43 (2010) 320–329.
[6] M. Belin, M. Kakizawa, E. Rigaud, J.M. Martin, Dual characterization of boundary friction thanks to the harmonic tribometer: identification of viscous and solid friction contributions, J. Phys. Conf. Ser. 258 (2010) 012008.
[7] F. Majdoub, J.L. Perret-Liaudet, M. Belin, J.M. Martin, Decaying law for the free oscillating response with a pseudo-polynomial friction law: analysis of a superlow lubricated friction test, J. Sound Vib. 348 (2015) 263–281.
[8] A. Hriouech, M. Belin, J. Perret-Liaudet, M.I. De Barros, M. Diaby, An original tribometer coupling a free-oscillation dynamic tribotest with a conventional linear reciprocating tribometer, in: Proceedings of the World Tribology Conference 2017, WTC'2017, Beijing, China, September 18–22, 2017; Poster ST-A-019.
[9] M. Belin, H. Arafune, T. Kamijo, J. Perret-Liaudet, T. Morinaga, S. Honma, et al., Low friction, lubricity, and durability of polymer brush coatings, characterized using the relaxation tribometer technique, Lubricants 6 (2018) 52.
[10] F. Majdoub, M. Belin, J.-M. Martin, J. Perret-Liaudet, M. Kano, K. Yoshida, Exploring low friction of lubricated DLC coatings in no-wear conditions with a new relaxation tribometer, Tribol. Int. 65 (2013) 278–285.
[11] F. Majdoub, J.-M. Martin, M. Belin, J. Perret-Liaudet, R. Iovine, Effect of temperature on lubricated steel/steel systems with or without fatty acids additives using an oscillating dynamic tribometer, Tribol. Lett. 54 (2014) 171–181.
[12] M. Belin, P.E. Mazeran, Exploring the lubrication regimes thanks to the relaxation tribometer, COST Nanotribo, Istambul Turkey, June 22–26, 2015.
[13] E. Rigaud, D. Mazuyer, J. Cayer-Barrioz, An interfacial friction law for a circular EHL contact under free sliding oscillating motion, Tribol. Lett. 51 (2013) 419–430.
[14] A. Le Bot, J. Scheibert, A.A. Vasko, O.M. Braun, Relaxation tribometry: a generic method to identify the nature of contact forces, Tribol. Lett. 67 (2019) 53.
[15] P.H. Cornuault, E. Rigaud, X. Roizard, M. Belin, J.M. Melot, Original free sliding oscillating ball-on-flat tribotests reveal some features of the frictional behaviour of alkylphosphonic acid self-assembled molecules, 45th Leeds-Lyon Symposium on Tribology, Leeds, September 4–7, 2018.

Chapter 24

Exploration of molecular behaviors in liquid superlubricity

Liran Ma and Jianbin Luo
State Key laboratory of Tribology, Tsinghua University, Beijing, P.R. China

24.1 Introduction

Superlubricity is intimately connected to many chemical, physical, and biological fields, thus it would be very useful to improve our understanding of the mechanism of friction and discover more materials with less wear. It has been extensively investigated as an ultralow friction state, which refers to a friction coefficient lower than 0.01 regardless of friction conditions [1]. Since the concept of superlubricity was proposed by Hirano et al. in the 1990s [2], various researches have emerged in recent decades to explore the underlying microscopic mechanism. In recent years, studies of liquid superlubricity were much more complex on account of an ambiguous microscopic mechanism and a lack of detection methods. A few kinds of liquid lubricants were found to possess superlubricity properties, such as polyelectrolytes with polymer brush appearances [3], some kinds of acid solutions [4,5], *Brasenia schreberi* mucilage [6,7] and confined salt solutions between mica surfaces [8]. Although reasonable explanations were proposed to interpret the underlying mechanism of liquid superlubricity, these theories diverged from each other. This resulted from no theoretical models that could be applied for all situations in regard to the mechanism of liquid superlubricity, and this was because several conditions and states were different in the proposed mechanisms that needed to be explained by different models, such as boundary lubrication, hydration lubrication, etc. Recently, the superlubricity of acid aqueous solution between an Si_3N_4 ball and an SiO_2 disc was reported by Li et al. [9,10], as shown in Fig. 24.1. This phenomenon was explained by the effect of a strong hydrogen bond network formed by phosphoric acid and water molecules [10]. This report speculated that the ultralow friction was attributable mainly to hydrogen bond effects and somewhat to dipole−dipole effects. However there were no direct detection methods to demonstrate a detailed and accurate molecular

Superlubricity. DOI: https://doi.org/10.1016/B978-0-444-64313-1.00024-7

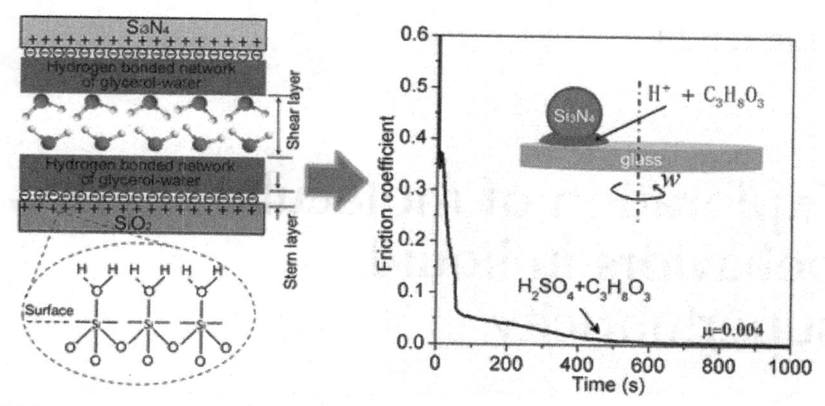

FIGURE 24.1 Superlubricity achieved with mixtures of acids and glycerol. *Reported by J.J. Li, C.H. Zhang, J.B. Luo, Superlubricity behavior with phosphoric acid–water network induced by rubbing, Langmuir 27(15) (2011) 9413–9417. Copyright 2013, American Chemical Society.*

structure of the phosphoric acid interface during the rubbing process, which was the most crucial aspect and urgently needed to be determined.

Superlubricity is a lubrication state with ultralow friction coefficient, which is different from either normal elastohydrodynamic lubrication or boundary lubrication. In the ball-disk experiments, superlubricity state has a nanoscale lubricant film in thickness. According to the theory of thin-film lubrication (TFL) proposed by Luo et al. [11,12], when the film thickness is at nanoscale, the lubricant molecules near the solid surfaces of the friction pair tend to be arranged due to the presence of a surface force and the orientation inducing effect of adsorbed molecules to liquid molecules nearby will induces the formation of the ordered layer, but the molecules farther from the solid surface tend to be more free liquid molecules. These orderly arranged molecules enhance thickness of the film effectively and improve the lubricant effects. However due to the limit of the research methods, the research on TFL is often focused on the lubrication properties and the measurement of mechanical quantities. The detailed molecular structure of lubricants cannot be directly detected by most microscopic or analytical tools but they can be interrogated through comparative analyses and simulations in which the molecular structure or changes can be predicted in the lubricant film during shearing. Therefore in order to clearly reveal the mechanism of fluid film lubrication, it is necessary to directly observe the effects of molecular arrangement and vibration mode on friction and lubrication.

24.2 Advances in analytical methods for molecular behavior of lubricants during sliding

The friction process itself is a superposition of multiple interactions, which leads to the complexity of research in friction mechanism. In recent decades, the calculations, analyses, and experiments in tribology have all improved a lot. More and more ways have been developed to study the physical and chemical process involved in friction. They can be mostly divided into direct dynamic observation and indirect chemical analysis. Among them, the direct dynamic observation is the method using scanning electron microscope (SEM), X-ray electron energy spectrum, and other technologies combined with a friction and wear tester, dynamically observing the process. However these techniques often require an ultravacuum environment, which is far from the environmental conditions of aqueous lubrication. Indirect chemical analysis uses all kinds of spectrums to measure the chemical changes of the surface after test. None of these methods can reflect the changes in the friction process in real time. In addition, tribological characteristics of the solid-liquid interface are closely related to the lubricant molecular structure and the solid materials, and it is difficult to eliminate the interference of bulk molecules with existing methods. Therefore it is necessary to study the surficial and interfacial molecular structure of materials in situ at the molecular level.

Nowadays, with the development of science and technology, experimental methods are changing with each passing day, and the research on molecular arrangement in nanoscale lubricant films has also made great progress. At present, the experimental methods to study the molecular structure and the interaction between lubricant molecules and surfaces can also be divided into indirect measurement and direct measurement. The advantages of direct measurement are various means and information. They are mostly used to measure physical quantities directly linked to tribological characteristics. For example, SEM and atomic force microscope (AFM) are used to measure morphologies, and X-ray photoelectron spectroscopy (XPS) and energy dispersive spectrum are used to analyze surface elements. By measuring this information, changes in molecular structure can be speculated. The surface force apparatus can accurately measure the nanogap between the two mica surfaces and the normal and lateral forces during friction, so as to predict the orderly arrangement of liquid molecules in the restricted lubricant film [13–15]. In addition, AFM scans the sample surface with nanoscale tips to obtain the fluctuation of the sample surface and the stress of the tip, so as to analyze the viscosity change of the adsorbed film on the sample surface [16,17]. Philippon et al. [18] used XPS to study phosphite and found iron phosphite compounds in the wear mark region of fresh metal surface, suggesting that adsorption was a prerequisite for the formation of tribofilm. Hakala et al. [19] used quartz crystal microbalance (QCM) to test the adsorption of hydrophobic proteins in water on the steel surface, and proved that the more water molecules are contained in the adsorption layer, and the smaller the friction coefficient was. Kato et al. [20] studied the process of nanooxide particles that form a sintered film during the sliding process on the surface of stainless steel. The results showed that if the oxide had a higher oxygen diffusion coefficient, it was easier to form a tribofilm. However these research methods have their own shortcomings under the requirements of in situ, real time, and atmospheric pressure, etc. and the experimental conditions are too different from which of macroscopic friction, so the molecular change information of lubrication friction cannot be reflected in situ.

The film thickness-measuring instrument based on optical interference can measure the friction process in situ. Domestic and international scholars [11,12,21] studied the film thickness in TFL regime, and have found that it gradually increases to a stable value with shear time. Therefore it is speculated that a layer of orderly arranged lubricant molecules is adsorbed on the solid surface, leading to the increase of film thickness. Boschkova et al. [22,23] also observed the orderly arrangement of surfactant molecules adsorbed on the surface of steel balls in the triboprocess by means of film thickness measuring instrument and QCM, and the arrangement structure could be immediately reformed after being destroyed by shear action, showing the characteristics of liquid crystal phase in general. All these experimental phenomena support the existence of ordered fluid film in TFL state. It can be seen that these indirect research methods can measure various mechanical information in tribology, but the lack of direct observation of molecular structure makes these theories to be further demonstrated.

The arrangement and structure of molecules are directly related to the atomic vibration or electron cloud distribution of the molecules themselves. Therefore the direct method to measure the information is called the direct method when studying the molecular structure, such as X-ray diffraction, fluorescence spectrum, and molecular vibration spectrum. Generally speaking, X-ray diffraction is most suitable for studying samples in static open space. For lubricant molecules in a confined space, x-rays must first pass through the gap between the two solid surfaces, making diffraction images difficult to obtain. Fluorescence spectrum is to observe the intensity changes of fluorescence signals in different polarization directions during the shearing process by using fluorescence produced by some molecules as research objects, and the changes of fluorescence molecular orientation with shearing action can be obtained [24,25]. However the shortcoming of the fluorescence spectrum method is that it can only measure the information of fluorescence molecules, which is only applicable to part of the research objects and cannot directly study commonly used lubricant molecules.

Molecular vibration spectrum measures the vibration frequency of molecules to directly characterize the molecular structure. The main methods include infrared (IR) absorption spectrum, Raman scattering spectrum and newly developed sum-frequency generation (SFG) spectrum. Cann et al. [26] measured the changes of two vibration peaks on the carbon chain of lubricant molecules with IR spectroscopy, and observed the relationship between film thickness changes and the orientation of lubricant molecules during friction.

As shown in Fig. 24.2, the CH_2 group on the lubricating oil molecular chain has three vibration peaks, among which the symmetric stretching vibration direction represented by 2853 cm^{-1} is perpendicular to the molecular chain, while the antisymmetric vibration direction represented by 2924 cm^{-1} is parallel to the molecular chain. By comparing the ratio of the two peak strengths, it is concluded that the carbon chains in the lubricated molecules are arranged more parallel to the surface of the sliding pair as the shear velocity increases.

Molecular vibration spectrum is a way to characterize the electron energy levels in molecular vibration and rotation. Different vibrational energy levels are closely related to the chemical bonds of molecules and their vibrating modes. Therefore information such as molecular arrangements, chemical states, and orientations of functional groups can be obtained by molecular vibration spectrum. Many kinds of molecular vibration spectra have been developed, and Raman spectra, with its high-sensitivity, high-resolution, and low environmental requirements, meets the needs of in situ, real-time, and atmospheric conditions for the study of molecular arrangement structure in lubrication.

Zhang et al. [27] designed a polarized Raman spectroscopy that measures an ordering process of nematic liquid crystal molecules of the contact area in situ. By changing the polarization direction of the laser in the plane of the contact area, the signal intensity distribution of liquid crystal molecules in the contact area within 360 degrees was obtained, as shown in Fig. 24.3. The results showed that the liquid crystal molecules are arranged along the direction of friction due to shear action.

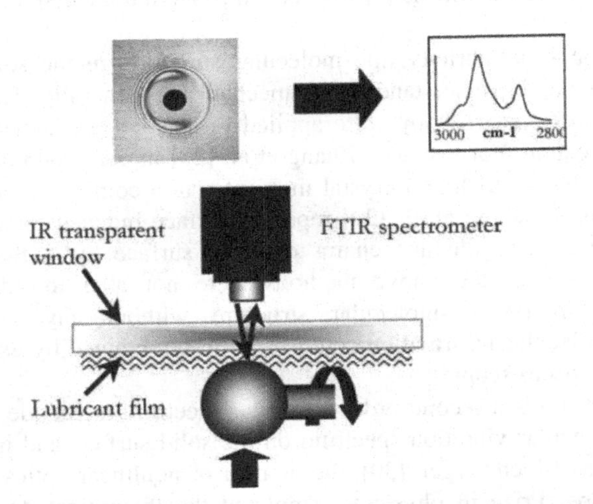

FIGURE 24.2 Infrared spectroscopy observation of the arrangement of lubricant molecules in the friction process. *From P.M. Cann, H.A. Spikes, In-contact IR spectroscopy of hydrocarbon lubricants, Tribol. Lett. 19(4) (2005) 289–297. Copyright 2005, Springer Science + Business Media, Inc.*

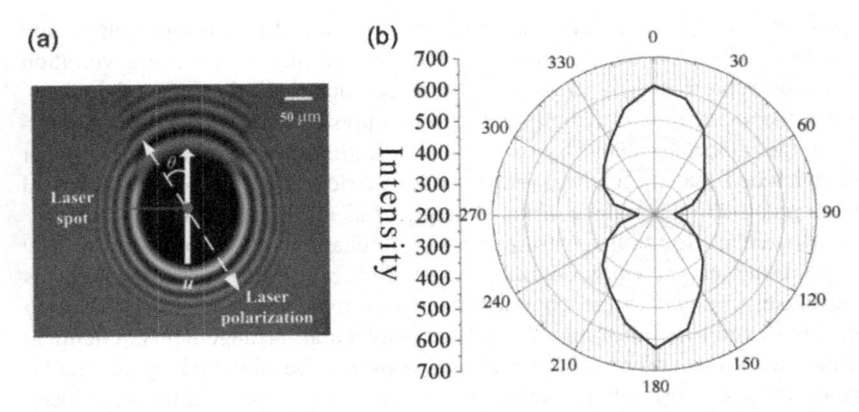

FIGURE 24.3 An ordering process of nematic liquid crystal molecules is observed by in situ polarized Raman spectroscopy of the contact area. *From S.H. Zhang, Y.H. Liu, J.B. Luo, In situ observation of the molecular ordering in the lubricating point contact area, J. Appl. Phys. 116 (1) (2014) 014302. Copyright 2014, AIP Publishing LLC.*

It can be seen that molecular vibration spectrum is an effective method to study molecular structure in situ. However these do not distinguish the effects of elevation and azimuth. Meanwhile, previous studies mostly focused on the effect of friction on molecular arrangement, but lack of research on the effect of molecular arrangement on friction. The relationship between specific molecular orientation and friction coefficient needs to be further studied, which will help to further improve the understanding of TFL mechanism.

During the superlubricity, the molecular structure on the solid surfaces was fundamental in understanding its mechanism. Generally, Raman spectroscopy and IR spectroscopy were applied to characterize molecular structure in great extent. For instance, Zhang et al. [28] in our group observed the orientation behavior of liquid crystal molecules in a confined area by use of Raman scattering. Jiang et al. [29] reported surface-induced torsional alignment of polydimethylsiloxane chains on mica surface by confocal Raman scattering. However the above methods were not able to purely detect solid–liquid interfacial molecular structure without any bulk signal. Interfacial molecular information could be entirely obtained by using nonlinear interface spectroscopy.

SFG spectrum is a second order nonlinear spectrum technique which provides the molecular vibration spectrum on the solid surface and interface. In 1962, Nicolaas Bloembergen [30], the pioneer of nonlinear optics (winner of the 1981 Nobel Prize in physics), completed the theoretical description of SFG spectrum. The most important characteristics of SFG spectrum are its unique interface selectivity and interface monolayer sensitivity. For isotropic bulk materials, such as the interior of substances other than crystals and

TABLE 24.1 Main advantages and limitations of sum-frequency generation spectrum.

Advantages	Limitations
In situ and in real time, no vacuum conditions are required	The sample must have infrared and Raman activity
Interfacial selectivity, monolayer molecules can be measured	The molecular structure of the bulk phase must be centrally symmetric
No need for molecular markers	Molecules on a surface or interface need a certain degree of order
Information at the molecular level	Light needs to reach the surface or interface studied
Molecular orientation at the interface	The sample cannot be damaged by laser

liquid crystal, the arrangement of molecules is usually isotropic. With similar electric dipole, the second-order polarizability is zero, and no sum-frequency signals are generated. However on the surface or interface, the two sides are different from each other, and the second-order nonlinear response may not be zero due to the broken symmetry of the molecules, so the surface and interface molecules can generate sum-frequency signals.

It was not until 1987 that Shen's research group realized the sum-frequency vibration spectrum's experimental technology [31,32]. Nowadays, sum-frequency vibration spectroscopy is one of the few effective methods to study the physical and chemical phenomena and processes on the surface or interface in real time in situ and at the molecular level. It is a surface characterization method that can be operated under ordinary pressure. The basic principle is that two beams of light, one IR and one visible (Vis) or ultraviolet, act on the surface or interface simultaneously in time and space, and then generate the third beam, called sum-frequency light. As a transition process of three photons (two incident light and one outgoing light), SFG spectrum can be regarded as a combination process of IR vibration excitation and anti-Stokes Raman emission. This requires that the molecules studied have both IR and Raman activity. The main advantages and limitations of SFG spectrum are shown in Table 24.1.

24.3 Direct observation of the liquid crystal molecules near the solid surface

In the recent decades, TFL theory has been used to explain the molecular behavior of lubricating films thinner than 100 nm. Surface-enhanced Raman spectroscopy has successfully distinguished the molecular ordering state affected by shear effect and surface effect, successfully showing the accumulation and orientation of liquid molecules in the TFL regime, and directly verifying the TFL model through experiments [33]. On the other hand, since the TFL molecular model was proposed by Luo et al. [11,12] in 1996, there has not been any experimental proof.

This work [33] is a new application of the TFL model. Here a thin layer of Ag nanorod film was deposited on a plano-convex surface as a SERS substrate, which improved the Raman intensity of interfacial molecules in the nanoconfined film, thereby significantly changed the alignment of the adsorbed molecules.

24.3.1 Materials and apparatus for directly observing the molecules

At both macro and nanoscales, 6CB (*p*-*n*-hexyl-*p*′-cyanobiphenyl, $C_6H_{13}(C_6H_6)_2CN$) possesses ordering and fluidity at the same time [34–38]. Therefore 6CB was chosen to be the testing lubricants on a convex-on-disk friction test platform equipped with a Raman spectrometer. Fig. 24.4 illustrates the experimental device. Between the upper SiO_2 plate and the lower lens the rolling point contact condition could be established.

Fig. 24.4 shows the lower substrate is a K9 plano-convex lens, and Fig. 24.5B shows the prepared Ag SERS substrate. Oblique angle deposition (OAD) technique was used to deposited slanted Ag nanorods in the electron beam system. According to OAD technique theory, Ag nanorods would grow perpendicular to the boundary. Fig. 24.5D shows an SEM image of Ag nanorods.

According to Hamrock-Dowson formula [39], the thickness of 6CB film in the contact center has a linear relationship with the shearing rate. A nanoscaled 6CB absorption film was formed between lens and SiO_2 plate, and with the method of relative optical interference intensity [11,40], the thickness of absorption film was measured. A single 6CB molecule could be

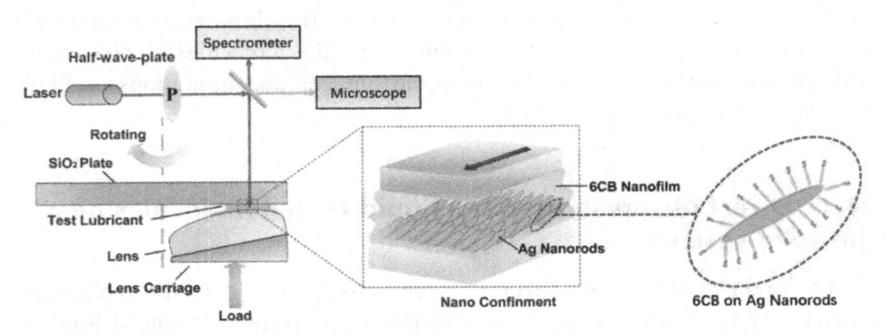

FIGURE 24.4 Illustration of the experimental apparatus with 6CB in the nanorestrain.

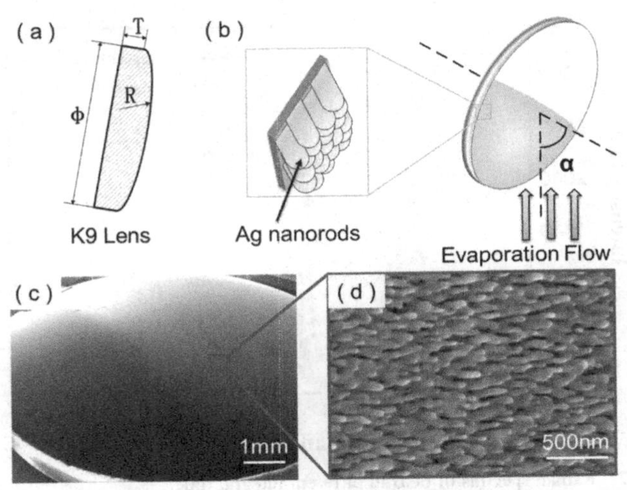

FIGURE 24.5 (A) Illustration of K9 plano-convex lens. (B) Schematic diagram of OAD technique. (C) SEM image of K9 plano-convex covered with Ti and Ag film. (D) SEM image of Ag nanorods [33].

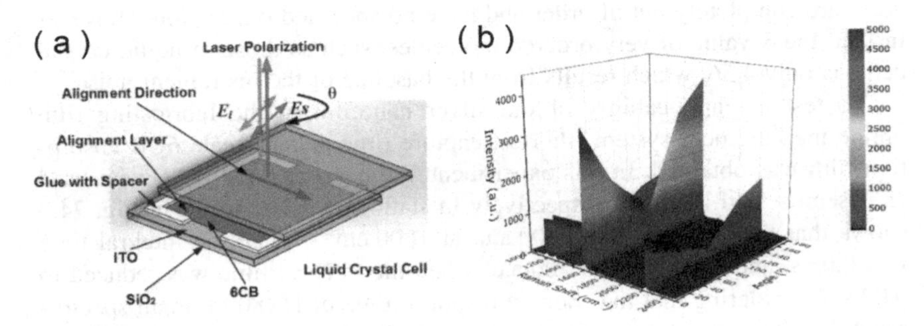

FIGURE 24.6 (A) Schematic diagram of the liquid crystal cell. (B) 6CB Raman spectra recorded with a laser polarized at different angles [33].

regarded as a structure with an uniaxial and rod-like shape, reported by Zhang et al. [27]. A liquid crystal cell can detect the relationship between the angle of the laser and the intensity of Raman Spectra at 1600 cm^{-1}, as shown in Fig. 24.6A. The angle between the direction of the alignment layer and the polarized light was defined as θ. Fig. 24.6B shows that when $\theta = 90$ degrees, the intensity at 1606 cm^{-1} peak reaches a minimum. In addition, when $\theta = 0/180$ degrees, it reaches a maximum. Therefore by calculating the polarization anisotropy of the C—C aromatic vibration mode of 6CB at 1606 cm^{-1}, the orientation of 6CB could be obtained. On the other hand, the ratio of the maximum intensity to the minimum intensity at a certain wavenumber can be defined as S (orderedness index), which represents the degree

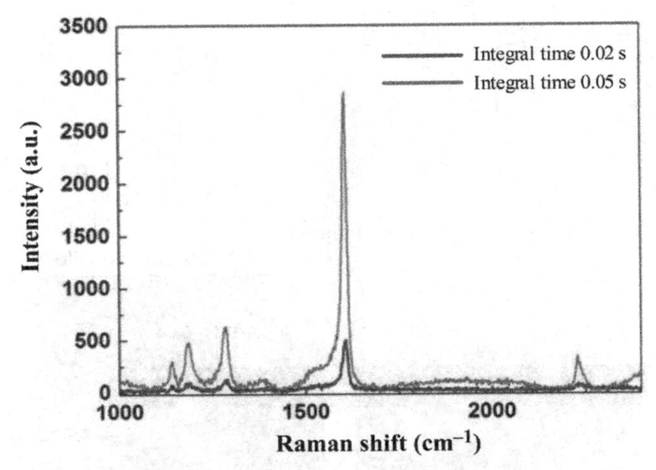

FIGURE 24.7 Raman spectras of 6CB in different integral time.

of order. Theoretically, the value of *S* lies between 1 and infinity. The larger the value of *S* is, the more ordered the molecules are. When *S* is 1, the molecules are completely out of order and have no specified orientation. However in fact, the S value of very ordered molecules, such as those in liquid crystal cells, is only 4.56, which results from the baseline of the instrument noise.

To test the applicability of the silver nanofilm in the lubricating film online measurement system, shortest capture time of nanoscale 6CB absorption film was obtained. In this experiment, Ag nanorods and 6CB were used as basement and lubricant respectively in static state. The result in Fig. 24.7 shows that the intensity was 2700 a.u. at 1600 cm^{-1} when the integral time was 0.05 s, and it reduced to 500 a.u. when the integral time was reduced to 0.02 s. Considering that the shortest response time of HR800 Raman spectrograph is 0.02 s, the shortest capture time of 6CB Raman scattering signal has reached the limit after the enhancement by Ag nanorods. According to the theory that the intensity of Raman signal is proportion to integral time, the shortest capture time can reach millisecond level.

24.3.2 Orientation of LC molecules adsorbed on Ag nanorods during lubrication

Different Raman signals excited by different polarized lasers were acquired and the orientation of molecules was calculated to analyze the influence of topography of Ag nanorods to absorbed 6CB molecules in static state. Fig. 24.8A–C shows the placing directions of the K9 planoconvex lens with Ag nanorods, and the arrows show the growth orientation of Ag nanorods. Fig. 24.8D–F are the radar maps of degree of molecular orientations of

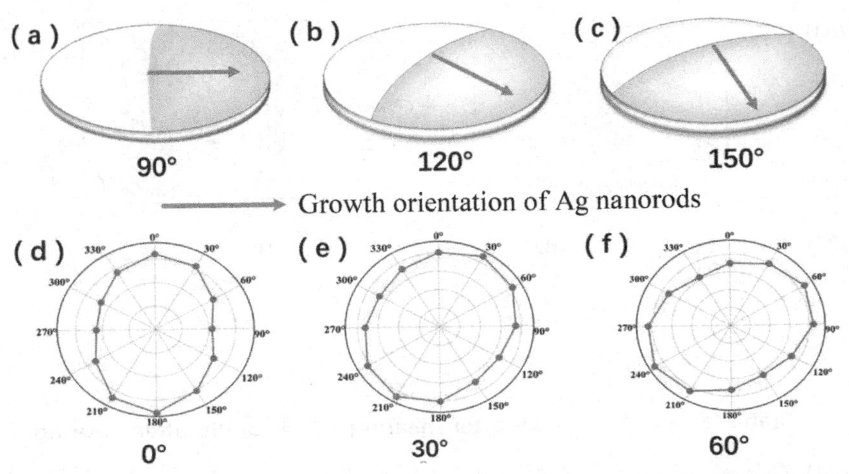

FIGURE 24.8 orientation of 6CB molecules absorbed on Ag nanorods with different incline angles in static state [33].

6CB. The result shows that when the orientations of Ag nanorods are 90, 120, and 150 degrees, respectively, the degree of molecular orientations of 6CB become 0, 30, and 60 degrees, respectively.

The aforementioned results reveals that in static state, the orientations of absorbed 6CB molecules and Ag nanorods are perpendicular to each other in the $X-Y$ plane, which means 6CB molecules are absorbed perpendicularly to the Ag nanorods along its principal axis.

As shown in Fig. 24.9A1, the statistics were measured on the $X-Y$ plane parallel to the friction pairs. The projection of the deposited Ag nanorods was along the X-axis direction, while the Y-axis was represented as 0 degree. Fig. 24.9A2 uses a laser, the measuring point, to collect the signal. It shows the initial microscopic image of point contact restrained between the film and silica plate. Fig. 24.9B1–B3 illustrates the Raman relative intensity of the adsorbed 6CB molecules at 1606 cm^{-1}. The Raman results show that under static conditions, the maximum Raman intensity can be obtained when the laser polarization is perpendicular to the growth direction of the Ag nanorods, and the minimum is obtained when the laser polarization is parallel to the growth direction. Next, when the silica plate is rotating at a speed of 100 mm/s, the shear-flow field is along the Y-axis. After 30 s of shearing, the silver film appeared to be slightly worn. Fig. 24.9B2 shows the relative intensity of Raman at this moment. The radar graph clearly showed now that S was around 1, and that the 6CB molecules were in a disorder state. The diagrams also show that the silver film reached a steady state after rotating the plate for approximately 400 s. Interestingly, Fig. 24.9B3 shows that when S is 2.23, the orientation of the measured here of 6CB molecules was in the

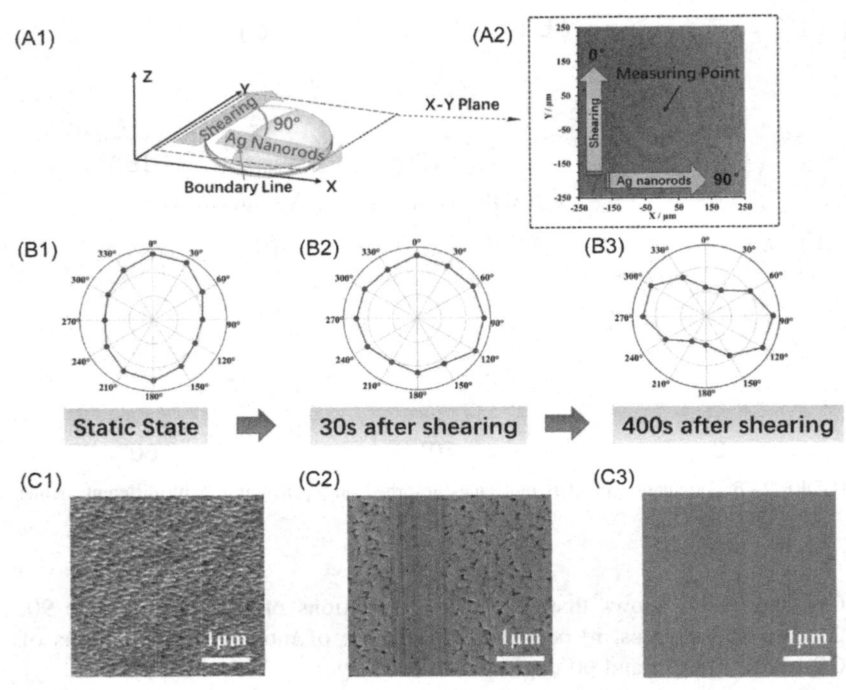

FIGURE 24.9 (A1) Illustrations of K9 plano-convex lens substrate covered with silver nanorod film under shearing. (A2) Microscopic image of the initial microscopic image of point contact restrained between the film and silica plate. (B) The radar graph indicating Raman relative intensity at different θ [the center point in (A2) is measuring point], relative intensity of 1600 cm^{-1} in corresponding Raman spectra. (C) SEM images of the silver film around the measuring point. Among these (B1) and (C1) are the initial state, (B2) and (C2) are after 30 s of shearing, (B3) and (C3) are after 400 s of shearing [33].

90-degree direction instead of the flow direction. The results show that the ordering state of the 6CB molecules may be greatly affected by the morphology of the Ag nanorods.

The SEM images clearly show that the morphologies of the measuring points in Fig. 24.9B1–B3 are very different, which indicates that the applied flow field has less influence, while the surface structure of the silver film mainly which keeps molecules in order.

24.3.3 The influence of different substrates on the orientation of LC molecules

Now we need to find out whether this phenomenon appears on the Ag nanorod surface specifically. The results are shown in Fig. 24.10. The bases were

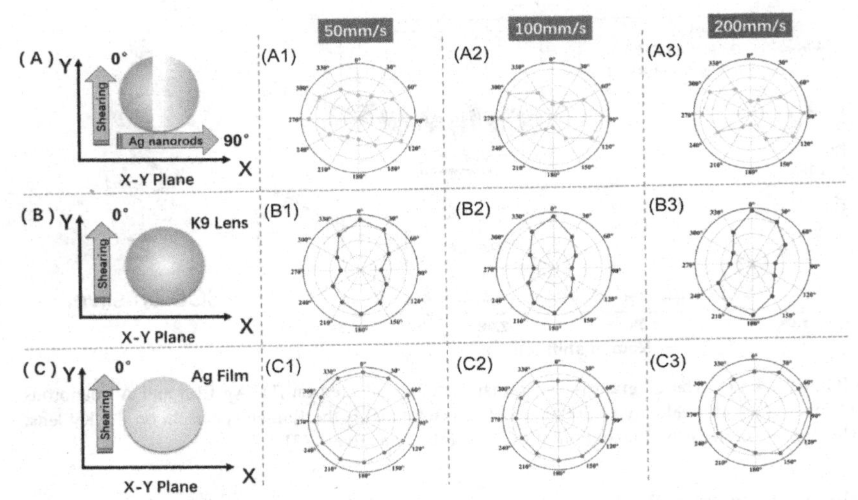

FIGURE 24.10 (A–C) Schematic diagrams of the substrate. (A1)–(C3) are Raman relative intensity graphs of 6CB on different substrates under shearing. The shear-flow speed is marked above the pictures. (A1–A3) Performed on Ag nanorod film bases; (B1–B3) performed on K9 plano-convex lens; (C1–C3) performed on flat Ag film bases [33].

a normal K9 plano-convex lens and another covered with silver. Surprisingly, the shape of the three sets of Raman radar charts varies greatly.

It is generally believed that when a flow field is applied, a fluid film will form in the gap between two solid surfaces, and since the liquid crystal (LC) molecules act as nanofilms, they are aligned along the flow direction under shear action [41]. However it should be noted that due to the SERS effect [42,43], most of the Raman signal received on the silver surface comes from the 6CB molecules adsorbed on or near it, which is true in both the Ag nanorods and flat Ag film. This conclusion can be drawn from the comparison shown in the chart. The enhancement is due to the SERS effect on the silver nanostructure, which is reported to be approximately 5 nm near the surface [44]. Therefore even though a fluid film is formed, only the information of the layer of adsorbed molecules on the solid surface and the molecules nearby can be obtained by Raman spectroscopy.

24.3.4 Model of molecular orientation on different part of lubricant film

Fig. 24.11 shows that the 2230 cm^{-1} peak in the Raman spectra broadened, which indicates that the cyano group (C–N) was affected by silver [45]. According to previous research, as shown in the inset panel in Fig. 24.11, 6CB molecules tend to use C–N bonds to adsorb vertically on the silver surface [46]. Thus the alignment of the initially static 6CB molecules would

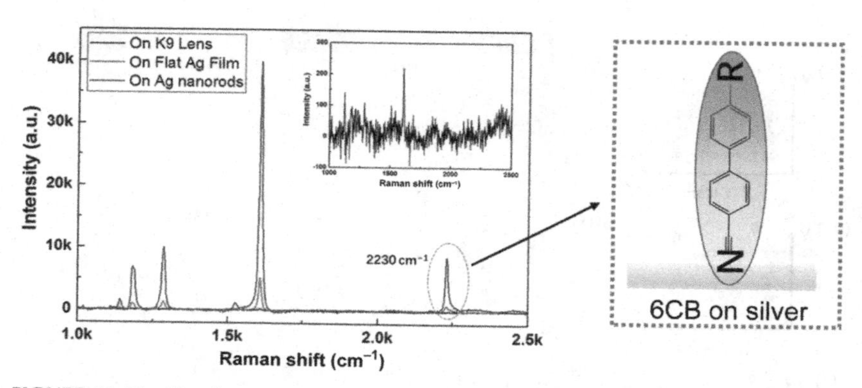

FIGURE 24.11 The diagram is the Raman spectra collected on flat Ag film and Ag nanorods when shearing at a velocity of 200 mm/s. The inset panel is the Raman spectrum on the K9 lens. The red rectangle is the illustration of 6CB adsorbed on silver [33].

always be perpendicular to the projection of the Ag nanorods, thus perpendicular to the $X-Y$ plane. We assumed that the Ag nanorods were redirected in the shearing direction due to the shearing process. As a result, regardless of the initial states of Ag nanorods, the 6CB molecules were adsorbed onto the Ag nanorods, which were aligned perpendicularly to the shearing direction.

A layered molecular model with orientation in a TFL system nanofilm is showed in Fig. 24.12. The first layer I is a thin adsorption layer formed near the surface. In this case, due to the influence of surface adsorption the 6CB molecules were perpendicular to the silver surface. The fluid layer in the middle of the nanofilm is marked as Layer III. From the results collected on K9 lens it could be determined that the 6CB in the fluid layer was arranged along the flow direction. We may assume that between layer I and layer III there is an ordered-molecule layer (marked as layer II), in which the molecules were oriented by inductive force. Therefore the conclusion can be drawn that the ordering behavior is significantly affected by the surface adsorption for near-surface or surface-based molecules. The orientation of molecules in the flow field inside the nanocell may be different. On the other hand, shearing forces can also influence the arrangement of molecules in middle level fluid. This conclusion agrees with the TFL model proposed by Luo et al. [11].

24.4 Detection of the water molecules on the liquid superlubricity interfaces

According to previous studies, superlubricity has been already achieved with phosphoric acid aqueous solutions. However due to the limitations of methods when detecting the interfacial molecules, there is no direct evidence to

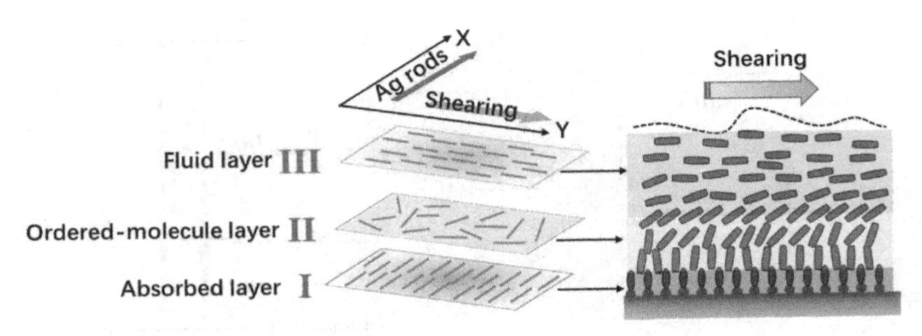

FIGURE 24.12 Illustrations of 6CB alignment under a shear-flow field [33].

clearly explain the underlying mechanism. With the development of SFG spectroscopy, we are able to explore the underlying microscopic mechanism for superlubricity in a phosphoric acid system. By capturing aqueous phosphoric solution between the surface of silicon nitride and silica, we found that phosphoric acid could greatly affect the ordered structure of interfacial water molecules, leading to SFG spectral changes the broad O—H bond stretching region [47], which provided a new idea for revealing the microscopic mechanism of liquid superlubricity.

24.4.1 Basic theory

Generally speaking, nonlinear optical interaction is a very weak optical process, which is difficult to detect. However when the vibration frequency of the incident light is coupled with the vibration frequency of the molecule itself, resonance will happen, greatly enhancing the polarization intensity and the output light intensity. Therefore the vibration mode of the molecule can be matched by changing the frequency of an incident light, and the SFG spectrum can be obtained.

We know that the vibrational energy levels of molecules are generally in the IR range, so the molecules will absorb the corresponding IR light and form the infrared absorption spectrum. Therefore in the sum-frequency vibrational spectrum, we use a beam of IR light and a beam of Vis light. The SFG process follows the conservation of energy, so the frequency of the outgoing sum frequency signals is the sum of the frequencies of the two incident lights. This is precisely the origin of the name sum-frequency generation. Therefore in the process of adjusting the frequency of IR light, the resonance of the sum-frequency process with the molecular vibration level is shown in Fig. 24.13.

When the energy of the incident light cannot match any vibrational level, only a few molecules can participate in the sum frequency process and the sum frequency signal is very small. However if the energy of IR light matches a certain vibrational energy level of the molecule, the molecule will

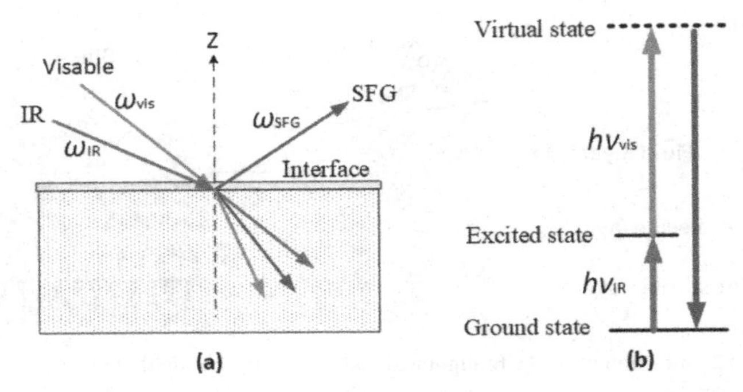

FIGURE 24.13 (A) Sum-frequency generation process and (B) schematic diagram of molecular vibration energy level.

absorb the energy and be excited to a more stable excited state, leading to a large increase in the number of molecules involved in the frequency process. Therefore the resonance enhancement of the sum-frequency process is achieved, and the energy of the sum-frequency signal also increases significantly. According to the energy level diagram, we can think the sum-frequency process in resonance as a superposition of an infrared absorption process and an antistokes Raman process. The relationship between the intensity of sum-frequency signal and the intensity of incident light is as follows:

$$I_{SFG}(\omega_{SFG}) \propto \left|\chi^{(2)}\right|^2 I_{IR}(\omega_{IR}) I_{Vis}(\omega_{Vis}) \tag{24.1}$$

where

$$\omega_{SFG} = \omega_{IR} + \omega_{Vis} \tag{24.2}$$

Therefore the SFG spectrum is actually how the square of $\chi^{(2)}$, a second-order nonlinear polarization, changes with respect to infrared frequency ω_{IR}. Usually you only need to know the energy of two incident beams and the sum-frequency signal energy to get the SFG spectrum.

However due to the use of femtosecond pulse length laser in the system, the width of monopulse IR light in the frequency domain is about $1000\ cm^{-1}$, so the intensity of IR light at a specific frequency cannot be accurately obtained. Therefore gallium arsenide should be used for calibration in the measurement process. Using the property that the second-order nonlinear polarization coefficient of gallium arsenide is a constant, the equation of gallium arsenide's sum-frequency signal can be obtained by:

$$I_{GaAS}(\omega_3) \propto I_{IR}(\omega_1) I_{Vis}(\omega_2) \tag{24.3}$$

Therefore in the experiment, the sum-frequency signal of the sample is measured first, and then the sum-frequency signal of gallium arsenide is measured in the same state. The final SFG spectrum is the ratio of the two sum-frequency signals:

$$\left|\chi^{(2)}\right|^2 \propto \frac{I_{SFG}(\omega_3)}{I_{GaAs}(\omega_3)} \tag{24.4}$$

Through the aforementioned description, it can be seen that frequency molecular vibration spectrum method is an excellent interface detection method. It not only retains the advantages of traditional molecular vibration spectrum, which can measure in situ, but also has the characteristics of interface selectivity. Therefore sum-frequency vibration spectrum is very suitable for the study of surface interface science, and will have great potential in the field of tribology in the future.

24.4.2 Experimental

Tribological experiments were conducted on a rotary sliding tester (UMT-3, CETR, as shown in Fig. 24.14). Diluted phosphoric acid solutions were selected as lubricants, whose concentrations were 0.1646 mol/L (pH value: 1.5). The constant load was 300 g, which indicated Hertzian contact pressure could reach 700 MPa. The spinning speed was maintained at 56 mm/s.

24.4.3 Detection of the water molecules on solid surfaces

Friction coefficient-time curve was shown in Fig. 24.15A, and ultralow friction coefficient occurred with the lubrication of phosphoric acid solution. Key measurement points of the Si_3N_4 ball and SiO_2 disc were demonstrated in Fig. 24.15B. Point A represented the edge of the track, and point B stood for the center of the track, in which the infrared beam and the Vis beam pass through the silicone disc and finally reach the same place simultaneously. SFG spectra of the stretching region of the O-H bond at the two measuring points were shown in Fig. 24.15C. It could be concluded that the edge of track reached the peak of 3200 cm^{-1} in "ice-like water" [48], whereas the SFG signal of the residues in track center almost entirely disappeared after tribological experiment. To verify the existence of phosphoric acid solution in the center of the friction track, the spectrum of the residues in the 1000–1200 cm^{-1} region was recorded in Fig. 24.15D. There is a peak at 1025 cm^{-1}, and another peak at 1150 cm^{-1}. These two peaks are respectively P=OH asymmetric (1010 cm^{-1}) and P=O symmetric (1170 cm^{-1}) stretching vibrations [49,50].

Molecular arrangement was relatively ordered in deionized (DI) water. However in phosphoric acid solution, the interaction of phosphoric acid with water molecules played an important role in disrupting the ordered molecular arrangement [51].

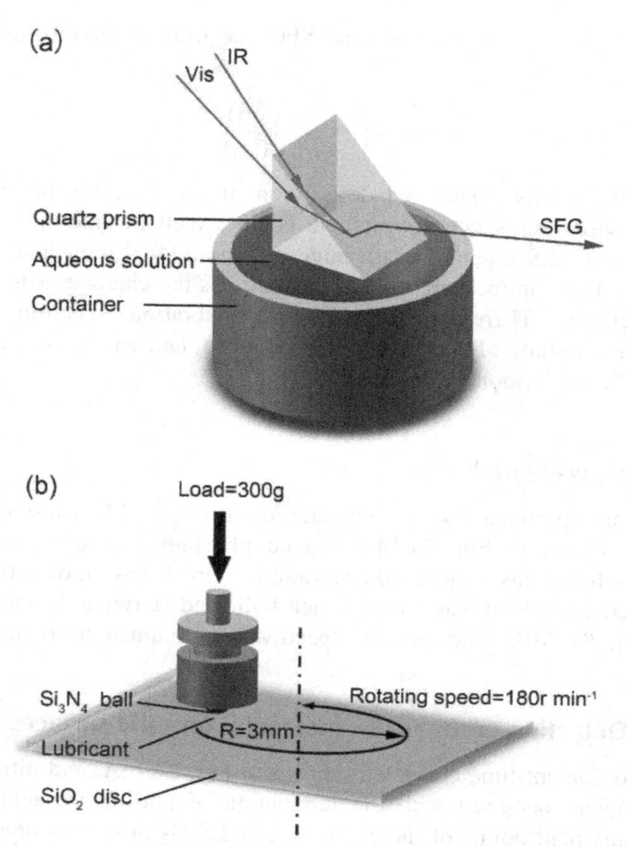

FIGURE 24.14 (A) A diagram of sum-frequency generation device and (B) schematic of the ball-on-disc friction tester. *From Y. Gao, L. Ma, Y. Liang, B.H. Li, J.B. Luo, water molecules liquid superlubricity interfaces achieved phosphoric acid solution, Biosurface Biotribol. 4(3) (2018) 94–98. Copyright 2019, Institution of Engineering and Technology.*

Using the method in Fig. 24.14A, the SFG spectra of various typical acid solutions absorbed on SiO_2 prisms were measured. The SFG spectra of DI water and three acid solutions in Fig. 24.16. (The data of phosphoric acid in Fig. 24.16 differed from those of Fig. 24.15 due to the varying degrees of plasma-treating [52].) The results demonstrated that the ratio of peak intensity (I1) at 3200 cm^{-1} to peak intensity (I2) at 3400 cm^{-1} was irrelevant to the value of pH. The (I1/I2) ratio of water, hydrochloric acid, and hyaluronic acid were at the same level, whereas the ratio of phosphoric acid was significantly lower than others. The interaction between phosphoric acid molecules and water molecules made it an ice-like structure rather than liquid-like structure, which could be the cause of this phenomenon.

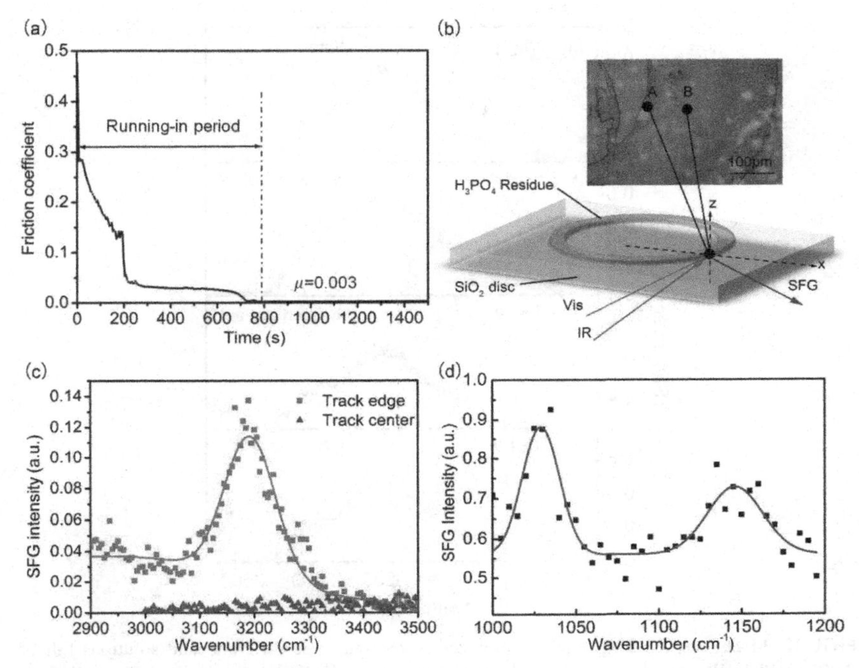

FIGURE 24.15 (A) Friction coefficient-time curve of phosphoric acid (pH = 1.5, concentration 0.1646 mol/L); (B) a schematic diagram of the sum-frequency generation from phosphoric acid solution residue in the contact region and the optical image of residues after friction (A: track edge, B: track center); (C) SFG spectra of track edge and track center for phosphoric acid solution residue in the O—H stretching region; (D) SFG spectrum of track center for phosphoric acid solution in the 1000—1200 cm^{-1} region. *From Y. Gao, L. Ma, Y. Liang, B.H. Li, J.B. Luo, water molecules liquid superlubricity interfaces achieved phosphoric acid solution, Biosurface Biotribol. 4(3) (2018) 94—98. Copyright 2019, Institution of Engineering and Technology.*

The above experimental results showed that the phosphoric acid solution can greatly affect the molecular arrangement, especially after friction. As shown in Fig. 24.17A, water molecules without phosphoric acid exhibited ordered arrangement at the liquid/solid surface, which corresponding to strong O—H stretching peaks in SFG spectra. With the addition of phosphoric acid, hydration changes the features of the solutions. From Fig. 24.17B and C, it could be concluded that the combination of phosphoric acid molecules and water molecules and would gradually break the original structure of interfacial water molecules, leading to lower SFG peak intensity. Higher degree of hydration caused greater level of molecular disorder. Molecular arrangement eventually reached the completely disorder state at the concentration reached 13.3 mol/L of acid. In that case, there will be no SFG peak in the O—H stretching region. Moreover, when phosphoric acid dissolved, hydrogen would be ionized, which would change the surface charge of SiO_2.

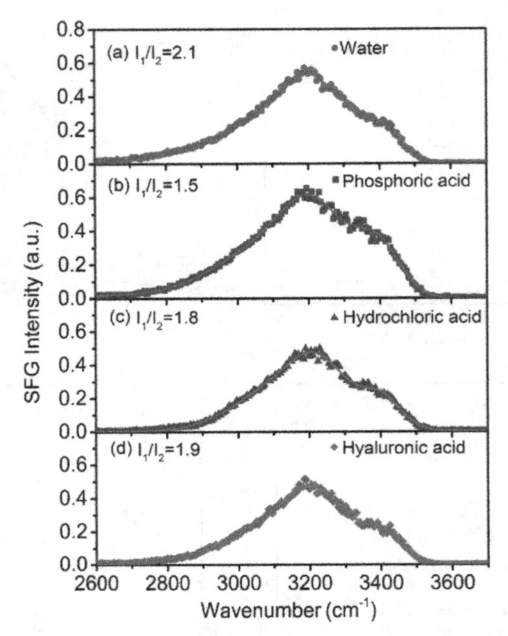

FIGURE 24.16 (A) SFG spectra of deionized water and three different acid solutions (all of 0.001 mol/L), (B) phosphoric acid, (C) hydrochloric acid, and (D) hyaluronic acid. *From Y. Gao, L. Ma, Y. Liang, B.H. Li, J.B. Luo, water molecules liquid superlubricity interfaces achieved phosphoric acid solution, Biosurface Biotribol. 4(3) (2018) 94–98. Copyright 2019, Institution of Engineering and Technology.*

Superlubricity could only be realized when the pH value of original phosphoric acid solution is between 1 and 1.5 [4,10]. In particular, the positively charged SiO_2 surface would attract oxygen atoms instead of hydrogen atoms if the pH value was less than pH_{pzc} (a pH value of zero surface charge), ~2 for SiO_2 [53]. In other words, surface electrical properties were changed with the rising concentration of phosphoric acid. The ordered molecular arrangement would also be disrupted. Combing all these experimental results, the chemical transformation of superlubricity could be revealed. The hydration of phosphoric acid might be faster in friction process, and the disappearance of the SFG intensity of the phosphoric acid residue could be attributed to this reason. In Fig. 24.15C, the concentration of phosphoric acid solution in the two measured points were nearly the same after equal time of evaporation. As a result, vanished SFG signal could not be explained by the evaporation of water during superlubricity. The strong hydration effect of phosphoric acid molecules might be the reason of the ultralow friction coefficients.

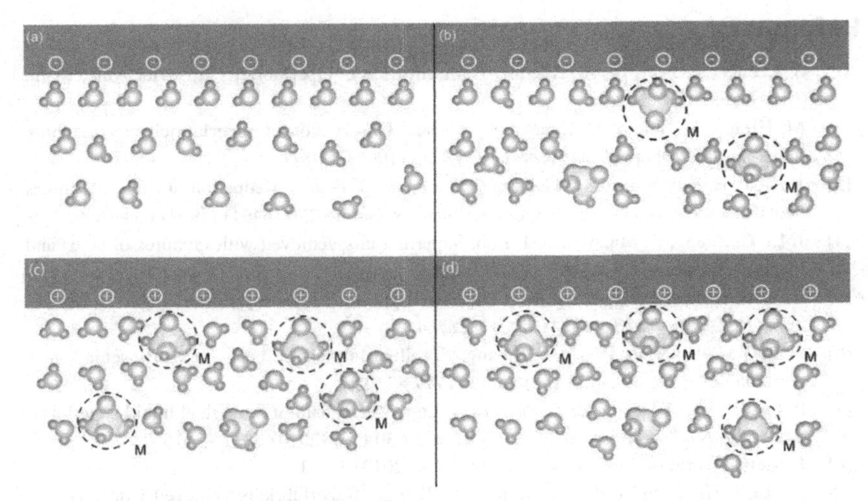

FIGURE 24.17 Schematic diagrams of the potential structure at four different solutions. (A) Water, (B) diluted phosphoric acid solution, (C) relatively concentrated phosphoric acid solution (pH = 1.5)/silica surface, and (D) phosphoric acid residue/silica surface after friction. The atoms from the biggest to the smallest stands for phosphorus, oxygen and hydrogen, respectively. M stands for a phosphoric acid molecule. *From Y. Gao, L. Ma, Y. Liang, B.H. Li, J.B. Luo, water molecules liquid superlubricity interfaces achieved phosphoric acid solution, Biosurface Biotribol. 4(3) (2018) 94–98. Copyright 2019, Institution of Engineering and Technology.*

24.5 Conclusion

In conclusion, based on some new real-time and online spectroscopic approaches, the molecular origins of liquid superlubricity have been able to be investigated, which helps to understand the underlying mechanism of the superlubricity process, even at the level of a nanolubricating film. The molecular structure, orientation, and degree of order of the molecules were able to be detected. Raman spectroscopy can be used to illuminate the orientation and ordering degree of molecules in the bulk film, while the SERS approach can be used to distinguish the difference between the bulk and interfacial molecules, based on those, the TFL molecular model was successfully demonstrated. Furthermore, SFG technique was used to show the water molecules on the superlubricity interface after rubbing via phosphoric acid solutions. All those work have successfully enabled the study of liquid superlubricity down to the molecular level.

Acknowledgments

This work was financially supported by National Natural Science Foundation of China (grant numbers 51527901).

References

[1] S. Lee, N. Spencer, in: A. Erdemir, J. Martin (Eds.), Superlubricity, Elsevier Amsterdam, The Netherlands, 2007.

[2] M. Hirano, K. Shinjo, R. Kaneko, Y. Murata, Observation of superlubricity by scanning tunneling microscopy, Phys. Rev. Lett. 78 (8) (1997) 1448.

[3] J. Klein, E. Kumacheva, D. Mahalu, D. Perahia, L.J. Fetters, Reduction of frictional forces between solid surfaces bearing polymer brushes, Nature 370 (6491) (1994) 634–636.

[4] J. Li, C. Zhang, L. Ma, Y. Liu, J. Luo, Superlubricity achieved with mixtures of acids and glycerol, Langmuir 29 (1) (2012) 271–275.

[5] Y. Gao, L. Ma, D. Guo, J. Luo, Ultra-low friction achieved by diluted lactic acid solutions, RSC Adv. 4 (55) (2014) 28860–28864.

[6] J. Li, Y. Liu, J. Luo, P. Liu, C. Zhang, Excellent lubricating behavior of Brasenia schreberi mucilage, Langmuir 28 (20) (2012) 7797–7802.

[7] P. Liu, Y. Liu, Y. Yang, Z. Chen, J. Li, J. Luo, Mechanism of biological liquid superlubricity of *Brasenia schreberi* mucilage, Langmuir 30 (13) (2014) 3811–3816.

[8] J. Klein, Hydration lubrication, Friction 1 (1) (2013) 1–23.

[9] J.J. Li, C.H. Zhang, L.R. Ma, Y.H. Liu, J.B. Luo, Superlubricity achieved with mixtures of acids and glycerol, Langmuir 29 (1) (2013) 271–275.

[10] J.J. Li, C.H. Zhang, J.B. Luo, Superlubricity behavior with phosphoric acid–water network induced by rubbing, Langmuir 27 (15) (2011) 9413–9417.

[11] J.B. Luo, S.Z. Wen, P. Huang, Thin film lubrication.1. Study on the transition between EHL and thin film lubrication using a relative optical interference intensity technique, Wear 194 (1–2) (1996) 107–115.

[12] J.B. Luo, M.W. Shen, S.Z. Wen, Tribological properties of nanoliquid film under an external electric field, J. Appl. Phys. 96 (11) (2004) 6733–6738.

[13] J. Yu, X. Banquy, G.W. Greene, D.D. Lowrey, J.N. Israelachvili, The boundary lubrication of chemically grafted and cross-linked hyaluronic acid in phosphate buffered saline and lipid solutions measured by the surface forces apparatus, Langmuir 28 (4) (2012) 2244–2250.

[14] S. Granick, Motions and relaxations of confined liquids, Science 253 (5026) (1991) 1374–1379.

[15] J. Klein, E. Kumacheva, Confinement-induced phase-transitions in simple liquids, Science 269 (5225) (1991) 816–819.

[16] K.S.K. Karuppiah, Y.B. Zhou, L.K. Woo, S. Sundararajan, Nanoscale friction switches: friction modulation of monomolecular assemblies using external electric fields, Langmuir 25 (20) (2009) 12114–12119.

[17] S.J. O'Shea, M.E. Welland, Atomic force microscopy at solid-liquid interfaces, Langmuir 14 (15) (1998) 4186–4197.

[18] D. Philippon, M.I. De Barros-Bouchet, T. Le Mogne, O. Lerasle, A. Bouffet, J.M. Martin, Role of nascent metallic surfaces on the tribochemistry of phosphite lubricant additives, Tribol. Int. 44 (6) (2011) 684–691.

[19] T.J. Hakala, P. Laaksonen, V. Saikko, T. Ahlroos, A. Helle, R. Mahlberg, et al., Adhesion and tribological properties of hydrophobin proteins in aqueous lubrication on stainless steel surfaces, RSC Adv. 2 (26) (2012) 9867–9872.

[20] H. Kato, K. Komai, Tribofilm formation and mild wear by tribo-sintering of nanometer-sized oxide particles on rubbing steel surfaces, Wear 262 (1–2) (2007) 36–41.

[21] M. Smeeth, H. Spikes, S. Gunsel, Boundary film formation by viscosity index improvers, Tribol. Trans. 39 (3) (1996) 726–734.

[22] K. Boschkova, J. Elvesjo, B. Kronberg, Frictional properties of lyotropic liquid crystalline mesophases at surfaces, Colloid. Surface A 166 (1–3) (2000) 67–77.

[23] K. Boschkova, B. Kronberg, J.J.R. Stalgren, K. Persson, M.R. Salagean, Lubrication in aqueous solutions using cationic surfactants - a study of static and dynamic forces, Langmuir 18 (5) (2002) 1680–1687.

[24] S.C. Bae, Z.Q. Lin, S. Granick, Conjugated polymers confined and sheared: photoluminescence and absorption dichroism in a surface forces apparatus, Macromolecules 38 (22) (2005) 9275–9279.

[25] A. Mukhopadhyay, S.C. Bae, J. Zhao, S. Granick, How confined lubricants diffuse during shear, Phys. Rev. Lett. 93 (23) (2004) 236105.

[26] P.M. Cann, H.A. Spikes, In-contact IR spectroscopy of hydrocarbon lubricants, Tribol. Lett. 19 (4) (2005) 289–297.

[27] S.H. Zhang, Y.H. Liu, J.B. Luo, In situ observation of the molecular ordering in the lubricating point contact area, J. Appl. Phys. 116 (1) (2014) 014302.

[28] S.H. Zhang, Y.J. Qiao, Y.H. Liu, L.R. Ma, J.B. Luo, Molecular behaviors in thin film lubrication-Part one: film formation for different polarities of molecules, Friction 7 (4) (2019) 372–387.

[29] S. Jiang, S.C. Bae, S. Granick, PDMS melts on mica studied by confocal Raman scattering, Langmuir 24 (4) (2008) 1489–1494.

[30] J.A. Armstrong, N. Bloembergen, J. Ducuing, P.S. Pershan, Interactions between light waves in a nonlinear dielectric, Phys. Rev. 127 (6) (1962) 1918.

[31] P. Guyotsionnest, J.H. Hunt, Y.R. Shen, Sum-frequency vibrational spectroscopy of a Langmuir film - study of molecular-orientation of a two-dimensional system, Phys. Rev. Lett. 59 (14) (1987) 1597–1600.

[32] J. Hunt, P. Guyot-Sionnest, Y.R. Shen, Observation C-H stretch vibrations of monolayers of molecules optical sum-frequency generation, Chem. Phys. Lett. 133 (3) (1987) 189–192.

[33] M. Gao, H.Y. Li, L.R. Ma, Y. Gao, L.W. Ma, J.B. Luo, Molecular behaviors in thin film lubrication-Part two: direct observation of the molecular orientation near the solid surface, Friction 7 (5) (2019) 479–488.

[34] J.P.F. Lagerwall, G. Scalia, A new era for liquid crystal research: applications of liquid crystals in soft matter nano-, bio- and microtechnology, Curr. Appl. Phys. 12 (6) (2012) 1387–1412.

[35] S.J. Woltman, G.D. Jay, G.P. Crawford, Liquid-crystal materials find a new order in biomedical applications, Nat. Mater. 6 (12) (2007) 929–938.

[36] W.M. Gibbons, P.J. Shannon, S.T. Sun, B.J. Swetlin, Surface-mediated alignment of nematic liquid-crystals with polarized laser-light, Nature 351 (6321) (1991) 49–50.

[37] H. Korner, A. Shiota, T.J. Bunning, C.K. Ober, Orientation-on-demand thin films: curing of liquid crystalline networks in ac electric fields, Science 272 (5259) (1996) 252–255.

[38] S. Kundu, M.H. Lee, S.H. Lee, S.W. Kang, In situ homeotropic alignment of nematic liquid crystals based on photoisomerization of azo-dye, physical adsorption of aggregates, and consequent topographical modification, Adv. Mater. 25 (24) (2013) 3365–3370.

[39] L. Berthe, A. Adams-Chaves, A.A. Lubrecht, Friction measurement indicating the transition between fully flooded and starved regimes in elasto-hydrodynamic lubrication, Proc. Inst. Mech. Eng. Part. J. 228 (12) (2014) 1403–1409.

[40] L.R. Ma, C.H. Zhang, Discussion on the technique of relative optical interference intensity for the measurement of lubricant film thickness, Tribol. Lett. 36 (3) (2009) 239–245.

[41] C. Gahwiller, Temperature dependence of flow alignment in nematic liquid-crystals, Phys. Rev. Lett. 28 (24) (1972) 1554.

[42] M.I. Stockman, Spasers explained, Nat. Photonics 2 (6) (2008) 327–329.

[43] W.L. Barnes, A. Dereux, T.W. Ebbesen, Surface plasmon subwavelength optics, Nature 424 (6950) (2003) 824–830.

[44] E. Hao, G.C. Schatz, Electromagnetic fields around silver nanoparticles and dimers, J. Chem. Phys. 120 (1) (2004) 357–366.

[45] M. Perrot, J.M. De Zen, W.G.J.Jo.Rs Rothschild, Mid- and low-frequency Raman spectra stable and metastable crystalline states of the 4-n-alkyl-4′-cyanobiphenyl (n = 9, 11, 12) liquid crystals, J. Raman Spectr. 23 (11) (1992) 633–636.

[46] G.W. Gray, A.J.M.C. Mosley, L. Crystals, The Raman spectra of 4-cyano-4′-pentylbiphenyl and 4-cyano-4′-pentyl-d_{11}-biphenyl, J. Mol. Cryst. Liq. Cryst. 35 (1–2) (1976) 71–81.

[47] Y. Gao, L. Ma, Y. Liang, B.H. Li, J.B. Luo, Water molecules liquid superlubricity interfaces achieved phosphoric acid solution, Biosurface Biotribol. 4 (3) (2018) 94–98.

[48] Q. Du, R. Superfine, E. Freysz, Y. Shen, Vibrational spectroscopy of water at the vapor/water interface, Phys. Rev. Lett. 70 (15) (1993) 2313.

[49] M. Cherif, A. Mgaidi, N. Ammar, G. Vallée, W. Fürst, A new investigation of aqueous orthophosphoric acid speciation using Raman spectroscopy, J. Solut. Chem. 29 (3) (2000) 255–269.

[50] R.F. Brandão, R.L. Quirino, V.M. Mello, A.P. Tavares, A.C. Peres, F. Guinhos, et al., Synthesis, characterization and use of Nb2O5 based catalysts in producing biofuels by transesterification, esterification and pyrolysis, J. Braz. Chem. Soc. 20 (5) (2009) 954–966.

[51] G.L. Richmond, Molecular bonding and interactions at aqueous surfaces as probed by vibrational sum frequency spectroscopy, Chem. Rev. 102 (8) (2002) 2693–2724.

[52] M. Kasuya, M. Hino, H. Yamada, M. Mizukami, H. Mori, S. Kajita, et al., Characterization of water confined between silica surfaces using the resonance shear measurement, J. Phys. Chem. C 117 (26) (2013) 13540–13546.

[53] D. Lis, E.H. Backus, J. Hunger, S.H. Parekh, M. Bonn, Liquid flow along a solid surface reversibly alters interfacial chemistry, Science 344 (6188) (2014) 1138–1142.

Chapter 25

Spatiotemporal manipulation of boundary lubrication by electro-charging and electrochemical methods

Yonggang Meng and Chenxu Liu
State Key Laboratory of Tribology, Tsinghua University, Beijing, P.R. China

25.1 Introduction

Friction and wear are of significance in performance, efficiency, reliability, and endurance of machinery, and have great impacts on human life and environment. Traditionally, friction and wear in a mechanical system are controlled through delicate combinations of the following engineering technologies. The first one is machine design and manufacturing, such as optimizations of load and pressure distributions, kinetics and dynamics of components, surface profiles, and topography. The second one is selection and development of materials, surface modification, and coatings. The third one is lubrication technology, including fluid and solid lubricants, additives, lubricant supply control, and maintenance. Along with the growing demands in reductions of energy consumption and emissions while increasing the power density, intelligence, and work life of mechanical systems, the traditional passive technologies mentioned earlier for the purpose of friction and wear control are not enough to meet the challenges. For instance, development and adoption of lower viscosity base oils for engine lubrication is a trend for lifting fuel economy of passenger cars. However experimental results [1] show that this approach does not always work well, especially when engine frequently runs at low speeds or in the durations of start-stop, often encountered in hybrid cars.

Active control of friction and wear is a goal that tribologists have been striving to achieve for the past decades. In literature, besides the term of active control of friction, there are several different terminologies coined to express the features of the emerging technology, such as tribotronics [2], active tribology [2], electromagnetic tuning of friction [3], potential-controlled boundary

Superlubricity. DOI: https://doi.org/10.1016/B978-0-444-64313-1.00025-9

lubrication [4], smart surfaces [5], etc. Although there is a lack of rigorous definition of the so-called active tribology, the objective of such a technology is to realize in-process fine-tuning, timely, and locally, of surface interactions by providing a supplementary external stimuli to while without changing the operation conditions of a mechanical system. The external stimuli could be mechanical, acoustic, photonic, thermal, electrostatic, electronic, electromagnetic, electrochemical, or a combination of these different ways. Among them, effects of an external electric or magnetic field on tribological behavior of materials have been mostly investigated in various lubrication conditions, and the mechanisms of the effects have been explored and discussed in depth.

This article presents a survey on the progress in manipulation of boundary lubrication by electro-charging and electrochemical methods in liquid lubrication systems, which is a subset of the active tribology researches. At first, potential control of boundary lubrication in pure water or acids will be described, followed by the work in aqueous surfactant solutions. Then, the study will be moved to controllability of boundary lubrication of polar base oils with different kinds of additives, including ionic liquids (ILs), nanoparticles, and zinc dialkyldithiophosphates (ZDDP) respectively. Finally, a summary of the findings and prospects for industrial application will be provided. The major part of the survey covers the research work done in the author's group.

25.2 Electrical potential manipulation of boundary lubrication in pure water or acids

In the early of 1950s, Bowden and Young [6] firstly introduced the potentiostatic technique of electrochemistry to the field of tribology. They found that the static friction of platinum on platinum in pure dilute sulfuric acid changed remarkably as the surface electric potential of the platinum wire was varied, and they attributed the reduction in friction to the adsorption of the hydrogen or oxygen gas film generated in electrolysis of water at the presence of a large negative or positive electric potential, which was considered to play the role of boundary lubrication. In 1969, Bockris and Argade [7] explained the effect of electric potential on friction in terms of the repulsive force between electrical double layers (EDLs) on the rubbing surfaces. The EDL repulsive effect on friction was supported by the experiments done in 1990s by Zhu et al. [8] with iron/iron and iron oxide/alumina rubbing contacts in Na_2SO_4 solution. The explanation of the EDL repulsive interaction effect has also been adopted to the findings of superlubricity of phosphoric acid-lubricated ceramic contacts in recent years [9]. In 2012, Valtiner et al. [10] investigated the effect of electrostatic surface potential as well as surface roughness and anodic oxide growth on interaction forces and friction in a nitric acid solution by using the electrochemical surface forces apparatus. They found that both surface forces and friction force of the acid solution at a nanogap could be affected by imposing an electric potential. By comparison with the Derjaguin–Landau–Verwey–Overbeek theory, it is found that not only the EDL force but also the other two exponentially repulsive interactions, steric and hydration forces, should be accounted.

A more profound effect of external electric field on water than EDL force is the phase transformation of water under the assistance of a strong electrostatic field. It has been argued for a long time that liquid water could transform to an icelike structure in a confined condition, and the presence of an electric field could change the effective viscosity as well as confined structure of water greatly. Pashazanusi et al. [11] demonstrated the anomalous potential dependence of friction of a 0.1 M NaCl solution on Au(111) surface by using a sharp or a colloidal atomic force microscopy (AFM) tip. When the gold surface is positively charged (+0.6 V vs Ag), coefficient of friction is as high as 3.1 for the sharp AFM tip and 2.1 for the colloidal AFM tip, respectively. When the charging state of the gold surface is switched to negative (−0.6 V vs Ag), the coefficient of friction drops to 0.12 and 0.16 for the sharp and colloidal AFM tips respectively. They attributed the remarkable and reversible change in friction to the transformation of water layer on the gold surface from the icelike structure at the positive potential to the liquid structure at the negative potential. Li et al. [12] showed that a remarkable increase in surface interaction due to such a transformation of water layer was observed only for a contact between the gold surfaces with a hydrophilic colloid such as SiO_2. For the hydrophobic polystyrene (PS)/gold contact, surface interaction was independent of the applied electric potential. By using an AFM with a SiO_2 microsphere tip sliding against a gold substrate in a 0.1 M NaCl solution and an electrochemical cell as shown in Fig. 25.1A reversible switching between superlubricity and high friction was realized by controlling the electrical potential (see Fig. 25.1B). The coefficient of friction at the electrical potential of −0.6 V versus Ag is as low as 0.004 (see Fig. 25.1C).

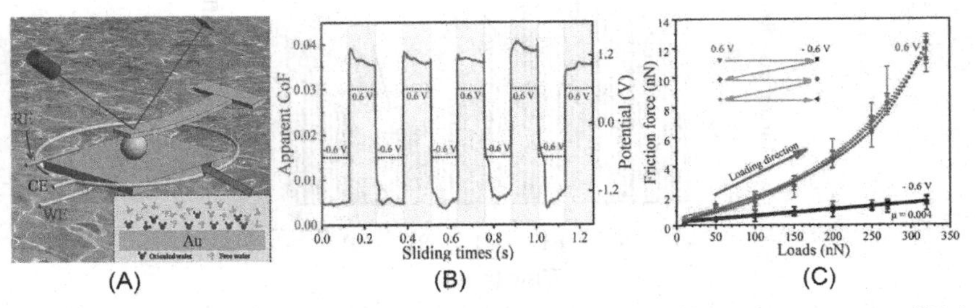

FIGURE 25.1 Reversible switching between superlubricity and high friction between a SiO_2/gold contact in aqueous solution of 0.1 M NaCl.

25.3 Electrical potential manipulation of boundary lubrication in aqueous surfactant solutions

For macroscale rough surface contacts, with the increase in normal load asperity contact gradually becomes profound because of the limited load carrying capacity of pure water, acidic and salt solution layers, and thus the effect of interfacial potential on friction gets weaker. Addition of some kinds of long chain organic molecules can effectively improve boundary lubrication of aqueous solutions in the cases. Jiang et al. [13] in 1998 found that external voltages could remarkably increase the friction coefficient of Al_2O_3/Cu sliding pairs in the emulsion of 1 wt.% zinc stearate dispersed into deionized water. The friction coefficient was about 0.15 in the absence of external voltages and about 0.45 in the presence of the voltage of +20 V when the lower specimen was used as cathode. They coined the name of potential controlled friction (PCF) as the expression of the phenomenon. Later, Jiang, and Meng [14–16] systematically studied the influences of electric potential on frictional behaviors of several kinds of ceramic/metal couples. By applying the theory of electrochemistry on the experimental findings, they proposed that the electric potential affected the friction of ceramic/metal couples by changing both the surface interactions and the state of boundary films, which were influenced by the double layer and the electrode current respectively in a tribochemical system. Meanwhile, a feedback control system was established to testify that the friction coefficient could be adjusted to vary in a predesigned way. Results of the studies had shown that the output could fit to the target values of friction coefficient well when control parameters were selected properly. One of the representative results was plotted in Fig. 25.2. Essential agreement between the controlled friction coefficient with its preassigned values was accomplished [15].

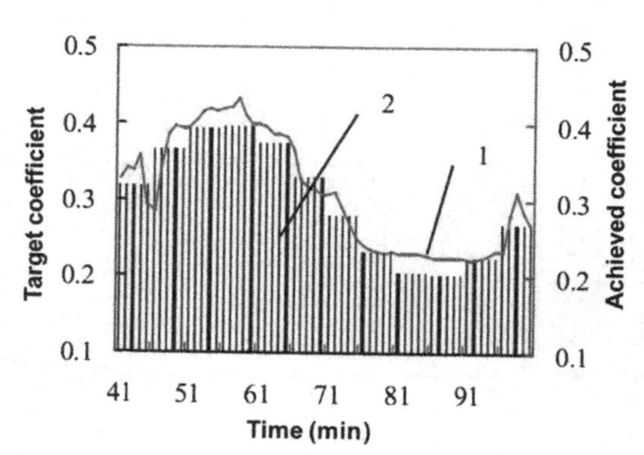

FIGURE 25.2 Experimental result of the controlled sinusoidal variation of friction coefficient: (1) target coefficient, (2) achieved coefficient [15].

In 2002, Chang et al. [17] found that the electrolysis of water might be the key trigger to the abrupt change in friction coefficient in the presence of a sufficiently negative electric potential, because the transition from a low to a high friction in their test conditions was coincident with that of the electrolysis of the solution. They attributed the observed change in friction to the cleaning action of the local high pH value due to the electrolysis of water under severely negative interfacial potentials for the laurylsulfonate solution. In addition, they developed the lubricant systems from the zinc stearate emulsion to aqueous solutions containing long-chain or short-chain anions. Though the applied potentials in their experiments were still tens of volts high (much higher than the voltage of water electrolysis), they creatively introduced the electrochemical characterization techniques in the PCF tests. In 2005, static cyclic voltammetry (CV) was used by Hu et al. [18] to seek for the possible electrochemical reactions on the metal surface, and experiments which combined CV with rubbing experiments were undertaken to study the actual electrochemical reactions when the PCF phenomenon was taking place. Moreover, the Micro Raman analysis was used to examine the adsorptive films on the metal surface under different voltages or other conditions. Their results had indicated that the PCF phenomenon was the results of electrochemical reactions and boundary film changes because of the lubricant adsorption or desorption on the metal surface.

Based on the aforementioned explorations, an innovative friction clutch based on PCF phenomenon was designed and fabricated to examine the clutch characteristics and the response of the friction torque of a brass/silicon nitride rubbing pair to external potentials. It was found that the output torque and friction coefficient of the frictional pair was changed by a factor of 2 upon applying of 20 V external voltages. The response time of the friction clutch to an external square-wave voltage was about $2\sim5$ s at the rising stage and about $3\sim8$ s at the descent stage [19].

In the aforementioned studies, the PCF phenomenon was influenced by many factors owing that the applied potentials were higher than the hydrolysis potential of water. Hence, the mechanism of the PCF had not been comprehensively understood and some misunderstanding about the mechanism still existed.

As a key step in understanding the PCF, He and Meng [20,21] researched the friction behaviors within the electrochemical window in 2009. They found that the PCF effect could be achieved within electrochemical window, where the hydrogen evolution reaction on the cathode and the friction pair oxidation on the anode could be eliminated, as long as the concentration of the surfactant solution was much lower than the critical micellar concentration.

In their experiments, sodium dodecyl sulfate (SDS) in the aqueous solution was used as the responsive ingredient of the PCF system. The effect of the SDS concentrations on the friction coefficient under different applied potentials is presented in Fig. 25.3A. The friction coefficients versus potential curves were similar in trend, and could be classified into a low concentration group (SDS concentration <5 mM) and a high concentration group (SDS concentration >5 mM), according to the different aggregate morphology of the SDS surfactant on stainless steel surfaces. Quartz crystal microbalance (QCM), electrochemical spectroscopy, AFM and lateral force microscopy were used to investigate the effects of electrode potential on adsorption and desorption of SDS surfactant. They found that the charge accumulation on the metal surface of the friction pair was a key factor, which determined the adsorption/desorption of surfactant ions, resulting in the falling or rising of friction coefficient.

Moreover, the effects of a few typical cationic surfactant solutions on friction have been studied for the purpose of comparison with anionic surfactant solutions, and the PCF phenomena of the silicon/ceramic and metal/metal contacts had been studied in the surfactant solutions. Through a number of convincing results in their experiments, the mechanisms of the PCF were clarified and attributed to the surfactant adsorption/desorption driven by the electric potential. Fig. 25.3B shows an example of the test results with an intentional control sequence of the "on-off" and "positive-negative" voltage applications. Friction coefficient is about 0.1 at the initial stage of open circuit state and then jumps up to around 0.5 when 2.5 V positive voltage is switched on. Friction coefficient can quickly decrease to the low value of 0.1 as a voltage of -0.2 V is switched to, and the experimental result is repeatable. The response times could be decreased to 0.2 ~ 1.5 s for the friction coefficient increasing and 0.5 ~ 2 s for the decreasing, depending on the magnitudes of the potential and the SDS concentration [20].

Based on the aforementioned findings in SDS aqueous solutions, Zhang and Meng [22,23] further studied the relationships among the adsorption of surfactant on solid-liquid interface, the boundary lubrication properties of the

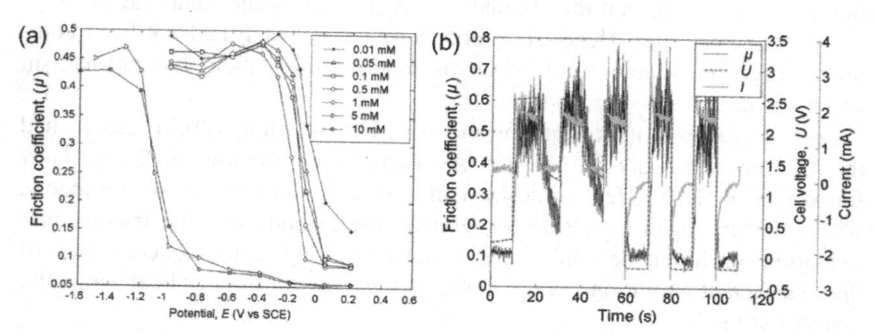

FIGURE 25.3 (A) Friction coefficient versus electrode potential in SDS solutions of different concentrations [21]. (B) Variation of friction coefficient under multiple and different stepwise voltage applications [20].

adsorption film and the control of boundary lubrication effect by applying potentials in 2015. The adsorption characteristics of SDS on stainless steel surfaces in aqueous solutions as well as the effect of the added $NaClO_4$ on adsorption were investigated using quart crystal microbalance with dissipation monitoring (QCM-D) [23]. To compare with the results in nanoscales, the boundary lubrication properties of SDS adsorption film as well as the control of boundary lubrication were investigated with a sharp AFM tip sliding on a smooth stainless steel sample. Their results showed that the adsorbed mass for achieving good boundary lubrication effect and the stick-slip behaviors of boundary lubrication in macroscopic scale and nanoscale friction process were similar in tendency, but different in the critical value of electrical potential which controls the transition between the low and high friction, as a result of the different boundary lubrication state between multi-asperity contact and single asperity contact.

Moreover, they introduced a novel method to facilitate rapid control of friction distribution on stainless steel surface by the means of bipolar electrochemistry [22]. As their results shown in Fig. 25.4, the friction coefficient at the cathodic pole is higher than that at the anode, as the adsorbed SDS molecules are repelled by electrostatic repulsion. The distribution profile of friction can be easily altered by adjusting the applied current in solution. It was also expected that the innovative potential-controlled friction method could find its applications where differential lubrication are needed, not only in metal-forming processes but also in other smart friction control areas.

The root of the potential controlled boundary lubrication behavior shown earlier lies in the change of Gibbs free energy of the tribosystem, accompanying with the distribution and structural changes of the responsive constituents, such as the surfactants in the aforementioned cases, under different electrical potentials. The concentration changes of the SDS surfactant on the metal surface under different charging states have been detected with the

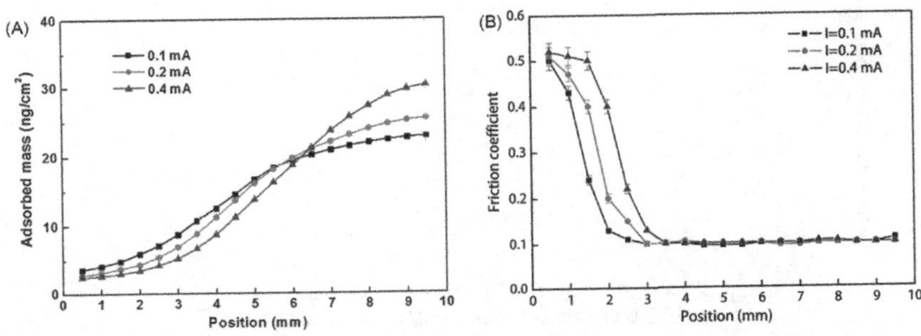

FIGURE 25.4 (A) Adsorbed mass of SDS on stainless steel surface for different positions along the bipolar electrode at different applied currents. (B) Distribution of friction coefficient for different positions along the surface of the bipolar electrode under the load of 3 N at different applied currents [22].

electrochemical QCM-D measurements at a high mass resolution of sub-ng/cm^2 [23]. The structural changes under different surface excesses of the responsive molecular have also been revealed with the electrochemical AFM measurements [21]. These molecular level observations are essentially consistent with molecular dynamics simulations (see Fig. 25.5).

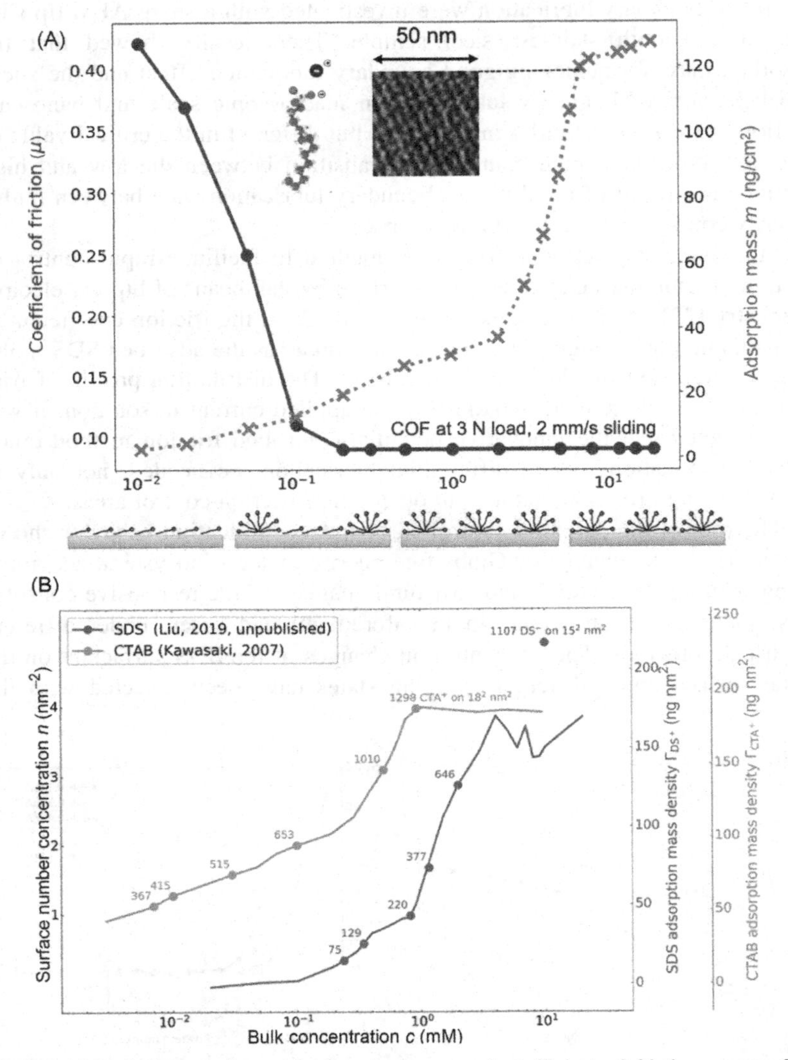

FIGURE 25.5 (A) Dependence of adsorption mass and coefficient of friction measured in aqueous SDS solutions on bulk concentration of SDS surfactant [21,24]. (B) Changes of morphology of adsorbed SDS *(the lower line)* and CTAB *(the upper line)* surfactants with bulk concentration [25]. *CTAB*, Cetyltrimethylammonium bromide; *SDS*, Sodium dodecyl sulfate.

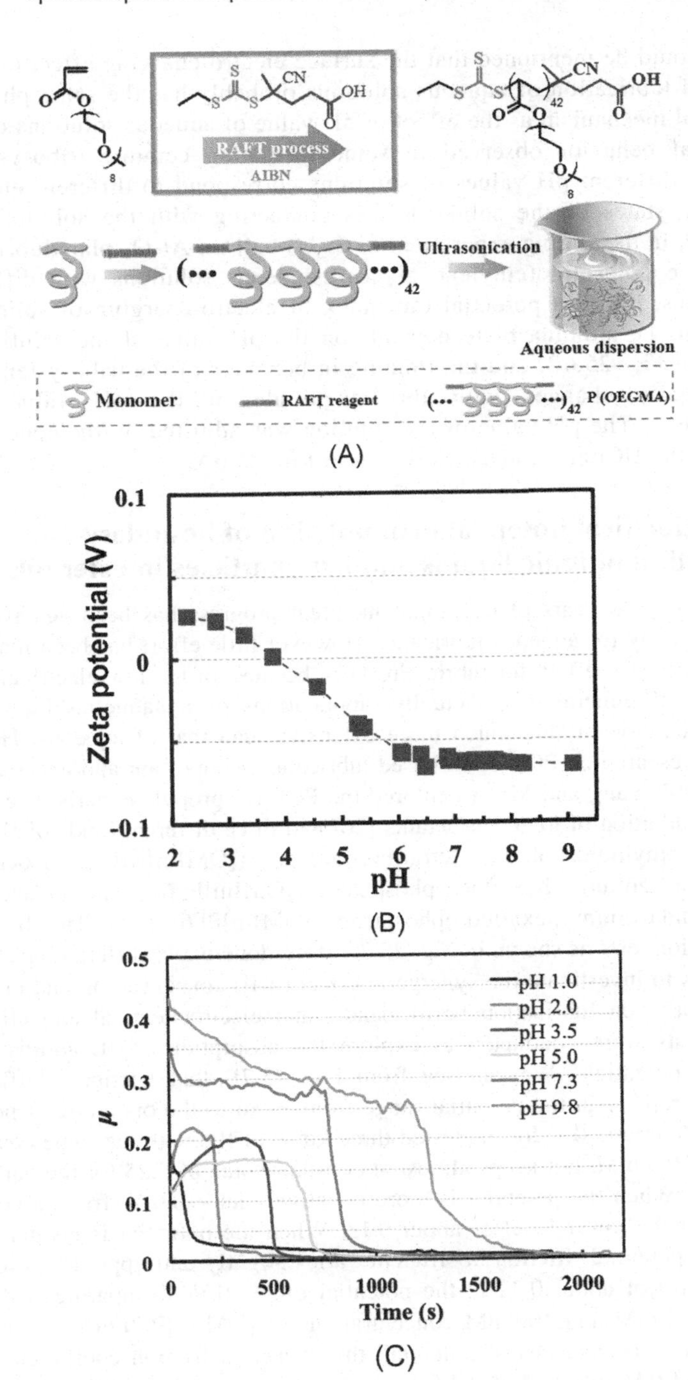

FIGURE 25.6 (A) Preparation of P(OEGMA) aqueous solution. (B) Dependence of zeta potential of alumina plate on the pH value of P(OEGMA) solution. (C) Effect of the pH value of P (OEGMA) solution on running-in behavior. *P(OEGMA)*, poly(oligo(ethylene glycol)).

It should be mentioned that the surface electro-charging effect on friction and lubrication of aqueous solutions probably has the same physical chemical mechanism as the effect of pH value of aqueous solutions on tribological behavior observed in some insulating ceramic tribosystems, because different pH values of solutions correspond to different electro-charging states of the solid surfaces contacting with the solution. For instance, in the sliding experiment of Si_3N_4 ball-on-Al_2O_3 plate lubricated under the poly(oligo(ethylene glycol)) aqueous solutions with different pH values, the zeta potential (an index of electro-charging of solid surfaces) of the alumina plate depends on the pH value of the solution as shown in Fig. 25.6B, and the running-in behavior of the tribosystem also substantially changes with the magnitude of the pH value (see Fig. 25.6C). The pH value of the solution was adjusted by the concentrations of the HCl and NaOH, as shown in Fig. 25.6A.

25.4 Electrical potential manipulation of boundary lubrication of ionic liquids and nanoparticles in ester oils

After nearly 20 years of development, great progress has been achieved in the PCF study for aqueous lubricants. However little effort has been made to explore the PCF effect for lubricating oils, because of the low electrical conductivity of mineral oils. Actually, applications of nonaqueous liquids as industrial lubricants are much more extensive than that of aqueous. Hence, detailed research on PCF in oil-based lubricants is important and necessary.

In 2014, Yang and Meng explored the PCF for propylene carbonate (PC) with the addition of ionic surfactants [26] and three different kinds of ILs: 1-octyl-3-methylimidazolium tetrafluoroborate ([OMIm]BF4), 1-octyl-3-methylimidazolium hexafluorophosphate ([OMIm]PF6), and 1-decyl-3-methylimidazolium hexafluorophosphate ([DMIm]PF6) [27]. The ball-on-disk friction tests as shown in Fig. 25.7A were done under different electrical potentials to investigate the synergetic effect of IL concentration and electrical potential on lubrication performance, and electrochemical and ellipsometric tests were conducted to explore the adsorption of IL additives at different potentials. We can see from Fig. 25.7B that friction coefficient increases rapidly in the potential range from -0.6 to the open circuit potential (OCP) around 0 V for the tested three kinds of ILs with the same concentration of 0.5 mM, but keeps almost at constant value of 0.25 for the bare PC lubricant. When the potential is more negative than -0.6 V, friction coefficient is at the lower level of about 0.13. When the potential is greater than the OCP potential, friction coefficient varies slightly and approaches to the higher level of about 0.22 at the potential of $+1.0$ V. Comparing with the results of 1 mM and 100 mM concentrations of [DMIm]PF6/PC, it is found that at the two extremes of potential, the measured friction coefficients are 0.12 at -1.0 V and 0.22 at $+1.0$ V respectively, regardless of the 200 times

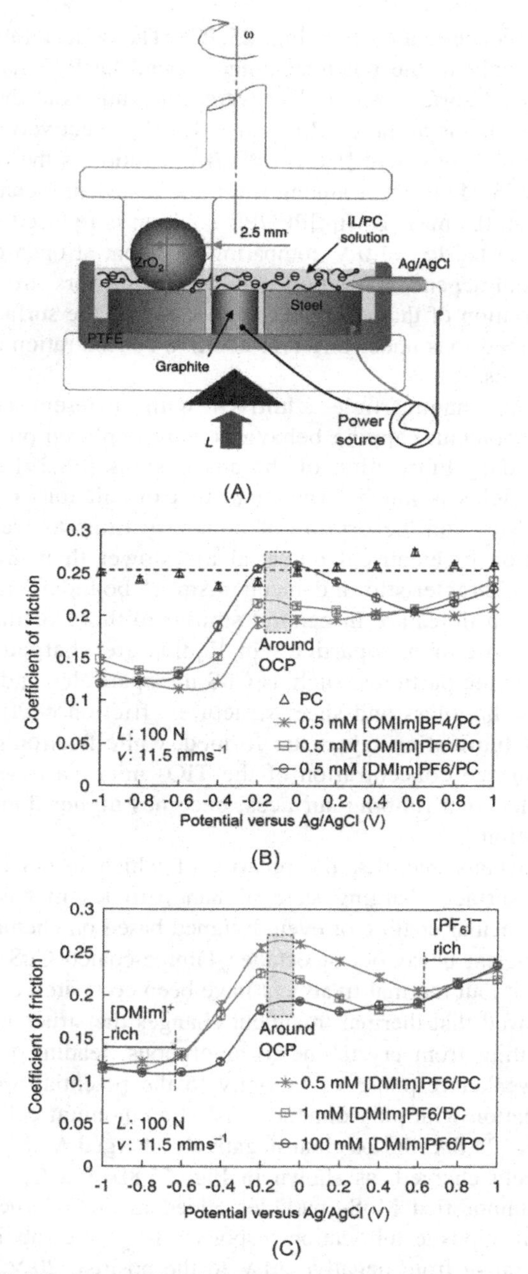

FIGURE 25.7 (A) Schematic of tribotest rig. The normal load is 100 N, and the relative speed is 20 rpm. (B) Potential dependence of coefficient of friction for AISI 4340 disk in base PC oil (empty triangles), 0.5 mM [OMIm]BF4/PC (asterisks), [OMIm]PF6/PC (empty squares), and [DMIm]PF6/PC (empty circles) solutions. (C) Potential dependence of coefficient of friction for AISI 4340 disk in 0.5 mM (asterisks), 1 mM (empty squares), and 100 mM (empty circles) [DMIm]PF6/PC solutions [27].

difference in IL concentration (see Fig. 25.7C). The concentration effect on friction appears only in the potential range around OCP. This implies that the AISI 4340 steel surface was fully covered the cations at the potential of -1.0 V, and the anions at the potential of $+1.0$ V respectively. The electrochemical test results show that [DMIm]PF6/PC solution is the lowest in corrosion against AISI 4340 steel among the three tested lubricants. The wear of steel surface in 0.5 mM [DMIm]PF6/PC solution is reduced when electrical potential is shifted to -1.0 V comparing with that at open-circuit potential. The potential-dependent friction and wear behaviors are explained in terms of the variation of the adsorbed ion species and the surface concentration of the adsorbed ions under different additive concentration and electrical potential conditions.

Most recently, nanoparticle additives with different compositions, structures, or surfactant capping behaviors were explored on the potential controlled boundary lubrication of the suspensions [28,29]. Because the size of nanoparticles is much larger than that of salt ions or surfactants, responses of adsorption, desorption of boundary films as well as friction with the variation of electrical potential are slower than those of molecule. However characteristic and mechanism of boundary lubrication of responsive nanoparticles are in general similar to those of the ionic additives, because most of nanoparticles in liquids are charged naturally in some extent. For the particles, such as CuS nanoparticles and MoS_2 nanoplates, with low hardness and sheet structures, friction coefficient is low when boundary lubrication layers are formed, while friction coefficient is higher when surface concentration of the TiO_2 microparticles, which are rough and harder than rubbing surfaces, becomes higher due to the field-assisted adsorption.

Unlike surfactants and ILs, the polarity of which is inherent and non-adjustable, the surface charging state of nanoparticles in base oil can be modified by molecular coating or even designed based on chemical engineering. The tribological behaviors of octadecylamine-coated CuS nanoparticles [29] with and without thermal treatment have been compared experimentally. The results showed that thermal treatment changes the structure of the octadecylamine coating from crystalline to amorphous, leading to less surface charging and weaker response of lubricity to the potential application. By surface modification with poly dimethyl disllyl ammonium chloride (PDDA) as shown in Fig. 25.8A, the original negatively charged MoS_2 nanoparticles become positively charged, as shown in Fig. 25.8B and C, and hence the modified and unmodified MoS_2 particles added in diethyl succinate (DES) ester oil present opposite lubrication responses when the rubbing surface is charged in the range from negative 20 V to the positive 20 V. This implies that there are more freedoms of selection and surface charge modification for nanoparticles to meet the specific requirement of active control of friction in practical applications.

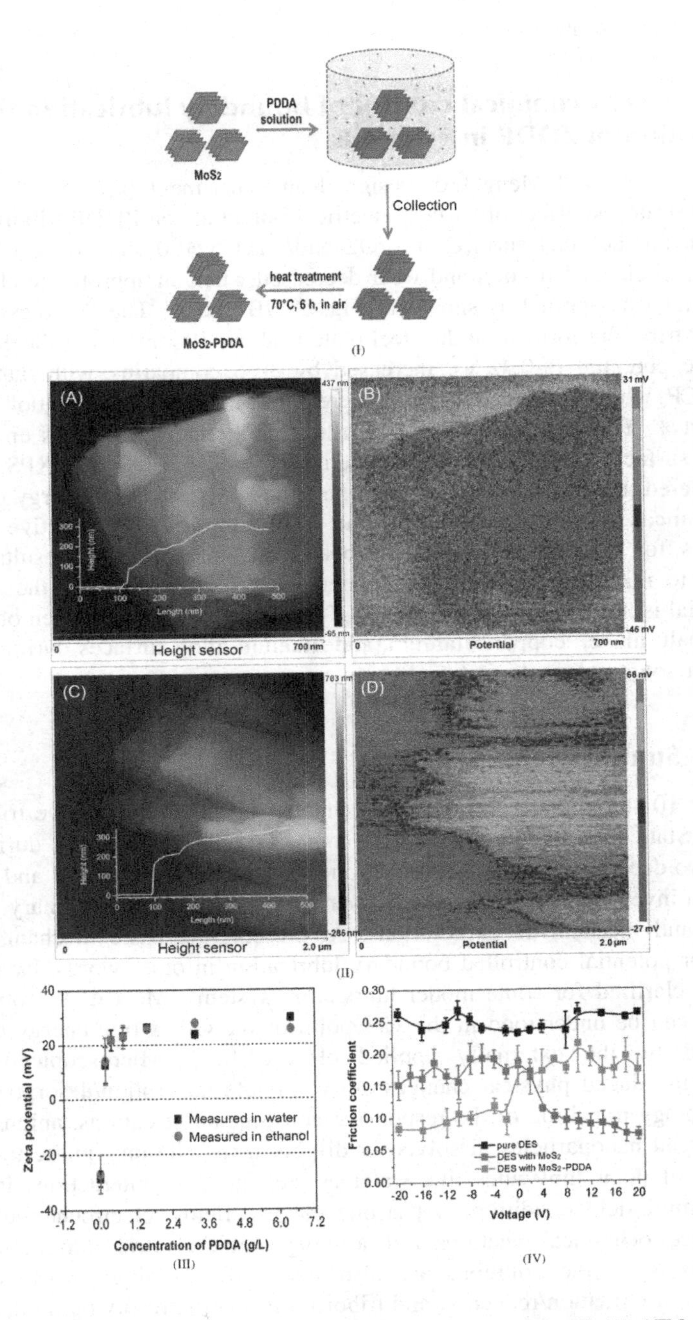

FIGURE 25.8 (I) Surface modification of MoS_2 particles with PDDA. (II) (A) AFM topography and (B) Surface potential image of MoS_2 particle; (C) AFM topography and (D) surface potential image of PDDA-treated MoS_2 particle. (III) Zeta potential of the MoS_2 particles and PDDA-treated MoS_2 particles dispersed in water or ethanol. (IV) Average friction coefficients measured for pure DES lubricant and for the DES suspensions with MoS_2 or PDDA-treated MoS_2 under different applied potentials.

25.5 Electrochemical control of boundary lubrication film formation of ZDDP in ester oils

In 2018, Cao and Meng [30] designed an experiment (see Fig. 25.9A) to demonstrate the effect of surface electrical potentials on ZDDP tribofilm formation on the steel surface of ZrO2 ball/AISI 52100 steel plate tribopair. The coefficient of friction and wear decreased when an appropriate electrical potential was applied as shown in Fig. 25.9B and C. The thickness of the ZDDP tribofilm formed on the steel plate is about 100 nm under the effective surface potential (+0.42 V), increased by 30% comparing with that under the OCP, which is equivalent to an increment of ZDDP concentration from 2 to 5 wt.%. The difference of ZDDP tribofilm chemical structures under different surface potentials were characterized by TEM, EDS, and XPS. It was considered that the effects of applied potentials on activation energy of electrochemical reactions as well as on the adsorption of ZDDP additive are the reasons for the observed electro-tribochemical experimental results. It is worth to note that the tribo-electrochemical effect induced by the applied potential is not only applicable to ZDDP additive, but also has been observed on cobalt, nickel, copper, titanium, and titanium alloy surfaces during tribotests for some polar oils such as PC.

25.6 Summary and prospects

Fig. 25.10 summarizes the progress achieved in the area of active tribology at the State Key Laboratory of Tribology, Tsinghua University, during the past two decades. The study started from macroscale, phenomenal and experimental investigations and gradually went on toward multidisciplinary, multiscale, and mechanistic explorations. At present, the basic mechanisms of PCF, or potential controlled boundary lubrication in other words, have been almost clarified for some model lubrication systems. Most of the observed results can be understood in the viewpoint of the Gibbs free energy change induced by the externally applied electric field. Microscopically, the potential-induced physical changes are reversible adsorption/desorption and morphology transition of the responsive constituents of cations, anions, and/or charged nanoparticles dissolved or dispersed in solutions, phase transformation of polar molecules like water as well as EDL interactions in relatively mild field conditions. At strong electric fields, electrochemical and tribo-electrochemical reactions with a nonignorable electric current, such as electrolysis of the solution, are also involved, resulting in irreversible changes of oxidation/reduction and tribofilm formation/resolving on the electrodes as well as decomposition of the solution. The lubrication systems encompassed various aqueous and nonaqueous solutions with or without additives of various kinds of surfactants, ILs, and nanoparticles. The tribomaterials are extended from metal/ceramic, metal/metal to ceramic/ceramic.

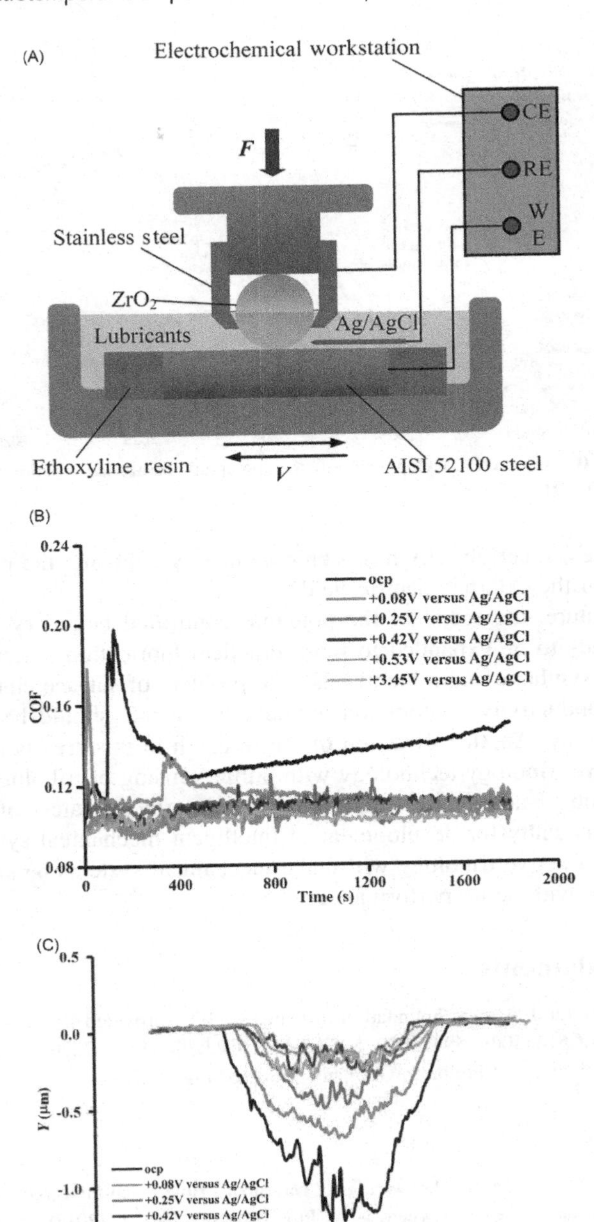

FIGURE 25.9 (A) Schematic of the electrochemical friction test system. (B) COF variations versus sliding time under different surface potentials. (C) Comparison of cross sections of wear scars under different surface potentials [30].

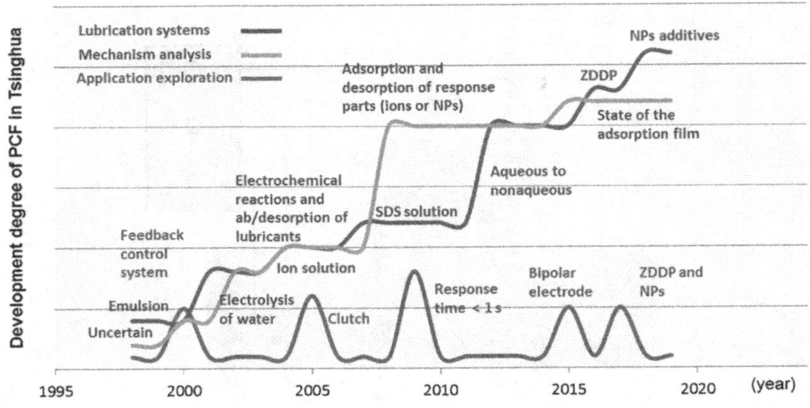

FIGURE 25.10 Research development of the potential controlled friction in Tsinghua University up to 2019.

In some cases, superlubricity has been obtained by adjusting the polarity and magnitude of the external electric field.

In the future, the scope of the potential controlled boundary lubrication systems needs to be expanded to more practical lubrication systems including popular synthetic and mineral oils. The problem of surface charging with the weak conductivity of common nonpolar lubricants should be overcome to open the way. Another direction of the research in this area is to incorporate the active tribology technology with online sensing of tribological working conditions and in-situ monitoring of interface states of machine elements, especially for development of intelligent mechanical systems. The advantages of active tribology will make mechanical systems be smarter and more reliable with better performance.

Acknowledgments

The National Natural Science Foundation of China (NSFC) provided financial support by the grant nos. of 51635009, 59575050, 59575048, 50823003, and 50721004. Mr. Johannes Hormann in University of Freiburg, Germany, prepared Fig. 25.5.

References

[1] R. Taylor, N. Morgan, R. Mainwaring, T. Davenport, How much mixed/boundary friction is there in an engine—and where is it? Proc. IMechE. Part. J. (2019). Available from: https://doi.org/10.1177/1350650119875316.

[2] S. Glavatskih, E. Hoglund, Tribotronics—towards active tribology, Tribol. Int. 41 (2008) 934–939.

[3] J. Krim, Controlling friction with external electric or magnetic fields: 25 examples, Front. Mech. Eng. (2019). Available from: https://doi.org/10.3389/fmech.2019.00022.

[4] J. Zhang, Y. Meng, Boundary lubrication by adsorption film, Friction 3 (2) (2015) 115–147.

[5] N. Nath, A. Chilkoti, Creating "smart" surfaces using stimuli responsive polymers, Adv. Mater. 14 (17) (2002) 1243–1247.
[6] F.P. Bowden, L. Young, Influence of interfacial potential on friction and surface damage, Research 3 (1950) 235–237.
[7] J.O. Bockris, S.D. Argade, Dependence of friction at wet contacts upon interfacial potential, J. Chem. Phys. 50 (4) (1969) 1622–1623.
[8] Y.Y. Zhu, G.H. Kelsall, H.A. Spikes, The influence of electrochemical potentials on the friction and wear of iron and iron oxides in aqueous systems, Tribol. Trans. 37 (4) (1994) 811–819.
[9] M. Deng, J. Li, C. Zhang, J. Ren, N. Zhou, J. Luo, Investigation of running-in process in water-based lubrication aimed at achieving super-low friction, Tribol. Int. 102 (2016) 257–264.
[10] M. Valtiner, A. Banquy, K. Kristiansen, G.W. Greene, J.N. Israelachvili, The electrochemical surface forces apparatus: the effect of surface roughness, electrostatic surface potentials, and anodic oxide growth on interaction forces, and friction between dissimilar surfaces in aqueous solutions, Langmuir 28 (2012) 13080–13093.
[11] L. Pashazanusi, M. Oguntoye, S. Oak, J. Albert, L. Pratt, N. Pesika, Anomalous potential-dependent friction on Au(111) measured by AFM, Langmuir 34 (2018) 801–806.
[12] S. Li, P. Bai, Y. Li, C. Chen, Y. Meng, Y. Tian, Electric potential-controlled interfacial interaction between gold and hydrophilic/hydrophobic surfaces in aqueous solutions, J. Phys. Chem. C 122 (2018) 22549–22555.
[13] H. Jiang, Y. Meng, S. Wen, Effects of external D.C. electric fields on friction and wear behavior of alumina-brass sliding pairs, Sci. China Ser. E 41 (6) (1998) 617–625.
[14] H. Jiang, Y. Meng, S. Wen, P. Wong, Effects of external electric fields on frictional behaviors of three kinds of ceramic/metal rubbing couples, Tribol. Int. 32 (3) (1999) 161–166.
[15] Y. Meng, H. Jiang, P.L. Wong, An Experimental study on voltage-controlled friction of alumina/brass couples in zinc stearate/water suspension, Tribol. Trans. 44 (4) (2001) 567–574.
[16] Y. Meng, H. Jiang, Q. Cheng, S. Wen, Modeling of the voltage-controlled friction effect, Sci. China Ser. A Math. 45 (9) (2002) 1219–1224.
[17] Q. Chang, Y. Meng, S. Wen, Influence of interfacial potential on the tribological behavior of brass/silicon dioxide rubbing couple, Appl. Surf. Sci. 202 (1) (2002) 120–125.
[18] B. Hu, Y. Meng, S. Wen, Influence of external voltage on the tribological behavior of brass/silicon nitride rubbing couple, J. Wuhan Univ. Technol. Mat. Sci. Edit. 19 (suppl) (2004) 37–40.
[19] B. Hu, Y. Meng, S. Wen, A preliminary experimental study on voltage-controlled friction clutch, Tribology 24 (1) (2004) 46–50 (in Chinese).
[20] S. He, Y. Meng, Y. Tian, Response characteristics of the potential-controlled friction of ZrO_2/stainless steel tribopairs in sodium dodecyl sulfate aqueous solutions, Tribol. Lett. 38 (2) (2010) 169–178.
[21] S. He, Y. Meng, Y. Tian, Correlation between adsorption/desorption of surfactant and change in friction of stainless steel in aqueous solutions under different electrode potentials, Tribol. Lett. 41 (2011) 485–494.
[22] J. Zhang, Y. Meng, Y. Tian, Control of friction distribution on stainless steel surface in sodium dodecyl sulfate aqueous solution by bipolar electrochemistry, Tribol. Lett. 59 (3) (2015) 43.

[23] J. Zhang, Y. Meng, Y. Tian, Effect of concentration and addition of ions on the adsorption of sodium dodecyl sulfate on stainless steel surface in aqueous solutions, Colloid. Surface A 484 (2015) 408–415.

[24] J. Zhang, Y. Meng, Stick-slip friction of stainless steel in sodium dodecyl sulfate aqueous solution in the boundary lubrication regime, Tribol. Lett. 56 (2014) 543–552.

[25] H. Kawasaki, K. Nishimura, R. Arakawa, Influence of the counterions of cetyltrimetylammonium salts on the surfactant adsorption onto gold surfaces and the formation of gold nanoparticles, J. Phys. Chem. C 111 (6) (2007) 2683–2690.

[26] X. Yang, Y. Meng, Y. Tian, Potential-controlled boundary lubrication of stainless steels in non-aqueous sodium dodecyl sulfate solutions, Tribol. Lett. 53 (2014) 17–26.

[27] X. Yang, Y. Meng, Y. Tian, Effect of imidazolium ionic liquid additives on lubrication performance of propylene carbonate under different electrical potentials, Tribol. Lett. 56 (2015) 161–169.

[28] C. Liu, O. Friedman, Y. Meng, Y. Golan, CuS nanoparticle additives for enhanced ester lubricant performance, ACS Appl. Nano Mater. 1 (12) (2018) 7060–7065.

[29] C. Liu, O. Friedman, Y. Li, Y. Meng, Y. Golan, Electric response of CuS nanoparticle lubricant additives: the effect of crystalline and amorphous octadecylamine surfactant capping layers, Langmuir 35 (48) (2019) 15825–15833.

[30] H. Cao, Y. Meng, Electrochemical effect on boundary lubrication of ZDDP additive blended in propylene carbonate/diethylsuccinate, Tribol. Int. 126 (2018) 229–239.

Chapter 26

Superlubricity of NiTi alloys

Qunfeng Zeng
Key Laboratory of Education Ministry for Modern Design and Rotor-Bearing System, Xi'an Jiaotong University, Xi'an, P.R. China

26.1 Introduction

60NiTi alloy is an intermetallic nickel-titanium alloy, exhibiting excellent engineering properties such as high electrical conductivity, high resistant to corrosion and galling, lower density than steel, and a broad combination of the superior mechanical and chemical properties [1]. Based on these special characteristics, 60NiTi alloy has been considered as a promising candidate material for oil-lubricated bearing and other types of moving mechanical applications. DellaCorte et al. [2] showed that 60NiTi alloy displays good tribological properties under synthetic hydrocarbon oil lubrication. However the coefficient of friction (CoF) of 60NiTi alloy is about 0.3 at the stage, which is still too high for roller and sliding bearing applications. Moreover, it is well known that titanium-based alloy is prone to mechanical degradation during sliding. Therefore it is necessary and urgent to select appropriate lubricants to reduce friction and wear of 60NiTi alloy for efficient and reliable industrial applications. The reduction in friction and wear under oil lubrication is crucial and important to improve the lubrication performances of the moving components and the energy conservation in many engineering fields [3,4]. Superlubricity has attracted increasing interests in recent years from both the research and industrial communities [5–8]. Researchers have tried to achieve superlubricity with polyhydroxyl group lubricants. Matta et al. [9] reported superlubricity of diamond-like carbon (DLC) films in presence of hydroxyl alcohols and the superlubricity mechanism is due to the degradation of glycerol and gliding on the triboformed OH-terminated surfaces [10]. Li et al. [11–13] found that the mixtures of acid and polyhydroxyl group lubricants exhibit superlubricity and the possible superlubricity mechanism is attributed to the hydrogen-bonded networks from hydroxyl groups in lubricants. So far, there are a number of liquid lubricants having potential superlubricity behaviors, even including water.

Superlubricity. DOI: https://doi.org/10.1016/B978-0-444-64313-1.00026-0

Tomizawa et al. [14] found that CoF of Si_3N_4 against itself under water lubrication is less than 0.002 after a running-in process, which is the first time to report that water when used as a lubricant can achieve superlubricity. Other ceramics such as SiC is also found to exhibit superlubricity with friction coefficients of less than 0.01 under water lubrication [15]. These achievements show that in the running-in process the tribochemical reactions of ceramics with water and the electric double layers of the water films play an important role in superlubricity. It is also found from the aforementioned studies that hydroxyl group of lubricants plays an important role in superlubricity. The previous work has shown that the lubrication performance of 60NiTi alloy can be strongly improved by the use of appropriate lubricant oils [1,16].

Therefore it is necessary to further investigate the influence of the chemical microstructures of lubricants, the friction conditions and the tribochemical process of lubricants reacting with the contact surface of the friction pair during sliding on the friction system, and achieve superlubricity and discuss the possible superlubricity mechanisms of 60NiTi alloy. In the present work, the oil lubrication tests were conducted using a pin-on-disc type configuration.

26.2 Experiment details

26.2.1 Preparation of Nitinol alloys

60NiTi alloy is an intermetallic nickel-titanium alloy with 60 wt.% nickel and 40 wt.% titanium. 60NiTi alloy was manufactured by Xi'an Saite Company and the physical properties of 60NiTi alloy were shown in Table 26.1 [17]. 60NiTi alloy was produced with the forging and hot rolling techniques by consumable electrode melting furnace to ensure compositional homogeneity. Firstly, the ingot of 60NiTi alloy was prepared through the processes of vacuum melting method in vacuum medium frequency induction melting furnace using appropriate amounts of Ni (electrolytic nickel, 99.97 wt.% of purity) and Ti (sponge titanium, 99.99 wt.% of purity) in a graphite crucible under vacuum of 10^{-1} Pa. The ratio of nickel in the starting material was 61.5 wt.%. A rod with 110 mm in diameter was obtained from the ingot. The rolled bar of the ingot was treated uniformly by electric

TABLE 26.1 The physical property of 60NiTi alloy.

Performance	Density (g/cm^3)	Hardness (MPa)	Thermal conductivity [W/(m K)]	Thermal expansion coefficient ($°C^{-1}$)	Elastic modulus (GPa)
60NiTi alloy	6.7	660	18	10×10^{-6}	~114

furnace at 1000 °C for 6 h. The smelting current and voltage was 100 A and 300 V, respectively. Prior to melting, the furnace was purged with Ar of high purity. All specimens were cooled after the treatment in ambient air. Then the specimen was forged by finish forging equipment at 850°C and kept at this temperature for 2 h appropriately. The diameter of 60NiTi alloy was forged from Φ110 to Φ50 mm. The rod was rolled by rolling mill at 820°C and kept at this temperature for 1 h. The diameter of 60NiTi alloy sample was then machined to Φ6 mm.

26.2.2 Tribological tests

The tribotests were carried out using a tribometer system (UMT-2 from Bruker Corp.) with a pin-on-disc configuration and unidirectional sliding to characterize the lubrication performance of lubricants in ambient air at room temperature. 60NiTi alloy pin sliding against high-speed tool steel disc are used as the friction pair. The sizes of 60NiTi alloy cylinder pin are 6 mm in diameter and 15 mm in length and the size of the disc of the high-speed tool steel is 30 mm in diameter and 5 mm in thickness. The surface roughness of pin and disc are around 0.200 and 0.006 μm, respectively. The tribotests were performed under a normal load of 30 N and a sliding velocity of 0.02 m/s. A normal load is 30 N corresponds to a calculated maximum contact pressure of 280 MPa. The relative humidity was about 50% RH in the laboratory in the process of tribotests. Before starting the tribotests, the specimens were cleaned ultrasonically by the acetone. Castor oil, glycerol, and other lubricants were used to lubricate the friction pair, respectively. Few drops of the tribotested lubricants were added to the surface of the steel disc before the tribotests and it was made sure that the tribotests were conducted under boundary lubrication condition. The lubrication status was estimated to be in the boundary lubrication regime, which demonstrates that the friction is not occurred in the hydrodynamic lubrication regime. A good initial indication of the lubrication mode is given by lambda ratio (λ), defined by Eq. (26.1), representing the ratio of theoretical minimum film thickness from smooth surface to the composite surface roughness of the friction pairs:

$$\lambda = \frac{h_{\min}}{(R_a{}^2 + R_b{}^2)^{\frac{1}{2}}} \tag{26.1}$$

where, λ—lambda ratio, R_a—roughness of pin/μm ($\sim$0.22), R_b—roughness of disc/μm ($\sim$0.006), $h_{\min}$—the minimum thickness of oil film/μm, which is determined as shown in Eq. (26.2). For the present lubrication case, a theoretical prediction of the oil film thickness can be obtained from the regression equations of Dowson and Higginson [18]. For the lubricated engineering components, known as lambda ratio λ, often provides a good initial indication of fluid film ($\lambda > 3$), mixed ($1 < \lambda < 3$), or boundary ($\lambda < 1$) lubrication. The central film thickness is expressed as regression equation consisting of two nondimensional parameters [18],

$$H_{min} = 2.8U^{0.65}W^{-0.21} \tag{26.2}$$

where, H_{min}—dimensionless minimum film thickness = h_{min}/R, R—the radius of the pin (45 mm), U—dimensionless speed = $\eta u/(E^*R)$, $E_{a,b}$—elastic moduli of pin and flat respectively, E^*—effective elastic modulus, $1/E^* = 1/2[1 - v_1{}^2/E_1 + 1 - v_2{}^2/E_2]$, ($E_1 = 114$ GPa, $E_2 = 210$ GPa, $v_1 = 0.33$, $v_2 = 0.3$; therefore we can calculate $E^* = 164.61$ GPa), W—dimensionless load = $F/(E^*R^2)$, F—normal load (30 N), u—the entrainment speed (0.034 m/s), η—dynamic viscosity under atmospheric conditions (674.32 Pa · s). We calculated $H_{min} = 2.79 \times 10^{-6}$. Therefore we obtained $h_{min} = 0.126$ μm. The calculation gave the lambda ratio well under unity (0.57), which means that the lubrication testing was run in the boundary lubrication regime and not the mixed lubrication and elastohydrodynamic lubrication in the present case. CoF is continuously recorded during the friction and wear test.

26.2.3 Surface characterization

The following facilities permit the observations of the worn surface and chemical characterizations of the surface within the contact area during sliding. After tests, samples were cleaned ultrasonically in acetone, and then the tribofilm on 60NiTi alloy pin and the steel disc were investigated by optical microscope, SEM and XPS. The worn surfaces were observed by a scanning electron microscopy (SEM, S-3000N, Hitachi Co. Japan).

26.3 Results and discussion

26.3.1 Selection of lubricants

Fig. 26.1A and B show the CoF of 60NiTi alloy pin sliding against the steel disc versus sliding time under dry friction and PAO oil lubrication

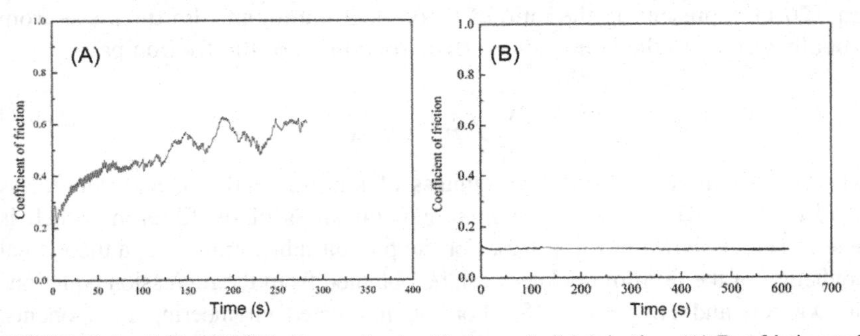

FIGURE 26.1 CoF of 60NiTi alloy under dry friction and oil lubrication. (A) Dry friction and (B) oil lubrication.

conditions, respectively. It is found that the tendency of CoF with sliding time is completely different. The CoF is low at initial step and then increases to around 0.6 with sliding time increasing under dry friction. However the CoF is constant around 0.11 under PAO oil lubrication irrespectively of sliding time. The frictional response of 60NiTi alloy is remarkable under PAO oil lubrication. Apparently, lubricant oil is beneficial to improve the tribological properties of 60NiTi alloy.

It is well known that the tribological properties are strongly dependent on the lubricant types and molecular microstructures. Therefore we select few lubricant oils with different molecular structures for analyses and comparison studies. The aim is to achieve the stable superlubricity with high wear resistance of 60NiTi alloy under oil lubrication. Fig. 26.2 shows the CoF as a function of sliding times under lubrication by castor oil, glycerol, glucose, and sorbitol respectively.

As is shown, the CoF increases rapidly at the initial stage, but then decreases with the sliding time for different lubricants. For example, CoF is around 0.05 at the initial stage, but then increases to around 0.2, and finally decreases to 0.12 at the stable stage under glycerol lubricant lubrication. However for castor oil, the CoF is around 0.07 at the initial stage and decreases steadily to 0.005 at the steady-state regime. It is concluded that castor oil is beneficial to the lubrication performances of 60NiTi alloy and only castor oil exhibits superlubricity for the 60NiTi alloy pin/steel disc friction pair.

After the tribotest, the worn surface morphology was examined in order to understand the formation of the morphological features and discuss the wear

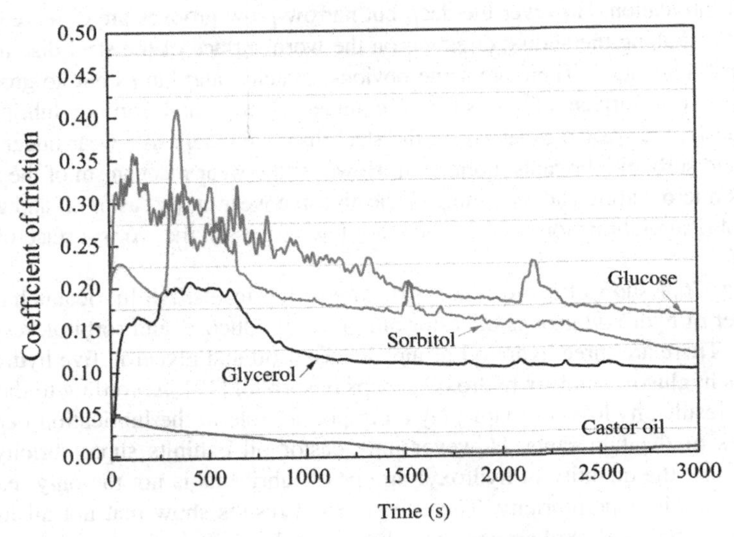

FIGURE 26.2 CoF of the 60NiTi test pair under different lubricants.

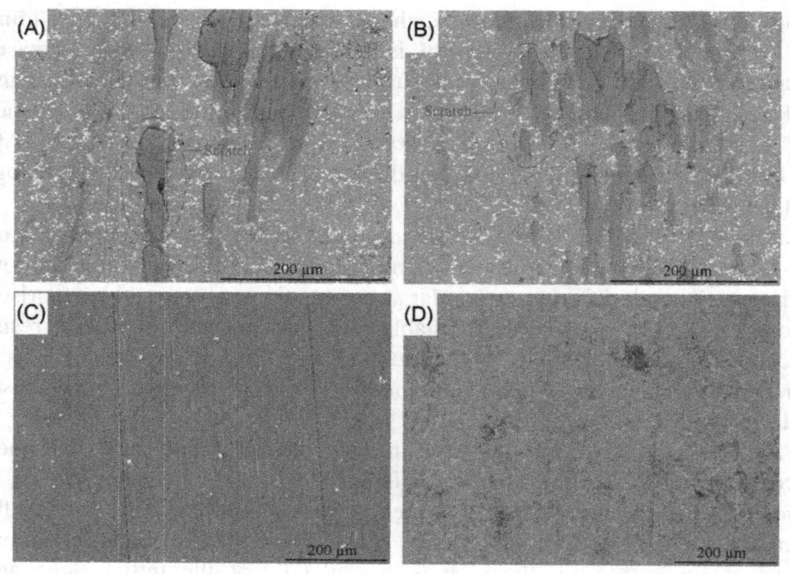

FIGURE 26.3 SEM images of the worn surface of disc under different polyhydroxyl group lubricants lubrication: (A) glucose; (B) sorbitol; (C) glycerol and; (D) castor oil.

mechanism. Fig. 26.3 shows the SEM images of the worn surface of the steel disc. It is found from the SEM observation that there is almost no visible trace of the wear and the shallow stripes are detected on the surface of the steel disc under castor oil lubrication. However the deep but narrow plow grooves are observed and distributed along the sliding direction on the worn surface of the steel disc under glycerol lubrication. There are some obvious scratches and lumps but no grooves on the worn surface of the steel disc under glucose and sorbitol lubricants. Obviously, the wear mechanism of the steel disc is the adhesive wear under glucose and sorbitol lubricants lubrication. However the wear mechanism of the steel disc is micro cutting and ploughing of the abrasive wear under castor oil and glycerol lubricant lubrication because there are few groves on the worn surface of the steel disc.

Fig. 26.4 shows the molecular structure of lubricants. All lubricants have a number of hydroxyl groups in castor oil, glycerol, glucose, and sorbitol, respectively. There are three hydroxyl groups in castor oil and glycerol, five hydroxyl groups in glucose, and six hydroxyl groups in sorbitol [19]. According to the tribotest results, hydroxyl groups play an important role in the lubrication performances of the lubricants. However only castor oil exhibits superlubricity. It seems that the quantity of hydroxyl groups of lubricants is not the only reason that results in superlubricity. The experimental results show that not all lubricants with polyhydroxyl groups can exhibit superlubricity.

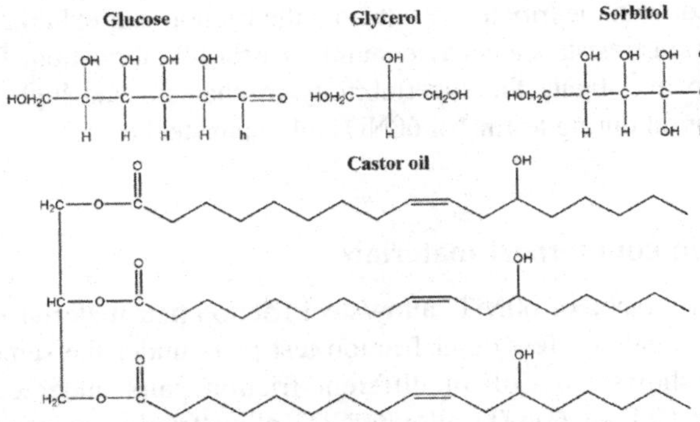

FIGURE 26.4 The chemical structure of polyhydroxyl group lubricants.

The lubrication performances of polyhydroxyl group lubricants are most likely involved in the chemical structures of these lubricants with the functional groups and their ability to trigger tribochemical reactions with the metal contact surfaces during sliding. Therefore the lubrication mechanisms of polyhydroxyl group lubricants are investigated and discussed from the views of the chemical structures and tribooxidations of lubricants and the tribochemical reactions of the lubricants with the metal surfaces simultaneously. There are a few of influencing factors, but the predominant factor is the chemical microstructure of lubricants that determine the polar attraction of the lubricant molecule for the active reaction sites on the contact interfaces and how they interact with the metal surfaces. Sorbitol ($C_6H_{14}O_6$) is a nonvolatile polyhydric sugar alcohol, which is usually used as lubricants. It is not easily to be oxidized into organic acid because it has high heat resistance and high thermal stability even under high temperature (above 200°C), thus the CoF is high under sorbitol lubrication because it is not oxidized into organic acid. Glucose is a monosaccharide with the formula of $C_6H_{12}O_6$. The hydroxyl groups are also difficult to be oxidized like sorbitol. Glycerol has three hydroxyl groups with the formula of $C_3H_8O_3$. The chemical microstructure of glycerol is that each carbon atom is bonded to an OH group. Castor oil exhibits the unique characteristics in the chemical microstructure for all polyhydroxyl group lubricants in the present work. Castor oil and its derivatives are often used as lubricants. It has a triglyceride with a very high content (~90%) of an 18-carbon monounsaturated fatty acid, ricinoleic acid, containing three ester groups, double bonds and a hydroxyl group with high viscosity. The ester linkage, double bonds, and hydroxyl groups in castor oil provide the active reaction sites in the tribochemical reaction. There are ester groups and double bonds besides hydroxyl groups in castor oil, which means that the chemical reactivity of castor oil is higher than that of other lubricants.

It is concluded from the tribotest results that the friction pair exhibits superlubricity and high wear-resistance behaviors under castor oil lubrication. Therefore it is necessary to investigate the superlubricity mechanism and high antiwear behavior of castor oil during testing of 60NiTi alloy pin/steel disc.

26.3.2 Friction counterpart materials

To verify the importance of 60NiTi alloy/steel friction pair material in superlubicity, we investigated a few other friction test pairs under the same conditions. Fig. 26.5 shows the CoF of different friction pairs under castor oil lubrications. The CoF of 60NiTi alloy/60NiTi alloy friction pair is relative stable at the initial stage, as shown in Fig. 26.5A, and after around 1200 s the CoF increases to about 0.3 and keeps on fluctuations. No superlubicity is observed in this case because CoF becomes high after a running-in period of 1200 s with sliding time. The CoF of 60NiTi alloy/copper flat friction pair is relatively stable throughout the sliding time. CoFs of GCr15 steel/HSS flat and 60NiTi alloy/HSS friction pair are relatively low, when comparing to 60NiTi alloy/Cu friction pair. The additional experimental results show that all the friction pairs could achieve low CoF below 0.1 in the steady state; however only 60NiTi alloy/HSS steel friction pair exhibits super-low friction below 0.01, as shown in Fig. 26.5A. 60NiTi alloy/HSS steel friction pair could achieve super-low friction; however the CoF becomes high when the HSS steel flat is replaced by 60NiTi alloy and copper flat. It means that the material of the friction pair should be appropriate to obtain super-low friction due to the tribochemical reactions between oil and the solid surface. This is the unique combination of the presence of Ni in the pin, the presence of iron in steel and the presence of branched OH in the backbone of the castor oil molecules that permit the intercalated Fe/Ni nanostructures to be formed and this provides the low friction. On the another hand, this kind of nanostructures are very similar to the ones described in details in the case of Ni/Fe batteries.

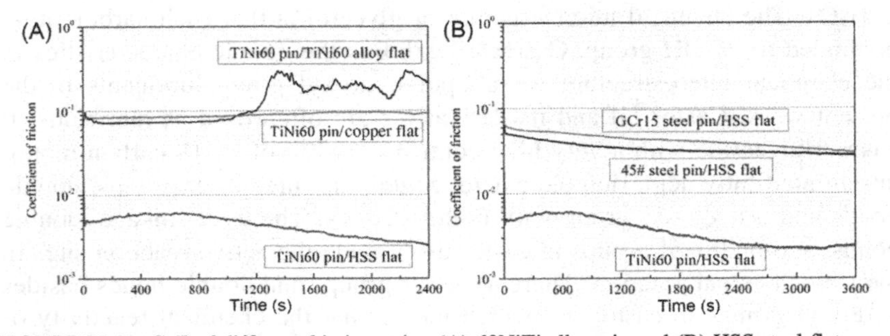

FIGURE 26.5 CoF of different friction pairs: (A) 60NiTi alloy pin and (B) HSS steel flat.

26.3.3 Friction parameters

Fig. 26.6 shows CoFs of 60NiTi alloy under different loads. It is clear that CoF is different at the initial and steady or stable stages. At the initial stage, CoF is 0.022 under 10 N load but increases from 0.015 to 0.054 between the loads from 20 to 50 N. However CoF decreases from 0.054 to 0.027 between the loads from 50 to 80 N. At the stable stage, CoF is unstable below 50 N, but CoF is super low, that is, 0.002 under the load of 40 N. CoF decreases from 0.005 to 0.003 with the increase of loads from 50 to 80 N. The time that took CoF to reach the stable value is 7, 7, 7, 7, 10, 9, 9, and 9 s with the increase of loads. That is, 60NiTi alloy takes longer time to stabilize at higher load.

Fig. 26.7 shows CoFs of 60NiTi alloy under castor oil lubrication and different speeds. Fig. 26.7A shows the Stribeck curve representing the evolution of CoF versus the sliding speed in log scale and the Lambda ratio under castor oil lubrication. The speed range is from 3 mm/s to 0.67 m/s, and the applied load is 30 N. Each value of the CoF in the curve in Fig. 26.7A corresponds to a unidirectional test of 1000 s and the CoF value is recorded in the steady state at the end of the test. For example, Fig. 26.7B represents the tribotest at the speed of 3 mm/s, the lowest sliding speed investigated here. First, we observed the amazingly low values of CoF at low speed. Second, it is very surprising to observe an inverted shape of this Stribeck curve compared with the typical Stribeck curves where friction increases sharply at the lowest sliding speeds. It is found that the CoF increases from 0.0005 to 0.03 with the increase of sliding speeds. The lowest CoF is only about sixtieth of the highest value. The lambda ratio is also provided at different speeds in Fig. 26.7A. In the present case, the CoF has a super low value of about 0.0005 under boundary lubrication, as shown in Fig. 26.7B. CoF decreases slowly to around 0.0005 after about 1000 s. Super-low friction force is kept during 1600 s. Finally, the CoF increases slightly to around 0.004 with increasing sliding time. It is concluded from the experimental results that there is huge difference between standard Stribeck curve described in the literature and our s-shape Stribeck curve under castor oil lubrication.

Fig. 26.8 shows optical and SEM images of the worn surfaces after the tribotest. It is seen that the contacting region of 60NiTi alloy pin is almost not worn after running for 5000 s and the circular Hertzian zone is hardly visible. Only some darker areas on the apparent contact area of the top surface of pin can be observed and roughness of pin surface is apparently not affected during the friction test. The SEM image of the wear track on the steel disc does not show any visible damage but only polishing inside the wear track. A magnification of SEM image, as shown in Fig. 26.8C, any specific scratch in the direction of sliding could not be found. From these observations, it can be deduced that only a colored very thin tribofilm is built on the pin surface and seems responsible for the amazing tribological behavior of the present lubricating system.

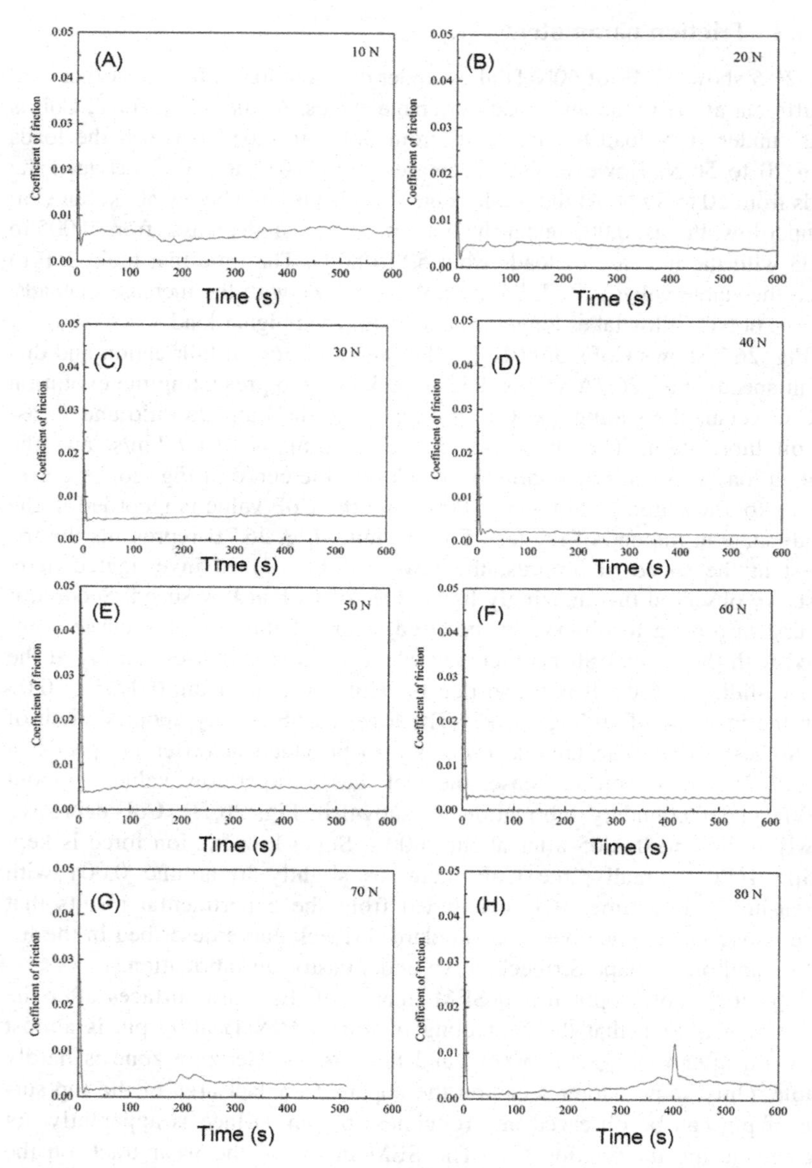

FIGURE 26.6 CoFs of 60NiTi alloy under different loads: (A) 10 N; (B) 20 N; (C) 30 N; (D) 40 N; (E) 50 N; (F) 60 N; (G) 70 N, and (H) 80 N.

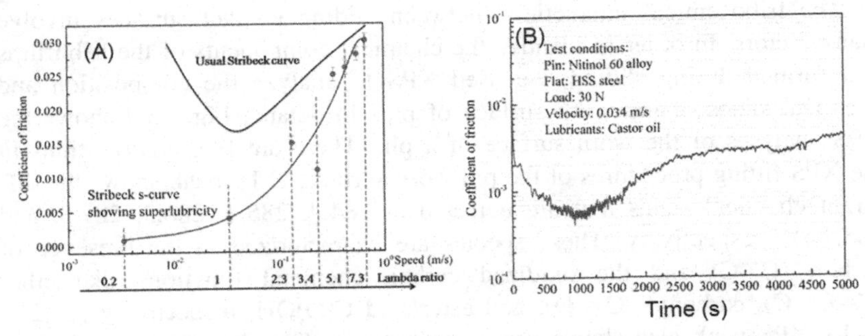

FIGURE 26.7 CoF of 60NiTi alloy under castor oil lubrication and at different speeds: (A) Stribeck curve usual shape and Stribeck curve with s-shape and (B) evolution of CoF.

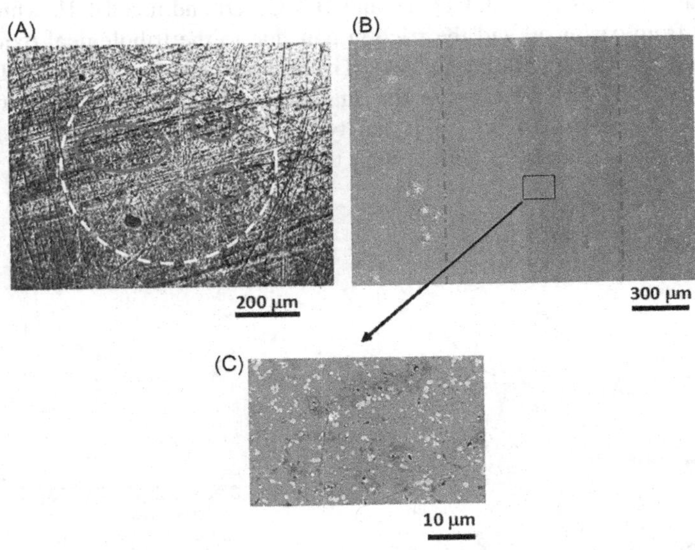

FIGURE 26.8 Worn surface images of the friction pair. (A) Optical image of pin, (B) the SEM image of disc, and (C) the worn surface of disc at higher magnification.

26.3.4 Superlubrictiy mechanism

To investigate the superlubricity mechanism of 60NiTi alloy/steel under castor oil lubrication, it is necessary to determine which kind of thin tribofilm is present on the friction surface. The presence of a tribofilm that is a result of the tribochemical reactions on the contact surface of pin is very difficult to be observed optically. We have speculated that this thin tribofilm results in superlubricity and high antiwear behavior under castor oil lubrication.

The tribocemical interactions between sliding contact surfaces involve many factors. In order to identify the chemical components of the tribofilms that formed during sliding, we used XPS to analyze the composition and chemical states of the worn surface of pins in details. Fig. 26.9 shows the XPS analyses of the worn surface of a pin. There are five contributions in the XPS fitting procedures of the pin worn surface. C 1 s peak shows the different chemical states that are detected at 284.4, 285.4, 286.3, 287.5, and 288.4 eV, respectively. These species are characterized in the presence of carbon (C—C) and the oxidized carbon chemical functions like ether (C—O—C), carbonyl (C=O), and ester/acid C(O)OH, respectively [20,21]. O 1 s XPS peak also shows five contributions. The characteristic peaks at 530.2 and 530.7 eV are generally attributed to the nickel oxide forms (Ni—O) or NiOOH and titanium dioxide form (Ti—O), respectively [22]. There are also other peaks at 531.8, 532.8, and 533.6 eV. These species are the characteristic peaks in the presence of oxygen and the oxidized carbon chemical functions (C=O, COOH, and O=C—O) and metal-OH, which are derived from castor oil and the friction pair due to the tribological oxidation and chemistry during sliding [23,24]. Ti 2p photopeaks at 455.2 eV (TiO or Ti_2O_3) and 465.2 eV (TiO_2) are the characteristic peaks of titanium oxides species. The peak of 459.5 eV is the typical peak of TiO_2. The measured peak intensity of Ni 2p at 853.2 eV in the tribofilms is strong, which is the

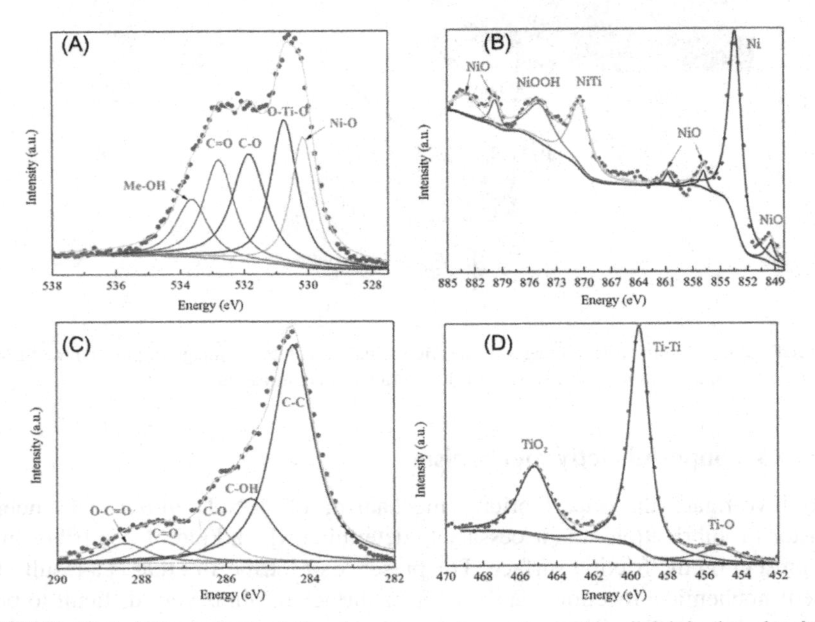

FIGURE 26.9 XPS analysis of the pin worn surface under castor oil lubrication showing different states of: (A) oxygen; (B) nickel; (C) carbon and (D) titanium.

typical peak of nickel oxide. According to Ni 2p spectrum data, it is concluded that a large amount of nickel is in the metallic form and others are nickel oxide and/or oxyhydroxide (NiOOH) [25]. There are Ni, NiO, and others compounds including NiOOH. The existence of NiO is also confirmed by O 1 s peak at 530.2 eV. Under high friction heating nickel can be subjected to the oxidation by the oxygen and water molecular in ambient air and oxidized further to NiOOH that is present in Ni 2p spectrum, as shown in Eqs. (26.3) and (26.4). These findings allow us to conclude that castor oil is the most suitable for the much enhanced lubrication performance of the 60NiTi alloy and steel friction pair in the present work.

$$2Ni + O_2 + 2H_2O \rightarrow 2Ni(OH)_2 \tag{26.3}$$

$$4Ni(OH)_2 + O_2 \rightarrow 4NiOOH + 2H_2O \tag{26.4}$$

In light of the XPS analyses obtained, oxidation of castor oil is due to the presence of ester group, double bonds, and hydroxyl group in molecules and this offers a high level of reactivity with oxygen, especially when it is placed in contact with humid air under frictional heat and high contact pressure simultaneously. The oxidation mechanism of castor oil can be explained as follows: (1) adsorption of castor oil molecules on metallic surface, and (2) hydrolysis of castor oil inside contact region under the combined effects of pressure, heat, and shear. The triacylglycerol group is firstly decomposed to diacylglycerol and ricinoleate, and then glycerol and ricinoleate. (3) Oxidation reactions. The hydroxyl group can oxidize into carbonyl group in ricinoleate. The presence of nascent metallic nickel can strongly enhance the oxidation reaction of castor oil. The stability of carbonyl group is lower than that of C=C bond; therefore C=C bond can be oxidized, and cleave C=C bond in ricinoleate under oil lubrication condition. Therefore ricinoleate can be degraded into hexanoic acid $CH_3(CH_2)_4COOH$. The XPS spectrum of carbon on the pin worn surface (Fig. 26.10) is in agreement with the

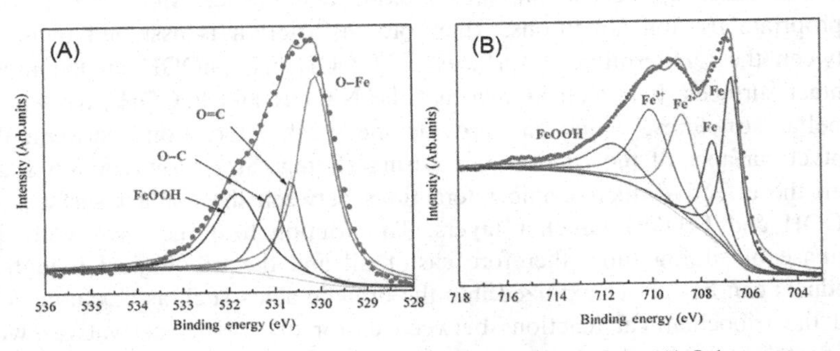

FIGURE 26.10 O 1 s and Fe 2p XPS spectra on the disc worn surface: (A) O 1 s spectrum and (B) Fe 2p spectrum.

formation of ethers (C–O), hydroxyl (C–OH) or carbonyl (C=O), and acid C(O)OH.

Fig. 26.10 shows the resolved XPS spectra of iron and oxygen obtained on the disc worn surface. No nickel is found on the steel disc surface and the thickness of tribofilm is only a few nanometers thick because metal iron is easily detected underneath the oxide contribution. It is clear that some oxide/hydroxide species are present in the tribofilms. The peaks in the region 530–531 eV correspond to O–C and O=C bonds originating from ether or carboxylic acid or esters and O–Fe as shown in Fig. 26.10A. O 1 s photopeak at 531.8 eV and Fe 2p at 711.5 eV are easier to identify and are characteristic of FeOOH.

Superlubricity is divided in two kinds of mechanisms depending on the use of solid or liquid lubricants. The superlubricity mechanism using solid lubricants such as DLC, graphite, grephene, and molybdenum disulfide is very different from that of liquid lubricants such as polymer brushes, ceramic materials with water and phosphoric acid, steel with glycerol solution, acid, and polyhydric alcohol. In the present cases, the superlubricity of castor oil and 60NiTi alloy/steel system is different from aforementioned mechanisms. Therefore it is necessary that we propose a new superlubricity mechanism model coupling the mechanisms of solid and liquid superlubricity to explain the superlubricity behavior of this friction system of 60NiTi alloy under castor oil lubrication.

In ambient air, FeOOH layers are naturally grown on the iron surface of the steel disc due to high chemical reactive behavior of iron with oxygen and water molecular and high relative humidity. This thin FeOOH layer should be kept during sliding in order to achieve super low and stable friction of the friction system. The CoF at the initial stage is high for other lubricants and this thin layer on the contact surface of the steel disc is destroyed at the initial stage of the tribotest, which explains the differences of high CoF obtained by glycerol, glucose, and sorbitol. At the initial stage, nickel on the top surface of 60NiTi alloy pin is easily oxidized by the oxygen and water molecules into nickel oxide and further NiOOH under the appropriate friction conditions. Therefore the friction is assumed to occur between the OH-terminated surfaces of FeOOH and NiOOH on the metal contact surfaces. It is well known that the NiOOH and FeOOH layers have lamellar structures, which can provide the weak interactions between the contact surfaces of the friction pair during sliding. Super-low friction starts from the effective microscopic interactions between the contact surfaces of NiOOH and FeOOH lamellar layers. The friction heat increases with the increase of sliding time; therefore castor oil and the tribological oxidation products can be further oxidized into the organic acids such as hexanoic acid and the tribochemical reactions between castor oil and its derivatives with the metal contact surfaces are unavoidable during sliding. Hexanoic acid not only is reacted with the metal contact surface to generate the terminated surfaces and but also could be intercalated through forming the coordinate bonds with NiOOH, which means the formation of two OH-terminated surfaces providing the electron-electron Coulomb repulsion force and van der Waals attraction [26]. According to the aforementioned analyses and results, we propose a new superlubricity model [27]. Fig. 26.11 shows the schematic showing the contact surfaces of NiOOH coordinating with the hexanoic acid and FeOOH on the solid interface. This superlubricity model is well supported by the tribotest results and XPS. Ni 2p peak at 856.8 and 874.7 eV in XPS and O 1 s peaks at 530.2 and 530.7 eV are consistent with the characteristic peaks of NiOOH after the tribotests. The hexanoic acid of the tribological oxidation products derived from castor oil has the possibility of the coordinative interaction of NiOOH. The expansion of hexanoic acid coordinational groups of NiOOH of the pin surfaces is preceded in the subsequent tribotests. The solid surfaces are separated by the OH-terminated layers and the coordinational hexanoic acid and the gaps between the solid surfaces become large, which results in superlubricity. The dipole−dipole effects between the hydrogen atoms of hexanoic acid or NiOOH and FeOOH form an interfacial Coulomb repulsion force and make a significant contribution to superlubricity of the friction system. The interfacial repulsion force is occurred between the interfaces of the positively charged hydrogen shielding carbon in the coordinating bond hexanoic acid on the top surface of 60NiTi alloy pin and hydrogen shielding oxygen in FeOOH on the top surface of the steel disc, which achieves super-low CoF of 0.005. Therefore based on the tribotest results and theoretical analysis, superlubricity of castor oil is attributed to easy shear between the triboformed OH-terminated surfaces and the shielding surfaces of hydrogen terminated carbon atoms.

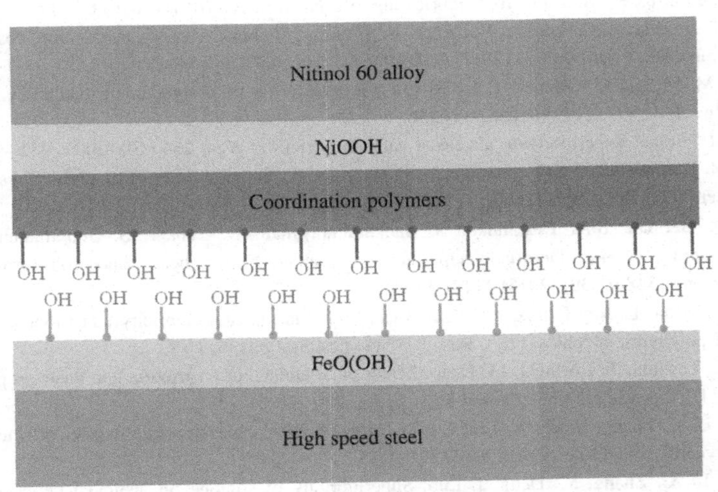

FIGURE 26.11 A schematic showing how superlubricity sliding between two contact surfaces.

26.4 Conclusions

The lubrication performances of 60NiTi alloy/steel friction pair are investigated by pin-on-disc tribometer under oil lubrication. Only castor oil with ester groups, double bonds and hydroxyl groups exhibits superlubricity with a CoF of 0.005, which indicates that the functional groups branched in castor oil play an important role in superlubricity of the friction system. Based on the tribotest results and theoretical analyses, superlubricity is attributed to the frictions occurring between the OH-terminated surfaces with hexanoic acid coordinated nickel oxyhydroxide and iron oxyhydroxide formed from the friction-induced oxidation of green lubricant and the tribochemical reactions and the positively charged surfaces, which produces the repulsive electrostatic forces and achieves superlubricity. This is beneficial to promote the energy efficiency, environment protection, and has tremendous potential for use in the industrial sector.

Acknowledgments

The present work is financially supported by the National Natural Science Foundation of China (51305331 and 51675409).

References

[1] C. Dellacorte, S. Pepper, R. Noebe, G. Glennon, Intermetallic nickel-titanium alloys for oil-lubricated bearing applications, Power Transm. Eng. 8 (2009) 26–35.

[2] C. DellaCorte, L. Moore, J. Clifton, Static indentation load capacity of the superelastic 60NiTi for rolling element bearings, NASA/TM-2012: 216016-1-11.

[3] A. Erdemir, G. Ramirez, O.L. Eryilmaz, B. Narayanan, Y. Liao, G. Kamath, et al., Carbon-based tribofilms from lubricating oils, Nature 536 (7614) (2016) 67–71.

[4] K. Holmberg, A. Erdemir, Influence of tribology on global energy consumption, costs and emissions, Friction 5 (3) (2017) 263–284.

[5] J.M. Martin, C. Donnet, T. Le Mogne, Superlubricity of molybdenum disulphide, Phys. Rev. B 48 (14) (1993) 10583–10586.

[6] M. Hirano, Superlubricity: a state of vanishing friction, Wear 254 (10) (2003) 932–940.

[7] M. Dienwiebel, G.S. Verhoeven, N. Pradeep, et al., Superlubricity of graphite, Phys. Rev. Lett. 92 (12) (2004). 126101–126104.

[8] D. Berman, S.A. Deshmukh, S. Sankaranarayanan, D. Sanket, S. Subramanian, A. Erdemir, et al., Macroscale superlubricity enabled by graphene nanoscroll formation, Science 348 (6239) (2015) 1118–1122.

[9] C. Matta, L. Joly-Pottuz, M.I.D.B. Bouchet, J. Martin, Superlubricity and tribochemistry of polyhydric alcohols, Phys. Rev. B 78 (8) (2008). 085436, 1–8.

[10] K. Yoshida, T. Horiuchi, M. Kano, Effect of organic acid on friction and wear properties of DLC coating, Tribol. Online 3 (3) (2008) 200–204.

[11] J. Li, C. Zhang, L. Ma, Y. Liu, J. Luo, Superlubricity achieved with mixtures of acids and glycerol, Langmuir 29 (1) (2012) 271–275.

[12] J. Li, C. Zhang, M. Deng, J. Luo, Superlubricity of silicone oil achieved between two surfaces by running-in with acid solution, RSC Adv. 5 (39) (2015) 30861–30868.

[13] J. Li, C. Zhang, J. Luo, Superlubricity achieved with mixtures of polyhydroxy alcohols and acids, Langmuir 29 (17) (2013) 5239–5245.

[14] H. Tomizawa, T.E. Fischer, Friction and wear of silicon nitride and silicon carbide in water: hydrodynamic lubrication at low sliding speed obtained by tribochemical wear, ASLE Trans. 30 (1) (1987) 41–46.

[15] F. Zhou, K. Adachi, K. Kato, Sliding friction and wear property of a-C and a-CN_x coatings against SiC balls in water, Thin Solid Films 514 (1) (2006) 231–239.

[16] C. DellaCorte, Nickel-titanium alloys for oil-lubricated bearing and mechanical component applications, in: ASME/STLE 2009 International Joint Tribology Conference, American Society of Mechanical Engineers Digital Collection, 2009, 225–227.

[17] Q. Zeng, G. Dong, Influence of load and sliding speed on super-low friction of 60NiTi alloy under castor oil lubrication, Tribol. Lett. 52 (1) (2013) 47–55.

[18] D. Dowson, G.R. Higginson, Elasto-Hydrodynamic Lubrication: International Series on Materials Science and Technology, Elsevier, 2014.

[19] Q. Zeng, Understanding the lubrication mechanism between the polyhydroxyl group lubricants and metal surfaces, J. Adhes. Sci. Technol. 32 (17) (2018) 1911–1924.

[20] A. Fujimoto, Y. Yamada, M. Koinuma, S. Sato, Origins of sp^3C peaks in C1s X-ray photoelectron spectra of carbon materials, Anal. Chem. 88 (12) (2016) 6110–6114.

[21] J. De Jesus, I. González, A. Quevedo, T. Puerta, Thermal decomposition of nickel acetate tetrahydrate: an integrated study by TGA, QMS and XPS techniques, J. Mol. Catal. A: Chem. 228 (1) (2005) 283–291.

[22] M. Kim, M. Kanatzidis, A. Facchetti, T. Marks, Low-temperature fabrication of high-performance metal oxide thin-film electronics via combustion processing, Nat. Mater. 10 (2011) 382–388.

[23] M. Acik, C. Mattevi, C. Gong, G. Lee, K. Cho, M. Chhowalla, et al., The role of intercalated water in multilayered graphene oxide, ACS Nano 4 (10) (2010) 5861–5868.

[24] T. Hayakawa, M. Yoshinari, K. Nemoto, Characterization and protein-adsorption behavior of deposited organic thin film onto titanium by plasma polymerization with hexamethyldisiloxane, Biomaterials 25 (1) (2004) 119–127.

[25] M.C. Biesinger, B.P. Payne, L.W.M. Lau, A. Gerson, R. Smart, X-ray photoelectron spectroscopic chemical state quantification of mixed nickel metal, oxide and hydroxide systems, Surf. Interface Anal. 41 (4) (2009) 324–332.

[26] A.A. Feiler, L. Bergström, M.W. Rutland, Superlubricity using repulsive van der Waals forces, Langmuir 24 (6) (2008) 2274–2276.

[27] Q. Zeng, G. Dong, J.M. Martin, Green superlubricity of Nitinol 60 alloy against steel in presence of castor oil, Sci. Rep. 6 (2016) 29992.

Index

Note: Page numbers followed by "*f*" and "*t*" refer to figures and tables, respectively.

D

E

F

G

H

I

M

N

O

P

Q

Printed in the United States
By Bookmasters